T0399149

HANDBOOK OF NONWOVENS

The Textile Institute Book Series

HANDBOOK OF NONWOVENS

SECOND EDITION

S.J. RUSSELL

University of Leeds, Leeds, United Kingdom

WP
WOODHEAD
PUBLISHING

ELSEVIER   An imprint of Elsevier

Woodhead Publishing is an imprint of Elsevier
50 Hampshire Street, 5th Floor, Cambridge, MA 02139, United States
The Boulevard, Langford Lane, Kidlington, OX5 1GB, United Kingdom

Notices
Knowledge and best practice in this field are constantly changing. As new research and experience broaden our understanding, changes in research methods, professional practices, or medical treatment may become necessary.

Practitioners and researchers must always rely on their own experience and knowledge in evaluating and using any information, methods, compounds, or experiments described herein. In using such information or methods they should be mindful of their own safety and the safety of others, including parties for whom they have a professional responsibility.

To the fullest extent of the law, neither the Publisher nor the authors, contributors, or editors, assume any liability for any injury and/or damage to persons or property as a matter of products liability, negligence or otherwise, or from any use or operation of any methods, products, instructions, or ideas contained in the material herein.

ISBN: 978-0-12-818912-2 (print)
ISBN: 978-0-12-818913-9 (online)

For information on all Woodhead publications
visit our website at https://www.elsevier.com/books-and-journals

Publisher: Matthew Deans
Acquisitions Editor: Brian Guerin
Editorial Project Manager: Mariana L. Kuhl
Production Project Manager: Manju Paramasivam
Cover Designer: Vicky Pearson Esser

Typeset by STRAIVE, India

Contents

5. Wetlaid web formation

A.A. Koukoulas

6. Spunbond and meltblown web formation

G.S. Bhat, S.R. Malkan, and S. Islam

7. Nanofibre and submicron fibre web formation

E. Stojanovska, S.J. Russell, and A. Kilic

8. Mechanical bonding

S.C. Anand, D. Brunnschweiler, G. Swarbrick, and S.J. Russell

9. Chemical bonding

R.A. Chapman, M. Molinari, S. Rana, and P. Goswami

10. Thermal bonding

A. Pourmohammadi

11. Finishing of nonwoven fabrics

M.J. Tipper and R. Ward

12. Characterisation, testing, and modelling of nonwoven fabrics

N. Mao, S.J. Russell, and B. Pourdeyhimi

Contributors

S.C. Anand University of Bolton, Bolton, United Kingdom

G.S. Bhat University of Georgia, Athens, GA, United States

D. Brunnschweiler[†]

A.G. Brydon Garnett Group of Associated Companies, Bradford & Hawick, United Kingdom

R.A. Chapman 3 The Wardens, Kenilworth, United Kingdom

R. Gharaei University of Leeds, Leeds, United Kingdom

P. Goswami Technical Textiles Research Centre, University of Huddersfield, Huddersfield, United Kingdom

S. Islam University of Georgia, Athens, GA, United States; Dhaka University of Engineering & Technology, Gazipur, Bangladesh

A. Kilic TEMAG Labs, Istanbul Technical University, Istanbul, Turkey

A.A. Koukoulas A2K Consultants LLC, Savannah, GA, United States

S.R. Malkan Hunter Douglas, Broomfield, CO, United States

N. Mao University of Leeds, Leeds, United Kingdom

M. Molinari Department of Chemical Sciences, School of Applied Sciences, University of Huddersfield, Queensgate, Huddersfield, United Kingdom

B. Pourdeyhimi The Nonwovens Institute (NWI), North Carolina State University, Raleigh, NC, United States

A. Pourmohammadi Department of Mechanical Engineering, University of Payam Noor, Tehran, Iran

S. Rana Technical Textiles Research Centre, University of Huddersfield, Huddersfield, United Kingdom

S.J. Russell University of Leeds, Leeds, United Kingdom

E. Stojanovska TEMAG Labs, Istanbul Technical University, Istanbul, Turkey

G. Swarbrick Groz-Beckert UK Ltd., Wigston, United Kingdom

M.J. Tipper Nonwovens Innovation & Research Institute Ltd (NIRI), Leeds, United Kingdom

G. Tronci University of Leeds, Leeds, United Kingdom

R. Ward Nonwovens Innovation & Research Institute Ltd (NIRI), Leeds, United Kingdom

A. Wilson Sustainable Nonwovens, Wakefield, United Kingdom

[†]Deceased

Acknowledgements

Special thanks are due to Dr. Andrew Hewitt of the Technical Textiles Research Centre, University of Huddersfield, United Kingdom, for his expert assistance in the editing and preparation of various diagrams and figures. The administrative support of Natasha Francis at the University of Leeds, United Kingdom, is also gratefully acknowledged. Thanks are also due to Matt Pealing at the University of Leeds for his assistance with figure images.

CHAPTER

1

Development of the nonwovens industry

A. Wilson

Sustainable Nonwovens, Wakefield, United Kingdom

1.1 Definition and classification

In defining what a nonwoven is, there is always at least one exception that breaks the rule. This is perhaps fitting, since whilst having been recognised in its own right for many decades, the nonwovens industry has drawn on the practices and know-how of many other fields of materials manufacturing with a piratical disregard and an eye to the most diverse range of end-use products. For this reason, it is possible for companies with almost nothing in common – with vastly different structures, raw materials and technologies, areas of research and development and finally, customers – to be grouped together under the nonwovens 'umbrella'. Many would define themselves by the customers they serve, as being in the disposable consumer products, medical, automotive or civil engineering industries, for example.

The term 'nonwoven' arises from well over half a century ago when nonwovens were often regarded as low-price substitutes for traditional textiles and were generally made from carded, staple ('dry') fibres on converted textile processing machinery. The yarn spinning stage is omitted in the nonwoven processing of staple fibres, with bonding (consolidation) of the web by various methods – chemical, mechanical or thermal – replacing the weaving (or knitting) together of the yarns in traditional textiles.

However, even then the process of stitchbonding, which originated in Eastern Europe in the 1950s, employed both layered and consolidating yarns, and the parallel developments from the paper and plastics fields, which have been crucial in shaping the multibillion-dollar nonwovens industry as it is today, had only tenuous links with textiles in the first place.

The term 'nonwoven' then – describing something that a product is not, as opposed to what it actually is – has never accurately represented its industry, but any attempts to replace it over the years have floundered. The illusion created by this misnomer has been of commodity manufacturing, when the opposite is often true. The nonwovens industry is highly profitable

Handbook of Nonwovens
https://doi.org/10.1016/B978-0-12-818912-2.00004-5

and very sophisticated, with healthy annual growth rates – sometimes in double digits in certain sectors and parts of the world. It is perhaps one of the most intensive in investing in new technology, and also in research and development.

Nonwovens are defined by ISO standard 9092 and CEN EN 29092. These two documents, identical in their content, are the internationally acknowledged definition of nonwovens. As industry, trade and technology have evolved since their first publication in 1988, these standards are regularly being updated by experts to better reflect the present understanding of nonwovens. The following text, prepared by the leading nonwovens trade associations, is the current definition of a nonwoven:

> An engineered fibrous assembly, primarily planar, which has been given a designed level of structural integrity by physical and/or chemical means, excluding weaving, knitting or paper making.

As such, nonwovens can be disposable – for single use, or with a limited life – or durable, with a much longer life, depending on their application. Their key functions include, but are not limited to

- Absorbency
- Bacterial barrier
- Cushioning
- Filtering
- Flame retardancy
- Liquid repellency
- Resilience
- Softness
- Sterility
- Strength
- Stretch
- Washability

A number of these properties are frequently combined to create fabrics suited for specific tasks, whilst achieving a profitable balance between product use-life and cost. They can also approximate the appearance, texture and strength of a woven fabric and can be as bulky as the thickest paddings. In combination with other materials, they provide a spectrum of products with diverse properties. Some of the most common products made with nonwovens include

- Agricultural coverings and seed strips
- Apparel interlinings
- Automotive headliners and upholstery
- Battery separators for electric vehicles
- Carpeting and upholstery fabrics, padding and backing
- Civil engineering fabrics/geosynthetics
- Disposable diapers
- Envelopes
- Filters
- House wraps
- Insulation

- Labels
- Laundry aids (fabric dryer-sheets)
- Reinforcements for composites
- Roofing products
- Sanitary napkins and tampons
- Sterile wraps, caps, gowns, masks and drapes used in the medical field
- Tags
- Wall coverings
- Wipes (household, personal, medical and industrial)

Disposable absorbent hygiene products (AHPs), consisting of baby diapers, femcare products (sanitary pads, panty liners, etc.) and those for coping with adult incontinence (AI) are by far the biggest grouping of nonwoven products. Whilst estimates for the retail value of the AHPs market vary significantly, they currently even out at around $90 billion annually, consisting of

- Baby diapers, worth $49 billion.
- Femcare products, worth $32 billion.
- Adult incontinence (AI) products, worth $9 billion.

Regionally, North America accounts for 36% of total AHP sales in 2019, Europe 25%, Asia Pacific 23% and the rest of the world 16%. In recent years, the Asia Pacific region has been achieving the highest growth rates of an average 7% across its markets, whilst growth in the United States and Europe has been only around 1% due to the already very high market penetration levels. The AI market will show the highest growth in AHP products going forward, due to people living longer around the world, and sadly due to the increase in obesity.

The general effect of higher disposable incomes is also that more people prioritise hygiene, once food and housing needs have been, or are in the process of being, satisfied. The explosive growth of AHPs in China over the past 20 years is the most striking example of this. Disposable wipes have meanwhile been one of the fastest growing markets for nonwovens over the past decade and can be considered an add-on to the general AHP sector. They now represent a retail market worth around a further $13 billion annually.

1.2 Dry, wet and spunlaid nonwovens

In dividing nonwovens into three major categories – drylaid, wetlaid and spunlaid – it can generally be said that drylaid materials have their origins in textiles, wetlaid materials in papermaking and spunlaid products in polymer extrusion and plastics (remembering that there is always at least one exception to the rule). Generally, nonwoven fabrics are made by forming a web from either fibres or continuous filaments, followed by bonding to stabilise the structure (Fig. 1.1). This may be, but is not always, followed by fabric finishing to engineer fabric properties. Finally, converting refers to a series of processes that convert nonwoven roll good fabrics into a multitude of finished products.

FIG. 1.1 Overview of major nonwoven manufacturing technologies. *Courtesy of D Brunnschweiler.*

Web Formation

Paper Technology Fibres 2-15 mm		Textile Technology Fibres 10-200 mm					Polymer extrusion Technology Filaments			
Wetlaid	Airlaid	Garnett	Worker-stripper cards	Cotton cards	Hybrid-airlay cards	Other hybrid cards	Spunbond	Meltblown	Flash spun	Fibrillated film

Web manipulation

Cross-lapping · Vertical-lapping · Scrambling · Drafting · Spreading

Bonding

Chemical (adhesive)	Solvent	Thermal	Spray	Needlepunching	Hydroentanglement (spunlace)	Stitch bonding	Ultrasonics
Calender/mangle/foulard		Calender (plain or embossed)		Pre-needling (tacking)	Plain	Fibre/web stitching	
Spray		Through-air (drum or lattice)		Plain and finish needling	Apertured	Yarn stitching	
Foam		Powder		Structuring	Embossed		
Print							
Lick-coat/kiss-roll							
Powder							

Fabric Finishing

Micro-corrugation · Singeing · Coating · Impregnation · Printing/embossing · Plasma · Lamination

Converting

Cutting/perforating · Folding/layering/placing/stacking · Impregnation/printing/sprinkling/coating/injection · Lamination/gluing/pressing · Thermoforming/moulding · Cleaning/sterilisation · Wrapping/packing

1.2.1 Drylaid nonwovens

The nonwoven technologies originating from the textile industry manipulate fibres in the dry state. In the most common drylaid process, the fibres are carded to make a web and then bonded by one or more of a number of methods – needlepunching, thermobonding, chemical bonding or hydroentanglement with high-pressure water jets – all of which are described in detail later in this book. The first drylaid systems owed much to the basic felting process known since medieval times. Airlaying technology or airlaid web formation also involves fibres that are manipulated in their dry state but is a process technology which originated from the paper industry. It involves converting cellulosic wood pulp or short cut synthetic fibres – or most often a combination of the two – into random-laid, low-density absorbent webs, using air instead of water to transport the fibres. Some airlaying systems are capable of handling longer natural or synthetic fibres.

1.2.2 Wetlaid nonwovens

Paper-based nonwoven fabrics are manufactured with machinery designed to manipulate short fibres suspended in water and are referred to as 'wetlaid'. To distinguish wetlaid nonwovens from wetlaid papers, a material is regarded by EDANA (the European Disposables and Nonwovens Association) as a nonwoven if more than 50% by mass of its fibrous content is made up of fibres (excluding chemically-digested vegetable fibres) with a length to diameter ratio greater than 300, or more than 30% fibre content for materials with a density less than $0.40 \, g/cm^3$. This definition excludes many constructions which others would class as nonwovens. The use of the wetlaid process is confined to a relatively small number of companies, being extremely capital intensive and involving substantial volumes of water which can be very effectively recirculated. The current leading manufacturers of drylaid and wetlaid nonwovens, in order of total nonwovens sales, are shown in Table 1.1.

1.2.3 Spunlaid nonwovens

Spunlaid or 'spunmelt' nonwovens – incorporating spunbond, meltblown and the many layered combinations of these products, e.g., SMS – are manufactured with machinery developed from the extrusion of plastics, with the nonwoven structures simultaneously formed from molten filaments and manipulated. In the basic spunbonding system, sheets of synthetic filaments are extruded from polymer onto a conveyor to make a randomly-oriented web, in a continuous polymer-to-fabric operation.

Most of the first spunbonding systems originated with fibre producers working with proprietary technologies such as DuPont in the United States, Rhone-Poulenc in France and Freudenberg in Germany. DuPont is regarded as the first to successfully commercialise spunbonding with its Typar product, initially launched as a tufted carpet backing fabric in the mid-1960s. The first commercial spunbonding system to be offered was the Docan system developed by the Lurgi engineering group in the 1960s and licensed to Corovin in Germany, Sodoca in France, Chemie Linz in Austria and Crown Zellerbach in the United States – all of which were later absorbed into other corporations, notably Berry Global – as well as Kimberly-Clark in the United States.

TABLE 1.1 The 20 leading manufacturers of drylaid and wetlaid nonwovens in 2020 by overall turnover.

Company	Headquartered
Berry Global	United States
Freudenberg	Germany
Ahlstrom-Munksjö	Finland
DuPont	United States
Kimberly-Clark	United States
Fitesa	Brazil
Glatfelter	United States
Lydall	United States
Suominen	Finland
TWE Group	Germany
Hollingsworth & Vose	United States
Zhejiang Kingsafe	China
Sandler	Germany
Toray Advanced Materials	South Korea/Japan
Jacob Holm	Switzerland
Georgia-Pacific	United States
Nan Liu	Taiwan
Fibertex Nonwovens	Denmark
Union Industries	Italy
Hassan Group	Turkey

Source: Reproduced with permission from Nonwovens Industry magazine.

The next major step towards the global commercialisation of the spunbond process was with the introduction of Reifenhäuser's Reicofil system in 1984. Since the introduction of the first Reicofil system, by Germany's Reifenhäuser in 1984, it has been continuously improved and more or less become the industry standard for the high-volume production of spunmelt fabrics for the disposable hygiene market. The staggering increase in the productivity of Reicofil machines is shown in Table 1.2.

1.2.3.1 Meltblown nonwovens

Meltblowing also produces fibrous webs directly from polymers using high-velocity air attenuation to form the filaments. The process is unique because it is used almost exclusively to produce microfibres rather than fibres the size of typical textile fibres. Meltblown microfibres generally have diameters in the range of 2 to $4\,\mu m$ which enhances the softness, cover, opacity and solid surface area of the webs obtained.

TABLE 1.2   Advances in Reicofil spunbonding production (1986–2017).

	Year	kg/h/m of beam
Reicofil 1 system		
	1986	50
	1992	100
Reicofil 2 system		
	1992	125
	1995	145
Reicofil 3 system		
	1995	150
	2002	195
Reicofil 4 system		
	2002	225
Reicofil 5 system		
	2017	270

Source: Based on data from Reifenhauser Reicofil.

The basic technology for producing microfibres was first developed in the 1950s by the US Naval Research Laboratory. The commercial significance of the work was recognised by oil and gas giant Exxon, which subsequently developed the technology further. Researchers at Exxon extended the basic design and first demonstrated the production of meltblown microfibres on a commercial scale by modifying sheet die technology. The meltblown process is a relatively complex manufacturing process. The processing window for successfully meltblowing polymers is very limited, and in order to produce webs with acceptable quality, tight control over processing parameters and raw materials specifications is required. Table 1.3 gives a basic comparison of spunbond and meltblown technologies. In terms of throughput, where the speed of the latest Reicofil 5 line for spunbonding is 270 kg/h/m of beam, for meltblowing, it is just 70 kg/h/m of beam.

1.2.3.2 Spunbond–meltblown systems

The spunbonding and meltblowing processes are now very often combined on multiple-beam machines, such as in popular SMS (spunbond–meltblown–spunbond) composite layer configurations, to exploit the properties of both types of these webs. To demonstrate the full extent of possibilities, for example, the Reifenhäuser Reicofil technology centre in Troisdorf, Germany, houses a seven-beam SSMMMSS line for both hygiene and medical products. As a result, the industry now collectively refers to spunbond and meltblown materials as 'spunmelts'. The current leading manufacturers of spunmelt nonwovens for hygiene, in order of overall turnover, are shown in Table 1.4.

TABLE 1.3   Comparison of spunbonding and meltblown technologies.

Parameter	Spunbond filter media	Meltblown filter media
Raw materials	Polymer resins	Polymer resins
Process equipment	Extruders	Extruders
Process air	Small quantity and lower temperature than extruded polymer fibre	Very high quantity of air is used. Temperature is close to that of extruded polymer fibre
Attenuation	Fibres are cooled and attenuated over a large distance between the die and the conveyor	Fibres are attenuated with strong air currents near the die or spinneret
Fibre diameters	Microfibre production is challenging	Microfibre production is relatively easy

Source: Data from Sustainable Nonwovens.

TABLE 1.4   The 20 leading manufacturers of spunmelt nonwovens in 2020 by overall turnover.

Company	Headquartered
Berry Global	United States
Freudenberg	Germany
Ahlstrom-Munksjö	Finland
Kimberly-Clark	United States
DuPont	United States
Fitesa	Brazil
PFNonwovens	Czech Republic
Zhejiang Kingsafe	China
Avgol	Israel
Toray Advanced Materials	South Korea/Japan
Fibertex Personal Care	Denmark
Asahi Kasei	Japan
Mitsui Chemicals	Japan
Gulsan	Turkey
Union Industries	Italy
Hassan Group	Turkey
Xingtai Nonwovens	China
Dalian Ruiguang	China
Toyobo	Japan
Halyard Health	United States

Source: Data from Sustainable Nonwovens.

1.3 Web forming

In all nonwoven manufacturing systems, the fibre (staple or filament) is deposited or laid on a forming or conveying surface, and the physical environment at this stage can be dry, wet or air quenched – drylaid, wetlaid or spunlaid. The drylaid and wetlaid processes make webs from staple fibres, whilst the spunlaid processes produce webs from continuous filaments.

The web formation stage in nonwoven manufacturing processes transforms previously prepared/formed fibres or filaments into layered, loosely-arranged fibre networks (webs, batts, mats or sheets). In operating web formation processes using machinery adapted from the textile, paper or extrusion industries, there are means to achieve the preferred fibre orientation in the web. This is important because fibre orientation influences the properties of the final fabric in different directions. Other critical fabric parameters established at the web formation stage are the unfinished product weight and the manufactured width. Each web-forming system is used for specific types of fibres or products, although one exception here is with highloft nonwoven production, which employs either cards and crosslappers or airlaid web formation systems.

1.4 Bonding

Web consolidation or nonwoven bonding processes can be by mechanical, chemical, solvent and/or thermal means. The degree of bonding is a primary factor in determining fabric integrity (strength), porosity, flexibility, softness and density (loft, thickness). Bonding may be carried out as a separate and distinct operation but is generally carried out 'in-line' with web formation so that the entire nonwoven production process is continuous. In some fabric constructions, more than one bonding process is used in sequence. Mechanical bonding methods include needlepunching, stitchbonding and hydroentangling via water jets. The latter method has grown considerably in popularity over the past decades, especially for applications such as disposable wipes employing cellulosic fibres. Chemical bonding methods involve applying adhesive binders to webs by saturating, spraying, printing or foaming techniques. Solvent bonding involves softening or partially dissolving fibres with a solvent to provide a self-bonded structure. Where thermoplastic material is present in the web, thermal bonding involving the use of heat and often pressure is used to fuse or weld fibres together and can also be applied after other bonding processes to achieve smoother fabric surfaces.

1.5 Raw materials

At present, man-made materials completely dominate nonwovens production, accounting for as much as 99% of total output. Man-made fibres fall into three classes – those made from natural polymers, those made from synthetic polymers and those made from inorganic materials.

Polypropylene (PP) is the most common polymer employed for the production of nonwovens, followed by polyester, regenerated cellulosics, acrylic, polyamide, cotton and other speciality fibres. As a thermoplastic polymer, PP's basic hydrocarbon chain consists of propylene-building blocks, and it is a hydrophobic material. PP fibre has very good tensile and abrasion-resistance properties and good resistance to acids, although their stability in alkaline conditions is limited. Being a polyolefin, PP has a maximum operating temperature of around 63°C. Both humid and dry heating conditions are detrimental to PP fibres above this temperature, even though the melting temperature is considerably higher. PP fibres as the basis of nonwovens have numerous advantages including

- Light weight – PP has one of the lowest specific gravity values of all fibres.
- The resulting fabrics are soft and provide comfort to the wearer.
- The fabrics are melt-processable and can be thermally bonded.
- They provide good bulk and cover in AHPs, in addition to strength, low static, abrasion resistance and resistance to deterioration from chemicals, mildew, perspiration, rot and weather, as well as staining and soiling.

This unique combination of properties provides the manufacturers of nonwovens with a valuable high-performance polymer or fibre at a very competitive price. Cellulosics however, and notably regenerated cellulose fibres such as lyocell, based on wood pulp, along with other natural fibres, are increasingly in demand as pressure continues to grow on the use of oil-based synthetics. Given such natural polymer materials are not suitable for melt processing, the development currently favours drylaid staple fibre manufacturing over spunlaid thermoplastic polymer-based methods. Given the increasing demand for more sustainable nonwoven products, there has also been a significant rise in the utilisation of recycled synthetic polymers, particularly polyester (PET), where a large proportion of the PET nonwovens made today are produced from recycled plastic bottle waste.

1.6 Market structure and development

Up until the 1990s, much of the world's nonwovens industry was based in the areas where the manufacturing technologies were conceived and developed – the United States, Europe and Japan. Carded webs, with bonding by needlepunching or heated calenders, could be small-scale enterprises, often part of textile companies, but at the same time, large companies, such as Freudenberg, Kimberly-Clark, DuPont and Asahi, invented much more sophisticated nonwoven technologies and nurtured them to commercial scale. Such large-scale production facilities were highly capital intensive, making it too risky for smaller companies to set up production, certainly if based on spunlaid, wetlaid, airlaid pulp and hydroentangling.

The industry can still be regarded as capital-intensive today, when considering that, according to the latest estimates, some 40 companies are responsible for 90% of total global nonwovens sales, and dedicated new spunmelt lines, which each add an average annual capacity of 15,000 tons of new product, involve capital investments of between $20 and 40 million. However, when machinery builders began to produce 'turn-key' production lines capable of making high-quality spunmelt nonwoven materials at competitive costs from the 1980s, the result was not only continued strong growth in the original three regions of the United States, Europe and Japan. It also sparked the beginning of global expansion, with many new local producers emerging, especially in Asia, which now accounts for 44% of all nonwovens production. According to a number of forecasts, including those from EDANA and INDA, around 11.8 million tons of nonwovens are now produced annually, with approximately 58% being drylaid or wetlaid (6.8 million tons) and 42% made by spunbond and meltblown technologies (5 million tons). The growth in spunmelt production in particular – as in the consumption of AHPs – has been most notable in China, having climbed from virtually zero in 1994 to close to an annual 2 million tons today.

Overview of nonwoven product applications

R. Gharaei and S.J. Russell
University of Leeds, Leeds, United Kingdom

2.1 Introduction

Given the enormous variety of different products containing nonwoven fabrics, an entire textbook could be dedicated to the subject of applications, and so this chapter is necessarily limited to a brief overview. End-use applications span a remarkably diverse range of consumer, industrial and healthcare markets, not least because of the unique combination of performance and value that nonwovens provide compared to traditional textiles. Without nonwovens, countless products that we take for granted in society would not exist, function properly, or be available at an affordable price.

Nonwovens are frequently associated with high volume, relatively inexpensive, single-use products such as wipes and baby diapers, where ease of use and convenience have driven remarkable growth in global consumer demand. While such products continue to represent a substantial proportion of the industry's output, nonwovens are important in many other markets. Today, the industry serves a full spectrum of price sensitive to high-value applications, encompassing short, multiple-use, and long-life products. In some markets, nonwoven fabrics embody the final product itself, while in many others, they are combined with films, coatings, adhesives, particulate materials, and other components. The extent to which nonwovens are ubiquitous is quite often overlooked because they are so frequently out of view, embedded within the interior of final products.

Irrespective of price-point, the fitness for purpose of nonwovens, as well as their cost-effectiveness usually entails commitment to innovation. It is therefore a mistake to think of all low-cost or single-use nonwoven products as being unsophisticated in terms of the underlying technology. Commercially, most of the nonwoven products in use today are the result of progressive technical developments in materials, manufacturing processes and product design, motivated largely by the need to remain competitive and to meet evolving consumer requirements. The large body of patent literature associated with nonwovens is testament to the intensive R&D activity underpinning many of the products in use, and the progressive attitude of companies that constitute the industry. Commitment to innovation continues to be fundamental to progress, and external forces are making this even more important. For example, an increasingly challenging regulatory landscape, particularly related to environmental sustainability, together with stricter product performance requirements means the design and specification of nonwoven products can never stand still.

The high production speeds and relatively low costs associated with nonwoven fabric manufacturing processes stem in part from the reliance on fewer manufacturing steps, compared to knitted or woven fabrics made from yarns. This means it is possible to convert raw materials in the form of polymer resins or fibres directly into fabrics as part of high-speed, integrated production processes. For example, modern spunbond lines can produce at up to 1200 m/min on the conveyor, with outputs up to ~270 kg/h/m. Nonwoven process technology is also inherently versatile such that it is possible to serve several different markets with similar manufacturing equipment. There are also fundamental differences in the structure and physical properties of nonwoven fabrics compared to conventional textiles, which means they are suited to different markets.

2.2 The service life of nonwovens

Nonwoven product applications can be categorised in terms of their service life. Historically, the terms 'disposable' and 'durable' have been used to describe this, but referring to products as single-use (short life), multiple-use, or long-life is arguably more accurate. Of course, this method of classification still oversimplifies reality as there are nearly always exceptions to the rule, depending on the specific product being considered. To represent the approximate service life, a temporal scale of seconds, minutes, hours, or years can be imagined as a range, as indicated by examples in Fig. 2.1 The service life in this context refers only to the use-phase and therefore should not be confused with the cradle-to-grave life cycle, which is used to assess the overall environmental impact of individual products.

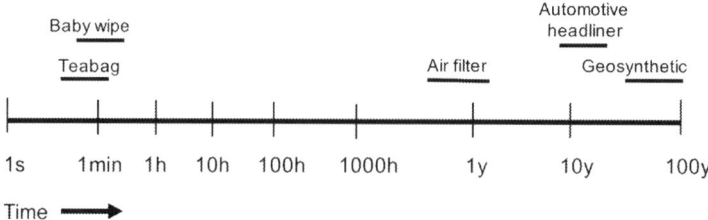

FIG. 2.1   Approximate service life of nonwoven products (seconds, minutes, hours, and years).

2.2.1 Single-use

The service life of single-use nonwoven products typically varies roughly from a few seconds to a few hours, prior to disposal by the user. The share of nonwoven output that falls into the single-use category varies in Europe, the United States and Asia, and generally accounts for a large proportion. Single-use applications are dominated by the hygiene and wipes sectors, including diaper, continence management and femcare products as well as baby wipes, but products are extremely diverse, covering many other items such as teabags, coffee filters, and pods, as well as personal protective equipment (PPE) to name just a few.

2.2.2 Multiple-use

The service life of multiple-use nonwoven products typically ranges from weeks to months. This encompasses nonwoven products that are used in an intermittent fashion, before being discarded. Clearly, this usually means they remain in use for longer than a single-use application. Examples include nonwoven products that are designed to be washed, laundered, or otherwise cleaned to 'regenerate' or reinvigorate them. Heavy-weight industrial wipes and mops for hard surface and floor cleaning applications, allergy protection covers used in bedding, and filter media for vacuum cleaners, are just a few examples.

2.2.3 Long-life

Long-life nonwovens typically have a service life measured in years, extending to many decades in some cases. The long-life category includes some of the oldest applications for nonwovens, developed during the early days of the industry. Such products include clothing interlinings, artificial leather, carpet underlay and floorcoverings, thermal insulation, roofing membranes, geosynthetics and automotive interior trim, including headliners. Long-life products are typically, but not exclusively, heavier than those intended for single-use, which means they also tend to be more costly to manufacture.

2.3 General categories of end-use application

Nonwoven product applications can be conveniently divided into three broad categories, although there are some crossovers: (i) consumer products, (ii) industrial products, and (iii) medical and healthcare products.

2.3.1 Consumer products

Consumer products in general can be classified as follows [1]:

1. *Convenience products* are purchased most frequently, at a low price and are readily available.
2. *Shopping products* are purchased less frequently, at a medium price-point, and attributes such as quality, style, and price influence purchasing decisions.

3. *Specialty products* are purchased less frequently, have distinctive characteristics or brand perception, are usually expensive and are often only available from certain vendors.

4. *Unsought products* are not thought to be necessary purchases until their need becomes apparent.

Nonwovens are relevant to all of four of these consumer product categories, which include applications in fast-moving consumer goods (FMCG), also known as, Consumer Packaged Goods (CPG). One of the largest consumer product markets is hygiene, which includes single-use baby diapers and wipes for personal care. According to EDANA [2], consumer products for hygiene and wipes for personal care, the majority of which are single-use, account for over 40% of manufactured nonwoven roll good output in Europe. In the convenience product category, single-use nonwovens generally compete on price and small margins, but high-volume consumption, makes production attractive and profitable. Nonwoven fabrics are also important as highly durable components in the manufacture of many long-life consumer products such as clothing (e.g. interlinings, artificial leather) and footwear, wallcoverings, as well as key components in consumer durables such as vehicles, and a large range of household appliances.

2.3.2 Industrial products

Industrial applications for nonwovens are extremely broad with a global CAGR of 7.2% projected by 2025 [3]. Nonwovens are used either as components in finished industrial products, as consumables to aid manufacturing, or to facilitate effective and safe operation of production processes. This covers a multitude of different applications, including for example, filters, fire barriers, abrasives, tapes, sorbents (liquid and gas), cable insulation, reinforcement of plastics, packaging, conveyor belts, papermaker's felts, and sound insulation. Industries making use of nonwovens include aerospace, automotive, chemical/pharmaceuticals, building and construction (including geosynthetics), defence, electric power, electronics, energy, food, meat packing, mining, petroleum, telecommunications, and the water industry.

2.3.3 Medical and healthcare products

The medical and healthcare sector accounts for less than 5% (by weight) of the total consumption of nonwovens in Europe, but it is an increasingly important area of development [4]. Nonwovens are routinely used by nursing and clinical practitioners both in hospital and increasingly in community care settings. Nonwovens also form critical performance components in a variety of regulated medical devices. The ageing population, together with the increased prevalence of chronic and degenerative disease, are amongst the major drivers of growth, and products range from low-cost single-use items routinely used for infection control, to high value, advanced wound care, extra corporeal and implantable devices.

2.4 Absorbent hygiene products

The majority of absorbent hygiene products (AHPs) are single-use, wearable items that are designed to manage body fluids, such as urine and menses, prevent leakage and remain discreet. They include as broad categories (Fig. 2.2):

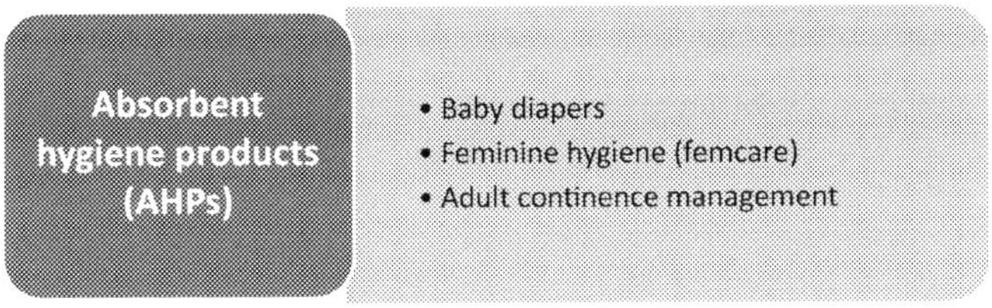

Absorbent hygiene products (AHPs)

- Baby diapers
- Feminine hygiene (femcare)
- Adult continence management

FIG. 2.2 Summary of major AHP products [5].

- Baby diaper and related products, which are predominantly single use.
- Other continence management products, primarily adult incontinence, which are predominantly single use.
- Feminine hygiene (femcare) products, e.g. single-use sanitary napkins, panty liners, and tampons.

The long-standing commercial success of AHPs is attributable to their convenience of use, performance, low-cost and widespread availability. There has been considerable incremental volume growth in AHPs for many years. Historically, the growth in AHP markets has mainly been fuelled by demand in developed countries but prospects for growth in some categories, notably baby diapers are greater in developing countries. For example, the baby diaper market in the United States and Europe is now considered to be mature and is characterised by slow growth, due in part to relatively low birth rates, and products that do not need changing as regularly. However, growth prospects in Africa and Asia are significantly higher. Adult continence care has been a particularly important area of growth and development in AHP markets. Remarkable innovation in AHPs has taken place over the years, not least in the design and performance of baby diapers, evidenced by the enormous body of patent literature charting progressive developments in product designs, innovative uses of materials and methods of manufacture. This has led to marked reductions in the weight and bulk of diaper products by up to 40% in the last 20 years [6], without compromising performance, as well as improvements in overall functionality, including fit and comfort.

Generally, AHPs are sophisticated multicomponent, multilayered products that are designed to conform to the body, manage fluids of different viscosities and compositions (e.g. menses and urine), suppress odour and skin irritation and provide thermophysical comfort. In terms of fluid management, the objectives are normally a rapid inlet time of fluid into the product, a prescribed absorption rate, high absorbent capacity and retention, as well as minimal wetback to the surface under pressure to ensure the topsheet remains dry. Simply stated, single-use baby diapers consist of the following main components, all of which can rely on nonwoven technology [7]:

- *Topsheet*—also referred to as coverstock, is the upper-most layer in contact with the skin, and which first comes into contact with the liquid 'insult' composed of urine. The primary purpose is to allow rapid liquid penetration through the topsheet into the underlying AHP structure, such that no residual liquid is retained by the surface. As short an 'inlet' time as possible is therefore key, which means the topsheet should not normally retain liquid as this can lead to skin wetness, as well as impede rapid inlet of subsequent liquid insults. Coupled with this is the need to prevent 'wetback', i.e. liquid that has penetrated the topsheet should not be able to strike back through the topsheet after inlet. The majority of nonwoven topsheets are therefore made of hydrophobic polymers, typically PP, but receive a hydrophilic surface treatment to reduce surface tension, minimising the liquid inlet time. Topsheets are traditionally made of very light-weight PP spunbond nonwovens (ca. 7.5–20 g/m^2), or a thin perforated film, depending on the manufacturer or brand, but other fabrics such as apertured hydroentangled substrates are also used in some other AHP products.

- *Acquisition-distribution layer (ADL)*—after initial inlet, the ADL is intended to promote lateral spreading of the liquid to ensure effective utilisation of the underlying absorbent core, inhibiting localised saturation, as well as preventing wetback to the overlying topsheet. Traditionally, ADLs are made of carded-through-air thermally bonded nonwovens in the range of 20–90 g/m^2. Normally, to function correctly, the ADL needs to be in direct contact with the topsheet and absorbent core. In practice, the lateral liquid spreading performance of some ADLs is quite limited, partly because of their low density and the competing requirement to minimise any potential for wetback.
- *Absorbent core*—the absorbent core is responsible for liquid absorption and retention, and therefore contains hydrophilic material, usually either wood pulp (fluff pulp), superabsorbent polymer (SAP) or mixtures of the two. Other components such as binders, short fibres and functional additives are included depending on the manufacturer. Reducing the pulp and increasing the SAP content has been instrumental in enabling thinner absorbent cores to be made. Key challenges relate to fixing the SAP in position preventing migration, avoiding gel blocking and addressing the suppressive effect of dissolved salts on the absorbent capacity of SAPs. Note that in some continence management products, a nonwoven core wrap encapsulates the absorbent core, improving its stability and preventing migration.
- *Backsheet*—the purpose of the backsheet is to prevent liquid leakage and strike-through from the absorbent core, while ideally, enabling water vapour transmission to the external environment through evaporation, which acts to reduce the overall water content in the core. Water vapour–permeable PE films, PE spunbonds, SMS fabrics and composite nonwoven-film fabrics are amongst the various substrates relevant to backsheets, as well as in some cases impermeable PE films [8]. For product differentiation and to enhance appearance, the ability to print the backsheet is increasingly important.

Leakage and comfort in wearable AHPs are influenced by the quality of fit and the performance of other components that contain nonwovens in their construction, such as elastic fittings, tabs, closures and fastenings. In these components, nonwovens often function as a backing or support for other layers, e.g. polymeric films that control aspects of physical properties or which improve aesthetics. A highly simplified diaper construction consists of a topsheet, acquisition and distribution layer (ADL), backsheet, absorbent core, fasteners and elastic fittings [9].

In contrast to the recent, relatively slow growth in the baby diaper market in developed countries, there has been progressively increasing demand for adult continence management products, attributed in part to a demographic shift associated with an ageing population. According to the Global Forum on Incontinence (GFI), 4%–8% of the global population is affected by incontinence, which equates to ca. 400 million people worldwide [10]. Urinary adult incontinence, specifically light incontinence is most common, and millions of sufferers remain clinically undiagnosed. Particularly, for adult continence management, recent developments have been aimed at combining the excellent liquid management properties of single-use products, with the superior aesthetics of reusables. Some companies have also extended their femcare range for managing menstruation, to address adult continence management.

The compositional complexity of diaper products can make understanding the materials they contain difficult to appreciate. A stewardship scheme overseen by EDANA now requires producers to monitor the presence of certain trace chemicals in AHPs using standardised test methods, making sure agreed guidance values are not exceeded and that details about product compositions are published.

One of the common criticisms of single-use AHPs relates to their end-of-life impact on the environment given that billions are disposed of each year. Depending on location, 2%–4% of municipal solid waste may consist of used AHPs. Such concerns have led to the development of new recycling technologies for used AHPs, including diapers, that enable the retrieval of valuable component materials such as the cellulose and mixed plastics from the waste, enabling their use as raw materials for other purposes. Numerous approaches and process technologies for handling AHP waste have been developed, and the recycling system of FaterSMART is just one example [11].

2.5 Wipes

Nonwoven wipes account for about 15% of all nonwovens produced, with significant growth in the last two decades due to their convenience, performance, and relatively low cost [12]. Although not exclusively, most nonwoven wipes are intended to be single-use. The progressive increase in the consumption of wipes is linked to rising disposable incomes, along with growth in institutional market demand from hotels, hospitals, schools, and universities, as well as organisations operating in industrial and healthcare settings. However, wipes are still considered a discretionary purchase and pricing remains influential for both consumers and retailers especially in developing markets.

The global nonwoven wipes market is segmented into two main categories: (1) dry wipes, and (2) wet or premoistened wipes, which includes flushable (dispersible) variants (Fig. 2.3). In wet or premoistened wipes, the nonwoven substrate is impregnated with an aqueous formulation or 'lotion', which contains auxiliaries such as surfactants to aid cleaning, alcohols, fragrances, viscosity modifiers, stabilisers and preservatives, according to the required

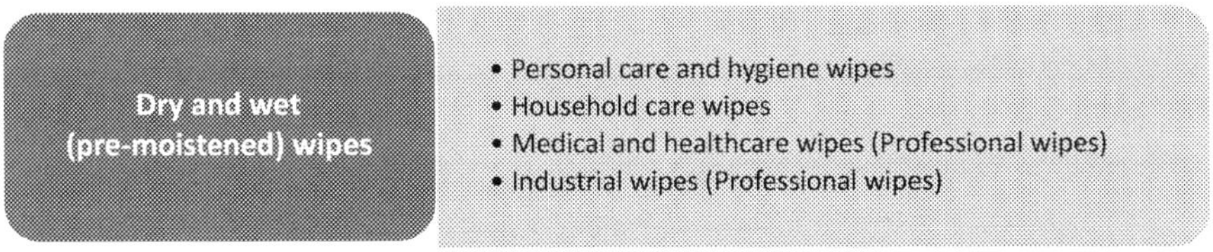

Dry and wet (pre-moistened) wipes	• Personal care and hygiene wipes • Household care wipes • Medical and healthcare wipes (Professional wipes) • Industrial wipes (Professional wipes)

FIG. 2.3 Main markets for wipes [5].

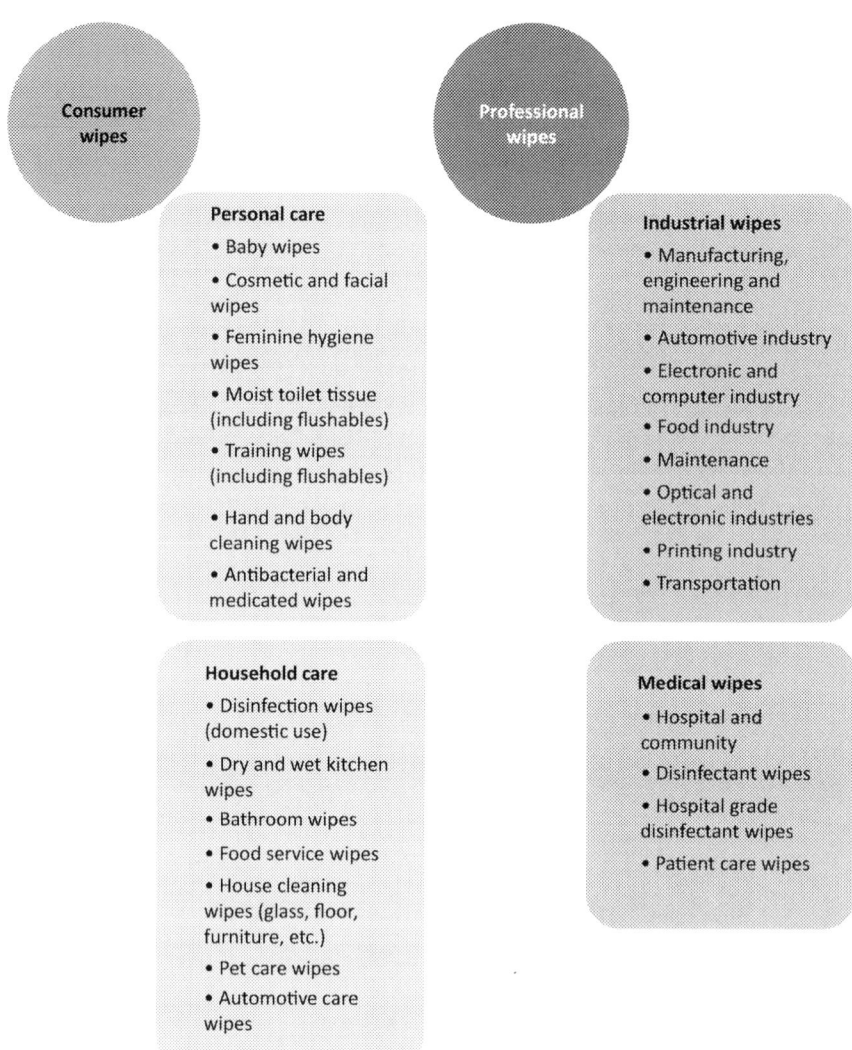

FIG. 2.4   Nonwoven wipes categories and examples based on end-use market [5].

function. Depending on the absorbent capacity of the nonwoven, liquid add-on levels are typically in the range of 80%–450%, based on the dry weight of the fabric.

For consumer product applications, wipes can be divided into two main categories (Fig. 2.4):

- *Personal care*: includes baby and intimate care wipes, moist toilet tissues (including a subcategory of those that are intended to be dispersible in the toilet, i.e., they are designed to be flushable by disintegrating following disposal), toddler training wipes (some of which are designed to be flushable), toddler face cleaning wipes, cosmetic and facial wipes, hand and body cleansing wipes, as well as antibacterial wipes for personal use. Most applications tend to be for uses involving skin contact.

- *Household care*: includes all purpose cleaning wipes, bathroom cleaning wipes, disinfection wipes for domestic use, floor wipes, glass window and screen cleaning wipes, kitchen wipes, and spectacle cleaning wipes. Most applications tend to be for use on hard surfaces.

For professional, nondomestic settings, wipes can be divided into the following two categories:

- *Industrial wipes:* include those used by the automotive, electronics, computer, food service and processing, optical, printing and transport industries, as well as for janitorial and maintenance applications.
- *Medical and healthcare* wipes: include those used for surface disinfection and for the prevention of healthcare associated infections, patient care and surgical wipes.

Nonwoven wipes are intended to remove and retain solids or liquids from hard, soft, organic, or inorganic surfaces by exploiting frictional, shear, adhesive, liquid capillary, or electrostatic forces, as well as the porosity, solid surface area and chemistry of the nonwoven substrate. Liquid and particle removal efficiency and retention are key to ensuring wiping is efficient, as well as to preventing recontamination of cleaned surfaces as wiping continues. The chemical composition of the wipe substrate depends on the required performance requirements and price-point. Wipe substrates are composed of either hydrophobic or hydrophilic fibres, either alone or in blends depending on the intended purpose. Commonly used fibres include polypropylene, polyester, PLA, viscose, lyocell, cotton, wood pulp, and blends thereof. Increasingly, FSC- or PEFC-certified nonwovens are offered for wipes applications. Where high aqueous liquid absorbency, coupled with high wet strength is required, blending hydrophilic and hydrophobic fibres, or chemical bonding of a hydrophilic fibre web with a hydrophobic latex binder are common. Where electrostatic forces are required to enhance dry (dust) particle collection efficiency, an entirely hydrophobic fibre composition is preferable. Virtually all the main web formation and bonding technologies are used to make nonwoven wipes depending on the specific application.

2.5.1 Personal care wipes

Wipes for the personal care sector account for a large share of the entire nonwoven market in terms of volume, with baby wipes being the dominant application [2]. A diversity of additional personal care wipes has emerged over the last 30 years or more, including intimate care wipes, moist toilet tissues, toddler face cleaning wipes as well as those for cosmetic and facial wipes. Such wipes are either designed for cleaning biological material away from the body, or to apply or remove cosmetics or cleaning agents from the skin. As such, the wipes must possess sufficient strength to remain intact, as well as resist linting during use. The perceived softness and conformability of the nonwoven are also important, as are liquid management properties. Personal care wipes may be dry or premoistened with a lotion. The latter are impregnated with either aqueous, alcohol in water, or oil in water formulations prior to use, to aid cleaning or cosmetic treatment of the skin, e.g. by moisturising and the addition of fragrances. Dry personal care wipes such as make-up removal pads, are used in conjunction with separately packaged cleaning agents or cosmetic lotions. Cosmetic premoistened wet wipes for facial use are commonly used to remove impurities, sweat and make up from face. Other wipes for personal use, include feminine hygiene wipes, moist toilet tissues, hand wipes, and body wash cloths. Medicated wipes that are used to provide pain relief, reduce itching, or provide protection to a localised area are also available. In fact, the number of different products has proliferated.

Nonwoven wipes for personal care normally involve skin contact and therefore the aesthetics, softness, conformability, and biocompatibility are important. Cellulosic staple fibres such as viscose, lyocell, cotton and wood pulp are important raw materials, particularly where liquid handling is required. Baby wipes and facial wipes are most commonly made from long (>25 mm) regenerated cellulose fibres, to ensure reasonable wet strength, whereas for flushable premoistened wipes, blends of shorter wood pulp and regenerated cellulose fibres (<10 mm long) are selected to balance the competing requirements of sufficient wet strength during use and rapid dispersibility following disposal in the toilet. One of the biggest challenges in making dispersible premoistened wipes capable of meeting industry guidelines (GD4) for flushability, is ensuring they have enough wet strength to be converted, and resist breaking when removed from the packaging by the consumer. Carded, airlaid and wetlaid substrates predominate in the overall personal care sector, combined with hydroentanglement or through-air thermal bonding to deliver the softness that is normally required for a personal care wipe product. Hydroentangling also facilitates surface texturing, as well as apertures and embossed features to be introduced during manufacturing to aid product differentiation.

2.5.1.1 Flushable wipes

Flushable, or more precisely dispersible nonwoven wipes, are a subcategory of personal care and are subject to specific industry guidelines developed by EDANA and INDA in relation to performance and labelling. Most nonwoven wipes should not be flushed in the toilet, as highlighted by the 'Do Not Flush' symbol on wipes packaging. When the nonwoven is contaminated with human waste during its intended use, the toilet is arguably a convenient and safe disposal route, provided it does not create issues for the sewer system and water management companies, or negatively impact the environment. This means flushable wipes are specifically designed and engineered to meet strict guidelines and performance criteria. The first industry guidelines, known as GD1 for Assessing the Flushability of Nonwoven Products were launched in 2008. Progressive development of the guidelines and the associated test methods has taken place since that time, with the current GD4 guidelines being introduced in 2018. Engineering dispersible, premoistened wipe products with sufficient strength to remain intact during use, and which rapidly disperse during disposal in the toilet is not a trivial challenge, but it is essential to facilitate ease of transit through waste stream pipelines, pumps, filters, and the entire wastewater treatment system. The patent literature details a multitude of different approaches for the design and manufacture of flushable wipe substrates, including producing fabrics entirely from blends of wood pulp and short cut regenerated cellulosic fibres (4–12 mm) followed by a reversible bonding step. Examples include blends of 70:30 or 80:20 wood pulp:lyocell or viscose that are wetlaid or airlaid to produce the web, followed by hydroentangling, or chemical bonding with a water soluble binder. In the case of hydroentangled substrates, it is possible to create sufficient wet strength through fibre-fibre friction to enable the wipe to remain intact during use, but with fibre entanglement low enough to facilitate rapid disintegration during disposal. In the case of chemical bonding using an ion-sensitive water soluble binder, the binder solubility can be temporarily reduced by salts added to the lotion during the wet wipe converting process.

2.5.2 Household care wipes

This includes wipes for cleaning and dusting interior and exterior household surfaces, in for example kitchens and bathrooms, as well as windows, floors and household appliances, where the cleaning of stains, liquid spillages, dust, and other particulate material is required. Generally, the focus is on cleaning hard polymeric, glass, ceramic, or metal surfaces. Dry and wet multisurface floor sweeping, and mopping products are included in this category. Carded as well as spunbond nonwovens are most common in the household care wipes sector.

For cleaning in the presence of liquids, fabric wet strength is important. One approach involves drylaid web formation and blends of hydrophilic (e.g. viscose) and hydrophobic fibres (e.g. PP or PET), followed by thermal or chemical bonding. Another approach to balancing liquid absorbency and good wet strength is chemical bonding of drylaid webs containing hydrophilic fibres, using a hydrophobic binder. The hardness of the binder (which is related to its glass transition temperature, or T_g), as well as the inclusion of particulate additives, provide a means to modulate the stiffness and frictional properties of the resulting wipe. Although aqueous liquid holding capacity is limited, PP spunbonds are also used in some cleaning applications, where resistance to linting and low purchase price is required, or where removal of grease or oily materials is needed and an oleophilic material aids performance.

Cleaning efficiency, nonwoven substrate strength and durability, the ability to be combined with a variety of different household cleaning agents and the management of liquids are therefore important. Both dry and premoistened wipes are produced depending on the intended purpose, the latter being attractive in terms of convenience.

Some nonwoven wipes are designed to be washable and are therefore made in slightly higher basis weights, e.g. $>100 \, \text{g/m}^2$ to improve their durability and ability to be reused. Floor wipes include long-life nonwoven products designed to replace traditional mops, such as for example P&G's Swiffer and other floor wipes containing microfibres. In wipes intended for dusting, electrostatic forces can be exploited to remove dry particles from household surfaces, and therefore, there is advantage in harnessing hydrophobic PP and PET nonwoven substrates (for example) capable of generating electrostatically charged fibre surfaces. In another example, for cleaning glass lenses, synthetic microfibre nonwovens resist linting during the cleaning process, and enable high cleaning efficiency because of the high specific surface area of the fibres.

2.5.3 Industrial wipes

Wipes used in industrial environments serve a diversity of purposes, from mopping up oil, fuel, paint and other petrochemical-based spillages (in contrast to aqueous liquids), cleaning tools and work surfaces, through to removing small particles from metal body panels during automotive manufacture and ensuring a high-quality paint finish. Key industrial sectors are automotive, aerospace, manufacturing, printing, janitorial and food. Industrial wipes include products specifically designed to polish and reduce static electricity on surfaces, as well as clean them. A high resistance to particle shedding or linting from the wipe is also important, particularly in the precision cleaning of surfaces, machinery and tools to minimise residual contamination.

In the automotive sector, 'tack' cloths are used for removing dust and other small particles from metal surfaces after they have been sanded or smoothed, before they are painted to produce a smooth finish. These fabrics commonly consist of a hydroentangled nonwoven treated with a formulation containing a tackifier agent (and in some cases a pressure sensitive adhesive), which reduces slippage between the wipe and the surface to be cleaned, assisting with the removal and retention of small particles by the wipe. Another application is sealant wipes for removing excess or residual sealant from metal, glass, or composite parts. Nonwovens for this purpose include hydroentangled apertured substrates made of PET/viscose blends, amongst others. Solvent or spirit wipes are used by the automotive sector for cleaning automotive parts before painting or the application of graphics/transfers as well as to remove any unwanted underbody coatings or paint from the surface. Clearly, in this application, solvent stability is critical, and therefore, PP meltblown fabrics are amongst the nonwoven substrates used for this application, with basis weights $>100\,\mathrm{g/m^2}$ to ensure sufficient durability.

Similarly, in other manufacturing environments the ability to wipe up spillages, particularly, grease, oils, and other petrochemical-based materials is necessary, as well as for the cleaning or tools and equipment. In the aerospace sector, there is greater emphasis on precision-cleaning, which means high cleaning efficiency, combined with a high resistance to linting. Hydroentangled fabrics are particularly valuable in this regard when based on carded or spunlaid webs, because of their long fibre or filamentous and highly entangled fibrous structure, which yields good durability and resistance to disintegration during wiping. Freedom from chemical binders in hydroentangled substrates also helps to minimise potential for unwanted abrasion leading to scratches during precision cleaning operations. Resistance to linting is also critical in the cleaning of machine parts in the printing industry.

For janitorial applications, the objectives in terms of cleaning wet and dry contamination are similar to household care, but more robust, heavier basis weight nonwoven fabrics are commonly encountered. This includes nonwovens containing microfibres for use in wet and dry mops. Rotary floor cleaners, utilise relatively thick nonwoven pads that allow debris to collect inside the fabric, due to their porosity, and away from the surface being cleaned, allowing the pad to operate for longer. In one example, such fabrics are made by carding, cross-lapping and spray bonding, wherein the resin binder applied to the fabric can contain abrasive additives of carefully selected dimensions, to improve cleaning power. By modulating the properties of the chemical binder and additives, such fabrics can be used for cleaning, polishing, burnishing, or sealing the floor as well as other surfaces.

Additionally, in food service and food preparation environments, nonwoven wipes are widely used for the cleaning and disinfection of surfaces, with products ranging from general purpose absorbent fabrics, to heavier basis weight variants for increased durability. Fabrics made from microfibres can reduce smearing and improve bacterial removal efficiency. Again, plain and apertured, hydroentangled substrates are frequently encountered in this sector.

2.5.4 Medical wipes

Medical wipes include dry and premoistened products for healthcare surface cleaning (detergent effect) and disinfection (pathogen killing effect), patient body/skin disinfection and care (including bathing wash cloths and drying), antiseptic wipes, preinjection swabs, and continence care. Surfaces in healthcare environments can be a source of pathogenic microorganism transmission, due to body fluid or aerosol contamination, or through poor hand hygiene. Nonwoven wipes are therefore used to remove contamination from frequently touched surfaces beside other measures to reduce the risk of Healthcare Associated Infections (HCAIs) in clinical settings. HCAIs account for significant morbidity and mortality, with more than 37,000 deaths per annum in Europe, but 20%–30% are potentially preventable via appropriate hygiene and control programmes [13]. Similarly, disinfecting the skin prior to surgery potentially reduces the risk of surgical site infections (SSIs). Premoistened nonwoven wipes loaded with biocidal chemistry can therefore be effective and convenient means of cleaning surfaces through a dynamic physicochemical mechanism. The specific biocidal chemistry and concentration in this context is crucial to consider, given bactericidal formulations, particularly those based on alcohol are not necessarily sporicidal or viricidal. Basic formulations and concentrations for disinfection depend on the specific application and the surface to be cleaned. Formulations such as 70% alcohol (typically ethanol) or 70% IPA (isopropanol) are common for solid surface cleaning, whereas 2% chlorhexidine is common for skin wipes. An example of a sporicidal formulation for surface disinfection is GAMA Healthcare's Clinell wipe, which uses hydrogen peroxide and water-activated peracetic acid. Wipes containing peracetic acid are also used to disinfect medical devices, such as transesophageal echocardiogram (TEE) and endovaginal probes. Dynamic wiping removes significantly more bacteria than adpression for all bacteria and wipe types [14]. The surface energies of the wipe fibres, surface roughness of the fibre and the contact surface can all influence wiping efficiency, depending on the type of bacterium [15]. Following the COVID-19 pandemic, demand for antibacterial, viricidal, disinfecting and healthcare wipes, is likely to remain at an elevated level for several years, and it is estimated that the global nonwoven wipes market will grow at an annual growth rate (CAGR) of 5.8% by 2025, exceeding previous projections [3,16].

For single-use disinfection wipes, spunbond fabrics are attractive because of their relatively low cost. The majority of spunbonds for medical wipe applications have low basis weights ($<30\,g/m^2$) and are made of PP and so have limited absorbent capacity. Despite this, it is not unusual to find PP spunbond wipes also being used manage liquid spillages in healthcare settings due to their low cost, but to do this effectively multiple sheets may be required to compensate for the lack of water absorbency, which is obviously wasteful. Drylaid nonwovens containing viscose or lyocell are important for patient body/skin disinfection and care, including bathing wash cloths and drying. For continence care wipes, cleansing, moisturising, antibacterial, antipruritic, barrier protection, and deodorising functions are important, and maceratable wipe variants are also available, often based on heavyweight airlaid fabrics with high absorbent capacity.

2.5.5 Environmental considerations

Environmental regulation poses a significant challenge in the design and development of certain AHPs (specifically femcare products) and wet wipes, particularly in respect of the polymer materials that can be used to manufacture them. The EU's Single-Use Plastic Directive (SUPD) as set out in Directive (EU) 2019/904 (SUPD) is intended to reduce the impact of marine litter and is particularly challenging for the industry because the definition of plastics only exempts 'natural polymers that have not been chemically modified' inhibiting the use of synthetic man-made fibres, as well as some man-made biobased and recycled materials. Existing and potential future developments in regulation, including extended producer responsibility (EPR) will continue to necessitate innovation in the design of nonwoven products.

2.6 Food and beverage

The food and beverage industry uses nonwoven fabrics in the processing of ingredients during manufacturing, as well as for packaging final products (Fig. 2.5). Materials also have to meet food safety regulatory requirements, the specifics of which depend on country location, and nonwoven manufacturers produce specific food grade fabrics for this purpose.

Nonwoven filter media are important for safe and hygienic processing of ingredients during food manufacturing. For example, liquid filter media incorporate wetlaid or spunbond membrane supports for filtration of milk, soft drinks, and alcoholic beverages. For dry filtration, nonwoven filter media, often in the form of bag filters are used as part of drying, roasting and milling processes where the management of dust (composed of flour and other substances) is required. Given the inherent flammability of some food ingredients, electrically conductive fibres are often incorporated into such filter media, as well as being made from materials with good thermal stability. Bag filters typically consist of needlepunched fabrics with and without scrim reinforcement. In commercial catering and cooking environments, in addition to wound paper substrates, nonwovens can be used for hot edible frying oil filters to remove particles, typically in the range of 10–50 µm. The fabrics are usually based on wetlaid nonwovens made of wood pulp or blended with regenerated cellulose fibres.

Food and drinks packaging is another major application. Appropriate selection of nonwoven fabrics used in food packaging can increase the distances over which perishable food and drinks can be transported as well as benefit shelf-life and storage times. Absorbent nonwoven pads, or 'soaker' pads, based on airlaid fabrics containing cellulose fibre/pulp or superabsorbent polymers (SAPs) are found in perishable meat, seafood and other refrigerated food packaging, principally to prevent the build-up of free liquid in food trays. In so doing, bacterial growth or changes in appearance or odour can be significantly delayed. While the absorbent capacity is provided by the airlaid fabric, laminated polyolefin–based spunbonds

FIG. 2.5   Summary of the major food and beverage markets [5].

or moisture-permeable films may be used as contact layers. SAP can also be laminated between two spunbonds to form a pouch, which provides the same function. Depending on the absorbent capacity required, the weight of food pads can reach >600 g/m^2. Heat-formable single-use food packaging trays and containers are also produced from spunbond substrates, sometimes in combination with films.

Perhaps one of the most established applications of nonwovens is in the production of teabags. Many are based on cellulosic wetlaid fabrics of about 12–18 g/m^2, traditionally containing a proportion of thermoplastic polymer such as PP or PLA to facilitate ultrasonic or heat sealing. Tea bags made of compostable PLA-based spunbonds are also available, compliant to EN13432. Wetlaid and spunbond nonwovens are used as filter media in vending machines for hot and cold drinks, as well as in the associated pouches or sachets, to package dry drink formulations prior to use. Another growing application for nonwovens is in biodegradable coffee pods and other formats for drinks dispensers.

2.7 Household and home furnishings

In addition to wipes, nonwovens are made into numerous single-use products for domestic applications including scouring pads, tablecloths, placemats, and napkins, but there are numerous long-life household products and home furnishings containing nonwovens, including floorcoverings, floor mats and wallcoverings, window blinds, mattresses and furniture components, including certain types of upholstery (Fig. 2.6).

Nonwoven carpet underlays are made in various constructions, but one of the most common is based on fabrics composed of mechanically recycled textile waste, produced by carding (or garnetting), cross-lapping and needling. In the manufacture of tufted carpets, spunbond carpet backings play a critical role in providing the support for the pile yarn, and contribute to the overall dimensional stability. Needlepunched nonwovens are used to make contract carpeting, floor tiles and mats, with the majority being based on carded, cross-lapped and needled fabrics (plain and structured) containing high denier (~17 dtex)

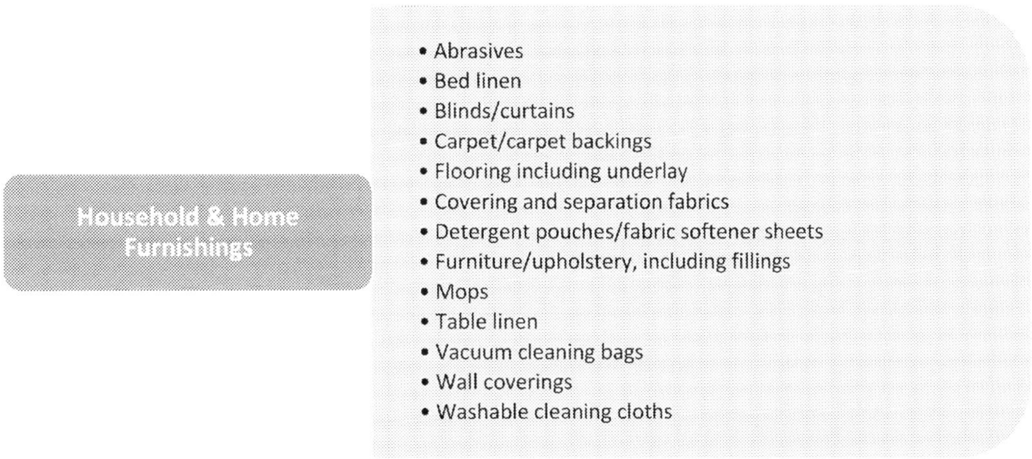

Household & Home Furnishings

- Abrasives
- Bed linen
- Blinds/curtains
- Carpet/carpet backings
- Flooring including underlay
- Covering and separation fabrics
- Detergent pouches/fabric softener sheets
- Furniture/upholstery, including fillings
- Mops
- Table linen
- Vacuum cleaning bags
- Wall coverings
- Washable cleaning cloths

FIG. 2.6   Summary of the major household and home furnishing markets [5].

dope dyed PP or PET fibres. Once made, the needled fabric is foam latex backed to improve dimensional stability and durability in use.

Wallcoverings are traditionally made of paper, but since the 1980s wetlaid nonwoven wallcoverings have progressively gained market share, particularly in Europe. Their big advantage is substantially higher wet strength compared to paper. As a result, nonwoven wallcoverings are more easily manipulated when wet during decorating, as well as more easily removed from the wall compared to paper-based substrates without tearing. It is also possible to 'paste the wall' rather than having to load the wallcovering with paste before application, making decorating more convenient. Most nonwoven wallcoverings are based on wetlaid fabrics containing cellulosic and thermoplastic synthetic fibre. Depending on the thermoplastic fibre content, attractive embossed effects created by calendering can obviate the need for traditional vinyl coating when used as facings. In addition to paintable or printable facings, nonwovens are also important as backings for flat or expanded vinyl wallcoverings, substantially increasing the ease of removal from walls without tearing. Owing to the enhanced opacity, increasing the basis weight of the nonwoven enables the manufacture of wall-linings to cover small imperfections such as uneven surfaces and cracks. A further growing market is in customised wallcoverings using the nonwoven as a substrate for inkjet printing.

Their relatively light-weight, ability to be cut and formed into pleats without fraying, has enabled nonwovens to build market share in the manufacture of window blinds, particularly for cellular and pleated blinds, which exploit the inherent stiffness of the fabrics. Various nonwoven substrates are used, either alone or in laminates as a backing support, with most based on 100% PET, including PET spunbonds as well as laminates containing drylaid and wetlaid fabrics with textile facings. Hydroentangled PET/PA microfibre nonwovens also find applications in curtains and roller blinds.

Traditionally, the construction of sprung mattresses uses heavy-weight nonwoven fabrics for the high-density insulator pad (or spring wrap) that covers the spring unit, and for the comfort layers that are placed above, beneath the outer ticking. Insulator pads include, but are not limited to, drylaid-needled fabrics containing mechanically recycled mixed fibres, including a proportion of wool to meet flame retardancy requirements. This is one of the oldest markets for postconsumer waste clothing in the United Kingdom. By contrast, comfort layers make greater use of virgin fibre and generally comprise of voluminous airlaid or carded webs followed by through-air thermal bonding or needling. Depending on the price-point of the mattress, drylaid nonwoven comfort layers are made of PET, wool, hemp and even cashmere, as well as various blends. Nonwovens are also used as mattress protectors. Some outer tickings in mattresses are made from stitch-bonded nonwovens, instead of conventional woven fabrics. Flame retardancy is a key issue in mattresses and FR-treated nonwoven barrier fabrics may need to be incorporated (e.g. based on stitchbonds) depending on the mattress design and prevailing regulatory requirements. In other forms of furniture, spunbond fabrics are frequently encountered in the linings of cushions, quilt backings, cushion ticking and pillowcases.

2.8 Clothing and footwear

Nonwovens have important applications in the manufacture of long-life clothing, shoes and clothing accessories, as well as in single-use clothing, some of which is categorised as personal protective equipment (PPE), as summarised in Fig. 2.7.

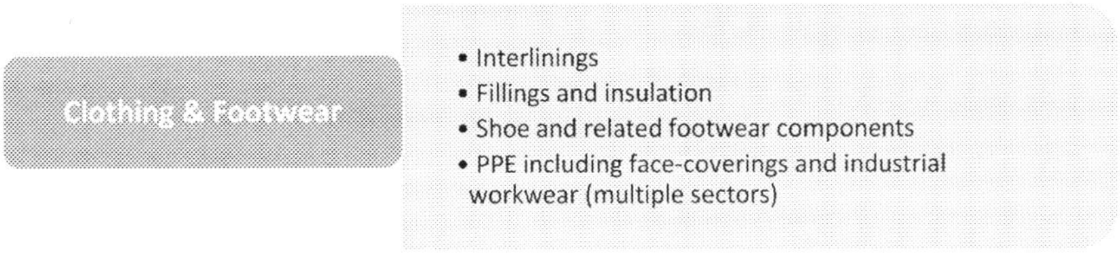

Clothing & Footwear

- Interlinings
- Fillings and insulation
- Shoe and related footwear components
- PPE including face-coverings and industrial workwear (multiple sectors)

FIG. 2.7   Summary of the major clothing and footwear markets [5].

Interlinings (interfacings) for long-life outerwear garments have been made from nonwovens since the 1950s, and it remains an important market. Both heat fusible and sew-in interlining variants are available. Generally, their function is to maintain the shape, fit and appearance of the garment, and so not surprisingly they are made in a variety of fabric constructions, basis weights, and compositions to provide the necessary mechanical properties and durability to washing and wearing, when in situ. Some are also designed to be water repellent. Nonwoven interlinings are mostly based on drylaid thermally bonded, drylaid chemically bonded and drylaid hydroentangled fabrics, as well as wetlaid chemically bonded, wetlaid thermally bonded and spunbond fabrics. PET and PA are most common in drylaid nonwoven interlinings, whereas wetlaids usually consist of wood pulp or viscose. Use of wetlaids is mainly focused on supports for embroidery backing, whereas the strength of spunbonds is useful in applications such as belt loop supports. In workwear and protective clothing, nonwoven linings and drop linings are incorporated to provide thermal barrier or other protective functions. Flame-retardant hydroentangled aramid fabrics for fire-fighter's jacket linings is just one amongst many examples.

Alcantara, Ultrasuede and similar synthetic leathers, are amongst the few nonwoven fabrics to have been accepted as face fabrics in long-life outerwear, in place of traditional woven and knitted textiles, due to their highly attractive physical and aesthetic properties. As an alternative to natural leather and suede, full garments, trims, accessories, and footwear for the fashion and sportswear industry are the primary applications. Traditionally, suede-like, synthetic leather fabrics are produced from island-in-the-sea bicomponent fibres using drylaid web formation, needlepunching, a dissolution step to remove the sea component and polyurethane impregnation. Approximate basis weights range from 30 to 350 g/m^2 and they are available in many different colourways. Other microfibre and microfilament nonwoven fabrics (the latter based on spunlaid-hydroentangled segmented pie bicomponent filaments) are used in long-life clothing as leather-replacements or 'micro-suedes'. High-loft nonwoven fillings or waddings are important for insulation in outdoor performance garments and sportswear, particularly in jackets, gloves, and sportswear. The fabrics typically comprise low-density drylaid staple fibre fabrics, meltblown fabrics, or combinations thereof. Examples include 3M's Thinsulate thermal insulation fabrics. Also, for outdoor clothing, nanofibre nonwovens are also beginning to penetrate the non-PFC (per- and poly-fluorinated compounds) waterproof, windproof and breathable (WWB) membrane market, replacing e-PTFE, where high water resistance is required, including after washing. One example is the XPore Brand F membrane, which is based on electrospun polyurethane (PU).

A related area to clothing is in the manufacture of long-life footwear (including workwear), where nonwoven fabrics are found in components such as, shoe linings (including Strobel types), counter linings, vamp linings, heel grips, and eyelet reinforcement. Given the variety of performance requirements demanded to enable shoe manufacture as well as performance during use, it is not surprising to find a large variety of different nonwoven constructions being used to modulate physical properties. In addition to fibre composition, methods of non-woven manufacture and dimensional properties, the addition of latex binders and chemical finishes such as antimicrobial agents are common. For example, latex-binder impregnation of carded, cross-lapped and needled fabrics enables modulation of stiffness and durability, as well as a vehicle for adding pigments or functional chemistry. Artificial (synthetic or 'faux') leathers used for the outer construction of shoes are made from nonwoven coating substrates impregnated or coated with for example, PU or PVC. Nonwovens are frequently used as coating substrates for film coatings, which can find use in clothing accessories, including in hand-bags and other leather goods.

Nonwoven fabrics are also widely used for single-use protective clothing, where the primary purpose is safety, either protecting users, or preventing contamination of sensitive working environments. This includes regulated products categorised as personal protective equipment (PPE). Such products are important in healthcare environments where there is a risk of cross-infection, as well as working environments where users need to be protected against exposure to respirable particles or fine dusts, aerosols, gases, liquids, biological and chemical agents. Sensitive working environments include clean rooms, food preparation, laboratories and in the electronics industry to protect components from contamination. The design, manufacturing and marketing of PPE is regulated and governed by international standards, e.g. European Standards (designated as EN). PPE is required to meet strict performance requirements and is subject to obligatory testing and specifications to ensure safety in use. The level of protection that the fabric is required to deliver is determined by the type of hazard or hazardous substances. The challenge in making nonwovens for PPE is to find a balance between the protective performance, which is defined in the associated test standards, and thermophysical comfort. The growth of single-use PPE made of nonwovens compared to woven alternatives is partly due to the lower purchase cost as well as the convenience of not having to handle contaminated items for laundering and reuse. Nonwoven single-use protective clothing is primarily based on PP or PE spunbonds, meltblowns or flash-spun fabrics, but hydroentangled substrates are also important in some markets where value is placed on enhanced softness and comfort.

2.9 Building and construction

Architectural practice, selection of building materials, construction methods as well as building regulations vary geographically, and therefore, consumption of nonwovens for different applications varies somewhat from country to country (Fig. 2.8).

Particularly in Europe, building regulations as well as governmental policies aimed at addressing climate change and reducing emissions are driving demands for more thermally efficient buildings and increased demand for nonwoven insulation. Nonwoven insulation relates to both thermal and acoustic insulation made of inorganic (e.g. mineral wool) and organic materials, including naturally derived (e.g. sheep's wool) or mechanically recycled fibre materials (e.g. recycled cotton or denim). In roofing, wetlaid glass fibre fabrics or needled heavy-weight PET spunbonds are used as supports to reinforce bituminous membranes

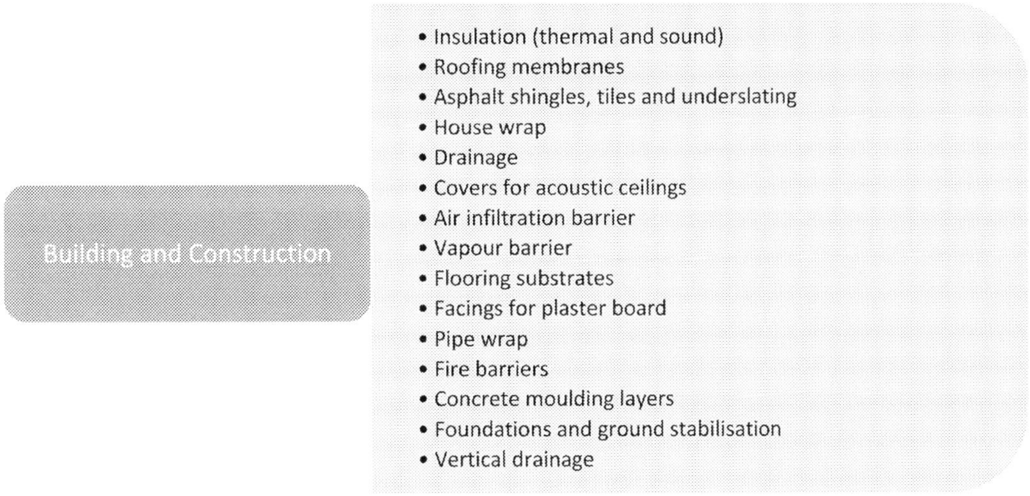

- Insulation (thermal and sound)
- Roofing membranes
- Asphalt shingles, tiles and underslating
- House wrap
- Drainage
- Covers for acoustic ceilings
- Air infiltration barrier
- Vapour barrier
- Flooring substrates
- Facings for plaster board
- Pipe wrap
- Fire barriers
- Concrete moulding layers
- Foundations and ground stabilisation
- Vertical drainage

Building and Construction

FIG. 2.8 Summary of major building and construction markets [5].

for sealing and waterproofing flat roofs. Similarly, glass fibre nonwovens are used ensure adequate tear strength, stiffness, and dimensional stability in asphalt roofing shingles. Bituminous membranes are also used to seal the rooves of tunnels.

In addition to traditional bitumen-backed roofing felts and underlay, reinforced by glass fibre, nonwoven bitumen-free vapour-permeable (breathable) roofing membranes are fitted under pitched roofs below the tiles. Many of these 'breather membranes' are based on PP or PE spunbond or PE flash-spun fabrics, frequently laminated to one or more other layers, e.g. meltblowns or films to modulate air and moisture vapour permeability. Facilitating air and vapour permeation through the roof space serves to prevent interstitial condensation. Being hydrophobic and relatively dense, such membranes also provide additional protection against the penetration of wind and rain into the building. By appropriate selection of the air permeability of the spunbond laminate, and/or by laminating with reflective aluminium foils, heat transfer in the building can also be appropriately managed. Similarly, moisture vapour–permeable spunbond and flash-spun (i.e. Tyvek) fabrics are used in housewrap applications (roof and wall) both in timber frame wall systems and in steel frame and concrete wall construction. Such fabrics may also be installed in suspended timber floors to prevent condensation.

Nonwoven insulation in the context of buildings refers to both thermal and acoustic insulation products, available in sheet form. Rock (mineral or stone) wool nonwoven fabrics comprise calcium–magnesium–aluminium silicate fibres directly extruded into the form of thick batts. Glass can be converted into batts using a similar melt-extrusion approach. Rock and glass-based insulation fabrics have inherently high thermal resistance compared to organic insulation materials. Drylaid alternatives include batts comprising of PET, wood fibre, hemp, wool, or FR-treated recycled cotton (e.g. mechanically recycled denim) made by carding and cross-lapping, or more usually air-laying to make the batt. A proportion of PET bicomponent fibre (up to ~25% by weight) is commonly added to natural fibre blends to enable consolidation through-air thermal bonding and production fabrics with appropriate compression-recovery behaviour. In acoustic nonwoven insulation, one of the biggest challenges is achieving adequate sound absorption without recourse to thick fabrics as there may be insufficient volumetric space to accommodate it, e.g. in acoustic ceiling panels. One example of a nonwoven used in suspended perforated metal ceilings to improve sound absorption is Freudenberg's 0.27 mm thick, 63 g/m^2 Soundtex product for airport interiors. Ceilings, wall panels, acoustic screens and baffles containing nonwoven fabrics, in conjunction with other components, are all intended to manage unwanted sound transmission or reflectance within the interior space. Nonwovens used in these applications are commonly drylaid and needled or thermally bonded FR-treated PET fabrics in basis weight ranging from about 75 to 500 g/m^2, depending on the required absorption coefficient [17]. Similarly, fabrics can be placed between concrete slabs and cast concrete floors during building construction to reduce noise transmission. Low frequency sound absorption continues to be most challenging for nonwoven media, unless used in combination with other materials. Incorporating multiple layers of electrospun nanofibre webs can substantially increase sound absorption without greatly increasing the thickness or basis weight of the underlying substrate. Nonwoven sound absorption can further be enhanced by impregnation with other porous materials such as activated carbon or metal–organic frameworks (MOFs) [18].

Related to general building and construction applications, a further use for nonwovens is in cured-in-place pipe (CIPP) sleeves to repair the inside walls of underground pipes, instead of having to dig up and replace large sections. The CIPP process involves inserting an uncured resin-impregnated fabric sleeve into the pipe, inflating it so that it conforms to the walls of the damaged pipe and finally, curing the resin to provide a new internal pipe surface.

2.10 Geosynthetics

Geosynthetics include geotextiles (woven and nonwoven), geogrids, geomembranes, and geocomposites. Nonwoven geosynthetics are important in a diverse range of applications. The civil engineering industry uses nonwovens in the building of roads, pavements, car parks, railways, harbours, sport fields, landfill sites, coastal and river defences, green roofs, and urban drainage systems (Fig. 2.9).

Root barriers that prevent damage to buildings by invasive plants is another example. Geosynthetics fulfil six main functions, specifically:

– Separation: to prevent the mixing of sand, soil and aggregate, while still permitting permeation of liquids and gases.
– Filtration: to retain soil in position, as liquids or gases pass through it.
– Drainage: to enable in-plane transmission of liquids and gases.
– Protection: to prevent physical damage to another layer positioned above or below.
– Reinforcement: to strengthen soil structures, which may be inclined or unstable.
– Containment: to provide an impermeable layer between different materials.

Nonwoven geosynthetics are predominantly made from PP carded, cross-lapped and needled fabrics, or heavy-weight PP spunbonds, with some PET-based products also being available. Drylaid processing is also leveraged to produce biobased alternatives, mainly from bast fibres. Nonwovens are also combined with nontextile materials including clays,

- Road and rail building
- Dam
- Canal and pond lining
- Hydraulic works
- Sewer lines
- Soil stabilisation and reinforcement
- Soil separation
- Drainage
- Landfill membrane protection
- Filtration
- Sedimentation and erosion control
- Weed control
- Root barriers
- Sport surfaces
- Asphalt overlay
- Impregnation base
- Drainage channel liners

Geosynthetics

FIG. 2.9 Summary of nonwoven geosynthetics markets [5].

thermoformed sheets and membranes, grids, nets and films, depending on the intended purpose to form geocomposites. One example is a geocuspate consisting of a vacuum formed 3D plastic sheet with projecting studs or cups, which when adhered to the nonwoven provides internal void space for the free passage of liquids or gases. Similarly, geosynthetic clay liners (GCLs) combine one or more layers of nonwoven with sodium bentonite clay particles, the latter conferring a valuable barrier function when the liner is exposed to water during use, because of the swelling of the clay. The inherent chemical resistance of PP renders it particularly valuable as a raw material in geosynthetics, and to improve ultraviolet light (UV) resistance, staple fibre PP for drylaid-needled geosynthetic fabrics is doped with ~1% carbon black.

Needled geosynthetics typically range from 3.5 to 17 mm in thickness with the primary applications being soil filtration, including filtration and separation in the prevention of coastal erosion and membrane liner protection for landfill sites. When bonded to a cuspate, the resulting geocomposite is suitable for horizontal drainage and green roof installations as well as for promoting gas venting. Owing to their lighter-weight and thickness, spunbonds are mainly directed at separation, stabilisation, and filtration functions, unless incorporated into geocomposites.

2.11 Automotive and transportation

Automotive represents a major industrial sector for the application for nonwovens [2] (Fig. 2.10).

Nonwovens are widely used in the automotive sector, both to assist with the manufacturing of vehicles, as well as to form parts of the finished vehicle itself. For example, in a typical

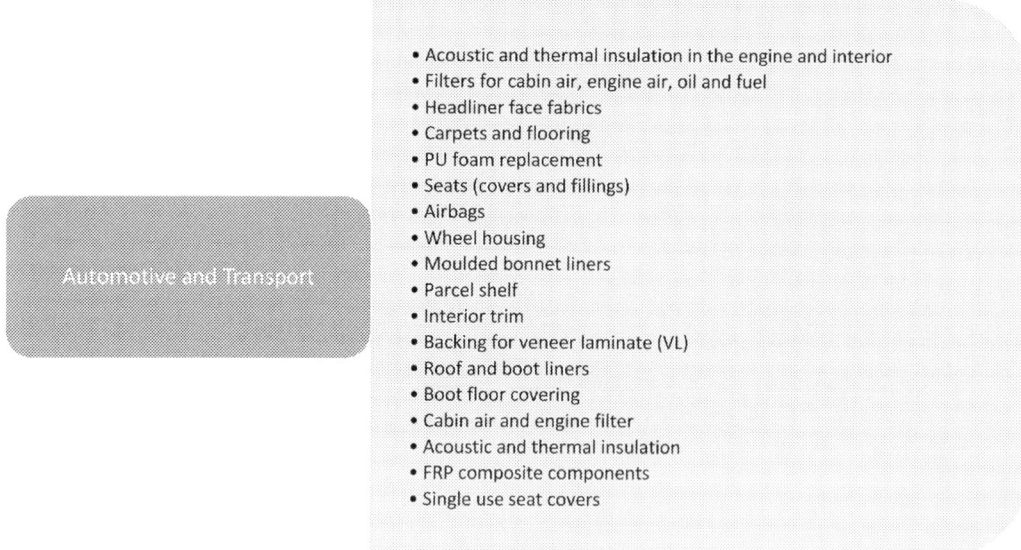

FIG. 2.10   Summary of automotive and transportation markets [5].

car, more than 40 component parts are made of nonwovens, accounting for more than 11% of the total fabrics used in the vehicle [19,20]. Major applications for nonwovens include (i) the interior trim, e.g. the boot liner, interior carpets, headliner, and parcel shelf, depending on the category of vehicle, (ii) management of noise, vibration, and harshness (NVH), (iii) air and liquid filtration, e.g. cabin air and engine filters, (iv) thermal insulation, which is often combined with NVH management and increasingly, (v) moulded internal parts and outer bodywork composites. Key drivers that have accelerated utilisation of non-wovens are vehicle weight reduction to improve fuel economy and gas emissions. According to a case study by Denkstatt, an average passenger car using nonwovens in all possible applications has the potential to save 55 kg CO_2 equivalents over its lifetime and generate ~30% less impact on the environment [21]. The progressive shift to electric vehicles means weight-saving continues to be of major importance, as a means to improve performance and vehicular range.

For tufted carpets in automotive applications, spunbond primary and secondary backings enable thermoforming or deep draw moulding of automotive carpets, while maintaining pile stability. For example, spunlaid nonwovens made of core-sheath bicomponent spunlaid filaments (PET/PA or PET/PP) are used extensively for this purpose. For carpets with a non-woven rather than a tufted facing, as well as interior trim such as headliners and boot liners, carded, cross-lapped and needled dope dyed staple fibre made of PET, PP, or PA predominate. Finished parts can comprise of a needlepunched face layer fabric laminated to a needlepunched thermoformable backing, e.g. for parcel shelves a PET facing, and PP backing is common but there are a number of different variations. PET spunbonds and glass wetlaid fabrics are used as carriers, or lamination support layers, in multilayer headliners that are faced with a range of other materials, e.g. knitted tricot fabrics. Nonwoven-based synthetic leathers are also used as automotive interior trim, often at a relatively high price premium.

Leading manufacturers in this area include, but are not limited to Alcantara (Italy), Dinamica Miko (Italy), Clarino by Kuraray (Japan) and E-Leather (United Kingdom) [22]. In E-Leather technology, mechanically recycled shredded waste leather shavings are used as part of a blend to make the nonwoven substrate. Webs are airlaid and consolidated by hydroentangling, eliminating the need for adhesives, and products are claimed to be 40% lighter than conventional leather [23].

Carded, cross-lapped and needled fabrics made of PET are heat moulded to make a variety of thermally stable NVH management products, including encapsulations for engines and electric motors (often in combination with foam) capable of operating up to 150–200°C, bonnet (or hood) liners, outer dashes, engine top covers, and outer boot floor insulators. Such products manage both noise and heat proximal to the source, e.g. around the engine bay. Inside the vehicle, beyond the carpets, needled fabrics extend to the floor mats, inner dash, inner wheel arch insulators, and inner trunk floor insulators. Because the thermal resistance requirements are lower than in the engine bay, needlepunched fabrics for inner dashes and floor insulators are available made of 50%–75% mechanically recycled cotton blended with thermoplastic fibre to render the fabric thermoformable. Fibre reinforced plastic (FRP) composites are another important area, particularly for structural parts in automotive interiors, based on for example, needled fabrics made of glass or bast fibres, blended with PP, to provide thermoforming capability. Other composites are made from needled or stitchbonded fabrics containing virgin or recycled carbon fibre.

2.12 Filtration

Nonwoven filter fabrics account for one of the largest markets amongst the industrial applications for nonwovens, and continue to be one of the most diverse in terms of the range of different products available (Fig. 2.11).

Nonwoven filter media are extensively used in liquid, air, and gas filtration in both domestic and industrial sectors owing to their ability to meet key requirements such as energy-efficiency, cost-effectiveness, and ease of scalable fabrication [24]. The global liquid filtration market is expected to grow at a CAGR of 12.89% during the forecast period of 2017–26 [25], while the filtration and air purification market is expected to grow at a CAGR of 8.2% over the period 2020–25 owing to factors such as growing health consciousness amongst consumers and concerns about airborne disease transmission [26]. Regulation, together with stricter environmental, health and safety standards, also have a major role in driving increased consumption of filter media across multiple market segments.

Nonwoven media are found in surface, depth, cake, and crossflow filtration applications. Drylaid, wetlaid, spunbond, meltblown, and nanofibre nonwovens are used for different specific filtration purposes, sometimes in multilayered combinations. Of course, fabric selection must consider the dimensions and properties of the materials to be filtered, flow rate requirements and the intended filtration mechanism. Depending on the particle or droplet size and type of carrier fluid (liquid or air/gas), the principal filtration mechanisms are straining, direct impaction, interception, diffusion, and electrostatic attraction. The latter, which relates to nonwoven electret filters, requires electrostatic charging of the fibres or additives in the fabric (usually by triboelectric charging, hydrocharging, charged particle bombardment, or corona discharge techniques).

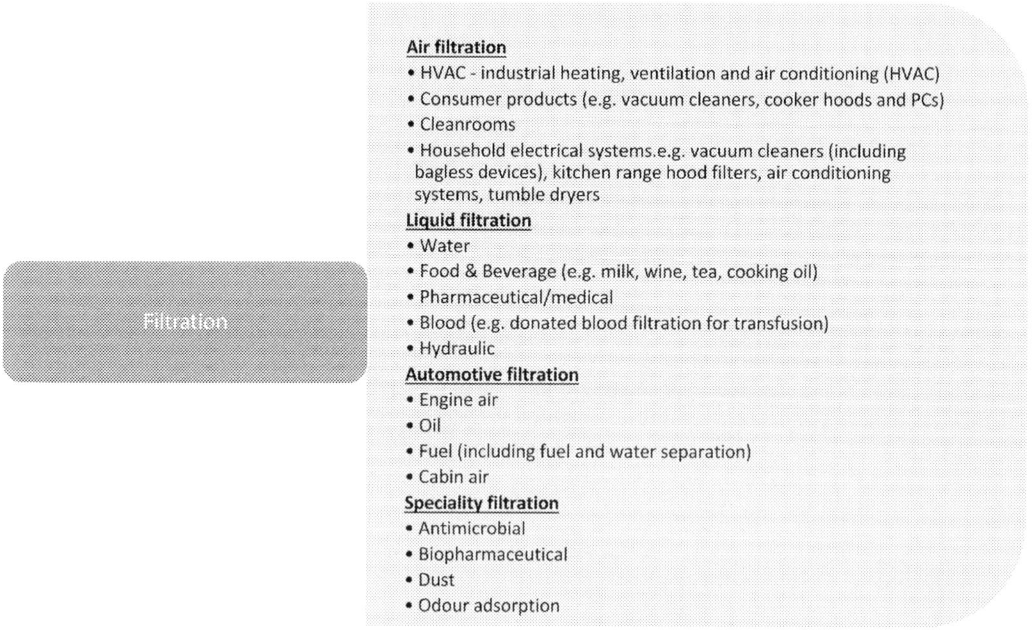

Air filtration
• HVAC - industrial heating, ventilation and air conditioning (HVAC)
• Consumer products (e.g. vacuum cleaners, cooker hoods and PCs)
• Cleanrooms
• Household electrical systems.e.g. vacuum cleaners (including
  bagless devices), kitchen range hood filters, air conditioning
  systems, tumble dryers

Liquid filtration
• Water
• Food & Beverage (e.g. milk, wine, tea, cooking oil)
• Pharmaceutical/medical
• Blood (e.g. donated blood filtration for transfusion)
• Hydraulic

Automotive filtration
• Engine air
• Oil
• Fuel (including fuel and water separation)
• Cabin air

Speciality filtration
• Antimicrobial
• Biopharmaceutical
• Dust
• Odour adsorption

Filtration

FIG. 2.11   Summary of major filtration markets [5].

Filtration efficiency and pressure drop (relating to flow resistance) are major concerns in most applications, and satisfactory performance in both usually requires a trade-off. Consequently, the porosity (or solidity), pore-size distribution, permeability, thickness, and specific surface area of the constituent fibres (related to fibre diameter) in the nonwoven are important considerations. To a large extent these factors dictate the choice of nonwoven process used to make the fabric, as each process technology results in a different structural architecture and has limitations in terms of the distribution of fibre diameters that can be present in resultant fabrics. For example, in general terms, drylaid nonwovens being relatively thick and highly porous ($P > 80\%$–99%) provide high particle holding capacity through depth filtration, as well as a low pressure drop, but there are limitations in terms of delivering narrow pore-size distributions, as well as incorporating microfibres below a certain diameter. In contrast, wetlaid nonwovens are much thinner and less porous, but pore-size distribution and basis weight are less variable and inclusion of microfibres, including inorganics is less problematic. Meltblown nonwoven filter media are widely used for fine particulate matter (PM) filtration which has been a substantial challenge for public health, climate and ecosystems in the past few years. The benefits of meltblown nonwovens as air filters include their high surface area per unit weight, high porosity, relatively narrow pore-size distribution and high filter efficiency. Nanofibre nonwovens provide a large surface area for particle capture, but thick nanoscale fibre layers can lead to high flow resistance and pressure drop [27].

In nonwoven filter media, filtration efficiency, and pressure drop are influenced by fibre diameter, porosity, pore size, tortuosity, fabric thickness, as well as other factors relating to the engineering design of the fabric. Drylaid, wetlaid, spunbond, meltblown, and nanofibre nonwovens all have characteristic differences in their structure and physical properties that govern their suitability as filter media for different applications. Partly for this reason, and to enhance performance in use, filter media are also combined in layers, including as 'supports', or separators. One example is the use of bicomponent spunbonds to confer mechanical support and enable pleating of delicate wetlaid, meltblown, nanofibre, and membrane filter media.

A wide range of commodity and high-performance polymers, as well as inorganic materials (notably glass and metal) are used as raw materials for nonwoven filter media, depending on the required chemical, thermal, and biological stability. Common synthetic polymers used in nonwoven filter media include PP, PA, PET, and PBT [28]. Higher performance materials such as PPS and aramid find applications in cartridge and bag filters where greater chemical and thermal resistance is required, usually in the form of drylaid-needled fabrics. Additives, surface coatings and chemical treatments are also important in some applications to modulate the properties of the component fibres, without recourse to changing the fibre itself, thereby achieving higher performance without prohibitively increasing cost. One example is in the manufacture of nonwoven coalescing filters to increase the removal efficiency of small water droplets from oil or fuel by increasing the hydrophilicity of PBT meltblown fibre surfaces. A wide variety of other materials are incorporated into filter fabrics, either to improve filter efficiency or provide additional functions. For example, in pleated meltblown and spunbond cabin air filter media, an activated carbon layer is provided to adsorb potentially harmful gases, e.g. VOCs and reduce odour in the vehicle interior. Another example relates to the challenge of maintaining charge stability in electret filters, where the addition of dielectric materials such as barium titanate [29] or magnesium stearate (MgSt) [30] in meltblowing enable maintenance of filtration efficiencies above 99% coupled with a low pressure drop.

2.13 Agriculture and horticulture

In these nonwoven applications the primary objectives are the protection, irrigation and general cultivation of plants (including fruit and vegetables), both in commercial and domestic settings (Fig. 2.12).

Protection in this context refers to preventing damage by invertebrates, particularly flying and crawling insects, and adverse environmental factors such as frost, hail, strong winds, UV light, elevated temperatures (particularly in greenhouses) and the control of weeds and roots.

Agriculture and Horticulture

- Crop cover
- Plant protection and mulching
- Seed blankets
- Weed control fabrics
- Greenhouse shading
- Root control bags
- Biodegradable plant pots
- Capillary matting
- Landscape fabric

FIG. 2.12 Summary of agriculture and horticulture markets [5].

Management of the microclimate around plants, particularly in terms of moisture vapour pressure (humidity) and temperature are important. Promoting cultivation using nonwovens improves yields and can also facilitate organic growing conditions, by reducing dependency on pesticides. Nonwovens are also used to protect and package valuable crops and other harvested materials as well as prevent crops from being damaged during storage.

Amongst the various applications, crop cover is one of the most well-established and generally relies on UV-stabilised PP spunbond fabrics in the range 15–50 g/m² depending on the strength of the fabric required. Of course, such fabrics are inherently porous, air and moisture vapour permeable.

Mulch or 'mulching' fabrics are intended to suppress weeds, control evaporative moisture loss and temperature in the soil around plants, as well as prevent damage due to invertebrates such as molluscs, e.g. snails and slugs, as well as insects. Again, PP spunbonds are used for this purpose (often mass pigmented brown, green, or black), but to enable aggressive weed suppression higher basis weights (compared to crop cover) up to about 100–150 g/m² may be required. Commercially, weed control mats are important for fruit and vegetable cultivation, particularly around the stems of plants, e.g. for newly planted figs, grapevines, and tomato crops. Degradable mulch fabrics produced from drylaid needlepunched fabrics made of natural fibres such as coir, wool, hemp, or jute are also important. Some incorporate film or scrim backings to increase dimensional stability. In commercial greenhouses, particularly in very warm climates, nonwovens have been used as evaporative cooling media instead of paper, enabling internal greenhouse temperature and humidity to be modulated by flowing ambient air over prewetted fabric. Such growing environments also enable valuable food crops to be grown in normally unsuitable geographic and climatic regions. Also, in greenhouse settings, PET spunbonds are widely applied as screens and curtains for UV protection as well as to prevent insect ingress.

In horticulture, capillary mattings and hanging basket linings are intended to spread and distribute water to the roots of plants by exploiting capillary pressure within the fabric's pore structure. Such products are made from both synthetic (e.g. PP), natural fibres and mechanically recycled waste fibre, and the majority are carded, cross-lapped and needled fabrics. Spunbond backings improve dimensional stability, where required. Nonwoven fabrics are also important as porous plant growing media for soilless and hydroponic systems, made from mineral wool (e.g. rockwool), coir and recycled fibre. Drylaid and spunlaid nonwoven pollination control bags can be used as alternatives for perforated films and paper-based substrates for plant breeding to control self and cross-pollination, as required. In another application, nonwoven biodegradable seed planting or propagation bags are filled with soil or compost and allow seedlings to be conveniently contained as well as replanted while allowing roots to grow through the fabric.

2.14 Healthcare and medical

Applications for nonwovens in healthcare include both regulated medical devices and nonregulated products. As in many other sectors, there is a diverse range of applications, encompassing single-use, multiple-use, and long-life products.

An important role of nonwovens in healthcare is to assist with the prevention and control of cross-contamination and infection in medical or surgical environments, specifically for preventing healthcare associated infections (HAIs), including surgical site infections (SSIs). According to the World Health Organisation (WHO), ca. 10% of patients per annum acquire an infection while receiving care and HAIs affect hundreds of millions of patients and occupational staff worldwide. New pathogens and multidrug-resistant strains of bacteria and viruses means effective infection prevention and control is more important than ever, but so too is compliance in the appropriate use of infection control products by users.

Barrier fabrics capable of preventing the transmission of liquids such as blood or other fluids containing potentially pathogenic species are designed to protect clinicians, patients as well as proximal surfaces from cross-contamination. This is particularly important in critical care areas, surgical preparation, and operating theatres, where surgical clothing, sheets or covers as well as bedding and drapes can provide a protective function. Compared to traditional woven cotton products such as gowns that are designed to be laundered and reused, spunbond and meltblown nonwovens are more resistant to linting, and being single-use, obviate the risk of progressively reduced barrier function due to repeated laundering and use. Developments in this market continue to be driven by strict regulation in terms of performance requirements as well as the cost to healthcare providers [31].

2.14.1 Gowns

In Europe, garments that are claimed to protect a user from hazards are regarded as personal protective equipment (PPE) and are therefore required to meet strict performance requirements. Important standards relating to regulated garments such as surgical gowns, drapes, and clean air suits include EN 13795:2011 + A1:2013, EN 14126:2003 and (EU) 2016/425). In the United States, surgical gowns are regulated by the FDA, normally as Class II medical devices, with four classification levels (1–4), according to the degree of critical risk, as indicated in Fig. 2.13 [32].

General test requirements for medical gowns include liquid barrier properties, permeability, tensile strength, linting propensity, and tear resistance. Furthermore, surgical gowns are sterilised before sale and since they are skin surface-contacting devices biocompatibility should also be considered (ISO 10993-5 cytotoxicity evaluation).

Manufacturing of gowns most commonly relies on thermally bonded spunbonds (S) and meltblowns (M), usually in combination (SMS, or with additional M-layers) or specific types of hydroentangled fabric, with or without surface treatments. Raw materials are mainly, but not limited to PP, PE, and PET [33]. For lower-risk applications, PP spunbonds as well as hydroentangled PET/wood pulp fabrics are important. Hydroentangled PET/wood pulp fabrics with basis weights of $\sim$68 g/m^2 possess excellent aesthetics and moisture vapour transmission, but limitations in terms of liquid penetration resistance and repellency necessitate chemical treatment. For higher levels of protection, four- or five-layer SMMS or SMMMS layered nonwovens in the range of 35–70 g/m^2 are produced, where increasing the basis weight of the M-layer, improves the level of liquid barrier performance. Further enhancement is possible by incorporating additives into the masterbatch prior to extrusion or chemical after-treatment of the fabric. Another approach is the use of trilaminates of about

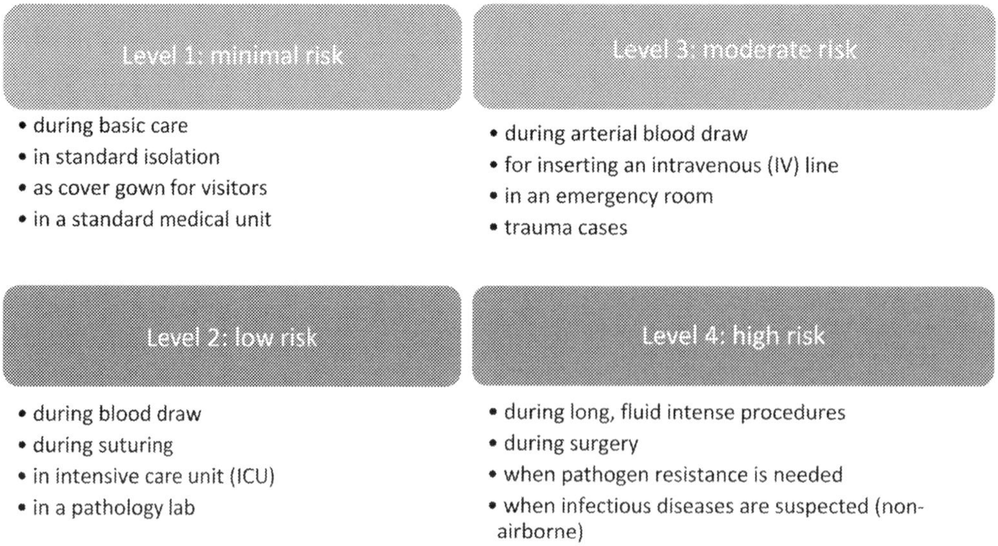

Level 1: minimal risk	Level 3: moderate risk
• during basic care • in standard isolation • as cover gown for visitors • in a standard medical unit	• during arterial blood draw • for inserting an intravenous (IV) line • in an emergency room • trauma cases
Level 2: low risk	Level 4: high risk
• during blood draw • during suturing • in intensive care unit (ICU) • in a pathology lab	• during long, fluid intense procedures • during surgery • when pathogen resistance is needed • when infectious diseases are suspected (non-airborne)

FIG. 2.13　Surgical gowns classifications and their intended use in healthcare facilities.

$60\,g/m^2$ that consist of spunbond (S) or SMS outer layers and a film membrane in the middle. PE film coated nonwovens are a further important type of substrate [31].

2.14.2 Face masks

Surgical face masks are medical devices, which in accordance with European Standard (BS EN 14683:2014) are single-use and are classified in terms of their bacterial filtration efficiency (BFE) as follows:

- Type I—BFE of $\geq 95\%$ for use by patients and other persons to reduce the risk of spread of infections particularly in epidemic or pandemic situations. This type is not suitable for use by healthcare professionals in an operating theatre or in other medical settings.
- Type II—BFE of $\geq 98\%$ and suitable for healthcare professionals during surgical procedures in operating theatres and other medical settings.
- Type IIR, in addition to a BFE of $\geq 98\%$, requires a splash resistance pressure equal to or greater than $16.0\,kPa$ for use in circumstances where there is a risk of exposure to potentially contaminated liquids and particles.

In 2020, control of respiratory droplet transmission and potential infection because of the COVID-19 global pandemic, created unprecedented demand for PPE-grade face masks [34].

Apart from BFE (and splash resistance), which are central performance requirements, airflow or breathing resistance (conformance to a specific differential pressure), biocompatibility and microbial cleanliness test (Bioburden) are also required. Furthermore, fluid resistance and flammability requirements also need to be considered [35]. Depending on specification, face masks in Europe are subject to requirements in the EN 14683 and EN149 standards, and they typically comprise of at least three layers: an inner comfort layer, a middle filtration layer and an outer protection layer. This is normally a PP spunbond–meltblown–spunbond fabric, wherein the BFE is mainly governed by the basis weight and mean fibre diameter of the meltblown filtration layer. The three layers are combined by calendering to form the mask substrate [36].

A respirator mask or filtering face piece (FFP) is classified as personal protective equipment (PPE) and is designed to provide protection against inhalation of airborne particles including aerosols of different sizes. They are widely used in industrial and healthcare settings, including by healthcare workers during aerosol-generating procedures [37]. Respirator masks are designed to protect the wearer from the inhalation of droplets and particles rather than preventing the release of exhaled respiratory particles from the wearer into the environment and are mostly not splash-resistant, especially if they have an exhalation valve [38]. When in addition to secondary aerosolisation, there is also a risk of splashing from bodily fluid, either a splash-resistant respirator or combined use of a respirator and a splash-resistant surgical face mask IIR may be advised. European standard EN 149:2001 + A1:2009 defines respirator masks (also known as half masks) in three classes, according to airborne filter efficiency and the total inward leakage under laboratory conditions, as follows (note that the total inward leakage relates to face seal leakage, filter penetration, and exhalation valve leakage):

FFP1: efficiency of at least 80% with maximum 22% total inward leakage.
FFP2: efficiency of at least 94% with maximum 8% total inward leakage. This class is closest to the USA's (NIOSH) N95 standard, which requires 95% efficiency and is sufficient for first-contact precautions, transport, visits, and supervision tasks.
FFP3: efficiency of at least 99% with maximum 2% total inward leakage, equivalent to the USA's N99 standard. Only this class is suitable for invasive patient care.

The filter efficiency test standard uses a NaCl test aerosol with a mean particle diameter of 0.6 μm. The performance in use of single-use respirator masks depends on effective sealing to the users face, and therefore, qualitative fit-tests are required. The majority of FFP classified respirators are made from layered, PP spunbond–meltblown–spunbond (SMS) media, wherein meltblown electrets (electrostatically charged to enable submicron particle capture) are employed to obtain high filter efficiency, without compromising airflow (breathing) resistance. Electrostatic charging may be carried during or after web formation. Methods include hydro-charging extruded filaments during meltblown fabric production, electrical charge bombardment, and corona discharge techniques.

To achieve the necessary FFP3 performance in terms of efficiency and charge stability, the melt flow index (MFI) and composition of the polymer resin are critical, as well as the mean fibre diameter and mode of electrostatic charging. Although more commonly associated with reusable respirators than single-use, triboelectrically charged electret nonwovens are also important in healthcare settings. For example, such electrets are used to prevent cross-contamination between patients and artificial respirator devices found in critical care settings. Commercially, most long-lasting tribo-charged media are made by carding, cross-lapping and needlepunching, and consist of two or more blended synthetic fibre types, or historically, fibres intermixed with solid resin particles. Selection of the components is based on their electrical insulation properties, understanding of the resultant charge signs (negative or positive) on the fibre surfaces, after agitation, and the relative position of materials in the triboelectric series. Examples of this type of tribo-charged electret include fibre blends of PP/modacrylic

(Technostat – from Hollingsworth & Vose) and PP/polymetaphenylene isophthalamide (Tribo-from Texel).

2.14.3 Wraps and packaging

Nonwovens are used as components in a variety of other healthcare products, some of which are briefly mentioned.

Nonwoven sterilisation wraps are used in conjunction with instrument trays. When surgical instruments are sterilised, the packaging must allow ingress of the sterilising fluid, while preventing penetration of solid particle or biological contamination. Nonwoven sterilisation wraps are produced in single or multiple layers as well as laminates, with PP SMS fabrics being particularly suited for this application.

Another rapidly developing area is in the packaging of pharmaceuticals, and the ability to deliver them direct to the end-user's premises. Many pharmaceuticals are temperature sensitive and require insulative packaging to enable them to be safely transported or stored and ensure their ultimate efficacy. In practice, this can require packaging materials to be capable of restricting the interior temperature from 2°C to 8°C for at least 24h, and up to >100h, depending on circumstances. The distribution and transport of controlled room temperature (CRT) pharmaceutical products has traditionally depended on incorporating ice packs, bulky foam fillings, and phase change materials (PCMs), but nonwoven insulation products have also penetrated this growing market. Most are based on drylaid, needlepunched fabrics. Two examples of such nonwoven medical packaging material are Woolcool (based on drylaid nonwoven wool fabrics) and Swiftpak (PLA needled fabric with a PBAT outer film layer).

2.14.4 Other medical devices

Medical nonwoven devices can be classified as implantable, nonimplantable and extra-corporeal [39], each encompassing a variety of products (Fig. 2.14).

In respect of nonimplantables, extensive applications for nonwovens exist in wound care for primary dressings in conventional and advanced products, including negative pressure wound therapy (NPWT). Wound management is a complex process, which increasingly involves prevention, diagnostics, and monitoring, as well as protecting and managing the wound environment to promote the healing process. Nonwovens are particularly important in the management of chronic wounds, where signs of healing fail to appear within at least 4 weeks and long-term, a significant amount of tissue loss can take place, e.g. through ulceration, mainly because of venous, arterial, diabetic foot, and pressure ulcers. The principal objectives of wound healing can be summarised as (a) protecting the wound, (b) preventing infection (including biofilms), (c) exudate management (including preventing maceration), (d) provision of a moist wound-healing environment, (e) promoting autolytic debridement, (f) reducing inflammation, and (g) managing other factors associated with morbidity such as pain and odour. Besides chronic wound management, nonwovens are used in dressings for surgical and acute wounds, and for management of intravenous and subcutaneous catheter sites.

Wound dressing selection is driven by many factors, not least the type of wound and the stage of wound healing, which can be thought of in at least three, nonlinear, overlapping

Non-implantable:
• Wound care products - dressings including absorbent pad, wound
  contact layer and other components
• Bandages
• Orthopaedic
• Plasters
• Transdermal drug delivery

Implantable:
• Regenerative medicine (e.g. soft tissue, including skin)
• In situ (implanted) drug delivery
• Hernia repair patches
• Staple line reinforcement buttresses
• Skin substitutes (burns dressings and cosmetic surgery)
• Periodontal and gingival repair
• Adhesion barriers
• Tendon and ligament repair materials

Extra-corporeal
• Blood filter and apheresis media

Medical

FIG. 2.14   Summary of major medical markets [5].

phases: (i) inflammation, (ii) proliferation, and (iii) maturation (sometimes referred to as remodelling). Often, chronic wounds do not progress beyond the inflammation phase and so fail to contract and accumulate collagenous tissue.

Whether delivered as a single layer, or more usually in the form of a composite or layered structure, nonwovens are used as primary dressings, i.e. they are used to directly cover the wound. Basic, passive wound contact dressings include nonwoven gauze-like fabrics produced from drylaid, apertured hydroentangled fabrics. Basic low adherent and absorbent nonwoven dressings (i.e. low pealing force when removed from the wound) tend to consist of a smooth perforated film or gauze wound contact layer and a cellulosic (e.g. viscose) drylaid nonwoven absorbent layer, with or without an adhesive border.

Advanced wound dressings are extremely diverse in construction and include those made from hydrogels, hydrocolloids, and other materials designed to enhance critical functions such as exudate management, odour adsorption, or conformance to the wound bed. Many are composite, multilayer constructs, and some utilise nonwovens as carriers for nonfibrous materials such as hydrocolloid particles. Some hydrogel sheet dressings are produced by impregnating hydrogels into a preformed nonwoven to provide absorbency and low adherent behaviour. Entirely fibrous hydrocolloid dressings such as Aquacel (Convatec) are produced from drylaid-needlepunched fabrics containing sodium carboxymethylcellulose (CMC), or other gel forming fibres, with or without an additional foam layer or adhesive border. Such dressings can manage heavily exuding wounds because of their high absorbent capacity and conform well to the irregular contour of the wound bed because of swelling through gelation. Similarly, drylaid calcium alginate dressings have been produced for years to provide haemostatic and exudate management properties. Other drylaid nonwoven products include blends of alginate and carboxymethylcellulose fibres, and composite dressings with an alginate contact layer and a viscose secondary layer. Another strategy for dealing with exudate management is to use a vapour-permeable film backing to promote evaporation from the back of the dressing, while preventing contamination from liquid or microorganism ingress. Composite nonwoven fabrics of this type normally consist of a hydrophilic fabric, e.g. a viscose rayon drylaid nonwoven pad, and a vapour-permeable adhesive film. Odour absorbent dressings incorporate activated carbon, often in the form of a nonwoven sheet, used in combination with a primary dressing layer, or combined in a composite structure with the drylaid viscose or alginate nonwoven primary dressing. In one example, a knitted activated charcoal fabric is encapsulated in a polyamide spunbonded fabric.

Antimicrobial nonwoven dressings are a further important category. Not least because of regulatory considerations, the objective is not necessarily to release antimicrobials into the wound, but rather to act on the exudate that is absorbed into the body of the dressing. Nonwoven dressings including those containing alginate or CMC hydrocolloid fibres are available containing silver in elemental nanocrystalline, organic compound or organic complex formats. In some dressings, silver-coated polyamide fibres are blended with alginate fibre and CMC. Medical grade honey, e.g. manuka is impregnated into nonwoven dressings including some drylaid products made of calcium alginate. Other antimicrobials associated with wound dressing products more generally include polihexanide (polyhexamethylene biguanide) and dialkylcarbamoyl chloride, iodine, glucose oxidase, and lactoperoxidase, but their use depends on the presence or not of infection and various contraindications [40].

Negative pressure wound therapy (NPWT) involves vacuum assisted management of wounds, particularly chronic, heavily exuding wounds such as pressure ulcers and bedsores and nonwoven dressings can also be used as fluid tight contact layers around the wound.

In stoma care, the bags used by ostomates to collect human waste, represent an important application area for nonwoven fabrics. This covers colostomy, ileostomy, and urostomy bag products. Depending on the manufacturer, one and two-piece (pouch-in-pouch) products are available, with nonwovens forming the basis of the inner and outer bags, as well as the integrated air vent. The inner layer is usually film coated to render it liquid impermeable, but flushable inner bags have also been developed more recently using water-soluble films and nonwoven substrates (or binders), where the solubility is a function of temperature. Ostomy pouches with a detachable, flushable nonwoven inner bag provide much needed convenience for ostomates. The integrated air vent in ostomy bags typically incorporates an activated carbon nonwoven layer to manage odour. In transdermal drug delivery systems, light-weight nonwovens are used both as backings laminated to thin films and as drug-loaded fabrics. The latter can be produced by electrospinning, one example of which is the transdermal drug delivery of contraceptives using polycaprolactone electrospun fabrics in skin patches.

A long-standing application for plasma treated PBT meltblown nonwovens is in gravity-fed leukoreduction filters (removal of white blood cells) as part of donated blood processing, to remove cells via filtration and adhesion to fibre surfaces. Extra-corporeal, therapeutic apheresis is another developing area, where devices, normally in the form of columns, have utilised a range of different nonwoven media. This includes whole blood apheresis, such as leukocytapheresis, where hydrophilic-treated PET or PP nonwovens have been used for treating conditions such as inflammatory bowel disease.

In relation to implantable devices, nonwovens are used to repair, enhance healing or restore the function of damaged or diseased tissues. In respect of hernia repair, although other types of textile mesh such as warp knits predominate, nonwoven PP fabrics have also been used in this application, including for laparoscopic ventral hernia repair. In the field of regenerative medicine, nonwoven scaffolds and barrier membranes made of bioresorbable polymers have been a major focus of development for the regeneration and restoration of function of for example, skin, cartilage, spinal cord, liver, and hard tissues. In the main, nonwovens, particularly those based on nanofibre webs, have attracted significant interest because of their high porosity and pore interconnectivity, promoting cell attachment and migration, although cell infiltration of nanofibre scaffolds remains a challenge in the engineering of thick, three-dimensional tissues. One of the challenges in this area, which is heavily regulated, is achieving the required reproducibility in terms of structural and dimensional properties.

2.15 Energy storage, fuel cells, and electronics

Another important application, particularly for wetlaid fabrics, is in energy storage devices notably for batteries, where a corrosion-resistant nonwoven battery separator is placed between the positive and negative electrodes to facilitate ion transfer and prevent short circuits. Extremely thin, wetlaid fabrics made of glass or other inorganic fibres, with basis weights as low as $2 g/m^2$ are produced for such applications. Wetlaid battery separators are also made from materials such as polyolefins, PET, and PVA. General applications for nonwoven fabrics include lead acid, nickel-metal hydride (NiMH), alkaline manganese, and lithium battery separators. Wetlaid nonwovens made of carbon fibre are also manufactured for gas diffusion layers (GDLs) in fuel cells, including proton electrolyte membrane (PEMFC), phosphoric acid (PAFC), and direct methanol (DMFC) fuel cells. Applications of electrically conductive nonwovens in electronics, many of which comprise metal coated fibres, include fabrics for electromagnetic interference (EMI) shielding, resistive heating fabrics and static charge dissipation.

References

[1] Manjulata M., Classification of products [Internet]. Economics Discussion. Available from: https://www.economicsdiscussion.net/marketing-2/classification-of-products/31799.
[2] EDANA, Nonwovens Markets, Facts and Figures [Internet]. EDANA Association. https://www.edana.org/nw-related-industry/nonwovens-markets.
[3] Smithers, The Future of Global Nonwoven Wipes to 2025 [Internet], Smithers, 2021. Available from: https://www.smithers.com/services/market-reports/nonwovens/the-future-of-global-nonwoven-wipes-to-2025.
[4] EDANA, Nonwoven Statistics 2020 [Internet]. EDANA Association. https://www.edana.org/docs/default-source/press-corner/nonwovens-statistics-2019.pdf?sfvrsn=dbad6d4c_6.
[5] EDANA, Nonwovens in daily life—Versatile products for modern life [Internet], EDANA Association, Available from https://www.edana.org/nw-related-industry/nonwovens-in-daily-life.
[6] M. Cordella, I. Bauer, A. Lehmann, M. Schulz, O. Wolf, Evolution of disposable baby diapers in Europe: life cycle assessment of environmental impacts and identification of key areas of improvement, J. Clean. Prod. 95 (2015) 322–331.
[7] EDANA, Sustainability Report 2007–2008, Absorbent Hygiene Products [Internet]. EDANA Association. Available from: https://www.edana.org/docs/default-source/sustainability/edana-sustainability-report—2007.pdf?sfvrsn=33a7d9b3_2.
[8] G. Kellie (Ed.), Advances in Technical Nonwovens, Woodhead Publishing, 2016.

[9] EDANA, Baby Diapers and Nappies Infographic [Internet]. EDANA Association. Available from: https://www.edana.org/nw-related-industry/nonwovens-in-daily-life/absorbent-hygiene-products/baby-diapers.

[10] FGI, About Incontinence [Internet]. Global Forum on Incontinence. Available from: https://www.gfiforum.com/incontinence.

[11] Interreg Europe, Good Practice: Fater SMART—Diaper Recycling [Internet]. Interreg Europe. Available from: https://www.interregeurope.eu/policylearning/good-practices/item/1902/fater-smart-diaper-recycling/.

[12] D. Zhang, Nonwovens for consumer and industrial wipes, in: Applications of Nonwovens in Technical Textiles, Woodhead Publishing, 2010, pp. 103–119.

[13] ECDC, Healthcare-Associated Infections [Internet], European Centre for Disease Prevention and Control, 2015. Available from: http://ecdc.europa.eu/en/healthtopics/Healthcare-associated_infections/Pages/index.aspx.

[14] S.A. Sattar, C. Bradley, R. Kibbee, R. Wesgate, M.A. Wilkinson, T. Sharpe, J.Y. Maillard, Disinfectant wipes are appropriate to control microbial bioburden from surfaces: use of a new ASTM standard test protocol to demonstrate efficacy, J. Hosp. Infect. 91 (4) (2015) 319–325.

[15] N.W. Edwards, E.L. Best, S.D. Connell, P. Goswami, C.M. Carr, M.H. Wilcox, S.J. Russell, Role of surface energy and nano-roughness in the removal efficiency of bacterial contamination by nonwoven wipes from frequently touched surfaces, Sci. Technol. Adv. Mater. 18 (1) (2017) 197–209.

[16] Smithers, The Future of Global Nonwoven Wipes to 2023 [Internet], Smithers, 2019. Available from: https://www.smithers.com/Services/market-reports/Materials/the-future-of-global-nonwoven-wipes-to-2023.

[17] M.D. Segura-Alcaraz, J.G. Segura Alcaraz, I. Montava-Seguí, M. Bonet-Aracil, The use of fabrics to improve the acoustic absorption: influence of the woven fabric thread density over a nonwoven, Autex Res. J. 18 (3) (2018) 269–280.

[18] Y. Chen, N. Jiang, Carbonized and activated non-wovens as high-performance acoustic materials: part I noise absorption, Text. Res. J. 77 (10) (2007) 785–791.

[19] J.Y. Chen, Nonwoven textiles in automotive interiors, in: Applications of Nonwovens in Technical Textiles, Woodhead Publishing, 2010, pp. 184–201.

[20] EDANA, Why Use Nonwovens in Automotive and Transportation? [Internet]. EDANA Association. Available from: https://www.edana.org/nw-related-industry/nonwovens-in-daily-life/automotive.

[21] Denkstatt, A Case Study on Automotive Nonwovens [Internet]. Denkstatt Available from: https://www.index17.ch/media/document/0/automotive-lightweight-fabrics.pdf.

[22] INDEX, Even Better Than the Real Thing [Internet], 2016, Available from: https://www.index17.ch/en/news/even-better-than-the-real-thing-599.

[23] C.G. Bevan, inventor, E Leather Ltd, assignee, Formation of Leather Sheet Material Using Hydroentanglement, United States Patent US 7,731,814, 2010.

[24] G. Mouret, D. Thomas, S. Chazelet, J.C. Appert-Collin, D. Bemer, Penetration of nanoparticles through fibrous filters perforated with defined pinholes, J. Aerosol Sci. 40 (9) (2009) 762–775.

[25] TMR Research, Liquid Filtration Market: Rising Allocations on Analysis and Innovation [Internet], Farming Sector, 2020. Available from: https://farmingsector.co.uk/uncategorized/152173/liquid-filtration-market-rising-allocations-on-analysis-and-innovation/.

[26] Modor Intelligence, Air Purifier Market—Growth, Trends, and Forecast (2020–2025) [Internet], Modor Intelligence, 2020. Available from: https://www.mordorintelligence.com/industry-reports/air-purifier-market.

[27] B.Y. Yeom, E. Shim, B. Pourdeyhimi, Boehmite nanoparticles incorporated electrospun nylon-6 nanofiber web for new electret filter media, Macromol. Res. 18 (9) (2010) 884–890.

[28] I.M. Hutten, Handbook of Nonwoven Filter Media, Elsevier, 2007.

[29] A. Kilic, E. Shim, B. Pourdeyhimi, Electrostatic capture efficiency enhancement of polypropylene electret filters with barium titanate, Aerosol Sci. Tech. 49 (8) (2015) 666–673.

[30] H. Zhang, J. Liu, X. Zhang, C. Huang, X. Jin, Design of electret polypropylene melt blown air filtration material containing nucleating agent for effective PM2. 5 capture, RSC Adv. 8 (15) (2018) 7932–7941.

[31] J.R. Ajmeri, C.J. Ajmeri, Nonwoven materials and technologies for medical applications, in: Handbook of Medical Textiles, Woodhead Publishing, 2011, pp. 106–131.

[32] FDA, Personal Protective Equipment for Infection Control; Medical Gowns [Internet], Food and Drug Administration, 2021. Available from: https://www.fda.gov/medical-devices/personal-protective-equipment-infection-control/medical-gowns.

[33] F.S. Kilinc, A review of isolation gowns in healthcare: fabric and gown properties, J. Eng. Fibers Fabr. 10 (3) (2015) 180–190.

[34] ECDC, Using Face Masks in the Community [Internet], European Centre for Disease Prevention and Control, 2020. Available from: https://www.ecdc.europa.eu/sites/default/files/documents/COVID-19-use-face-masks-community.pdf.

[35] T. Oberg, L.M. Brosseau, Surgical mask filter and fit performance, Am. J. Infect. Control 36 (4) (2008) 276–282.

[36] EDANA, How Medical Masks Are Made [Internet]. EDANA Association. Available from: https://www.edana.org/docs/default-source/publications/how-medical-masks-are-made-bottlenecks-and-solutions.pdf?sfvrsn=a8fb647c_7.

[37] National Institute for Occupational Safety and Health (NIOSH), Use of Respirators and Surgical Masks for Protection Against Healthcare Hazards [Internet], Centre for Disease Control and Prevention, 2018. Available from: https://www.cdc.gov/niosh/topics/healthcarehsps/respiratory.html.

[38] ECDC, Safe Use of Personal Protective Equipment in the Treatment of Infectious Diseases of High Consequence [Internet], European Centre for Disease Prevention and Control, 2014. Available from: https://www.ecdc.europa.eu/sites/default/files/media/en/publications/Publications/safe-use-of-ppe.pdf.

[39] A.J. Rigby, S.C. Anand, A.R. Horrocks, Textile materials for medical and healthcare applications, J. Text. Inst. 88 (3) (1997) 83–93.

[40] National Institute for Health and Care Excellence, Soft Polymer Dressing [Internet]. BNF. Available from: https://bnf.nice.org.uk/wound-management/soft-polymer-dressings.html.

Raw materials and polymer science for nonwovens

G. Tronci and S.J. Russell

University of Leeds, Leeds, United Kingdom

3.1 Introduction

Nonwovens are engineered fabrics that address the performance requirements of an increasingly wide range of industrial applications. In contrast to traditional textiles, nonwovens are manufactured from webs of staple fibres or filaments that are bonded by chemical, mechanical, or thermal means, omitting classic yarn spinning, weaving, or knitting processes. The chemical composition and molecular characteristics of the various materials used to make nonwovens can be thought of as customisable building blocks that influence the cost, processing characteristics, environmental impact, and overall performance of final products. By careful consideration and selection of these various raw material components, nonwovens continue to find a multitude of applications across industrial, healthcare, and consumer product markets. In this chapter, the major raw materials used in nonwoven fabric manufacture are discussed, paying particular attention to the role that polymer systems play in controlling nonwoven properties and functionality.

The breadth and versatility of manufacturing processes used by the nonwovens industry mean that nearly any fibre-forming material, whether natural or man-made in origin, can be made into a nonwoven fabric. Of course, the compatibility of a particular polymer or fibre material with individual nonwoven processes varies considerably, so they have to be properly matched to the intended manufacturing route. In attempting to introduce new fibres or alternative materials into the nonwovens industry, commercial feasibility is dependent not only on process compatibility, but also on the cost, volume of supply, as well as the precise combination of physical properties being offered.

Fibre-forming capability and physical properties are determined by the molecular characteristics of the raw materials [1]. From the molecular scale up to the nanoscale, this means the chemical sequence of the polymer backbone and the assembly of macromolecular chains or inorganic crystals are important, as are manufacturing conditions, where the chemistry and structure of the fibre may be controlled. It is therefore important to understand process–structure–property relationships to enable target cost and performance requirements to be met, as well as to enable rational tuning of raw material formulations and fibre microstructure.

3.2 Fundamentals of polymer science

Various materials, mainly polymers and organic molecules of low and high molar mass, as well as glass, ceramics, and metals have proven instrumental building blocks for the development of engineered nonwovens. Polymer systems from either synthetic or natural origin play a very special role in this context because they can be customised in terms of macroscopic properties and formats, according to small variations in their organic composition. Polymers are the most commonly used raw materials in the manufacture of nonwovens. Depending on the molecular structure, rheological, and thermal properties, polymers can be spun into fibrewebs, filaments or staple fibres and assembled into nonwoven assemblies by a range of processes.

On a molecular scale, a polymer is obtained by the polymerisation of monomers, and consists of monomer-repeating units that are connected by chemical bonds to form long molecular chains. The sequence and connectivity of repeating units can be easily tailored by, for example, the application of co-monomers or the introduction of specific chemical functions in the monomer, resulting in a wider range of fibre customisation possibilities. Although complex polymer architectures, e.g., branched or multiarm, have been reported, linear polymers are mainly used for the manufacture of fibres and nonwovens. Hence, three main characteristics describe the molecular organisation of polymers: polydispersity, chemical sequence, and polymer morphology (Fig. 3.1).

Owing to the statistical nature of the polymerisation reaction, polymers result in a mixture of macromolecules of different lengths with a specific distribution in *molecular weight*, in

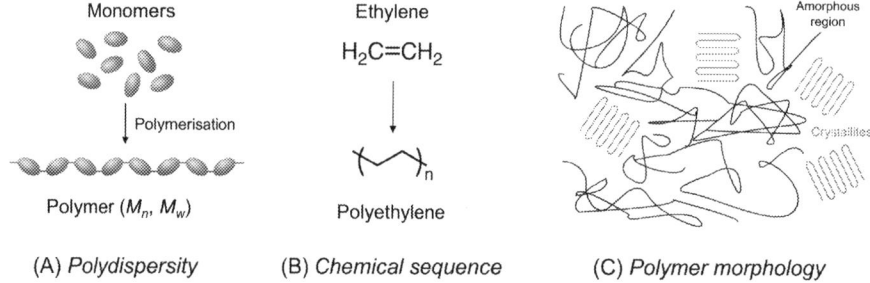

(A) *Polydispersity*   (B) *Chemical sequence*   (C) *Polymer morphology*

FIG. 3.1 (A) Polymers consist of a distribution of long chains with different molecular weights (M_n, M_w), resulting in a polydisperse system. (B) The chemical sequence of the polymer reflects the chemical composition, connectivity, and structure of the monomer. (C) The resulting polymer can display either regular (crystalline) or randomly (amorphous) nanoscale domains, depending on its macromolecular features.

contrast to common organic molecules. A characteristic feature of polymers is therefore their *polydispersity* (Fig. 3.1A), which is quantified as the ratio between the weight averaged molecular weight (M_w) and the number average molecular weight (M_n), and describes the breadth of the molecular weight distribution of the polymer chains. Typical polymerisation reactions lead to polydispersity indexes of about 2.0, although lower values of up to 1.1 can also be achieved, indicating a narrow distribution of polymer chains. Polymers with a polydispersity in the range of 1.0 to 2.0 are expected to generate reproducible fibre spinning and tensile properties. Other than the polydispersity, linear polymers with increased molecular weights are appealing when fibres with high tenacity are required [2], although increasing the molecular weight above a certain threshold makes continuous spinning of aligned fibres challenging due to the level of entanglement between polymer chains [3,4].

Other than the polydispersity and chain length, the *chemical sequence* of the repeating units, described by its composition, connectivity (linear, branched) and spatial configuration (*tacticity*), is a key parameter that can be varied to tailor macroscopic properties of synthetic polymers (Fig. 3.1B). For example, polypropylene (PP) is a thermoplastic polyolefin obtained by the polymerisation of propylene monomers, whilst viscose rayon is a natural polysaccharide derived from the partial degradation of cellulose and consists of hundreds of glucose repeating units. In light of the different monomers employed, polymers have significantly different melting (T_m) and glass transition (T_g) temperatures (Table 3.1) as well as mechanical properties, such as fibre tenacity ($< 0.1\,\mathrm{cN\,dtex^{-1}}$ (PP); $< 4.4\,\mathrm{cN\,dtex^{-1}}$ (rayon)). Together with

TABLE 3.1 Chemical and thermal properties of a selection of fibre-forming polymers used in the manufacture of nonwoven fabrics.

Fibre-forming polymer	Chemical structure	Class	$T_g/°C$	$T_m/°C$
Poly(lactide) (PLA)		Aliphatic polyester	50–80	130–180
Polyethylene terephthalate (PET)		Aromatic polyester	78–120	260–270
Polyethylene (PE)		Polyolefin	−110	105–180
Polypropylene (PP)		Polyolefin	−20–0	130–171
Acrylonitrile copolymer	x > 85 wt.%	Acrylics	100	300
Poly(hexamethylene adipamide) (Nylon 66)		Polyamide	55–58	260
Poly(p-phenylene terephthalamide) (Kevlar)		Aramid	300	500
Viscose rayon		Polysaccharide	20	> 150 (degrades)

the chemical composition of the monomers, polymers can also be tailored depending on their tacticity, which describes the spatial configuration of the monomer repeating units depending on the arrangement of a specific substituent. If the polymer chain contains carbon atoms linked to two different substituents, three structures of the same polymers can be achieved, i.e., isotactic (same substituent orientation), syndiotactic (alternating substituent orientation), and atactic (random orientation). For example, by controlling the spatial arrangement of the methyl group, isotactic PP was found to display a much higher tensile strength at break ($\sigma \sim 200\text{--}400\,\text{MPa}$) with respect to the syndiotactic variant ($\sigma \sim 130\text{--}280\,\text{MPa}$) [5]. Depending on the chemical substituents in the monomer repeating unit and the spatial configuration of the polymer backbone, polymer chains can mediate noncovalent interactions (e.g. hydrophobic, electrostatic, hydrogen bonding) and display either regular (*crystalline*) or random (*amorphous*) *morphology* (Fig. 3.1C).

The presence of crystallites is key to achieving the formation of mechanically stable, continuous filaments and fibres. On the one hand, the presence of crystallites enables compliance with solvent-free fibre manufacturing routes, e.g., melt spinning and spunmelt nonwoven production [6]. On the other hand, the presence of crystallites act as physical crosslinks between polymer chains, so that spinning of polymer filaments with increased mechanical strength and decreased water-induced swelling capability can be achieved. In light of these considerations, semicrystalline rather than amorphous synthetic polymers are mostly employed for the manufacture of nonwovens.

3.3 Synthetic polymers for nonwovens

Synthetic polymers provide enormous capacity for customisation of fabric properties, even though in practice a relatively small number of polymer types make up the bulk of industrial production. At present, the largest proportion of spunlaid nonwovens is made from polyolefins, notably polypropylene, whereas for drylaid fabrics, polyester terephthalate (PET) predominates. The physical and mechanical properties of synthetic fibres can be altered by manipulating the molecular organisation of the polymer, as well as the use of additives, fillers or via incorporation of other materials. Drawing and postspinning chemical or thermal treatments can also be employed to modulate molecular order and physical properties. The manufacture of fibres and filaments from synthetic polymers relies on the spinning of long-chain polymers usually by means of wet, dry or melt spinning. In the case of fibres, continuous filaments are cut into defined lengths, before being assembled into webs in a separate step by drylaid or wetlaid web forming methods. In spunbonding, meltblowing, forcespinning, and electrospinning processes, extruded filaments are directly assembled into webs, usually in one integrated process [7]. Web bonding exploits the basic physical and thermal properties of the constituent materials, e.g., thermoplastic polymer content is required for thermal bonding, whereas low fibre bending modulus and high extension at break assist with mechanical bonding. In the following, the main classes of synthetic polymers relevant to the nonwovens industry are briefly reviewed and certain process–structure–property relationships are discussed.

3.3.1 Polyesters

Polyesters are a family of polymers that are important building blocks for the manufacture of a variety of nonwoven fabrics. Polyesters can be processed in the form of fibrewebs, continuous filaments or as staple fibre and are characterised by relatively high tensile modulus and tenacity, dimensional stability, weather resistance, melt-processability, and recyclability. The presence of the ester bond between repeating units also equips respective polyester fibres with hydrolytic degradability (Fig. 3.2).

This characteristic has been widely exploited for applications in pharmaceuticals and medicine [8], resulting in fibres with controlled drug delivery capability, dental membranes with controlled degradability and regenerative devices. Although polyesters can be susceptible to degradation in water, the fibres are typically hydrophobic in the nondegraded state and have very low moisture absorption [9]. The most widely used fibre-forming polyester is polyethylene terephthalate (PET) (Table 3.1).

Polyethylene terephthalate (PET) is a petroleum-based polyester primarily used in the manufacture of meltspun staple fibres for drylaid web forming processes particularly carding, or as a polymer feedstock for spunmelt nonwoven manufacture, particularly where final products require greater temperature resistance than is possible with polypropylene (PP) [10]. At the other end of the volume manufacturing spectrum, electrospun PET nonwovens are also produced for applications in areas such as liquid filtration [11]. In comparison to PLA, PET has higher glass transition ($T_g \sim 78$–$120°C$) and melting ($T_m \sim 260$–$270°C$) temperatures, which reflect the presence of an aromatic ring in the PET repeating unit (Table 3.1) and the varying degree of crystallinity. The chemical sequence of this polymer has also remarkable impact at the macroscopic level, whereby PET fibres are mechanically strong, with a fibre tenacity of up to $8\,cN\,dtex^{-1}$, depending on the molecular weight of the polymer. PET fibres can also display a remarkably high elastic recovery (70%–100% following low elongation) and resistance to abrasion, and weak acidic and alkaline conditions, whilst it degrades at low and high pH. As a result of their durability, environmental stability and relatively low cost, PET fibres have been used for a wide range of nonwoven applications, from industrial and consumer goods (e.g. filter media, reusable hard surface wipes, furniture components, waddings and insulation) to high value products (battery separators, inkjet printable display media and automotive interiors, e.g. headliners).

Polybutylene terephthalate (PBT) is a semicrystalline thermoplastic aromatic polyester with similar properties and composition compared to PET (Fig. 3.3).

FIG. 3.2 Hydrolysis of a polyester chain generates two partially degraded polymer chains with either carboxylic or hydroxyl termination.

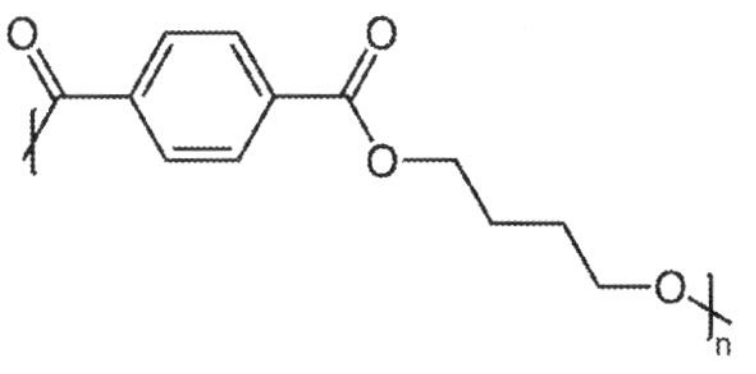

FIG. 3.3 Chemical structure of PBT, which is typically synthesised via polycondensation of terephthalic acid with 1,4-butanediol.

With the flexible butylene chain present between ester groups, PBT can crystallise much faster than PET, and that is why it is often blended to enhance the processability of the latter [12]. In comparison to PET, PBT also shows lower melting ($T_m = 227°C$) and glass ($T_g = 40°C$) transition temperatures, making it compatible with spunbond and meltblown processes [13]. Despite that, resulting nonwovens typically have good impact strength at temperatures as low as $-40°C$ and resistance to moisture, creep, fire, fats, and oils. For these reasons, PBT nonwovens have gained increasing attention for a wide range of industrial applications, including automotive, electronics, and medical devices [7,8]. For biomedical applications, PBT nonwovens are employed as degradable scaffolds for the cultivation of human cells [14], and PBT meltblowns are widely used in blood filtration products, e.g., leukocytes. Given the relatively inert surface, chemical modification of PBT fibres is often carried out, via for example, plasma treatment, or grafting with either synthetic polymers [15] or peptidomimetic molecules [16], aiming to improve the separation properties and leukocyte adhesion and filtration, as well as the surface reactivity towards biochemical assays. Owing to their fast crystallisation, PBT chains cannot be as conveniently oriented as PET, so that the macroscopic properties of the highly crystalline PBT (as much as 60% crystallinity) are fairly similar to the ones of unoriented crystalline PET. Nevertheless, PBT offers high tenacity ($\sim 1\,g\,den^{-1}$) and elongation at break (up to 300%), excellent electrical properties and acid/base chemical resistance, rapid moulding cycles, and excellent reproducible mould shrinkage. Owing to low moisture absorption rates ($< 0.1\%$ after 24h), PBT also exhibits excellent wet state dimensional stability, although prolonged incubation in water results in hydrolytic degradation of the fibre due to the lower crystallinity, compared to, e.g., PET.

Copolyesters are employed to expand the applicability and the fibre properties of native polyesters, e.g., with regards to melting temperature, control of thermal shrinkage, improved dyeability and processability, as well as introducing specific properties that are inherent to natural fibres. With regards to PET, introduction of a comonomer during the polymer synthesis effectively breaks down the regularity of the chemical sequence of PET, so that polymer morphology and thermo-mechanical properties can be customised according to the co/ monomer ratio. This is achieved by introducing a second acid or diol component in the synthesis of PET, e.g., diethylene isophthalate, diethylene orthophthalate, *p*-hydroxybenzoic acid, and diethylene adipate [17]. Corresponding copolyesters normally show intermediate thermal properties, with their T_m or T_g falling in between those of the corresponding aliphatic or aromatic polyesters. Incorporation of aliphatic segments into the copolyester sequence typically results in decreased T_m or T_g, so that increased melt processability and fibre's hydrolytic degradability are accomplished. Resulting fibres also present widely customisable shrinkage (27%–70%), depending on the composition of the starting comonomers, and the manufacturing conditions applied during fibre drawing and heat treatment. A typical example of copolyester fibre is poly(tetramethylene adipate-*co*-terephthalate) (PTAT, Fig. 3.4), which is commercially traded as Eastar Bio [18].

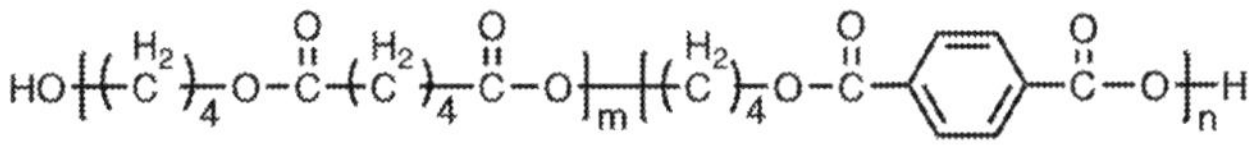

FIG. 3.4 Chemical structure of PTAT as derived from 1,4-butanediol and adipic and terephthalic acids.

PTAT exhibits higher hydrophilicity and better processability than other biodegradable polyesters, whilst its processability is comparable to that of low density polyethylene (LDPE) but with reduced environmental impact after disposal. Respective PTAT nonwovens are semicrystalline and enable finer fibres, higher throughputs and higher spinning speeds (> 4500 min^{-1}) [19], combined with relevant drapability, soft hand, and elastic properties. In light of their relatively inert chemical composition, these fabrics are compatible with gamma irradiation sterilisation, radio frequency bonding and ultrasonic sealing, which make them appealing for medical applications, e.g., for the design of hospital surgical packs, wipes, bandages, and face masks. Because of the ability to reduce the melting temperature compared to PET, co-polyesters are also used as sheath components in core-sheath and side by side binder fibres.

Poly(lactide) (PLA) is an aliphatic polyester (Table 3.1), which derives from the polymerisation of lactide, a cyclic diester extracted from fully renewable resources such as corn and rice. Consequent to lactic acid chirality, polymeric lactide exists as three isomers: L-lactide, D-lactide, and *meso*-lactide. Therefore, four different types of PLA are available, i.e., poly (L-lactide), poly(D-lactide), poly(DL-lactide), and *meso*(polylactide). Poly(L-lactide) (PLLA) is mostly used industrially due to its degree of crystallinity ($X_c \sim 30\%$–35%), which provides the polymer with relatively high glass transition temperature ($T_g \sim 57$–$60°C$) and tensile modulus ($E \sim 4.8$ GPa), and low strain at break ($\varepsilon \sim 6\%$). In light of its crystallinity, PLA fibres display low water absorption capacity, with a moisture regain of 0.4% to 0.6%. Given these features and the well-known hydrolytic degradability, PLA finds uses in a diversity of nonwoven products, notably in food packaging applications such as heat sealable tea bags (made of ~ 18 gsm spunbond), mulch mattings for agriculture (~ 100 gsm spunbond fabric) and meltblown and electrospun medical devices. Industrially, PLA fibres for nonwovens are typically processed by melt extrusion, followed by drawing and cooling so that the polymer chains are organised in an axially aligned state. It is available both as staple fibre for drylaid and wetlaid processing as well as in the form of resin/masterbatch formats for making spunmelt nonwovens. Depending on the molecular weight and spinning conditions, the tensile strength of PLA fibres is about 35 cN tex^{-1} and the melting point is somewhat lower than PET at ~ 155–$170°C$.

3.3.2 Polyolefins

The key polyolefins used by the nonwovens industry are polypropylene and polyethylene. These consist of long aliphatic hydrocarbon chains, whose chemical elements are carbon and hydrogen. In view of their chemical composition, polyolefin fibres are highly hydrophobic, display minimal moisture regain, and are highly resistant to common chemical agents, particularly acids and organic solvents. The density of polyolefin fibres is also very low, i.e., lower than water, and is measured in the range of 0.90 to 0.96 g cm^{-3}, depending on the presence of an alkyl side group (X) in the monomer repeating unit [($-CH_2-CHX-$)$_n$] and the crystallinity of the polymer. The size and nature of the side group have direct impact on polymer crystallinity and chain alignment in the fibre, thermal properties and elasticity [20]. Despite their excellent chemical stability, one of the main limitations of polyolefin fibres is their susceptibility to degradation by oxidation. Oxidative reactions can be initiated via either thermal or photochemical mechanisms and can be mitigated by the introduction of suitable additives in the fibre. Oxidation of polyolefin fibres leads to cleavage of polymer chains, a deterioration in fibre mechanical properties and undesirable discoloration. Thermal oxidation initiates at elevated temperature via a complex reaction cascade, mostly involving the abstraction of hydrogen atoms from the polyolefin (P) chain by singlet oxygen molecules, resulting in the generation of free radicals and ultimately in chain cleavage (Eq. 3.1):

$$PH + {}^1O_2 \longrightarrow P\bullet + \bullet OOH \tag{3.1}$$

In light of this mechanism, risks of thermally induced oxidative degradation are present during the extrusion of polyolefin melts, given that high temperature are required to achieve suitable melt flow properties. As a result, an increase in melt flow index can be observed in the case of thermal oxidation. Oxidation of polyolefin fibres can also be triggered by ultra-violet radiation ($\lambda \sim 300$–400 nm), mostly due the presence of hydroperoxide residues produced during fibre manufacture. Hydroperoxides are photochemically cleaved into free radicals (Eq. 3.2):

$$POOH \xrightarrow{h\upsilon} PO\bullet + \bullet OH \tag{3.2}$$

The generation of free radicals leads to the propagation of the photo-oxidation reaction, ultimately resulting in chain scission and polymer degradation. To preserve the chemical composition of polyolefins during melt spinning and to ensure stability in the final nonwoven product, antioxidant stabilisers are introduced in the polymer melt to minimise the risk of thermal oxidation (e.g. phenols) and to confer stability against ultra-violet (UV) radiation (e.g. thioesters).

Polypropylene (PP) is the most widely used polyolefin building block for the production of fibres and spunmelt nonwovens. Structurally similar to polyethylene (PE), it contains an additional methyl group in its repeating unit (Table 3.1). The presence of the methyl group leads to improved mechanical properties and thermal resistance, although the chemical resistance decreases compared to PE. Whilst there is a range depending on the polymer grade (Table 3.1), the melting point for PP nonwovens is typically in the range of 160–170°C. In addition to the molecular weight and polydispersity, the properties of PP depend on the specific polymer tacticity. In isotactic PP, the methyl groups are oriented on one side of the polymer backbone, generating a greater degree of crystallinity and material stiffness than both atactic polypropylene and PE. This is why most of the PP that is industrially used is in the isotactic configuration, whereby stereospecific catalysts are used during the polymerisation reaction to obtain PP products with high isotactic index [21]. PP fibres and filaments are usually obtained via melt spinning. In melt processing, the tensile strength of fibres is typically directly related to the molecular weight of the fibre-forming polymer [22]. Like all thermoplastic polymers used in the production of nonwovens, the viscosity of the polymer melt is highly important both in terms of processing performance, but also ultimate fabric properties. A melt flow index in the range of either 18 to 35 or 100 to 2000 is typically required for the preparation of spunbond and meltblown PP fibre webs, respectively.

Polyethylene (PE) is used to make staple fibres as well as spunmelt nonwovens. PE fibres are generally produced via melt extrusion or gel spinning, the latter method being mostly applied to ultra-high molecular weight polyethylene (UHMWPE). In both cases, the lack or minimal presence of side chain branching along the polymer backbone results in fibres with increased oxidation stability. The relatively low melting temperature of PE has led to its use as a sheath polymer in core-sheath and side by side bicomponent fibres for thermal bonding. It can also yield extensible, relatively soft handling fabrics, which is valued in skin-contact applications. In contrast to high-volume commodity applications of PE, UHMWPE has a molecular weight in the range of 2 to 6×10^6 g mol^{-1} and can therefore generate highly oriented fibres with significantly higher crystallinity (95%–99%) with respect to conventional melt spun PE fibres, resulting in superior UV resistance, tensile modulus and strength at break. Such high-performance UHMWPE fibres include Dyneema and Spectra, which are used to make non-woven fabrics for ballistic or high velocity projectile and fragment protection.

3.3.3 Acrylics

Acrylic fibres are made from synthetic copolymers containing at least 85 wt% of acrylonitrile monomer (Table 3.1). If the acrylonitrile content is in the range of 35–85 wt%, the fibre is referred to as modacrylic. In both cases, the vinyl comonomer is added to the polymerisation system to improve the processability of the acrylic homopolymer to yield polyacrylonitrile (PAN). This is because of the extensive hydrogen bonding and strong van der Waals interactions between α-hydrogens and electronegative nitrile groups, resulting in tight packing of the PAN chains.

Typical monomers used for the manufacture of acrylic fibres include vinyl chloride, vinyl acetate, vinyl alcohol, vinylidene chloride, acrylic acid, methacrylic acid, and methacrylate esters. Acrylic fibres are mostly produced via either wet spinning or dry spinning and possess high chemical stability at ambient temperature, whilst undergoing thermal transitions at elevated temperatures. Their physical properties are largely affected by the type and content of comonomer used during the synthesis of the fibre-forming polymer and their crystallinity. For example, acrylics fibres containing vinyl alcohol repeating units display increased adsorption capacity in light of the hydrogen bonding capability of the comonomer. Also, structural irregularity can occur in the fibre depending on the comonomer used during the polymerisation reaction, which result in increased fibre accessibility to moisture. At the macroscale, acrylic fibres exhibit excellent acid and sunlight resistance. They have therefore been used extensively in applications where resistance to environmental degradation is important. From a mechanical point of view, the tenacity of acrylic fibres can be up to 3.5 cN dtex^{-1} depending on the moisture content, whilst up to 50% elongation at break can be achieved. Acrylics fibres also display good elastic recovery based on 2% elongation, and 50% to 95% recovery up to 5%.

3.3.4 Polyamides

Polyamide (PA) fibres consist of monomer repeating units connected by amide bonds (Table 3.1). The most industrially relevant polyamide fibres include nylon 6 (PA6) and nylon 66 (PA66), which are generally tough, strong and durable. The name of the specific nylon is directly related to the number of carbon atoms in the repeating unit of the polymer; i.e., nylon

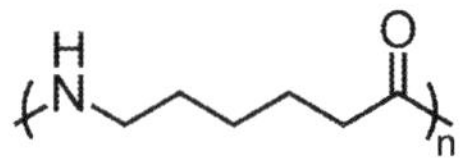

FIG. 3.5   Chemical structure of nylon 6, displaying six carbon atoms in the repeating unit.

6 is formed from repeating units with six carbon atoms (Fig. 3.5), whilst 12 carbons are present in the repeating unit of nylon 66.

Although both polymers are classified as polyamide, two different polymerisation mechanisms are used for the synthesis of the two above-mentioned nylons. Nylon 6 is synthesised via ring-opening chain growth polymerisation of caprolactam in the presence of water vapour and an acid catalyst at the melt. After product purification, the polymer is typically melt spun into fibres at 250–260°C. On the other hand, nylon 66 is synthesised by step growth polymerisation of hexamethylene diamine and adipic acid, prior to melt spinning at 280–290°C. With a degree of polymerisation in the range of 100 to 250 repeating units, strong hydrogen bonds are established between the polymer chains of both nylons due to the presence of amide linkages along the backbone, resulting in a 'pleated sheet' structure.

The significant hydrogen bonding capability of these polyamides plays an important role in fibre spinning and explains the increased melting temperature, molecular rigidity and mechanical properties of melt spun nylons with respect to, e.g., PP and PE fibres (Table 3.1). Both nylon 6 and 66 fibres are strong and display a dry and wet tenacity of up to $\sim$80 and $\sim$70 g tex^{-1}, respectively, whereby the former group has increased melting temperature and decreased dyeing capability, likely due the decreased molecular weight of the nylon 6 repeating unit. Dry nylon fibres have elongations at break of 15%–50%, which increase somewhat on wetting, whilst an elastic recovery of 99% is observed from elongations up to 10% [23]. The mechanical properties of melt spun nylon fibres can be significantly tailored via postspinning fibre drawing with respect to the case of the semicrystalline polymer, due to the influence that the draw ratio and temperature have on polymer chain alignment, fibre crystallinity and fibre elasticity. For example, drawing of nylon 66 fibres above T_g can lead to significantly increased crystallinity and only marginal changes in polymer chain orientation. When the same drawing is applied below the T_g, nylon 66 fibres may not show a significant increase in crystallinity, despite the polymer chain orientation being strongly affected [24]. Polyamides are conventionally based on petrochemical-derived feedstock. However, bio-sourced polyamide materials that can be made from renewable resources have been known since the 1940s. One example is PA11 (e.g. Rilsan, Arkema), based on castor oil. The material is also utilised in thermoplastic block copolymer elastomers, where PA11 forms the rigid block in combination with soft, polyether blocks (Pebax Rnew, Arkema). Both variants can be processed into nonwoven fabrics, including by means of melt blowing.

3.3.5   Aramids

Aramid fibres are manufactured from long-chain synthetic polyamide in which at least 85% of the amide linkages are attached directly to the two aromatic rings (Table 3.1). Aramids first entered the market in the early 1960s as *m*-aramid fibres (developed by DuPont) under the trade name Nomex, and later *p*-aramid fibres as Kevlar (Table 3.1). Other examples of *p*-aramids include Twaron and Technora (Teijin) as well as Teijinconex (*m*-aramid).

The chemical structure of the polymer chains is such that the bonds are aligned along the fibre axis, resulting in outstanding strength, flexibility, and abrasion tolerance. The presence of aromatic rings and hydrogen bond-mediating amide groups in the fibre-forming polymer backbone leads to high performance fibres, with remarkable mechanical properties and tenacity, as well as superior resistance to abrasion, cutting, organic solvents and heat. In contrast to other synthetic fibres, aramid fibres do not melt but decompose at ~500°C, retain about 50% of their strength up to 300°C, and display negligible shrinkage at high temperature. Other than their heat resistance, aramids are sensitive to acids, bases and certain salts, and possess poor resistance to UV radiation, as evidenced by discolouration and a significant deterioration in mechanical properties. Due to their inert nature, aramid fibres are particularly useful in body and vehicle armour, protective clothing, and a variety of applications where impact resistance is important, including the manufacture of military helmets, protective gloves, and suits for fire-fighters.

From a manufacturing perspective, aramid fibres are prepared via wet or dry spinning of a hot concentrated polymer solution supplemented with inorganic salt. In the case of Kevlar, the fibre-forming polymer is synthesised via low-temperature polycondensation reaction between p-phenylene diamine (pPD) and terephthaloyl chloride (TCI) in an alkyl amide solvent, e.g., dimethyl acetamide (Fig. 3.6).

The polymer product, i.e., poly(p-phenylene terephthalamide), is isolated by precipitation with water, neutralised and subsequently washed and dried. The solvents used to prepare the wet spinning polymer solution are very strong chemicals, such as 100% sulphuric acid, and they produce fibres that are generally resistant to less aggressive media. Fibre wet spinning is carried out through a spinneret into weak acid or water, whereby inorganic salts can be washed away from the fibre-forming polymer. Unlike wet spinning, the salts are more difficult to remove in the dry spinning process and that is why this process is mainly used to produce the weaker m-aramid fibres. Following fibre formation, a series of stretching, solvent removal and heat-treatment stages is then applied, which induces orientation and alignment of the liquid crystalline domains in the flow direction. Resulting Kevlar fibres exhibit anisotropic properties, with higher strength and modulus in the fibre longitudinal direction than in the axial direction. Due to the chain alignment in the fibres, Kevlar displays low density ($\sim$1.38 1.47 g cm^{-3}), high tensile modulus ($\sim$125 GPa) and high ultimate tensile stress ($\sim$ 3.6 GPa), which make them significantly stronger than steel, glass fibre and nylon.

The alignment of polymer chains generates strong anisotropy of the fibre properties, i.e., high tensile strength and stiffness along the fibre axis and low strength and stiffness in the cross direction of the fibre. Furthermore, the compressive strength is low both along and

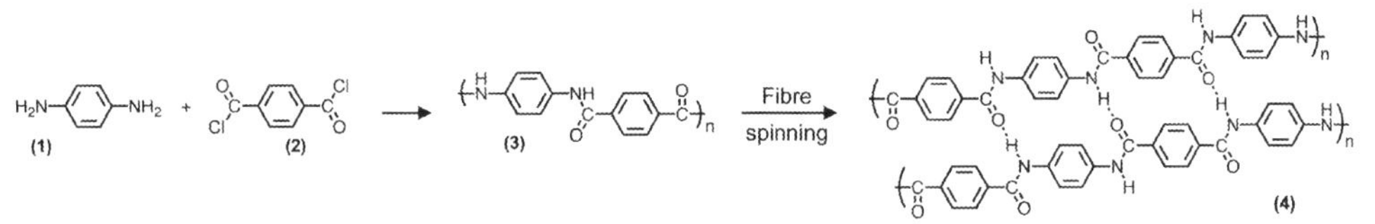

FIG. 3.6 Polycondensation between pPD (1) and TCI (2) generates poly(p-phenylene terephthalamide) (PPTA, 3). Wet spinning of PPTA leads to highly oriented Kevlar fibres (4) made of stiff polymer chains containing aromatic and hydrogen bond-mediating amide groups.

across the fibre axis, posing challenges in the applicability of aramid fibres in many applications where wear resistance and compressive strength are required. Copolymers and block copolymers containing aromatic polyamides have therefore been studied with respect to both the polymer synthesis and the fibre-forming capability of resulting products. These efforts led Teijin Aramid to develop and commercialise Technora, a copolyamide fibre made via condensation polymerisation of TCI with a mixture of pPD and 3,4′-diaminodiphenylether (3,4′-ODA). This fibre is closely related to Teijin Aramid's Twaron, with the only difference being that the latter is derived from PPD alone and not from a mixture of pPD and 3,4′-ODA (as in the former Technora). Together with the superior fibre properties, the relatively simple synthesis of the copolyamide in only one amide solvent allows fibre spinning directly after polymer production, so that increased manufacturing yields can be realised.

3.3.6 Thermoplastic elastomers

Thermoplastic elastomers (TPEs) relate to polymers and copolymers that have excellent elastic recovery after extension. Generally, they consist of block copolymers, where hard and soft blocks (or segments) are connected in specific ratios and sequences. Most commonly in nonwovens, TPEs are based on polyurethane (PU or PUR), but others include styrenic block copolymers, polyamide block copolymers (COPA) including polyether block amides (PEBA), copolyesters (COPE) or polyether ester block copolymers and polymer blends, notably thermoplastic blends containing polyolefins (TPO elastomers).

Polyurethanes represent a family of elastomeric block copolymer materials that are traditionally formed by reacting a 'soft' elastic polyol with either a 'hard' diisocyanate or an isocyanate polymer in the presence of an appropriate catalyst. Therefore, by varying the sequence and concentrations of the soft and hard blocks, as well as the molecular weight of each of these segments, PUs can be made with a variety of starting materials to modulate mechanical and other physical properties. PUs can also be produced using bio-based PA or PET polyester polyol building blocks, such as Priplast (Croda). Thermoplastic TPUs are used in meltblowns and spunbonds for applications where low modulus, excellent multidirectional extensibility, elastic recovery and tear resistance are required.

In styrenic block copolymers (SBS), the hard, thermoplastic blocks are comprised of styrene and the soft blocks, of butadiene. The heat and chemical stability of SBS-based TPEs can be greatly increased by hydrogenation to produce SEBs by removal of carbon double bonds in the butadiene blocks. Melt blowing of styrenic block copolymer materials is possible, but like many TPEs mixing with other materials can improve processing performance.

COPAs are copolymers made of hard thermoplastic polyamide blocks (e.g. PA6, 66, 11, or 12) and soft polyester, polyether-ester or polyether blocks. For example, PEBA's consist of polyether and polyamide soft and hard segments with ester linkages, wherein the specific polyether and polyamide blocks are selectable, depending on the processing performance and bulk properties required. These TPEs have good solvent and thermal resistance, and water vapour permeability is adjustable depending on the hydrophilicity of the block copolymer composition.

Similarly, the chemical resistance, mechanical and thermal properties of COPEs, which consist of hard polyester blocks and soft polyether blocks, can be controlled by careful selection of the composition and proportion of each block in the copolymer structure.

Bulk material properties in elastomeric TPOs can be achieved by blending low cost thermoplastic polyolefins such as PP and elastomers such as ethylene propylene rubber (EPR) or noncrosslinked ethylene propylene diene (EPDM). In another approach, (low or high-density) PE is blended with ethylene copolymers and monomers such as ethylene vinyl acetate (EVA), ethylene methyl acrylate (EMA) or ethylene ethyacrylate (EEA). Vistamaxx (ExxonMobil) copolymer is an example of a semicrystalline propylene-based elastomer (PBE) made of isotactic propylene repeat units with random ethylene distribution that when blended with PP and PE is used in spunbond and meltblown nonwoven production. Resulting fabrics have applications in wearable skin contact hygiene products, e.g., diapers, where elasticity and softness are required.

3.3.7 Other high performance synthetic fibres

In addition to *m*-aramids and *p*-aramids, discussed in Section 3.3.5, other speciality materials are used in nonwovens where for example, excellent thermal or chemical stability is required.

Poly(p-phenylene-2-6-benzobisoxazole) (PBO). PBO fibres exhibit a combination of extraordinary characteristics, such as excellent thermal stability, antiabrasion and mechanical properties, as well as resistance to creep, chemical degradation and flame retardancy. Commercialised as Zylon, these fibres present a unique extended rigid-rod molecular configuration, resulting in low polarity, low surface energy, and low coefficient of friction, making them excellent antifriction and wear-resistant material. PBO fibres display the highest tensile strength (~6 GPa, i.e. ~1.6-fold higher than Kevlar) and tensile modulus (270 GPa, 1200 g den^{-1}) amongst high-performance fibres (Table 3.2).

They also exhibit superior thermal stability (decomposition temperature: > 650°C) compared to any other heat-resistant polymers, such as polyimides (decomposition temperature: ~600°C) and Kevlar (decomposition temperature: ~500°C). These characteristics have

TABLE 3.2 Overview of the macroscopic properties of common high performance fibres (HPFs).

HPF	PBO	PBI	PPS	PF	MF	PAI	PEEK
Breaking tenacity (g den^{-1})	42	2.7	4.3–5.2	1.5	2	4	6–7
Specific gravity	1.54	1.4	1.37	1.27	1.4	1.67	1.3
Elongation at break (%)	3.5	27	25–35	10–50	11	19	15–50
Tensile modulus (N den^{-1})	1200	32	31–42	29–40	55	27–36	57
Moisture regain (%)[a]	2.0	15	0.6	6	5	4	0.1
Shrinkage (%)[b]	< 1.0	< 1.0	0–5 [c]	< 0.1	< 1	0	1.1

[a] *At 55% RH.*
[b] *In dry air at 177°C for 30 min.*
[c] *In boiling water.*

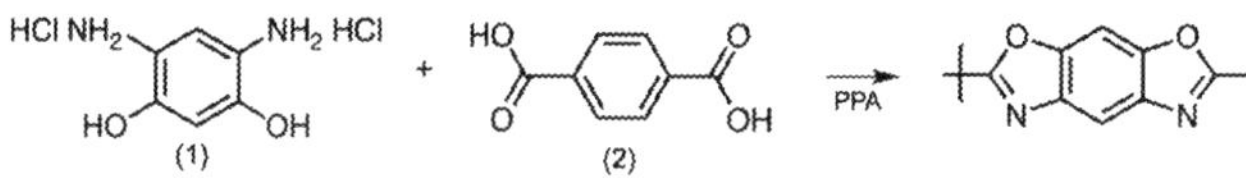

FIG. 3.7   Synthesis of PBO from DADBO (1) and TA (2). The polymer is then dry jet spun to yield high performance fibres.

therefore made them the material of choice for high-temperature application of military installations, aerospace crafts, and common industry.

PBO is synthesised from 4,6-diamino-1,3-benzenediol dihydrochloride (DABDO) and terephthalic acid (TA) in 77% of poly(phosphoric acid) (PPA) (Fig. 3.7).

The temperature is increased to 60–80°C to enable dehydrochlorination of the amine monomer and then lowered to 50°C, whereby phosphorus pentoxide is added (~70–80 wt.%) to induce polymerisation. Due to the rigid molecular structure of the polymer product, its insolubility in organic solvents and the lack of a glass transition, wet spinning and melt spinning cannot be applied. PBO fibres can therefore be produced via dry jet spinning of the polymer solution in PPA. Fibre extrusion is carried out under heat through an air gap into a coagulating bath, i.e., water, followed by washing, drying, and heat treatment (500–700°C).

Poly(benzimidazole) (PBI). PBI is the building block of another class of extremely heat-resistant synthetic fibres. Structurally, PBI is a completely aromatic polyamide, whereby recurring imidazole groups are present as one of the main structural repeat units. The only commercial PBI fibre is poly(2,2'-*m*-phenylene-5,5'-bibenzimidazole) [25], which is synthesised from an aromatic tetraamine and an aromatic dicarboxylic acid or a derivative of it (Fig. 3.8). Due to the fully aromatic and rigid structure, the lack of a melting transition and the presence of a very high glass transition temperature ($T_g = 425$°C), the formation of PBI fibres is typically expensive and difficult.

Dry spinning is employed with dimethyl acetamide as the solvent, generating fibres with outstanding heat stability and chemical resistance against alcohols, hydrocarbons, chlorinated solvents, hydrogen sulphide, weak acids and bases, and many other chemicals. PBI fibres are also flexible, have low tenacity and moisture regain (Table 3.2), and present an outstanding strength-to-weight ratio, making them attractive for a wide range of high-temperature applications. They are often used for safety and heat resistant garments like safety gloves, and fire-fighter uniforms, as well as plastic reinforcements, for heat and chemical resistant filters, and for various civil engineering applications. Blends of PBI with other high-performance fibres, e.g., Kevlar, are also common in order to achieve product customisation and improved manufacturability.

FIG. 3.8 Synthesis of PBI from tetraaminobiphenyl (TAB, 1) and diphenylisophtalate (DPIP, 2).

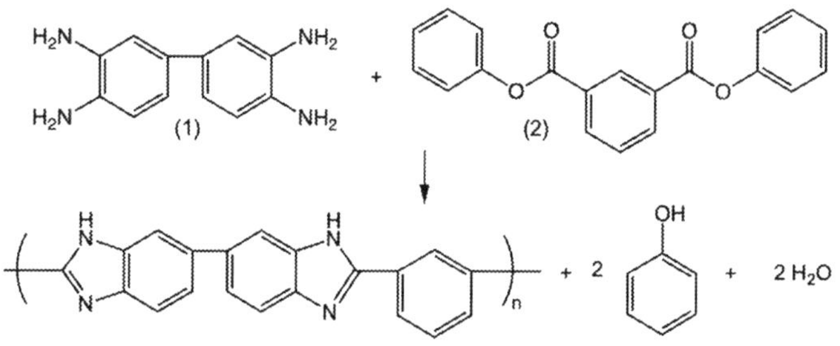

Poly(phenylene sulphide) (PPS). PPS is a synthetic polymer consisting of aromatic rings linked with sulphides. Due to its molecular structure, PPS generates high-performance, semi-crystalline fibres with a crystallinity in the range of 50%–65% and with exceptional chemical and heat resistance, high dimensional stability, low moisture absorption, and high resistance to dyeing and hydrolysis (in either acidic or alkaline conditions), organic solvents and to nearly all common chemicals. Because of their high melting transition temperature ($T_m = 285°C$), PPS fibres withstand continuous use up to 190°C, whilst fibre formation can be accomplished via melt spinning with high reproducibility and tolerance. Aforementioned thermal properties also explain the excellent creep and property retention of PPS fibres at both high and low temperatures, with typical tensile strength at break, tensile modulus and elongation at break (of staples) of up to ~130 MPa, 31 to 42 g den^{-1}, and 35% (Table 3.2), respectively, depending on whether a linear, cured, branched or filled configuration is present. An example of commercial PPS-based nonwoven fabrics is TORCON, that is used in a wide variety of industrial applications, including bag filters, in the aerospace industry as a replacement of aluminium for structural components, and in the automotive industry for actuator housings and for the gears used for power transfer to breaks, due to its dimensional stability over a wide temperature range, impact resistance and chemical resistivity.

Phenolic resins. These are synthetic polymers obtained from the reaction of phenol (or substituted phenol) with formaldehyde. Phenolic fibres (PFs) were invented by Dr. James Economy and coworkers at the Carborundum Company in the late 1960s and early 1970s and have been commercialised under the trademark Kynol. On the molecular level, they exhibit a highly aromatic, nonoriented, crosslinked structure, which makes them infusible, insoluble, and generally intractable (Fig. 3.9). In light of these characteristics and the presence of a crosslinked network, PFs cannot be realised using conventional spinning processes. Fibre formation is firstly carried out with a precursor polymer (novolac), which is then treated chemically with formaldehyde and heat to achieve crosslinking in the fibre state.

Variation of the formaldehyde/phenol ratio therefore enables changes in crosslink density and fibre macroscopic properties. As made of thermoset polymers, PFs have no melting point but only a decomposing point starting from 220°C, a characteristic that has been exploited for making weather and boil proof plywood. Due to the crosslinked structure and high degree of aromaticity, these fibres display outstanding heat resistance, good UV stability, very low flammability, excellent electrical and thermal insulation properties, and excellent chemical resistance to most organic solvents as well as acids and alkalines. On the other hand, they

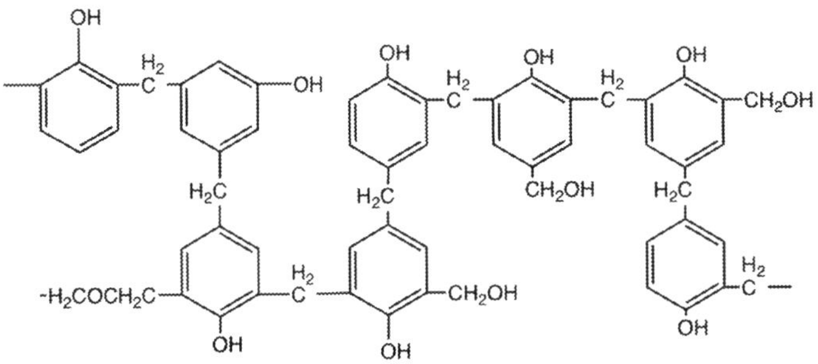

FIG. 3.9 Chemical structure of typical PFs, exhibiting a covalent network with a high degree of aromaticity.

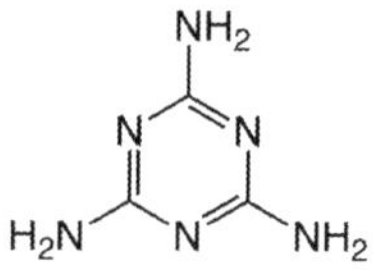

FIG. 3.10 Chemical structure of melamine. The primary amino groups react with formaldehyde to generate a crosslinked network. The crosslink density can be controlled by addition of bifunctional melamine derivatives, e.g., benzoguanamine and acetoguanamine.

have poor abrasion resistance and only low to fair tenacity and lower tensile modulus compared to other heat resistant high-performance fibres such as Kevlar and PBO (Table 3.2). Nevertheless, PFs find applications as flame resistant fabrics, such as seat linings, thermal insulators for ventilation ducts, and military vehicles, electric arc protection wrapping tapes, spark and metal splash protection (welding), and chemical resistant garments and gloves. Various mechanical applications have also been met, e.g., gaskets, disk brake pads, brake linings, and reinforcement for composites.

Melamine. Melamine can be used in the presence of formaldehyde to generate thermoset fibres, whereby melamine's primary amino groups undergo crosslinking reaction with formaldehyde, generating methylene and dimethylene ether linkages (Fig. 3.10). As in the case of PFs, melamine fibres (MFs) undergo continuous polymerisation and crosslinking under heat, whereby modified melamine derivatives are often used to improve fibre's breaking elongation (Table 3.2), e.g., in Basofil.

MFs are commonly used as a cost effective heat resistant material, whereby the presence of crosslinked network enables continuous operation at a temperature of 200°C. MFs have flame resistance and self-extinguishing properties due to the release of nitrogen gas when burned or charred, whilst displaying outstanding heat and dimensional stability, durability and resistance to chemicals and ultraviolet light. The most outstanding physical properties of MFs are therefore the high limiting oxygen index (LOI), low thermal conductivity, and the fact that they do not shrink, melt, or drip when exposed to a flame.

Polyamide-imides (PAIs). PAIs are used to generate strong and heat-resistant synthetic fibres, which are made via polycondensation of a diisocyanate, e.g., phenylene diisocyanate (MDI), with an acid anhydride, e.g., trimetallic anhydride (TMA). As with other aromatic-rich fibres (Fig. 3.11), PAI fibres possess outstanding heat resistance, low flammability, and good chemical resistance to aliphatic and aromatic hydrocarbons as well as chlorinated and fluorinated solvents.

In comparison to aramid fibres, PAI fibres have outstanding resistance to UV radiation and twofold lower thermal conductivity, as well as decreased mechanical properties due to the

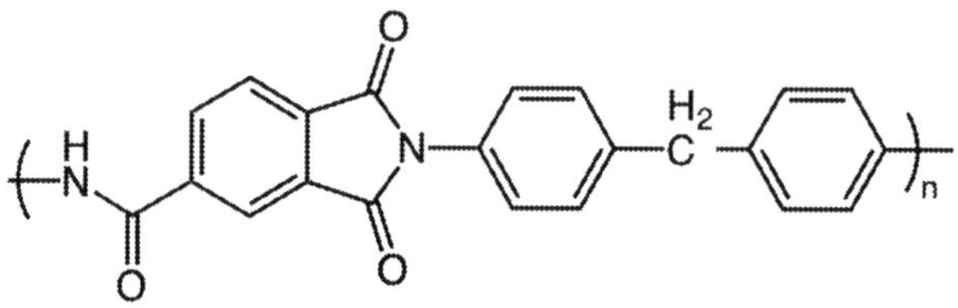

FIG. 3.11 Chemical structure of PAI obtained via polycondensation reaction of trimetallic anhydride chloride with methylenedianiline.

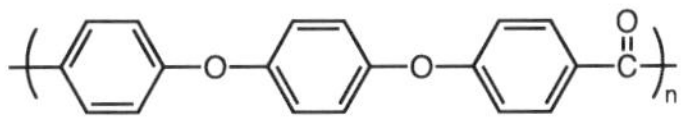

FIG. 3.12   Chemical structure of PEEK.

lower content of aromatic structures. Other than that, they also display very high resistance to abrasion due to its amorphous structure, high elongation at break, good elasticity and dimensional stability, and medium to high tenacity over a wide temperature range (Table 3.2). PAI fibres are therefore mainly used for protective clothing such heat resistant safety gloves, flight suits, tank crew coveralls, fire-fighting uniforms and industrial workwear for personnel working in hazardous industries. PAI textiles are also used for some technical applications such as hot gas filtration.

Polyetheretherketone (PEEK). PEEK is a semicrystalline and aromatic thermoplastic polymer, which belongs to the family of polyetherketones (PEKs). Its thermoplasticity ($T_g = 143°C$; $T_m = 334°C$) makes it compliant with high-temperature meltspinning, so that an extensive range of monofilaments and fibres can be realised with no need of additives. Whilst it was first commercialised by ICI Advanced Materials in the late 1970s (Victrex PEEK) for injection moulding processing, subsequent work carried out by ICI and the University of Leeds showed the polymer's potential for the manufacture of high-specification fibres. From the chemical point of view, PEEK's repeat units contain one ketone and two ether groups, so that polymer chains with a linear, fully aromatic and highly stable structure are accomplished (Fig. 3.12).

This molecular configuration is responsible for PEEK's excellent thermal behaviour, which ensures continuous operation at temperature as high as 260°C, and a nonbrittle mechanical response at temperature as low as −60°C. PEEK is also unaffected by high-temperature steam and most fluids and chemical reagents, although it is soluble in concentrated sulphuric acid (>50%) and degraded by strong oxidising agents such as nitric acid. Besides that, PEEK fibres are tough and present low friction, low wear, cut-resistant surface and abrasion resistance at elevated temperatures and relatively high surface speeds. From a mechanical perspective, PEEK fibres highlight low creep and low shrinkage (Table 3.2), especially below its T_g, as well as excellent dynamic recovery and flex fatigue performance. Because of their excellent purity, they are generally inert in the biological environment, as also evidenced by the fact that they have been approved by the FDA for use in medical devices, e.g., in the context of dental implants. Together with their self-extinguishing properties (LOI = 35%), PEEK fibres are also appealing from a sustainable point of view, since the recovery and recycling of PEEK can be accomplished with nearly preserved physical properties.

Polyvinyl alcohol (PVA, PVOH, or PVAL). PVA is derived from polyvinyl acetate following hydrolysis. Depending on the degree of polymerisation and hydrolysis, fibres can be produced that are either insoluble or soluble in water. Solubility can also be modulated by temperature, and warm/hot water-soluble variants are in the range 30°C to >100°C. In nonwovens, PVA is considered a speciality material with quite specific applications. Its ability to be dissolved in water, even at low temperature, is distinctive amongst synthetic polymers. Examples of PVA staple fibres include Kuralon and Kuralon K-II (Kuraray), which differ slightly in their manufacturing methods, and Vinylon or Mewlon (Unitika). Both water soluble and insoluble PVA grades are used for papermaking and wetlaid nonwoven production, where fibres ranging from 0.3 to 3 denier are short cut to 2 to 8mm. Given the temperature-sensitive dissolution behaviour that is possible in water-soluble grades, PVA fibres are incorporated as a binder component to increase the network strength of wetlaid fabrics made from inorganic materials such as glass. Existing applications for wetlaid nonwovens containing short cut PVA fibres include battery separator fabrics and filter media. PVA staple fibres are also cut to longer lengths to enable drylaid web formation, and melt-processable PVA resins for spunmelt manufacture, such as melt blowing are also produced. Nonwovens containing PVA fibres alone or in blends, also include sacrificial, water soluble backing fabrics used for embroidery and cold-water soluble ostomy bag liners. The latter facilitates convenient disposal of used ostomy bags in the toilet following use.

3.3.8 Biodegradable and bio-based polyesters

In addition to materials such as PVA and PLA, other biodegradable polymers for nonwoven production include petroleum-based poly(ε-caprolactone) (PCL) and polybutylene adipate terephthalate (PBAT), as well as bio-based polybutylene succinate (PBS), polyhydroxyalkanoates (PHAs) and polyethylene furanoate (PEF) [26]. PCL nano-to-microscale fibres can be made via either melt spinning or wet spinning, whilst corresponding webs are typically produced by electrospinning or melt blowing. Electrospun PCL webs have been reported to display an averaged tensile modulus of up to 60 MPa [27–29] and have found wide applicability in the biomedical industry, whereby their hydrolysis makes them appealing for the manufacture of, e.g., degradable sutures, controlled drug delivery devices, and degradable implants. Another well-known biodegradable aliphatic polyester is PBS, which is a crystalline thermoplastic polymer typically synthesised via polycondensation of 1,4-butaindiol and succinic acid [30]. These monomers can be obtained from renewable sources, e.g., by fermenting lignocellulosic sugars (succinic acid) and by hydrocracking starches and sugars (butanediol). PBS can be degraded by enzymes, bacteria, and fungi as well as in soil, offering opportunities in circular economy. PBS displays a tensile strength ($\sigma = 20$–30 MPa) in between the ones of polyethylene (PE) and polypropylene (PP), whilst its melting transition temperature is comparable to that of low-density polyethylene (LDPE). Due to its excellent biodegradability, thermal and chemical resistivity, PBS has found wide applicability as degradable plastic for the manufacture of, e.g., films, sheets foam and bottles. On the other hand, its relatively low elastic modulus, tensile strength, and rate of degradation, often restricts its manufacturability so that PBS is often blended with other polymers to generate composites [31]. To overcome the manufacturing limitations of PBS, PBAT has also been employed. PBAT is an aliphatic-aromatic copolyester, which is synthesised by polycondensation between butane diol (BDO), adipic acid (AA), and terephthalic acid (PTA). Due to presence of aliphatic and aromatic units in the polymer backbone, PBAT presents a unique combination of relevant tensile properties ($\sigma = 20$ MPa, $\varepsilon > 650\%$) and hydrolytic biodegradability [32]. PBAT has therefore been converted into multiple products, such as Ecoflex, Ecopond, and Origo-Bi, which have been applied into a variety of industrial applications, including shopping bags, waste bags, cutlery, and mulch film. Other than PBAT, PHAs are a family of biocompatible polyesters produced by bacteria that can safely be degraded in the human body and the environment. They therefore hold promise in overcoming both petroleum reliance and plastic pollution, although the low production yield and restricted process-structure relations has so far limited their industrial uptake. Tailoring of macroscopic properties and manufacturability can be achieved by varying the repeat unit chain length, the side chain functionalities, and comonomer composition [33]. While PHAs typically undergo faster hydrolysis than PLA, short length chain (Slc) variants, e.g., poly(3-hydroxybutyrate, PHB), display very high crystallinity, causing brittleness and poor fibre manufacturability. Medium length chain (Mlc) PHAs typically show attractive low crystallinity ($T_m \sim 60°C$), whereby the increase in side-chain length may impact on polymer hydrophobicity and hydrolytic degradability. PHA fibres have been produced by electrospinning and wet spinning [34] and are approved for use in certain medical devices. The applicability of PHAs in healthcare as well as single use hygiene products is still in its infancy, due to challenging polymer processability (brittleness) and low quantity of bacterial production. Other than biodegradable polyesters, polyethylene furanoate (PEF) has recently emerged as a promising high-performance plastic similar to PET. The increased gas diffusion barrier, tensile strength, and glass transition temperature make PEF suitable for applications in, e.g., long-lasting packaging materials. It can be synthesised via polycondensation of the bio-derived monomers monoethylene glycol and 2,5-furandicarboxylic acid.

3.4 Regenerated cellulosics and man-made fibres derived from natural polymers

The British Plastics Federation (BPF) identifies five different feedstocks on which bio-based polymers are typically based, as follows:

- Cellulose (mainly but not exclusively obtained from wood pulp).
- Lignin (commonly extracted from pulp waste).
- Starch (principally obtained from maize, potatoes, and cassava).
- Oils (e.g. from castor beans, soya beans and oilseed rape, for use as starting materials to make bio-based PA, PBTs, and PUs).
- Proteins (e.g. casein, soy, and maize).
- Xylans (e.g. extracted from cereal grains).

With the obvious exception of cellulose, which is a very important source of raw material in the nonwoven industry, and to a lesser extent PLA, based on corn starch (or sugar cane), consumption of bio-based polymers in the nonwoven industry is relatively small. However, progressive growth in the utilisation of bio-based materials is taking place as material supply issues, cost, polymer and fibre processing challenges, as well as regulatory requirements are addressed (see also Section 3.3.8).

In the nonwovens industry, naturally abundant polymers, such as polysaccharides are made into man-made fibres via dissolution in an appropriate solvent and then extrusion of the polymer solution into a coagulation bath, or through solvent evaporation, i.e., wet spinning, dry spinning, or derivatives thereof. Industrially, wet spinning is the most common route. Cellulose, normally in the form of wood pulp is the usual starting material and is the basis of viscose rayon, cellulose acetate and lyocell fibre manufacture. These regenerated cellulosic man-made fibres are produced from the same polysaccharides that make up classical natural fibres such as cotton, hemp, flax, and the structural fibres found in wood (i.e. wood pulp). However, regenerated cellulosic fibres are formed by chemical processing of the cellulosic feedstock, before being spun into functional cellulose-derived fibres.

In contrast to cellulose acetate, both viscose rayon and lyocell have the same chemical composition as cellulose following purification and spinning and they have emerged as the most important regenerated cellulose fibres for the manufacture of nonwovens. The importance of such cellulosic fibres has been brought into sharper focus by the Single-Use Plastics Directive (SUPD) in Europe, which excludes 'natural polymers that not been chemically modified' from the formal definition of a plastic.

3.4.1 Viscose rayon

Viscose rayon is a polysaccharide-based fibre made of cellulose, a linear polymer consisting of β-D-glucose units (Table 3.1) that is present in all plants. The cellulose is traditionally derived from wood pulp, which has an average cellulose content of 40%, but other raw materials can include cotton linter pulp and bamboo. Like naturally occurring cellulosic fibres such as cotton, viscose rayon is hygroscopic, and fibre properties therefore differ in the dry and wet states. The tenacity of viscose rayon ranges from 3 to 4.5 and 2 to $3.6\,g\,dtex^{-1}$ in the dry and wet state [35], depending on the grade as low and high modulus (modal) variants are available. Although sharing a similar chemical composition, rayon differs from cotton with regards to its molecular weight and crystallinity, being about one-fifth and one-half that of cotton, respectively. These differences make the traditional or standard viscose rayon relatively weaker and more extensible (ε: 8%–10%), but about twice as absorbent as cotton (ε: 7%–9%). Thus, under standard conditions (65% RH and 20°C), the fibre absorbs about 12% to 14% moisture, and when soaked, can swell in and absorb almost $70\,wt\%$ water, resulting in a significantly weaker fibre. In comparison to cotton, viscose rayon also has a greater number of free hydroxyl groups, so that specific compounds can be readily grafted onto the polysaccharide backbone in order to tailor macroscopic properties and introduce specific product functionalities. Globally, approaches to the management of sustainable forests, emissions levels and water impacts in the viscose rayon manufacturing supply chain varies between manufacturers such that is not possible to make generalised comments about the environmental sustainability of the fibre.

The manufacture of regenerated cellulosic fibres, such as viscose rayon, converts the purified nonwater-soluble cellulose material into a soluble compound via chemical modification. The solution of the cellulose derivative is passed through a spinneret yielding multiple soft filaments that are then converted or 'regenerated' into almost pure cellulose via a subsequent chemical modification step. To manufacture viscose rayon, wood pulp is treated with aqueous sodium hydroxide (typically $16-19\,wt\%$) to form 'alkali cellulose' (Fig. 3.13). During this step, the cellulose chains are cleaved and allowed to depolymerise depending on the temperature and the presence of inorganic additives, such as metal oxides and hydroxides, and oxygen. The alkali cellulose is then treated with carbon disulphide to form sodium cellulose xanthate (Fig. 3.13). The higher the ratio of cellulose to combined sulphur, the lower the solubility of the cellulose xanthate. The xanthate is dissolved in a mineral acid solution, such as sulphuric acid. In this step, the xanthate groups are hydrolysed to

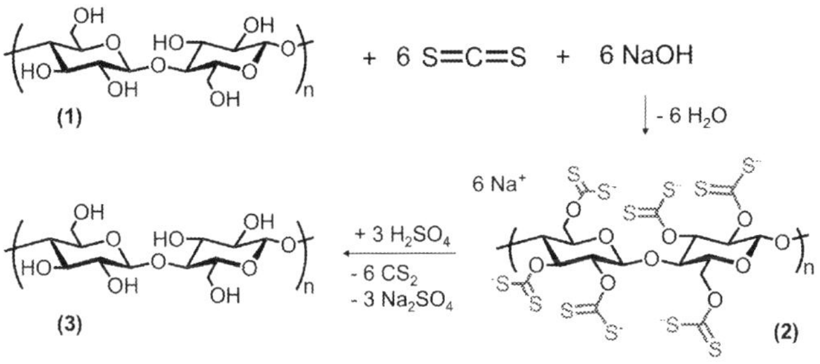

FIG. 3.13 Manufacture of viscose rayon. Wood pulp is incubated in aqueous sodium hydroxide to generate depolymerised alkali cellulose with decreased molecular weight (1). Treatment of alkali cellulose in carbon disulphide leads to sodium cellulose xanthate (2). This product is dissolved in sulphuric acid, whereby xanthate groups are hydrolysed to regenerate the cellulose (3).

regenerate cellulose and release dithiocarbonic acid that later decomposes to carbon disulphide and water (Fig. 3.13).

Aside from regenerated cellulose, acidification by-products include hydrogen sulphide, sulphur and carbon disulphide. Acid residues are removed by washing of the regenerated cellulose thread, whilst a sodium sulphide solution is added to remove sulphur; other impurities are oxidised by bleaching with sodium hypochlorite solution or hydrogen peroxide solution. The use of highly toxic compounds, e.g., carbon dioxide, has historically raised health and safety issues during the production of viscose rayon. These issues have partially been mitigated by the implementation of more stringent processes based on European Standards, which allow recovery and reuse of chemicals and other natural resources, reducing the exposure to workers as well as making the overall process much safer. Other viscose fibres include higher tenacity variants (modal), delustred and spun-dyed products, and those with a trilobal (star-shaped) fibre cross-section (Viscostar). The latter has important applications in tampons for liquid management.

3.4.2 Cellulose acetate

Cellulose acetate fibre is a man-made cellulose-derived fibre. Its chemical composition is similar to that of cellulose, with the main difference being that most of the hydroxyl groups are derivatised into acetate esters. Cellulose acetate is therefore no longer considered a regenerated cellulose fibre because the polymer composition is chemically different from that of cellulose. Cellulose acetate fibres have relatively low strength values in the range of 1 to $1.3\,g\,dtex^{-1}$ (dry) and 0.8 to $0.9\,g\,dtex^{-1}$ (wet). The dry state extension at break can be as high as 35%, whilst it increases slightly, up to 45% with wet fibres. Cellulose acetate fibres have also a moderate water absorption capability with a standard moisture regain of 6.5% (close to that of cotton) and good dimensional stability with some stretch and resistance to shrinkage.

Cellulose acetate was firstly synthesised by Paul Schützenberger in 1865, in an attempt to dissolve cotton in acetic anhydride under heat. When this solution was poured in water, white amorphous flakes, called cellulose triacetate, were precipitated. Several contributions to the science of cellulose derivatisation followed, resulting in a series of key publications and patents. Despite the cellulosic product that could be obtained, the formation of cellulose acetate fibres itself did not directly lead to acetate fibre spinning until the method of converting triacetate into secondary acetate (diacetate) was discovered (Fig. 3.14) by George Miles in 1904 [36]. The cellulose diacetate product was demonstrated to be soluble in relatively cheap and

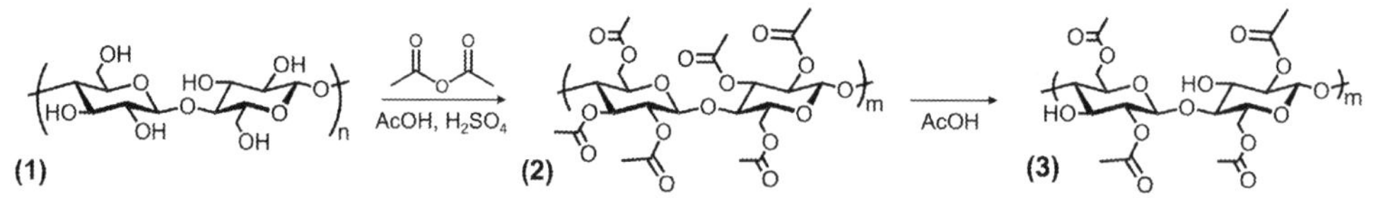

FIG. 3.14 Manufacture of cellulose acetate. Either cotton or wood pulp fibres (1) are incubated in a mixture of acetic anhydride, acetic acid and sulphuric acid, yielding the viscous gelatinous product of cellulose triacetate (2). Incubation of this product in acetic acid generates cellulose diacetate via acid hydrolysis and the random regeneration of acetate esters into hydroxyl groups (3).

nontoxic solvents such as acetone, in contrast to the use of challenging solvents, such as chloroform, nitrobenzene and epichlorohydrin, for the pristine acetylated product. This development solved the problem of directly dissolving cellulose triacetate for fibre spinning.

The first commercial production of acetate filament was undertaken by the British Cellulose Co. Ltd. in 1921. One year later, Camille Dreyfus (one of the British Cellulose Co. Ltd. owners) launched acetate fibre production in the United States. Despite these early advancements, the application of cellulose acetate fibres was rapidly affected by the expansion and relatively low price of synthetic fibres. Cellulose acetate is used to make a variety of nonwoven fabrics, for example, Eastman Vestera staple fibres produced in a near closed-loop fashion are suitable for drylaid or wetlaid web formation, and high loft Celaire multilayer nonwovens from Celanese, which contain a continuous filament core, are used in woundcare, absorbent core and insulation products.

3.4.3 Lyocell

Lyocell is a regenerated cellulose fibre produced by dissolving wood pulp or cotton cellulose in N-methylmorpholine N-oxide (NMMO) by heating to make a viscous spinning solution. Following filtration of this solution, the following manufacturing steps are applied:

(1) The filtered solution is extruded through a spinneret via dry wet spinning against a diluted amine oxide coagulation batch, yielding the solidified filaments.

(2) Following washing with demineralised water to remove any NMMO residue, the lyocell fibres pass through a drying area to remove the water, followed by a finishing step, whereby a lubricant, e.g., soap, silicone, or other agents, is applied to the fibre, aiming to facilitate the future steps of web formation and bonding. The dried, finished fibres at this stage in the form of the so-called *tow*, which is a large untwisted bundle of continuous length filaments.

(3) The bundles of tow can then be crimped. Fibre crimping improves compatibility with subsequent carded web formation, improving interfibre cohesion and enabling high speed production. Following crimping, staple fibres are formed by cutting the tow and then can be baled prior to shipment to a nonwoven roll good manufacturer. The entire manufacturing process, from unrolling the raw cellulose to baling the fibre, takes only about 2 h.

(4) The amine oxide used to dissolve the cellulose and dry wet spin the fibre is recovered (~99%) and reused in the manufacturing process. The dilute solution is evaporated to remove the water, and the amine oxide reused to dissolve the wood pulp or cotton cellulose.

In contrast to viscose rayon, water absorption has less impact on the fibre strength of lyocell, even though the fibre is comparatively hygroscopic. The dry tenacity of lyocell fibres ranges from 38 to 42 cN tex^{-1} compared to a wet tenacity of 34 to 38 cN tex^{-1}. Lyocell fibres have a characteristically fibrillar structure, which means the fibre can be mechanically refined like wood pulp to make a suspension in water and then made in to paper substrates or wetlaid fabrics. High-pressure hydroentanglement is also capable of inducing fibrillation as well as fibre entanglement, allowing the opacity and surface texture of the fabric to modified during bonding. Lyocell is produced in different grades for nonwovens, bearing in mind the specific requirements of different high-speed web formation processes and end-products.

Hygiene is an important nonwoven market for lyocell fibres. For wetlaid and airlaid fabrics, short cut lyocell (fibre length <10 mm) is blended with wood pulp to increase wet strength in the manufacture of cost-effective single use, premoistened dispersible wipes. Typical blend proportions are 80:20 and 70:30 wood pulp:lyocell. Given the need for liquid absorbency and high wet strength during use and biodegradability following disposal, wipes have become a primary application for lyocell. Whereas lyocell for nonwovens is predominantly manufactured as staple fibre, Lenzing Web Technology enables direct formation of lyocell fabrics made from continuous filaments.

3.4.4 Bacterial cellulose

Cellulose fibres are generally isolated by chemical treatments from the cell walls of wood and plants. With the development of molecular biology and in vitro models, the mechanism underlying the biosynthesis of cellulose has been extensively explored over the past 30 years. These efforts have aimed at renewable natural polymers to meet current societal constraints determined by the growing population, energy crisis and environment pollution.

Bacterial cellulose (also known as microbial cellulose) is a promising natural polymer synthesised by specific bacteria, e.g., *Gluconacetobacter*, *Sarcina*, and *Gluconacetobacter xylinus*. Bacterial cellulose is biocompatible and highly hydrophilic as the plant-derived variant, whilst it produces three-dimensional fibre networks with decreased fibre diameters of approximately 10 to 100 nm [37], i.e., significantly smaller than typical man-made fibres (Fig. 3.15). In contrast to plant-derived cellulose, this product displays high purity and no hemicellulose and lignin species, making it attractive in light of its simple production and purification process.

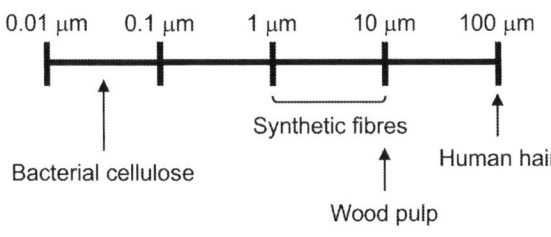

FIG. 3.15 Diameters of bacterial cellulose, synthetic and natural fibres.

The presence of inter- and intra-fibrillar hydrogen bonds and the high degree of polymerisation (up to 8000) at the molecular scale generate materials with increased crystallinity (up to 90%) and fibrillation, superior Young's modulus ($\sim$15–35 GPa) and tensile strength ($\sim$200–300 MPa) and high water content. Because of its unique structural and mechanical properties, bacterial cellulose is expected to become a commodity material in various fields, with its reinforcing effect already being exploited during paper-making to improve the physical properties of papers.

Improved quality-controlled, scalable and cost-effective processes of manufacture and recycling are still needed to realise full exploitation of bacterial cellulose at industrial scale [38]. To address this challenges, bespoke in vitro culture systems have been investigated to tailor bacterial cellulose growth for textile and garment applications [39]. Tailor-shaped cultivation techniques proved to generate bacterial cellulose growth within garment panels according to sustainable design practice. The whole production cycle was successfully demonstrated in environment-friendly conditions, whereby no cutting step was required, and no material waste generated, with the potential for significant reduction in labour, time and energy costs.

3.4.5 Cuprammonium rayon

In nonwovens, cuprammonium rayon (or cupro/cupra) fabric production is noteworthy because it uses a wet-spunlaid process to produce nonwoven fabrics made of regenerated cellulose continuous filaments rather than staple fibres. Derived from the historical Bemberg process, Bemliese nonwoven fabrics (Asahi Kasei), are industrially made from cotton linters. Cuprammonium solution is spun into filaments which are solidified to produce a coherent web and nonwoven structure. Traditionally, cuprammonium solution is produced by dissolving cotton linters in a solution of ammonia and copper hydroxide or an ammonia solution containing copper sulphate with added alkali, which produce highly stable solutions. This is followed by filament deposition to form a web, coagulation and bonding, normally by means of hydroentangling.

3.4.6 Other man-made fibres derived from natural polymers

From the early days of the man-made fibres industry, various methods were developed to produce regenerated fibres from a variety of naturally derived polymers, such as casein (protein in milk), soybean, groundnuts, zein (maize protein), and collagen but following the rapid growth of synthetic petrochemical-based fibre production, many were discontinued. Today, increasing demand for nonwovens based on renewable materials has reinvigorated interest in naturally derived polymers for making fibres as well as coatings and binders for nonwovens. A brief overview of some of the materials being used by the nonwovens industry, albeit for specialist or small-scale applications, is worthwhile.

3.4.6.1 Alginate

Alginate is a linear, water soluble polysaccharide that exists widely in many species of brown seaweed. It consists of (1-4)-linked units of α-D-mannuronate (M) and β-L guluronate

(G), whereby its chemical composition and sequence of the M and G residues depend on the biological source and the state of maturation of the plant. Alginate fibres can be prepared by extruding solutions of sodium alginate into an aqueous calcium chloride bath via simple wet spinning, so that the presence of calcium ions enables the development of ionic crosslinks with the G units of the fibre-forming alginate [40]. Resulting alginate fibres exhibit a density, tenacity and elongation of $1.78\,g\,cm^{-3}$, 11 to $18\,cN\,tex^{-1}$, and 5% (at 21°C and 65% RH), respectively, whilst a water absorbency and moisture sorption in the range of $18\,g \times 100\,cm^{-2}$ and 20% to 35% (at 20°C and 65% RH), respectively, are common. Together with the well-known biocompatibility, aforementioned characteristics have therefore made alginate fibres particularly appealing for wound dressing applications. In situ, calcium-crosslinked alginate fibres absorb the exudate released from the wound, so that a water-swollen material is formed due to the ion exchange between the fibre-crosslinking calcium ions and the sodium ions in the exudates. The gel-forming property of alginate supports the development of a moist wound healing environment, which is key for chronic wound care, as well as facilitating low adherence to the wound site, permitting the removal of the dressing with minimal pain and tissue injury. The ability of alginate to interact with divalent ions has also been exploited to generate bacteriostatic wound dressings, via replacement of sodium ions with zinc, or even silver, ions during fibre formation. Alginate dressings are therefore manufactured to treat acute and chronic wounds, whereby alginate is often blended with other polymer phases to accomplish antimicrobial and antiinflammation functionalities for wound healing. Most commonly, to make nonwoven wound dressings, staple fibre alginate is carded, cross-lapped and needlepunched. The fibres are generally quite brittle and so minimisation of fibre breakage during fibre opening and carding operations is crucial to maximise yield.

3.4.6.2 Chitin

Chitin is the second most abundant biopolymer that is found widely in cell walls of fungi, moulds, and yeasts, and in the cuticular and exoskeletons of invertebrates, such as crabs, shrimps, and insects. Its chemical sequence consists of long polysaccharide chains of N-acetylglucosamine units, which form ordered crystalline microfibrils with structural function similar to the one of collagen in animals, and cellulose in terrestrial plants [41]. Chitin microfibrils are held together via hydrogen bonds between the carbonyl and amine groups along the polysaccharide chains, so that three different fibril configurations (α, β and γ) can be found depending on the packing and polarities of adjacent chains. This tight network structure therefore brings inherent challenges with respect to the solubilisation of chitin and its processability into fibres. Alteration and breakdown of chitin crystalline domains has been pursued via chemical modification of the amino functionalities, such as etherification, esterification, crosslinking, and graft copolymerisation. Other than supporting fibre formation and improve physical properties, these chemical strategies have also been demonstrated to introduce relevant fibre functionalities, e.g., antibacterial, antifungal, antiviral, antiacid, nonallergenic, biocompatibility, and biodegradability. Dibutyryl chitin (DBCH) has been obtained by esterification with butyric anhydride in the presence of perchloric acid, so solubilisation in ethyl alcohol and dry wet spinning could generate DBCH fibres. Wet spinning of the DBCH solution (14.5%) in dimethylformamide was also performed to accomplish filaments, whereby subsequent fibre treatment in alkaline condition was also carried out to induce chitin regeneration. Resulting products therefore exhibited a gradual increase of crystallinity, density, tensile strength, and average elongation at rupture, as well as a decrease of fibre diameter.

Other than chemical modification, strong organic solvents commonly used for cellulose spinning, such as strong acids, fluoroalcohols, and chloroalcohols, have also been applied to chitin, aiming to break down the intra- and intermolecular hydrogen bonds [42]. For example, a solvent system of dichloroacetic acid, formic acid, and diisopropyl ether was applied, although the corrosive nature of the former was found to induce uncontrollable degradation of the polymer chains, so that mechanically deficient fibres with low wet state tenacities were formed. Other than that, solubilisation of chitin in trichlroacetic acid proved to generate high tensile strength wet spun fibres, although large fibre diameters (Ø ~0.25 mm) were recorded, despite fibre drawing postspinning.

To overcome aforementioned limitations in mechanical properties, the incorporation of chitin in composite fibres has been demonstrated, aiming to integrate the stiff chitin phase with soft polymers in single fibres. Composite fibres of alginate and chitin derivatives have been prepared by spinning of the polymer blend against a coagulating bath supplemented with calcium chloride. Here, the strong interactions originating from intermolecular hydrogen bonds and electrostatic complexation were leveraged to ensure miscibility between the two polymer phases. Enhanced tensile strength and elongation at break were obtained at low chitin content (≤ 30 wt%), whilst decreasing wet-state mechanical properties were recorded at increased concentrations. Other than that, the introduction of chitin in the composite fibre could also be pursued to improve the water-retention properties of alginate fibres, whereby bespoke chemical modifications could be introduced to further improve fibre properties and functions.

3.4.6.3 Chitosan

Chitosan is derived from the partial deacetylation of chitin and represents the only linear cationic polysaccharide. With a degree of deacetylation typically higher than 70%, chitosan is a copolymer comprising glucosamine and N-acetylglucosamine units, so that it mimics the chemical composition of the glycosaminoglycans of the extracellular matrix of biological tissues. It presents competitive biodegradability, biocompatibility, immunological, antibacterial and wound healing properties, as well as metal-chelating, film- and fibre-forming properties [43]. It has therefore been widely applied in the biomedical field as wound dressing, haemostat, regenerative scaffold, and for the controlled delivery of drugs and genes [44]. The degree of deacetylation has a striking effect on the solubility of chitosan, so that clear polysaccharide solutions can be prepared in acidic environment. In contrast to the case of chitin, the formation of chitosan fibres is much more straightforward, whereby the polymer is dissolved in acetic acid solution and then extruded through the spinneret into a caustic coagulation bath to obtain a regenerated fibre. Due to the presence of primary amino groups along the polysaccharide chains and the decreased polymer crystallinity, resulting fibres swell in water and present poor wet-state strength (tenacity 2.0 g den^{-1}). Although the fibre properties are affected by spinning conditions, such as spin–stretch ratio, coagulation bath concentration, and drying conditions, postspinning chemical modification, e.g., crosslinking [45] and reacetylation [46], offers an additional dimension to improve thermal stability, dry and wet strengths, and cytocompatibility. Fibre tenacity can also be improved (up to

$4.4\,g\,den^{-1}$) via incorporation of surfactants into the coagulation bath, whereby respective products display antimicrobial, antithrombogenic, hemostatic, deodorising, moisture controlling, and nonallergenic properties. As in the case of chitin, wet spun composite fibres made of chitosan and, e.g., cellulose can be accomplished with increased moisture retention compared to cellulosic fibres, whilst presenting dyeability towards direct and reactive dyes. Due to their skin hydration capability and minimal irritation risks, chitosan fibres can find applicability for clothing for people with sensitive skin, sutures, antimicrobial wound dressings, bandages, scaffolds for cell culture, as well as for the design of bone-like degradable implants.

3.5 Natural fibres for nonwovens

Globally, natural fibres account for a relatively small proportion of total nonwoven production compared to man-made fibres, but they are important in a variety of both single use, as well as durable products. In Europe, wood pulp accounts for about 18% by weight of total staple fibre nonwovens production, with conventional natural fibres accounting for less than 10%. Given the extensive literature that is available on the science and technology of natural fibres only a brief overview is given herein, specifically related to applications in nonwovens.

3.5.1 Wood pulp

A large proportion of nonwoven products, such as those used in single use hygiene and wipes applications, are required to readily absorb aqueous liquids, which means hygroscopic material is required. Compared to regenerated cellulose fibres, wood pulp provides a low-cost alternative. Owing to the very short length of wood pulp (generally <5mm) and the associated low fabric strength when wet, blending with man-made fibres enables nonwoven fabrics containing pulp to remain intact when saturated. Note that a large variety of different wood pulp grades exist, based on softwoods and hardwoods, and all are characterised by marked differences in fibre dimensions, as well as physical properties. Wood pulp is an important feedstock in both short fibre airlaying (specifically fluff pulp) and wetlaying systems, where the pulp is normally blended with short-cut man-made fibres, and also in coform products, where the pulp is intermixed with continuous filaments in spunlaid processes. For airlaid nonwovens, fluff pulp made from softwoods containing long tracheid fibres (2–5mm) as well as much shorter fibres (or fines) is used to make the voluminous absorbent cores of various single use hygiene products including diapers, feminine napkins and adult continence management products. There are also applications in single use food pads, food napkins, table covers, and medical products.

3.5.2 Cotton

The fibre consists mainly of cellulose (80%–90%) and varies in length (10–50mm) and diameter, as well as mechanical properties. Technically, nonwovens can be made from long fibre cotton grades as well as linters, which are shorter in length. A major application for nonwovens containing purified or bleached cotton fibres is in the manufacture of single use cosmetic pads using drylaid web formation followed by either hydroentangling or needlepunching. Commercially, wipes made of 100% cotton or blends are also produced by carding followed by hydroentanglement, and such fabrics are notable for their inherently high wet strength when saturated. Note that cotton is one of the very few hygroscopic fibres that can be stronger wet than dry. Mechanically recycled cotton fibres derived from clothing such as denim are also used in drylaid nonwoven production to make acoustic and thermal insulation.

3.5.3 Bast fibres

Flax, hemp, jute, and kenaf fibres are cellulosic fibres extracted from the stems of plants. Compared to cotton, these fibres contain a higher proportion of noncellulose impurities such as lignin, and because of the multicellular structure, ultimate fibre dimensions both in terms of diameter and length depends on the degree to which this is progressively broken down. Fibre dimensions and mechanical properties are therefore heavily dependent on the physiochemical processes used to extract the fibres from the plant stem, as well as subsequent fibre processing. The resulting fibre properties govern compatibility with specific nonwoven processes, as well as influence physical properties. Generally, bast fibres are characterised by relatively high modulus and tensile strength and have been adopted as a lower density alternative to glass reinforcements in natural fibre reinforced plastic (NFRP) composites. Normally, the fibres are converted into needlepunched drylaid nonwovens containing a proportion of thermoplastic fibre, or a suitable matrix polymer (thermoplastic or thermoset), to enable, for example, vacuum forming or compression moulding of three-dimensional automotive composites. Other applications for heavy-weight fabrics containing bast fibres include mulch and erosion control mattings, growing medium for horticulture and underlay. Drylaid hemp nonwovens are also used as luxury mattress fillings. Wall coverings, acoustic ceiling panels and room dividers have also been made based on carded-hydroentangled flax or hemp fabrics, blended with fibres such as viscose. Nonwoven wallpapers have also been made. Cottonisation of flax and hemp yields much finer fibres, with potential for wider applications in light-weight nonwoven fabrics, particularly after blending. Following intensive mechanical beating to reduce the fibre dimensions (<8 mm in length), bast fibre variants such as manila hemp are extensively used both in papermaking and the manufacture of wetlaid nonwovens, e.g., for tea bag fabrics, where lustre and high wet strength is needed.

3.5.4 Other plant-based natural fibres

Numerous other natural fibres derived commonly as a by-product of food production, e.g., banana, pineapple, coconut (coir), and agave (sisal) have been exploited as raw materials for nonwoven fabric manufacture. Because of the relatively large fibre diameters associated with such materials following extraction, drylaid web formation followed by needlepunching is the most commonly adopted manufacturing route, with products finding applications in Natural Fibre Reinforced Plastics (NFRPs), geosynthetics, floorcoverings and mattings, erosion control fabrics, and carpet backings. Efforts have also been made to develop higher value nonwoven products from these plant-based feedstocks. For example, Pinatex (Ananas Anam) fabric has been developed to provide an alternative to leather for luxury fashion, as well as interior spaces, including the automotive sector. Degummed pineapple waste leaf fibre is blended with 20% PLA and converted into a drylaid nonwoven before being coated with a water-based PU resin as part of a separate finishing process, yielding a variety of aesthetically pleasing fabrics.

3.5.5 Protein-based natural fibres

Wool is a mechanically resilient and inherently flame-retardant keratinous polypeptide fibre and as such it has been used for years as a component in the manufacture of drylaid nonwoven insulator pads as well as luxury comfort layer fillings for mattresses. Traditionally, for insulator pads next to the spring unit, mixed fibre blends containing a proportion of mechanically recycled postindustrial or postconsumer wool have been used as raw materials. For the comfort layers below the outer ticking fabric, some high-quality mattresses contain a high proportion of drylaid virgin wool and sometimes even cashmere. Drylaid nonwoven battings containing coarse virgin wool combined with an insecticide treatment are produced for thermal insulation used in buildings. Waste leather is also exploited by the nonwovens industry. In the manufacture of E-Leather, mechanically shredded postindustrial (unused) scrap leather that would normally be sent to landfill is the main feedstock in a unique airlaid hydroentangled nonwoven fabric manufacturing process. The resulting high value, engineered leather fabrics resemble natural leather but have the benefit of being produced as roll goods in uniform weights, thicknesses, widths and colourways, for use by multiple airlines and railways as seat coverings, amongst other applications.

3.5.6 Mycelium

In complete contrast to the conventional processes for making nonwoven fabrics, the harnessing of biotechnology provides an intriguing approach for producing nonwoven fabrics. Although the basic technical approach has been known for many years, industrially scalable mycelium-based bio-fabrication, is now being harnessed as a means to address the aim of producing fabrics resembling leather and other high value fibrous substrates using sustainable production methods. One example is Mylo (Bolt), a supple, relatively high strength fabric made from mycelium cells that is intended to provide an alternative to natural and synthetic leather. Another is example is Reishi (MycoWorks). Mycelium is the branching underground structure of mushrooms that grows as tiny threads, called *hyphae*. The largest part of these hyphae usually grows underground, generating fungal colonies in and on the soil and in many other substrates. Within the right conditions, from a single mycelium network multiple fruiting bodies called Basidiocarps, generally known as mushrooms, are formed above the surface. Mycelium-based nonwoven materials can therefore be produced by inoculating an individual strain of fungi in a porous substrate, which functions as the growing medium for the secretion of organic mycelium fibres. Depending on selected fungi, e.g., The Oyster mushroom (Pleurotus Ostreatus), dense mycelium networks can be obtained, which grow fast and are not easily infected by competing organisms. The quality and properties of the fibres is also determined by the characteristics of the substrate and the growing environment. The growing process can be stopped at will by drying or heating the material. When it is dried, the mycelium stays in a 'hibernated' state, which can be overcome under appropriate environmental conditions so that mycelium growth can be resumed. Other than drying, heating kills the mycelium and stops its growth irreversibly. Heated mycelium fibres exhibit similar properties to polystyrene or other foams, although further heating under pressure will make the material denser and stronger, similar to natural materials like wood. Other than macroscopic properties, mycelium fibres also display significantly increased biodegradability with respect to conventional building materials. Whilst they can be formed in about 6-week exposure to the environmental ground and display a lifespan of ca. 20 years if maintained under stable and favourable conditions, mycelium fibres can be returned to the ecosystem as a useful nutrient for plants. Furthermore, the energy consumption and inherent emissions of mycelium-based components are significantly lower compared to conventional building materials. Whilst the production of 1 m^3 of polystyrene costs more than 4.5 MJ and 462 kg of CO_2 emission, only 652 MJ are needed to produce the same volume of mycelium fibres with only 31 kg of CO_2 emission generated.

3.6 Inorganic fibres for nonwovens

Other than polymers, nonwoven fabrics are also be made from glass, ceramic, metallic and carbon fibres. Each of these fibres has a distinct combination of properties, and typically they are employed to manufacture durable, high-temperature, or chemical-resistant nonwoven products across a range of sectors, including building and construction, filtration, and transport, to name just a few.

3.6.1 Glass fibres

Glass fibres are typically used in nonwoven reinforcements, backings or coating substrates, where temperature stability, dimensional stability and relatively low cost are required. There are also important applications in air and liquid filtration. Most glass fibres are based on silica (SiO_2) and can be supplemented with oxides of calcium, boron, sodium, iron and aluminium. As per their composition and morphology, these glasses consist of what is a hard, amorphous and brittle material, made by fusing components and rapid cooling to prevent crystallisation. Typical glass fibre compositions are E, C and S. The E-glass variant is mainly used for electrical applications, since it draws well and has good tensile strength ($\sigma = 3.45$ GPa), Young's modulus ($E = 76$ GPa), and electrical and weathering properties. C-glass has excellent corrosion resistance, e.g., with respect to E-glass, but a lower tensile strength ($\sigma = 3.30$ GPa). Finally, S-glass is very strong, with the highest values of Young's modulus ($E = 85.5$ GPa), tensile strength ($\sigma = 4.60$ GPa) and temperature resistance ($T_{max} = 650°C$) [47].

Silica-based glasses consist primarily of covalently bonded tetrahedra, with silicon at the centre and oxygen at the corners. The oxygen atoms are shared between tetrahedra, leading to a rigid three-dimensional network. This network configuration is responsible for the macroscopic tensile strength and modulus of the glass fibres. Introduction of low-valency elements, such as Ca, Na, and K, tends to break up this network by forming ionic bonds with oxygen atoms, which then no longer bond the tetrahedra together. Addition of these elements therefore leads to glasses with decreased stiffness but improved formability.

Glass fibres are produced by melting the raw materials, whereby fibre diameters in the range of 8–12 μm are typically obtained. In nozzle-drawing, the glass is melted in a heated melt tub at 1250–1400°C and passed through nozzle holes of 1- to 25-μm diameter. Resulting melted filaments are drawn mechanically downwards, solidified, cut and assembled into webs. Other than nozzle-drawing, nozzle-blowing operates in a similar fashion with the only difference that the melted filaments are attenuated by compressed air, so that decreased fibre diameters in the range of 6–10 μm are generated. Ultimately, rod-drawing involves the collection of the melted filaments via a rotating drum so that a dry web is obtained at the end of the process.

Numerous industrial applications have been established for glass fibre nonwovens. Traditionally, the roofing industry employs wetlaid-latex bonded glass fibre fabrics or tissues, where the high-temperature resistance is needed for thermal protection and in fire walls. Wetlaid glass fibre nonwovens in construction also include bituminous roofing substrates, roofing shingles and foam and mineral insulation facers. Wetlaid glass nonwovens are used as reinforcement in vinyl floor tiles and HVAC filtration systems are amongst the applications for airlaid glass fabrics. Glass microfibre nonwovens are extensively used as filter media both for air purification and liquid filtration. Another area is in composites, where glass veils are used to improve the surface finish of structural components such as windmill blades. Wetlaid fabrics made of C-Glass microfibres are also used to manufacture cryogenic insulation products. Silica and basalt fibres are used in similar applications to glass where high-temperature insulation is required.

3.6.2 Ceramic fibres

Nonwovens made of ceramic fibres are important in high-temperature (>1000°C) insulation for refractory-type applications and the material is characterised by a polycrystalline structure. According to their composition, these materials can be classified as nonoxide fibres, i.e., made of silicon carbide (SiC), or oxide fibres, based on the alumina–silica (Al_2O_3–SiO_2) system and on α-alumina (α-Al_2O_3). Ceramic fibre manufacture requires an organic or mineral precursor fibre, which is then heat-treated and pyrolysed for a very short time. Large-diameter (~100–150 μm) fibres are made by chemical vapour deposition (CVD) onto a fine (~10–30 μm diameter) tungsten wire. The wire is fed continuously through a reaction chamber in the presence of an organosilicon gas compound, such as methyltrichlorosilane (CH_3SiCl_3). At elevated temperature, the gas dissociates at the wire surface to deposit the SiC. Surface, e.g., graphitic, layers are often deposited in a second reactor to improve the compatibility between the two phases and fibre resistance to handling. This process is carried out industrially and respective products display an elastic modulus of up to 200 GPa, whilst values of fibre tenacity and density are in the range of ~3 GPa and ~2 to 3 g cm^{-3}, respectively. Other than using tungsten wires, a similar process has been developed using either poly(acrylonitrile) or polycarbosilane as fibre precursor. Polycarbosilane is produced via the reaction of dichlorodimethylsilane with sodium in an autoclave under heating.

This product is spun into fibres, which are pyrolised at up to 1300°C. The resulting fibre composition includes a substantial proportion of silicon dioxide (SiO_2), free carbon as well as silicon carbide.

Industrially, alumina fibre nonwovens, e.g., Saffil, are made by a proprietary solution extrusion process, which directly forms heavy-weight ceramic fibre webs on a moving conveyor, before needlepunching to improve consolidation. The alumina fibres can be combined with either a sacrificial PP carrier, or an acrylic binder that degrades at >400°C, to improve the fabric's stability during handling and installation. Such alumina fabrics find applications in for example, expansion gap fillings, as well as kiln and industrial furnace linings, where thermal stability up to 1600°C is required.

3.6.3 Carbon fibres

Carbon fibres consist of long thin strands (~0.005–0.010 mm in diameter) mainly composed of carbon atoms. The carbon atoms are bonded together and configured in microscopic crystals, which are aligned parallel to the longitudinal axis of the fibre. The crystal alignment is responsible for the superior mechanical properties of these fibres. With a tensile strength and tensile modulus exceeding 6 and 600 GPa, respectively, and a low density (1.8–2.0 g cm^{-3}), carbon fibres are the strongest fibres currently available to reinforce polymeric matrices. The manufacture of carbon fibre relies on thermal treatment of poly(acrylonitrile) (Fig. 3.16).

The bulk polymer is drawn down to a fibre and stretched to induce chain alignment. Following the polymer heating, the pendant nitrile groups react to produce a ladder polymer, consisting of a row of six-membered rings. High-temperature oxidative treatment is ultimately induced resulting in polymer carbonisation and chains joining in graphite planes.

Other than poly(acrylonitrile), carbon fibres can also be obtained from mesophase pitch. Pitch is a complex mixture of thousands of different species of hydrocarbon and heterocyclic molecules. Following temperature raise above 350°C, condensation reactions occur, leading to large, flat molecules. This product is called 'mesophase pitch' and consists of a viscous liquid exhibiting local molecular alignment (i.e. a liquid crystal). This liquid is rapidly extruded through a multihole spinneret, generating aligned filaments. The product is then oxidised at temperatures below the softening point and converted thermally (at about 2000°C) into a graphitic fibre with high degree of axial orientation. The graphitic structure provides pitch-based

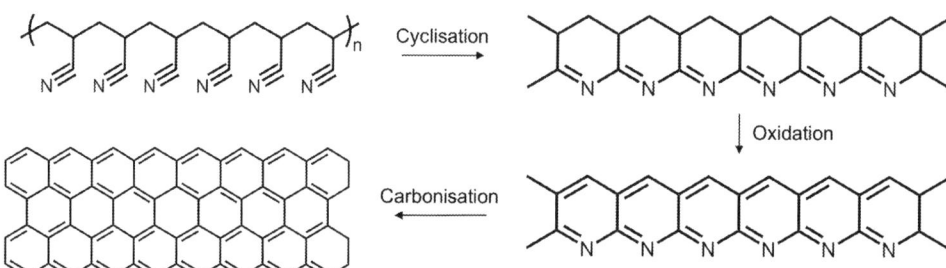

FIG. 3.16   Synthesis of carbon fibre from poly(acrylonitrile). Low-temperature treatment induces cyclisation of nitrile groups, followed by oxidative treatment at high temperature. Resulting product is mainly composed of carbon, whereby polymer chains are joint into graphite planes.

fibres with very high thermal conductivities ($\sim 1000\,\mathrm{W\,m^{-1}\,K^{-1}}$), which are much higher than the ones measured in PAN-based fibres and copper.

Lightweight wetlaid carbon fibre nonwovens of $<20\,\mathrm{g\,m^{-2}}$ are produced for surfacing veils for composites, resistive heaters and EMI shielding and are often adhesively bonded using for example PVA to improve fabric strength. Heavier-weight, carbon fibre nonwoven manufacture has been mainly directed at applications in the automotive industry, harnessing mainly postindustrial recycled carbon fibre as a feedstock, or fibre recovered from recycled carbon fibre reinforced polymer (CFRP) parts extracted by thermolysis or solvolysis. Carding and cross-lapping is a common platform for carbon fibre nonwoven production, making fabrics in the range of 100–$600\,\mathrm{g\,m^{-2}}$. In addition to the manufacture of pre-pregs, blending carbon fibre with a proportion of thermoplastic fibre both increases processing efficiency and enables thermoformable carbon fibre composite parts to be produced.

3.7 Metal fibres

Stainless steel and other metal alloy fibres have been available for many years, made by melt spinning or related processes. Staple fibres are formed by cutting or stretch breaking of metallic filament tow. They have found a variety of applications in the nonwoven industry, notably for the purpose of static elimination in durable products such air filter media and shoe insoles made from carded blends of stainless steel and polymers such as PET. A minimum of about 5% by weight of the fibre is required in the blend to yield a functional effect. Bekinox stainless steel staple fibres of ca. 12 μm and 60 mm cut length have been used in such blends to make carded and needlepunched products. Filter media where high thermal and chemical stability is required is another important area. In addition to needlepunched bag filter media, wetlaid fabrics made from metal fibres are found in air and liquid filtration and may also be sintered to improve dimensional stability.

3.8 Recycled materials

For many years, a section of the nonwovens industry has specialised in the mechanical recycling of postindustrial and postconsumer textile waste, including used clothing, to make new, long-lasting products such as insulator pads to cover the spring unit in the construction of mattresses and sound absorption fabrics in vehicles. Textile waste from various sources is mechanically shredded or 'pulled' back into fibre, enabling it to be airlaid, garnetted, or otherwise drylaid into batts and then consolidated, usually by needlepunching and/or thermal bonding (provided there is thermoplastic content). Mechanical recycling processes are extremely versatile, handling mixed textile waste in various forms, regardless of whether it is of synthetic or natural origin.

Recycled PET (rPET) based on plastic bottle chip is extensively used for fibre production in the nonwovens industry and in recent years, rising global demand has diminished the price differential with virgin PET. Historically, other sources of postconsumer polymer waste has been more difficult to recycle into feedstock suitable for cost-effective, high quality fibre production, but significant progress continues to be made. Solvent-based physical separation, or chemical recycling processes are amongst the methods being harnessed for recycling of waste products containing polyolefins, polyamides, polyester, cellulose and other polymers.

Development of scalable waste recycling processes for petroleum-based plastics such as PP is exemplified by industry initiatives such as P&G's PureCycle process, Sabic's Tru-Circle, BASF's ChemCycling process, and the collaborative work of OMV and Borealis (amongst others). Some examples of chemical recycling approaches specifically aimed at providing polymer for fibre production are summarised below.

3.8.1 Chemical recycling using ionic liquids

Driven mainly by a desire to establish closed-loop recycling of postconsumer clothing waste in the textile industry, notable progress has been made using ionic fluids to separate and recycle PET and cotton. These two fibrous components are frequently blended together in clothing and despite the two individual components being recyclable, mechanical techniques cannot cost effectively separate them. Solvent-induced separation of polyester/cotton blends has therefore been attempted, aiming to avoid the use of environmentally damaging solvents, such as hexafluoroisopropanol and phenol/tetrachloroethane, for the solubilisation of polyester fibres. Ionic liquids, e.g., 1-allyl-3-methylimidazolium chloride (AMIMCI), have therefore been introduced as alternative processing solvents for the chemical separation of polyester/cotton blends. Rather than targeting the polyester phase, AMIMCI selectively dissolves the cotton component, so that polyester fibres can be recovered with high yield. In the process developed by De Silva et al. [48], the textile material is added to AMIMCI at 80°C at 10 wt% concentration and incubated for 6 h to enable dissolution of the cotton fibres. The undissolved polyester component is then collected and rinsed with water. Whilst AMIMCI can be recycled and reused, the dissolved cotton phase can be regenerated into various forms, such as wet spun fibres or cast films. Moreover, the material properties can now be sufficiently well preserved in the recovered products to enable subsequent processing into fabrics. In light of these remarkable results, ionic liquid-based recycling is in commercial development, with technologies such as Ioncell (Finland) and Worn Again Technologies (United Kingdom).

3.8.2 Chemical recycling of PET using liquid state polycondensation

More recently, processes have been developed to enable PET waste generated during extrusion, nonwoven roll good production or converting to be recovered by chemical recycling, based on methods such as liquid state polycondensation (LSP). The LSP process enables a rapid and controlled increase in the intrinsic viscosity (IV) of the waste PET, enabling it to be harnessed for high quality fibre extrusion as part of a continuous process. By exploiting the inherent tendency of PET to condense under vacuum in the melt (liquid) phase, LSP results in an extremely quick increase in the IV value, so that good control of molecular weight and fibre mechanical properties can be accomplished. Application of strong vacuum in the process facilitates decontamination of the material from harmful chemicals, ensuring 100% food-safe material applicability. An example of a LSP system is P:REACT (NGR) technology.

The melt phase enters multiple vertical strands of the reactor, so that large surface area to volume ratios can be realised. Contamination molecules can therefore migrate to the surface quickly, where they are removed in the vacuum system and collected in a waste container. The decontaminated material can then be either extruded into recycled pellets or transferred directly to downstream fibre or spunmelt production. Another example of a PET chemical recycling system that does not rely on solid state polymerisation (SSP) and enables boosting of IV as part of an integrated process is JUMP technology (Gneuss).

3.8.3 Chemical recycling of cellulosic waste

Lenzing Refibra is a lyocell fibre produced partly from postindustrial and postconsumer cotton waste, including discarded clothing. The cellulose waste is first converted into pulp and up to 30% is blended with wood pulp sourced from sustainably managed forests to produce new fibre using the N-methylmorpholine N-oxide (NMMO) solvent-based lyocell fibre production process. Postproduction cotton textile waste is mixed with wood dissolving pulp and dissolved in an NMMO solution, followed by spinning of lyocell-type cellulosic fibres [49]. This lyocell process benefits from a closed loop system, which negates environmental issues arising from chemical discharge. Resulting fibres exhibit low fibrillation and have good moisture management capabilities. A target recycled content of up to 50% is in development aiming to minimise reliance on the pure wood pulp content. Given that many waste clothing fabrics containing cotton are made of blends with polyester, ongoing and future developments also aim to integrate the cotton/polyester separation step with chemical recycling.

3.9 Other polymer materials for nonwovens

A nonwoven is a fabric in which the fibres are held together by physical or chemical bonding [50]. Beyond fibres and filaments, a large variety of other polymer materials are used by the nonwovens industry both as raw materials to control physical properties, as well as to enable bonding of the fabric itself. Polymer additives may be incorporated within the fibres themselves or applied afterwards in the form of binders, coatings, powders or particles, to enhance web consolidation or the macroscopic properties of the fabric. Depending on their physical state, polymer additives can be classified as either dry, e.g., powders or wet, e.g., solutions, dispersions and emulsions, to generate an integral or sacrificial component of the resulting nonwoven fabric.

3.9.1 Polymer-based wet binders

Polymer-based wet binders can be widely customised depending on the specific polymer and solvent composition. They are often synthesised from vinyl co/monomers by emulsion polymerisation in the presence of a surfactant and an initiator molecule [51]. Typical vinyl co/monomers include ethylene styrene, vinyl acetate, vinyl chloride acrylic esters, acrylonitrile and methyl methacrylate. Given the variety of co/monomers and surfactant formulations, the range of commercially available wet binders and respective properties are constantly

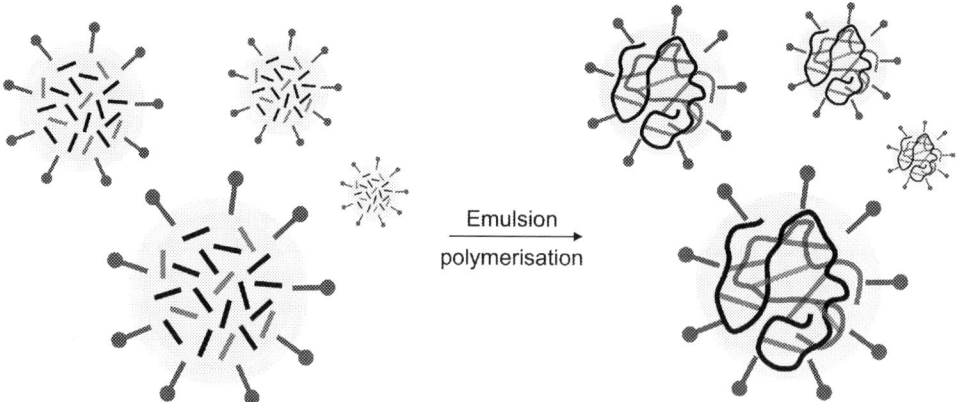

FIG. 3.17   Water-in-oil emulsion containing a surfactant (—●), vinyl monomer (—) and crosslinker (----). Latex particles are obtained via polymerisation occurs within the emulsion droplets.

expanding. It is therefore key to control the glass transition temperature and wettability of the emulsion polymer according to requirements of the selected spinning method and web substrate, given their impact on polymer stiffness, and kinetics and yield of emulsion polymerisation, respectively.

The delivery of the polymerising emulsion to the fabric should be considered carefully in order to minimise evaporation whilst migration into the structure occurs. Control of the emulsion storing temperature and time is also key to minimise the risk of binding agent coagulation, which can inherently reduce the adhesivity of the system. To address these challenges, crosslinking agents are incorporated in the polymerising emulsion to further enhance binder adhesivity as well as fibre and nonwoven properties (Fig. 3.17).

Crosslinking agents are also important when controlled water and solvent resistance are required, since they mediate the synthesis of a covalently crosslinked polymer networks. Current crosslinking agents include acrylic acid, methacrylic acid, acrylamide, *n*-methylol acrylamide; they have replaced aldehyde-based binders to minimise environmental impact and reduce toxicity levels. Employment of the above-mentioned crosslinking agents has also proven effective to enable wet binder compliance with more stringent safety standards, requiring the absence of volatile residual monomers, volatile adhesives, phenols or heavy metals, as well as complete biodegradability in water and landfills.

3.9.2 Dry polymer binders

Dry polymer binders consist of adhesive polymers in the form of powders, particles and fibres that can be used to consolidate nonwoven structures. The binding mechanism stems from their solubility and wetting characteristics and/or thermal properties, by which adhesive bonds amongst fibres can be established. Depending on the binding mechanism, dry polymer binders can therefore be classified as either soluble, which become sticky under the influence of a solvent and can be removed following fibre adhesion, or thermoplastic, which leads to fibre bonding via heating or upon melting. Dry polymer binders are typically applied by mixing with the fibres during web formation or scattering them over dry webs or prebonded nonwoven fabrics. Hot melt application is another route wherein, the liquid polymer is applied on to the substrate after melting. These approaches therefore require the polymer binder and the fibre-forming polymer in the substrate to be compatible such that the binder is able to induce good interfacial bonding between the two materials, and that the distribution of the binder across the surface and through-thickness is appropriately controlled to yield the required consolidation. Dry polymer binders available in different forms from pellets, powders to fibres, are made of semicrystalline polymers, such as polypropylene, polyethylene, co-polyamides, and co-polyesters, with melting transitions typically in the range of 85–150°C, although some powders based on polymers such as polyetherketoneketone (PEKK) require higher temperatures.

Although the majority of water-based binder systems applied to webs and nonwovens are latex based, i.e., the polymer binder is not applied after solubilisation, there are instances where water-soluble binders are used. Poly vinyl alcohol (PVA) has been widely employed as soluble binder, in light of its compatibility with the aqueous environment. Following incubation in water, PVA swells and dissolves when temperature is increased to 60–90°C [52]. Once applied to the substrate, swelling of the PVA attached to the fibres varies depending on the temperature and the PVA's molecular weight.

3.9.3 Superabsorbent polymers

Superabsorbent polymers (SAPs) consist of covalently crosslinked water swollen polymer networks, i.e., hydrogels, which are capable of retaining liquids even under pressure. Despite storing large amount of water, these polymers are characterised by an elastic behaviour, which is explained by the presence of covalent bonds between the polymer chains [53]. Although not employed as nonwoven building blocks, superabsorbent polymers are inherently linked to the manufacture of nonwoven fabrics, especially where water absorption and retention are key, e.g., in the context of hygiene products. At a molecular scale, superabsorbent polymers are made of acrylate co/polymers; the carboxyl groups present in the repeating units mediate the water uptake and swelling behaviour of the polymer in water, whereby the electrostatic repulsion between side groups is responsible for the large water absorption. SAP in the nonwovens industry is predominantly based on sodium polyacrylate and is mainly utilised in the form of powders and granules. These SAPs are designed to absorb between 200 and 400 times their own weight in water, but liquid absorption is negatively affected at higher ionic strength, i.e., salt content of the liquid, which affects the absorptive capacity of body fluids, e.g., urine, menses and wound exudate. In contrast to powders and granules, Super Absorbent Fibre (SAF) materials are also available, and can be directly made into nonwovens, by for example drylaid web formation.

Superabsorbent polymers are mostly manufactured in solution via radical polymerisation reaction. The acrylic monomer is derivatised into sodium acrylate via partial neutralisation and then polymerised in the presence of a comonomer, so that a covalently crosslinked hydrogel is generated. The radical generation mechanism as well as the removal of oxygen and the temperature play a major role in the polymerisation reaction, crosslink density and swelling behaviour of resulting superabsorbent polymer. Once formed, the hydrogel product is dried and converted into powders by grinding and sieving. The same mechanism of radical polymerisation can also be carried out in suspension, where the sodium acrylate derivative is dispersed in a hydrophobic organic medium prior to polymerisation with a polyfunctional crosslinker. As the polymerisation reaction occurs, covalent networks in the form of porous droplets are formed in the dispersion. Other mechanisms for the manufacture of superabsorbent polymers include surface crosslinking, where reactive species are introduced on the surface of the polymer to generate an additional crosslinked network, or in situ polymerisation, where the superabsorbent polymer is directly formed on the nonwoven. The characterisation of the water uptake and retention capability of the superabsorbent polymers is also key. Typical tests are carried out to quantify the centrifuge retention capacity and the absorption against pressure. In the former case, the maximum retention of a physiological sodium solution is measured in the hydrogel following centrifugation. In the latter case, the polymer swelling is measured over time following application of a gravimetric load of either 0.3, 0.7, or 0.9 psi (as representative weight of the wearer).

3.10 Outlook

Significant steps are being made in the industry to improve resource utilisation and reduce environmental impacts, as well as meet the continually developing demands of consumer, industrial and healthcare markets. Increasing environmental regulation linked to, for example, air quality and emissions, plastic waste disposal, and sound pollution provides the nonwoven industry with both major opportunities and challenges. Notable developments include increased quantities of recycled polymer being used in nonwoven materials and the development of new circular economies, involving closed loop recycling of postindustrial and postconsumer waste streams. The industry currently utilises large quantities of nonrenewable, thermoplastic synthetic polymers, but switching to renewable biobased or recycled alternatives is challenging. Particularly in relation to spunlaid processes, where thermoplastic polymers and melt extrusion predominate, there are fundamental incompatibilities between the properties of many renewable naturally derived materials and the requirements of high-speed processing. Where there is process compatibility, reduced production speeds, higher costs and differences in physical properties need to be addressed. Significant ongoing developments in polymer science, process innovation and nonwoven product design are aiming to addressing these challenges.

References

[1] A.J. Pennings, J. Smook, Mechanical properties of ultra-high molecular weight polyethylene fibres in relation to structural changes and chain scissioning upon spinning and hot-drawing, J. Mater. Sci. 19 (1984) 3443–3450.

[2] G. Tronci, R.S. Kanuparti, M.T. Arafat, J. Yin, D.J. Wood, S.J. Russell, Wet-spinnability and crosslinked fibre properties of two collagen polypeptides with varied molecular weight, Int. J. Biol. Macromol. 81 (2015) 112–120.

[3] Q. Mao, T.P. Wyatt, A.-T. Chien, J. Chen, D. Yao, Melt spinning of high-strength fiber from low-molecular-weight polypropylene, Polym. Eng. Sci. 56 (2016) 233–239.

[4] S.L. Shenoy, W.D. Bates, H.L. Frisch, G.E. Wnek, Role of chain entanglements on fiber formation during electrospinning of polymer solutions: good solvent, non-specific polymer–polymer interaction limit, Polymer 46 (2005) 3372–3384.

[5] H. Uehara, Y. Yamazaki, T. Kanamoto, Tensile properties of highly syndiotactic polypropylene, Polymer 37 (1996) 57–64.

[6] L. Fambri, A. Pegoretti, R. Fenner, S.D. Incardona, C. Migliaresi, Biodegradable fibres of poly(L-lactic acid) produced by melt spinning, Polymer 38 (1997) 79–85.

[7] M.A. Narter, S.K. Batra, D.R. Buchanan, Micromechanics of three-dimensional fibrewebs: constitutive equations, Proc. Royal Soc. A 455 (1999) 3543–3563.

[8] M.B. Bazbouz, H. Liang, G. Tronci, A UV-cured nanofibrous membrane of vinylbenzylated gelatin-poly (ε-caprolactone) dimethacrylate co-network by scalable free surface electrospinning, Mater. Sci. Eng. C 91 (2018) 541–555.

[9] L.N. Woodard, M.A. Grunlan, Hydrolytic degradation and erosion of polyester biomaterials, ACS Macro Lett. 7 (2018) 976–982.

[10] C.-Y. Chen, J.A. Cuculo, P.A. Tucker, Effects of spinning conditions on morphology and properties of polyethylene terephthalate fibers spun at high speeds, J. Appl. Polym. Sci. 44 (1992) 447–458.

[11] A. Hadjizadeh, A. Ajji, M.N. Bureau, Nano/micro electro-spun polyethylene terephthalate fibrous mat preparation and characterization, J. Mech. Behav. Biomed. Mater. 4 (2011) 340–351.

[12] M. Nofar, H. Oğuz, Development of PBT/recycled-PET blends and the influence of using chain extender, J. Polym. Environ. (27) (2019) 1404–1417.

[13] D.V. Rosato, D.V. Rosato, M.V. Rosato, Plastic property, in: D. Rosato, D. Rosato, M. Rosato (Eds.), Plastic Product Material and Process Selection Handbook, Elsevier, 2004, pp. 40–129.

[14] S. Grande, P. Cools, M. Asadian, J. Van Guyse, I. Onyshchenko, H. Declercq, R. Morent, R. Hoogenboom, N. De Geyter, Fabrication of PEOT/PBT nanofibers by atmospheric pressure plasma jet treatment of electrospinning solutions for tissue engineering, Macromol. Biosci. 18 (2018) 1800309.

[15] M. Heller, Q. Li, K. Esinhart, B. Pourdeyhimi, C. Boi, R.G. Carbonell, Heat induced grafting of poly(glycidyl methacrylate) on polybutylene terephthalate nonwovens for bioseparations, Ind. Eng. Chem. Res. 59 (2020) 5371–5380.

[16] C. Salvagnini, A. Roback, M. Momtaz, V. Pourcelle, J. Marchand-Brynaert, Surface functionalization of a poly(butylene terephthalate) (PBT) melt-blown filtration membrane by wet chemistry and photo-grafting, J. Biomater. Sci. Polym. Ed. 18 (2007) 1491–1516.

[17] L.A. Anan'eva, V.P. Petrov, L.P. Sharova, Preparation of polyester fibres with improved consumer properties, Fibre Chem. 23 (1992) 410–414.

[18] T.-Y. Liu, W.-C. Lin, M.-C. Yang, S.-Y. Chen, Miscibility, thermal characterization and crystallization of poly(L-lactide) and poly(tetramethylene adipate-co-terephthalate) blend membranes, Polymer 46 (2005) 12586–12594.

[19] G. Bhat, D.V. Parikh, Biodegradable materials for nonwovens, in: Applications of Nonwovens in Technical Textiles, Woodhead Publishing Series in Textiles, 2010, pp. 46–62.

[20] R.R. Mather, The structural and chemical properties of polyolefin fibres, in: S.C.O. Ugbolue (Ed.), Polyolefin Fibres, Woodhead Publishing, 2009, pp. 35–56.

[21] C. Duplay, B. Monasse, J.-M. Haudin, J.-L. Costa, Shear-induced crystallization of polypropylene: influence of molecular structure, Polym. Int. 48 (1999) 320–326.

[22] R. Nayak, I.L. Kyratzis, Y.B. Truong, R. Padhye, L. Arnold, Structural and mechanical properties of polypropylene nanofibres fabricated by meltblowing, J. Text. Inst. 106 (2015) 629–640.

[23] V. Moody, H.L. Needles, Major fibers and their properties, in: V. Moody, H.L. Needles (Eds.), Tufted Carpet—Textile Fibers, Dyes, Finishes, and Processes, William Andrew, 2004, pp. 35–59 (Chapter 3).

[24] N. Vasanthan, S.B. Ruetsch, D.R. Salem, Structure development of polyamide 66 fibers by X-ray diffraction and FTIR spectroscopy, J. Polym. Sci. Polym. Phys. Ed. 40 (2002) 1940–1948.

[25] R.B. Sandor, PBI (polybenzimidazole): synthesis, properties and applications, High Perform. Polym. 2 (1990) 25–37.

[26] J.-G. Rosenboom, R. Langer, Bioplastics for a circular economy, Nat. Rev. 7 (2022) 117.

[27] A. Contreras, M.J. Raxworthy, S. Wood, J.D. Schiffman, G. Tronci, Photodynamically active electrospun fibers for antibiotic-free infection control, ACS Appl. Bio Mater. 2 (2019) 4258–4270.

[28] S.-F. Chou, K.A. Woodrow, Relationships between mechanical properties and drug release from electrospun fibers of PCL and PLGA blends, J. Mech. Behav. Biomed. Mater. 65 (2017) 724–733.

[29] S.R. Baker, S. Banerjee, K. Bonin, M. Guthold, Determining the mechanical properties of electrospun poly-ε-caprolactone (PCL) nanofibers using AFM and a novel fiber anchoring technique, Mater. Sci. Eng. C 59 (2016) 203–212.

[30] N. Soatthiyanon, C. Aumnate, K. Srikulkit, Rheological, tensile, and thermal properties of poly(butylene succinate) composites filled with two types of cellulose (kenaf cellulose fiber and commercial cellulose), Polym. Compos. 41 (2020) 2777–2791.

[31] W. Jia, R.H. Gong, C. Soutis, P.J. Hogg, Biodegradable fibre reinforced composites composed of polylactic acid and polybutylene succinate, Plast. Rubber Compos. 43 (2014) 82–88.

[32] J. Jian, Z. Xiangbin, H. Xianbo, An overview on synthesis, properties and applications ofpoly(butylene-adipate-co-terephthalate)—PBAT, Adv. Ind. Eng. Polym. Res. 3 (2020) 19–26.

[33] R. Muthuraj, O. Valerio, T.H. Mekonnen, Recent developments in short- and medium-chain- length Polyhydroxyalkanoates: Production, properties, and applications, Int. J. Biol. Macromol. 187 (2021) 422–440.

[34] B. Singhi, E.N. Ford, M.W. King, The effect of wet spinning conditions on the structure and properties of poly-4-hydroxybutyrate fibers, J. Biomed. Mater. Res. 109 (2021) 982–989.

[35] H.L. Hergerth, G.C. Daul, Rayon—a fiber with a future, ACS Symp. Ser. 58 (1977) 3–11.

[36] J.W.S. Hearle, C. Woodings, Fibres related to cellulose, in: Regenerated Cellulose Fibers, Woodhouse Publishing Series in Textiles, 2001, pp. 56–173 (6).

[37] I. Reiniati, A.N. Hrymak, A. Margaritis, Recent developments in the production and applications of bacterial cellulose fibers and nanocrystals, Crit. Rev. Biotechnol. 37 (2017) 510–524.

[38] W. Czaja, A. Krystynowicz, S. Bielecki, R.M. Brown, Microbial cellulose—the natural power to heal wounds, Biomaterials 27 (2006) 145–151.

[39] Z. Xiang, J. Zhang, Q. Liu, Y. Chen, J. Li, F. Lu, Improved dispersion of bacterial cellulose fibers for the reinforcement of paper made from recycled fibers, Nanomaterials 9 (2019) 58, https://doi.org/10.3390/nano9010058.

[40] R. Russo, M. Malinconico, G. Santagata, Effect of cross-linking with calcium ions on the physical properties of alginate films, Biomacromolecules 8 (2007) 3193–3197.

[41] C.K.S. Pillai, W. Paul, C.P. Sharma, Chitin and chitosan polymers: chemistry, solubility and fiber forma, Prog. Polym. Sci. 34 (2009) 641–678.

[42] O.C. Agboh, Y. Qin, Chitin and chitosan fibers, Polym. Adv. Technol. 8 (1996) 355–365.

[43] G. Salihu, P. Goswami, S. Russell, Hybrid electrospun nonwovens from chitosan/cellulose acetate, Cellulose 19 (2012) 739–749.

[44] G. Tronci, H. Ajiro, S.J. Russell, D.J. Wood, M. Akashi, Tunable drug-loading capability of chitosan hydrogels with varied network architectures, Acta Biomater. 10 (2014) 821–830.

[45] G. Tronci, P. Buiga, A. Alhilou, T. Do, S.J. Russell, D.J. Wood, Hydrolytic and lysozymic degradability of chitosan systems with heparin-mimicking pendant groups, Mater. Lett. 188 (2017) 359–363.

[46] Q.C. East, Y. Qin, Wet spinning of chitosan and the acetylation of chitosan fibres, J. Appl. Polym. Sci. 50 (1993) 1773–1779.

[47] D. Hull, T. Clyne, Fibres and matrices, in: An Introduction to Composite Materials, Cambridge University Press, 1996, pp. 9–38, https://doi.org/10.1017/CBO9781139170130.004.

[48] R. De Silva, X. Wang, N. Byrne, Recycling textiles: the use of ionic liquids in the separation of cotton polyester blends, RSC Adv. 4 (2014) 29094.

[49] H. Wedin, M. Lopes, H. Sixta, M. Hummel, Evaluation of post-consumer cellulosic textile waste for chemical recycling based on cellulose degree of polymerization and molar mass distribution, Text. Res. J. 89 (2019) 5067–5075.

[50] N.H. Sherwood, Binders for nonwoven fabrics, Ind. Eng. Chem. 51 (1959) 907–910.

[51] M. Chana, A. Almutairi, Nanogels as imaging agents for modalities spanning the electromagnetic spectrum, Mater. Horiz. 3 (2016) 21–40.

[52] S.E. Bulman, P. Goswami, G. Tronci, S.J. Russell, C. Carr, Investigation into the potential use of poly(vinyl alcohol)/methylglyoxal fibres as antibacterial wound dressing components, J. Biomater. Appl. 29 (2015) 1193–1200.

[53] R. Holmes, X.B. Yang, A. Dunne, L. Florea, D. Wood, G. Tronci, Thiol-ene photo-click collagen-PEG hydrogels: impact of water-soluble photoinitiators on cell viability, gelation kinetics and rheological properties, Polymers 9 (2017) 226, https://doi.org/10.3390/polym9060226.

CHAPTER

4

Drylaid web formation

A.G. Brydon[a], A. Pourmohammadi[b], and S.J. Russell[c]

[a]Garnett Group of Associated Companies, Bradford & Hawick, United Kingdom [b]Department of Mechanical Engineering, University of Payam Noor, Tehran, Iran [c]University of Leeds, Leeds, United Kingdom

4.1 Introduction

The drylaid nonwoven sector utilises carding, garnetting, airlaying, and, in certain specialist applications, direct feed batt formation processes to convert staple fibres into a web or batt structure that is uniform in weight per unit area.

4.2 Selection of fibres for carding

Virtually, any fibre that can be carded can be, and probably already is, used in nonwovens including polymeric, glass, and ceramic materials. Table 4.1 gives a general overview of some of the fibres that are carded either alone or in blends. Man-made fibres account for the majority of raw materials consumed by the nonwoven carding sector, with polyester (PET) being the most prevalent. This is principally because of PET's suitability for multiple product applications and comparatively low cost. Polypropylene (PP) is also important in carding, particularly for manufacturing heavyweight needlepunched fabrics for durable products such as floorcoverings and geosynthetics, as well as needlepunched filter media and lightweight thermal bonded fabrics for hygiene disposables. Carding of viscose rayon is important in the hygiene, wipes, and medical sectors, principally because of its high-moisture absorption. Lyocell combines moisture absorption with good wet strength, which is particularly useful in products such as wipes. Thermoplastic core-sheath bicomponent fibres are widely used to aid thermal bonding following carded web formation, and carding of island in the sea or segmented pie bicomponent fibres is a key step in the manufacture of some synthetic leather fabrics made of microfibres. A large variety of other fibre materials, including, but

TABLE 4.1 Summary of fibre properties.

		Physical properties of textile fibres					
Fibre type	Name	Range of diameter (µ)	Density (g/cm^3)	Tenacity (gf/tex)	Breaking extension (%)	Moisture regain 65% r.h. (%)	Melting point (°C)
Natural	Cotton	11–22	1.52	35	7	7	–
	Flax	5–40	1.52	55	3	7	–
	Jute	8–30	1.52	50	2	12	–
	Wool	18–44	1.31	12	40	14	–
	Silk	10–15	1.34	40	23	10	–
Regenerated	Viscose rayon	12+	1.46–1.54	20	20	13	–
	Acetate	15+	1.32	13	24	6	230
	Triacetate	15+	1.32	12	30	4	230
Synthetic	Nylon 6	14+	1.14	32–65	30–55	2.8–5	225
	Nylon 6.6	14+	1.14	32–65	16–66	2.8–5	250
	Polyester	12+	1.34	25–54	12–55	0.4	250
	Acrylic	12+	1.16	20–30	20–28	1.5	Sticks at 235
	Polypropylene	–	0.91	60	20	0.1	165
	Spandex (Lycra)	–	1.21	6–8	444–555	1.3	230
Inorganic	Glass	5+	2.54	76	2–5	0	800
	Asbestos	0.01–0.30	2.5	–	–	1	1500

not limited to alginate, aramids, PTFE, PVA, carbon, silica, and glass, are also carded for niche applications.

Natural fibres such as cotton and wool have been carded for as long as cards have been in existence, and together with flax, hemp, and other bast fibres, they find specific applications in drylaid nonwoven web formation. However, man-made fibres are by far the chief feed-stock for high-speed nonwoven carding systems.

Fundamental to the suitability of a particular fibre for drylaid processing is its machine compatibility, as well as its influence on fabric properties. There are numerous examples of new fibre developments that have been slow to progress because of processing problems, particularly during carding. Common problems are uncontrolled static electricity, low fibre-to-fibre cohesion, and insufficient fibre extension (the minimum required is 2%–5%), leading to fibre breakage and poor yield.

Whilst exceptions do exist, the general range of fibre dimensions suitable for carding can be given approximately as 1 to 300 dtex fibre linear density and 15 to 250 mm mean fibre length.

In practice, such a wide range of fibre dimensions could not be satisfactorily processed on one card, without modifying the card roller configuration and layout, process settings, and the card wire. Blending extends the range of fibre lengths and fineness that can be processed, and in certain sectors of the industry, carrier fibres are used to aid processing of short, stiff, or low surface friction materials.

Cotton and other short-staple fibres of <60mm fibre length are used in the short-staple spinning industry, where traditionally, a modular sequence of processes has been developed to prepare, card, and spin the fibre into yarn. Man-made fibres of similar diameter to cotton are therefore cut to a similar length so that they can be processed on the same machinery, either in 100% or blended form, depending on end-use requirements. Fibres are commonly square cut to one length prior to processing. This gives a different fibre length distribution from natural fibres, which typically have a trapezoidal-shaped distribution. It should also be understood that the mean fibre length and the fibre length distribution as measured, before and after any carding process, are substantially different, due to breakage or permanent elongation of fibres during the process.

Historically, cotton cards have been used by the nonwovens industry to make webs destined for use in feminine hygiene and some absorbent medical products, made of short-staple cotton or viscose rayon fibres of $c.$ 28 to 45mm mean fibre length. However, the use of short-staple or cotton 'flat' cards in the nonwovens industry is not extensive, because the revolving flats limit the maximum card width to about 1.5m, and the mixing power is significantly lower than a worker-stripper card. The majority of carded webs for nonwovens are manufactured from fibres with a mean length in the range of 38 to 100mm, although in some specialist applications, fibres outside this range are used. Accordingly, the nonwovens industry relies mainly on worker-stripper (roller top) cards, which are well suited to a wider range of fibre lengths and are capable of operating over larger working widths of up to 4m, or more.

Fibre characteristics have a big influence on carding performance. Web cohesion, fibre breakage, nep formation and web weight uniformity are key quality parameters, and are influenced by fibre diameter, fibre length, fibre tensile properties, fibre finish and crimp. During the production of man-made fibres, crimp is introduced to increase frictional resistance and web cohesion during carding, bulk and sometimes elastic recovery. However, the crimp can decay during carding due to the applied forces and temperatures that are generated, and regenerated cellulosic fibres are particularly prone to this. The applied forces and fibre deformation taking place in carding also give rise to fibre breakage that modifies the original fibre length distribution, and some temperature sensitive materials such as PVC can be subject to thermal shrinkage during the process.

Fibre finish applied to the fibre prior to carding, modifies both fibre to fibre friction (cohesion) and fibre to metal friction (holding power of the wire), and can reduce propensity for fibre breakage. The finish also influences the extent to which static electricity is generated during carding. Fibre finish is normally topically applied before carding. Both the static and dynamic friction are important, fibre to fibre, and fibre to metal. The ability of a fibre finish to increase fibre cohesion whilst at the same time reducing friction is an example of frictional 'stick–slip' behaviour. A useful analogy is to imagine two sheets of glass coated by a thin film of lubricant. Placed together, the glass sheets easily slide over each other, but it is not so easy to prise the sheets apart. In carding, the fibres should readily slide against each other, but in a controlled manner. Fibre finishes also contain anti-static agents, the effectiveness of which is particularly important when carding hydrophobic fibres such as polypropylene. Issues with static electricity are less likely with hygroscopic fibres having a relatively high-moisture content or regain.

Other additives can be present in the fibre finish, either to improve downstream processing efficiency, or to address the end-use requirements of finished fabrics. Accordingly, finish additives are available to improve fibre wetting by modifying surface energy, or to reduce foaming during hydroentangling of carded webs. Other finishes address food contact approval regulations or have been formulated to be biodegradable.

4.3 Opening of fibres

It is traditionally said that 'well opened is half carded' and this old adage remains remarkably true. The more mechanical work that has to be done on the card to break down and disentangle large tufts of fibres, the more likely are fibre breakage, nep formation and web weight variation.

Traditionally, the need for opening and blending emerged to tackle the inconsistency within and between batches of fibres, particularly as fibres used in carding at that time were predominantly natural and hence their physical properties were affected by seasonal and environmental factors that were beyond the control of industrial processors. Natural fibres also require mechanical cleaning to remove impurities.

In the majority of nonwoven carding operations, the most important raw material is man-made fibre. Such fibre is cut to length, or to a defined length distribution, and has a fibre finish pre-applied to aid processing. The fibre, which is supplied in dense press-packed bales, usually needs to be opened prior to carding. In other words, large tufts of fibre need to be removed, progressively reduced in size and disentangled, whilst minimising fibre damage. The reduction in tuft size promotes consistent fibre feeding to the card, which is particularly important in high production operations.

Although the general rule is to open the fibre as uniformly as possible, there is a risk of opening a batch of fibres too well, with the result that it is so voluminous that problems are experienced in card feeding. This is a particular problem in high production carding, where a low-density fibre flow can restrict the maximum production rate of the card feeder. The volume of opened fibre also depends on the fibre type, fibre fineness, crimp and stiffness. Partly for this reason, fibres having different specifications cannot be expected to be processed using the same machine settings. If the tuft density is too low, fibres may roll in the feed hopper leading to nep formation and entanglement. This can be overcome by providing a suitable feed rate differential between the feed entering the hopper and the stream leaving the hopper. This maintains a reservoir so that consolidation can be achieved by gravitational compaction, either in the hopper chamber or in the volumetric feed chute.

The opening process converts large, densely packed tufts from a bale into smaller tufts, or in some cases individual fibres. Intensive fibre opening is not problematic provided the production rate of the opened fibre is matched with the rest of the processing line. The goal is to maximise the degree of opening at a particular opening device, whilst minimising the associated fibre damage, particularly fibre breakage. In practice, this is achieved by selecting appropriate throughput rates for a given fibre type and tuft density and selection of appropriate types of opening system. Variations in feed rate through an opening machine will tend to lead to variations in the degree of fibre opening and the associated fibre damage. The sequencing of different opening processes also influences the ability to achieve progressive opening, as well as the extent of fibre damage. A suitable dwell time within a condenser or hopper chamber or other suitable buffer zone is required to ensure that a sufficient mass of fibre passes through the feeder to evenly distribute the batch and to achieve the desired production rate. Opened fibre is sometimes re-packed into loose bales or held in blending bins prior to carding, so it can be conveniently stored prior to carding.

4.3.1 Bale breakers

Commonly, bale breaking hopper feeders are the first point of entry to a nonwoven process line (see Fig. 4.1). A bale breaker is similar to a conventional hopper feeder in its construction

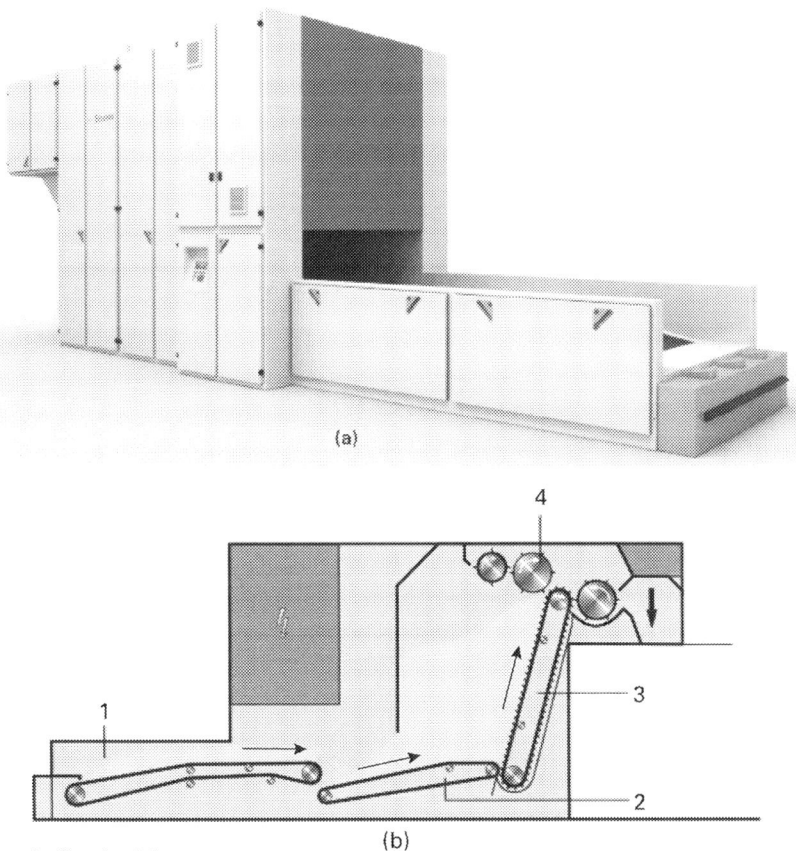

1. Feed table
2. Feed lattice
3. Spiked (inclined) lattice
4. Evener roller

FIG. 4.1 (A) and (B) Bale opener. *Courtesy of Trützschler Nonwovens.*

but is usually more robust, particularly the design of the spiked lattice and the beater rolls. An extended floor apron, often constructed from a chain of steel rolls rather than a conveyor belt, accepts bales directly from a fork-lift truck. Where the width of the bale opener is relatively narrow (e.g. 1500 mm) each feeder accommodates a single line of bales.

Alternatively, a wide bale opener is used that can accommodate several bales side by side. The individual bales can consist of the same raw material or several different ones to make up the blend. Such bales are very dense and the purpose of the bale breaker, as its name suggests, is simply to break it down into manageable clumps and to pneumatically feed these to the opening machine at a relatively consistent rate.

4.3.2 Bale pickers

Although blending large batches of fibre for nonwoven manufacturing mainly involves bale breakers, bale pickers used in the cotton spinning industry can also be utilised. In such systems, rows of bales are positioned in line formation (usually adjacent), and a mechanical picking device traverses across the top of the bales, progressively removing small tufts from each, in the correct proportions (see Fig. 4.2). Rotating spiked rollers set on a pivot arm inside the bale picker head remove the tufts as they run across the top of the bales. Because of the small tufts, a well-distributed mix can be produced in the blending bin. Systems have been introduced that use a variation of this concept for long staple fibres.

4.3.3 Fibre openers

Fibre openers used by the nonwovens industry have partly evolved from traditional machines that were originally designed for cotton and wool processing. Heavy-duty fibre openers such as Fearnoughts (see Fig. 4.3) are found in some sectors of the nonwoven industry, particularly those processing fibres of >50 mm and colour blends. Multi-roll openers, pickers or fine openers commonly suffice for other applications. Fine openers provide efficient in-line opening of fibres up to about 100 mm length. Such machines are arranged horizontally or vertically and are incorporated in feeding units and chute feeds, as well as blending hoppers (see Figs 4.3 and 4.4). In chute feed systems, a pair of feed rollers presents fibre to a revolving opening roller that is clothed with either pins or coarse card wire.

FIG. 4.2   Bale picker (automatic bale opener). *Courtesy of Trützschler Nonwovens.*

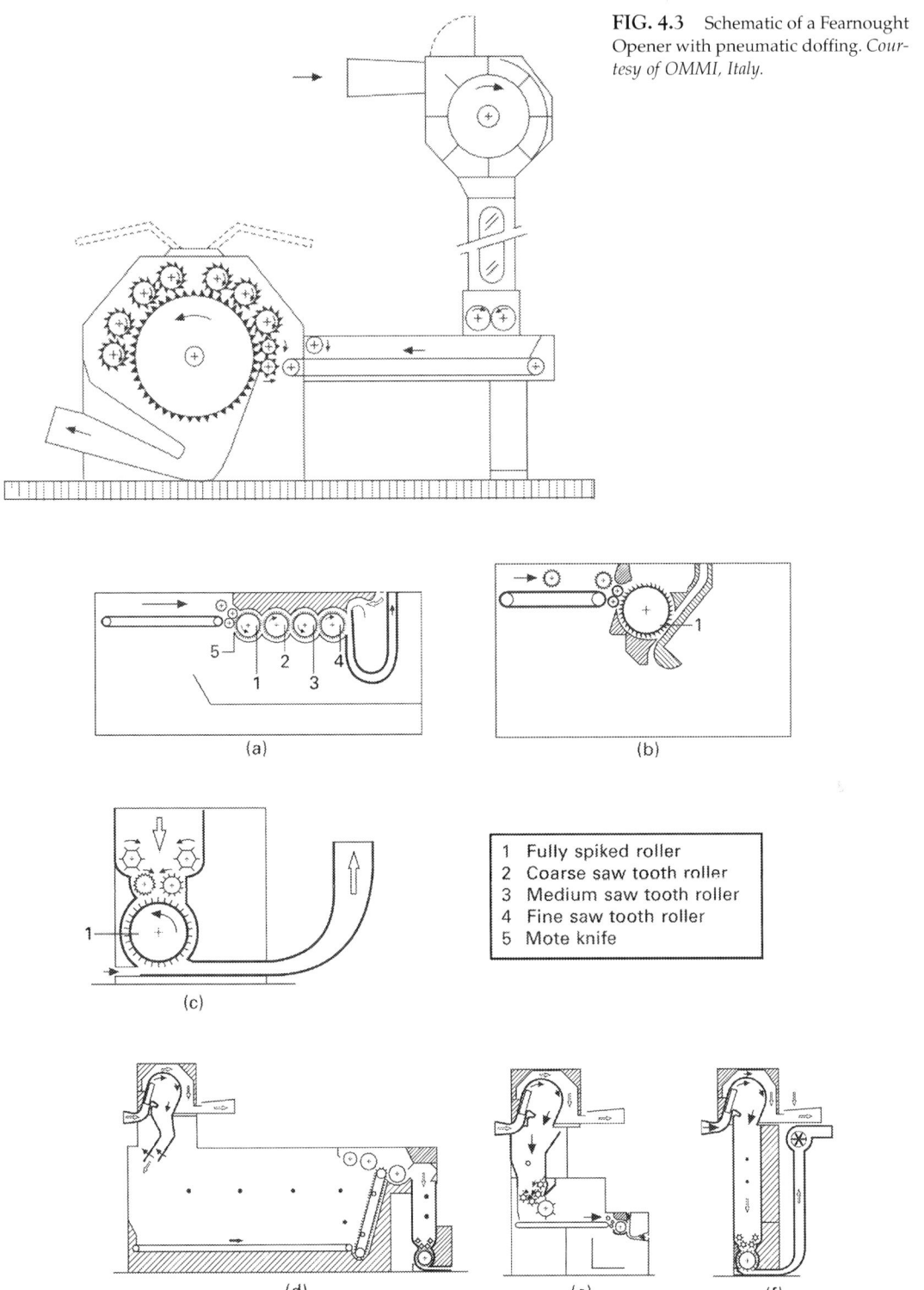

FIG. 4.3 Schematic of a Fearnought Opener with pneumatic doffing. *Courtesy of OMMI, Italy.*

1	Fully spiked roller
2	Coarse saw tooth roller
3	Medium saw tooth roller
4	Fine saw tooth roller
5	Mote knife

(a)

(b)

(c)

(d)

(e)

(f)

FIG. 4.4 Opening machine variants and integrated feeding and fibre opening units; (A) multi-roll opener; (B) single roll opener; (C) universal opener; (D) blending hopper with universal opener; (E) feeding unit with single roll opener; (F) feed trunk with universal opener. *Courtesy of Trützschler Nonwovens.*

A secondary chute with delivery rollers that feed a finely pinned opening roller operating with a high surface speed follows this. Examples are shown in Fig. 4.4.

Single roll openers are suitable for opening fibres such as polyester, whereas a multi-roll opener may be employed to open bleached cotton or viscose rayon where the tufts are more heavily entangled. One of the most important considerations in opening is the state of the incoming fibre, in terms of tuft density and fibre entanglement.

Fibre entanglement is generally reduced at the expense of fibre breakage, and to minimise such fibre damage, gradual opening using a sequence of opening units (rather than one single unit) is required to progressively reduce the tuft size. Based on this stepwise approach, a theoretical optimum opening curve has been proposed (see example in Fig. 4.5). As well as the design of the feed system and the number of opening rollers used, the type of clothing, pin density or blade frequency, gauge settings and surface speeds are also varied according to the degree of fibre opening required, and the incoming tuft size. The most intensive opening is generally achieved by presenting fibre to the opener roller (or beater) via a pair of clamped feed rollers rather than by an airstream. The theoretical tuft size after each stage of opening can be estimated based on the opening roller design, feed rate and fibre linear density.

It is important to recognise that decreasing the average tuft size by progressive fibre separation also promotes homogeneous mixing of the different fibre components because the tufts are smaller. Also, as the tufts are reduced in size, any solid particles or impurities that may be present are more likely to be liberated from the fibre. Clearly, it is advantageous to remove such impurities before carding, if possible, to maximise the life of the card clothing and yield.

4.3.4 Disc opener

The disc opener shown in Fig. 4.6 is remarkable in that it has only one moving part. Fibre is drawn through the system under negative pressure. As the fibre enters the expansion chamber, it makes contact with a high-speed rotating disc that is studded with stainless steel pins. The pins drag the fibre across a stationary, pinned plate and the opening takes place between the pins on the plate and those on the disc. Fibre then continues within the airflow, and is transported out of the machine via the exit chamber.

4.4 Mixing and blending

Different fibre types, grades or dimensions are mixed or blended either to obtain a particular combination of physical properties in the final fabric or for economic reasons to minimise cost.

Although most fibres utilised in nonwovens are not dyed, adequate blending is still important because of the fibre variation within bales as well as bale-to-bale. Visual assessment of blending is not reliable because most of the blends appear white. Bale-to-bale variations occur in respect of crimp frequency (crimps/cm), fibre finish application level and fibre entanglement. Fused, co-terminus ends and cutting problems experienced by the fibre producer are sometimes evident in bales, which can impact fibre processing performance, as well as the

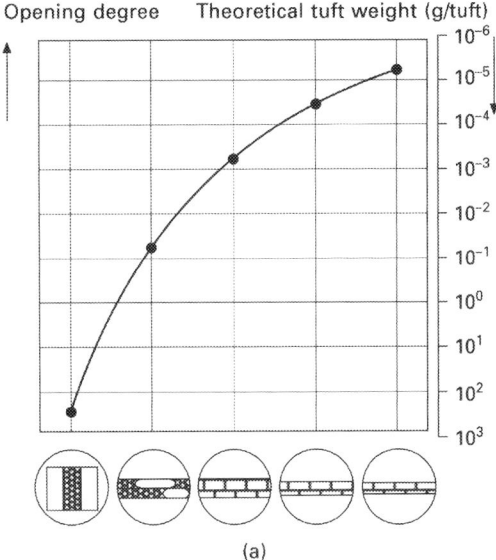

For pinned rollers:

$$N = \frac{\text{number of fibres per minute (F/min)}}{\text{speed of roller (rpm)} \times \text{surface area of roller (cm}^2) \times \text{points/cm}^2}$$

For beater rollers:

$$N = \frac{\text{number of fibres per minute (F/min)}}{\text{blows per minute (B/min)}}$$

$$(\text{F/min}) = \frac{\text{mass of fibre per minute (mg/min)} \times 10^5}{\text{fibre linear density (mtex)} \times \text{average fibre length (cm)}}$$

B/min = number of blades or pins on roller × roller speed (rpm)

Examples: calculation of N for different rollers

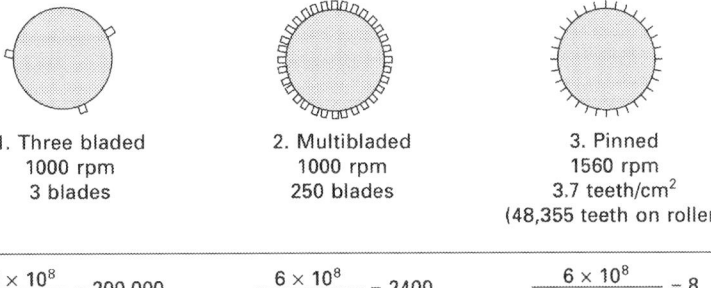

1. Three bladed
1000 rpm
3 blades

2. Multibladed
1000 rpm
250 blades

3. Pinned
1560 rpm
3.7 teeth/cm^2
(48,355 teeth on roller)

$$\frac{6 \times 10^8}{3 \times 1000} = 200{,}000 \qquad \frac{6 \times 10^8}{250 \times 1000} = 2400 \qquad \frac{6 \times 10^8}{1560 \times 48{,}355} = 8$$

(assuming a feed rate of 6 × 10^8 fibres per minute)

(b)

FIG. 4.5 Opening sequence for fibres; (A) progressive fibre opening across a series of fibre opening units; (B) fibres per blade or tooth calculation. *Courtesy of Trützschler Nonwovens.*

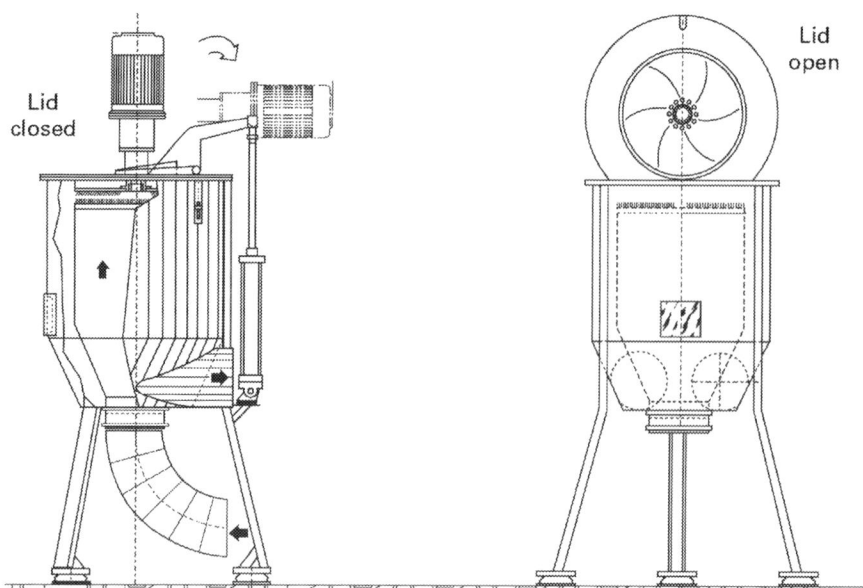

FIG. 4.6   Disc opener. *Courtesy of J Stummer Konstruktion, Germany.*

visual appearance of the finished product. In some sectors, such as the manufacture of needlepunched floorcoverings, stock dyed or spun dyed fibres are blended to create specific colour and shade effects. Clearly, the intimacy of blending in such blends must be consistent throughout the entire batch to minimise shade variations.

Fundamentally, the properties of a nonwoven fabric depend on the fibre composition, and it is therefore important that the blend components are consistently in proportion to minimise variations and to ensure product specifications are achieved. Poor blending leads to various processing and quality problems, some of which only become apparent after downstream processing. When one fibre component constitutes a small proportion of the total blend, for example <10% by weight, ensuring its uniform distribution can be particularly problematic.

Microprocessor controlled dosing systems assist in this regard. Nevertheless, where a particular blend component is a very small proportion of the total, for example in some thermal bonding applications using fusible fibres, pre-blending of that component with one or more of the other components, is sometimes carried out. The pre-blended, sub-component can then be treated as a single component during processing, resulting in a more thorough distribution throughout the entire blend. Where small component sizes are standard, sophisticated blending systems are available that claim uniform integration of a particular component down to as low as 1% of the total blend. Manual feed weigh-blenders are found in small blending rooms, or as a preparatory blending step to pre-mix small blend components.

Weigh-blenders incorporate a weigh conveyor onto which blend components are manually layered one on top of the other by the operator, in the correct weight proportions to produce a sandwich blend. A continuous succession of such blend sandwiches is then transported on the conveyor to a hopper or opener, before being intimately mixed within a blending system.

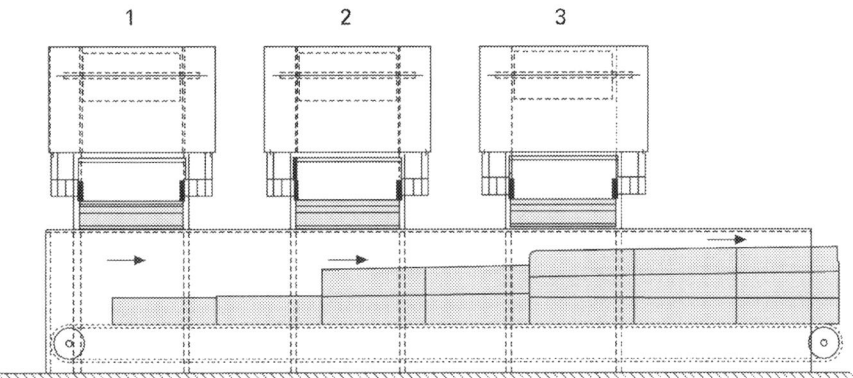

FIG. 4.7 Multiple hopper feed. *Courtesy of Garnett Controls Ltd.*

4.4.1 Multi-hopper systems

A multiple hopper arrangement is commonly used when between two and six blend components are to be combined in particular proportions, or where at least one component must be uniformly mixed in a low proportion (<10%). The hoppers can either be conventional in design, being fed from a single or a series of bale breakers or may be constructed as bale breakers with a weighing device attached to each (see Fig. 4.7). The hoppers typically weigh the fibre and drop the dosed weights onto a cross conveyor that runs under the hopper feeders. In some cases, where production is particularly high, volumetric chute-feed hoppers are used. Increasingly, multi-hopper systems utilise microprocessor control to feed a predetermined weight of fibre to the cross conveyor. The individual weights of fibre are synchronised to ensure that the cross conveyor receives a correctly proportioned blend. The conveyor then feeds the fibre to the next machine in line, which is usually an opener. In some cases, multiple weigh systems have weigh pans that are fed by silos rather than individual hoppers.

4.4.2 Metal detection

Sometimes, accidental contamination of the fibre occurs from a variety of common metal objects such as wire, screws, broken card clothing, small machine parts, spikes from conveyors, and any number of unusual objects. Consequently, it is necessary to incorporate metal-detection devices in blending systems as well as at the feeding section of cards. The consequences of allowing metal to enter a card in terms of damage and lost production are serious. The simplest form of metal detection is the 'magnetic hump' which is fitted with powerful magnets to catch stray metal objects. The unit is fitted in-line within the ductwork that conveys the fibre to the machine. In-line diverting devices detect metal within ductwork by electronic means. When metal is detected a signal is generated by a microprocessor, which opens the duct to divert the contaminated fibre to a holding bin where the metal can be manually removed and the fibre recovered. Many hopper feeders have a row of magnets mounted above the spiked apron. Metal-detection devices can also be fitted above the card feed apron or integrated into the card feed rollers. Because of the fine gauge setting between the feed rollers, any metal that passes though the feed rollers comes into contact with both rollers and completes a circuit that triggers an emergency stop either to the feed section or the entire card.

4.4.3 Fibre lubrication and spray systems

Although fibre lubrication is not universally used by the nonwovens industry because fibres are delivered ready for processing by the man-made fibre supplier, it is sometimes desirable to apply additional formulations. This may be a lubricant in the case of natural fibres, or an auxiliary such as an anti-static agent. Spray systems accurately dose and apply such additives directly onto the fibre. In some cases, the addition of water alone can be an effective processing aid. Water can be sprayed during the blending and opening stages, or on-line using an atomiser to apply a fine mist prior to carding.

4.4.4 The influence of moisture content

Moisture is important both in respect of fibre processing performance and the properties of the final fabric. Some hydrophobic fibres, particularly polyester and polypropylene are prone to static electricity during carding, which becomes most evident when the relative humidity is low. This can lead to problems in handling lightweight webs. The fibre breaking strength of cotton increases as the moisture content increases, whereas for viscose rayon and most other hygroscopic fibres the reverse is true. The extension at break of many hygroscopic fibres as well as frictional properties is also affected by changes in moisture content. Hydrophilic fibre finishes are applied to hydrophobic man-made fibres, to improve wetting out during hydroentanglement as well as to control static in carding. There are also important economic considerations in relation to moisture content, particularly where hygroscopic fibres are to be converted into medical, hygiene or wipe products because of the change in mass. Moisture measurement and control systems continuously monitor the moisture content of fibres, batts, and fabrics on-line. These systems either simply measure and report, giving alarms when the moisture content drifts beyond the pre-set limits, or they can affect automatic control.

Automatic re-hydration is achieved by water sprays or control of the throughput speed. The moisture content can also be controlled by automatic temperature adjustment in the dryer. In processing, the benefits of controlling the moisture content can be seen in the productivity of the process, in terms of both higher throughput and the reduction of static. Where a product is sold by weight, it is clearly important to ensure that the correct moisture content is maintained. For example, hygiene products such as cotton wool are produced and packed to a specific invoice weight. This packed weight is therefore made up of the fibre as well as an allowable amount of absorbed moisture. Areas of the industry where moisture control is utilised include feminine hygiene products, medical products, absorbent wipes, backing for floorcoverings, hydroentanglement installations, and some thermal bonding applications.

4.4.5 Blending hoppers and self-emptying bins

The traditional method of batch blending involves successive horizontal layering of the entire blend (composed of many bales) to form a block or 'stack' and then vertical slicing to

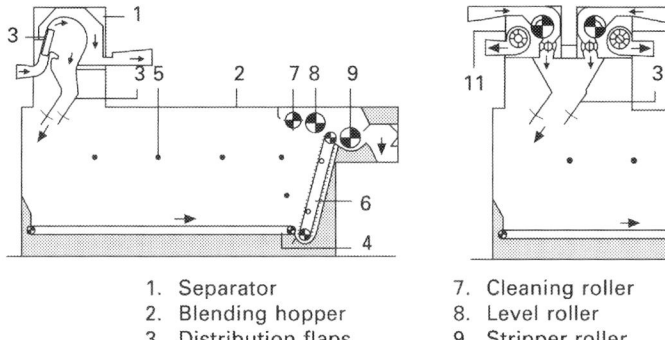

1. Separator
2. Blending hopper
3. Distribution flaps
4. Feed lattice
5. Light barriers
6. Spiked lattice
7. Cleaning roller
8. Level roller
9. Stripper roller
10. Recycled fibre (e.g. edge trim)
11. Condenser

FIG. 4.8   Blending hoppers. *Courtesy of Trützschler Nonwovens.*

produce small tufts and a homogeneous blend. The same principle is used in automatic blending bins. Semi-continuous and continuous blending are also common in the nonwoven industry, where the production line allows blending between only a few bales (<1t), rather than the entire blend (1–>10t). Blending hoppers allow continuous mixing. Fibre is fed into the machine either from a telescopic or fixed rotary distributor and is deposited into horizontal layers.

In modern small to medium capacity blending bins, a moving floor conveys the fibre in the direction of a revolving spiked lattice, which takes vertical slices from the fibre and discharges it pneumatically to the next machine (Fig. 4.8). The movement of the floor is electronically controlled to optimise throughput, preventing feeding variations. Large capacity automatic self- emptying bins work on a similar principle to blending hoppers except that they allow a much larger blend to be assembled and are intended mainly for batch rather than continuous blending. In this case, the bin emptier moves within the bin in a controlled manner. Fibre is deposited into the rectangular bin using a telescopic duct and rotary spreader to form horizontal layers and a spiked apron removes the fibre in vertical slices. A spiked inclined apron forms part of the emptying unit that moves progressively into the bin to remove the fibre.

4.4.6 Multimixers

An example of a multimixer is shown in Fig. 4.9, which incorporates a series of individual silos into which fibre is alternately fed. The size of the silos is variable depending on the desired production capacity and the number is usually six or ten. In such systems, horizontal slicing of the blend is performed by simultaneous removal of fibre from the bottom of each silo using either air or mechanical conveyors.

4.4.7 Buffer zones

To ensure uniform and continuous flow through the system, buffer zones may be required to provide interim storage. These usually take the form of silos with delivery rolls at the base.

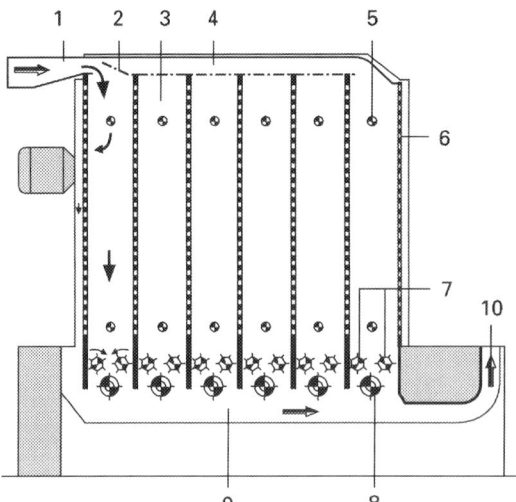

1. Fibre feed funnel
2. Closing flap
3. Mixing chamber
4. Feed duct
5. Light barrier to control level
6. Perforated plate
7. Delivery rollers
8. Opening rollers
9. Blending duct
10. Suction funnel

FIG. 4.9   Multimixer. *Courtesy of Trützschler Nonwovens.*

Such storage is also common in fully automated factories not only within the blending system but also between blending and carding to ensure a continuous supply of fibre to each carding machine. Where a number of cards are running the same fibre blend, it is common to utilise automatic feeding at every stage of manufacturing from bale to final fabric. A nonwoven line is usually continuous and of course must be fed with a continuous supply of fibre. The hopper feeding system that feeds the card utilises an optical sensor or similar device to call for additional supply when the level of fibre in its reservoir falls below a specific level.

In compact plants such a supply may come direct from the blending and opening line, with a diverter or distribution device directing the fibre flow to an appropriate card line. In high-production manufacturing plants, there may be specifically designated, self-emptying storage chambers that take the form of self-emptying bins or silos with delivery rolls. Such reservoirs supply fibre to the card lines on demand and in turn are supplied by the blending and opening equipment. The advantage of such systems in large operations is that the storage capacity can be utilised for a particular blend type to ensure an adequate supply for a given number of cards, whilst the opening equipment can be utilised for preparing other blends for different lines.

4.4.8 Cleaning systems

With some exceptions such as the processing of hemp, flax and similar natural fibres in the production of airlaid waddings, fibre cleaning is not universally used by the nonwovens industry. Natural fibres such as cotton are generally purchased in a pre-cleaned form, including for medical applications. Most opening machines can perform a mechanical cleaning function if the continuous scroll under the opening rollers or beaters is replaced with a perforated or slotted grid so that solid particles can be separated from the fibre by gravitation or centrifugal force. Additional cleaning apparatus is available to combat impurities such as discoloured

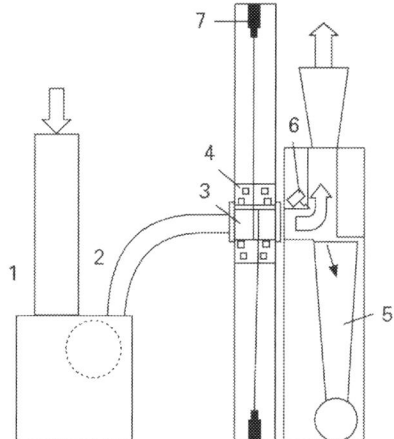

1. Chute from a fine opening machine
2. Rectangular fibre transport duct
3. Inspection shaft
4. Upper and lower illumination tubes
5. Contaminants collection chute
6. Air valves
7. Colour cameras

FIG. 4.10 SCAN-e-JET system for the automatic detection and removal of foreign matter in fibre processing lines. *Courtesy of H. Hergeth GmbH.*

fibre, or bale wrap material. Colour cameras are positioned at each side of a glass channel through which the fibre is directed. This is ideally situated directly after an opening machine when the fibre is mixed with air and is in an open state. A series of air nozzles operating across the channel are used to remove contaminants and fibre is directed through a separate exit duct. Such systems are able to record the position of the impurity and because multiple nozzles are used across the channel, only those nozzles in proximity to the contaminated fibre can be activated. This minimises the amount of good fibre that is lost in the discharge. An example of such a detection and removal system is shown in Fig. 4.10.

4.5 Carding

The purpose of carding is to disentangle and mix fibres, as well as to form a web that is uniform in weight per unit area. Carding relies on a series of fibre disentangling and layering actions accomplished by the interaction of toothed rollers situated throughout the carding machine. The terminology used to describe specific operations and machine parts varies in different countries and sectors of the industry, but the fundamental process is universal. The subject of carding is one that is widely debated, often misunderstood, and which has traditionally been viewed as more of an art than a science. There is no doubt that significant skill and experience is required to successfully produce the perfect web, if indeed such an icon exists, yet the basic principles of carding are few, and well worth the time it takes to understand them. For the purpose of this section, carding is broken down to its basic fundamental principles in order to explain and illustrate the interactions that take place within a card.

Essentially, the principles of carding can be largely explained in just two basic actions. The first and most basic principle of carding is 'working' and the second is 'stripping'. An understanding of these two core principles will lead to an understanding of the entire process, which is essentially a succession of 'working' and 'stripping' actions linked by incidental actions that are derivatives of the fundamental principles.

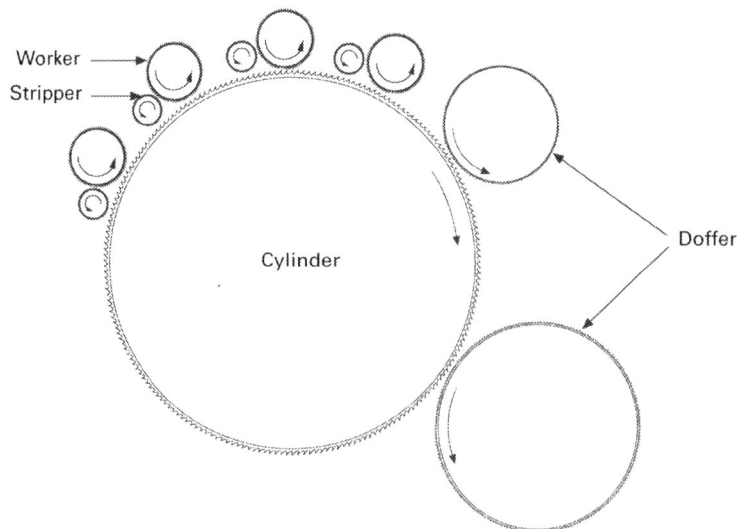

FIG. 4.11   Arrangement of rollers within a basic carding machine.

Whilst in the nonwovens industry there is no such thing as a standard carding machine, every roller card has a central cylinder (or swift) that is normally the largest roller. Smaller satellite rollers, called workers and strippers, which normally operate in pairs, are situated around the cylinder and these carry out the basic functions of working and stripping. Many cards have more than one cylinder, each with its own satellite rollers but to explain the basic principles of carding, consider a simple card configuration as shown in Fig. 4.11.

The cylinder is the heart of the carding machine and is the central distributor of fibres during the process. The worker-stripper pairings around the perimeter of the cylinder have both a carding and mixing function. A proportion of the fibre passing through the machine is disentangled or 'worked' between the cylinder wire and the worker wire, and some is 'delayed' as it revolves on the surface of the workers and strippers before being returned to the cylinder. The doffer rollers condense and remove fibre from the cylinder in the form of a continuous web. A proportion of the circulating fibre is not removed by the doffer and is recycled by the cylinder to be combined with the fresh fibre that is continuously fed to the cylinder. Therefore, during carding both fresh (uncarded) fibre and recycled (carded) fibre circulates on the cylinder in various proportions depending on machine configuration and settings. This contributes to the mixing power of the card.

4.5.1 Principles of working

The points of the teeth on a worker roller directly oppose the points of the cylinder teeth in a point-to-point relationship (Fig. 4.12). The worker revolves in the opposite rotary direction to that of the cylinder but because the bottom of the worker is set closely to the top of the cylinder teeth, the effect is that teeth on each roller travel in the same lateral direction at their point of interaction. The cylinder, being the main distributor of fibre, conveys fibres towards the worker and as the fibre passes the worker teeth, some is trapped on the worker teeth

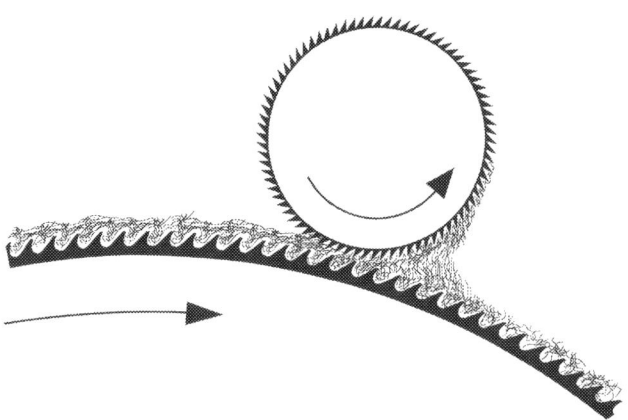

FIG. 4.12   Diagram of worker to cylinder action.

whose surface speed is slower than that of the cylinder. Since part of the fibre tuft that is trapped on the worker is also trapped on the cylinder, a separation or carding (working) action takes place as each roller revolves.

The efficiency of a worker tooth to collect fibre from the cylinder is highest when the tooth is empty. When an empty worker tooth enters the arc of contact with the cylinder, fibre from the cylinder is immediately caught and there is little resistance preventing the fibre being taken by the worker. As the tooth fills with fibre, resistance increases due to the build-up of fibre between the teeth and consequently, the efficiency with which the teeth can accumulate fibre is reduced. When worker rotation is slow, a high loading of fibre on the workers is clearly visible. Because of this, it is often wrongly assumed that slow workers promote better carding, i.e. if the worker has collected so much fibre it must be very efficient. In fact, the opposite is true. Assuming the cylinder speed is constant, the efficiency of a worker generally increases as its speed increases. This is because a higher worker speed results in more empty teeth being presented to the cylinder at the arc of contact per unit time, and these empty teeth are free to collect fibre from the cylinder. Slow workers, on the other hand, take longer to recirculate the fibre back onto the cylinder and therefore spread the fibre they have caught over a greater area. This results in more efficient mixing. In summary, faster workers promote more efficient carding, whilst slower workers provide better mixing. Usually, a balance of the two factors is required, which is feasible given the multiple worker-stripper pairings available on a full-size nonwoven card.

4.5.2 Principles of stripping

Once trapped by the worker, the fibre is carried around as the roller rotates. Before the worker teeth are re-presented to the cylinder, the fibre must be removed, otherwise the worker will continue to collect fibre until it becomes full and can no longer 'work'. The role of the stripper is to remove fibre from the worker and to re-present that fibre back onto the cylinder (see Fig. 4.13). To do this efficiently the teeth of the stripper must interact with the back of those on the worker. In turn, the fibre is removed from the stripper by the

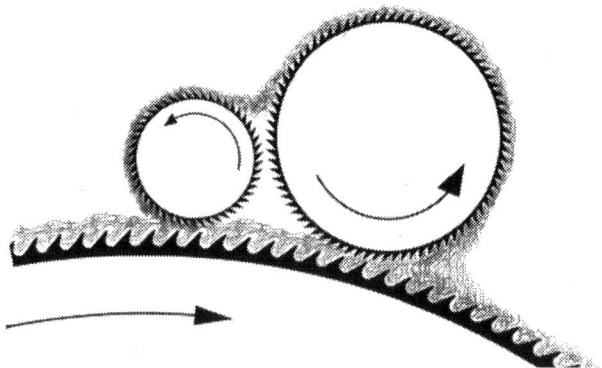

FIG. 4.13   Diagram of worker and stripper.

teeth on the cylinder whose points interact with the back of those on the stripper. Therefore, there is an initial 'working' action between the worker and the cylinder, then a stripping action between the stripper and the worker, followed by a further stripping action between the cylinder and the stripper. This series of actions represent the fundamental operational function of a carding machine. It is by a succession of such interactions that fibre is progressively, 'worked' and 'stripped' within a carding machine until the fibres are so uniformly distributed and individualised that a homogeneous web can be formed.

4.5.3 Interaction between card rollers

As shown in Fig. 4.13, fibre presented to a worker must first pass beneath the corresponding stripper. The reason the stripper is not placed at the other side of the worker is because this would adversely affect fibre separation, particularly when carding longer fibres. If the stripper is situated behind the worker, rather than in front of it, the transfer of fibre from cylinder to worker, then from worker to stripper and subsequently back to the cylinder would still be carried out, and a degree of working would indeed take place. However, because of the short linear distance between the pick-up points of the worker and stripper, fibres longer than this distance could be trapped by both rollers at the same time, leading to fibre breakage or rolling, which results in nep formation.

By positioning the stripper at the other side of the worker, the worker is able to draw the full length of a fibre away from the cylinder, thus separating the incoming fibre bundles, and straightening the fibres before re- presenting the fibre back to the cylinder via the stripper. A further argument is that by changing the relative positions of the worker and stripper, and at the same time changing the direction of the worker, a satisfactory arrangement may be achieved that would carry out the same carding function, but without re-presenting fibre to the same worker. However, in such a case, the fibre is re-presented to the next worker in line and although overall, less re-presentation of fibre to the same workers takes place, this also leads to reduced fibre working. Moreover, in that situation, the teeth of the worker travel in an opposing direction to those on the cylinder and the worker needs to be driven against the opposing force of the cylinder. This leads to higher energy expenditure because of the large mechanical forces placed on the machine drives, rollers and card clothing. This is particularly disadvantageous in the early stages of carding where the fibre bundles are entangled. By rotating the teeth away from those on the cylinder, the worker effectively yields to the force of the cylinder. More importantly, the force between the two rollers, can be controlled by adjusting the worker speed.

As an example, where fibre loadings are to be reduced by increasing the worker speed and hence an increase in the number of fresh empty worker points presented to the cylinder; when operating in the reverse direction these points are presented at increased opposing speed and the applied forces increase. This can result in fibre breakage and increased fibre packing between the teeth. In contrast, when the worker revolves in the normal direction, more fresh points are presented by increasing the speed in the same direction as the cylinder, creating reduced opposing forces. This results in a gentler action and more effective transfer of fibre. Carding efficiency also increases because of the increased ratio of empty teeth to incoming fibre.

It is also important to consider the following point, which is one that is commonly misunderstood. In Fig. 4.13, fibre is carried by the cylinder and trapped by the worker. The fibre is carried around the worker until it is removed by the stripper and then in turn is removed from the stripper and is re-taken by the cylinder. At this point, the fibre is again presented to the same worker, and it is easy to speculate as to why it does not simply continue to be trapped by the same worker and be perpetually carried around the same set of rollers. In fact, only a proportion of the fibres are re-circulated around the same rollers. Because the surface speed of the worker is much slower than that of the cylinder, the worker catches only a proportion of the in-coming fibre from the cylinder. Moreover, at the point of worker-cylinder interaction, part of the fibre tuft caught by the worker is simultaneously held by the cylinder. Thus, each time fibre is presented to the worker by the cylinder, separation takes place. It is the many successive separations of fibre by these rollers that break down the fibre tufts until the fibres are in such a disentangled state that a homogeneous web can be formed. Fibre that is not trapped by the worker continues to be trapped by the cylinder and is carried to the next worker, where the same process takes place. This progressive separating and layering of fibres leads to efficient fibre mixing. The process described above is the essence of carding. The following elements are additional and serve to prepare, transport, manipulate or consolidate the fibre.

4.6 Roller operations

The function of a doffer is to remove or 'doff' (remove) fibre from the cylinder so as to produce a continuous web (see Fig. 4.14). The easiest way to understand the doffing action is to consider it as a large-diameter worker. The tooth direction as well as the direction of rotation is the same as that of a worker.

But, whereas the function of a worker is to 'work' the fibre, to break down the fibre bundles, and return it via the stripper to the same cylinder, the function of a doffer is to consolidate it into a web structure so that it can be removed in the form of a web. The essential difference between a doffer and a worker is that the doffer accepts fibre from the cylinder and conveys it away, without re-circulating the fibre to the cylinder using a stripper. The doffer teeth are

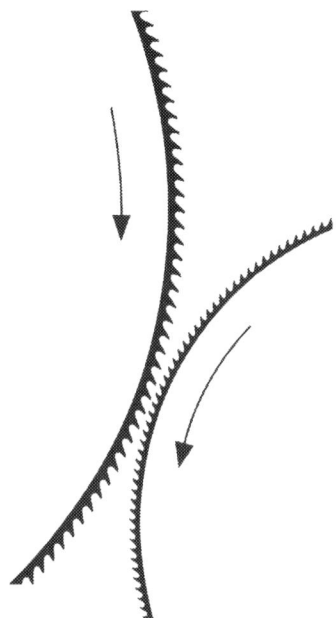

FIG. 4.14   Diagram of cylinder and doffer.

designed to accept fibre and to hold it efficiently and the doffer is larger in diameter than a worker creating a greater arc of contact between the doffer and the cylinder. This allows more efficient fibre transfer as well enabling the formation of fibres into a web. Generally, as an individual tooth becomes full, its ability to hold more fibre decreases. Consequently, the faster a doffer travels, the emptier are the teeth it presents to the cylinder, and therefore, more fibres leave the cylinder. As more fibres leave via the doffer, fewer re-circulate around the cylinder. This explains why increased doffer speeds result in lower cylinder loadings.

It is known that increasing the throughput, with the consequent increase in cylinder loading, tends to negatively influence fibre breakage, nep formation and general web quality. Adjustment of the doffer speed or specifically, the swift-doffer surface speed ratio, is one of the most efficient methods of controlling cylinder loading and hence web quality. The doffer wire design is also important and is discussed later in this Chapter. A further consideration is the doffer to cylinder setting. Although decreasing the setting gap theoretically promotes increased transfer efficiency, in practice, with a suitable doffer wire design and doffer speed, efficient fibre transfer can be achieved without tight settings. This is helpful in minimising wear of the card wire.

4.6.1 Multiple doffers

Nonwoven cards commonly utilise a double-doffer arrangement and some cards have been designed with up to four separate doffers or web take-off devices. As previously noted, the fibre transfer efficiency from cylinder to the doffer is maximised when the teeth are empty. When card throughput is high, the transfer efficiency of a doffer decreases as it fills up and a greater proportion of the incoming fibre recirculates on the cylinder leading to cylinder loading. Quality problems such as nep formation then arise, particularly with fine fibres. Whilst increasing the doffer speed reduces the loading, it may also lead to problems removing the web, particularly when a fly-comb is used. Additionally, an upper delivery speed limitation may be imposed by subsequent processes, such as a cross-lapper, in the production line. Double doffers tend to increase the fibre transfer efficiency from the cylinder allowing higher production rates to be achieved without overloading the cylinder or necessitating the use of very high doffer speeds.

However, the structure and uniformity of the two webs (top and bottom) may be different. The top doffer has 'first bite' at the fibre and unless adjustments are made, the web from the top doffer tends to be heavier than that produced by the bottom doffer. Structural differences between the two webs may affect the respective web tensions as well as the physical properties of the resulting fabric. In practice, the proportion of fibre taken by each doffer is balanced by adopting different top and bottom doffer to cylinder gauge settings and the use of different doffer diameters. The top doffer is often smaller to provide a smaller arc of contact with the cylinder. The tooth design on each roller can also be varied to alter the relative pickup of each roller. Despite the difference in roller diameter, the surface speed of each doffer must match the line speed. Slight adjustments in surface speed between the two doffers may be used to control different web tensions.

4.6.2 Transfer rollers

Where a carding machine has more than one cylinder, the fibre must be transported from one section to another. Sometimes this is carried out by a 'middle doffer', which allows reconsolidation of the incoming fibre stream and provides an additional working point. However, more usually in nonwoven carding, a transfer roller is used. Whilst a doffer operates like a worker in terms of its relative tooth interaction with the cylinder, a transfer roller operates like a stripper. The fundamental difference between a doffer and a transfer roller is that a transfer roller takes fibre from the preceding cylinder, whereas a doffer has the fibre put onto it by the cylinder. The surface speed of a doffer is slower than that of a cylinder and the points are opposing, so that as the cylinder passes the doffer, fibre is deposited into its teeth by the cylinder. The surface speed of a transfer roller is greater than the cylinder from which it receives fibre and the teeth of the transfer roller act upon the back of the cylinder teeth in the same way that a stripper interacts with a worker. The faster moving transfer roller therefore 'strips' fibre off the cylinder and conveys it to the next cylinder, which in turn 'strips' the transfer roller by virtue of its even greater surface speed. A transfer roller is commonly used to convey fibre between the initial 'breaker' or 'breast' section of a carding machine (see Fig. 4.15) to the main cylinder.

On the initial breaker section, because the fibre is in a pre-opened state, the cylinder to worker settings may be comparatively open with a low tooth population. If a doffer is used in such circumstances, with a similarly open setting, the fibre transfer efficiency to the next section would be low and fibres would re-circulate on the cylinder, increasing fibre loading and the likelihood of neps and fibre breakage. Alternatively, if a closely set doffer is used, the fibres would be subjected to forces that could induce fibre breakage or damage the card clothing. Alternatively, a transfer roller facilitates a progressive increase in surface speed between

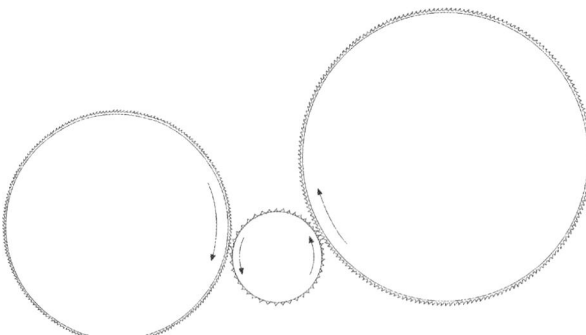

FIG. 4.15   Diagram of cylinder to transfer.

two card sections and entangled fibre tufts are not subjected to the large forces associated with large difference in relative roller surface speeds. This tends to minimise fibre breakage and the formation of neps.

4.6.3 Operation of card feed rollers

Although the feed roller section is clearly the first section of the card, the interaction of these rollers is more readily appreciated once the fundamental principles of carding are considered. Obviously, the purpose of the feed rollers is to feed fibre into the machine. There is another important principle involved. Fig. 4.16 shows a typical feed-roller arrangement. Clearly, due to the direction of roller rotation, the rollers would feed fibre to the licker-in even

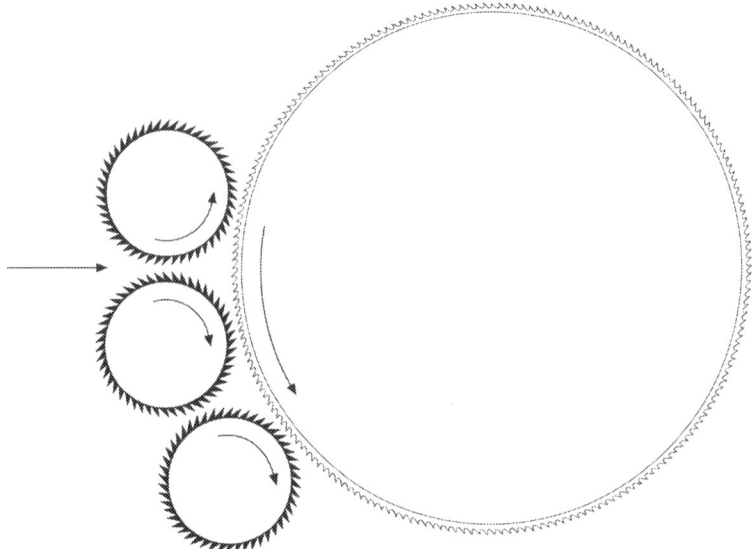

FIG. 4.16   Diagram of feed rollers.

if the surfaces were not covered with teeth. The function of the teeth, however, is not merely to help feed the fibre forward into the card but more importantly, to clamp the fibre fringe, which is created between the feed roller and the licker-in, thus preventing snatching of the tufts by the licker- in (sometimes called taker-in because it 'takes' fibre from the feed roller and carries it into the card). The licker-in takes fibre from the slower feed rollers using the tooth points, in just the same way that a stripper takes fibre from a worker. This is a 'stripping' action. The feed roller teeth, which face in the opposite direction to the direction of travel, trap fibres and prevent larger tufts from being intermittently snatched. This facilitates a uniform release of fibre to the licker as the feed rollers rotate. The licker conveys the fibre to the next roller in line, which in turn strips the fibre from the licker.

Down-striking and up-striking licker-in arrangements are used. Whereas a down-striking licker-in assists in the gravitational removal of impurities, the choice of configuration generally depends on the roller configuration for a given card design, the fibre length and the direction of rotation of the next roller in line.

For example, a short card requiring minimum carding of short fibres may be designed to take fibres from the feed rollers using a down-striking licker straight to the cylinder. In contrast, a card designed for long fibres that requires more progressive carding, may use an up-striking licker followed by a transfer to a breaker cylinder, before reaching the cylinder. The position of the clearer roller is important. Where the licker-in is up-striking, the clearer is situated above the feed rollers and where the licker-in is down-striking, the clearer is below the feed rollers. For both licker-in configurations, once the licker-in starts to take fibre in the direction of rotation, it is confronted with a point-to-point relationship between its own teeth and those of one of the feed rollers. In effect, there is a crude carding action, which is minimised by continued rotation of the feed roller and the utilisation of teeth that are designed to release the fibre. The function of the clearer roll is simply to prevent the accumulation of fibre on the feed roller and to move it in the direction of the licker-in rotation.

4.6.4 Dish feed

A dish feed arrangement (see Fig. 4.17) is sometimes utilised in place of feed rollers. Dish feeds are available in both down-striking and up-striking arrangements. In nonwovens, their use is mainly confined to short fibres that are less prone to being snatched and dragged into the licker-in. A dish feed arrangement utilises a single roller that operates against a feed plate, limiting the access of fibre to the licker-in and feeding by rotation of a single roll.

The design and setting of the dish with respect to the licker-in influence the degree of opening, waste extraction, removal of good fibres, and fibre breakage.

4.6.5 Web doffing and web structure modification: Randomisers, scramblers, and condensing rollers

The nonwovens industry is distinguished in the world of carding by the variety of different web geometries and structures that may be produced by mechanical manipulation of the web before it leaves the card. Generally, using different roller arrangements, the aim is to improve web isotropy, which is often expressed in terms of the machine direction to cross direction (or

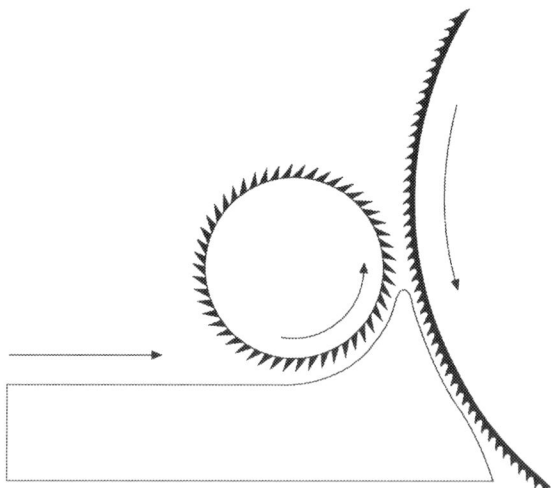

FIG. 4.17 Diagram of dish feed arrangement.

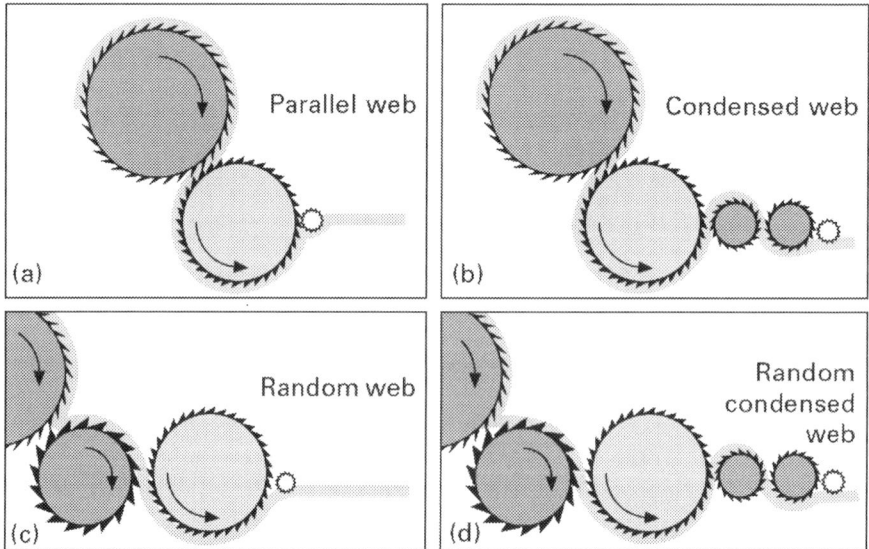

FIG. 4.18 Production of (A) parallel-laid (conventional), (B) condensed, (C) random and (D) combined random/condensed webs in carding. *Courtesy of ANDRITZ Asselin-Thibeau.*

MD/CD) ratio of properties, typically the tensile strength. Fig. 4.18A shows a conventional cylinder-doffer arrangement (parallel or parallel-laid web), which produces an anisotropic web with a comparatively high MD/CD ratio. The MD/CD may be >5:1 depending on card configuration. Under normal circumstances, a card web is significantly stronger in the machine direction than in the cross direction because successive working and stripping imparts a degree of fibre parallelism. A condenser roller (see Fig. 4.18B) gives scope to reduce the preferential fibre orientation in the MD and to set the optimum doffer speed at the same time as modifying the structure of web. The condenser roller surface speed is lower than that of the doffer, which means that fibre leaving the doffer is suddenly decelerated so that fibre is condensed or piled into the condenser rollers. The fibres bunch up vertically within the teeth resulting in a three-dimensional 'condensed' structure. Since the condenser roller surface speed is lower than the doffer, it adversely affects the output speed of the carding machine. To overcome this, the surface speeds of the condenser rollers are matched to the required card line speed and the doffer is operated at a higher surface speed than would normally be the case. In such an arrangement, it is the condenser rollers and not the doffer whose speed is governed by the overall line speed, or the next machine in line (e.g. the cross-lapper).

Randomiser or scrambling rollers as they are also known (see Fig. 4.18C) can be operated between the main cylinder and the final doffer to create 'randomised' webs having an MD/CD ratio nearer to one. The randomising or scrambling rollers disrupt the preferential orientation of fibres and redistribute fibres into a 'randomised' or 'scrambled' web.

Randomisers work by instantaneously changing the flow direction of fibres moving at high velocity by introducing an opposing carding action that modifies fibre orientation partly by air turbulence developed between the cylinder and randomising roller. The card wire is designed to accommodate fibre rearrangement whilst still allowing the release of fibre to form a web. Randomisers are often used in 'straight-through' carding applications where the web is bonded immediately after leaving the card and is not formed into a cross-lapped batt. Thermal calender bonding is a typical example. Some nonwoven producers utilise randomisers even when a cross-lapper is in use, for example when lightweight products are being manufactured using a minimal number of laps to form the batt or when the application requires an isotropic structure, for example in the production of geosynthetic fabrics.

Scramblers and condensing rollers can be used in combination (Fig. 4.18D) to achieve MD/CD ratios approaching one. Also, double doffer cards are frequently run with one doffer operating with a condenser roller and the other without. The two webs are then combined at the end of the card.

4.6.6 Web removal systems

Traditionally, a high-speed fly-comb is used to remove the web from a nonwoven card. Such combs are capable of operating up to 3200 rpm (see Fig. 4.19). However, as production speeds have increased, roller take-off has become standard on high-speed cards. Rollers and combs are also be used in combination. Roller doffing systems include the Dofmaster system (Hollingsworth), which incorporates a fluted roller in conjunction with a snap-in blade and the LDS arrangement (ANDRITZ Asselin-Thibeau), which utilises a fluted roller in conjunction with a suction system to control web transfer to the apron (see Fig. 4.19).

4.7 Card clothing

The card clothing or 'wire' covering is critical. Without it the card is totally ineffective and the choice of card clothing for particular rollers within the machine is most important. Metallic wire is most commonly used as opposed to the original 'flexible' card clothing, which

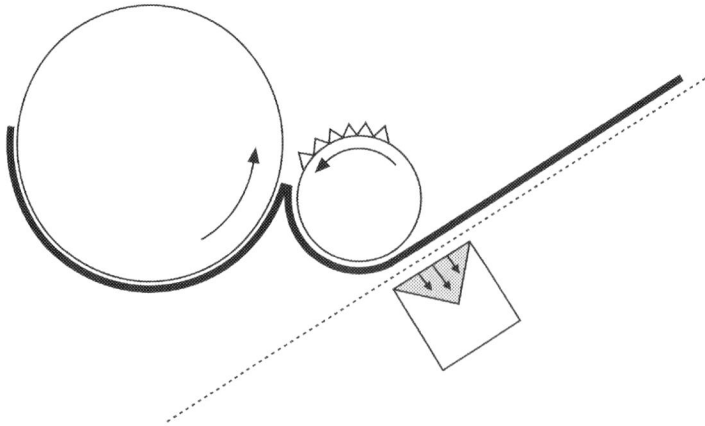

FIG. 4.19 Example of web detachment system – suction-assisted roller take-off. *Courtesy of ANDRITZ Asselin-Thibeau.*

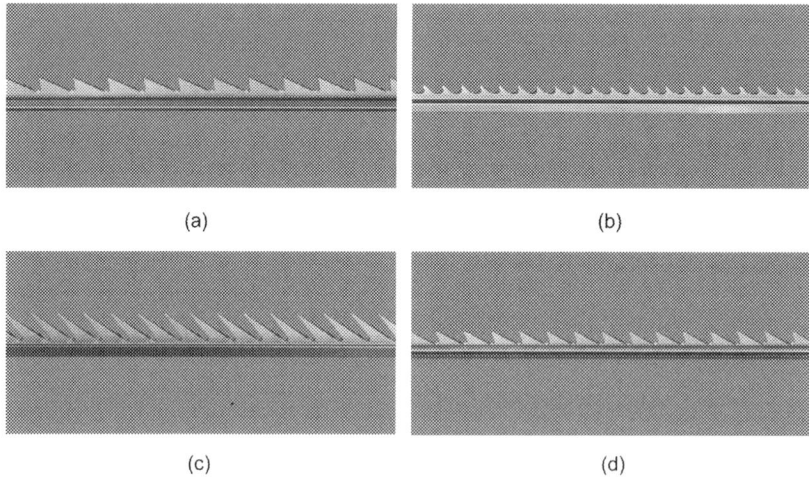

FIG. 4.20 Metallic card wire, with examples for each card roller: (A) feed roller; (B) cylinder; (C) worker; (D) stripper. *Courtesy of Groz-Beckert.*

consisted of fine metal wires mounted in composite fabric foundations. Metallic wire is manufactured in continuous strips with a shaped profile that is stamped to create a single row of teeth. The 'wire' is wound around the card rollers under tension from one side to the other until the surface is covered with teeth of specific point density and geometry. Since flexible card clothing is now largely confined to the carding of natural fibres in the traditional textile industry and is rarely used in nonwovens, all references to card clothing in this section will be to metallic wire unless specifically stated otherwise.

Since the introduction of metallic card clothing (Fig. 4.20), thousands of different wire designs aimed at improving the carding process have been devised. Research and development

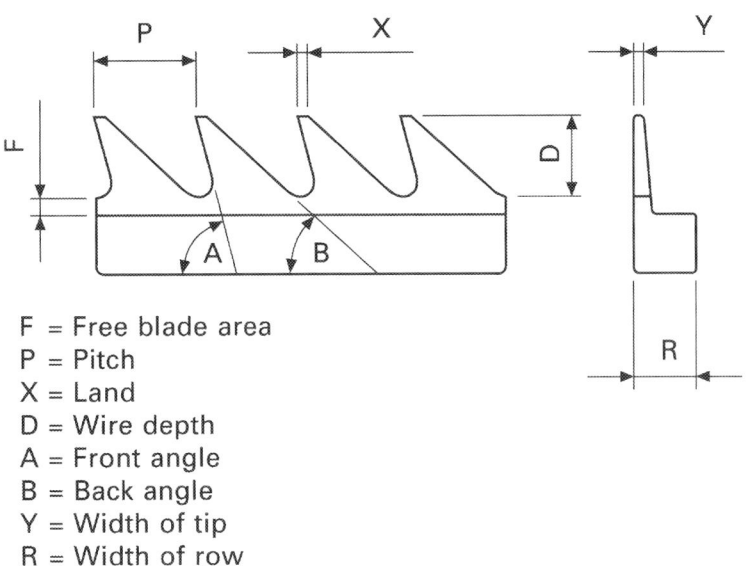

F = Free blade area
P = Pitch
X = Land
D = Wire depth
A = Front angle
B = Back angle
Y = Width of tip
R = Width of row

FIG. 4.21   Characterisation of metallic wire features.

on new wire design is continuing to respond to the introduction of new fibres, higher card production rates and the need for improved web quality. The choice of clothing depends on fibre properties and dimensions, the card configuration, settings and production rate. Fig. 4.21 indicates the basic design of metallic card wire and illustrates how the tooth profile and the cross-section are characterised.

4.7.1 Tooth depth

Fig. 4.21 illustrates the two dimensions (D and F) that determine tooth depth. Dimension D is the distance from the tip of the point to the bottom of the 'mouth'. This is the working depth of the tooth and determines the holding capacity of a particular wire. The working depth affects the loading capacity of the roller onto which the wire is wound. Therefore, it follows that rollers requiring high fibre loading are clothed with wire having a comparatively large tooth depth. Such rollers are principally workers and doffers, which must readily accept fibre from a faster moving roller, i.e. the cylinder, as they carry out their respective functions within the machine. Conversely, cylinder wire whose function is to interact with the worker must not overload and hence low depth wire is generally selected. Similarly, stripper wire, whose function is merely to transfer fibre from one roller to another, does not require high wire depth.

The selection of tooth depth is also influenced by fibre diameter and fibre length. For example, long and coarse fibres, which are invariably utilised for heavyweight products such as needled floorcoverings at high throughput, require greater wire depth in order to accommodate the volume of fibre that needs to pass through the card. Fine fibres, which are typically converted into lightweight webs, require shallow depths to prevent overloading and nep formation and to keep the fibre near to the tops of the teeth to permit transfer to subsequent rollers, notably the doffer. When carding fine fibres (particularly those below 1.7 dtex), if the tooth depth is too high, fibre is trapped within the teeth and fibres near the top of the teeth are processed in preference to those packed into the base.

Dimension F is the free-blade area, which is the space between the bottom of the mouth and the surface of the shoulder. This space affects the freedom of movement of fibres in the wire as well as the aerodynamics of the process. The setting gaps between card rollers typically range from about 0.27 mm to 0.70 mm. The work done within those spaces is considerable, as are the speeds of the various rollers, particularly the cylinder, whose surface speed may be in excess of 1500 m/min. The dynamics are such that a significant movement of air occurs within the card, which in a confined and enclosed space creates air velocities that result in turbulence and large pressure differentials. This influences the movement of fibre within the card and may lead to uncontrolled fibre migration in the direction of the highest air pressure. In order to control such effects, the free-fibre depth can be used to alter the aerodynamic properties of the carding elements. Since the card wires across the width of the machine are relatively closely spaced and a single fibre can bridge several teeth, it will be seen how much of the fibre can be kept away from the shoulder and how the space beneath the mouth of the tooth can allow air to dissipate.

Of course, since many of the fibres are aligned in the longitudinal direction, they will enter the free-space area. However, it is unlikely that a fibre will be so perfectly straight that it will sit between adjacent rows of wire. In practice, such fibres are quickly caught by one or more teeth at some point along their length and fibres are transferred between the available teeth. In this state, the fibre within the card can be described as 'fluid'. It is when such fluidity is interrupted that the majority of quality problems occur. With some types of fine, short fibres such as cotton, the free blade depth is reduced in order to minimise cylinder loading.

4.7.2 Wire angle

Referring to Fig. 4.21 there are two key angles to consider in the design of card wire. The first is the 'front' or leading angle and the other is the 'back' or trailing angle.

4.7.2.1 Front angle

The front angle influences the degree to which a tooth captures fibre from the roller with which it interacts. This controls the 'appetite' of a particular wire. The front angle also determines the ability of a tooth to hold and carry fibre and this is particularly important in high-speed carding of fine fibres. Centrifugal force increases as roller speed increases. Consequently, the front angle influences the degree to which fibre will 'lift' from the tooth as a result of that centrifugal force and equally, the degree to which it will be held when subject to mechanical forces. Again, the function of the roller in question is a key consideration. If we consider the cylinder, for example, the function of this roller is to carry the fibres to the workers and hold part of them as they are presented to the worker so that there is a force of separation which disentangles the tufts, yet at the same time it must progressively release the fibre preferentially to the worker, allowing the worker to take a significant share and thus avoiding overloading of the cylinder. Conversely, a low cylinder wire angle will tend to hold the fibre too well, resulting in re-circulation of fibre and overloading, with a consequent increase in nep formation and web patchiness. Typically, the front angles used on cylinder wires are between 70 and 80 degrees.

In comparison, consider a worker that interacts with the cylinder described above. A typical front angle used on a worker wire is between 50 and 60 degrees producing a more, 'hungry' tooth than that of the cylinder. Given that the function of the worker is to capture fibres from the cylinder and hold them securely so that they are 'worked' as the rollers interact, the captured fibres must then continue to be held as the worker rotates. Accordingly, the angle on the worker must be more 'hungry' than that of the cylinder so that fibre on the surface of the cylinder that comes into contact with the worker, is more likely to be released by the cylinder and retained by the worker yet not so readily released that effective carding is not achieved. The front angle of the worker teeth therefore influences carding power. Consequently, the relationship between the front angles of interacting cylinders and workers, as well as between cylinders and doffers (which as previously discussed have a similar function), is one of the most critical on a carding machine. This is demonstrated in Fig. 4.22, which illustrates the relative interaction of a worker and cylinder teeth.

In Fig. 4.22 there are two different worker front angles interacting with a cylinder wire having a front angle of 80 degrees. The one on the left shows a worker wire (on top) that has a similar front angle to the cylinder. Compare this with the one on the right, which has a front angle of 50 degrees and the effect can be clearly observed. The lower angle of the worker on the right of Fig. 4.22, results in a more efficient transfer of fibre from cylinder to worker. Progressively decreasing the angle eventually closes the mouth of the worker teeth and begins to have a negative effect. A low-fibre transfer efficiency from cylinder to worker results in excessive fibre loading on the cylinder, which leads to web quality problems. Conversely, if the worker wire is too efficient in capturing fibre from the cylinder, the fibre will transfer too easily and will not be effectively 'worked', which in turn will negatively influence the web quality.

It is also worth noting the effect of roller diameter. In the examples shown, the tooth angle is taken from the horizontal plane. Clearly, rollers are not flat but cylindrical and the effective angle of a tooth is influenced by the roller diameter around which the card wire is wound. A card that has small diameter workers will generally require a different worker tooth angle compared to a card with relatively large diameter workers if the same tooth point relationship

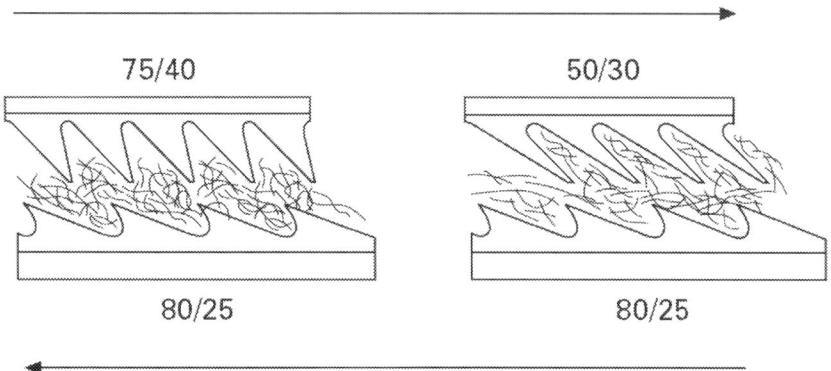

FIG. 4.22   Representation of different fibre to wire interactions using various wire angles.

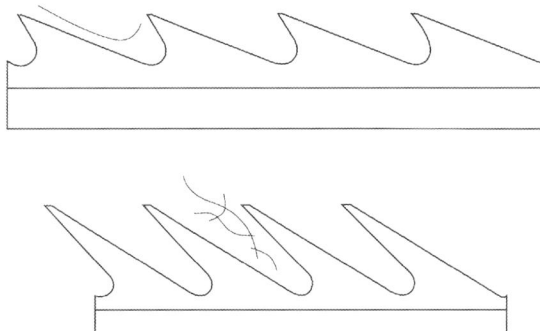

FIG. 4.23 Comparison of wires with different back angles.

is to be achieved with the cylinder wire. It is the effective wound wire angle relationship that is important, not the theoretical relationship before the wire is wrapped around the rollers.

4.7.2.2 Back angle

The 'back angle' of a tooth influences the card wire loading properties as well as the overall strength of the tooth. A high back angle promotes increased loading capacity and is therefore employed on rollers such as workers and doffers. Cylinders and strippers, which must not operate with high fibre loadings, utilise lower back angles, which minimise trapping of fibres in the mouth of the teeth (Fig. 4.23).

4.7.2.3 Point density

The choice of tooth population (or point density) for a particular roller is largely determined by the fibre type and fibre dimensions and is commonly expressed as the number of teeth per unit area (points/in^2). Normally, the tooth population increases as the fibre diameter decreases. The point density is determined by the pitch (P) and the number of tooth points across the card (R). Referring to Fig. 4.21, the 'pitch' (P) is the number of points along a one inch (25.4 mm), length of the wire. R indicates how many tooth points are present in one inch across the roller after spiral winding, which is governed by the width of the wire, or more specifically by its shoulder, and the way that each row of wire joins with the previous one. Some wires are positioned such that the rows butt against each other, some are fitted within grooves on the surface of the roller, whilst others utilise an interlocking arrangement. The wire mounting arrangements are discussed later. The point density of a roller is calculated by multiplying the pitch by the number of rows.

4.7.2.4 Point profile

The tooth point design influences such factors as point penetration into the fibre and fibre to metal friction as well as tooth strength and resilience. To appreciate the effect of point design on the function of the tooth, consider the point of a needle. A sharp, symmetrical, uniformly shaped point provides good penetration into the fibre, whilst allowing free release. A needlepoint also minimises frictional contact between the tooth and fibre minimising wire wear. However, a needlepoint tooth is relatively weak at its tip. Additionally, the penetration angle of the tip is largely governed by the overall tooth angle. Because of the way metallic card clothing is made, teeth can be stamped out of the wire in specific shapes allowing different dimensions and shape features to be included for specific purposes. Fig. 4.21 shows the two fundamental dimensions of a tooth point. Dimension Y is the thickness of the tooth at its point, whilst X is the length. This latter dimension is often called the land area.

Increasing the land area behind the effective front edge of the tooth creates support and added strength. However, the land area also affects the metal to fibre contact area and hence frictional characteristics. The sides of the tooth can also be engineered to include tapers, grooves or striations. These surface features influence fibre to metal friction, strength, and the available free space. Examples of different point profiles and wire specifications for nonwoven cards produced by one wire manufacturer are shown in Fig. 4.24. A combination of design elements is used including special surface finishes, point shapes, and surface features.

4.7.2.5 Wire foundation

The original Garnett wire was fitted into grooves cut into the surface of the rollers. Some heavy feed roller wires are still mounted in this way to provide additional stability in demanding areas.

Licker-in

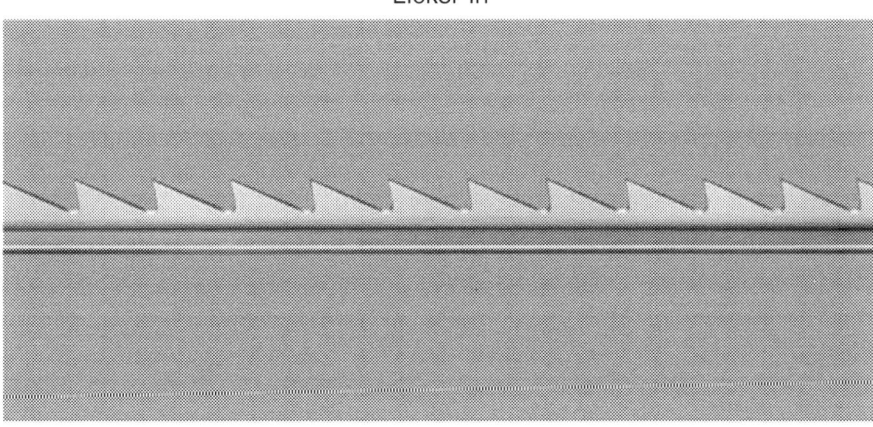

Product	Height (mm)	Rib (mm)	Pitch (mm)	Working angle	Front angle	PPSI
VA8/650/70	5.50	3.17	6.50	20°	70°	31
VA8/650/80	5.50	3.17	6.50	10°	80°	31
VE08/550/95	5.50	3.17	5.50	−5°	95°	37
VA10/650/70	4.70	2.54	6.50	20°	70°	40
VA10/550/70	4.70	2.54	5.50	20°	70°	48
VC10/400/75	4.70	2.54	4.00	15°	75°	65

FIG. 4.24 Wire profiles. *Courtesy of Groz-Beckert.*

(Continued)

Breast & Main Cylinder

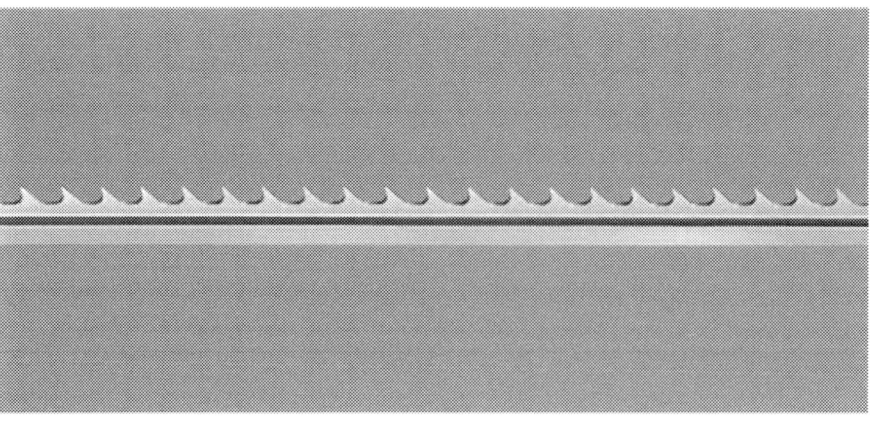

Product	Height (mm)	Rib (mm)	Pitch (mm)	Working angle	Front angle	PPSI
VA12/400/75	4.70	2.12	4.00	15°	75°	76
VA16/400/75	4.50	1.59	4.00	15°	75°	102
VA16/425/75	3.80	1.59	4.25	15°	75°	96
VA20/300/80	3.80	1.59	3.20	10°	80°	159
VA20/300/70	3.80	1.59	3.00	20°	70°	169
VA24/300/75	3.80	1.06	3.00	15°	75°	203
VF28/320/80	3.80	0.90	3.20	10°	80°	222
VA28/300/75	3.80	0.90	3.00	15°	75°	240
VF30/270/75	3.80	0.85	2.70	15°	75°	282
VF30/180/75	3.30	0.85	1.80	15°	75°	422
P090/320/70	3.20	0.90	3.20	20°	70°	226
P090/270/75	3.20	0.90	2.70	15°	75°	270
P090/130/75	3.20	0.90	1.30	15°	75°	551
P065/180/75	2.80	0.65	1.80	15°	75°	551
P090/180/75	2.50	0.90	1.80	15°	75°	400
P090/160/75	2.50	0.90	1.60	15°	75°	448
P050/280/78	2.50	0.50	2.80	12°	78°	461

FIG. 4.24—CONT'D

(Continued)

Workers

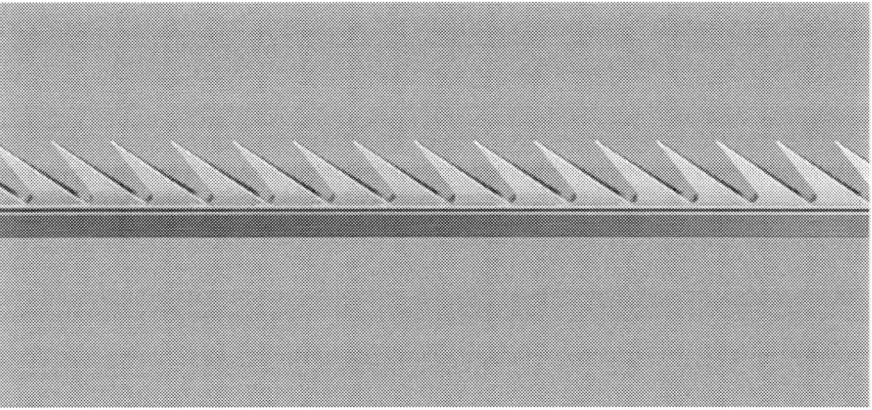

Product	Height (mm)	Rib (mm)	Pitch (mm)	Working angle	Front angle	PPSI
VL20/360/50	5.30	1.27	3.60	40°	50°	14
VL20/360/50 EvoStep®	5.30	1.27	3.60	40°	50°	141
VL24/360/50	5.30	1.06	3.60	40°	50°	169
VL24/250/50	5.30	1.06	2.50	40°	50°	244
VH16/360/53	5.00	1.59	3.60	37°	53°	113
P100/360/50	5.00	1.00	3.60	40°	50°	179
P100/220/50	5.00	1.00	2.20	40°	50°	293
VA20/360/50	4.50	1.27	3.60	40°	50°	141
VA24/360/50	4.50	1.06	3.60	40°	50°	169
VA20/250/50	4.50	1.27	2.50	40°	50°	203
VF28/250/50	4.50	0.90	2.50	40°	50°	287
P095/250/50	4.00	0.95	2.50	40°	50°	272
P095/210/50	4.00	0.95	2.10	40°	50°	323

FIG. 4.24—CONT'D

(Continued)

Strippers

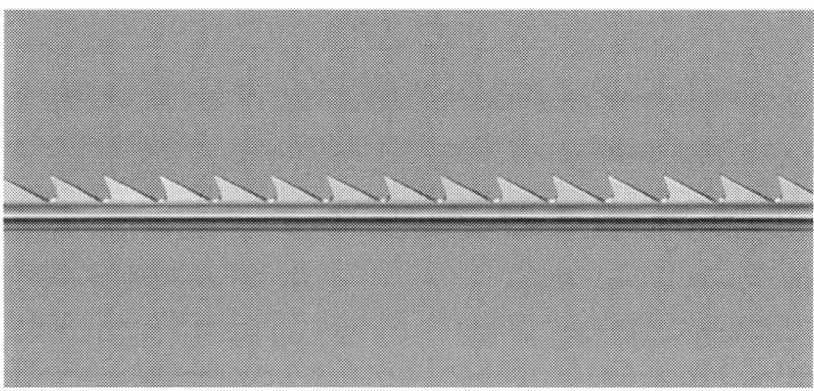

Product	Height (mm)	Rib (mm)	Pitch (mm)	Working angle	Front angle	PPSI
VA12/500/70	4.70	2.12	5.00	20°	70°	61
VA16/425/75	4.50	1.59	4.25	15°	75°	96
VA16/400/75	4.50	1.59	4.40	15°	75°	102
VA16/360/50	4.50	1.59	3.60	40°	50°	113
VA20/360/50	4.00	1.59	3.60	40°	50°	141
VA20/300/70	3.80	1.27	3.00	20°	70°	169

Transfer rollers

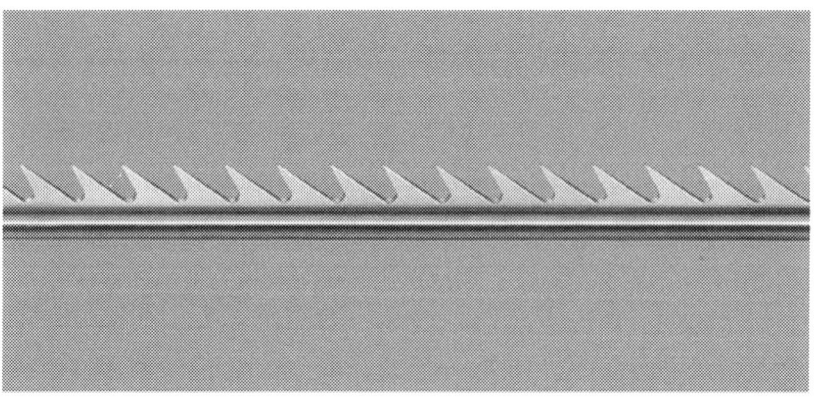

Product	Height (mm)	Rib (mm)	Pitch (mm)	Working angle	Front angle	PPSI
VA12/500/60	4.70	2.12	5.00	30°	60°	61
VA12/360/60	4.70	2.12	3.60	30°	60°	61
VA14/360/60	4.70	1.81	3.60	30°	60°	99
VA16/360/60	4.50	1.59	3.60	30°	60°	113
VA20/360/60	4.50	1.27	3.60	30°	60°	141

FIG. 4.24—CONT'D

(Continued)

Doffers

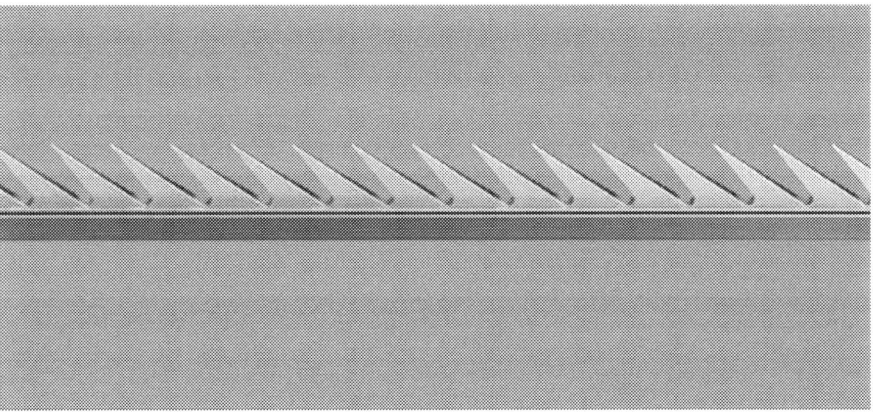

Product	Height (mm)	Rib (mm)	Pitch (mm)	Working angle	Front angle	PPSI
VH20/360/53	5.30	1.27	3.60	37°	53°	140
VL20/360/50	5.30	1.27	3.60	40°	50°	141
VL24/360/50	5.30	1.06	3.60	40°	50°	169
VL24/360/50 EvoStep®	5.30	1.06	3.60	40°	50°	169
VL24/250/50	5.30	1.06	2.50	40°	50°	244
VH16/360/53	5.00	1.59	3.60	37°	53°	113
P100/360/50	5.00	1.00	3.60	40°	50°	179
P100/220/50	5.00	1.00	2.20	40°	50°	293
VA20/360/50	4.50	1.27	3.60	40°	50°	141
VA24/360/50	4.50	1.06	3.60	40°	50°	169
VA20/250/50	4.50	1.27	2.50	40°	50°	203
Groz-Beckert SiroLock® VG28/250/50x SL Plattinium	4.50	0.90	2.50	40°	50°	284
VF28/250/50	4.50	0.90	2.50	40°	50°	287
P095/250/50	4.00	0.95	2.50	40°	50°	272
P095/210/50	4.00	0.95	2.10	40°	50°	323
Groz-Beckert InLine SiroLock® plus P080/250/40H40 SL+Plattinium	4.00	0.80	2.30	40°	50°	351

FIG. 4.24—CONT'D

Standard surface-mounted wire utilises a shoulder that butt-joins each row side by side. Interlocking wire utilises a shoulder whose shape is such that each row is 'locked' into the adjacent rows. Foreign objects such as pieces of metal cause most accidental damage in carding. With surface-mounted wire, this results not only in wire breakage, but as the wire unwinds on the roller, damage is caused to other rollers in close proximity. The result is catastrophic. With interlocking wire, damage is confined to the points of the teeth and generally the wire remains on the roller. Although the card clothing on the affected roller may be irreparable, the extent of damage to the card is limited.

4.7.2.6 Examples of different wire designs

Introduced by Garnett Wire, UK, random pitch wire was primarily developed to overcome the spiralling effect associated with conventional wire profiles. The defined spiralling patterns created by the distribution of teeth on a wound card roller are claimed to introduce irregularities in the web. Previously, two wires of different pitch were wound onto rollers to counter the effect. The tips of random pitch teeth have a flat surface and the land area of consecutive teeth differs (see Fig. 4.25). Although the tooth pitch varies along the wire, the working angle and mouth of each tooth remain the same.

The flat top tooth is claimed to aid fibre breakage as well as self-cleaning of the card during run-off. Card clothing manufacturers produce card wire with either single- or double-sided serrated edges, which aims to increase the holding power of the teeth particularly for low-friction fibres, which may otherwise slip off the tooth. Enhanced point wire (see Fig. 4.26) has a contoured point, which is claimed to pick up fibres efficiently from the cylinder. The manufacturers claim that more open settings can be used in comparison to conventional wire, thus reducing the risk of excessive card wire wear or damage. The profile is tapered to allow easy fibre release.

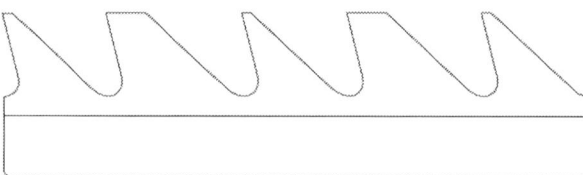

FIG. 4.25   Profile of random pitch wire.

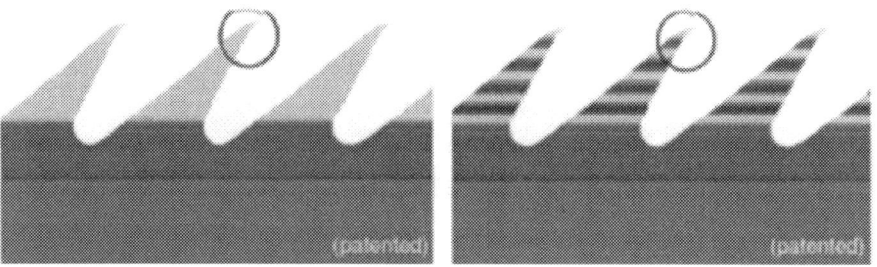

FIG. 4.26   Non-serrated and serrated enhanced point wire. *Courtesy of John D. Hollingsworth, USA.*

4.8 Carding machine configurations

Web formation in carding is traditionally referred to as either parallel-laid or cross-laid, depending on the specific arrangement of the process, and the way in which the web is handled prior to bonding. Fibres in a conventional carded web tend to exhibit a degree of preferential orientation in the machine direction. Many high-speed carding installations follow a parallel-laid, 'straight-through' process, where the web from the card is transported directly to the bonding stage. This includes situations where two or more carded webs are laid directly on top of each other in the direction of production, to form a single web, ready for bonding. Here, multiple cards are arranged in sequence, and the webs from each are laid onto a common conveyor to make the final parallel-laid web. Alternatively, the web leaving the card undergoes a separate lapping process in which the machine direction is turned through ninety degrees with respect to the exit of the card, i.e. cross-lapping, which produces cross-laid webs. Carding machines function around only a few basic principles that are repeated successively until a satisfactory web is produced. There are many diverse carded nonwoven products and a wide variety of raw materials are used in their manufacture. Consequently, a variety of different card configurations have come into use.

4.8.1 Garnett machines

In the USA, it is common to refer to some carding machines as Garnetts rather than cards. This is a term that is commonly misunderstood because garnetting in Europe means something quite different. The relatively recent introduction in the USA of large, high-production European-style carding lines to the nonwoven industry, has widened the apparent gulf in productivity between an American-style Garnett and a European card. In fact, both are carding machines in their true sense and the principles are the same. Garnetting in Europe refers to the process in which textile waste, either in yarn or fabric form, is mechanically recycled by first cutting and then tearing pieces in an extremely robust machine. Feed materials above about 50mm in length need to be cut prior to the process.

Garnett machines were originally manufactured by P&C Garnett Ltd. in the UK. The process is also called 'rag tearing', 'pulling' or 'rag grinding' and in traditional industries the resulting reclaimed material was called 'mungo' (from hard waste) and 'shoddy' (from soft waste). Nowadays, it is more commonly referred to as mechanical waste recycling and the resulting fibre is sold as flock, or is directly formed into products. Traditionally, Garnett machines have mechanically recycled waste such as used clothing, yarns and pieces into nonwoven batts and fabrics for carpet underlay, mattress components, acoustic and thermal insulation for automotive and other technical uses. Many machines are still in use. The recycling machinery was fundamentally similar to a card in that the point to point 'working' principle was utilised. However, the purpose of garnetting is to break down waste yarn and fabric and not specifically to produce a uniform fibre web. To do this, the machine is robustly built and utilises small diameter rollers. The workers and strippers are the same diameter but are situated in a different position from that of a card.

In traditional Garnett machines the worker is placed before the stripper because the main function is to break down the waste, then simply to move it on to the next worker.

Consequently, there is considerable fibre breakage in the process and limited fibre mixing. Many years ago, the association and later the amalgamation of P&C Garnett with the Bywater Machine Company, who developed some of the first needlepunching technology, resulted in Garnett-Bywater becoming a leading manufacturer of turn-key nonwoven lines. It is from this early history that the term 'Garnett' originated in respect to its use in the USA.

Although the compact carding machines produced by Garnett specifically for nonwovens were different from the larger roller cards used in traditional fibre processing, the carding principles were the same. The rollers were smaller, and the machines were shorter. This reflected the preference for processing short fibres and the limited amount of work that was required to produce a web for wadding at that time. Inevitably, as the popularity of compact cards for that purpose increased, other manufacturers, principally in the USA, began to produce their own versions of small, compact cards, specifically for nonwovens, and Garnett became a generic term.

4.8.2 Carding machines

Commercial nonwoven carding installations vary from small low-production machines used for manufacturing specialist products, for example, in the medical industry for wound dressing components, to high-speed units for the manufacture of lightweight fabrics suitable for single-use consumer products.

Depending on the machine configuration and the fibre type, carding machine production rates can be >500 kg/hr/m width. At high delivery speeds of >150 m/min, the control of web uniformity can be particularly challenging. Although now superseded by Dilo Group's VectorQuadroCard (VQC) and MultiCard, the Delta card (Fig. 4.27) is one example of a high-speed card with a double transfer system providing two separate intermediate streams of fibre between the first and second cylinders, designed to increase mixing power. The suction assisted multiple web take-off system allows delivery of multiple webs (up to four depending on machine design) directly onto an air-permeable conveyor. Suction-assisted take-off helps to maintain web uniformity and requisite isotropy by minimising unwanted tensioning during high-speed processing of lightweight webs. Suction-assisted web transfer on permeable belts or rollers can take place both as the web is transferred from the take-off roller to the conveyor, and also conveyor to conveyor, between the card and the next process in line.

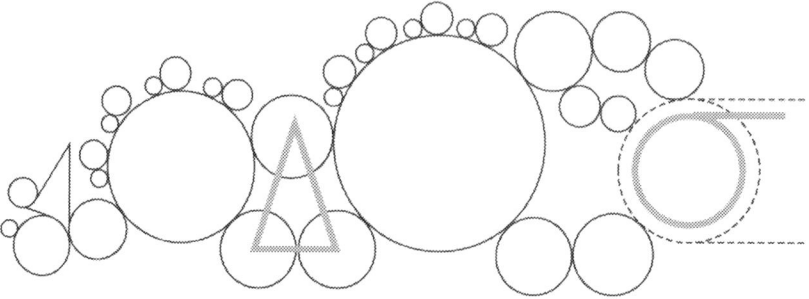

FIG. 4.27   Delta card. *Courtesy of Spinnbau, now Dilo Group.*

High-speed cards are effectively sealed and use internal suction to enable removal of fibre debris and other particles during carding, as well as to control air flows and air pressure distribution inside the machine, to prevent disturbance of fibres on the rollers. Internal humidity control within the card is also possible.

The airflow created by the rotation of the main cylinder in carding is harnessed by the Injection card (Fig. 4.28). A high-velocity Venturi is generated between the edge of a metal plate and the worker surface, causing fibres from the worker to be replaced on the cylinder without the need for a stripper roller. The absence of stripper rollers removes one of the limitations on maximum card width, which is the deflection of small-diameter rollers. Since fibres in the Injection card do not revolve around the worker rollers in the same manner as a conventional card, there is reduced circulation of fibres within the machine. Partly for this reason, overcarding of fine fibres and nep formation are claimed to be minimised. It has also been suggested that the elimination of stripper rollers leads to a more isotropic web. Subject to fibre specifications and card width, even high-speed cards launched twenty years ago, such as that shown in Fig. 4.29, were claimed to be capable of 400 m/min web output speed. Working widths of high-speed cards have now reached >4.5 m, and up to 6 m in parallel-laid format.

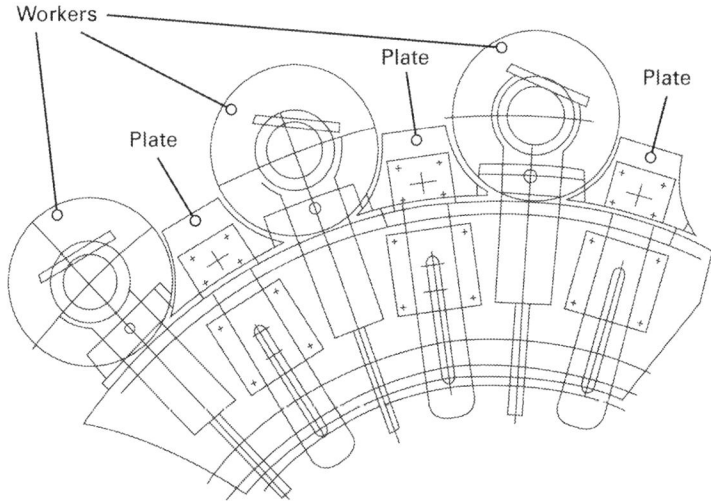

FIG. 4.28   Injection card. *Courtesy of FOR, now Autefa Solutions.*

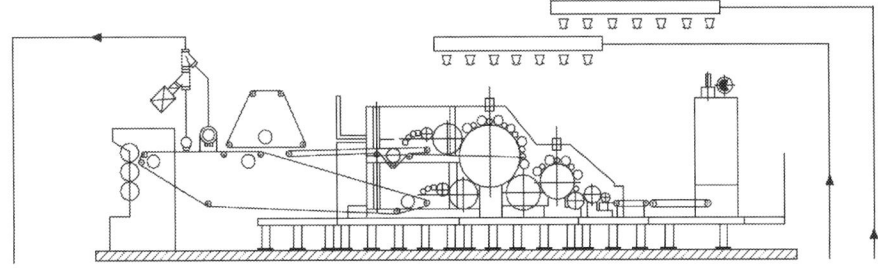

FIG. 4.29   Hyperspeed card. *Courtesy of Spinnbau, now Dilo Group.*

The maximum width may be lower if the card is followed by a cross-lapper. High production of lightweight webs on such wide cards makes minimising web weight variation critically important. For any web, the fibre fineness influences the web weight limits of the card. ANDRITZ Asselin-Thibeau has studied the dependency of web weight on fineness and has defined a coefficient K, which for a card producing a lightweight web for hygiene operating with condenser rollers varies between 12 and 25. The web weight (g/m^2) limits L_1 and L_2 as a function of fibre fineness (denier) n is then determined as follows,

$$L = K\sqrt{n} \tag{4.1}$$

Therefore, if $n = 2$ denier, then

$$
\begin{aligned}
L_1 &= 12\sqrt{2} = 17\,g/m^2 \\
L_2 &= 25\sqrt{2} = 35\,g/m^2
\end{aligned} \tag{4.2}
$$

The value of K depends on the composition of the card and the take-off arrangement. Over the years, a variety of unconventional card designs have emerged, and whilst some have been discontinued, it is instructive to be aware of their layout.

Roller train cards were developed many years ago and are distinguishable from conventional cards because of the absence of a main cylinder. Instead, a series of small rollers are arranged in such a way that multiple transfer actions take place in sequence. An example of this type of machine was the Turbo-Lofter (Fig. 4.30), now discontinued, which was developed for high-loft products up to $200\,g/m^2$. Webs with MD/CD ratios approaching 1:1 were claimed. The webs were also voluminous and characterised by partial 3D fibre orientation. The roller train approach embodied in the Turbo-Lofter card is re-emerging as part of a new generation of airlay-carding machine hybrids for the production of lightweight webs at high speed. It is important to appreciate that carding machines form part of integrated nonwoven production lines for different fabrics. A variety of card and bonding configurations used to make different products is shown in Fig. 4.31.

Generally, with reference to the card configuration, increasing the number of cylinders as well as the use of a breast section increases both the carding and mixing power of the machine and the ability to achieve progressive working and mixing actions. As the number of

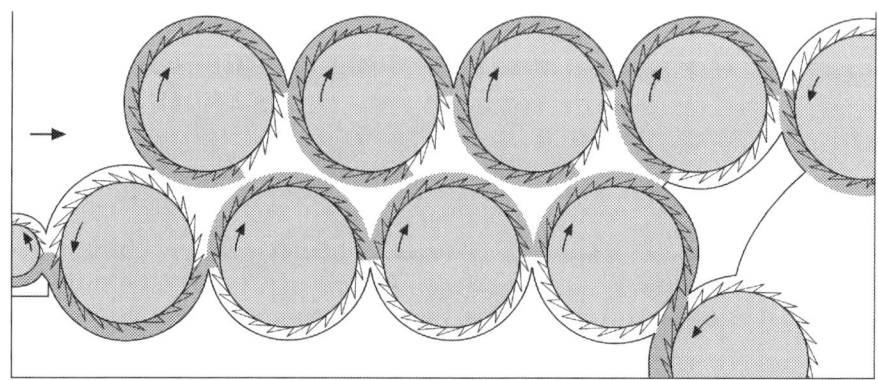

FIG. 4.30   Fibre passage through the Turbo-Lofter.

FIG. 4.31 Different card configurations. *Courtesy of Bonino .*

worker-stripper pairings increases, the greater is the available carding power and the degree of mixing that may be achieved. There is also the potential to gradually progress fibre disentanglement through the machine if a large number of worker-stripper pairings are available. Double doffers improve clearance of fibre from the cylinder. Condensing or condenser rollers on the card increase the web isotropy and in modern cards, they can be retracted to vary web isotropy as required for different products. Short cards are selected for processing brittle (e.g. ceramic) or low-friction fibres, where serious fibre damage can result from over-carding.

Examples of high-speed carding systems produced in different design configurations include, but are not limited to, the Web Master FUTURA range (Autefa Solutions), T-Web cards (Trützschler Nonwovens), aXcess and eXcelle cards (ANDRITZ), and the VectorQuadroCard (Dilo Group). The Bonino Turbo card can be considered a hybrid machine incorporating traditional worker rollers around a high-speed cylinder to provide efficient opening. But instead of web removal by means of a doffer, the fibre is discharged aerodynamically in the manner of an airlay machine. The technology was originally developed to process natural fibres and blends of natural fibres with synthetic. The resulting webs, in the range of $80\,g/m^2$ to $4000\,g/m^2$ are more voluminous than those created with traditional cards and cross-lappers and with a more balanced MD/CD ratio. The airflow through the machine is arranged in a closed loop, with the same air re-circulated and filtered. The closed nature of the system makes it particularly suited for use with bonding powder (for subsequent thermal bonding). It also handles dusty materials well and production rates can be up to $500\,kg/h/m$ width.

4.9 Card feed control, weight measurement, and other control systems

The inclusion of automatic feed control in a nonwoven card is standard. From a technical viewpoint, consideration of the various setting parameters on a carding machine highlights the need for consistent fibre feeding. Fibre loadings on the cylinders, workers and doffers affect not only web regularity, but also quality indicators such as nep formation and fibre breakage. The choice of settings and card wire design for a nonwoven card are governed not only by fibre properties but also by the card throughput and fibre loading.

Uncontrolled fibre feeding to a card can result in short- and long-term variations (including a long-term average drift in the weight per unit area) as well as undermining roller settings and favourable fibre-card wire interactions. The latest types of card clothing allow higher fibre throughput (usually in conjunction with very high cylinder speeds) and greater fibre holding capacity using comparatively low wire heights (to reduce loading and minimise air turbulence) and narrow roller settings. Since uniform fibre loading within the wire is critical to web quality in such high-production scenarios, variations in the feed rate can quickly lead to overloading, fibre damage and nep formation as well as increased waste and fly. An increase in production rate requires passing a larger volume of fibre through the volumetric chute feed and this effectively decreases the time available for fibre to consolidate effectively within the chute. The higher the throughput via a volumetric chute feed, the greater is the likelihood of significant feed variation. Some type of feed control system is therefore essential.

The blueprint for a nonwoven fabric is the web. Nonwoven manufacturers sell their product by measured length or area, rather than by weight per unit area. This is complicated by the implications of linear web weight variations. Using relatively short cards without control, such variations can be >10%.

Nonwoven products must normally meet a minimum technical specification. Since underweight fabric is considered more undesirable to the customer than overweight, it is common practice to err on the side of caution and produce webs slightly over the target weight. Typically, this means that the average weight is above the target weight. Since the product is then sold by length, roll number or area and not by weight, the manufacturer effectively gives away fibre for free. This by itself is a motivation for using automatic weight control systems. By controlling the weight of fibre fed into the card it is possible both to improve web weight uniformity and save a significant amount of fibre, which increases production cost efficiency.

4.9.1 Microprocessor controlled weigh-pan systems

Microprocessor controlled weigh-pan systems are extensively used by most sectors of the staple fibre processing industry. Whilst much of the carding technology adopted in the nonwovens industry originated from the traditional spinning sector, the 'Microweigh' was the first card feed control system to be developed specifically within the nonwovens sector before being offered to the traditional industries. Such systems (Fig. 4.32) can form part of a new hopper or are retrofitted to existing installations.

Initially, the feed to the weigh pan is regulated by controlling the speed of the spiked lattice to ensure a regular flow of fibre regardless of the amount of fibre in the hopper. This counteracts the well-known susceptibility of weigh-pan systems to hopper load variations and the consequent changes in weight of fibre against the spiked lattice, which results in uneven fibre flow. When the weigh pan achieves the pre-set weight, the spiked lattice is stopped, and trap doors are closed above the pan. At this point, some fibre is still in mid-air, on its way into the pan. This is called 'in-flight fibre' and in traditional hopper feeds is a common source of irregularity. The Microweigh controls in-flight fibre by constantly monitoring in-flight values

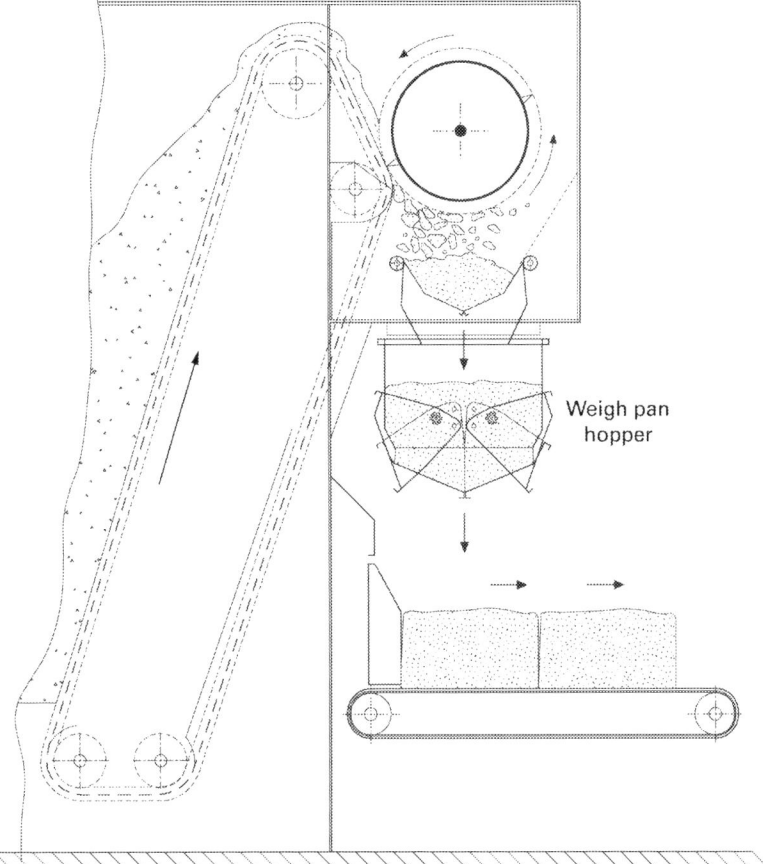

FIG. 4.32 Microweigh weigh pan hopper feeder. *Courtesy of Garnett Controls Ltd., UK.*

and adjusting the stop point in anticipation of a calculated weight of fibre falling into the pan when the lattice has stopped. Once the individual weight of fibre has been delivered to the weigh pan, a further control function carries out a quality check. If the weight in the pan does not exactly match the pre-set target weight, a correction is made before the fibre is allowed to enter the card. This is achieved in one of two ways.

The first, and more usual option is to allocate a specific space on the card in-feed sheet for that particular weigh. For example, if the weight is detected to be 1% heavier than the target weight, the drop point is automatically adjusted to allocate 1% more space on the card feed sheet. This is made possible by dropping the contents of the weigh pan onto the in-feed sheet according to counted electronic pulses from an encoder, as opposed to the cycle cam of a traditional hopper. A light weight is therefore dropped at a lower pulse count, whereas a heavier weigh is dropped at a higher pulse count. Since the speed of the card in-feed sheet is constant, the result is a uniform allocation of weight per unit area on the in-feed sheet, which produces a regular feed. This function is termed 'distance dropping'.

Alternatively, fine-tuning of the individual in-feed weight is achieved by controlling the card feed roller speed. A weigh, which is above or below the pre-set weight, is allowed to drop onto the card feed sheet and is transported to the card feed rollers. Again, the linear speed of the feed sheet is monitored by a signal from an encoder, which allows the microprocessor to determine the exact point where any particular weigh-pan drop will enter the feed rollers. A feed roller speed adjustment is made to correct any deviation from the target weight. Therefore, a weight that is 1% too heavy is corrected by a 1% reduction in the feed roller speed.

The Microweigh XLM system (Garnett Controls Ltd., UK) also incorporates a moisture control system. It is known that significant moisture loss occurs during carding. Since many fibres arriving at the card may contain combined or even interstitially held water (if they are lubricated prior to carding), any irregularities in the moisture content within a batch of fibre will lead to weight irregularities in the final product. This is particularly relevant when processing hygroscopic fibres, for example, in the production of medical products composed of polysaccharides or cellulosics. The Microweigh XLM incorporates a Streat Instruments (New Zealand) moisture measurement system within the weigh-pan assembly, which measures the moisture content of each consecutive weigh. The moisture value is subtracted from the total weight and the Microweigh bases its control functions on the dry weight of fibre thereby eliminating the effect of moisture variations on product weight regularity. Whilst Microweigh was one of the first weight-control systems to be widely adopted, and remains one of the most precise systems available, there is a limitation on fibre throughput. The production rate of modern nonwoven cards has surpassed the production capabilities of weigh-pan systems. Despite this, such systems are in operation throughout the nonwovens industry, particularly for processing speciality fibres such as calcium alginate where the fibre physical properties do not allow, or require, the use of high-production machinery, but when maximum feed weight accuracy is essential.

4.9.2 Volumetric chute feed systems

The majority of nonwoven cards use volumetric feeds. Basic systems consist of a vertical chute arrangement into which fibre, usually fed from a spiked apron, is deposited from the top (Fig. 4.33). Most volumetric chutes incorporate a vibrating or 'shaking' wall to encourage the fibre to compact, thus filling air pockets and generally distributing the fibres evenly to form a continuous stream of fibre and to assist in the movement of fibre though the chute. A pair of fluted feed rollers is normally situated at the exit of the chute to continuously move the fibre onto the card feed sheet. Some chute feeds incorporate a reserve chamber and a main chute to improve feed uniformity. Volumetric chute feeds were introduced to overcome the production limitations associated with traditional weigh-pan-type systems, even before the advent of weight-control systems. The volumetric chute was designed to produce an even distribution of fibre at the card feed section than was thought possible using weigh pans. However, whilst the fibre from a chute feed appears uniform and continuous, variations in tuft density and packing density within the chute still lead to web weight variations. There are also significant differences in the way different fibre types pack under gravity in chute feeds. For these reasons, volumetric chute feed control systems were developed.

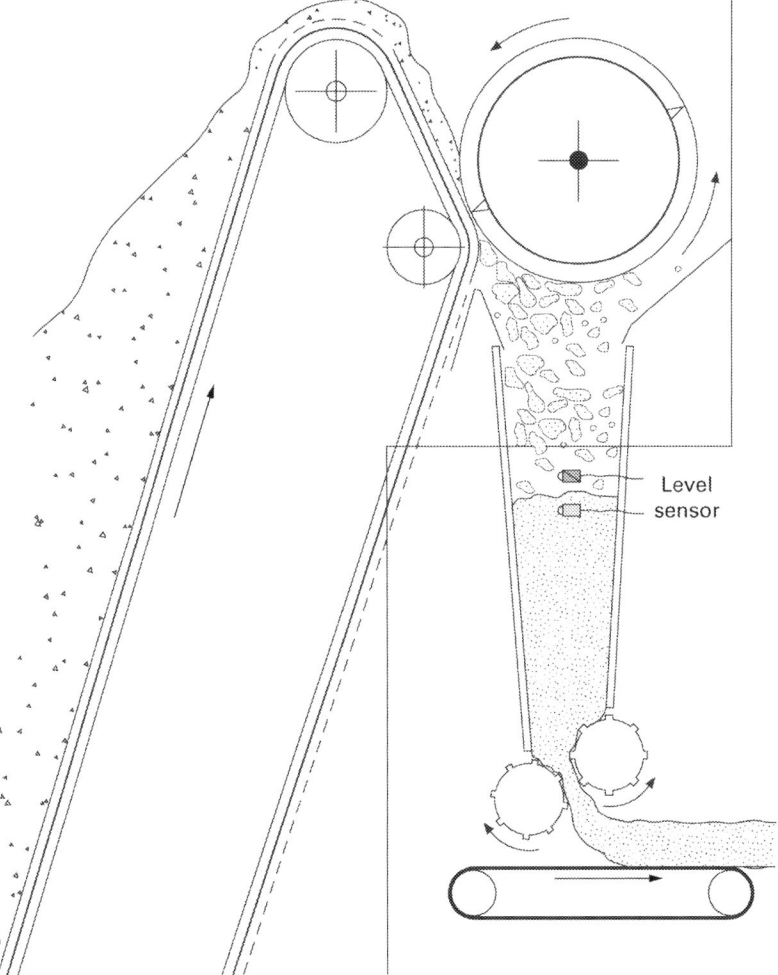

FIG. 4.33 Volumetric feed system. *Courtesy of Garnett Controls Ltd., UK.*

4.9.3 Long-term variation controllers

Controllers of long-term feed weight variations usually operate by maintaining a constant average feed into the card. Their operation often relies on continuously adjusting the card feed roller speed in response to measured short-term variations in the in-feed fibre stream.

4.9.4 Electromagnetic radiation systems

Servolap, originally developed by HDB (Belgium), was one of the first volumetric weight-control systems on the market. The system measures the mass of fibre on the card feed sheet by directing an isotopic ray through the material. Originally a radioactive source was used to produce the ray, which in recent years has been replaced by an X-ray generator. The emission ray from the source is directed through the full width of the fibre assembly on the card feed sheet from one side of the machine to the other. The residual radiation from the source is collected at the opposite side by a scintillation tube, which converts it to an electric signal that is inversely proportional to the density of the mass of fibre that is being conveyed. This signal is used to automatically regulate the speed of the card feed rollers to compensate for in-feed mass variations.

4.9.5 Weigh platforms

A stationary weigh platform is located between the delivery rollers of the chute and the card feed rollers. As fibre passes over the weigh platform, variations in weight are measured and recorded. The signal obtained from the measuring device is used to control the speed of the card feed roller speed. The Microchute (Garnett Controls Ltd., UK) measures over a relatively short distance and is able to detect and control short-term variation in the fibre feed. However, because of the short measuring distance, there is a degree of 'bridging' over the weigh platform due to the consolidation that takes place within the volumetric chute. This means that fibre on the weigh platform is to some extent supported by fibre before and after the platform. In practice, therefore, these systems provide comparative information with which to control the variation but do not provide absolute weight values.

4.9.6 Weigh-belt systems

Weigh belts were introduced as long-term weight controllers allowing the user to directly input the desired production rate. The weigh belt measures over a much longer distance to minimise the 'bridging' effect, which occurs with weigh-platform systems. Additionally, because the belt is moving at the same speed as the fibre moving into the card, weigh belts are able to measure both speed and distance, as well as weight. A disadvantage of this type of system arises because the measuring distance is long, and the system is unable to measure and control short-term variations occurring inside the weighing area. Depending on the card configuration and its mixing power, such variations introduced at the feed section can be amplified in the resultant web. Additionally, the entire assembly is comparatively heavy given the small mass of fibre that it needs to weigh. Consequently, the measuring precision of the weigh-belt system can be limited.

4.9.7 Roller weighing systems

Rollaweigh (Garnett Controls Ltd) was designed to incorporate the advantages and eliminate the disadvantages of short-term regulation systems and weigh- belt controllers. The original system has two distinct measuring zones (Fig. 4.34). The five-roller assembly of the Rollaweigh operates essentially like a weigh belt measuring over a longer distance, thereby minimising the 'bridging' effect. Moreover, because there is no belt, the weigh assembly is lighter than a weigh-belt system, which improves the resolution and weighing accuracy. The surface speed of the rollers is identical to that of the fibre as it moves into the card;

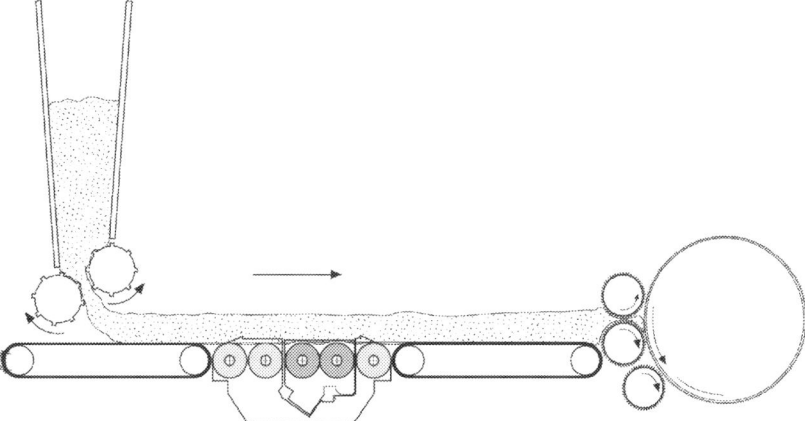

FIG. 4.34 Rollaweigh system. *Courtesy of Garnett Controls Ltd., UK.*

therefore, the system can accurately measure distance and speed, as well as weight. The user enters the desired production rate and the system automatically maintains the desired throughput, regardless of changes in fibre characteristics. To regulate short-term variation in the in-feed fibre stream, the system incorporates a secondary control loop within the overall five-roller system. In the secondary short-term zone, an additional measurement is taken over only two rollers. The short-term weight control system then adjusts the card feed roller speed in order to compensate for these shorter- term weight variations. The system may be installed as part of a new carding installation or retrofitted to existing machinery.

4.9.7.1 *On-line basis weight measurement*

Basis weight measurements of web, batt or final fabric weight can be determined using a variety of sensor technologies including gamma backscatter, near infra-red and beta transmission. Non-contact sensors measure the output of a nonwoven line by either traversing the width of the product or by a number of individual sensors situated across the width. The information from such systems is used to monitor product density in both the MD and CD and although the measurement is taken after the point at which automatic regulations are realistically possible, the data produced is useful for on-line quality- control purposes. Web and fabric scanners are incorporated into closed loop weight-control systems that integrate operations in the entire production line. Adjustments are made to the instantaneous speeds of the card feed rollers, the card and cross-lapper to regulate both along the card as well as across the batt and fabric weight variations.

4.9.7.2 *Cross-machine direction (CD) controllers*

It is possible to control the fibre density in both the long and cross-machine directions. The Scanfeed system (Fig. 4.35) incorporates an upper feed chute within which a feed roll and opening roll are situated at the base. A constant airflow is directed through the chute and allowed to escape through outlet vents situated at the front and behind the position where fibre accumulates at the base of the chute. Fibre distribution in the chute is regulated by

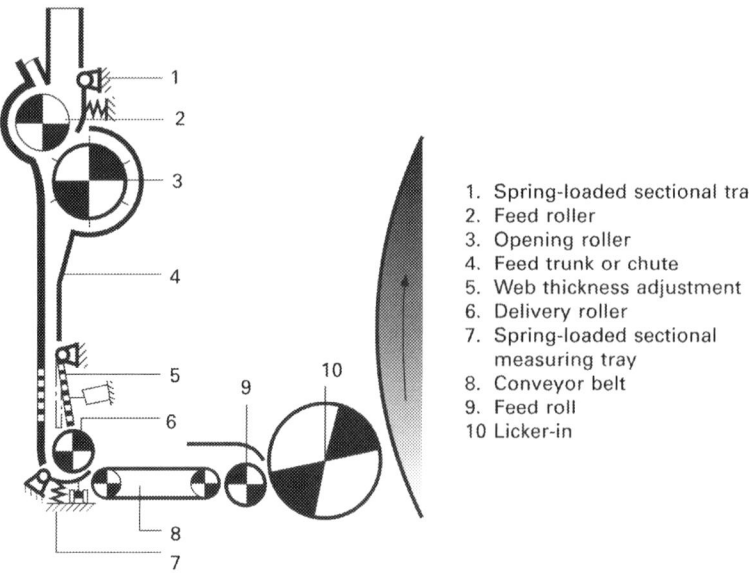

1. Spring-loaded sectional tray
2. Feed roller
3. Opening roller
4. Feed trunk or chute
5. Web thickness adjustment
6. Delivery roller
7. Spring-loaded sectional
   measuring tray
8. Conveyor belt
9. Feed roll
10 Licker-in

FIG. 4.35   Feed density control. *Scanfeed, courtesy of Trützschler Nonwovens.*

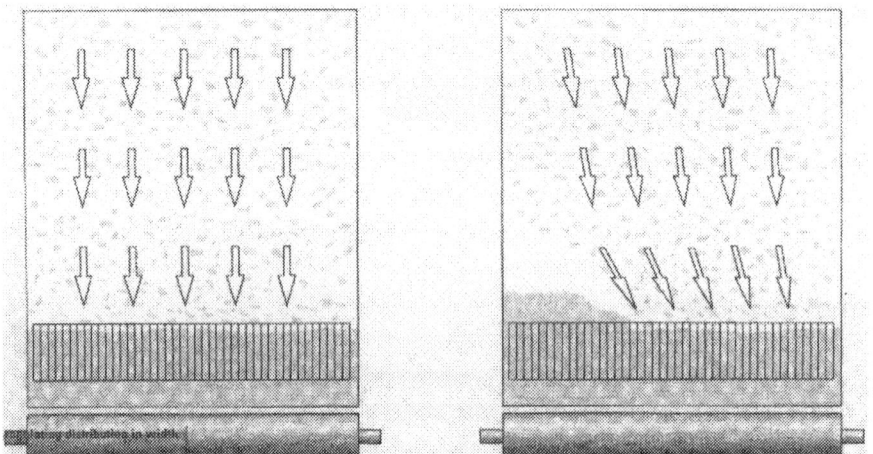

FIG. 4.36   Self-regulation of feed weight in width. *Scanfeed, courtesy of Trützschler Nonwovens.*

the airflow. If the fibre in the chute is denser at one side, or in one particular area, the air pressure correspondingly increases in this area because the packed fibre is less air permeable. Consequently, the differential pressure, which is created between areas of different density in the chute, preferentially directs the falling fibre to fill the lower density areas, which are more air permeable (Fig. 4.36). At the same time, a series of spring-loaded flaps across the width of the chute provide pressure regulation as they open and close according to the thickness of the fibre passing through.

Further uniformity of cross-card density is achieved by a web profile control system in the lower feed chute. At the base of the feed chute, the side wall is split into a series of profiled boxes having pivoting flaps that are automatically adjusted to increase or decrease the volume available for fibre to pass. Below this point, a chute feed delivery roller is mounted above a series of spring-loaded sectional trays, which are pivoted so they can open and close according to the fibre density passing through. As fibre exits the chute feed, the movement of the flaps is continuously measured and a signal is generated to control actuators on the flaps of the profile boxes. If a tray below the delivery roller opens as a result of higher fibre density passing through that section of the chute, a servomotor automatically closes the flap on the corresponding profile box. This reduces the volume available for fibre passage and reduces the effective air volume passing through at that point thereby causing fibre to preferentially flow into less densely packed areas of the chute. This automatic regulation of fibre flow is continuous and since the distance between the measurement point and the control point is relatively short, the system provides effective short-term control of both longitudinal and cross-machine feed uniformity.

The Scanfeed system can be integrated with a card in such a manner that the bottom delivery roller of the lower chute becomes the card feed roller, thus eliminating the traditional feed sheet. The measuring accuracy of the individual tray sections is claimed to be such that with the appropriate calibration, a direct correlation with web weight can be achieved without the need for an additional weight-control system. The potential for consolidating a batt produced by such a feed system directly (using, for example, hydroentanglement or needlepunching) without the use of a traditional web-forming step has also been proposed. For making heavyweight products, for example, quilted fabrics, the use of a volumetric hopper feeder to form a batt structure in place of the conventional card and cross-lapper has been demonstrated.

4.9.8 Localised fibre dosing system

Direct localised dosing of additional fibre onto the batt after it leaves the card feed system is also possible, allowing dynamic control of short-term weight variation. Such an approach can also be used to provide a uniform fibre feed for airlaid web formation systems, as well as direct batt formers, using volumetric chute feeds. IsoFeed (Dilo Group) technology involves the use of multiple, individual airlay units mounted across the full width of the conveyor, each with an operating width of about 33 mm. This allows the requisite weight of fibre to be precisely dosed onto the exact location of the batt in order to compensate for a lightweight area.

4.10 Cross-lapping

A cross-lapper (or cross-folder) is a continuous web transfer machine that normally follows a card or Garnett machine as part of an integrated web formation system. The web is layered from side to side onto a lower conveyor or bottom lattice, which runs perpendicular to the in-feed web to form a diagonally stratified batt (sometimes also called a wadding or fleece), which typically consists of 4 to >15 layers depending on requirements. Commercially, cross-lapped batt weights range from about $50\,\text{g}/\text{m}^2$ to over $1500\,\text{g}/\text{m}^2$ depending on fibre properties, the web weight per unit area and intended product application.

The original method of producing a cross-lapped batt used a system known as a camel back, so termed because of the shape of the web path used by the machine. In the camel back, a conveyor transports the emerging web from a card, upwards to a pivot point from which the conveyor system reciprocates to layer the web onto a cross conveyor. Such systems utilised simple harmonic motion to reciprocate the web layering conveyor, and as such produced heavy edges at the end of each traverse due to overfeed of the web as the mechanism decelerated and then accelerated at the sides.

Nowadays, horizontal cross-lappers predominate. They consist of a number of interacting conveyor aprons that operate in conjunction with traversing carriages and drive rollers. The carding machine delivers the web to the in-feed conveyor, which transports it onto the top sheet or belt assembly. The carriages reciprocate as the web is transported within the belts. The exiting web is then layered concertina-fashion via a short or long path, onto a lower conveyor which runs at ninety degrees to the web in-feed direction (Fig. 4.37). The relative speeds of the transporting and bottom lattice conveyors determine the number of layers in the batt and the angle of laydown. The number of layers is controllable and depends on the required basis weight and uniformity of the final nonwoven fabric.

The cross-lapping process results in a major change in the preferential orientation of fibres in the web. The fibres in the carded web entering the cross-lapper tend to have a preferential fibre orientation in the MD although they are by no means completely parallel to the web axis. Since during cross-lapping the cross-laid web (or batt) leaves the machine at ninety degrees to the card, the fibres in the batt have a preferential fibre orientation, which is nearer to the CD. In fact, two peak orientation angles can be measured in the batt after cross-lapping. The exact laydown angle depends on the speed of the in-feed web from the card v_1, and the lower conveyor speed of the cross-lapper v_2, and can be determined as follows:

$$\tan\emptyset = \frac{v_2}{v_1} \tag{4.3}$$

The linear production speed of the cross-lapper depends on both the web laydown width (w_2) on the lower conveyor and the number of layers required in the cross-lapped batt (l). If more layers are required and the speed of cross-lapper is already at its maximum, then the output speed of the cross-lapper, i.e. v_2 is reduced. Also, for a given web in-feed speed from the card (v_1), and cross-lapping speed, more time is needed to lay down a single layer of web

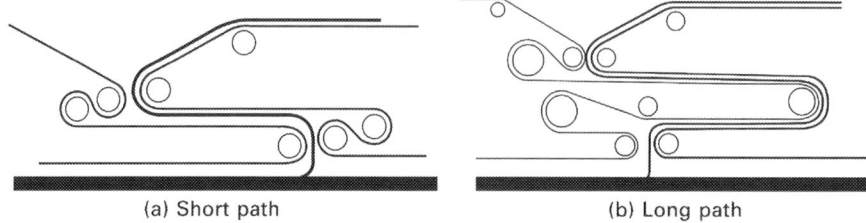

(a) Short path   (b) Long path

FIG. 4.37   Web path through a cross-lapper. *Courtesy of ANDRITZ Asselin-Thibeau.*

as the laydown width (w_2) increases. Based on the width of the incoming web from the card (w_1), the number of layers in the cross-lapped batt can be calculated as follows:

$$l = \left(\frac{w_1}{v_2}\right) \times \left(\frac{v_1}{w_2}\right) \tag{4.4}$$

In practice, the laydown width varies depending on product requirements and for special-ist applications such as papermakers' felts, it can reach >17 m. Therefore, cross-lapping enables the production of batts much wider than the initial web fed from the carding machine, which is limited to <5 m and is most usually about 2.5 m. For certain applications, such as the manufacture of filter media, reinforcing scrims (pre-formed nets, fabrics, meshes) or yarns, can be combined with the batt during cross-lapping to increase fabric dimensional stability. If required, more than one web can be lapped and combined from two different cards.

In operation, cross-lapping machines undergo significant shifts in mass at high speed. Air currents and inter-belt tension variations can produce web faults since the webs are ex-tremely delicate and are easily deformed by mechanical and aerodynamic forces. Basic cross-lappers of early design can introduce irregularities as the conveyors change direction at the end of the traverse and the surface speed changes due to the inertia of the conveyors. The resulting web overthrow at the edges is pronounced when older cross-lappers operate at high speed and the batt width tends to increase. The problem may be partly compensated by setting the laydown width slightly narrower than required. The production rate of basic horizontal cross-lappers is limited by the necessity to instantaneously reverse the conveyor mechanism, which develops a large momentum as the carriage changes direction at the end of its traverse.

Ideally, the motion of the carriage versus time produces a square sine wave to achieve per-fect reciprocation. In practice, as the carriage is decelerated, there is a momentary dwell whilst the direction is reversed and then the carriage is rapidly accelerated to the desired speed in the opposite direction. A time displacement curve for the standard carriage motion would show a dwell time at the end of each traverse as the direction of travel is reversed. Since the output of the carding machine is constant, so too is the output from the top section of the cross-lapper. Consequently, at the point where the carriage stops and reverses, the web overfeeds at the edges as it is fed onto the bottom conveyor resulting in heavy edges. To minimise the resulting variation, the carriage speed is often set to run slightly faster than the web delivery speed. Whilst this reduces the overfeed at the ends of the traverse, it also means the carriage is too fast in the centre of the traverse and consequently, drafting of the web takes place.

Clearly, a critical balance must be achieved. If the traverse speed is too high, the web will break, and if too slow web overfeed occurs creating folds and creases. A further consideration is the motion of the web itself as it exits the conveyor at speed, which may crease or fold caus-ing batt faults at the edges. To minimise this, a speed differential can be deliberately created between the upper and lower carriages, the top carriage being run slightly faster than the in-feed and the lower carriage travelling slightly faster than the top.

Tension in the card web during cross-lapping can lead to dimensional variations across the batt width as the fibres recover from extension after laydown. In subsequent needlepunching, width shrinkage frequently occurs as a result of the fibre reorientation induced by the needles as well as the applied take-up tension. In general, heavy edges are an inherent quality issue in cross-lapping; the weight profile across the width of the batt therefore 'smiles', or is said to exhibit a 'bath-tub' profile. Traditionally, the heavy edges of the batt are continuously trimmed by slitting after bonding. Typically, about 100 mm of the edge is trimmed and the waste fibre is then recycled back through the card. On a typical line, such edge trims could represent 2.2% of total production, which can equate to over 70 t of fibre unnecessarily recycled per line per year.

The original method of reversing the carriages involved a clutch mechanism, that relied on alternating connection of the driving clutch-plate with driven plates that ran in opposite directions. This arrangement places significant demands on the clutch, particularly at high operating speeds. In addition, the wider the cross-lapper, the higher is the load on the mechanism and the corresponding inertia. As speeds increased, the basic design of the cross- lapper evolved.

Double aprons are now used to sandwich the web and control its motion through the machine. Short web path cross-lappers allow higher lapping speeds and minimise the turning of the web in the conveyors. This minimises the introduction of web irregularities. The concept of a carriage has changed from that of a heavily constructed unit within a distinctive framework, to a lighter mechanism incorporating a relatively simple series of rollers in which the web may be turned only once during its passage through the cross-lapper.

Fibre reinforced composite rollers have replaced metal. External chain drives, a common source of mechanical problems, particularly when exposed to airborne fibres, have been eliminated in favour of integral drive systems where the aprons act as the main means of transmission as well as the web carrier. Reversing clutches have been replaced by gears and brakes, allowing controlled deceleration and acceleration of the apron carriage. Laydown rollers beneath the cross-lapper positively lay the web in a controlled manner, rather than relying on gravity, as was previously the norm.

The use of perforated conveyor aprons allows the evacuation of air to reduce undesirable air currents whilst anti-static aprons minimise problems such as fibre sticking to the aprons causing laps, web breaks, production stoppages and batt irregularities. The complex arrangement of conveyor belts within a cross-lapper, the high speeds, and the relative movement of respective roller positions make it important to control the lateral position of the conveyors. A cross-lapper apron is a consumable item that is expensive to replace. Early cross-lappers suffered from tracking problems and variations in tension caused gradual migration of the apron to one side leading to serious damage. Tracking control systems are used to prevent such events by ensuring that equal tension is maintained at both sides of each apron to prevent distortion. Should a deviation of the apron position occur, tension is applied to the appropriate side to force the apron back onto the correct path. In some systems, the apron edges are continuously monitored by optical sensors and automatic adjustments are made by a tracking control system.

Recent developments in cross-lapping technology have primarily focused on increasing speeds. The increased productivity has further highlighted the need to reconcile the problem of achieving high web uniformity at high production rates. The common practice of trimming over-weight edges produces significant waste and an alternative approach is preferred. To directly address the problem, modern cross-lappers include sophisticated drive control systems. The logistical aspects of the process are worthy of consideration. The input width of a cross-lapper is the same as the card, which feeds it and is typically 2.5 m wide. Since the cross-lapper must accommodate the web with room at each side to prevent roller laps, the width of the corresponding conveyor could be 2.7 m

The output of the cross-lapper is not restricted by the width of the card. Indeed, this is one of the advantages of having a cross-lapper in the line. Generally, the subsequent bonding process, which is commonly needlepunching, is capable of much lower linear speeds than a card. A card that is 2.5 m working width can be used to feed a bonding process that is at least twice that width, thus optimising the production capacity of the card and maximising product width in an economical manner. In practice, although machine widths are continuing to increase, most of the industry uses batt widths of 5 m and below.

4.10.1 Profiling cross-lappers

To counteract the batt weight variation across the width, or the 'smile' effect due to cross-lapping, profiling cross-lappers have been widely adopted. Although the term profiling has become somewhat generic, it was first introduced by Asselin (now ANDRITZ Asselin-Thibeau). Some manufacturers use the terms 'density control' or 'contour distribution'. Profiling cross-lappers manipulate the web between the conveyors by controlled drafting and condensation (or web storage), as well as control of web laydown to produce batts with specified cross-machine weight profiles. In practice, this is achieved by introducing controlled velocity variations in the conveyors using variable speed drives. Sandwiching the web between two aprons and controlling the velocity curve of the traversing web as well as its output velocity enable such control. The most productive, high-speed lappers incorporate a short path that requires only a single turn of the web within the cross-lapper. Input speeds of about 200 m/min are claimed based on a web input from a 2.5 m card.

Profiling considers not only the width-wise weight profile of the batt formed by the cross-lapper but also the width-wise weight profile after subsequent bonding, particularly needlepunching. Since needlepunching can accentuate the heavy-edge effect, a reverse smile profile is frequently pre-programmed so that the profiling cross-lapper produces a batt with lighter edges and a heavy middle section (Fig. 4.38). The laydown is controlled at every stage of each traverse. Variable speed drives fitted to sections of the cross-lapper allow the web to be laid at a specific proportional rate across a series of control zones. The web can be laid symmetrically with an induced bias in weight towards the centre of the batt, with a view to compensating for the increase in edge weight, which is associated with subsequent needlepunching. Indeed, such is the opportunity to control the batt density across the width of the laydown, that density biases can be deliberately introduced as required to suit specific production lines such as nonwoven production for subsequent 3D moulding.

(a)   (b)   (c)

FIG. 4.38 Transverse cross-sectional weight profiles of nonwoven fabrics (profiled and unprofiled). (A) Conventionally cross-lapped and bonded; (B) Profiled batt (before bonding); (C) Profiled fabric (after bonding). *Courtesy of ANDRITZ Asselin-Thibeau.*

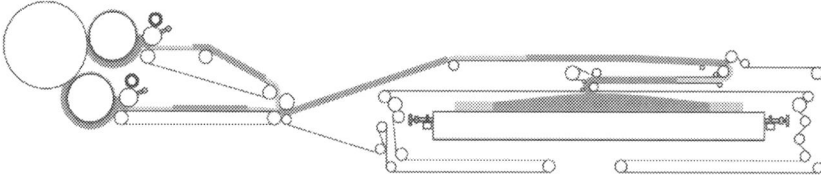

FIG. 4.39 ProDyn System. *Courtesy of ANDRITZ Asselin-Thibeau.*

Profiling can also be achieved by dynamically modulating the short-term web weight leaving the card via control of the doffer speeds. This effectively means profiling the periodic weight per unit area of the web before it arrives at the cross-lapper, reducing reliance on drafting or stretching of the web to achieve the required weight profile. Examples are the ANDRITZ ProDyn and ProWid systems (Fig. 4.39). The AUTEFA WebMax system uses a different approach, whereby the carded web weight is rapidly modulated by a separate unit placed between the card and the cross-lapper infeed so that a profiled batt can be produced. This system can also be retrofitted to existing lines. In some profiling systems, combining control of the carded web weight, with stretching within the cross-lapper, can also be used to provide even greater control over batt weight uniformity. Naturally, such systems rely on sophisticated control systems, capable of continuously scanning the output batt or needled fabric weight, and synchronisation of card feed rate, doffer speeds, cross-lapper speeds, and drafting speeds.

4.10.2 Camel-back cross lapping

Although less common than conventional horizontal cross-lappers, camel-back systems are worthy of mention. The web from the card is carried upwards onto a conveyor, which swings backwards and forwards like a pendulum over the top of the bottom lattice, laying the web down across the width. The web is supported during the process to avoid unwanted drafting, sometimes with the aid of suction. Dilo's Hyperlayer HLSC is an example of a camel-back cross-lapper. Camel-back cross-lappers are limited to maximum laydown widths of about 5 m, which is much less than a horizontal cross-lapper. However, with high infeed speeds, they allow for minimal unwanted drafting of lightweight webs, and arguably permit a more precise laydown.

4.11 Batt drafting

A batt drafter is used to increase the fibre orientation in the machine direction after the formation of a cross-lapped batt. This is particularly important in applications such as geotextiles where uniformly isotropic tensile properties and burst strength may be required.

Batt drafters normally consist of a series of nipped roller drafting units extending the full width of the machine, the surface speed of each set of rollers increases from the input to the output to control the draft and the maximum draft may be in the range of 30% to 260%. Needlepunched fabric drafters are also used after or between successive needle looms to manipulate fibre orientation in the fabric. Typically, drafts of 20% to 60% are applied to fabrics in widths up to 7.4 m. Whilst draft is principally applied to manipulate the MD/CD strength it also influences other structural features such as density, thickness and permeability. Although drafting can produce MD/CD ratios approximating to 1, it is frequently associated with an increase in weight variation.

4.12 Vertically lapped (perpendicular-laid) web formation

Vertically lapped (perpendicular-laid) nonwovens are gaining acceptance in an increasing number of applications. Such fabrics are used as foam replacement materials in the automotive industry, depth filtration media, thermal insulation, and in bedding products such as futons. Various methods of corrugating webs to form perpendicular-laid fabrics have been devised over the years and all produce a concertina-like, three-dimensional structure, which after bonding exhibits high recovery from compression. The Corweb, Struto, and V-lap systems are amongst the most well-known processes.

A carded web, which normally contains a proportion of thermoplastic fibre, typically a core-sheath or side by side bicomponent, is mechanically formed into a series of vertical folds that are stabilised by through-air thermal bonding. Blends may be composed of thermoplastic synthetic fibres, reclaimed waste materials and natural fibres such as cotton and wool. In addition to fibre composition, the fold frequency (which influences packing density) and the fold orientation affect fabric properties. The fold frequency and orientation are controlled by the choice of lapping device and the web overfeed setting.

The Struto system (Struto International Inc.), is an established perpendicular-lapping process (Fig. 4.40). The more recently developed V-lap system is similar but differs mechanically in respect of the drive system for the presser bar and comb. In the Struto and V-lap systems, a

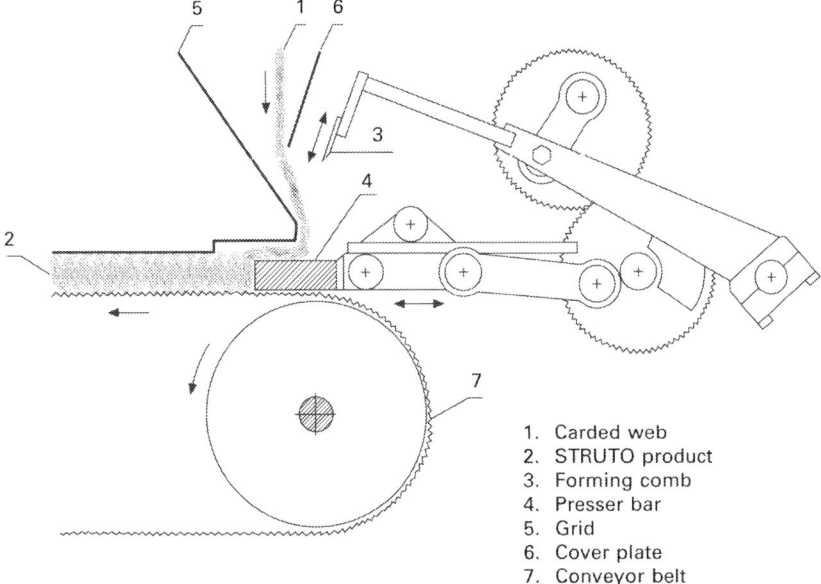

1. Carded web
2. STRUTO product
3. Forming comb
4. Presser bar
5. Grid
6. Cover plate
7. Conveyor belt

FIG. 4.40 Formation of perpendicular-laid webs. *Courtesy of Struto International Inc., USA.*

reciprocating lapping device is used to continuously consolidate the carded web into a vertically folded batt immediately prior to through-air bonding. A proportion of low-melt thermoplastic fibre in the blend enables thermal bonding of the structure either in its basic lapped form or in conjunction with a scrim or support fabric, which can be introduced before the oven.

Thereafter the fabric (Fig. 4.41) is cooled and subsequently wound. Slitting the fabric in a similar manner to woven carpets to make two separate materials produces thinner fabrics.

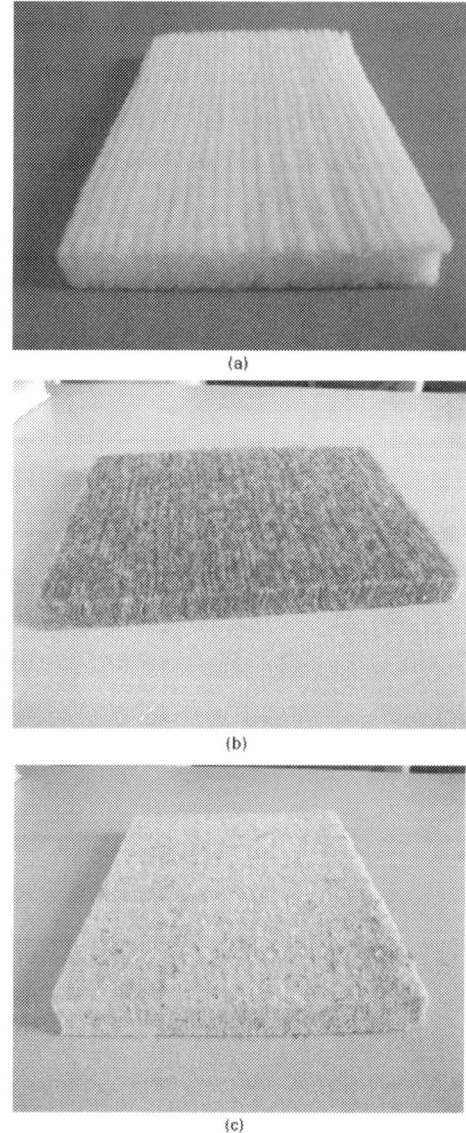

(a)

(b)

(c)

FIG. 4.41   Cross-section of various Struto fabrics; (A) PET (heavy web) $500 \, g/m^2$; (B) PET (standard) $500 \, g/m^2$; and (C) (shoddy) $1050 \, g/m^2$. *Courtesy of Struto International Inc., USA.*

The compression properties of the fabrics are strongly influenced by the proportion of thermoplastic bicomponent fibre present and the fibre diameter, which governs fibre rigidity. Stiff, board-like products are produced using a high proportion of relatively coarse bicomponent fibre (>5 dtex).

Depending on composition and fabric structure, the fabrics have higher resistance to compression and elastic recovery than comparable cross-lapped and high-loft airlaid fabrics (see Fig. 4.42). To maximise the resistance to compression-recovery properties, vertical orientation of the fibre in each web fold is usually preferred instead of a slightly inclined orientation.

The V-Lap system incorporates its own patented design which is structurally much heavier and robust and with fewer moving parts than older vertically lapping technologies. The machine is claimed to be better balanced with easy maintenance and well suited to continuous production. Changing from one product to another is simple, utilising a quick release change-over, which is claimed to reduce the product change time by up to two thirds. A specialised barbed needle configuration on the pusher bar allows improved control over difficult fibres such as ceramic and siliconised materials. The V-Lap has a broad product weight range from

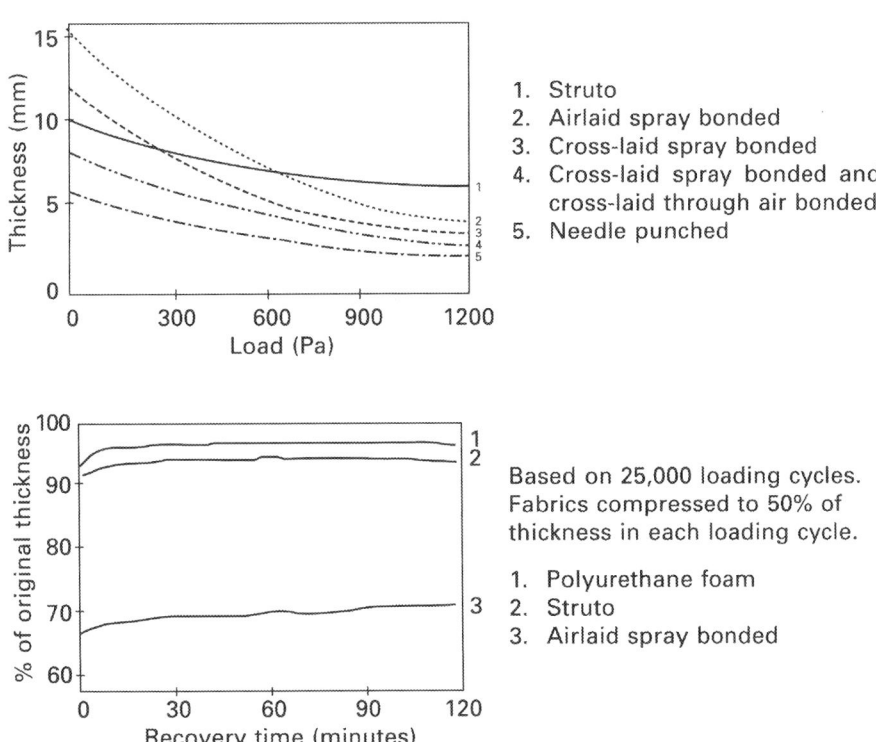

FIG. 4.42 Comparison of the load vs. thickness and elastic recovery behaviour of 150 g/m^2 Struto fabric compared with other nonwoven materials. *Courtesy of Struto International Inc., USA.*

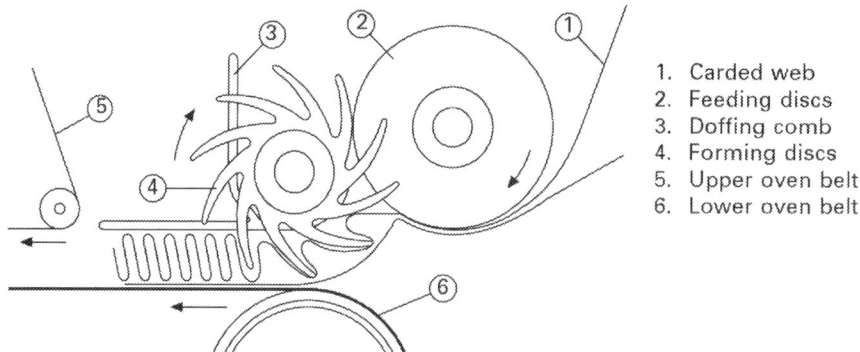

1. Carded web
2. Feeding discs
3. Doffing comb
4. Forming discs
5. Upper oven belt
6. Lower oven belt

FIG. 4.43   Rotary lapper. *Wavemaker, courtesy of Santex, Italy.*

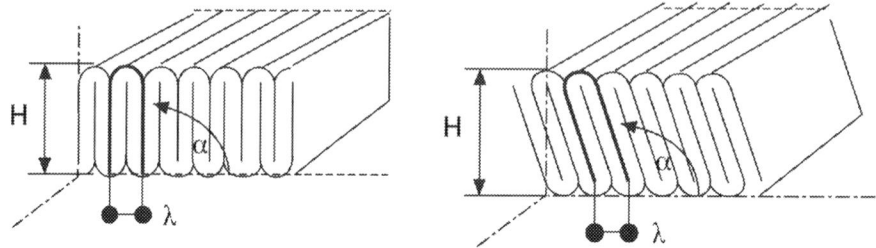

FIG. 4.44   Comparison of folded web cross-sections produced by perpendicular-laid web formation systems.

$50 \, g/m^2$ to $3000 \, g/m^2$ and can run at much higher speeds up to 1400 rpm and over a much wider thickness range of 10 mm to 80 mm compared to earlier vertical lapping devices.

In contrast to the Struto and V-Lap technologies, the Wavemaker system (Santex, Italy) utilises a rotary-forming disc to create the web folds (Figs 4.43 and 4.44). The first rotary and reciprocating lappers originated at the University of Liberec. Whilst the rotary lapper leads to significantly higher production rates than the reciprocating version used by the Struto and v-lap system the latter produces a more pronounced z-directional fold orientation, which is approximately perpendicular to the fabric plane. The fold structures produced by rotary lappers tend to slope slightly relative to the fabric plane, and therefore, the resistance to compression of fabrics produced by rotary and reciprocating lappers is different.

Through-air thermal bonding is used to stabilise the resulting structure. Alternatively, the Rotis system was developed for introducing fibre entanglements in Struto vertically lapped webs using revolving, barbless tubes arranged across the machine and perpendicular to the two sides of the web. These tubes are used to entangle surface fibres in discrete continuous longitudinal rows thereby connecting successive folds of web in-plane. Preformed scrims or fabrics may also be introduced from above and below to form a composite fabric structure in situ.

Applications for perpendicular-laid nonwovens include foam replacement and fillings for mattresses, bedding, seating, furniture, and some wearable healthcare products as well as sound absorption for vehicles and filter media.

4.13 Introduction to airlaid web formation

Airlaying (aerodynamic or airlaid web formation) refers to a family of drylaid web formation processes used in the manufacture of products containing pulp or short cut fibres (e.g. for wipes, absorbent cores and food packaging pads), and products containing longer fibres (e.g. for high-loft waddings, filtration media, interlinings, automotive components and mattress fillings).

Airlaid webs are characteristically isotropic. In contrast to carded webs, MD/CD ratios approaching one may be obtained depending on fibre specifications and machine parameters. Airlaid webs are therefore frequently referred to as 'random-laid'. Additionally, airlay processes are highly versatile in terms of their compatibility with different fibre types and specifications. This versatility partly arises from the principles of fibre transport and deposition used in airlaying as well as the variety of airlay machine designs available.

Airlaying, like other technologies, has certain benefits and limitations.

Amongst the benefits are:

- Isotropic web properties.
- Three-dimensional structure if the basis weight is above about $50\,g/m^2$ producing voluminous, high-loft structures with a very low density.
- Compatibility with a wide variety of generic fibre types including natural and synthetic polymer fibres, ceramics, metals including steel, carbon, melamine, aramids and other high-performance fibres.

The main limitations are:

- Fabric uniformity is highly dependent on fibre opening and individualisation prior to web forming.
- Air flow irregularity adjacent to the walls of the conduit can lead to variability across the web structure.
- Fibre entanglement in the airstream can lead to web faults.

Depending on fibre type and fineness, airlaying is claimed to be more efficient than carding in the production of webs greater than 150 to $200\,g/m^2$, where production rates of $250\,kg/h/m$ can be achieved [1].

4.13.1 Raw material specifications and fibre preparation for airlaying

A variety of fibres are used by the airlaying industry. Wood pulp (usually fluff pulp) continues to play a major role, often in blends with short cut man-made fibres. The characteristics of some typical pulps are now briefly discussed.

4.13.1.1 Wood pulp and natural fibres

Wood pulp can be produced by mechanical or chemical processes. Thermomechanical pulping (TMP) involves passing wood chips between rotating plates having raised bars at high temperature and pressure. The heating softens the lignin, which is a natural phenolic resin holding the cellulose fibres together, making it possible to separate the fibres. A yield of over

TABLE 4.2   Key properties of wood pulp fibres.

Pulp type	Main species	Fibre length (mm)	Fineness (mg/100 m)	Fibres/g (× 10⁶)
Southern US (Kraft)	Southern pine	2.70	45.6	2.6
Scandinavian (Kraft)	Spruce/pine	2.06	27.0	5.0
Northwest (Sulphite)	Spruce/fir	2.08	33.0	4.2
Cotton linter pulp		1.8	25.0	
Cold caustic extracted cellulose		1.8	34.0	
Cross-linked cellulose		2.3	40.0	

90% of wood fibres can be obtained. In contrast, chemical pulping (Kraft process) dissolves the lignin using suitable chemicals such as caustic soda and sulphur under heat and pressure. The chemical pulping process produces lower fibre yield than mechanical pulping, typically 50% to 60% [2]. Some typical wood pulp fibres currently used in airlaying are:

- Southern Softwood Kraft. Manufactured in the southeastern USA and used in products where absorbency, softness, cleanliness and brightness are required. Softwood fibres are used to provide strength and bulk. They tend to produce less dust and lint, providing a cleaner conversion process.
- Scandinavian Sulphate (Kraft). Fluffs are of shorter length and coarser than American southern pines (see Table 4.2).
- Northern Softwood Sulphite. Used on a smaller scale for speciality productswhere superior formation (low fibre entanglement), softness, and high brightness are required. They are mainly used in products such as airlaid tabletop covers and wipes as well as feminine hygiene pads [3].

The main critical parameters that characterise wood pulp fibres are: [4]

- Wood species
- Pulping process (mechanical or chemical process)
- Fibre length
- Fibre fineness
- Fibre stiffness
- Special treatments.

Table 4.2 summarises some of the key properties of wood pulp fibres relevant to the airlaying industry. Generally, the finer pulp fibres give rise to higher softness, wicking rate and printability. On the other hand, coarse and long fibres produce more resilient and bulkier fabric structures with better total absorption capacity and a higher porosity.

4.13.1.2 Man-made fibres

The man-made fibres used by the airlaying industry fall into two main categories, natural polymer-based fibres (e.g. regenerated cellulosic fibres, such as viscose rayon and lyocell) and synthetic polymer-based fibres (e.g. polyamide, polyester and the polyolefins). The regenerated cellulose fibres such as viscose rayon and lyocell are very hydrophilic and similar in their absorbency characteristics to wood pulp. They can hydrogen bond and are typically cut to fibre lengths of 3 to 12 mm. The short cut length makes them suitable for inclusion in airlaid products particularly in blends with wood pulp, to increase the strength of the airlaid fabric. Additionally, the inclusion of fibres in the blend contributes to higher abrasion resistance and often, a softer handle as compared to the shorter, stiff wood pulp fibres.

Synthetic polymer fibres such as PET, PA, PP and PE are hydrophobic and are particularly effective in maintaining the bulkiness of airlaid fabrics in wet conditions. Such fibres are used in blends with wood pulp and sometimes SAP in liquid acquisition layers for nappies as well as other absorptive products. Synthetic fibres have a high wet strength as compared to viscose rayon (which decreases in strength when wet) and can markedly increase the durability and strength of the fabric in use. The effect of fibre parameters (crimp level, fibre fineness, fibre length and fibre cross-sectional configuration) on the performance of a thermally bonded airlaid fabric was investigated by Gammelgard [5], using both the Dan-web and M&J short fibre airlaying systems. The main findings of the study may be summarised as follows:

- Finer fibres increase the tensile strength of the product. Changing from 3.3 dtex to 1.7 dtex fibres increased the tensile strength by up to 40%.
- The tensile strength of the airlaid web increased with decreasing crimp level. This may be attributed to the fewer bonding points available in crimped fibres. Also, it was pointed out that the lower the crimp level, the higher the fibre throughput. Therefore, crimp may be used to control the production capacity of an airlaid line and should be optimised depending on the type of web formation system employed (i.e. Dan-web or M&J system).
- The tensile strength varied with the proportion of PE in the PP/PE bicomponent (BICO) fibre. Up to a certain point the tensile strength increased with an increase in the proportion of PE. Further increases affected the PP core, which became weak and broke before the thermal bonding points failed. The optimum proportion of PP and PE in the bicomponent fibre was claimed to be 35/65 for a concentric sheath- core bicomponent fibre or 1.7 dtex (AL-Special–C).
- M&J and Dan-web lines perform differently regarding fibre length.
- It was concluded that in the M&J system, selection of 3 mm fibre length optimises the production capacity whereas, 4 mm fibre length optimises the fabric tensile strength. The Dan-web line was claimed to have greater flexibility regarding fibre length without affecting production capacity (6–8 mm fibre length) compared to the M&J line (3–4 mm fibre length) [5].

In long fibre airlaying systems, all types of synthetic fibres between 1.7 and 150 dtex linear density and staple lengths of 40 to 90 mm can be processed, as well as natural fibres such as cotton, wool, jute, flax, hemp, kenaf, recycled fibre materials, and specialist high-performance fibres such as P84 (polyimide fibre).

4.13.1.3 Superabsorbents

Superabsorbent polymers (SAPs), which are available as powders, granules, beads or more recently as fibres, can be used to augment the water holding capacity of airlaid webs containing fluff wood pulp and other short cut fibres. The capacity of superabsorbents (cross-linked hydrogels) to absorb aqueous liquids is several times higher than wood pulp fibres and their function is to immobilise as much fluid as possible without releasing it even when the fabric structure is compressed. Liquid absorption is reduced in saline conditions. The powder form is usually added to the airstream in which wood pulp fibres are suspended prior to airlaying. The fibre component, which is more expensive, can be blended or formed as an individual layer as part of a composite web. Superabsorbent fibres are designed to absorb fluids without losing their fibrous structure and therefore retain a proportion of the dry fibre strength. Typically, such fibres absorb 95% of their ultimate absorbent capacity in 15 s. One example is SAF fibre (Technical Absorbents, UK) which is normally cut to a staple length of 6 or 12 mm for use in pulp airlaying systems.

Typically, 10% to 40% of SAF is used in blends with woodpulp and/or staple fibres. Bonding of webs containing superabsorbent fibres is carried out using thermal bonding (assuming a thermoplastic fibre is also added to the blend) or latex (chemical) bonding. During processing of such fibres it is recommended that the relative humidity be kept below 60% and preferably 55% [6] to prevent unwanted gelatinisation of the fibre. Some of the advantages of superabsorbent fibres over superabsorbent powders are derived from their physical form, rather than their chemical nature. The advantages are summarised below:

- Fibres absorb fluids faster than powder with the same absorption capacity.
- Fibres are better integrated within the web structure.
- Fabrics containing fibres are flexible and soft in contrast to the powders that are abrasive and confer a rough and harsh handle to fabrics.
- Fibres are easier to incorporate into the airlaid structure and are less likely to migrate from the structure during subsequent bonding and in use.

4.13.1.4 Fibre preparation

In airlaying, it is important to introduce opened and preferably, individualised fibres to the airstream so that a uniform web without any tangled clumps or fibre flocks can be formed. It is important to note the difference in state between opened and individualised fibres. The term 'opened' fibres refers to a collection of fibres that is substantially free of clumps, tangles, knots, or similar dense entanglements, but there is still significant frictional interaction between the fibres. In contrast, 'individualised' fibres have no mechanical or frictional interaction with other fibres.

Various methods for fibre opening and separation have been designed for airlaying lines. The majority of opening systems are the same as those used prior to carding. In addition, hammer mills or customised openers are utilised. In general, opening and fibre separation can be accomplished using a clamped feeding unit consisting of a feed chamber equipped with a fine opener, a vibration chute feed with a weighing device followed by a further opening section composed of a pinned or saw-toothed roller with or without worker-stripper rollers. Typical examples of feed roller designs that can be used to separate fibres prior to airlaying are shown in Fig. 4.45. In pulp airlaying, the hammermill dominates fibre preparation procedures. A hammer mill disintegrates the feed material so it can be uniformly distributed through the forming heads. The increasing use of the Sunds defibrator has increased the importance of the disc refiner. The Sunds system incorporates the use of a bale shredder, screening equipment and a disc refiner.

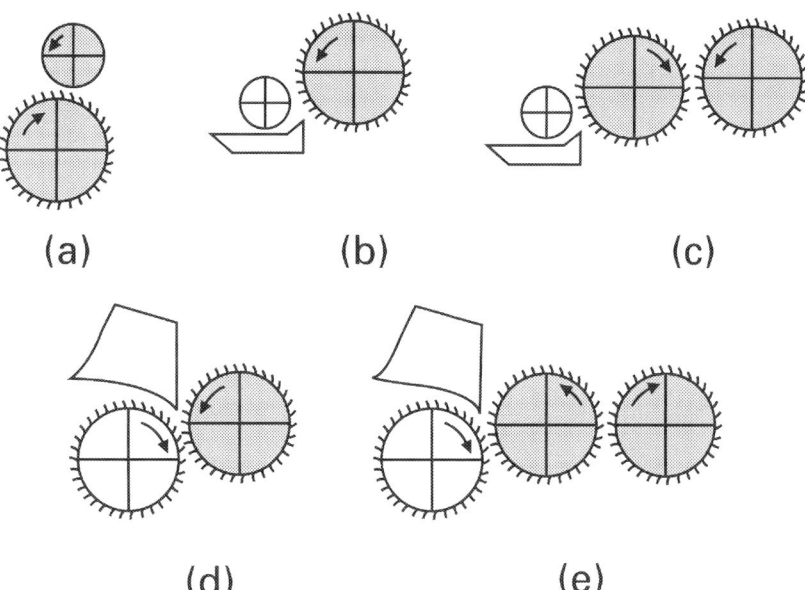

FIG. 4.45 Feeding systems used in airlaying.

4.14 Airlaying technology

Airlaying involves uniformly dispersing individualised fibres in an airstream and transporting this air-fibre mixture towards a permeable screen or conveyor where the air is separated, and the fibres are randomly deposited in the form of a web. Fibre separation is therefore an essential part of the airlaying process and strongly influences the global and local uniformity of the final web. In the formation of lightweight webs, it is essential to ensure that opened, individualised fibres, free from clumps and entanglements are introduced into the airstream.

The fibre orientation in the final web is mainly influenced by the dynamics of the airflow in the fibre transport chamber near the landing area. In practice, this can be strongly affected by the rotation of the opening or fibre dispersing unit above the transport chamber. The following methods are used to transport fibres from the opening unit to the web-forming section:

- Free fall.
- Compressed air.
- Air suction.
- Closed air circuit.
- A combination of compressed air and air suction systems.

The principle of airlaid web formation using a suction assisted landing area is shown schematically in Fig. 4.46. In this particular machine design, pre-opened fibres, which can be prepared using the feeding, mixing and opening systems described in Section 4.3, are fed to a pair of feed rollers which, in the same way as carding, are designed to grip the fibre and prevent large clumps from being drawn into the system. To ensure feed regularity, which is critical given that no long-term levelling of weight variation is possible within the airlaid-forming head, automatic feed control systems of the type used by the carding industry can be applied.

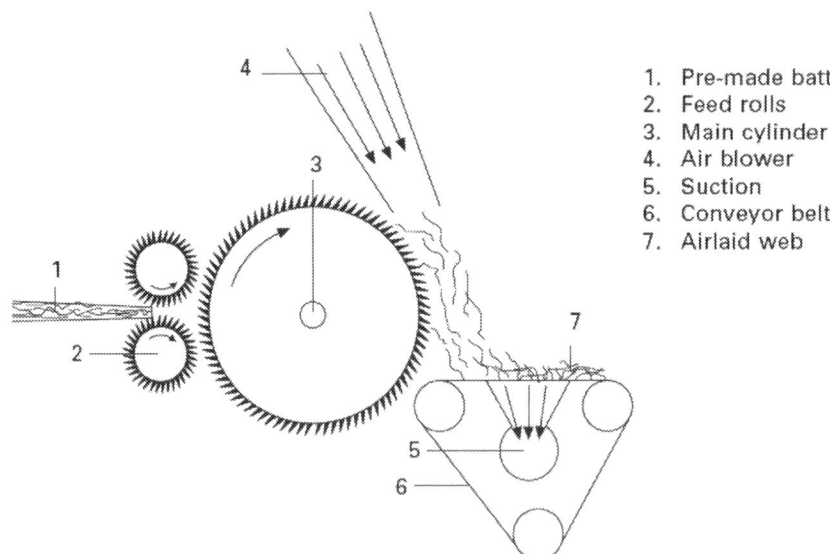

1. Pre-made batt
2. Feed rolls
3. Main cylinder
4. Air blower
5. Suction
6. Conveyor belt
7. Airlaid web

FIG. 4.46 Principle of web formation in a simple airlaying process.

The rotating drum or cylinder removes fibres from the fringe presented by the feed rollers. The fibres are transported by hooking around the wire teeth on the drum and are subsequently removed by a high-velocity airstream directed over the wire teeth surface. In this way, the fibres are mixed with air and transported with it to an air permeable conveyor where the air is separated and the fibres are deposited to form the web or batt structure.

Airlaying technology may be classified according to the raw materials used for processing. Using this form of classification there are two main types; the type that uses natural or man-made textile fibre (cut length >25 mm) and the type that employs short cut fibres (generally <25 mm) and wood pulp (1.5–6 mm). The importance of airlaying textile length fibres was recognised when the basic need for forming webs with a random fibre orientation was recognised. In contrast to carded webs, which are characterised by anisotropy due to the preferential orientation of fibres in the machine direction, airlaying allows webs with practically multi-dimensional (random) fibre orientation in a wide range of web thicknesses. Although work on airlaying was done by Kellner [7] in 1892, it is generally agreed that the first commercial airlaying process capable of processing textile fibres was pioneered by The Curlator Corporation (known as the Rando Machine Corporation) in the 1940s.

A form of airlaid batt is also formed by rag-pulling machines and pickers provided the opened material is condensed onto a suction screen or conveyor. Some commercial airlaid machines have evolved from this background in waste reclamation and mechanical recycling of textile materials, and they are also suited to processing natural fibres such as hemp, jute coir, sisal, and cotton. Specialist systems of this type are manufactured by companies such as Laroche (France), DOA (Austria) and Schirp (Germany). Carding machine hybrids have also been developed that possess worker-stripper rollers or fixed flats to intensively open the fibres prior to web forming. However, instead of using a conventional doffer to produce a web, the fibre from the card is airlaid onto a permeable conveyor. This approach is useful in the manufacture of lightweight isotropic webs at high output speeds. Much has been published on the development of airlaying technology, particularly in the patent literature. Some of the systems that have achieved some commercial acceptance are briefly reviewed.

4.14.1 Dual rotor systems

Various versions of dual rotor airlaying are referred to in US patents 3512218 [8] (1970), 3535187 [9] (1970), 3740797 [10] (1973), 3768118 [11] (1973), 3772739 [12] (1973), and 4018646 [13] (1977). The dual rotor airlaying system comprises a pair of contra/rotating rollers equipped with a fibre feeding device. The fibres are ejected from the rollers by a combination of centrifugal force and high-velocity airflow to a transfer duct. The doffed fibres are then deposited onto a moving conveyor belt downstream from the doffing point. It is claimed that an airlaid web of homogeneously blended short and long fibres can be produced by this system. This airlaying technology was used from the mid-1960s by Johnson and Johnson to produce nappy facings until the company decided to terminate its presence in this market.

4.14.2 Rando-webber

The Rando-webber (Rando Machine Corporation, formerly The Curlator Corporation) is one of the oldest aerodynamic web-forming methods and is still in use today. The design features of the machine are described in US patents 2451915 [14], 2700188 [15] and 2890497 [16]. Wood then made a number of improvements as described in US patents 3768119 [17] and 3972092 [18]. The Rando process normally consists of three units, (i) opening and blending, (ii) feeding, and (iii) web forming. The feeding section is similar to a hopper feeder unit with an inclined lattice, evening condenser and stripper roll. The web formation part is equipped with an input unit including a feed roll and a feed plate, opening unit (licker-in roll) and a conveying section. Fibres are pre-opened at an early stage of the process in the opening and blending section and then fed through the feeding unit to the web formation zone where they are further opened and individualised by the actions of the licker-in. The opened fibres are then removed from the licker-in to the transport duct by means of a high-velocity airstream and centrifugal forces generated due to the rotational speed of the licker-in. Finally, the fibres are deposited onto the cylindrical condenser to form an airlaid web.

The output speeds of the conventional cylindrical condensers are limited due to the effect of the centrifugal force created by the high surface speed of the cylinder type condenser and also the space available for air removal from the cylinder, which is important for a high fibre production rate. For these reasons, Rando-webbers normally have cylindrical condensers and relatively narrow widths up to about two metres (model A and B). The continuous screen condensers are recommended for high production rates and wider widths (model C), see Fig. 4.47. Rando-webbers can produce webs of 10 to $3000 \, \text{g/m}^2$ and can process virgin or recycled fibres for applications in filtration, home furnishings, automotive fabrics, insulation and some medical specialities [19]. A Rando-webber (model B) has been utilised to form webs from melamine fibre (Basofil, BASF) with very low to zero crimp, 2.5 denier and 50 mm length for application in industrial filtration and insulation [20].

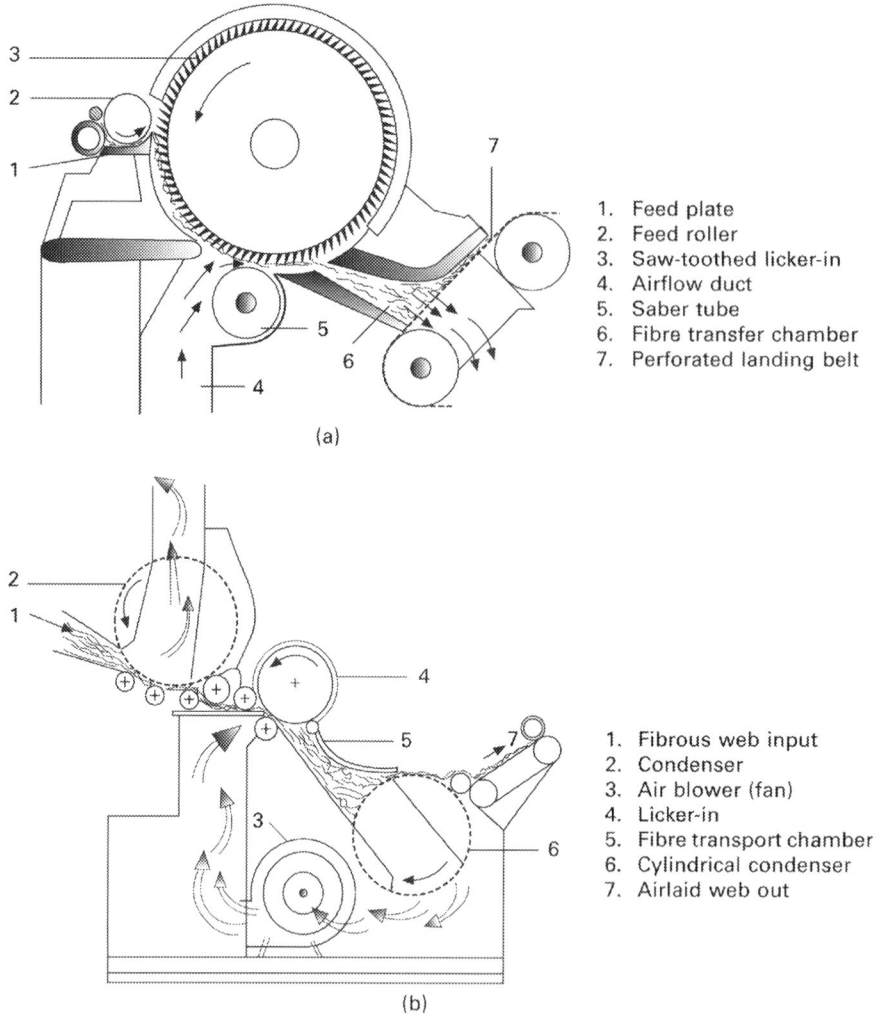

1. Feed plate
2. Feed roller
3. Saw-toothed licker-in
4. Airflow duct
5. Saber tube
6. Fibre transfer chamber
7. Perforated landing belt

(a)

1. Fibrous web input
2. Condenser
3. Air blower (fan)
4. Licker-in
5. Fibre transport chamber
6. Cylindrical condenser
7. Airlaid web out

(b)

FIG. 4.47   Rando-webber systems with (A) perforated screen and (B) cylindrical condensers.

4.14.3 Airlaid process for production of blended composite web structures

In US patent 3535187 [9], Wood described an apparatus for producing airlaid structures composed of two or more separated layers of different randomly orientated fibres. The airlaying apparatus included two licker-ins and rotary feed condenser assemblies. Individualised fibres from each licker-in were deposited as layers on separate cylindrical condenser screens. The two condenser screens were positioned adjacent to each other and the layers of the fibres on the condensers were compressed between the two condensers to form a composite nonwoven web having some blending of the fibres at the interface between the layers, see Fig. 4.48. In the process, the fibres were removed from their respective licker-ins by

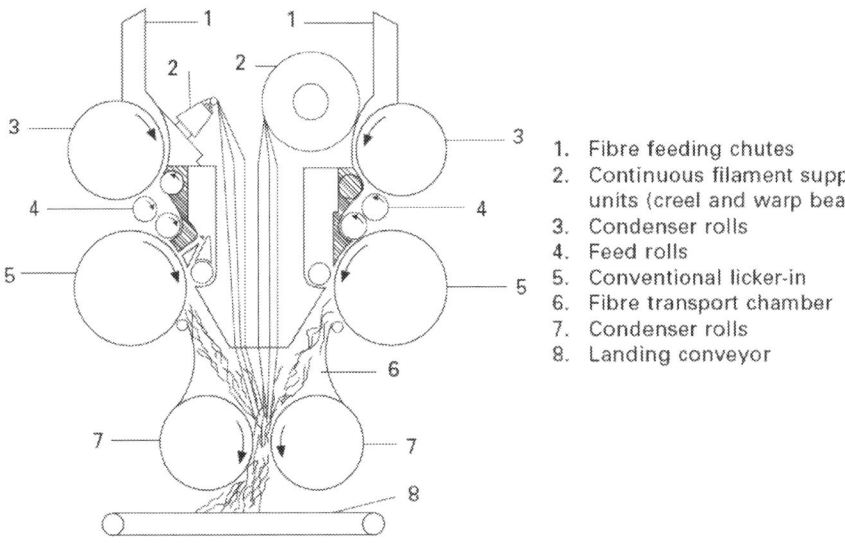

FIG. 4.48   Schematic view of the airlaying system explained in USP 3535187 [9].

1. Fibre feeding chutes
2. Continuous filament supply units (creel and warp beam)
3. Condenser rolls
4. Feed rolls
5. Conventional licker-in
6. Fibre transport chamber
7. Condenser rolls
8. Landing conveyor

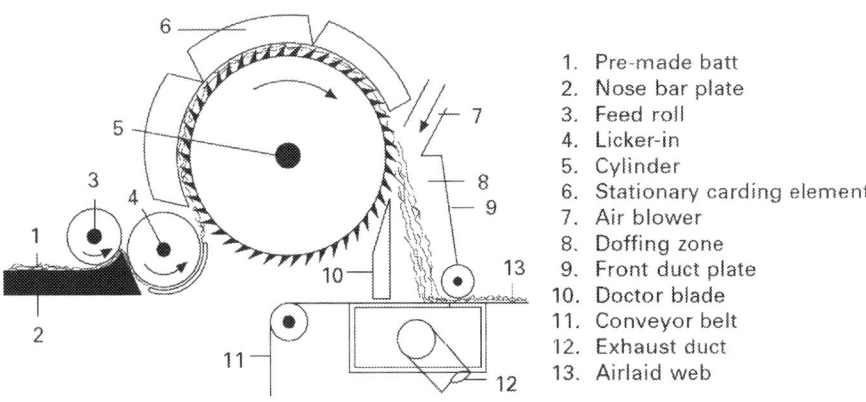

FIG. 4.49   Chicopee airlaying system based on USP 4475271 [21].

1. Pre-made batt
2. Nose bar plate
3. Feed roll
4. Licker-in
5. Cylinder
6. Stationary carding element
7. Air blower
8. Doffing zone
9. Front duct plate
10. Doctor blade
11. Conveyor belt
12. Exhaust duct
13. Airlaid web

high-speed turbulent air streams with higher velocities than the peripheral speeds of the licker-ins. The webs produced were composed of textile fibres, although processing of continuous filaments and wood pulp fibres was also suggested.

4.14.4 Chicopee system

Lovgren in US patent 4475271 [21] described a method for producing highly uniform webs at high speed (see Fig. 4.49), which incorporated (i) a feeding unit with a rotating toothed roll for opening the fibres; (ii) a main toothed cylinder with stationary carding elements to individualise the fibres; (iii) a transport duct where the individualised fibres are released from the toothed roll by centrifugal force and high-speed airflow passing tangentially to the surface of the cylinder; and (iv) a forming section where the fibres are condensed on a foraminous screen and form a randomly arranged fibrous web. Preferably, the airflow in the transport duct is turbulent which provides a nearly flat velocity profile except at the boundaries.

Also, in the airlaying machine, the air velocity (e.g. 140 m/s) should be substantially higher than the surface speed of the cylinder (e.g. 20–60 m/s) or the velocity of fibres coming off the cylinder, so that the fibres are kept under tension until they reach the landing area. In this manner, it is claimed that the fibres can be uniformly dispersed without any tendency to clump or condense. Staple fibres ranging from 13 to 75 mm can be used in this system.

4.14.5 Autefa airlay systems

The K12 random web-forming machine was developed in 1968 [22,23] to produce fabrics in the basic weight range of 20 to 2000 g/m^2, depending on fibre specifications. The airlay system was originally developed by Fehrer (now Autefa Solutions). In the K12, a laminar airflow is used to carry the fibres through the transport chamber; the airflow is produced by a patented transverse jet stream in an open system and is not separated from the surrounding air. The K12 is more particularly suited to coarse fibres (10–110 dtex) [23]. Fig. 4.50 shows a schematic view of the airlaying system.

Demand for lighter-weight airlaid webs of 10 to 100 g/m^2 led to development of the K21 high-performance random airlaying machine [24]. In contrast to the K12, which has only one cylinder, the K21 consisted of four carding cylinders each with a pair of worker-stripper rollers. A proportion of the total flow of fibres into the machine is airlaid by each cylinder onto a common conveyor belt. Airlaying is performed by a combination of centrifugal force created by the rapid rotation of the cylinders (30–40 m/sec) and suction under the conveyor belt. Fibres are deposited on the belt in four different positions [24], which allows levelling of local weight variations in the web by intensive doubling of the incoming fibres along the collection zone.

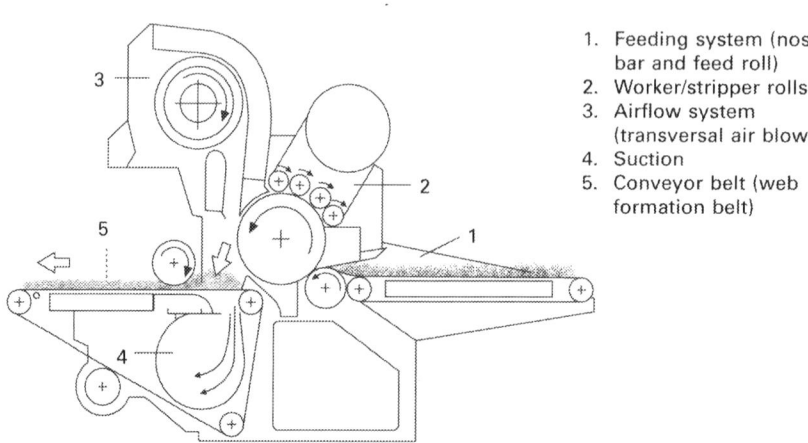

1. Feeding system (nose bar and feed roll)
2. Worker/stripper rolls
3. Airflow system (transversal air blower)
4. Suction
5. Conveyor belt (web formation belt)

FIG. 4.50   K12 airlaying system [23].

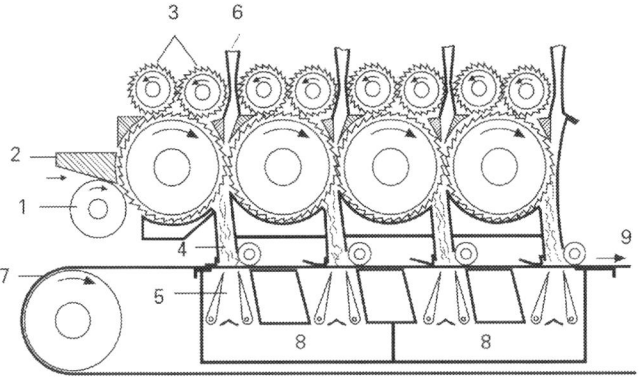

1. Feed roll
2. Nose bar
3. Worker/stripper rolls
4. Fibre transport chamber
5. Suction nozzle
6. Air passage
7. Collecting surface (perforated belt)
8. Suction
9. Airlaid web

FIG. 4.51   Schematic view of K21 airlaying system [25].

The design of the web-forming zone is different from the K12 which is a closed system. The K12 is an open system (see Fig. 4.51). It is claimed that this design allows production speeds up to 150 m/min. The K21 was designed for processing synthetic and viscose rayon fibres of 1.7 to 3.3 dtex with a throughput of up to 300 kg/h/m [23,24]. Today, a two-stage process arrangement is common, where a separate airlay unit (V12/R) operating without worker rollers, makes a uniform batt that is fed to the K12 system as part of an integrated processing line known as the V21/R-K12.

Autefa's V12/R-K12 aerodynamic web for forming system is particularly suited to the processing of mechanically recycled fibres as well as natural fibres such as coir, jute, sisal, cotton, hemp and wool in the weight range of 400 to 7000 g/m².

4.14.6 K12 high-loft system

High-loft waddings are low-density, highly three-dimensional batt structures that are usually bonded by through-air ovens or by spray bonding and curing. It is advantageous for a proportion of the fibres in such waddings to be oriented in the z-direction (thickness direction) so as to maximise resistance to compression. Whilst this is achieved to some extent in most heavy weight airlaid webs due to progressive blocking of the suction below the conveyor, a modification to the Fehrer (now Autefa Solutions) K12 airlaying machine consists of an upper rotating suction drum, see Fig. 4.52.

In this design adaptation, fibres are released from the cylinder and deposited partly onto the surface of the suction drum and partly onto the conveyor belt, whose directions of travel are away from each other. Therefore, the web is formed and fibres re-orientated between the two suctioned moving surfaces, which results in an increase in the vertically orientated fibres in the web producing bulkier structures [25]. The suction drum can be adjusted horizontally and vertically according to requirements. Using this attachment, the K12 High Loft is claimed to produce webs with 80% greater thickness compared to the conventional K12 machine. The high-loft device can be retrofitted to existing K12 airlaying machines [26].

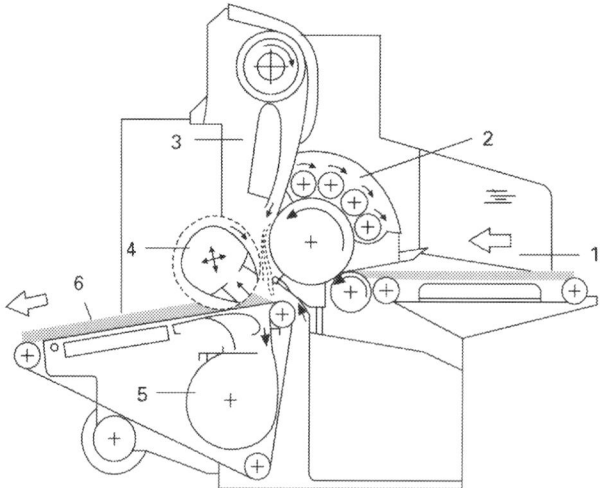

1. Feeding system (nose bar and feed roll)
2. Worker/stripper rolls
3. Airflow system (transversal air blower)
4. High-loft device
5. Suction
6. Airlaid web

FIG. 4.52   High-loft airlaying system.

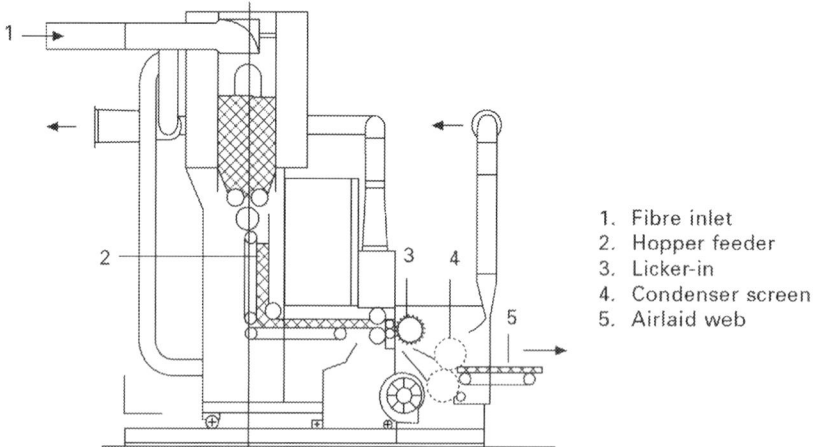

1. Fibre inlet
2. Hopper feeder
3. Licker-in
4. Condenser screen
5. Airlaid web

FIG. 4.53   Schematic view of the DOA airlaying system.

4.14.7 DOA system

The basic principle of this system is shown in Fig. 4.53. Opened fibre is passed through the feeding unit to an opening device (e.g. licker-in roller) to open and individualise fibres. The opened fibres are then released from the licker-in by means of an airstream and centrifugal forces created by the rotation of the opening roller. Fibres are finally transferred to a pair of condenser screens where the air is sucked away and a fibrous web is formed. It is known that the uniformity of airlaid webs can be improved by using two airlaying zones where fibres undertake more intensive opening actions (model 1044 and 1048). Screen drum diameters can be selected (40, 55 or 80 cm) for producing webs up to a thickness of 350 mm with the capability of introducing powders, foam or liquid additives with the fibres.

It is claimed that due to the air-blowing system used, an excellent random distribution of fibres can be obtained and that these fibres are entangled in the landing area very uniformly and consistently in all directions. A wide range of synthetic fibres as well as natural fibres are processed (such as cotton, rayon, jute, flax, coir fibre, sisal, wood, coconut and even straw), reprocessed wool, and also reclaimed raw materials (waste fibres). Applications include moulded products, needlepunched felts, insulation, automotive fabrics such as bodyshell insulation, high loft waddings, geotextiles, apparel components, furnishings, mattress components, carpet underlay, fibre glass batts, and filter fabrics amongst others.

4.14.8 Laroche system

The Laroche airlaying process, exemplified by the Flexiloft system, is intended to handle various types of fibres from short staple (e.g. cotton, synthetics including aramids, carbon, glass, and recycled materials) to long and coarse natural fibres (e.g. hemp, flax, sisal, and coir). Generally, the fibre length is in the range of 20 to 75 mm. The machine also processes recycled materials such as carpet waste and quilted bedspreads. The batts are normally needlepunched or thermally bonded to produce finished fabrics [27]. Typically, fibres are fed via a rotating condenser to a feed chamber and are then transported by a spiked lattice to a volumetric hopper feeder with a vibrating chute feed. Two pairs of feed rolls deliver a fibrous mat to a conveyor belt with a continuous weighing device. The mat is then fed to a pinned opening roller where the fibres are opened by the action of the clothed opener roller. The fibres are then directed onto a perforated belt to form an airlaid batt structure. The airlaid products produced are intended for use as mattress pads, carpet underlay, insulation, agricultural felts, automotive components (e.g. preformed panels from bast fibres) and geotextile substrates. The airlaid batt weights ranges from 200 to 5000 g/m^2 with a production speed of up to 10 to 15 m/min depending on the required web weight and the fibre type. The Laroche Resinfelt airlaid system is specifically intended for manufacturing semi-cured resonated fabrics to make moulded automotive components as well as other parts for consumer products.

4.14.9 Spinnbau hybrid system

A method for producing lightweight airlaid webs with high uniformity at a high production rate is described in US patent 5839166 [28] (1998). The Hollingsworth carding system (US patent 5007137 [29], 1991) is utilised to open the fibres and transfer them to the main saw-toothed cylinder (47–72 m/s), covered with stationary carding elements, where the fibres are intensively opened [30]. The fibres are then thrown onto the second cylinder in a random fashion, by centrifugal force. The surface speed of the second cylinder is 80% to 110% of the main cylinder. Due to the high surface speed of the second cylinder and after a very short residence time on the second cylinder, the fibres are thrown tangentially into the transport duct where they are transported by the entrained airflow generated by the high rotational speeds of the main and the second cylinder. Additional air may be drawn in through the

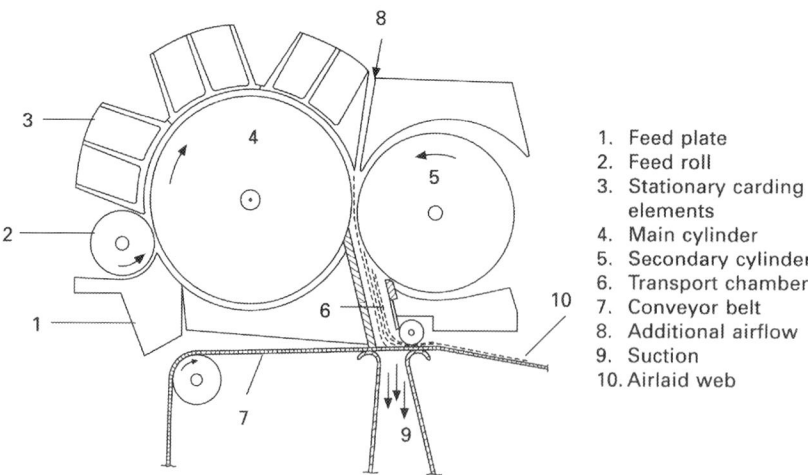

1. Feed plate
2. Feed roll
3. Stationary carding elements
4. Main cylinder
5. Secondary cylinder
6. Transport chamber
7. Conveyor belt
8. Additional airflow
9. Suction
10. Airlaid web

FIG. 4.54   Spinnbau airlaying system [28].

gap between the two cylinders (Fig. 4.54). The fibres are finally deposited onto the conveyor belt where the web is formed. The main technical features are summarised as follows:

- Fibre specifications 1.7–200 dtex, 30–60 mm length.
- Web weight c.16–250 g/m^2.
- Fibre throughput (depending on fibre fineness and fibre type) up to 200 kg/h/m.
- Working width 4 m.
- Web speed 20–200 m/min.

4.14.10 Other hybrid systems

A variety of hybrid carding systems have been developed. US patent 6195845 [31] is one example of a hybrid card-airlay machine comprising a conventional two-section carding machine (a breast and main cylinder), double doffers, an outlet cylinder to take up the two condensed webs removed by the doffers, an air blower and a perforated conveyor belt where the web is formed (Fig. 4.55). Elements of this approach have been commercialised, such as in the Airweb system (NSC France). The Airweb system, which has since been superseded was claimed to have the following features:

- Typical MD/CD ratio of 1.2–1.5:1.
- Production rate of 200–260 kg/h/m.
- Web weights of 35–200 g/m^2.
- Fibre types cotton, viscose rayon, PET, PP, PA.
- Fibre length of 10–40 mm [32].

The Trützschler TWF-NCA airlay card (T-web carding technologies) is another example, where a roller train card designed to process cotton and other short staple fibres is combined with an airlaying unit to produce random-laid webs of 30 to 400 g/m^2.

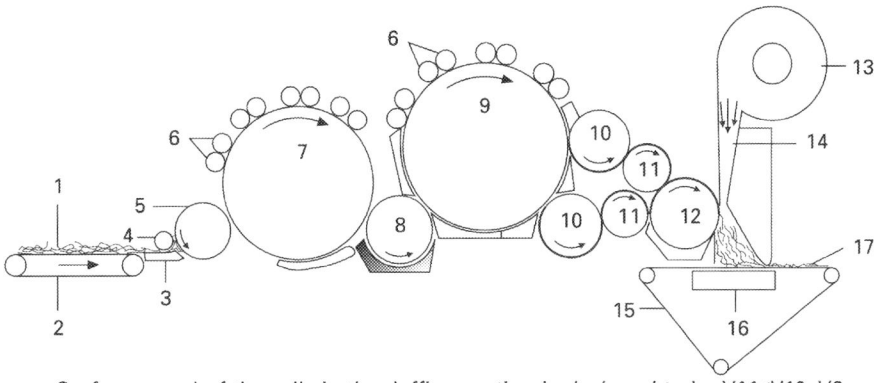

Surface speed of the rolls in the doffing section is designed to be V11≤V10<V9

1. Fibre feed
2. Conveyor belt
3. Feed plate
4. Feed roll
5. Licker-in
6. Worker/stripper rollers

7. First cylinder (breast)
8. Transfer roll
9. Main cylinder
10 Doffers
11. Condenser rolls
12. Take up roll

13. Air blower (fan)
14. Airflow channel
15. Perforated conveyor belt
16. Suction box
17. Airlaid web

FIG. 4.55   Thibeau hybrid card airlaying machine [31].

4.14.11 Airlaying technology for pulp and short cut fibres

Pulp fibre airlaying technology was essentially designed as a 'dry' alternative to conventional paper making. Wood pulp (normally fluff pulp) is formed into paper-like products either alone or in blends with short-cut fibres. The advantages over the wet laying process are (i) fabrics with greater softness and bulk and (ii) lower capital investment and low environmental impact. The early developments in pulp airlay technology were mainly concerned with paper products rather than textile or nonwoven fabrics. Since the focus was on web formation from cellulose pulp derived from trees, the fibre lengths used varied from about 3 to 20 mm. The highest production speeds were achieved with the shortest fibres (<10 mm) and longer fibres required utilisation of a higher air volume because a lower concentration of fibre in air helps to minimise fibre entanglement during the process.

At an early stage, this type of airlaying technology was limited by output speed, web uniformity and web weight limitations. Due to the uniformity problems, it was not practical to make isotropic webs lighter than 30 g/m^2. The manufacturers of pulp airlaying machinery include Oerlikon Nonwoven, Dan-Web, M&J Airlaid Products (now Mölnlycke), Anpap, Honshu, and Campen.

Historically, in the late 1950s, Hejtl (a Finnish engineer) issued a patent on dry-forming webs from cellulose pulp. He started joint work with Danish inventor Karl Kroyer and developed a process called the Combined Fibre Distributor. Before 1980, Kroyer's systems were sold to the American Can Company and the Fort Howard Paper Co. The American Can Co. was bought by James River Corporation. Later the James River and Ford Howard paper makers were merged together into one business called Fort James, one of the biggest airlaid suppliers of its time. In 2000, Fort James Corp. was acquired by Georgia-Pacific. Moller & Jochumsen Paper Division's (M&J) connection with Kroyer began in October 1981.

In late 1986, M&J entered into an agreement with Kroyer whereby M&J would market, licence and sell airlaid plants based on the Kroyer patents. Under the agreement M&J would use Kroyer's patent rights, both existing and future, within the field of dry-forming and wet forming [28]. Since the first patent, Kroyer has described several different apparatuses to modify and improve airlaying technology. Some of these are detailed in USP 3581706 [33], 1971, USP 4014635 [34], 1977, USP 4144619 [35], 1979, USP 4494278 [36], 1985, and USP 5471712 [37], 1995.

In 1975, John Mosgaard left Kroyer and set up a business based on drum former airlaying technology (Scan-Web, renamed Dan-Web in 1981). The basic difference between the Dan-Web and Kroyer process is the means by which fibres are distributed before deposition onto the perforated screen to form a web. Kroyer's process uses agitators to distribute fibres, whilst Dan-Web uses two rotating drum formers with brush rolls inside each drum. Independently, Honshu Paper (now part of Oji Paper) developed and commercialised a third process (USP 3781150, 1973) [38] that was kept for internal use, apart from licensing to Johnson & Johnson. In this system, pulp is first broken and then defibrillated in a rotator inside a stationary cylindrical screen. The fibres pass through the screen and a vacuum pulls them onto a moving wire where the web is made. The web is then spray bonded.

4.14.11.1 *Flat-bed forming*

This type of airlaying system is mainly directed at airlaying of pulp, short cut fibres normally <10 mm in length and blends thereof. Historically, it was the work of Karl Kroyer that sparked real interest in the development of short fibre airlay technology. In one of the US patents granted to Kroyer, USP 3581706 [33], an apparatus for the production of uniform fibre webs free from fibre entanglements or lumps was explained, see Fig. 4.56.

The system comprises a housing having a perforated bottom wall, an inlet for a stream of air containing suspended fibres and a stirring device having impellers mounted for rotation at a short distance above the perforated bottom wall. In this apparatus, disintegrated fibrous

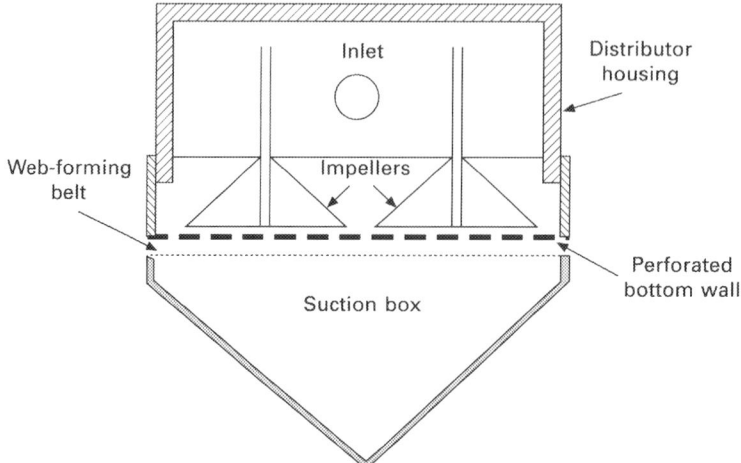

FIG. 4.56   Schematic view of airlay system in USP 3581706 [33].

materials suspended in an airstream are fed to the distributor box where they are subjected to an air current generated by a suction box located underneath the forming belt. Consequently, the individual fibres of suitable size pass through the screen and are deposited on the forming belt. The rotating impellers partly distribute the fibres uniformly across the screen and partly disintegrate the fibre entanglements or lumps carried into the housing or formed during the process. By suitably adjusting the size of the holes and the total free area of the perforated bottom wall as well as the distance between the impellers and the bottom wall, fibre passage through the bottom wall may be controlled to regulate the production rate. Other patents assigned to Kroyer describe methods of producing uniform webs using the same principle, for example, USP 4014635 [34], USP 4144619 [35] and USP 4494278 [36]. In USP 4144619 [35], it was explained that fibres pass through a vibrating screen faster than one that is static. Accordingly, an apparatus for producing a web using a vibratable screen was designed whereby a cylindrical brush roll is mounted within the fibre distribution chamber, which is in contact with the screen, and therefore, as it rotates, the screen vibrates, see Fig. 4.57.

In a typical example of this system a brush roll of 25 cm diameter, rotating at a speed of 700 rpm, was positioned within the distribution chamber to vibrate the screen. The passage of the fibres through the screen is believed to be influenced by:

- The vibration of the brush where the amplitude and frequency can be controlled.
- The rubbing of the brush roll on the screen.
- The aerodynamic effect of the brush roll.
- The positive net air pressure inside the chamber.
- The suction box.

The process was claimed to produce webs in the range of 10 to 300 g/m². Cellulosic fibres, as well as glass fibres could also be processed by this particular system. USP 4212607 [39] assigned to American Can Company described an improved apparatus for producing webs from short fibres. It was explained that the motion of the forming wire in the Kroyer machine

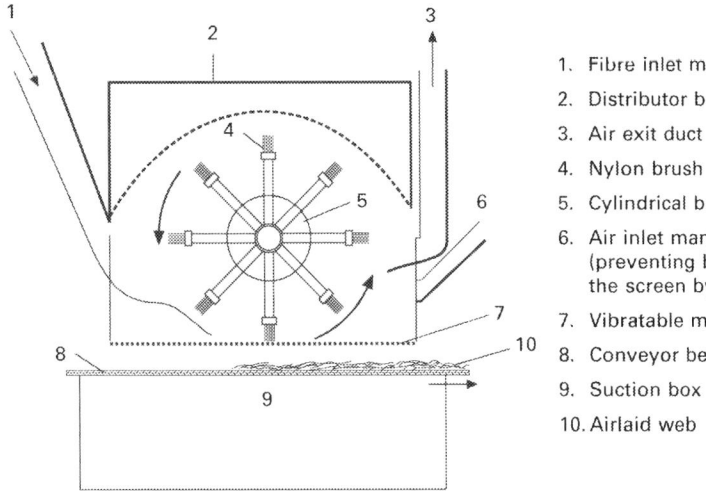

1. Fibre inlet manifold
2. Distributor box
3. Air exit duct
4. Nylon brush
5. Cylindrical brush roll
6. Air inlet manifold (preventing blockage of the screen by fibres)
7. Vibratable mesh screen
8. Conveyor belt
9. Suction box
10. Airlaid web

FIG. 4.57 Schematic view of the apparatus explained in USP 4144619 [31].

tends to cause fibres to orientate in the direction of travel, thereby forming a web that is stronger in one direction than in another. To deliver fibres to the forming wire in a random fashion, a continuously moving screen was provided for sifting fibres onto a forming wire that is moving at substantially the same velocity and in the same direction as the screen. In this way, it was argued that with no relative motion between the screen and the forming wire, fibres are not orientated primarily in the direction of motion of the forming wire but are randomly orientated. At an early stage of development, a defibrator such as a hammer mill was used to feed the finely divided pulp directly onto the forming surface. Later, a fibre distributor was placed between the defibrator and the forming surface. The purpose was both to distribute the fibre uniformly over the forming surface, and to reduce fibre clumps formed during the process so that a uniform fibrous layer free from clumps, or so-called 'fish-eyes', could be obtained.

The inclusion of a fibre distributor in the airlay machine was disclosed in USP 4494278 [31]. This modified airlaying system comprises a fibre distributor that includes at least two closely spaced rows of stirring devices, a suction box and a perforated forming wire. It was explained that this apparatus was capable of handling different fibrous materials so that a composite structure could be produced in a single step. In a further development, an adjustable screen for the fibre distribution section was described in USP 5471712 [37], which improved the quality of the products made from mixtures of fibres having different lengths at high production speed. The screen was characterised by a diagonally stretchable wire net comprising meshes in the form of parallelograms. Previously, the screen used for producing webs from cellulosic fibres had quadratic or rectangular mesh openings with dimensions of $2.5 \times 2.5 \, mm^2$. Using parallelogram-shaped openings made it possible for longer fibres (synthetics) to pass through without clumps. A typical screen of this type is shown in Fig. 4.58. The long side of the opening is preferably between 6 and 10 mm and the short side is between 1 and 4 mm.

4.14.11.2 Drum-forming technology

A diagram of a typical rotating forming drum airlaid system is shown in Fig. 4.59. Dan-Web has been one of the major developers of such machinery. The raw materials are delivered by one or more hammer mills or grinders to the forming drums which allow properly sized and distributed fibres to be deposited on a moving wire and formed into an airlaid web.

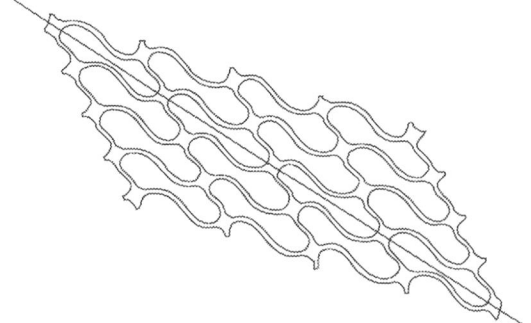

FIG. 4.58 Schematic view of the upper screen explained in USP 5471712 [37], 1995.

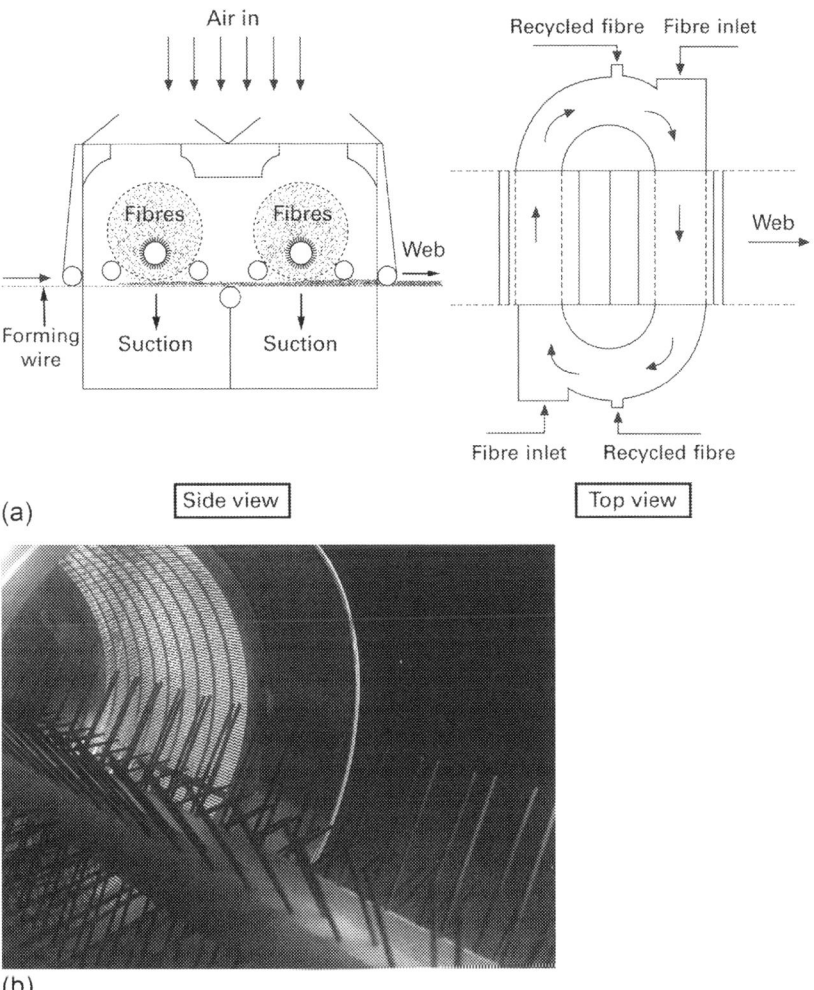

FIG. 4.59 (A) Dan-Web drum former; (B) Drum and internal brush.

Fibres that are not of the proper size are recycled for further processing and are subsequently deposited on the forming wire. As with most airlay systems of this type, a key consideration for the achievement of uniform web formation is the concentration of the air–fibre mixture, or dilution ratio. The advantages claimed for using rotating drums over agitators and flat beds are (i) greater flexibility in terms of the maximum fibre length (up to 15mm) and (ii) reduced fibre buildup in the system and a completely uniform distribution of fibres across the web.

The important feature of this particular airlaying technology is the design of the forming drum illustrated in Fig. 4.59A. It comprises two contra-rotating perforated drums situated transversely above the forming wire and connected to fixed pipes. Inside the drums and transverse to the forming 'wire' or conveyor belt is a rotating brush roll that removes fibres from the transport airstream and directs them through the perforated drums (Fig. 4.59B). The fibres are then deposited onto the wire by means of a vacuum located underneath the forming head. As with other types of pulp airlay machines, it is not uncommon to use multiple forming heads to allow higher production and the capability to produce multi-layer web structures from different blends at throughputs >400 kg/h/m and line speeds of 400 m/min or more. Although predominantly intended for pulp and short fibres, modified airlay drum formers have also been developed capable of handling longer fibres ranging from 4 to 40 mm. This makes the technology relevant to some web-forming applications that are currently dominated by carding. The main advantages of using longer fibres is the increased fabric tensile strength and lower fibre consumption for a given fabric weight. Normally, production capacity decreases in such systems as the fibre length increases. However, it is claimed that, at least on a small-scale drum former, the reduction in capacity due to the processing of longer fibres has been eliminated. The products are used for applications such as high-quality tabletop coverings (free from entanglements and lumps).

4.14.11.3 Honshu's TDS (totally dry system)

The principles of the Honshu TDS are explained in US Patents 3781150 [38] and 3,886,629 [39]. Also US Patent 3984898 [40] describes Honshu's method for producing multi-layer structures using short and long cut fibres. In Honshu's apparatus, pulp fibres from a shredding unit are delivered to a disintegrator device that is designed to produce finely separated fibres. It essentially comprises a plurality of disintegrating elements or blade runners as shown in Fig. 4.60, placed in the shaft inside a separating wall. This wall is 1.5 to 3 mm thick and is provided with openings (3–5 mm in diameter) uniformly distributed over its entire circumference. The total open area ranges from 30% to 50%. The sifting operation at the wall is related to the peripheral speed of the blades and to the diameter of the openings. For the above-specified diameters of openings, the peripheral speed of the blades is 60 to 80 m/s. In order to maintain the desired volumetric ratio of air to fibres, air intake valves are used. The opened fibres are screened through the wall and then deposited onto a perforated conveyor belt with the help of suction positioned at a distance of 150 to 300 mm from the lower end of the wall, underneath the belt. Using this approach, web weights of 20 to 200 g/m^2 and delivery speeds of 100 to 300 m/min are claimed to be achievable. A schematic illustration of this system is shown in Fig. 4.61. Using this method production of multilayer structures from short-fibre (2–5 mm) layers and long-fibre (20–50 mm) layers, simultaneously in a single stage, is claimed. Both layers are integrated by interfibre bonds at their interfaces. It is claimed that a relatively thin multi-layer structure can be produced by reducing the weight of the long fibre layer to about 5 g/m^2. This system has been mainly used in Japan.

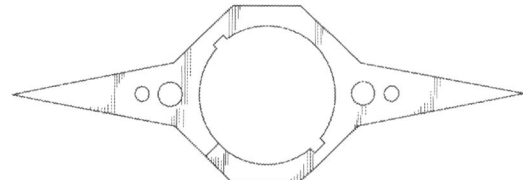

FIG. 4.60   Opening element (blade) explained in USP 3781150 [38].

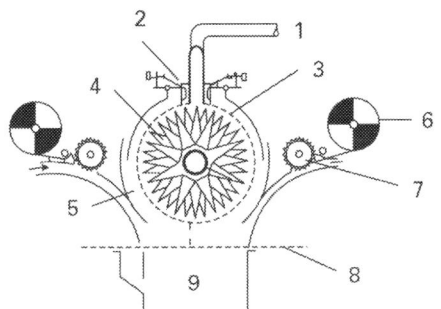

1. Wood pulp feeding
2. Air intake
3. Separating wall (dispersed fibres are screened through this wall)
4. Disintegrator blades
5. Dispersing chamber
6. Long fibre supply
7. Opening roll for long fibre
8. Conveyor belt
9. Suction

FIG. 4.61 Schematic view of Honshu system explained in USP 3781150 [38].

4.15 Developments in airlaying

Progressive developments have taken place in the airlaying industry that reflect the growing importance and versatility of the technology. Key developments include the ability to handle a wider range of raw materials, including longer natural fibres, recycled textile fibre, recycled paper and card, glass and ceramic fibres, as well as major reductions in energy and air consumption in the process.

4.15.1 Formfiber SPIKE

The flexibility of flatbed airlaid forming is exemplified by the SPIKE technology of Formfiber (Denmark). A tall rectangular chamber mounted over the conveyor belt contains rows of rapidly rotating spiked rollers, which create turbulent airflow in which the fibres are dispersed, mixed and transported. A large-fibre deposition area below facilitates uniform web formation on the conveyor. This technology is noteworthy because of the variety of materials that can be accommodated. As well as short natural and synthetic fibres, recycled fibre waste is compatible, as are fibres up to 75 mm in length. Whereas most flatbed-forming systems for pulp and short cut fibre focus on producing webs less than $100 \, \text{g/m}^2$, the Formfiber system claims an upper limit of about $5000 \, \text{g/m}^2$ with web thicknesses up to 400 mm.

4.15.2 Advance nonwoven CAFT2500

The CAFT (Carding Air-laid Fusion Technology) former from Advance Nonwoven (Denmark) launched at ITMA 2019 also utilises high-speed rotating rollers with protruding spikes within a tall distribution tower mounted over the conveyor. As such it is particularly well suited to heavier web weights ($150–15,000 \, \text{g/m}^2$) and the processing of mechanically recycled textiles and coarse natural fibres. Compatibility with fibres across a range of lengths from 1 to 100 mm is claimed. The technology has been used to make insulation fabrics from unusual raw materials such as waste paper and feathers.

4.15.3 Integrated forming and bonding (IFB)

A proprietary method referred to as IFB was developed by Metso. The system utilises an airlaying machine with the capability of simultaneously forming and bonding 100% synthetic fibres or a combination of natural and synthetic fibres. The IFB process is claimed to be particularly suited for producing high loft nonwovens and related products that demand high cross- directional tensile strength and uniform Z-directional bonding. It is claimed that a variety of natural fibres such as flax, wood fibres, cotton linters or other recycled fibres are compatible with the system. The main applications are building insulation materials, automotive parts, furniture, and construction boards.

4.15.4 Lap formair

This refers to two different types of airlay system by Cormatex (Lap Formair V and H), which produce minimum batt weights of ca. $150 \, g/m^2$. The systems enable fibre orientation in the batt to be entirely planar or have a degree of preferential Z-directional orientation. Through-air thermal bonding or needlepunching follow airlaid batt formation.

4.15.5 Star former

The Star Former (ME Consulting, Denmark) incorporates a drum-type forming head and is claimed to be capable of production rates of $1000 \, kg/h/m$. The system processes fibres of 4 to 50 mm length and a variety of raw materials are suitable including superabsorbent polymers, wood pulp and natural fibres such as flax and cotton. Basis weights of 25 to 3000 gsm [41] may be produced depending on fibre composition.

4.15.6 Campen drum-forming system

An energy efficient, versatile airlaying system intended for short fibres in the length range of 1.5 to 18 mm, and web weights from $10 \, g/m^2$ upwards has recently been developed by Campen (Denmark). The system operates with perforated drums and internal agitators, and is designed to process pulp, short cut regenerated cellulosic and synthetic fibres, as well as inorganic fibres and mechanically recycled paper materials. The working width is between 0.6 m and 6 m, with output speeds up to 250 m/min.

4.15.7 Combined airlay and hydroentanglement technology

Mainly in response to a significant growth in the use of single-use wipes as well as other hygiene fabrics, it has been recognised that integrated multi-layer composite nonwoven fabrics combining airlaid webs with other web types could bring improved product performance, particularly in respect of modulating liquid absorption and fabric wet strength. Carded webs and airlaid webs composed of wood pulp produced on the same line are combined and then hydroentangled to produce a multi-layer product.

Two claimed benefits of this approach are the capability to replace viscose rayon fibre with wood pulp, which is about one-third cheaper, whilst still retaining the required physical properties of the product and the potential to increase production capacity. It is claimed that to ensure good bonding and minimal loss of wood pulp, the carded web component must be pre-bonded, for example, by through-air thermal bonding and the pulp layer hydroentangled on a flat-wire section with 8 to 10 injectors. The composite is integrated by hydroentangling the individual layers. Even with these precautions, 4% to 5% of the wood pulp can be lost through the hydroentanglement machine conveyor and is removed and discarded from the water during recycling. The capital investment involved in such a process is comparatively high and is less flexible than a carding line. Therefore, in practice an airlaying machine of this type may need to be dedicated to similar product types [42,43]. Hydroentangled 100% wood pulp fabrics for use in distribution or absorption layers, or certain industrial applications have also been developed [44]. Airlaid webs are also combined with spunlaid webs prior to bonding to make high wet strength, water absorbent fabrics.

4.15.8 Multi-layer composite airlaid webs for diapers

Historically, the 'super-site' airlaying concept demonstrated how the manufacture of disposable diapers could be simplified by assembling various functional layers by means of individual airlay-forming heads operating over a common conveyor belt. In this way, the acquisition layer, absorbent core and retaining layers could be assembled before bonding. As a result, the conventional converting processes of unwinding, layering and joining of multiple preformed fabrics containing wood pulp, SAP and other components could potentially be simplified. The ability to make very lightweight layers within the multilayer structure was a further advantage.

4.15.9 3D web preforms and moulds

A method of airlaid web formation using an intensive fibre opening machine has been developed which features a large collection area to promote the formation of a uniform web as well as reduced airflow velocities (see Fig. 4.62). This process may be used to form preformed, contoured or profiled webs, within certain process boundaries [45]. Technically, shaped or preformed 3D-moulded webs can be produced directly on many airlaying systems using a three-dimensional, contoured forming surface and appropriate suction. The basic approach using a rotary conveyor has been described by Eldim (USA) and is relevant to other fibre or filament deposition processes.

4.15.10 Roller draft airlaying system

Lin et al. designed an experimental airlaying system intended to improve the web weight uniformity and fibre randomisation using mechanical attenuation and airflow [46]. The approach, referred to as 'roller drafting airlaying', basically consists of a roller drafting device, a

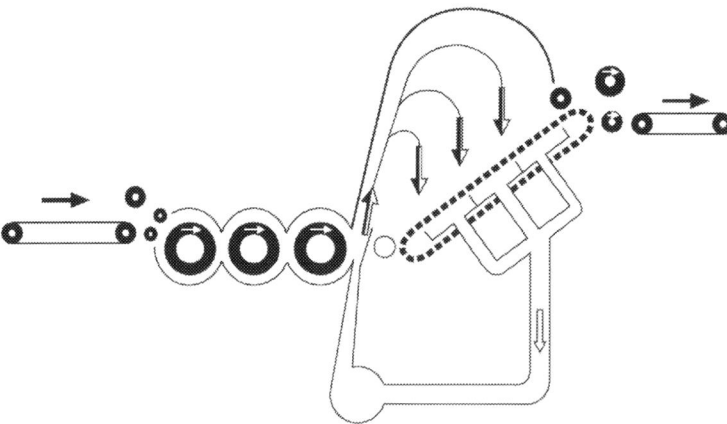

FIG. 4.62   Basic principle of the aerodynamic web formation process explained by Paschen [45].

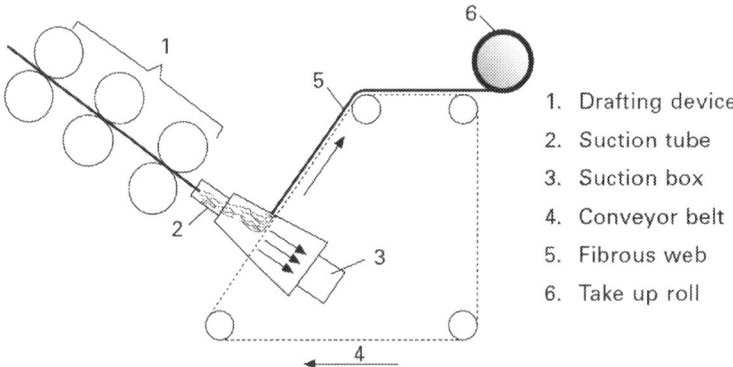

FIG. 4.63   Schematic diagram of roller drafting system [46].

1. Drafting device
2. Suction tube
3. Suction box
4. Conveyor belt
5. Fibrous web
6. Take up roll

web collection unit, a conveyor and suction box as shown in Fig. 4.63. A sliver is drafted by three pairs of rollers and converted into loose fibres. The airflow from the suction device transports the loose fibres into the suction tube at high speed and transports them towards the main chamber where their velocity is immediately reduced due to the geometry of the chamber. Finally, fibres are collected on the conveyor belt and form a web with random fibre orientations.

4.15.11   Inverted airlaying systems

Combining webs from multiple web formation systems as part of an integrated manufacturing process has also been demonstrated (Fig. 4.64). Historically, significant work on inverted airlaying has been done, including by Dan-Web.

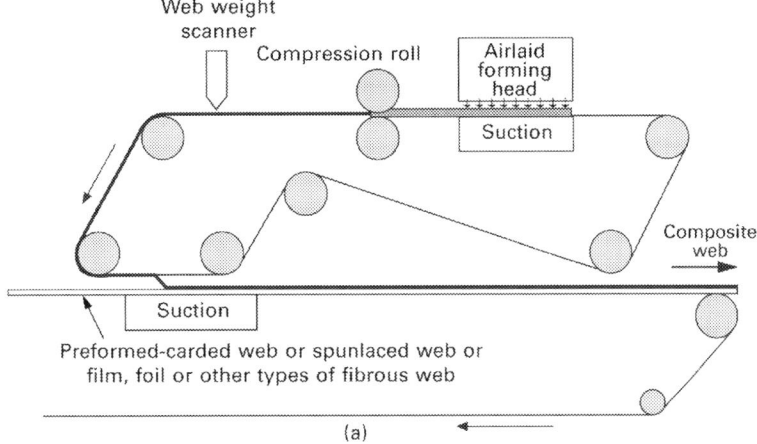

(a)

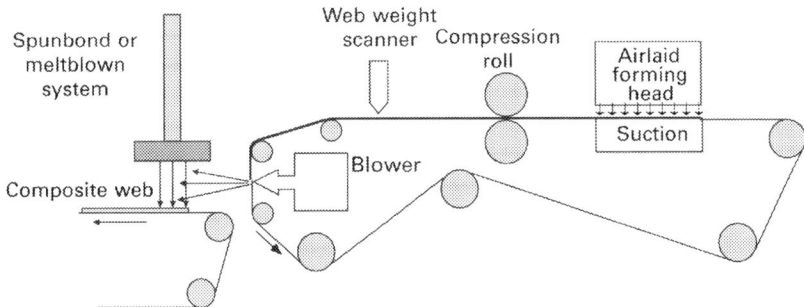

Using this technique fibres are blown directly into a spunbond or meltblown fibre stream.
(b)

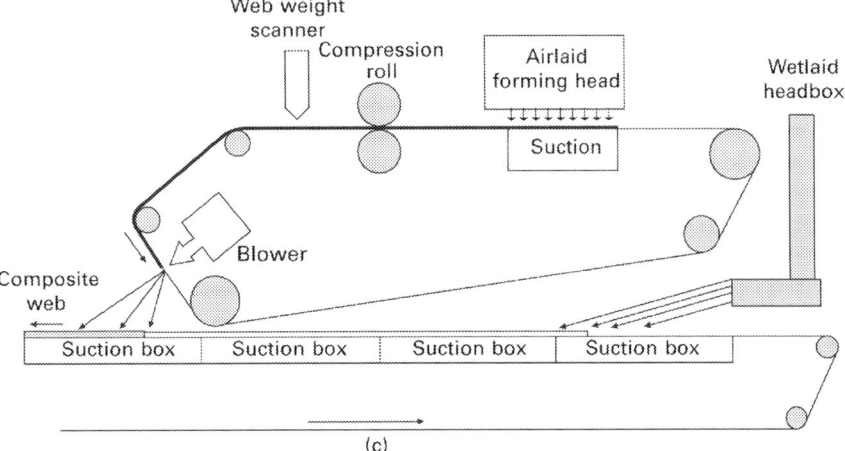

(c)

FIG. 4.64 Combination of airlaid with other web formation methods; (A) airlaid with a carded web, fabric, film, foil, or other structure; (B) airlaid with spunbond or meltblown; (C) airlaid with wetlaid.

4.16 Airflow and fibre dynamics in airlaying

In this section the importance of the airflow and fibre dynamics in the transport chamber of airlaying machines is discussed. Some manufacturers make use of turbulent flow in the transport chamber, whilst others consider turbulent flow to be detrimental to the process and seek to achieve a laminar flow. Either way, it is important to understand how the airflow affects the uniformity and fibre orientation in an airlaid web. Once fibre has been dispersed into the airflow, the air-fibre mixture generally passes through a duct or transport chamber before the fibres are deposited on the conveyor or screen.

4.16.1 Airlaying of textile fibres

USP 4475271 [21] explains the preference for turbulent airflow in the transport channel of an airlaying machine. The average airflow velocity is higher than the fibre velocity where fibres are under tension during their travel towards the landing area. Such turbulent flow, except for the narrow boundary edges at the sides of the channel, produces a relatively flat velocity profile, which encourages the formation of a uniform web across the width of the machine. In contrast, laminar flow produces a more curved velocity profile, which tends to deposit more fibres in the centre of the web than at the sides. An airlaying system using a straight and laminar airflow profile in the transport chamber is described in WO 9720976 [47].

The speed of airflow was less than 95% of the disperser roller. It is claimed that otherwise the airstream tends to blow fibre off the disperser roll, undermining the intended effect of centrifugally doffing the fibre. If the fibre is blown off the roller, it tends to come off in clumps and creates an unstable flow (i.e. more turbulence, larger eddies and vortices). The straightened airstream passing over the surface of the roller resulted in a smoother, more gentle delivery of fibres from the teeth on the roller to the straightened/laminar airstream. In a study by Pourmohammadi et al. [48], on the fibre trajectory in the transport channel of a small-scale airlaying system (of the Fehrer K12 type), the airflow in the transport channel was estimated to be turbulent with a relatively flat velocity profile across the width of the channel. The fibre trajectory in the transport channel was then theoretically modelled. The theoretical results were then compared with the experimental observations obtained by high-speed photography and reasonable agreements were obtained. This may suggest that the assumption of turbulent airflow is relevant. In this study, the effect of machine parameters on the fibre trajectory was also reported [48].

4.16.2 Airlaying of short fibres and pulp

The airflow dynamics in the transport chamber of an experimental pulp airlaying system have been studied using a Laser Doppler Anemometer [49]. The velocity varies along both the length and height of the transport chamber and the velocity variation is markedly affected and increased by rotation of the rotors in the fibre dispersing section above the transport chamber. Generally, the velocity increased from the top to the bottom of the transport chamber. In another attempt to improve understanding of the airflow behaviour in such airlaying systems, Bradean et al. used computational fluid dynamics (CFD) to model the effect of blade rotation on the airflow pattern in the region between the top grid and the conveyor belt. The results showed that the airflow patterns in this region are very dependent on the inlet velocity profile when the grid size is large and is independent of the geometry below the belt when the belt size is small. By decreasing the grid size, the flow between the grid and the belt becomes more uniform, but of course in practice, any decrease in the grid size is limited by the need to ensure fibre penetration. Therefore, depending on the rotational speed of the blades and the grid size, the flow in the region between the grid and the conveyor belt is either steady and uniformly downwards or unsteady and three-dimensional, in which case the rotation of the blades is the most important machine parameter affecting the airflow and fibre dynamics [50].

One of the limiting factors in airlaying is the tendency of fibres to entangle after they have been dispersed in air. The probability of fibre entanglement increases as the fibre length increases, and this can negatively impact web uniformity. Therefore, a high dilution factor, or a low concentration of fibre in air, is needed to keep fibres apart. A basic approach to address this is to apply the fibre stacking theory, first discussed by Wood [51]. Essentially, the amount of air required for each fibre to be conveyed without being entangled with its neighbour is equal to the volume of a sphere having a diameter equal to the fibre length. Wood's Law therefore calculates the volume of air required to encapsulate each fibre to enable fibre separation during airlaying. Based on this spherical volume theory, the required volume of air to convey the fibre depends on both, fibre dimensions and the number of fibres passing through the airlaying system per unit time. The latter is of course dictated by the amount of fibre being fed into the system per unit time. Based on Wood's Law, the air volume flow rate (Q) can be determined as follows [51]:

$$Q = 1.5 \times 10^6 \times \pi \times \frac{P \times l^2}{d} \ (\text{m}^3/\text{h})$$

where,

P = production rate (kg/h)
l = fibre length (m)
d = fibre linear density (denier)

Clearly, if the volume flow rate of air in the machine (Q_1) is lower than the calculated value of Q, then there will be greater fibre accumulation. The ratio, Q/Q_1 commonly referred to as the concentration factor (C) is therefore a useful means of evaluating the dilution of fibres in air, and the degree to which fibre separation during airlaying is possible. Practically, operating with high concentration factors may be necessary to achieve economically acceptable production rates, and for the same reason, the sensitivity of Q to fibre length means that most airlaying systems tend to favour short fibres.

The concentration is generally in the range of 10^4 fibres per litre. This gives a crowding factor, N, of less than 1.0. The crowding factor is defined as the number of fibres in a spherical volume with a diameter equal to the fibre length. A crowding factor value of <1 signifies that fibres should come into only occasional contact during the process [52]. The same principle has been applied in the wetlaying of pulp and short fibre, where fibre-to-fibre interactions in the liquid suspension are minimised by a low concentration of fibre in the flow.

Another important consideration particularly for airlaying of short fibres and pulp, is the ability of the material to pass through the slots or holes in the screen, prior to web formation. This affects the residence time of fibres in the machine and the rate at which the web thickness can be built up by a single airlaying head. Kumar et al. [53] discussed the main parameters influencing the passage of fibres through slots in pulp screening, and the basic principles are also highly relevant to the operation of airlaid web formers processing pulp and short fibre. The fibre passage efficiency increases with increasing velocity through the slot or aperture (due to the suction applied underneath) as well as the slot width. Fibre passage also increases with an increase in fibre flexibility and a decrease in fibre length.

4.16.2.1 Fibre dynamics in airlaying

A comprehensive study of the fibre dynamics in the transport chamber of an experimental Kroyer type airlaying system [54] processing short fibres has been conducted using high-speed photography. It was established that fibres move intensively in three dimensions at the top of the transport chamber and as they travel towards the landing area their motion becomes steadier and more uniform. Fibre landing behaviour was also studied by analysing sequential photographic images because landing behaviour has a direct effect on structure and therefore, the physical properties of the formed web. This experimental study (which used 5 mm length, uncrimped viscose fibre) showed that about 38% of fibres landed on end before falling down flat onto the conveyor belt (hit-fall configuration), about 30% landed flat, along their full length (flat landing configuration), and about 16% were curved in the middle and land on one side (mid curved land) [54]. A mathematical model has been developed of the fluid and fibre dynamics in the transport chamber of the same system [55]. The airflow was considered to be turbulent because of the magnitude of the air velocity and the geometry of the machine. The fibres were modelled as ellipsoids and the trajectories of the centre of mass of the fibres were obtained. Reasonable agreement was found between theoretical predictions and experimental findings.

4.17 Bonding of airlaid webs

Different bonding technologies are utilised to consolidate airlaid webs including latex or chemical bonding, thermal bonding, multi-bonding and mechanical bonding (mainly needlepunching and hydroentanglement). The choice of system depends on fibre dimensions (principally fibre length) in the web, the proportion of fluff pulp (if present), web weight, whether or not the web contains particulate materials such as SAP and the final fabric properties required [56,57]. In addition to hydrogen-bonded airlaid fabrics (HBAL) made by calendering webs containing fluff pulp, other options include LBAL/SBAL/BBAL, TBAL, and MBAL.

4.17.1 Latex bonding airlaying (LBAL)

Also referred to as SBAL or BBAL, this refers to spray or binder bonding of airlaid webs, most commonly containing fluff pulp. Latex bonding was one of the first bonding methods used by the airlaying industry. Twenty years ago, about 85% of pulp fibre based airlaid products were latex bonded, but since that time the importance of thermal and combined (multi-) bonding systems has increased, as well as bonding using other techniques such as hydroentanglement. The latex binders used are normally synthetic copolymers produced via emulsion polymerisation to form a stable emulsion or latex. A typical binder solution is sprayed onto the airlaid web, which is then cured in an oven. Latex bonded fabrics have a cloth-like appearance and feel and can be used in place of conventional tissue and synthetic woven fabrics. The major problem in latex bonding is the ability of the binder to penetrate thick structures. Since the binder is sprayed onto the surface of a web and is transported into the interior by the water carrier and the applied vacuum, an increase in web thickness and web weight reduces the ability of the binder to reach the centre of the web. Generally, the lower the glass transition temperature (T_g) the softer is the binder. One of the largest markets for LBAL has been in the feminine hygiene absorbent core and wiping sectors.

4.17.2 Thermal bonding airlaying (TBAL)

Thermal bonding involves forming a homogeneous airlaid web of the base fibre (e.g. fluff pulp) and the bonding component (i.e. a bicomponent or homofil thermoplastic fibre or particle), heating the web to the softening temperature of the fusible bonding element and then cooling the web. As the heated fibres or particles start to melt, they become adhesive in nature and create bonding points in the web. The thermal bonding process offers web design flexibility, since the web can be embossed with different patterns during thermal calender bonding. The utilisation of bicomponent fibres can produce a high-loft web with excellent bonding in the X, Y, and Z directions of the web [57]. High loft, thick airlaid fabric structures provide increased void areas within the web and consequently increase the liquid holding capacity. Thermally bonded airlaid products are generally used in absorbent core articles and medical disposable products.

This type of bonding is advantageous because it saves energy, there is less environmental pollution and the recycling of webs and fabric is possible. The main limitations of thermally bonded airlaid webs are, first, the dust generated during high-speed production, which necessitates frequent line stoppages for cleaning. The second concern is the low or irregular tensile strength of the final product particularly when low levels of binder fibre are used; therefore, the industry is being forced to modify and optimise the process to overcome these apparent problems.

4.17.3 Multi-bonding airlaying (MBAL)

The response of the industry to the challenge for products with lower dust levels (due to fluff pulp drop out) and higher strength has been the development of multi-bonded airlaid webs. In simple terms, MBAL is a combination of thermal and latex bonding technology. It produces a finished web that has high loft, low density, exceptional fluid penetration, high absorbent capacity, good tensile strength, a soft cloth-like handle and a very low lint or dust level.

4.17.4 Mechanical bonding

Needlepunching is not compatible with very short cut fibres of <10 mm in length, or fluff pulp and is used to bond heavier weight airlaid webs containing longer fibres. It is one of the oldest methods of bonding nonwoven fabrics. The principle of needling, or needlepunching, consists of mechanically interlocking fibres by oscillating barbed needles through the airlaid batt to induce entanglement. Typically, needling is used to consolidate the fibrous structure, to densify it, and/or to control the porosity. High-speed needle looms are capable of 3000 rpm, although for bonding airlaid webs, lower speed looms are capable of balancing production. Needlepunched airlaid products can be made from a large variety of textile-length fibres including Basofil, Miraflex [58], PET, PP, hemp, jute, flax and kenaf for applications such as protective apparel, filtration, geosynthetics, automotive panel components (e.g. car doors) and shoe linings. It is also a convenient means of bonding airlaid webs containing mechanically recycled textile fibre waste.

Hydroentanglement or spunlacing of lightweight airlaid webs containing fluff pulp, short cut fibre or blends of both, is important in the production of mainly single-use hygiene fabrics, typically wipes, and various systems are available as turnkey installations [59,60]. The airlaid-hydroentangling process gives a low raw material cost product, which it is claimed, can be tailor-made to suit any requirements. Composite fabric structures can also be produced from multiple webs using hydroentanglement for applications in the medical and sanitary, healthcare and cosmetics (e.g. applicators) sectors.

4.18 Physical properties and practical applications of airlaid fabrics

Airlaid webs are characterised by a more random fibre orientation than a carded web, but they are not truly isotropic particularly in the Z-direction. Theoretically, a random web would have an isotropic structure having the same properties in all directions. The web structure is also modified by the bonding process, so the properties of the resulting fabric are not entirely isotropic. The cross-section of the web as it builds up in the fibre landing area during pulp and short fibre airlaying tends to be wedge shaped [61]. Many of the incoming fibres continually land on the tapered end because the suction is stronger at this side and fewer fibres obscure the openings in the belt. The angle of the taper depends on the level of suction. Incoming fibres tend to fall against the taper wall, and thus, their orientation in the structure is partly governed by the angle of the taper. The web density is also influenced by the suction level, and increasing the suction tends to increase the web density. The property differences between fabrics produced from airlaid and carded webs, as well as composite structures have been compared [62].

Generally, airlaid fabrics exhibit lower anisotropy than parallel-laid carded webs. When made from short cut fibres, wood pulp or blends thereof, airlaid webs are usually more voluminous than wetlaid webs.

The physical properties of airlaid nonwoven fabrics are highly dependent on the physical properties of the constituent fibres, the blend ratio, the web geometry and the bonding process. The general properties of airlaid fabrics are:

- Near isotropic structure.
- High loft (if required).
- High porosity (95%–>99%).
- High absorbency and wicking rate (depending on fibre content).
- Soft handle (compared to wetlaids).
- Adequate tensile strength.
- Good resiliency (compression recovery).
- High thermal resistance.

Depending on the fibre composition, airlaid fabrics are claimed to have higher liquid absorption and faster liquid transfer (acquisition and distribution) compared to other nonwoven fabrics particularly those produced from carded webs [63]. Using multiple airlaying heads, multi-layer fabrics can be assembled to manage various liquid management functions. One such example is a five-layer composite fabric, originally developed for use in nappies [64]. The use of airlaid webs in the core of such liquid absorbent products improves the performance and allows a thinner product to be manufactured, which is advantageous in sanitary and incontinence applications. The acquisition rate depends on the fibre type and fibre dimensions (length, fineness, crimp level and cross-sectional shape). The surface chemistry is also important as this influences surface wetting during the introduction of liquid. Pulp and short fibre airlaid webs can produce more rigid fabric structures with lower air permeability than fabrics produced from carded webs.

4.18.1 Applications and markets for airlaid products

Depending on the fibre composition and the bonding method, a wide variety of airlaid products are manufactured. For heavyweight airlaid fabrics, there are important applications in long-life products such as high-loft waddings for the clothing, mattress and furniture industries, sound and thermal insulation, geosynthetics and roofing felts, filters, barrier materials, wall and floor coverings, moulded products and pre-formed automotive components. Airlaid fabrics are also produced from mechanically recycled fibre waste and bast fibres to make fabrics for the automotive industry. For light-weight, single-use airlaid fabrics, major applications include wipes (domestic and industrial), absorbent cores, acquisition and distribution layers for hygiene products (including femcare), as well as medical fabrics [65]. They are also widely used to produce food packaging, napkins, and tablecloths. Airlaid products can also be classified in terms of the different bonding routes used to make them:

- Chemical bonding: napkins, tablecloths and wipes.
- Thermal bonding: diaper components (absorbent core and the acquisition and distribution layer), feminine hygiene/incontinence products, food packaging, sound and thermal insulation.
- Hydroentangling: wet and dry wipes for consumer and industrial markets, medical products (including single-use gowns, curtains, woundcare dressings, bed sheets), food packaging and filter media.
- Needlepunching: interlinings and shoe linings, waddings, medical and hygiene products, geosynthetics and roofing felts, insulation felts, automotive components, filter media and industrial wipes.

The increase in airlaid capacity for various product applications during the early 2000s has been well documented [66]. Short fibre and pulp airlaid fabrics have become an important alternative to carded-hydroentangled products, particularly for single-use wipes, due to a softer, spongier handle and lower price [67]. The market for both wet and dry airlaid wipes has grown across multiple market segments [68]. Airlaid-hydroentangled fabrics consisting of >70% wood pulp blended with a small proportion (<30%) of short cut (4–8 mm) regenerated cellulose fibres enable a good balance of wet strength and dispersibility for flushable (dispersible) wet wipe substrates. Food packaging is another important market for airlaid fabrics made of hydrophilic fibres, including absorbent 'soaker' pads for meat, seafood, and fruit. In the medical sector, similar fabrics are used as absorbent cores in wound dressings, surgery pads, body bags, and medical packaging kits.

4.19 Direct feed batt formation

In certain applications, for example in the production of batts for quilting, it is possible to eliminate the need for web formation by carding and cross-lapping or airlaying altogether, and a highly compact direct batt-forming process can be adopted. Provided the fibre is sufficiently pre-opened, it is possible to form a comparatively uniform batt using a carefully constructed volumetric feed hopper or chute feed. The batt emerging from the feed system can then be directly quilted or otherwise bonded to make a finished fabric. The minimum batt weight is limited to about $250 \, g/m^2$ and the width is from about 1.6–3.2 m. The maximum fibre length that can be processed in this way is about 75 mm and such systems are claimed to be capable of up to 400 kg/h. One example is the chute feed system developed by Masias Maquinaria, which enables direct batt formation from waste materials, including blends of mechanically recycled fibre and foam particles.

4.20 Additive drylaid web formation

Generally, like most other nonwoven production systems, drylaid web formation is usually aimed at large-scale, high-speed manufacturing of webs that are uniform in terms of weight per unit area.

Recent developments have created more versatility in drylaid web-forming systems, which include potential for mass customisation of nonwoven products as well as reduced waste in subsequent bonding and converting processes. One example in the drylaid sector is Dilo Group's 3D-Lofter system, which allows highly localised dosing of fibres onto a preformed nonwoven web, batt, or pre-needled fabric, across the width of the machine. In the 3D-lofter, a series of individually programmable aerodynamic web-forming modules mounted on a gantry above a conveyor, allow narrow tracks of fibre to be laid down, across the width of the machine. Each module is individually fed with preformed slivers stored in cans. This enables precise dose weights of fibre to be additively placed at localised positions across the width of a moving preformed web, batt, or pre-needled fabric, carried on the conveyor below. A nonwoven can therefore be made with locally customised thickness and weight profiles, prior to being needlepunched or otherwise bonded, such that subsequent converting, cutting, or forming operations produce less waste, and can be more efficient. This additive manufacturing approach can be harnessed to produce shaped preforms for automotive interior components, composites, and other deep moulded 3D products, where slightly thicker areas can help to compensate for stretching of the fabric in the mould to avoid quality defects.

References

[1] J. Lunenschloss, W. Albrecht (Eds.), Non-woven Bonded Fabrics, Halstead Press, New York, 1985. translator, Janet Hock; translation editor, David Sharp.

[2] www.ppic.org.uk.

[3] www.nonwovens.com.

[4] D. Sens, Fluff fibre morphology, in: Paper Presented in Marketing Technology Service, MTS, Conference, Ottawa, Canada, 2002.

[5] E. Gammelgard, Fibres for airlaid thermal bonding, Nonwovens World 24 (1997) 81–85.

[6] R. Heath, P. Akers, High performance airlaid fabrics containing SA fibres, Nonwovens World 5 (1996) 66–69.

[7] C. Kellner, Art of Preparing Short Fibres for Spinning and Other Purposes, US patent 480588, 1892.

[8] H.H. Langdon, Machine for Forming Random Fibre Webs, US patent 3512218, 1970.

[9] D.E. Wood, Apparatus for Manufacturing Nonwoven Textile Articles, US patent 3535187, 1970.

[10] A.P. Farrington, Method of Forming Webs and Apparatus Therefore, US patent 3740797, 1973.

[11] A.P. Ruffo, P.K. Goyal, Web Forming Process, US patent 3768118, 1973.

[12] E.G. Lovgren, Web Forming Apparatus, US patent 3772739, 1973.

[13] A.P. Ruffo, et al., Nonwoven Fabric, US patent 4018646, 1977.

[14] F.M. Buresh, Machine and Method for Forming Fibre Webs, US patent 2451915, 1948.

[15] F.M. Buresh, et al., Fibre Web Forming Machine, US patent 2700188, 1955.

[16] H.H. Langdon, Machine for Forming Random Fibre Web, US patent 2890497, 1959.

[17] D.E. Wood, Machine for Forming Random Fibre Webs, US patent 3768119, 1973.

[18] D.E. Wood, Machine for Forming Fiber Webs, US patent 3972092, 1976.

[19] I. Pivko, Nonwovens Industry, 1991.

[20] A.M. Reader, Preparation of 100% Melamine Fibre Web Via Air-Laying Technology, Book of papers, INDA-Tec, Massachusetts, USA, 1997.

[21] E.G. Lovgren, et al., Process and Apparatus for Producing Uniform Fibrous Web at High Rate of Speed, US Patent 4475271, 1984. Chicopee.

[22] E. Fehrer, R. Fehrer, Improvement in or Relating to Apparatus for the Manufacture of a Hair or Fiber Web, GB patent 1090827, 1967.

[23] H. Jakob, Experience with random web technology, Melliand Textilberichte 70 (3) (1989) E76.

[24] E. Fehrer, Apparatus for Making a Fibrous Web, US patent 4583267, 1986.

[25] Lennox-Kerr, High Loft Offers More for Less, Nonwoven Report International, 1998, p. 21.

[26] H. Jakob, Aerodynamic web-forming and needle-punching, Text. Technol. Int. (1996) 79–80.

[27] P. Poillet, Text. World 150 (4) (2000) 27–28.

[28] H. Graute, Carding Machine and Process for Producing an Aerodynamic Card Web, US patent 5839166, Spinnbau, 1998.

[29] H. Graute, Carding Apparatus, US patent 5007137, Hergeth Hollingsworth GmbH, 1991.

[30] S. Bernhardt, Nonwovens World, 2001, p. 79 (June-July).

[31] J. DuPont, Method and an Installation for Forming a Fiber Web by the Airlay Technique, US patent 6195845, 2001. Thibeau.

[32] J. Kleppe, Air-laid: care and feeding of a growing market, Nonwovens World 22 (1990) 27–29.

[33] T.B. Rasmussen, Apparatus for Uniformly Distributing a Disintegrated Fibrous Material on a Fibre Layer Forming Surface, US patent 3581706, 1971.

[34] K.K.K. Kroyer, Apparatus for the Deposition of a Uniform Layer of Dry Fibres on a Foraminous Forming Surface, US patent 4014635, 1977.

[35] D.G.W. White, Dry-Laying a Web of Particulate or Fibrous Material, US patent 4144619, 1979.

[36] K.K.K. Kroyer, et al., Apparatus for the Production of a Fibrous Web, US patent 4494278, 1985.

[37] K.K.K. Kroyer, Adjustable Screen for a Distribution for Making a Sheet-Formed Fibrous Product, US patent 5471712, 1995.

[38] H. Matsumura, Apparatus for Producing Multilayer Fibrous Structures, US patent 3781150, 1973.

[39] S. Nakai, H. Matsumura, Apparatus for Producing Fibrous Mats, US patent 3886629, 1975.

[40] H. Matsumura, et al., Multilayer fibrous structures, US patent 3984898, 1976.

[41] J. Westphal, New air lay system for longer fibres, in: Paper Presented in Marketing Technology Service, MTS, Conference, Ottawa, Canada, 2002.

[42] J. Dobel, The future of airlaid forming, Nonwovens World (2002) 53 (June-July).

[43] D. Feroe, Two layer air-lace technology, in: Paper Presented in Marketing Technology Service, MTS, Conference, Ottawa, Canada, 2002.

[44] H.S. Jensen, Moving Air-Laid Forward-Hydroentangling 100% Woodpulp and M&J Novel Highloft Synthetics Technology, INSIGHT, 2002.

[45] A. Paschen, B. Wulfhorst, Aerodynamic web formation for the creation of new nonwoven structures, Tech. Text. 44 (2001) 13–16.

[46] F.-J. Lin, I.-S. Tsai, Configuration of PET fibre arrangement in roller drafting air-laid webs, Text. Res. J. 71 (1) (2001) 75–80.

[47] P.O. Staples, et al., Feeding Carded Fibre to an Airlay, WO patent 9720976, 1997.

[48] A. Pourmohammadi, Fibre Dynamics in Air-Laid Nonwoven Process, PhD thesis, University of Leeds, 1998.

[49] A. Pourmohammadi, et al., 'A study of the airflow and fibre dynamics in the transport chamber of a sifting air-laying system' part I, Int. Nonwovens J. (2000) 31–34 (Sept.).

[50] R. Bradean, et al., Modelling of an aerodynamic technique of forming web structures in textile industries: I. turbulent air flow past filters, Math. Eng. Ind. 8 (2) (2001) 137–160.

[51] D.E. Wood, Air-laid low density nonwovens, in: G.E. Cusick (Ed.), Nonwoven Conference Papers, Manchester UMIST, 1980.

[52] R.J. Kerekes, C.J. Schell, Characterization of fibre flocculation regimes by a crowding factor, J. Pulp Pap. Sci. 18 (1) (1992) 32–38.

[53] A. Kumar, W. Robert, J. Richard, Factors controlling the passage of fibres through slots, TAPPI J. 81 (5) (1998) 247–254.

[54] A. Pourmohammadi, et al., A study of the airflow and fibre dynamics in the transport chamber of a sifting air-laying system' part II, Int. Nonwovens J. os-9 (3) (2000) 22–26.

[55] D.B. Ingham, et al., Mathematical and computational modelling of fluid and fibre dynamics in a sifting air-laying process, Int. J. Appl. Mech. Eng. 6 (2) (2001) 287–306.

[56] P. Mango, Low density latex bonded airlaid, Nonwovens World (1998) 54–61 (Spring).

[57] J. Westphal, Bonding options for airlaid webs, Nonwovens World (1997) 63–69 (Spring).

[58] C.M. Kenney, et al., New-glass fibre geometry—a study of nonwovens processability, TAPPI J. 80 (3) (1997) 169–177.

[59] M. Ruckert, Fleissner AquaJet needling process for nonwovens in the technical and medical sphere, AVR Allg. Vliesst. Rep. 27 (3) (1999) 11.

[60] ICBT Perfojet, Nonwovens Ind. 30 (5) (1999) 88.

[61] A. Pourmohammadi, et al., Structurally Engineered Air Laid Webs, INDA-Tec, Atlanta, Georgia, USA, 1999.

[62] E.A. Vaughn, C.W. Jarvis, M.O. Arena, 13th Annual Technical Symposium, Boston, 1985, pp. 267–302.

[63] E.G. Gammelgaard, What are the Future Prospects for Carded, Thermal Bonded and Airlaid Nonwovens?, Nonwovens Report International, 1998, pp. 24–29.

[64] H.S. Jensen, The super site for ultra thin air-laid diaper production, in: Presented at Insight, 1999.

[65] J.T. Conley, Airlaid markets—the next wave, Nonwovens World (2002) 75–79 (June-July).

[66] B. Stevens, History I, in: Paper Presented in Marketing Technology Service, MTS, Conference, Ottawa, Canada, 2002.

[67] K. Bitz, Air laid: what's next? Nonwovens Ind. (2002) 30 (October).

[68] P. Hanna, Key Factors Driving Airlaid Demand, Supply and Profitability, INTC, 2002.

CHAPTER

5

Wetlaid web formation

A.A. *Koukoulas*

A2K Consultants LLC, Savannah, GA, United States

5.1 Introduction

Wetlaid nonwovens are engineered paper-like materials. Like paper, wetlaid nonwovens are made from a suspension of fibres, which is dewatered to produce a fibrous mat. However, unlike paper, wetlaid nonwovens can contain long fibres (up to 40mm) and achieve their structural integrity by physical, thermal, and/or chemical means other than hydrogen bonding [1]. Wetlaid nonwovens can also contain wood pulp, other natural fibres, pigments, colourants, thermoset resins, functional additives, and coatings that provide or enhance product attributes. As a result, the process of wetlaid nonwoven manufacturing yields a range of versatile products that are used in a variety of industrial applications as well as durable and disposable consumer goods.

Although considerable technical advances have been made that are specific to wetlaid nonwovens, the production of wetlaid nonwovens is largely based on scientific and engineering principles developed by the paper industry. Many of the unit operations associated with paper manufacturing are found in wetlaid nonwovens production. However, the man-made fibres used in wetlaid nonwovens have physical properties that are very different than the natural fibres used in papermaking. As a result, the production of wetlaid nonwovens relies on processes and equipment that are uniquely designed to compensate for these differences.

5.1.1 History of papermaking

The origins of papermaking can be traced back to 105 CE and Cai Lan (or Ts'ai Lun), who is attributed as being the first to develop the art of making hand-made papers [2]. Ts'ai Lun made his paper from a combination of bark from hemp, as well as waste fibres from discarded cloth and fishing nets. These materials were macerated with a mortar and wooden pestle to soften and liberate the fibres, and were then added to water to form a fibrous suspension. Following gentle mechanically dewatering across a screen made from parallel bamboo strips

FIG. 5.1 The jiaozi, developed under the Chinese Song Dynasty (969–1279 CE), was the first promissory note made from paper.

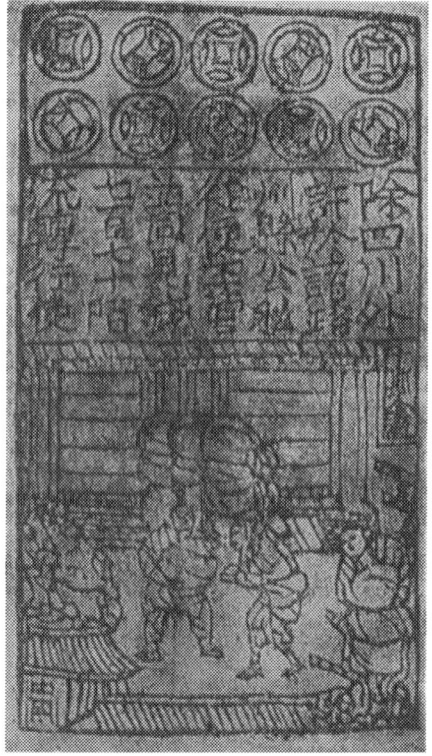

or reeds woven into a rollable mat, a fibrous web was formed and air-dried to produce a paper product. The result was a much-improved and lower-cost alternative to other writing substrates, specifically, bamboo strips, and silk fabrics (Fig. 5.1).

Hand-made papermaking would prevail until the turn of the 19th century in addition to the development of the flat-wire papermachine by Henry Fourdrinier, a machine that Herman Melville called 'a miracle of inscrutable intricacy', and the cylinder mould machine by John Dickenson [3]. Both machines automated the process of papermaking, reducing labour costs and making paper more available.

The flat-wire machine or eponymous Fourdrinier machine (Fig. 5.2) took an aqueous suspension of pulp, kept in a state of agitation, onto an endless wire cloth and between two deckles [4]. Once on the wire, the fibrous web would begin to dewater, forming a wet mat. This was then passed between felted *squeeze* or *couch* rolls as well as press rolls to further remove water from the web, which was then wound up on a reel. In contrast, the cylinder mould machine used a cylinder covered by a wire mesh, which was partially submerged in a vat containing a pulp suspension. By rotating the cylinder and keeping a difference in the levels of the pulp suspension or *stock* and the level of the back water inside the mould, a driving force was created depositing fibres on the wire. The resulting fibre mat would then be doctored-off, pressed, and dried into paper. A number of improvements would follow

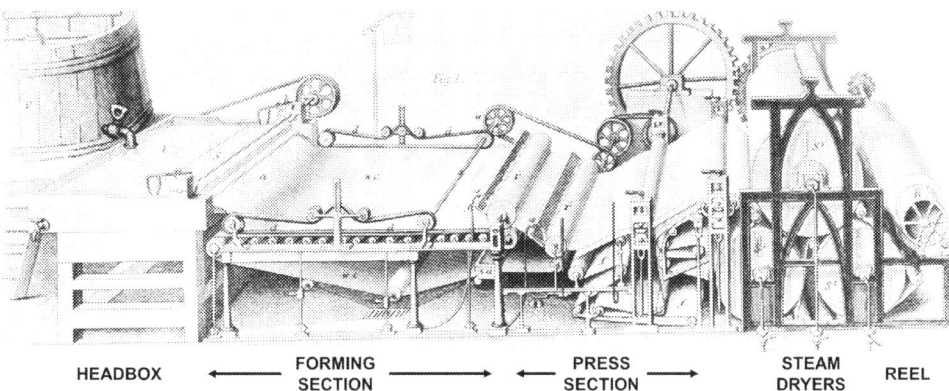

| HEADBOX | ← | FORMING SECTION | → | ← | PRESS SECTION | → | STEAM DRYERS | REEL |

FIG. 5.2 Early Fourdrinier paper machine showing the continuous production of paper. Dilute pulp delivered to the headbox is subjected to unit operations of forming, pressing, and drying, and collected as paper on the reel. *Adapted from G.E. Sellers, in: E.S. Ferguson (Ed.), Early Engineering Reminiscences (1815–40) of George Escol Sellers, Smithsonian Institution United States National Museum, Bulletin 238, 1965, p. 119.*

including the addition of steam drying cylinders by Crompton in 1823 and the introduction of vacuum boxes by Canson in 1826 to enhance dewatering along the wire.

With the advent of chemical (sulphate-based) kraft pulping in the mid-1800s and the rapid replacement of cotton and other nonwood fibres with wood pulps, modern methods of producing paper were developed around the use of relatively short wood-based fibres of 1–2 mm in length. As will be discussed, fibre length has a profound impact on wetlaid equipment design and operations. Arguably, fibre length has its greatest impact on water use and management. For example, the solids content or consistency of wood pulp dispersions used in papermaking is about 0.3%–0.5% by weight. In contrast, the consistency of long fibre dispersions used in nonwovens production can be up to 100 times less.

In the 1930s, C.H. Dexter & Sons developed a method to mass produce Japanese yoshino paper, which was traditionally hand-made using kozo fibre [5]. Kozo bast fibres, which are obtained from the bark of the mulberry tree, have fibre lengths of up to 12 mm. Dexter replaced expensive kozo fibres with relatively low-cost abaca fibres and used a consistency of 0.002%. To economically produce paper from these highly diluted suspensions, a more effective dewatering process was required than was possible with either the flat-wire or cylinder machines available during this time. To achieve a more aggressive drainage profile, the company introduced the first inclined wire machine (Fig. 5.3) [6]. In the mid-1940s, Dexter would use an inclined wire paper machine to produce one of the first nonwoven products, a filter paper made from a blend of natural and polyvinyl chloride (vinyon) man-made fibres [7].

5.1.2 Transition to man-made fibres

The basis of wetlaid nonwovens begins with development and commercialisation of man-made fibres and their addition to traditional papermaking furnishes. In time, their expanded use would result in paper-like materials made from 100% man-made fibres, as well as blends

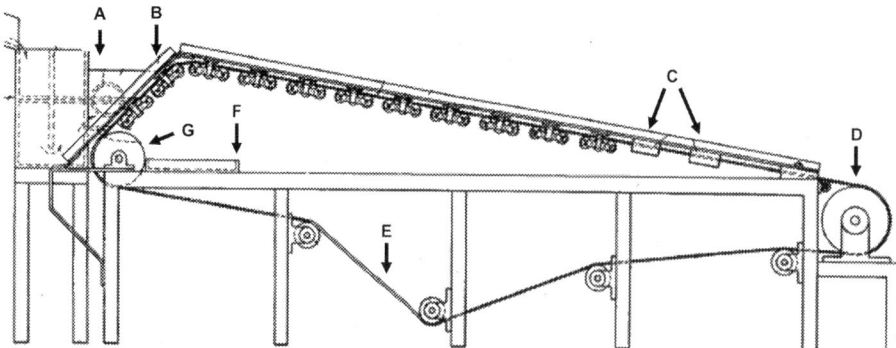

FIG. 5.3   An early example of an inclined-wire forming section showing headbox (A), inclined portion (B), suction boxes (C), couch roll (D), wire (E), saveall (F), and breast roll (G). *Adapted from F.H. Osborne, Process of Making Porous Long Fibered Nonhydrated Paper, U.S. Patent No. 2,045,095, 1936.*

of wood pulp with short cut man-made fibres. Driving the transition from natural to man-made fibres was the unprecedented advancements in polymer chemistry and new manufacturing methods developed to overcome some of the challenges posed by man-made fibres.

The earliest man-made fibres were in the form of acetate rayon, which is prepared through the solvent spinning of cellulose acetate and acetone. Upon the evaporation of the solvent, solid filaments are produced [8]. In 1892, Cross and Bevan would develop viscose rayon fibres via the cellulose xanthate intermediate, which resulted in fibres that were thicker and more pliable than previously achieved. Work on regenerated cellulose proved to be great success stories for companies like American Viscose, Courtaulds, and DuPont Rayon Company [9].

Research and development in man-made polymers in the early part of the 19th century resulted in a number of breakthrough materials. In 1907, Leo Baekland developed a synthetic phenol formaldehyde resin or Bakelite, which is considered to be the first plastic made from man-made components. A thermosetting phenol formaldehyde resin, Bakelite is formed from a condensation reaction of phenol with formaldehyde.[a] The development of synthetic rubber would soon follow in 1910. However, arguably the most commercially important man-made polymer discovery, i.e., polyamide or nylon 6,6 would not be developed until 1937.[b] Wallace Carothers' discovery of nylon 6,6 would see unprecedented commercial success first as a replacement for silk, followed by multiple applications in both the traditional textile and non-wovens industries. In 1939, vinylon, a man-made fibre made from polyvinyl alcohol, would also be discovered [10], followed by the progressive development of other new man-made fibres based on synthetic or naturally derived polymer building blocks.

The development of man-made fibres and their use in papermaking furnishes created a platform for designing novel industrial and consumer products. Man-made fibres would be used to impart a range of material attributes most notably strength, flexibility, softness,

[a]The IUPAC name for Bakelite is polyoxybenzylmethylenglycolanhydride.

[b]The IUPAC name for nylon 6,6 is poly[imino(1,6-dioxohexamethylene) iminohexamethylene].

and porosity to nonwoven products. Man-made fibres would also be used to impart unique properties and characteristics rendering the product suitable for a range of diversified applications.

5.1.3 Current state

Wetlaid nonwovens production represents the smallest segment of the array of available nonwovens production methods. Recent estimates by SmithersPira report global consumption of wetlaid nonwovens equal to 11.3 billion square metres or 460,900 tonnes, representing a market size of about $1.3 billion [11].

Products produced by wetlaid forming are found in a wide variety of consumer and industrial segments including but not limited to hygiene applications, filtration, and building and roofing materials. Fabric compositions range from blends of wood pulp and man-made fibres for single use wet-wipe products, to fabrics made entirely from high performance organic or inorganic man-made fibres such as aramids, glass or carbon for mechanical high temperature resistant reinforcements, composites or prepreg applications. Another important category of wetlaid fabrics includes those that are further processed to modify physical properties such as double re-crepe (DRC) laminated tissue.

Heightened environmental awareness by consumers particularly in relation to single use items such as wipes, coupled with government restrictions on the use of fossil-based plastics, are likely to further stimulate demand for wetlaid nonwovens made of sustainable materials. Sustainably-derived *green* products potentially favour natural fibres and certain regenerated cellulose fibres over petroleum-derived man-made fibres. Natural fibres, and the so-called biobased fibres, such as those made from polylactic acid (PLA), are recyclable, biodegradable, and industrially compostable. Often when recovered from industrial or consumer goods, materials made from bio-based fibres can be used as a recycled feedstock. Petroleum-derived man-made fibres, on the other hand, are difficult to economically recycle and in most cases, they are nonbiodegradable; the recovery rate of plastics is less than 16%. Plastic waste is having a profound impact on marine life and the effects of microplastics on marine life and bio-accumulation in humans are of growing concern [12]. Given the importance of wetlaid forming in the manufacture of water-dispersible and flushable single use wipes and related products, these environmental factors have the potential to increase demand for both natural and specific bio-based man-made fibres.

5.2 General web forming principles

In its most basic description, wetlaid forming takes a well-dispersed blend of fibres and additives in water and removes the water to form a uniform paper-like structure. The appearance and properties of this material will be influenced by the choice of fibre and its surface treatments, the types of additives, the procedures used during stock preparation, and the type of forming process used, including drying type. Each variable will impact to some degree the product and the process of production.

Wetlaid forming begins with the dispersion of fibres in water. Natural fibres, due to their surface chemistry, are relatively easy to disperse. Natural fibres have oxygen-rich surfaces that are easily solvated by water. Their hydrophilic properties facilitate dispersion. In contrast, the surface of petroleum-derived man-made fibres, made from polyolefins and other petroleum-derived polymers, are hydrophobic and do not easily interact with water [13].

Fibre properties, specifically length and flexibility, will also have a significant impact on the preparation of the dispersion. The tendency to form coherent networks or *flocs* increases with fibre length and conformability, making separation of fibres from water more difficult. Also, long fibres tend to have poor sedimentation or settling velocities [14]. As a result, higher levels of dilution are required and the use of surfactants, dispersants, and thickeners, may be needed to achieve adequate dispersion.

In the dewatering process, dispersed fibres along with other materials and chemical additives, known collectively as the pulp stock or *furnish,* are forced to impinge on an endless filtration medium or forming fabric, which acts as a conveyor. The forming fabric is also referred to as the 'wire' and in the early days of papermaking these were made from bronze wire. Today, forming fabrics are woven man-made polymer yarns. As shown in Fig. 5.4, depending on the impingement angle of the stock, the fluid flow will include drainage forces, oriented shear, and turbulence [15]. Control of these hydrodynamic forces will impact final product properties. Following forming of the wet web, the sheet is dried to produce a nonwoven product. The final product may contain additional materials and chemicals that impart specific properties to the web. The final product may also undergo additional processing steps, such as bonding by processes such as hydroentanglement, followed by drying as well as coating and calendering steps to achieve desired properties.

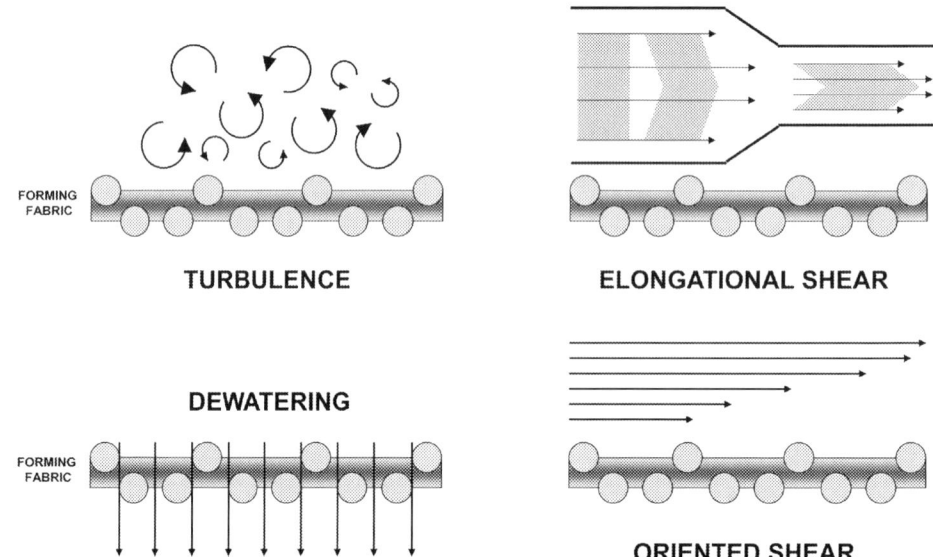

FIG. 5.4   *Adapted from B. Norman, Web forming, in: H. Paulapuro (Ed.), Papermaking Science and Technology, vol. 8(6), Fapet Oy, Helsinki, 2000, pp. 191–250.*

5.3 Fibre selection

The production of wetlaid nonwovens is highly dependent on the type of fibre or fibre blends selected. In addition to wood pulp, a wide range of fibre types can be used, originating from three basic categories: natural fibres, man-made fibres, including both synthetics and those made from regenerated natural polymers, and inorganic fibres. The length of fibres used in wetlaid forming can reach 40 mm but is generally 10–15 mm or less [16].

Wood pulp and man-made fibres are most commonly used in wetlaid forming but inorganic fibres, particularly glass is important across a variety of applications. Along with fibres wetlaid nonwovens may contain additives such as fillers and a variety of chemical additives from those that can improve operational performance, such as the dewatering rate of the stock, to additives that can modify fabric properties, such as flame resistance. Lastly, fibre uniformity, in terms of dimensions, physical properties and morphology, differs between the various fibre types. Man-made fibres are made using chemical reactions where the properties of the feedstocks are highly controlled and using extrusion processes that yield relatively consistent fibre dimensions. The properties of man-made fibres are therefore quite uniform, but defects can still be introduced during the processing of the polymer tow that negatively affect uniformity and processing performance. For example, cutting of the filament tow to make staple fibres can lead to fusing of fibre ends creating a defect known as *logs*. Nonuniform surface properties may also create unwanted defects [17]. In contrast, natural fibres show high degrees of nonuniformity, especially in terms of fibre length and fineness.

There are six basic types of natural fibres: bast, leaf, seed, core, grass, and reed; and all other types such as wood and roots [18]. Important bast fibres include flax, hemp, jute, and ramie. Abaca is a commonly used leaf fibre and cotton is a commonly used seed fibre. The differences in physical properties between these types are quite large. For example, following extraction, the fibre length for nonwood fibres averages about 8 mm but can be as long as 120 mm [19]. Hardwoods, such as Northern Bleached Hardwoods (NBH) or Bleached Eucalyptus Kraft (BEK), have fibre lengths ranging from 0.8 to 1.8 mm, whereas as Southern Bleached Softwood Kraft (SBSK) fibres can be as long as 4.2 mm. Northern Bleached Hardwood Kraft (NBHK) have fibre lengths that range from 2.8 to 3.4 mm. Natural fibres also exhibit relatively large variability within a fibre type. Variations in lignin content, fibre length, fibre fineness, degree of defibrillation, fines content and other mechanical properties will impact the final properties of the nonwoven. Commonly used natural fibres in wetlaid forming are listed in Table 5.1.

5.3.1 Natural fibres

Natural fibres comprise a range of biogenic fibres derived from both plant and woody species. These fibres are characterised by their morphology, chemical composition, and heterogeneity. Natural fibres, especially wood fibres, can be viewed as composite structures owing to their cell wall structure. As shown in Fig. 5.5, fibres from woody biomass are made from multiple structural elements that define their cell walls. The S2 layer makes up the greatest proportion of the wall thickness and is of greatest interest to nonwovens applications. The

TABLE 5.1 Typical fibre types used in wetlaid forming.

Fibre	Type	Length (mm)	Diameter (μm)	Aspect ratio
Wood	BEK	1.0	16.0	62
	NBHK	1.4	18.0	78
	NBSK	3.0	40.0	75
	SBSK	4.2	40.0	105
Nonwood	Abaca	6.0	20.0	300
	Sisal	3.0	20.0	150
Viscose rayon	1/4″/3d	6.4	16.0	397
	1/2″/1.5d	12.7	11.8	1080
	1/2″/3d	12.7	16.0	793
	1/2″/3d	12.7	18.0	706
	3/4″/20d	19.1	43.0	444
Polyester	1/4″/1.5d	6.4	12.4	512
	1/4″/3d	6.4	18.0	353
	1/2″/1.2d	12.7	11.1	1144
	1/2″/1.5d	12.7	12.4	1024
	3/4″/1.2d	19.1	11.1	1720
	3/4″/1.5d	19.1	12.4	1540
Glass	106 microfibre	1.0	0.63	1600
High performance	Aramid	6.0	12.0	500
	Carbon	25.0	7.0	3571

Key: *Aspect ratio = length (mm) ÷ diameter (μm) × 1000. BEK, Bleached Eucalyptus Kraft; NBHK, Northern Bleached Hardwood Kraft; NBSK, Northern Bleached Softwood Kraft; SBSK, Southern Bleached Softwood Kraft.*
From I.M. Hutten, The concerns of forming wet lay nonwovens from long fibres, in: TAPPI 1995 Nonwovens Conference, Atlanta, 1995, 161–172, with contributions by the author.

microfibril angle, which characterises the orientation of cellulose microfibrils in the S2 layer, determines the stiffness of the fibre [20]. The S2 layer is important as the pulping processes used to produce market pulps of importance to nonwovens applications remove the other layers, including the lignin-rich middle lamella and the primary wall or P layer.

In terms of chemical composition, natural fibres owe much of their chemical properties to cellulose. A structural material found in at least one third of the world's biogenic material, cellulose is found in both plants and woody biomass, and found in its nearly pure state attached to the seeds of cotton. Cellulose is a linear polymer of β-D-glucopyranose units linked by 1,4 glycosidic bonds. The polymer exhibits a wide range of molecular weights and degrees of polymerisation, which is highly dependent on its sample history.

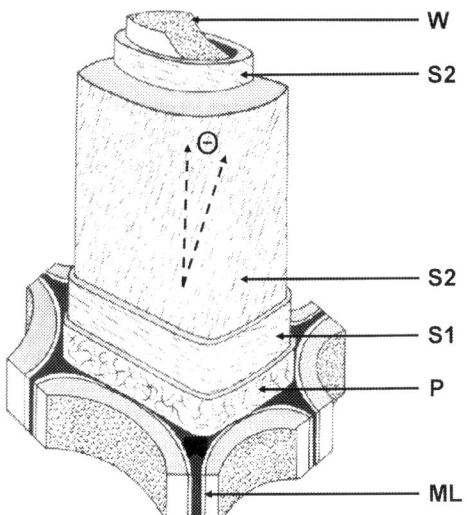

FIG. 5.5 Structural elements of the wood cell: middle lamella (ML); primary wall (P); outer (S1), middle (S2), and inner (S3) layers of the secondary wall; and the warty layer (W). The microfibril angle (Θ) is the angle between fibre axis and the microfibrils in the S2 layer. *Adapted from W.A. Côté Jr., Wood Ultrastructure, University of Washington Press, Seattle, Washington, 1967.*

In terms of its morphology, cellulose has several crystalline states as well as amorphous regions. Microfibrils, which constitute the basic structural units of the plant cell wall, are 10–30-nm wide with indefinite length. Cellulose forms strong hydrogen bonds due to its high concentration of free hydroxyl groups. As a result, it readily reacts with water and water vapour. Both the macro- and microstructure of cellulose is of commercial interest to nonwovens applications. The use of nanocellulose is also seeing considerable early-stage research interest [21].

5.3.1.1 *Wood pulp*

Wood pulp is arguably the most common natural fibre used in wetlaid forming representing about half of the total fibre used [22]. Wood pulp is also used in a variety of dry-forming processes, such as airlaid forming and the Coform meltblown process, which intermix polypropylene filaments and wood pulp to make the fabric [23]. Most wood pulps are made using the kraft pulping process, which combines caustic (NaOH) and sodium sulphide (Na_2S), along with temperature, pressure, and time, to break the chemical bonds between lignin, hemicellulose, and cellulose and thus liberate cellulose fibres from the lignocellulosic matrix. Wood pulp can come from either hardwood or softwood species. Hardwood pulps are generally shorter and finer than softwood pulps, which tend to be long and coarse. Wetlaid forming is ideally suited to wood pulp and other natural fibres because these fibres are easily dispersed in water and are capable of self-bonding, due to entanglement and the formation of hydrogen bonds between fibres during web formation.

5.3.1.2 *Nonwood fibres*

Nonwood fibres of relevance to wetlaid applications include those obtained from abaca, cotton, flax, and hemp, as well as agricultural wastes, such as wheat straw. Fibres from these sources can be found in a number of specialty products including wipes, tea bags, coffee filters, as well as number of other filtration products.

Cotton fibres have excellent wet strength and liquid absorption properties. However, as with many nonwood fibres, their long fibre length (up to 36 mm) requires that they be chopped to <18 mm before they can be dispersed in water [24]. The relatively high cost of cotton fibre has limited its use in nonwoven applications [25]. However, the sustainability goals of brand owners, such as H&M, IKEA, P&G, and Walmart, are driving expanded use of cotton in certain nonwovens applications.

5.3.2 Bio-based fibres

Bio-based fibres are man-made but are produced with biogenic raw materials, such as cellulose, starch, and sugar. Viscose rayon and lyocell are made from dissolving pulp: a specialty pulp in which the hemicellulose fraction is removed. In the production of rayon, dissolving pulp is reacted with carbon disulphide (CS_2) and caustic (NaOH) to produce the dope cellulose xanthate, which is then spun into fibres. The lyocell process uses the organic solvent N-methylmorpholine N-oxide (NNMO) to produce the dope. Unlike the production of rayon, the lyocell solvent is recovered making the process much more sustainable and environmentally friendly.

Other bio-based fibres include those made from bio-based aliphatic polyesters, such as polylactic acid (PLA) and polyhydroxyalkanoate (PHA). These materials are made through the bacterial fermentation of plant sugars, derived from corn, sugar cane, and other sugar-producing plants. In PLA synthesis, glucose derived from corn starch is enzymatically hydrolysed to dextrose and fermented to produce lactic acid [26]. A catalysis step follows to make lactide dimer, the intermediate to the polylactide polymer, which is spun into fibres. Still other bio-based fibres include those made from bio-based polyethylene terephthalate (bio-PET), polybutylene succinate (PBS), and polycaprolactone (PCL).

Bio-based fibres are generally used in applications wherein the product must biodegrade. However, not all bio-based materials are biodegradable. For example, bio-PET is not biodegradable as the name may imply but it is recyclable. Adding to the confusion, oxy-degradable plastics, such as poly(butylene adipate-*co*-terephthalate) (PBAt), are fossil-derived and biodegradable. Bio-based polymers are generally more expensive than their fossil-based alternatives. However, when analysing costs, the total cost of ownership, including time in use must be considered, as well as increasingly, the environmental impacts across the full life cycle of the product.

5.3.3 Petroleum-derived man-made fibres

Petroleum-derived man-made fibres represent a range of polymer types, from commodity polymers, such as polyethylene (PE) and polyethylene terephthalate (PET), to high-performance materials made from polyaramids, glass, carbon and metal coated carbon fibres.

Microfibres can also be used in the process. For example, PET microfibres less than 0.4 denier with round, ribbon or flat cross-sections and a cut length of 1.5 mm have been developed by Eastman (Cyphrex) specifically for wetlaid forming. Commodity staple fibres made from polyolefins, namely PE and PET, as well as PE/PET bicomponent fibres, represent about 20% of all fibres used in wetlaid nonwovens production [22]. In their unmodified form, polyolefin fibres are highly hydrophobic and require additional processing steps to ensure adequate dispersion [16]. Polyolefin fibres impart strength and dimensional stability to wetlaid nonwovens.

Man-made fibres made with relatively low melting points represent a unique class of fibres called melt bondable ('or low melt') staple fibres, which can be either mono- or bicomponent. Used in wetlaid forming, these meltable fibres impart sheet strength to the nonwoven after thermal bonding. Products containing a proportion of meltable fibres can also enable heat-sealing of fabrics during converting of final products, e.g., in the manufacture of tea bags, and be used to enhance interply bonding in laminated structures. The three most common polymers used to make low melt bicomponent fibres are PE, PP, and PET (often co-PET). In these fibres, the lowest melting point polymer, such as PE or PP is spun in combination with a higher melting point polymer, such as PET, to produce a melt-bondable fibre. These are produced in either sheath/core or side-by-side configurations. The advantage of bicomponent fibres is that when the sheath component melts, the fibre maintains its fibrous structure and imparts dimensional stability to the nonwoven by reducing curl and shrinkage [27]. A common application is the blending of bicomponent fibres with pulp to form a nonwoven web, which is then thermally bonded to enhance sheet strength [28]. Also available are bicomponent melt-bondable fibres made from biodegradable polymers derived from petroleum, such as polyvinyl alcohol (PVOH), or biogenic sources, such as polylactic acid (PLA) [29].

High performance fibres, such as polyaramid and carbon fibres, are used for demanding applications requiring high temperature and chemical resistance. For aramid-based nonwovens, staple fibres and fibrids, which are film-like polymer-based binders, are combined using papermaking processes to produce synthetic paper-like composites with high strength and thermal resistance [30]. Fibrids are produced by subjecting a polymer granule into an external shear field, which can be generated through shear precipitation or mechanical action, such as refining [31]. The specific surface area of fibrids generated by these methods can be as high as $60 \, m^2/kg$. Paper comprising man-made fibre and fibrids can be calendered at elevated pressures and temperatures above the second order transition temperature of the polymer to enhance internal bonding and increase paper strength. Fabrics made from aramid fibres are sold under the Kevlar, Nomex, and Twaron brands and are used in a variety of industrial applications including: personal protection wear and composite structures used in aerospace applications [32], as well as electrical insulation paper, printed circuit boards, and as separators for Li-ion batteries [33].

5.3.4 Inorganic fibres

Wetlaid forming allows the use of the broadest range of fibre types including inorganic fibres, which can be in the form of metallic or metallised fibres as well as mineral-based fibres. Metallic fibres can be both pure metal, such as aluminium, gold, iron, silver, tantalum, and zinc, or alloys, or in the form of metallised fibres. These fibres are used for a variety of specialised products including battery electrodes, friction material, and filter media. High porosity flow battery electrodes made from carbon fibres are also made using wetlaid forming [34]. Metal coated fibres used in wetlaid web formation include nickel or copper coated carbon.

Common mineral-based fibres include cut glass strand, microfibre glass, and ceramics. Natural mineral fibres from layered silicate minerals, such as mica and vermiculite, are also used. Nonwovens made from mineral-based fibres provide high temperature stability and flame retardancy as well as mechanical reinforcement because of their relatively high modulus. There are numerous applications for wetlaid glass fabrics including, bituminous roofing membranes and gypsum board facings as well as carpet tile and luxury vinyl tile (LVT) backings.

Wetlaid fabrics containing mineral fibres, an exfoliating graphite intumescent, and actives such as alumina trihydrate (ATH), together with a binder are used to make composite fire protection fabrics suitable for use in vehicles and buildings. Pitch and PAN-based carbon fibres, carbon whiskers and silicon carbide fibres are also processable using wetlaid forming.

5.3.5 Fibrillated fibres

With many fibres there is significant scope to modify their morphology during processing to influence properties such as cohesion, opacity, light reflectance, and handle. For example, with man-made fibres, cold drawing and washing out of plasticisers can be used to create a rough surface [35]. Wood pulp, natural cellulosic fibres, and certain man-made fibres, e.g., lyocell and aramids, can also be heavily modified through fibrillation prior to wetlaid forming.

Synthetic wood pulp (SWP) is produced by fibrillating susceptible man-made fibres using an external shear field to induce longitudinal splitting and generation of submicron fibrils of much smaller diameter and high surface area. The shear field ise created by any number of mechanical methods including by adjustment of the gap of a double-disk refiner. Depending on the mechanical preparation method, instead of discrete individual fibrils being formed, small fibrous branched structures of very low diameter are produced known as fibrids. Fibrids resemble pulp when dispersed in water, and when blended with man-made fibres the resulting wetlaid fabrics have been traditionally referred to as Textryls.

Early methods of producing SWP recommended refining in the presence of an organic dispersant, such as polyvinyl alcohol [36]. The fibrils created by the shearing action are distributed along the length of the fibre and have a length to width ratio of about 2 to about 100 with diameters less than 1 μm. Alternatively, high surface area polyolefin fibres can be produced by flash-spinning [37]. Fibrillated polyolefin fibres with physical and morphological characteristics similar to wood fibres were introduced in the late 1960s by Crown Zellerbach under the trademark SWP. Mitsui Chemicals now markets them under the Fybrel and SWP brands. The high surface area of fibrillated fibres enhances the mechanical strength of the nonwoven and decreases its pore size, which can improve filtration efficiency. Cellulose acetate fibrets are another example of fibrillated fibres that are easy to disperse in water and that can provide strength advantages relative to papermaking fibres [38]. Fibrillated lyocell fibres are also available from a number of fibre suppliers. Kevlar fibres, based on aromatic polyamide chemistry, can be prepared in a form that is highly fibrillated and easy to disperse [30].

5.3.6 Nanofibres

Over the last 20 years, the use of nanofibres has gained considerable attention. Nanofibres made from polyolefins [39], cellulose [40], and chitosan [41] have all sparked interest in the ability to create new and improved products from these materials. Of particular relevance to wetlaid forming is the use of nanocellulose, in the form of nanofibrillated and nanocrystalline morphologies, to impart enhanced mechanical and barrier properties to films and paper [42]. Although dimensions for nanofibres vary widely, reported diameters tend to be less than 50nm, with lengths less than 1μm, and high aspect ratios. Recent examples of nanofibres in wetlaid forming include such diverse applications as: lithium ion battery separators [43]; polymer laminates used as semistructural panels for aircraft interiors [44]; nanofibres as coating material for paper and paperboard [45]; and, the use of lyocell nanofibres to produce a disposable wipe [46]. Fig. 5.6 shows the nanostructure of fibrillated lyocell fibres.

5.4 Chemico-physical considerations

5.4.1 Fibre morphology

The properties of a wetlaid nonwoven product and the performance of the manufacturing process are intimately linked to the physical and mechanical properties of fibres; hence, the importance of fibre selection. Important factors to consider when selecting fibres are related to fibre length, diameter, and linear density; material properties, such as tensile strength or

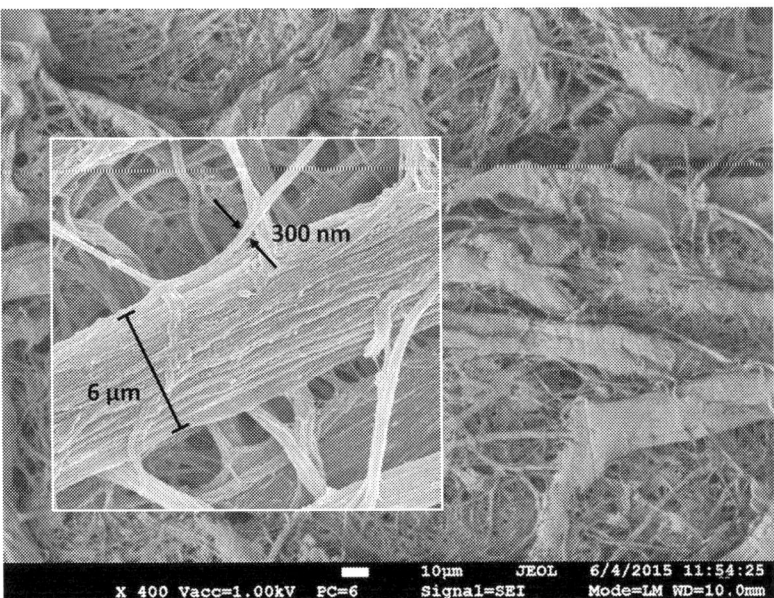

FIG. 5.6 Photomicrographs of a nonwoven mat containing fibrillated lyocell fibres. *Courtesy of the author.*

tenacity and elongation at break; and physico-chemical properties, such as the degree of hygroscopicity.

Fibre coarseness (C), also referred to as linear density or fineness, is directly related to the number of fibres per unit weight:

$$C = m/n\,l_n$$

where m is the weight of the fibre, n = total number of fibres, and l_n is the arithmetic mean length of the fibre [47]. For wood fibres, coarse fibres tend to have thicker cell walls and a lower degree of fibre collapse. Pulps made from long, flexible fibres of a low coarseness index, such as northern softwood species, are preferred in spunlace applications due their entanglement potential. Fig. 5.7 compares the fibre diameter and surface area for commonly fibres used in wetlaid applications.

5.4.2 Dispersion of fibres

Dispersing fibres in water requires that they first be wet-out. This can be a challenge with many man-made fibres as their surface energies are quite low. For example, untreated PE has a surface energy of $30\,\mathrm{kJ\,m^{-2}}$, and a contact angle of 100 degrees, which means that it is not wettable. In comparison, raw viscose rayon has a contact angle of 68.3 degrees, which indicates moderate wettability. Treatment of fibres through washing, chemical modification (i.e. by adding wetting surfactants), the addition of inorganic fillers, such as calcium carbonate, to the polymer matrix, and surface treatments, including corona and plasma treatments, will

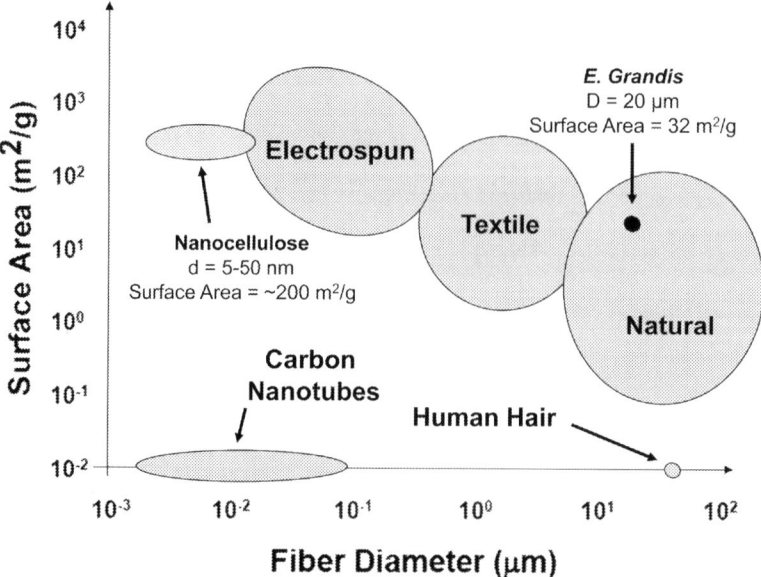

FIG. 5.7   Comparison of the surface area and fibre length of eucalyptus hardwood fibre (*E. grandis*) with other fibre types and nanocellulose. *Adapted from P.W. Gibson, H. Schreuder-Gibson, D. Rivin, Transport properties of porous membranes based on electrospun nanofibres, Colloids Surf. A Physicochem. Eng. Asp. 187–188 (2001) 469–481.*

increase the surface energy of the fibres making them more wettable and thus easier to disperse in water [48].

The degree of reactivity between fibres and water greatly influences the wetlaid manufacturing process as well as the properties of the final product. Wetlaid forming is a multiphase flow process wherein fibre length and flexibility determine product structure and define operational limits [49]. Entanglement among fibres in suspension will depend in part on the free volume of rotation, a property that is specific to the fibre. If a fibre in solution has enough volume from a large hydrodynamic radius between it and other fibres, it can rotate freely without collisions that would lead to flocculation. Mainly for this reason a high dilution ratio in water is needed to keep the fibres separated from each other prior to web formation.

The relationship between fibre hydrodynamics and dilution factor is clearly expressed in the concept of the crowding factor. Kerekes and Schell [50] showed that a crowding factor (N_C) can be defined as follows:

$$N_C = (2/3)\, C_v\, (L/d)^2$$

where C_v is the volume-based concentration of fibre material in the suspension, L is the fibre length, and d is the fibre diameter. The amount of dilution water required to keep the value of N_C below a defined limit will increase as the square of the ratio of L/d. A crowding factor $N_C < 1$ represents dilute fibre concentrations, wherein fibre-to-fibre contact is rare. On the other hand, $N_C > 60$ represents concentrated fibre stocks wherein contact between the fibres is continuous. The fibre's Reynolds number and force factor account for additional hydrodynamic forces influencing the behaviour of fibres in suspension [50]. Fig. 5.8 illustrates the

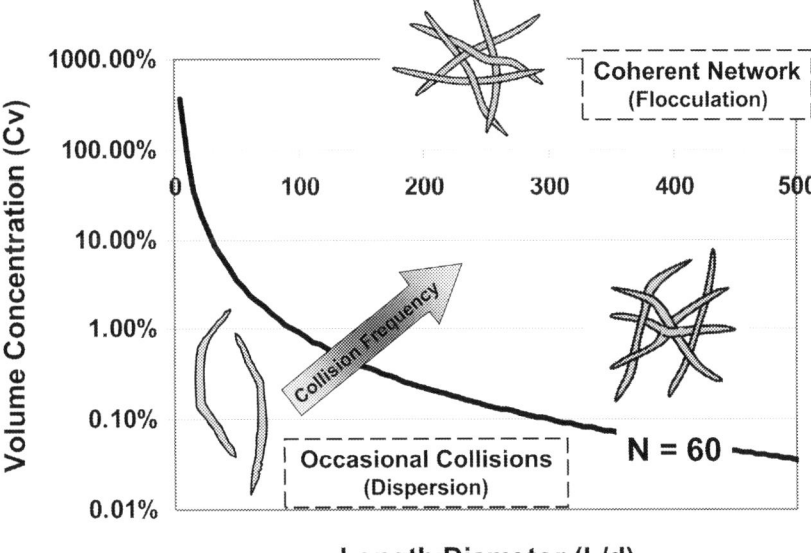

FIG. 5.8 The impact of L/d ratio on volume concentration and the collision frequency of fibres in suspension. *Adapted from M.A. Hubbe, A.A. Koukoulas, Wet-laid nonwovens – chemical approaches using synthetic and cellulosic fibres, BioResources 11(2) (2016) 5500–5552.*

concept of the crowding factor and the influence of aspect ratio (L/d) on volume concentration and the tendency for fibres to form networks or flocs.

All things being equal, a well-dispersed stock will result in a more uniform nonwoven web with a low coefficient of mass variation. Flocs generated in the stock, unless redispersed, will appear as regions of high mass density in the nonwoven web. Wetlaid manufacturers refer to this as *poor formation,* that is, high point-to-point mass variation. Poorly formed webs will also exhibit lower strength and higher porosity.

In addition to fibre length, the degree to which the stock is well-dispersed depends on the chemico-physical properties of the fibres, the properties of the dispersing medium, and the degree of hydrodynamic shear imparted to the stock. The chemico-physical properties of the fibres will depend on their surface morphology (smooth vs fibrillated) and surface chemistry (hydrophilic vs hydrophobic). Smooth, hydrophobic fibres, such as PE staple fibre, will experience a high degree of surface tension that will oppose wetting and dispersion.

Spin-finishes with anionic, cationic, and nonionic surfactants will promote wetting of the fibres and allow them to disperse more easily. Alternatively, surfactants can be added to the stock to promote dispersion. In the case of PE, nonionic surfactants, such as oxyalkylated fatty amines, in the amount of 1–200 ppm are known to be effective dispersing agents [51]. Other commonly used dispersants include: alkyl amine acetate, cationic and oxidised starch, cationic polyamide-polyamine, polyethylene glycol amine, and polyacrylates. In all cases, good adsorption of the surfactant to the surface of the fibre is critical to its effectiveness. Adjusting pH above the pK_a of the surfactant may be needed to promote adsorption onto the fibre.

Once dispersed, the stock requires constant agitation to ensure that the fibres remain dispersed. Variations in flow and pulsations will destabilise the suspension and create flocs. Optimising specific applied energy, shear rate, and agitation time to minimise defects, such as flocs or fibre bundles, is critical. Avoidance of turbulent flow and the creation of vortices, which can promote rope formation, are also recommended [52].

5.4.3 Dewatering

The nonwoven web is made through a continuous water removal process. Making handsheets in the laboratory, wherein stock is allowed to drain on a fine-mesh wire, is dominated by drainage. In wetlaid forming using commercial equipment, other hydrodynamic forces become important (see Fig. 5.4). As fibres in suspension form coherent networks, a fibre web is created and filtration of water across the structure becomes a dominant dewatering mechanism. The degree to which fibres swell and retain water will influence both dewatering and drying. Swollen fibres exhibiting a high degree of fibrillation, both on the fibre surface and within the cell wall, will show a larger resistance to filtration and tend to drain more slowly [53]. In contrast, most petroleum-derived man-made fibres are not reactive in water. As their solubility parameters are low, the interaction between the fibre and water is best described through relative differences in surface chemistry, electrophoretic potential, and the degree to which they are soluble in water (Table 5.2). All else being equal, fibres made from materials with higher concentrations of oxygen at their surfaces will tend to be more strongly bound to water. These differences will impact dewatering and will become more pronounced with the degree of fibrillation or specific surface of the fibre.

TABLE 5.2 Hildebrand solubility index for water and common polymer systems used to make staple fibres for wetlaid forming.

Material	δ_H (MPa$^{1/2}$)
Poly(propylene) (PP)	16.2
Poly(ethylene) (PE, LDPE, HDPE)	16.7
Poly(vinyl acetate) (PVA)	19.6
Polylactic acid (PLA)	20.2
Poly(ethylene terephthalate) (PET)	21.2
Poly(caprolactam) (nylon 6)	25.5
Poly(hexamethylene adipamide) (nylon 6,6)	26.1
Poly(vinyl alcohol) (PVOH)	30.5
Cellulose	40.1
Water	47.8

Fibre–water interactions are described by the degree of solvation, capillary forces within the internal pore structure, interstitial water contained between fibres. The degree of solvation is related to surface chemistry. Cellulose, which has three hydroxyl groups per glucose residue, is highly solvated, while polyethylene, which does not contain oxygen, is not (Fig. 5.9). Useful methods for characterising fibre–water interactions include the solubility indices, such as the Hildebrand solubility index, the fibre saturation point (FSP), and the bound water content of the fibre [53]. The FSP is a measure of the total volume of water within fibre pores both bound to cellulose and in associated solvation spheres. Similarly, the reactivity of hydrophilic man-made fibres, such as viscose rayon, modal, and lyocell fibres, will depend on the pore structure of the material [54]. Bound water content is a measure of water bound directly to cellulose and will vary with fibre type. For example, the bound water content for cotton (0.19 g/g) is lower than either fully-bleached softwood kraft (0.23 g/g) or fully-bleached hardwood kraft (0.29 g/g), which is due, in part, to the presence of sulphonated groups created during the pulping process and the higher degree of crystallinity in cotton [55].

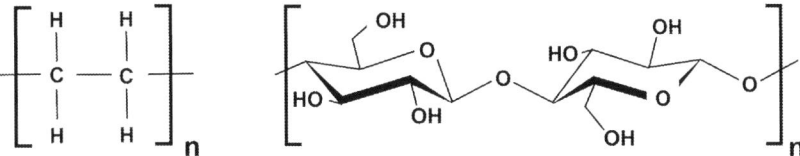

FIG. 5.9 Molecular structure of polyethylene (left) and cellulose (right). The hydroxyl groups on cellulose promote both solvation with water and the formation of intermolecular hydrogen bonds.

5.4.4 Web properties

Wetlaid nonwovens will appear to have varying degrees of uniformity, on length-scales no greater than about 10–15 mm. The visual impression of nonuniformity is known as *formation* or *look-through* [56]. A uniform sheet will be said to be 'well-formed'. Generally speaking, compared to other nonwoven web forming technologies, lightweight wetlaid webs and fabrics are among the most uniform in terms of weight per unit area.

Formation is closely related to the point-to-point mass variation of the sheet. All else being equal, products made from short, fine fibres will have better point-to-point mass uniformity and formation, smoother surfaces, and lower tear strength. Longer, coarse fibres will produce a product with relatively high tear strength and folding resistance.

The overall strength of the nonwoven will depend on the strength of component fibres and the network strength, which will come from the fibre bonding and the degree of entanglement. Natural fibres will benefit from hydrogen bonding to create network strength whereas petroleum-derived man-made fibres, which have much lower bonding potential, may require thermal bonding or the addition of bonding agents, such a latex binder. With man-made fibres, fibre length, diameter and crimp will also influence the degree of bonding.

Finite element analysis has been widely used to model the mechanical properties of thermally bonded nonwoven materials [57]. Page developed a semiempirical equation that is useful in modelling the tensile strength of paper [58]:

$$1/T = 9/8Z + 1/B$$

The Page equation can be viewed as the nonwovens equivalent of a simple electric circuit describing resistors in parallel. The first term governed by the zero-span tensile strength (Z) is dependent on the intrinsic strength of the fibre, whereas the second term is governed the contribution from interfibre bonding (B).

5.5 Machinery and equipment

5.5.1 Mill overview

Unlike dry-forming, the wetlaid forming manufacturing platform can be complicated because of the large water utilisation requirements. Wetlaid nonwoven manufacturing facilities require waste-water treatment and must be capable of monitoring the quality of wastewater to meet permissible limits. Drylaid manufacturers typically do not have co-located wetlaid capabilities and the necessary processing expertise is quite different, although manufacturers of spunlace nonwovens will be familiar with the waste-water handling needs of wetlaid forming. The added infrastructure requirements can also contribute to higher capital costs, as compared to drylaid nonwoven manufacturing.

A well-organised mill will be designed to maximise the efficiency of workflows (Fig. 5.10). Raw material handling and deliveries should be co-located. Stock supply should be up from the *wet-end* of the wetlaid machine. The *dry-end* of the machine houses all ancillary processing such as pressing, drying, and reel building. The wetlaid machine may also have in-line unit

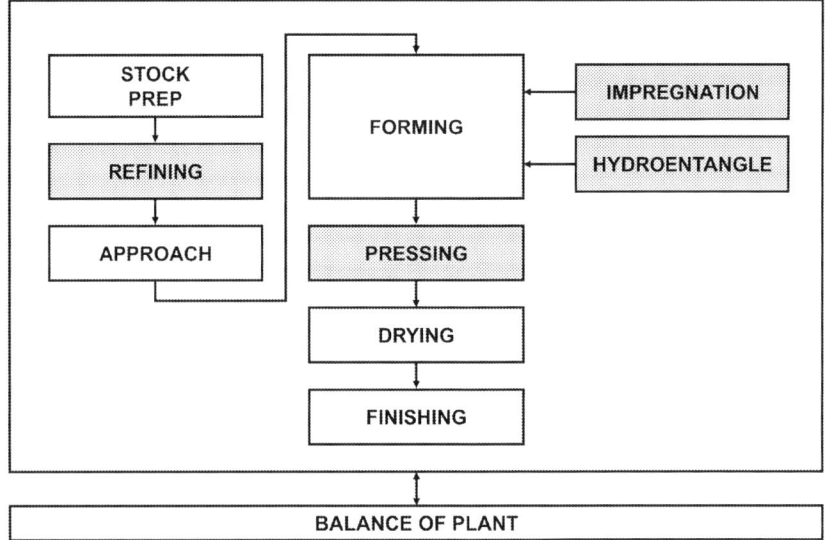

FIG. 5.10 Typical process flow for wetlaid nonwoven production. Optional unit operations for bonding the wetlaid web are highlighted in *blue* (*gray* in print version).

operations such as coaters, hydroentangling units, and calenders. Beyond the reel, the mill may have off-machine unit operations such as surface treaters, coaters, calenders, and slitting.

5.5.2 Stock preparation

Production of wetlaid nonwovens begins with stock preparation, where fibres or fibres blended with wood pulp and other additives are dispersed in water to produce a fibre slurry. As discussed earlier, wetlaid nonwoven manufacture is similar to a papermaking process. However, the longer length of fibres used in wetlaid nonwovens production has significant impact on water use and management associated with conventional papermaking. A typical papermaking process using wood pulp, which is very short in length, might use a dilution factor of about 0.5 g/L based on the dry weight of the fibre. This corresponds to a ratio of fibre to water or *consistency* of 0.5%. In other words, the initial water content constitutes 99.5% of the initial weight of the stock. In comparison, a wetlaid nonwovens process can necessitate a consistency as low as 0.001% to keep the longer fibres separated. In this case, every tonne of nonwoven produced would require about 10,000 tonnes of water. However, it is important to recognise that the water used in wetlaid forming is a recyclable transport medium, akin to air in dry-forming processes. The typical freshwater requirement in many wetlaid applications can be below 10 L/kg [59].

High-shear agitation is usually required to create a uniform dispersion of fibres and additives. A hydrapulper or similar high-shear mixing device, such as sloping bottom tanks with side entry impellers, are also commonly used. Hydrapulpers are well-suited for dispersing hydrophobic fibres such as PE. Side entry designs, which generally provide less severe agitation, are suitable only for readily dispersible fibres. In all cases, mixing tanks, also called make-down tanks, should be selected with smooth surfaces and designs that inhibit the creation of vortices to avoid air entrainment and fibre flocculation. Once the stock is well-dispersed, it should be held under constant agitation.

In applications that require additional tensile or internal bond strength, the stock may be refined to defibrillate the fibres and enhance their surface area. The additional surface area promotes interfibre bonding and develops strength in the wetlaid web. However, it also promotes flocculation before the web is formed, so care must be taken to avoid this. Although a number of refining types are available, disk refiners, either single- or double-disk are the most commonly used. Stock components are usually refined separately, especially if natural fibres are used, and then blended together and stored in a run-tank for delivery to the wetlaid former.

Following refining, in-line cleaning systems are commonly used to remove contaminants from the stock. Centrifugal cleaners remove heavy contaminants, such as sand, by driving them to the periphery of the cleaner and then down to the reject outlet by centrifugal force. Magnetic pulp-separators are used to remove metallic contaminants that may damage screens and fabrics. If high pressure screens are used, the distance between screen holes should be greater than twice the fibre length to avoid blocking.

The approach system refers to the pumps and piping needed to deliver the stock to the wetlaid former. In designing the approach system, care is taken to maintain appropriate stock flow rates and to use straight-through piping layouts to minimise dead zones that can cause fibres to flocculate and form deposits. Constant flows should be maintained within all piping and flushed and drained when not in use. The approach system can be as simple as a single pipe. However, modern approach systems are usually based on a tapered manifold design to ensure uniform flows across the width of the headbox. Alternatively, radial distributors can be used. The influence of web formation speed, stock flow rates, and stock uniformity, both dimensional and temporal, should be considered in any approach design.

Modern approach systems will have a number of in-line sensing equipment, including tank level controls and flow meters, as well as consistency, pH and conductivity sensors. These can be monitored manually or tied into mill-wide distributed control system (DCS). In all cases, care should be taken to ensure that meters and sensors are routinely calibrated.

Increasing stock consistency has been a long-standing goal of wetlaid forming because it would open the opportunity to lower water usage, increase operating throughput, and reduce energy costs, especially those related to drying. New wetlaid web forming methods that can increase stock consistency have been proposed. Foam forming is one example that has potential to reduce water use in wetlaid nonwovens. On the papermaking side, a method for tissue production using form forming and recycled wood pulp claims consistencies of up to 60% [60]. However, adoption of high consistency forming processes remain limited, expect for some used to produce paperboard, which can operate with headbox consistencies up to 5% [61].

In addition to fibres, a number of other materials and chemicals may be added to the stock (Table 5.3). These include binders, fillers, surfactants, thickeners, and defoamers, as well as specialty chemicals that impart a specific functionality to the nonwoven product or enhance web formation. Wet and dry strength additives may be used for products containing natural fibres. Lindström et al. [62] provide an excellent review of these additives. When considering the use of additives, care must be taken to optimise the order of addition as this can

TABLE 5.3 Common additives used in the production of wetlaid nonwovens.

Type	Purpose	Examples
Binders	Promote adhesion between fibres and enhance structural integrity	*Dispersions:* PVOH, starch *Emulsions:* SB Latexes *Solids:* PTFE, PVdF
Defoamers	Used to avoid excessive levels of entrained bubbles and/or floating foam	DF-122, TERGITOL L-64, Q2-5247
Surfactants	Increase the surface energy of hydrophobic (petroleum-derived) fibres to promote wetting and enhance dispersion	Aerosol OT, Milease T, Pluronic F108, Q2-5211, Triton X-114
Thickeners	Increase the slurry viscosity to promote fibre dispersion	Aqualon, Separan AP-30, Sumifloc

Adapted from M.A. Hubbe, A.A. Koukoulas, Wet-laid nonwovens – chemical approaches using synthetic and cellulosic fibres, BioResources 11(2) (2016) 5500–5552.

dramatically influence the effectiveness of the additive. Maximising the retention of the additive is also critical to optimising overall materials costs.

5.5.3 Web forming

The web forming section consists of the stock delivery system or headbox and the former, which in some cases can be combined in a single unit. Three dominant forming designs continue to remain in use: the cylinderical mould machine or cylinder machine, the flat-bed former, and the inclined former. Fig. 5.11 shows the basic elements of cylinder mould and inclined-wire formers. Relative to either Fourdrinier or inclined-wire formers, cylinder mould machines tend to be narrow (down to 0.5 m) and slow (as low as 10 m/min). However, their ability to handle high dilution levels make them ideally suited for specialty products made from high-performance fibres, such as ceramic fibres, glass microfibres, and polyaramids. Cylinder machines are extremely versatile and can produce webs from 15 to $1000 \, g/m^2$ on the same machine with very little adjustment. Of the many cylinder machines available, the Rotoformer by Allimand has one of the largest installed bases. The radial distributor system is commonly used on cylinder machines to improve the basis weight profile. The Octopus distributor by Kadant Inc. is the most well-known, although this can be sourced from many companies. With regards to flat-bed and inclined formers, leading suppliers include Andritz, Allimand, Kobayshi, and Voith. Other suppliers include Beston Machinery *Co.* Ltd., Daisho Tekkosho, Hardayal, Henan Fuyuan Machinery Manufacturing *Co*. Ltd., Zhengzhou Guangmao, Ltd. The neXformer by Andritz, the Hydrofibre by Allimand and the HydroFormer by Voith are the most commonly used inclined-wire machine used in wetlaid nonwovens.

Wetlaid forming requires significantly higher volumes of water than conventional papermaking. As a result, the pressure or head at the web forming area in wetlaid machines is typically much greater than it is in papermaking. In conventional papermaking, the forming area is outside the slice, whereas in wetlaid the forming area begins in the headbox. In contrast, the headbox pond level in wetlaid forming is >10 times that of conventional

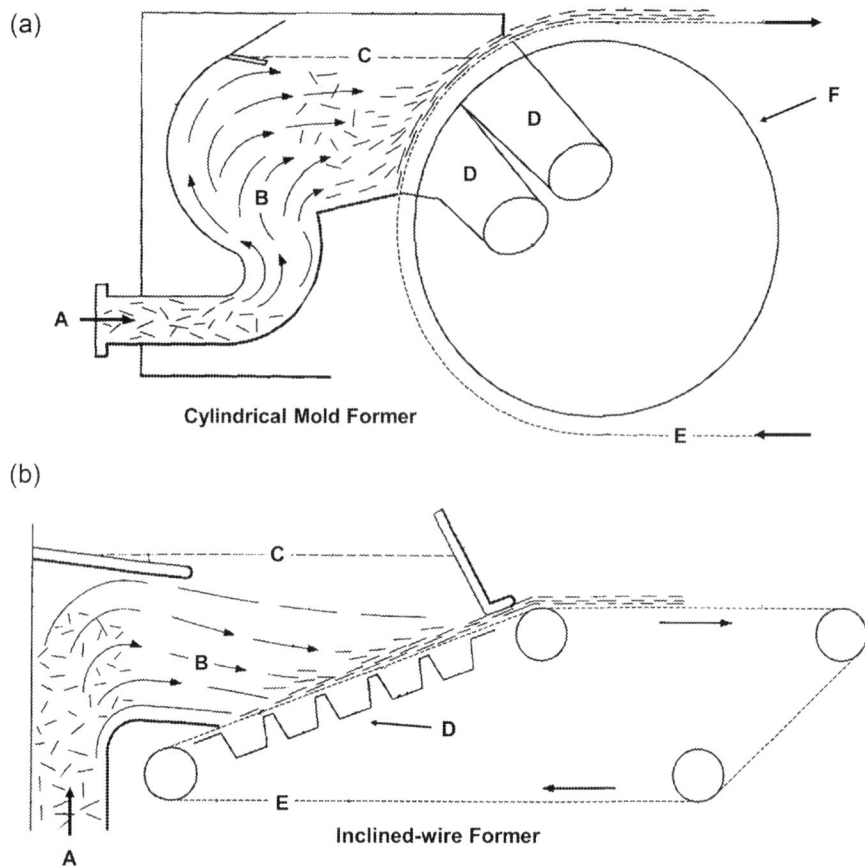

FIG. 5.11 Depiction of stock flows and fibre deposition in two common formers used in wetlaid manufacturing: (a) The cylindrical mould and (b) the inclined-wire former, showing stock inlet (A), stock flow (B), pond height (C), vacuum boxes (D), forming wire (E), and rotating forming roll (F). *Adapted from S.P. Scheinberg, Wetlay Process for Manufacture of Highly-Oriented Fibrous Mats, U.S. Patent No. 6,066,235, 2000.*

papermaking (75–100 cm) and considerably greater than this in closed (air-padded) or hydraulic headboxes [63]. Moreover, the enhanced water-handling capability of inclined formers results in a drastically reduced forming area, from tens of metres to 60–100 cm. As a result, stock flow velocities are also more easily controlled (Fig. 5.11).

The headbox is designed to uniformly deliver the stock across the width of the former. A properly designed headbox will provide controlled levels of turbulence, stock or jet impingement, and fibre orientation. An evolution of headbox design changes has led to significant improvements in processing efficiency and product quality. The earliest headbox designs, known as open headboxes, were operated under atmospheric pressure. In an open headbox, a differential stock height within the headbox provides a driving force or *head* to deliver stock to the slice. Rectifier rolls or other turbulence generators located in the headbox may be used to maintain stock uniformity across the slice.

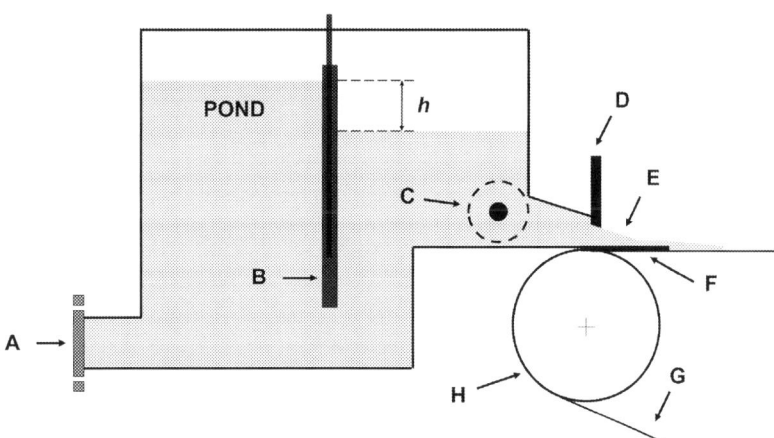

FIG. 5.12 Simple air-padded headbox used with flat-bed formers showing stock inlet (A), adjustable baffle (B), rectifier roll (C), adjustable slice (D), stock jet (E), forming apron (F), breast roll (H), and forming fabric or wire (G).

Open headboxes are used for speeds below 400 m/min. As headbox height controls the jet speed, open headboxes became impractical at high speed. The basic elements of a simple headbox used with flat-bed forming are shown in Fig. 5.12 and include a stock inlet; a partitioned interior created by an adjustable baffle to control stock levels that create two *ponds* with a differential head; an adjustable slice, which controls stock flow to the former; and an apron, which acts as a landing zone for the exiting stock or jet. The headbox contains at least one or more rectifier or *holey rolls*. Closed or *air-padded* headboxes overcame this limitation and dampened the level of pulsations observed with open headbox designs. These are designed for speeds up to 1000 m/min. Hydraulic headboxes use a tapered header design in combination with turbulence generators or tubes to deliver stock to the slice opening. These headboxes are used for speeds above 600 m/min. They are compact and can be used at any angle. The flexibility of the hydraulic headbox design has led to the development of a number of forming designs including: roll, blade, and shoe gap formers, as well as s-wrap, c-wrap, and crescent formers (Fig. 5.13).

In designing headboxes, a balance is sought between turbulence generation and length-scale knowing that the time to form a coherent floc is quite short. For softwood fibres, coherent floc can form in about 0.2 s. This corresponds to a distance of about 40 cm for a machine speed of 1200 m/min. Another important variable is the jet to wire ratio (J/W), which is the ratio between the stock or jet velocity and that of the moving wire or forming fabric. The J/W establishes the degree of turbulence or *activity* of the stock as it dewaters across the wire. It also influences the degree of fibre orientation, tensile ratio (MD/CD), and the quality of formation. Fibre orientation will also influence the anisotropy of the final wetlaid nonwoven fabric in terms of the directional in-plane transport of liquids [64]. A J/W ratio of between 0.9 and 1.1 is typical, and the best degree of formation is achieved with a J/W=1.0 [65].

The former is the main drainage element of the wetlaid forming machine. It consists of an endless wire or forming fabric, and dewatering elements, such as table rolls and foils, as well as vacuum boxes. These devices span the width of the machine and are placed along its length to remove water from the web. Table rolls and foils are placed in contact with and below the

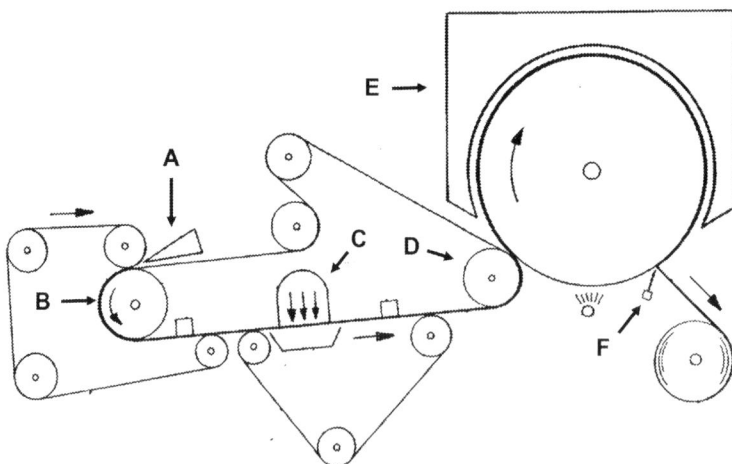

FIG. 5.13 Typical crescent former configuration showing hydraulic headbox (A), forming roll (B), air press (C), pressure roll (D), Yankee cylinder and drying hood (E), and creping doctor (F). *Adapted from F.-J. Chen, et al., Wet-Resilient Webs and Disposable Articles Made Therewith, U.S. Patent No. 6,808,790, 2004.*

forming fabric. Table rolls freely rotate in the machine direction, creating a vacuum at the out-going nip to remove water. The vacuum pulse associated with table rolls can also induce turbulence in the stock. Foils are tapered wing-like elements that create a milder vacuum than table rolls as they do not generate a pulse. Their landing and angles can be adjusted to create a range of vacuum profiles. During the forming process water is removed through a number of dewatering mechanisms. As the stock dewaters its consistency increases. Fast-draining stocks, made from petroleum-derived man-made fibres, will dewater quickly, whereas slow-draining stocks, such as those containing wood pulp, natural fibres or regenerated cellulosic man-made fibres will tend to dewater more slowly and will thus require more active methods of dewatering.

When the consistency is greater than about 2.5%, passive vacuum elements, such as table rolls and foils become ineffective. At this point, vacuum boxes are used. Vacuum boxes are typically placed near the so-called *dry line*. This is a point along the machine where fibres begin to disrupt the air-water interface whereupon the web transitions from a wet shiny surface to one that is matt in appearance. Except for evaporative losses, all water removed on the former is collected in the save-all. Most facilities will recycle this water and use it for dilution and shower water. However, in some demanding nonwoven grades, this water is not re-used. The recovery and reuse of process water or system closure is possible and has been demonstrated. Water treatment systems are typically more complex with grades based on natural fibres because organic material can enter the discharge water. This may require biological treatment to meet water quality standards for biological oxygen demand (BOD) and chemical oxygen demand (COD) prior to discharge [66].

5.5.3.1 Recent advances

Although wetlaid machine design has evolved considerably over many years, it is likely that the basic design elements and unit operations would be easily recognised by the likes of Fourdrinier and Dickenson. Innovation in wetlaid forming has been largely focused on maximising dewatering rates while achieving the highest levels of web uniformity. In addition, tissue manufacturers have helped drive the development of a number of innovative machine designs, to improve the strength, softness, and absorbency of tissue and towel grades.

As an alternative approach to conventional wetlaid forming, foam forming is process that uses stabilised foams as a transfer medium for fibres that can substantially reduce water use [67]. Foam forming has been used with a range of natural and man-made fibres and demonstrated using 6-mm-long Tencel fibres [68]. In the foam-laid technology, aqueous foam is used (instead of water alone) as a process fluid and flowing medium. The air content in the foam is controlled to be within 60%–70% of the total volume, containing small, spherical air bubbles with diameters below 100 μm. These foams allow the use of a wide variety of different raw materials, from nanoparticles up to 200 mm long fibres, and also other low density materials [69].

Multilayer wetlaid web forming is used to produce a variety of high performance products. The process provides additional flexibility in product design to achieve both enhanced performance and optimisation of material costs. Multilayer nonwovens are produced in series, by forming an additional layer on a previously formed layer; in parallel, by separately forming individual layers and laminating them together; or concurrently, using a multilayer headbox [70]. The layers may be discretely organised (i.e., the mixing or interaction between the adjacent layers is very weak), or they may be mixed (bound) together very efficiently. Multilayer forming options to manufacture such composite wetlaid nonwovens are numerous. One example is the multipond inclined former (Fig. 5.14).

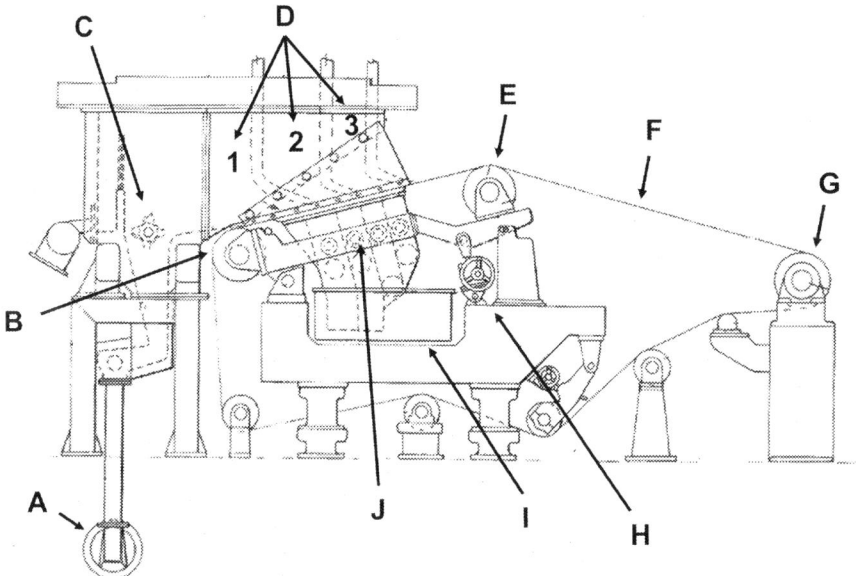

FIG. 5.14 Example of an inclined-wire former designed by Sandy Hill Corporation used to produce multilayer composites. The Deltaformer showing stock inlet (A), breast roll (B), rectifier roll (C), three ponds and slice elements (D), wire roll (E), forming wire (F), couch roll (G), table angle adjustment (H), save-all (I), and suction boxes (J). *Adapted from M.B. Keller, Inclined Multiplyformer, U.S. Patent No. 5,011,575, 1991.*

FIG. 5.15 Multi-layer headbox design showing three independent stock supply lines (A), headbox (B), mixing chambers (C), slice dividers (D), and breast rolls (E). *Adapted from H.W. Verseput, Multiple Ply Web Former with Divided Slice Chamber, U.S. Patent No. 3,923,593, 1975.*

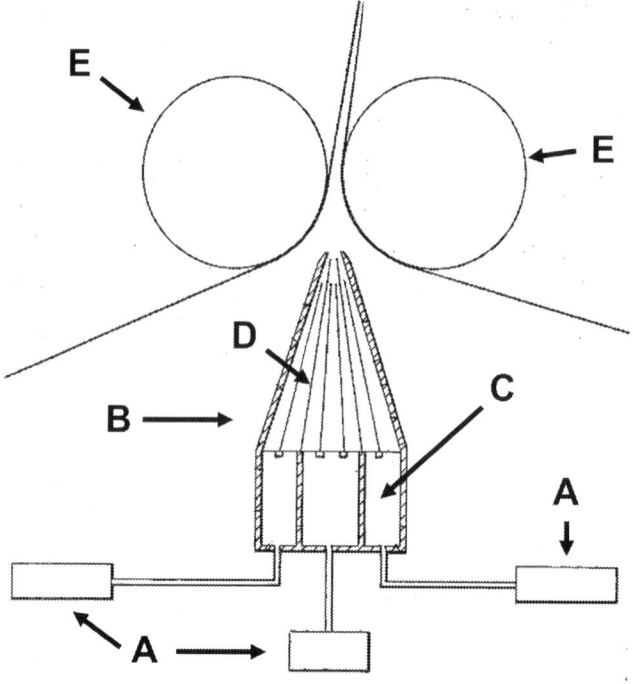

Multilayer or stratified headboxes use vane inserts in hydraulic headboxes to improve the velocity profile of the jet. The vanes are oriented parallel to the stock flow to create separate chambers in the headbox, which can be supplied by different stocks to create a multilayer jet and ultimately a multi-ply sheet. Beloit introduced the first stratified headbox, the Converflo, which was followed by the first multilayer headbox, the Strata Flo. The challenge of multiheadboxes has been mixing between layers. However, a number of improvements have been made to address this problem. For example, the MasterJet (Voith) uses lamella to separate the plies, whereas the OptiFlo headbox from Valmet introduces a water layer between the top- and bottom-slice layers (Fig. 5.15). Acoustic radiation has also been proposed as a means for creating a layered sheet [71].

5.5.4 Wet pressing

The newly formed still-wet nonwoven web contains water, which must be removed. The amount of water will depend on the nature of the fibres and additives used to prepare the stock. Petroleum-derived man-made fibres do not absorb water in any appreciable amount and will drain quickly. In contrast, natural fibres, wood pulp and regenerated cellulosic man-made fibres both swell in water and retain a significant amount of water in the pore structure of the web. Nonwovens made from natural fibres can contain upwards of 75% water by weight when they leave the forming section.

In order to efficiently remove this water, the still-wet nonwoven is passed through a 'press section'. In its simplest configuration, the press section comprises two rolls and one or two felts. The web entering the nip is compressed to allow water to transfer from the wet paper to the felt. Since most wetlaid nonwovens are made from furnishes that are free draining, most wetlaid machines will have a simple *straight-through* press section that is either single- or double-felted. Light press loads (<50 kN/m) are common. In some cases, such as the production of carbon or glass mats, press sections may not be used.

In the wet press nip, the nonwoven web will experience a pressure profile that is determined by the external load and nip length. Although the mechanics of dewatering is complex, the limits to wet pressing and hence water removal will be a function of whether the nip is compression controlled or flow controlled [72]. Compression-controlled nips occur with lightweight (<100 g/m^2) webs made from furnishes that are free draining. Here, dewatering is controlled by mechanical compression and care must be taken not to apply excessive nip loads that may *crush* the web. Crushing occurs when water flow in the nip is in the plane of the sheet. Nonwoven webs containing a high percentage of petroleum-derived man-made fibres will fall under this regime. In comparison, heavier webs made from furnishes that are difficult to drain will create a flow-controlled nip. In this case, the dwell-time in the nip will control the amount of water removed [73]. The solids content of the web exiting the press section will be >50% for most nonwoven grades.

5.5.5 Drying

The final process of water removal from the wet web is done in the dryer section. Depending on the furnish and product specification, drying of the wet sheet can be accomplished using several dryer technologies using either contact or noncontact driers. Additionally, for fast-draining fibres, such as carbon or glass fibres, an oven is commonly used. The dryer section will remove upwards of 50% of the remaining water in the sheet to achieve a targeted final moisture content, which is typically 5%–7% by weight. However, the target moisture content may be at or near 0% or *bone dry* if the sheet is then off-machine calendered or coated. Table 5.4 gives a list of common wetlaid products and the types of dryer used in each application.

TABLE 5.4 Drying options for common wetlaid nonwovens.

Products	Basis weights (gsm)	Dryer type
Battery separators, filtration media, high performance papers, wall coverings	4–200	Contact Through-air
Coffee and tea filters	10–50	Through-air
Disposable wipes, industrial wipes	40–100	Through-air
Carbon, ceramic, glass, and metal fibre high performance nonwovens	20–250	Oven

Contact driers include both steam-heated cylinders or *cans* and Yankee dryers [74]. Cylinders are typically made from cast-iron and are 1.8 m in diameter. These are placed in series and coupled with a dryer fabric, which is used to support the sheet and hold it tightly against the cylinder. Steam pressures are subatmospheric to 1000 kPa. Drying takes place by the transfer of heat from the steam through the dryer shell and into the web. Water is then evaporated from the web into the surrounding air during the draw between the cylinders. The dryer section is hooded to maximise ventilation and recover heat. Yankee dryers are large cast-iron cylinders up to 6.7 m in diameter with extremely high drying rates, typically 4–6× greater than conventional steam-heated cylinders. The use of hot air to accelerate drying rates, the so-called impingement drying, can be used in contact drying. High drying rates of up to 200 kg m^{-2} can be achieved. Impingement drying is typically used for higher basis weight products (>75 gsm) with high natural fibre content.

When maximum bulk preservation is required, as in the production of tissue grades, through-air dryers are used. The through-air dryer or TAD features a rotating perforated drying cylinder, a supporting fabric, and a hood. Typically, hot air is supplied from the drying cylinder to pass through the still-wet web and supporting fabric, with the resulting moist hot air vented by hood. If a TAD unit is used, care is taken to minimise wet pressing. In most cases, a press section is not used. The supporting TAD fabric may be embossed to impart a pattern to the web [75].

Preheating of the web prior to the oven with infrared heating units can also be used to activate thermal bonding if low melt fibres are present, and impart sensible heat to the wet web. Other noncontact drying options should be considered when using fibres with low melting points as these can stick to dryer cans. In addition, care must be taken when using these fibres as the web will stretch during drying and lose tension, which can cause poor uniformity in the sheet and create operational issues such as poor reel building.

5.5.6 Bonding of wetlaid webs

Depending on the fibre composition of the wetlaid web and the intended application of the final fabric, it may be consolidated by hydrogen bonding through pressing and drying, thermally bonded, mechanically bonded by hydroentangling, or chemically bonded using latex binders. Bonding can also involve combinations of processes used in sequence. For example, some lightweight wetlaid webs intended for hygiene applications are both hydroentangled and latex bonded, while others are hydroentangled and thermally bonded. These nonwoven bonding processes take place as part of one continuous manufacturing operation following wetlaid web forming.

5.5.6.1 *Hydrogen bonding*

In wet pressing and drying, wetlaid webs made of cellulose fibres including wood pulp can exploit hydroxyl groups to form hydrogen bonding and generate significant network strength if the fibres are fibrillated prior to web formation. Such wetlaid hydrogen bonded fabrics tend to have poor wet strength, so may be further processed to improve physical properties by, for example, coating.

5.5.6.2 Hydroentangling

Wetlaid webs comprising mixtures of wood pulp and man-made fibres are used to produce a variety of nonwovens products such as consumer and industrial wipes, and towels.

These products typically require a high-degree of wet-strength. Increased wet strength can be achieved by chemical bonding, mechanical bonding, or a combination of both. Chemical bonding includes the addition of wet-strength additives, such as cross-linking agents, binders, low-melt fibres, and superabsorbers [76]. In terms of mechanical bonding, hydroentangling is the most common for wetlaid webs.

Hydroentanglement (or spunlacing) is a process of using fluid forces to intertwine constituent fibres in the web, increasing interfibre friction and thereby the final fabric strength. Small-diameter water jets operating at high pressure concentrate hydraulic energy towards the surface of the moving web [77]. The jets entangle the fibres, which enhances network strength even in the absence of chemical bonding agents [16]. In addition to strength, hydroentanglement can also impart desirable physical properties, such as a soft handle and drapability.

Each water jet manifold or injector houses a jet strip with of 100–200 µm diameter holes arranged in rows having 30–80 holes per 25 mm [78]. Water pressures will depend on the application, but systems are typically designed for a maximum operating pressure of 40 MPa. Suction boxes may be placed under the conveyor to collect water and loose fibre that is released from the web. Operating conditions are chosen to optimise fabric properties and to minimise fibre losses, which is a concern when short fibres or wood pulp are present.

Hydroentangling units can be placed at various positions after the wetlaid web has been formed, depending on the needs of the final product. For most applications where the webs contain natural fibres, the hydroentangling unit will be placed at a position where the web solids are at least 15%. On wetlaid lines they are typically configured for single-side applications.

Wetlaid spunlacing or WLS has seen expanded use in the production of biodegradable (flushable) wipes. Flushable in this context means water dispersible according to INDA and EDANA's GD4 industry flushability standard [79]. For such applications, the product is typically made from wood pulp and short cut lyocell (<10 mm), although other man-made fibres can be used. Hydroentangling at high pressure can create sufficient dry and wet fabric strength, yet the fibres can be easily re-dispersed in toilet water during flushing. Wetlaid webs containing pulp, regenerated cellulosic fibres, and a very small proportion bicomponent fibres are also hydroentangled and thermally bonded to further modulate wet strength to make flushable wipes.

In selecting fibre types for hydroentangling, fibre morphology is an additional consideration as those with crenulated cross-sections can produce nonwoven fabrics with superior strength than those made from fibres with round or oval cross-section [80]. Hydroentangling can also be used to create patterns and microperforations in fabrics. Its ability to fibrillate fibres such as lyocell and wood pulp can be used to modify the interfibre pore size distribution of the fabric. Hydroentangling can also be used to manufacture nonwoven preforms for composites using blends of glass and low melt polyester or bicomponent sheath/core (polyester/polyethylene) [81].

5.5.6.3 Latex bonding

Latex bonding involves adhering fibres in wetlaid webs using water-based latex emulsions. The latex can be added either prior to web formation or afterwards. In contrast to the bonding of drylaid webs, where the latex is only applied after the web has been formed, in wetlaid processing, latex emulsions can also be mixed with the fibre suspension in water prior to web forming. The mixing process is facilitated by adjusting the pH or ionic concentration to destabilise the water-based emulsion, enabling improved intermixing with the fibre. Common latex emulsions are based on ethyl acetate, acrylic and butadiene styrene copolymer.

5.5.6.4 *Thermal bonding*

Low melt binder fibres can be included in wetlaid webs to facilitate thermal bonding, as well as to confer fabrics with the ability to be heat sealed, which may be essential for subsequent conversion into finished products. While homofil thermoplastic fibres are used, particularly if the web is to be heat calendered, copolyesters and bicomponents based on for example, PP/PE or PET/coPET, in sheath-core or side by side configurations are most common. Wetlaid webs containing such bicomponent fibres are usually through-air bonded. Although not strictly an example of thermal bonding, incorporation of certain hot water soluble PVA binders in the web that can be solubilised in the interstitial water during the drying process can also create adhesive bonding of fibres during subsequent cooling.

5.5.7 Impregnation and coating

A variety of unit operations can be used to transfer chemicals and materials to wetlaid webs to impart additional functionality to the nonwovens product. These materials are typically low-viscosity slurries made from natural or man-made polymers, such as starch and polyvinyl alcohol, or thermoset resins, such as styrene-butadiene latexes, or pigments. Solutions and slurries can be applied neat or formulated as a mixture. When activated, typically through heating, they impart valuable properties, such as liquid barrier resistance or substantially modified mechanical properties in the wetlaid nonwoven. For example, addition of nanofibrillated cellulose improves tear resistance and tensile strength [82], while addition of expandable microspheres can enhance thermal insulation and impact resistance [83].

Impregnation and coating can be done both in-line with wetlaid nonwovens production *on-machine* or off-machine. Common processing methods include dip coating, spray coating, curtain coating, and size-press application. Off-machine surface treatments include extrusion, lamination, plasma deposition of films, cascade coating and a variety of printing methods, such as gravure. Impregnation involves complete saturation of the substrate whereas dip coating transfers the coating to the moving web through direct immersion. Dip coating is the simplest coating method but coating weight or *pick-up* can be difficult to control as it is a nonmetering coating method. In contrast, impregnation units or impregnators are designed to deliver controlled amounts of the coating material to the substrate.

5.5.8 Finishing

Finishing refers to unit operations designed to impart added functionality to the wetlaid nonwoven web. These are typically mechanical operations, calendering and creping being the most commonly used.

5.5.8.1 *Calendering*

Calendering is the process of passing the wetlaid nonwoven through one or more nips formed by loaded calender rolls. Typically, the surface of at least one of the rolls is smooth. Application of heat and pressure imparts smoothness and increases gloss on the fabric surface. Calendering can irreversibly deform the thickness of the fabric causing it to densify. This can be used to improve cross-direction calliper uniformity and reel building, as well as be used to decrease air permeability. Calender rolls are usually made from steel and heated either by induction systems or hot oil. The rolls may be crowned or be of the variable crown type, which use narrow (150 mm) inductively heated control zones to adjust roll diameter across the length of the roll. Prior to entering the nip, steam may be applied to the surface of the nonwoven. The steam acts as a plasticiser, enhancing smoothness development and can be used to preserve bulk.

A simple calendering configuration is a single nip calender comprising one heated steel roll and one covered roll, with a synthetic rubber cover. The hardness of the soft roll determines the degree of compression under load. As the web enters the calender nip it experiences a shear field, which causes the web to conform to the surface of the steel roll. As a result, the surface topography of the steel roll is imparted to the web. The choice of calender rolls and calendering conditions will depend on the material properties of the nonwoven and the intended product application. Typical calendering conditions for different types of wetlaid nonwovens are listed in Table 5.5.

5.5.8.2 *Creping*

Nonwovens containing wood pulp or regenerated cellulose fibres can derive much of their dry strength from hydrogen bonding between fibres. In applications where softness is desirable, such as tissue grades, the degree of strength imparted by hydrogen bonding may be too high, causing the product to be too stiff. A common method of imparting softness, as well as a means to increase the bulk and/or extensibility of the fabric is through creping. In one common approach, the still-moist web is brought into intimate contact with the hot surface of a Yankee cylinder, which causes the sheet to adhere strongly it. Release agents and creping doctors are used to remove the web. The creping doctor weakens the internal bond strength of the web and creates microfolds in the sheet, which imparts bulk and softness. These characteristics are especially important in the production of tissue and towel [84].

By introducing man-made fibres, such as PE/PET bicomponent fibres, to the furnish it is possible to create permanent creping, which does not wash out. This has been used to produce a variety of products including medical gowns [85]. Other methods for introducing

TABLE 5.5 Calendering conditions for wetlaid nonwovens.

Product	Furnish	Line load (kN/m)	Temperature (°C)
Battery separator	Polyolefin fibres/cellulose fibres	400–600	120–150
High performance paper	Aramid fibres	150–250	100–400
Insulating paper	Polyolefin fibres/aramid fibres/kaolin	125–175	175–200
Ultrafiltration membrane	Polyolefin fibres/lyocell fibres	50–250	180–240

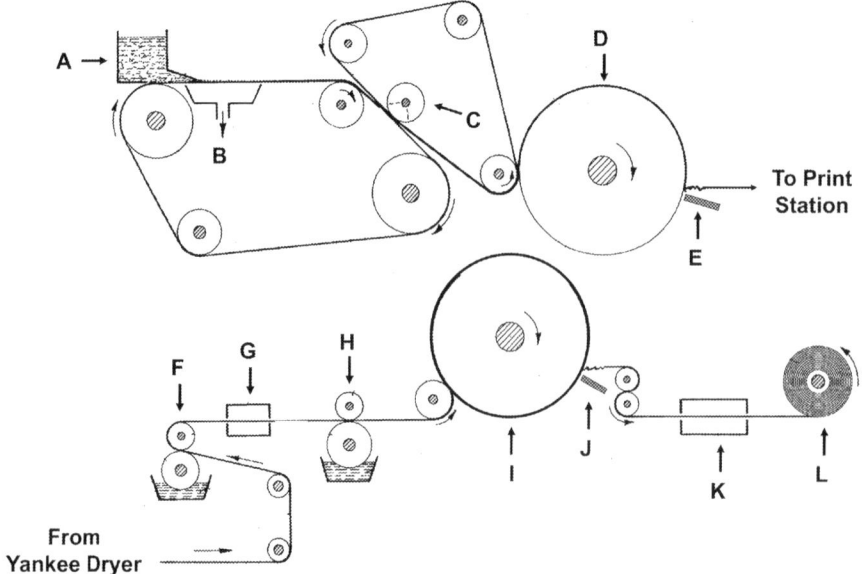

FIG. 5.16 The DRC process showing former (A), save-all (B), vacuum pick-up roll (C), Yankee dryers (D, I) and doctor blades (E, J), print stations (F, H), through-air dryers (G, K), and reel (L). *Adapted from V.R. Gentile, et al., Absorbent Unitary Laminate-Like Fibrous Webs and Method for Producing Them, U.S. Patent No. 3,879,257, 1975.*

mechanical (compressive) creping are by Sanforising, Micrex, or Clupak. In the Clupak process, a moist web is introduced into a nip formed between a smooth metal drum and a longitudinally prestretched elastic rubber blanket. The rubber is then allowed to contract back to its original dimensions, causing the web to shrink with it and slip across the surface of the metal roll. The resultant shrinkage is then dried into the sheet [86].

An important development in creping methods was the DRC (double re-crepe) process [87]. In the DRC process, shown in Fig. 5.16, the wet web is first printed on one side with an emulsion polymer, which imparts strength to the web. It is then creped, printed on the second size and re-creped. The resulting nonwoven product exhibits both softness and a high degree of durability and strength.

5.6 Summary

Although wetlaid forming accounts for less than one-quarter of the estimated $47 billion global nonwovens market, it represents a powerful manufacturing platform for producers of industrial and consumer nonwoven fabrics. Wetlaid products are found in a range of markets including personal care and hygiene, air and liquid filtration, building products, electronics, and aeronautics.

Wetlaid forming is ideally suited to the production of lightweight nonwoven fabrics with low variation in weight per unit area. It is particularly suited to making fabrics where mechanical properties are fully or at least partly derived from hydrogen bonding that occurs between fibres. As such, natural fibres, including wood pulp, natural occurring cellulosic fibres, such as cotton and regenerated cellulose fibres are ideal fibre types that are commonly used in wetlaid forming. Products made from cellulosic fibres, specifically hygiene products and wipes, are seeing strong demand from consumers for their ease of use, biodegradability, and perceived sustainability. Moreover, a hydrogen bonded product can be made to redisperse, which has opened new and growing market segments in flushable products.

Wetlaid forming is also ideal for fibres that are difficult to process by other web forming methods, when a uniform, lightweight fabric is required. As a result, wetlaid forming is used to produce a range of products containing specialty and high-performance fibres such as those made from glass, metals, polyaramids, and carbon fibres. In addition, the ability to formulate stocks containing a mixture of fibre types and additives allows the product developer to enjoy large degrees of freedom in terms of material selection. Additional process technologies, such as hydroentanglement, coating, and calendering, can be used to refine product design and impart enhanced properties that can achieve targeted product performance.

A major barrier to wetlaid forming is the fact that it is water-based. Water handling and management can indeed complicate the production process and increase capital costs relative to drylaid forming. Moreover, companies with core competencies in drylaid forming are not necessarily eager to adopt new manufacturing platforms based on water. On the other hand, hydroentanglement is water-based and well-established in the nonwovens industry. By displacing binders and petroleum-derived low-melt binder fibres, hydroentanglement has proven to be one of the most sustainable methods for bonding nonwoven webs. From this vantage point, moving from carded spunlace manufacturing to wetlaid is considered a low-barrier transition.

As discussed, wetlaid forming is uniquely suited for certain fibre types and products. The increasing pressure from all levels of society as well as changes in both national and global legislative frameworks is forcing many segments of the industry to develop sustainable end-of-life solutions, particularly when it comes to single use products. These secular changes are expected to place much higher emphasis on sustainability and biodegradability with respect to environmental impacts. As such, wetlaid forming will continue to be a vital and relevant component of nonwovens manufacturing, realizing increasing importance in the near future.

References

[1] International Organization for Standardization, Nonwovens Vocabulary, 2019, ISO Standard No. 9092:2019. Retrieved from: https://www.iso.org/obp/ui/#iso:std:iso:9092:ed-3:v1:en.

[2] M.A. Hubbe, C. Bowden, Handmade paper, review, BioResources 4 (4) (2009) 1736–1792.

[3] J.A. McGaw, Most Wonderful Machine: Mechanization and Social Change in Berkshire Paper Making, 1801–1885, Princeton University Press, 1987.

[4] D.W. Manson, Fourdrinier papermaking, in: B.A. Thorp, M.J. Kocurek (Eds.), Pulp and Paper Manufacture, Vol. 7, Paper Machine Operations, third ed., TAPPI Press, Atlanta, 1991, pp. 192–215.

[5] K. Masuda, Japanese paper and Hyōgu, Pap. Conserv. 9 (1) (1985) 32–41.

[6] F.H. Osborne, Process of Making Porous Long Fibered Nonhydrated Paper, 1936. US Pat. 2,045,095.

[7] F.H. Osborne, Thermoplastic Paper and Process of Preparing the Same, 1944. US Pat. 2,414,833.

[8] E. Sjöström, Wood Chemistry: Fundamentals and Applications, Academic Press, New York, 1981.

[9] A. Kinnane, DuPont: Form the Banks of the Brandywine to Miracles of Science, E.I. du Pont de Nemours and Company, Wilmington, DE, 2002.

[10] I. Sakurada, The Sakurada laboratory, in: The Commemoration Volume for the Silver Jubilee, Kyoto University, 1951, pp. 84–91.

[11] P. Mango, The Future of Wetlaid Nonwovens to 2023, SmithersPira, Surrey, United Kingdom, 2019.

[12] A.E. Johnson, We Need to Kick our Addiction to Plastic, Scientific American, 2017. Retrieved from: https://blogs.scientificamerican.com/observations/we-need-to-kick-our-addiction-to-plastic/.

[13] M. Ring, M.P. Godsay, R.S. Swenson, J.N. Kent, Method for Forming Wet-Laid Nonwoven Fibres, 1977. US. Pat. 4,007,083.

[14] E.J. Tozzi, D.M. Lavenson, M.J. McCarthy, R.L. Powell, Effect of fibre length, flow rate, and concentration on velocity profiles of cellulosic fibre suspensions, Acta Mech. 224 (10) (2013) 2301–2310.

[15] J.D. Parker, The Sheet Forming Process, TAPPI Press, Atlanta, GA, 1972.

[16] M.A. Hubbe, A.A. Koukoulas, Wet-laid nonwovens—chemical approaches using synthetic and cellulosic fibres, BioResources 11 (2) (2016) 5500–5552.

[17] J.M. Keith, Dispersion of synthetic fibre for wet-lay nonwovens, in: TAPPI 1994 Nonwovens Conference Proceedings, 1994, pp. 237–239.

[18] R.M. Rowell, Natural fibres: types and properties, in: Properties of Natural-Fibre Composites, Woodhead Publishing, 2014, pp. 3–66.

[19] J.S. Han, Properties of nonwood fibers, in: Proceedings of the Korean Society of Wood Science and Technology Annual Meeting, 1998.

[20] J.R. Barnett, V.A. Bonham, Cellulose microfibril angle in the cell wall of wood fibres, Biol. Rev. Camb. Philos. Soc. 79 (2) (2004) 461–472.

[21] C. Salas, M.A. Hubbe, O.J. Rojas, Nanocellulose applications in papermaking, in: Z. Fang, R. Smith Jr., X.F. Tian (Eds.), Production of Materials From Sustainable Biomass Resources. Biofuels and Biorefineries, vol. 9, Springer, Singapore, 2019.

[22] A.A. Koukoulas, Wet-laid nonwovens: where do we go from here? in: INDA RISE2017, Raleigh, NC, 2017.

[23] C.E. Everhart, D.O. Fischer, F.R. Radwanski, H. Skoog, High Pulp Content Nonwoven Composite Fabric, 1994. U.S. Pat. No. 5,284,703.

[24] R. Farer, S. Batra, T. Gilmore, Processing cotton and cotton/pulp blends for wet-laid nonwovens using novel dispersion technologies, in: 1998 Nonwovens Conf.: Proceedings, TAPPI Press, Atlanta, 1998, pp. 133–136.

[25] A.P.S. Sawhney, B.D. Condon, Future of cotton in nonwovens, Text. Asia (2008) 13–15.

[26] P.R. Gruber, E.S. Hall, J.J. Kolstad, M.L. Iwen, R.D. Benson, R.L. Borchardt, Continuous Process for the Manufacture of a Purified Lactide From Esters of Lactic Acid, 1993. U.S. Pat. No. 5,247,059.

[27] D.J. Haynes, Melt-Bondable Fibres for Use in Nonwoven Web, 1992. U.S. Pat. No. 5,082,720.

[28] S.F. Nielsen, B.L. Davies, Wet Laid Bonded Fibrous Web Containing Bicomponent Fibres Including Lldpe, 1992. U.S. Pat. No. 5,167,765.

[29] J.S. Dugan, Heat Bondable Biodegradable Fibres With Enhanced Adhesion, 2003. U.S. Pat. No. 6,509,092.

[30] E.A. Merriman, Kevlar® aramid pulp for paper making, in: Seminar Notes: 1981 Nonwoven Fibres and Binders, TAPPI Press, Atlanta, 1981, pp. 5–9.

[31] P.W. Morgan, Synthetic Polymer Fibrid Paper, 1961. U.S. Patent 2,999,788.

[32] Dupont, NOMEX®: A Family of Insulation Materials, 2016. http://www2.dupont.com/ReliatranV3/en_RU/products/Nomex/More/Nomex_family.html.

[33] Teijin, Twaron Paper, 2016, Retrieved from: http://www.teijinaramid.com/aramids/twaron/twaron-paper/.

[34] S.L. Sinsabaugh, G. Pensero, H. Liu, L.P. Hetzel, High Surface Area Flow Battery Electrodes, 2018. U.S. Pat. No. 9,893,363.

[35] W.A. Hare, Process for Making Rough Surface Filaments, 1954. U.S. Pat. No. 2,695,835.

[36] D.W. Lare, Self-Bonding Synthetic Wood Pulp and Paper-Like Films Thereof and Method for Production of Same, 1977. U.S. Pat. No. 4,049,492.

[37] H. Shin, R.K. Siemionko, Flash Spinning Process, 1997. U.S. Pat. No. 5,672,307.

[38] S.M. Kozak, Use of specialty cellulose fibres and modified cellulose fibres in nonwovens, in: TAPPI Nonwovens Conf. Proc., Marco Island, FL, 1991, pp. 37–48.

[39] Z.-M. Huang, Z. Zhang, M. Kotaki, S. Ramakrishna, A review on polymer nanofibres by electrospinning and their applications in nanocomposites, Compos. Sci. Technol. 63 (15) (2003) 2223–2253.

[40] J.H. Kim, et al., Review of nanocellulose for sustainable future materials, Int. J. Precis. Eng. Manuf. Green Technol. 2 (2) (2015) 197–213.

[41] M.Z. Elsabee, H.F. Naguib, R. Elsayed Morsi, Chitosan based nanofibres, review, Mater. Sci. Eng. C 32 (7) (2012) 1711–1726.

[42] M.A. Hubbe, A. Ferrer, P. Tyagi, Y. Yin, C. Salas, L. Pal, O.J. Rojas, Nanocellulose in thin films, coatings, and plies for packaging applications: a review, BioResources 12 (1) (2017) 2143–2233.

[43] B.G. Morin, Methods of Making Single-Layer Lithium Ion Battery Separators Having Nanofibre and Microfibre Components, 2012. U.S. Pat. No. 8,936,878.

[44] S. Sohn, S.L. Lawrence, E.O. Teutsch, Reinforced Polymer Laminate, 2019. U.S. Pat. No. Appl. 2019/0001650.

[45] V. Kumar, D.W. Bousfield, M. Toivakka, Slot die coating of nanocellulose on paperboard, TAPPI J. 7 (1) (2018) 9–17.

[46] D.W. Sumnicht, J.H. Miller, Disposable Cellulosic Wiper, 2012. U.S. Pat. No. 8,187,422.

[47] R.E. Mark, Mechanical properties of fibers, in: R.E. Mark (Ed.), Handbook of Physical Testing of Paper, second ed., vol. 1, Marcel Dekker, New York, 2002.

[48] S.K. Nemani, R.K. Annavarapu, B. Mohammadian, A. Raiyan, J. Heil, M.A. Haque, A. Abdelaal, H. Sojoudi, Surface modification of polymers: methods and applications, Adv. Mater. Interfaces (2018), 1801247.

[49] F. Lundell, D. Söderberg, H. Alfredsson, Fluid mechanics of papermaking, Annu. Rev. Fluid Mech. 43 (2011) 195–217.

[50] R.J. Kerekes, C.J. Schell, Characterization of fibre flocculation regimes by a crowding factor, J. Pulp Pap. Sci. 18 (1) (1992) J32–J38.

[51] E.J. Powers, S.F. Nielsen, J.E. Smith, T.S. Thornburg, Filtration Structures of Wet Laid, Bicomponent Fibre, 1996. U.S. Pat. No. 5,580,459.

[52] M.K. Ramasubramanian, D.A. Shiffler, A. Jayachandran, A computational fluid dynamics modeling and experimental study of the mixing process for dispersion of synthetic fibres in wet-lay forming, TAPPI J. (2010) 6–13.

[53] H. Corté, Cellulose-water interactions, in: H.F. Rance (Ed.), The Raw Materials and Processing of Papermaking, Elsevier, New York, 1980, pp. 1–90.

[54] K. Stana-Kleinscheka, T. Kreze, V. Ribitsch, S. Strnad, Reactivity and electrokinetical properties of different types of regenerated cellulose fibres, Colloids Surf. A Physicochem. Eng. Asp. 195 (1–3) (2001) 275–284.

[55] A.A. Koukoulas, Measurement of bound water content of pulp fibers using near-infrared diffuse reflectance spectroscopy, 1993 (Unpublished manuscript).

[56] B. Radvan, Forming the web of paper, in: H.F. Rance (Ed.), The Raw Materials and Processing of Papermaking, Elsevier, New York, 1980, p. 165.

[57] X. Hou, M. Acar, V.V. Silberschmidt, 2D finite element analysis of thermally bonded nonwoven materials: continuous and discontinuous models, Comput. Mater. Sci. 46 (3) (2009) 700–707.

[58] D.H. Page, A theory for the tensile strength of paper, TAPPI 57 (4) (1969) 678–681.

[59] K. Poehler, Personal Communication, 2019.

[60] M.A. Hermans, R.J. Makolin, K.A. Goerg, F.-J. Chen, Method of Treating Papermaking Fibres for Making Tissue, 1994. U.S. Pat. No. 5,348,620.

[61] J I Bergstrom, Method of Forming a Multi-Ply Web From Paper Stock, 1983. U.S. Pat. No. 4,376,012.

[62] T. Lindström, L. Wågberg, T. Larsson, On the nature of joint strength in paper—a review of dry and wet strength resins used in paper manufacturing, in: 13th Fundamental Research Symposium, Cambridge, 2005, pp. 1447–1464.

[63] I.M. Hutten, The concerns of forming wet lay nonwovens from long fibres, in: TAPPI 1995 Nonwovens Conference, Atlanta, 1995, pp. 161–172.

[64] C. Deng, J. Hou, X. Zhang, R.H. Gong, X. Jin, Controllable anisotropic properties of wet-laid hydroentangled nonwovens, TAPPI J. 18 (3) (2019) 173–180.

[65] O. Svensson, S. Schröder, The effect of jet/wire speed ratio on various paper characteristics, Svensk. Papperstidn. 68 (2) (1965) 25–33.

[66] U. Hamm, S. Schabel, Effluent-free papermaking: industrial experiences and latest developments in the German paper industry, Water Sci. Technol. 55 (6) (2007) 205–211.

[67] A.P.J. Gatward, Long-fibre developments in UK and Europe, Pap. Technol. Ind. (1973) 272–274.

[68] A. Koponen, et al., The effect of in-line foam generation on foam quality and sheet formation in foam forming, Nord. Pulp Pap. Res. J. 33(3) (2018) 482–495, https://doi.org/10.1515/npprj-2018-3051.

[69] K. Kinnunen, et al., Benefits of foam forming technology and its application in high MFC addition structures, in: Advances in Pulp and Paper Research 2013, FRC, Manchester, 2013, pp. 837–850.

[70] B.W. Attwood, Multi-ply forming, in: B.A. Thorp, M.J. Kocurek (Eds.), Pulp and Paper Manufacture, Vol. 7, Paper Machine Operations, third ed., TAPPI Press, Atlanta, 1991, pp. 237–259.

[71] C.S. Park, H. Xu, Method of Making a Stratified Paper, 2005. U.S. Pat. No. 6,902,650.

[72] R.J. Kerkekes, J.D. McDonald, A decreasing permeability model of wet pressing: theory, in: TAPPI Proceedings 1991 Engineering Conference, TAPPI Press, Atlanta, 1991, pp. 551–558.

[73] R.A. Reese, Pressing operations, in: B.A. Thorp, M.J. Kocurek (Eds.), Pulp and Paper Manufacture, Vol. 7, Paper Machine Operations, third ed., TAPPI Press, Atlanta, 1991, pp. 260–281.

[74] K.C. Hill, Paper drying, in: B.A. Thorp, M.J. Kocurek (Eds.), Pulp and Paper Manufacture, Vol. 7, Paper Machine Operations, third ed., TAPPI Press, Atlanta, 1991, pp. 282–305.

[75] A.P. Bakken, M.A. Burazin, J.D. Lindsay, Non-Woven Through Air Dryer and Transfer Fabrics for Tissue Making, 2004. U.S. Pat. No. 6,875,315.

[76] J.M. Euripides, L.C. Phillips, S.F. Nielsen, Process for Making a Non-Woven, Wet-Laid, Superabsorbent Polymer-Impregnated Structure, 1998. U.S. Pat. No. 5,795,439.

[77] H. Viazmensky, C.E. Richard, J.E. Williamson, Water Entanglement Process and Product, 1991. U.S. Pat. No. 5,009,747.

[78] C.F. White, Hydroentanglement technology applied to wet-formed and other precursor webs, TAPPI J. 73 (6) (1990) 187–192.

[79] INDA/EDANA, Guidelines for Assessing the Flushability of Disposable Nonwoven Products, 2018. https://www.edana.org/docs/default-source/product-stewardship/guidelines-for-assessing-the-flushability-of-disposable-nonwoven-products-ed-4-finalb76f3ccdd5286df88968ff0000bfc5c0.pdf?sfvrsn=34b4409b_2. (Accessed 3 July 2020).

[80] J.H. Manning, Hydroentangled Nonwoven Fabric Containing Synthetic Fibres Having a Ribbon-Shaped Crenulated Cross-Section and Method of Producing the Same, 1992. U.S. Pat. No. 5,106,457.

[81] N. Vaidya, B. Pourdeyhimi, D. Shiffler, M. Acar, The manufacturing of wet-laid hydroentangled glass fibre composites: preliminary results, Int. Nonwovens J. 12(4) (2003) 55–59.

[82] N. Cartier, M. Dufour, F. Mavrikos, S. Merlet, A. Vincent, Wet-Laid Nonwoven Comprising Nanofibrillar Cellulose and a Method of Manufacturing Such, 2014. Eur. Pat. No. 2,781,652.

[83] P.A. Geel, Process of Manufacturing a Wet-Laid Veil, 2002. U.S. Pat. No. 6,497,787.

[84] T. de Assis, L.W. Reisinger, L. Pal, J. Pawlak, H. Jameel, R.W. Gonzalez, Understanding the effect of machine technology and cellulosic fibres on tissue properties—a review, BioResources 13 (2) (2018) 4593–4629.

[85] J.H. Manning, I.M. Hutten, Synthetic Fibre Paper Having a Permanent Crepe, 1992. U.S. Pat. No. 5,094,717.

[86] D.P. Dumbleton, Longitudinal Compression of Individual Pulp Fibres, Institute of Paper Chemistry, Appleton, Wisconsin, 1971.

[87] V.R. Gentile, R.R. Hepford, N.A. Jappe Jr., C.J. Roberts, G.E. Steward, Absorbent Unitary Laminate-Like Fibrous Webs and Method for Producing Them, 1975. U.S. Pat. No. 3,879,257.

CHAPTER

6

Spunbond and meltblown web formation

G.S. Bhat[a], S.R. Malkan[b], and S. Islam[a,c]

[a]University of Georgia, Athens, GA, United States [b]Hunter Douglas, Broomfield, CO, United States [c]Dhaka University of Engineering & Technology, Gazipur, Bangladesh

6.1 Introduction

Spunlaid or polymer-laid nonwoven fabrics are produced by extrusion spinning processes, in which filaments are directly collected to form a web instead of being formed into tows or yarns as in conventional spinning. It encompasses two main processes, spunbond and meltblown, often operated in sequence as part of an integrated 'spunmelt' production line. As these processes eliminate intermediate steps, they provide opportunities for increasing production and cost reductions. In fact, melt spinning, which is the basis of the majority of commercial spunbond and meltblown processes, is one of the most cost-efficient methods of producing filaments.

Both spunbonding and meltblowing are similar in principle, but the technologies used are quite different. In both processes, molten polymer is extruded through a spinneret, cooled, and then collected on a moving belt or rotating drum. In the spunbonding process, a continuous web of extruded filaments is formed and subsequently bonded, whereas in the meltblowing process, molten filaments are attenuated and entangled, and self-bonded as they are deposited on a collector. Because of distinct differences in the structure and properties of the fabrics, these processes have grown in parallel since their inception in the late 1950s, and for some applications, they are also used in combination to produce composite nonwovens in the form of bilaminates or trilaminates, e.g., spunbond–meltblown–spunbond (SMS) and other multilayer fabrics. Advancements in polymer chemistry and extrusion technology have enabled an increasingly varied range of products to be developed based on spunlaid and meltblown technologies.

Some of the most important research commenced in the late 1950s, and an extensive number of processing and product patents have been reported over the years. There has been

Handbook of Nonwovens
https://doi.org/10.1016/B978-0-12-818912-2.00001-X

continuing research and development concerned with spunlaid systems, as well as increased acceptance of the fabrics in new product areas, making these one of the most heavily utilised types of nonwoven.

6.2 Historical context

Nonwoven fabrics as we know them today are comparatively new, with developments taking place early in the twentieth century. The concept of spunlaid nonwovens emerged in the 1950s. Since then it has been one of the fastest growing technologies to produce fabrics for various essential applications. The use of such nonwovens is continuing to increase, with progressive developments in the process technology, the resin feedstocks, and the ability to produce fabrics with desired properties at the lowest cost possible.

6.2.1 History of spunbond technology

The concept of spunbond fabric production (or spunlaid web formation) developed simultaneously in Europe and the USA in the late 1950s, but it was later, in the mid-1960s to the early 1970s, that the commercial potential of spunbond technology was fully recognised. Numerous patents on spunbond process design were filed during this period, and some were commercially adopted.

In the USA, DuPont developed and commercialised the first spunbond process in the late 1950s [1]. In 1965, research at DuPont resulted in the manufacture of a polyester spunbond product called Reemay. This was followed by the polypropylene (PP) spunbond fabric Typar and the flashspun polyethylene (PE) fabric Tyvek [2]. Although DuPont subsequently sold both the Reemay and Typar technologies to Reemay Inc., DuPont still remains a significant force in the spunbond market with Tyvek branded products.

In Europe, the German company Freudenberg filed several patents on the spunbond process in 1959. Six years later, Freudenberg introduced their first spunbond process, Lutravil and made a product called Viledon M, made from mixed polyamides [3]. The German company became the first European spunbond manufacturer, and in 1970, Freudenberg incorporated all its spunbond activities into the Carl Freudenberg Spunbond Division based in Kaiserslautern. At this facility, polyester, polyamide, and later PP spunbonded fabrics were produced. About the same time in 1970, Lurgi Kohle and Mineraloltechnik GmbH, Germany, introduced the Docan spunbond process and in 1971 started licensing the process in the USA. Many US and European spunbond roll goods manufacturers use the Docan process with proprietary modifications. The 1950s were an important time for the initiation of spunbond process commercialisation [4], with both Freudenberg and DuPont commercialising nonwovens made from synthetic polymers during the 1950s and 1960s [2]. Several companies including Lurgi Kohle and Mineral Öltechnik GmbH, Reifenhäuser, Amoco Fibres and Textiles, and Sodoca also began commercialisation based on similar technologies [4].

There were comparatively few major developments in the spunbonding process from 1971 to the early 1980s, until in 1984, Reifenhauser GmbH, Germany, introduced the Reicofil spunbond system. A few years before its introduction, Reifenhauser had purchased the rights to an East German patent for the production of polyamide spunbond webs [1]. Considerable research and development effort based on the original East German process resulted in the development of the Reicofil system for producing webs primarily from PP. [1] Other spunbond processes have since been developed. Amoco Fibres and Fabrics USA introduced the RFX system, and Sodoca, France, introduced the S-TEX system [5]. These processes were reported to make highly uniform webs even at low basis weights, thereby providing opportunities that the earlier spunbonding processes were unable to meet. Kobelco (Japan), Neumag-Ason, (Germany), and Nordson (USA) also operated in this market with new spunbonding systems, and in Italy, Meccaniche Moderne S.p.A also started offering spunbond equipment. Over the last 30 years or so, various other spunbond technologies have been developed or further refined (Table 6.1), which in some cases followed acquisition of existing process technology. Current turnkey spunbond machinery providers include Reifenhauser, Oerlikon Neumag, Andritz, Farè, SICAM, Ramina, YingYang, Runjuxiang Machinery, Zhejiang Yanpeng Nonwoven Machinery, Hills Inc., and others.

Technological developments in spunbond processing continued during the 1990s and 2000s, including the processing of biobased plastics such as PLA, and production of smaller diameter filaments. In 1998, Fiberweb developed fabrics containing multicomponent hollow fibres and J&M introduced their MultiFil Composite Filament System [6]. In the 2000s, Reifenhauser offered bicomponent spunbond and meltblown processes, further extending the versatility of this form of nonwoven production [4]. Kobelco, Nordson, Hills, Nippon Kodoshi, Sima, and others also developed complete spunbond line equipment, but some of this technology has since been acquired by other machinery suppliers. Recently, further technological advancements have occurred in spunbond technology. Morman et al. [7] developed elastic fabrics, and Degroot et al. [8] reported a spunbond process for producing nonwoven fabrics with staple or binder fibres from an ethylene-based polymer. Initially, spunbond production was limited to Western Europe, USA, and Japan, but this technology has now spread all over the world. The production of fabrics from a wider range of polymer

TABLE 6.1 Major spunbond equipment manufacturers (current and historical).

Company	Country	Process
Reifenhauser	Germany	Reicofil I, II, III, IV, and V
Zimmer	Germany	Docan/NST
STP Impianti	Italy	Modified Docan
NWT (Perfobond)	Italy/Switzerland	Multiple Slot
Kobelco	Japan	Kobelko (NKK)
Karl Fischer (Inventa)	Germany (Switzerland)	Karl Fischer
Nordson	USA	MicroFil
Oerlikon Neumag	Germany/USA	Ason
Andritz	Austria	Perfobond/Spunjet

types is also a major area of interest in spunlaid manufacturing systems, including the use of thermoplastic polyurethane (TPU) and a variety of melt-spinnable biobased plastics.

6.2.2 History of melt blowing technology

Van A. Wente of the Naval Research Laboratories was the first person who explained the concept of melt blowing process in 1954 [9]. Exxon Research established the first 10-in. melt blowing line in 1965 and also started the first large scale semiworks unit in 1970 [6]. In the following year, the USA started the first commercial meltblowing line and in 1972 meltblowing equipment was offered [6]. Ergon Nonwovens produced oil sorbents and PP wipes in the 1970s, by utilising the characteristically small pore sizes within meltblown fabrics [6]. Johnson & Johnson produced facemasks early in the 1970s to replace the use of glass fibre [6]. Toa Nenryo started to develop artificial leather produced by the meltblown process in the 1980s [6]. Exxon also established a collaboration with the University of Tennessee, Knoxville, in 1983 to enable continuous research and development of meltblown process technology [6].

Several technological advancements in the meltblown process and equipment occurred in the 1990s. Kimberly-Clark developed melt blowing apparatus to produce crimped or uncrimped fibres at reduced energy costs [6]. J and M Laboratories also worked on modified melt blowing dies with a large outlet and developed a method to provide a variable die length [4]. Fiberweb also contributed significantly to the development of the meltblown process. This included melt blowing of liquid crystals in 1995 for high-temperature filter media. They also developed a meltblown process to produce subdenier and microdenier fibres with enhanced strength and barrier properties, as well as modified dies to produce continuous, easily splittable hollow fibres in 1998 [6].

Luo et al. [10] developed a meltblown process in 2004, to produce filaments from cellulose, where a variable degree of mechanical attenuation could be applied to modulate the fineness. Hassan et al. [11] successfully fabricated nanofibre meltblown membranes directly from polymer resins with a fibre diameter ranging from 1 to 2 μm in 2013. In 2016, Wang et al. [12] reported an environmentally friendly water-extractable meltblown process, enabling nanofibre composite nonwovens of PBT to be produced with a fibre diameter of 66 nm. There are numerous suppliers of meltblowing equipment including, for example, Reifenhauser, Oerlikon, Catbridge Machinery, Biax-Fiberfilm Corp, Farè, Ramina, Zhejiang CL Nonwoven Machinery, Hills Inc., Fibre Extrusion Technology (FET), and Useon.

6.3 Polymer resins used in spunbonding and meltblowing processes

In general, high-molecular-weight and broad-molecular-weight distribution polymers such as polypropylene, polyester, and polyamide can be processed by spunbonding to produce uniform webs. Medium melt viscosity polymers, commonly used for production of fibres by melt spinning, are also used. In contrast, low-molecular-weight and relatively narrow-molecular-weight distribution polymers are preferred for meltblowing. Polypropylene has been the most commonly used polymer in spunbonding and meltblowing processes.

Today, other important polymers are polyesters (PET, PBT, PLA), PE, polystyrene, polyure-thane, polyamide, and polycarbonate.

6.3.1 Requirements of polymer resins for spunlaid nonwovens

The main resin characteristics affecting the extrusion and spinning processes during the production of spunlaid nonwovens are as follows: [13]

- **Melting point**. Most PP resins melt at around 165°C, whereas PE resins melt at around 120–140°C. The melting point directly affects the polymer processing temperature. The higher the melting point, the higher the energy requirements.
- **Thermal bonding**. Thermal bonding is an integral step in many spunbonding processes. It influences the structural integrity and drape characteristics of the finished product. PP has a broader thermal bonding window than PE, mainly due to a higher melting range and higher crystallinity. The thermal bonding range for PP is from 125°C to 155°C, and for PE, it ranges from 90°C to 110°C.
- **Molecular weight distribution (MWD)**. Both spunbonding and meltblowing require relatively narrow-molecular-weight resins. The narrow-molecular-weight distribution reduces the melt elasticity and melt strength of the resin so that the melt stream can be drawn into fine denier filaments without excessive draw force. A broad MWD increases melt elasticity and melt strength, which prohibits fibre drawing and can lead to filament breaks due to draw resonance (melt instability) phenomena.
- **Melt viscosity**. The melt viscosity is a function of the melt flow rate (MFR) or melt-mass flow rate (MFR) and melt temperature. The MFR is inversely proportional to (shear) viscosity and is usually expressed as the mass flow (g/10 min). The melt viscosity of the resin has to be appropriate in order to form fine filaments. The typical MFR range suitable for spunbonding is 20–80 and for melt blowing, 30–1500.
- **Resin cleanliness**. Owing to the fine capillary diameter of the spinneret utilised in spunbonding and meltblowing, it is important to have a resin with practically no contaminants. The contaminants block the spinneret holes during processing, causing inconsistency in the final product, as well as leading to filament breaks. Usually, the contaminants are removed using a two-step melt filtration system.

6.3.2 Polyolefin resins

The use of polyolefins, especially PP, has dominated the production of meltblown and spunbonded nonwovens. One of the main reasons for the importance of polyolefins in spunlaid nonwovens is that the raw materials are relatively inexpensive and are available throughout the world. Polyolefin resins offer a relatively attractive cost, combined with good value and ease of use when compared to resins such as polyesters and polyamides that are more popular for traditional textiles. Moreover, progressive advances in polyolefin fibre grade resins have strengthened their attractiveness in terms of the price-performance ratio.

Commercial polyolefin technologies over the last six decades have gone through significant changes with introduction, growth, and stabilisation or maturity phases [14]. The drive for technology evolution is reflected by the industry's desire to continuously improve control of the molecular architecture, which leads to improved polymer performance. Some of the key developments or milestones in polyolefin technology can be summarised as follows: [15]

- In the 1930s, ICI set a trend of making versatile plastics by introducing its high-pressure process for making PE resins.
- In the 1950s, the discovery of stereo-regular polyolefins and the incredibly rapid development of catalysts and processes led to commercialisation of crystalline isotactic PP and HDPE resins.
- In the 1970s and through the 1990s, the invention of the low-pressure, gas-phase process for making linear polyolefins industrially developed.
- In the 2000s, the introduction of a single site catalyst for making superior polyolefin resins set another technological trend.

Data in Table 6.2 shows the total usage of PP and PE in nonwovens [13]. It is estimated that, of this total usage, about 45%–55% is used in spunbond and meltblown nonwovens applications. Polypropylene has the major share of the disposable diapers, sanitary product, and medical apparel markets, and is the principal polymer used in the production of geosynthetics, nonwoven furniture construction sheeting, and carpet components. PP is also widely used in wet filtration applications. PE is used predominantly in the USA for industrial garments, house wrap, envelopes, and other paper-like products.

Commercially available polyolefin resins span a very wide range of molecular weights and comonomer contents ranging from extremely viscous high-molecular-weight resins to low-molecular-weight liquids; from highly crystalline, stiff materials to low modulus, amorphous polymers. The two main polyolefin resins used in spunbond and meltblowing processes are polypropylene (PP) and polyethylene (PE). Although both of these are members of the olefin family, they have significantly different processing requirements and performance variations [13].

Polypropylene, the most widely used resin, exists in three forms: isotactic, syndiotactic, and atactic. Commercially, isotactic PP is most common.

1. Isotactic PP is a stereo-specific polymer because the propylene units are added head to tail so that their methyl groups are all on the same side of the plane of the polymeric backbone. It crystallises in a helical form and exhibits good mechanical properties, such as stiffness and tensile strength. Isotactic PP is sold commercially in three basic types of products: homopolymer, random copolymer, and block copolymer. Homopolymer has the highest stiffness and melting point of the three types and is marketed in a wide range of MFRs.

TABLE 6.2   World consumption of polyolefin resins (in million lbs) [13].

Region	Polypropylene (PP)	Polyethylene (PE)
United States	815	75
Europe	760	40
Japan and the Far East	150	<5

2. Syndiotactic PP is made by inserting the monomer units in an alternating configuration. It lacks the stiffness of the isotactic form but has better impact and clarity.
3. Atactic PP is made via random insertion of the monomer. This form lacks the crystallinity of the other two. It is mainly used in roofing tars and adhesives applications.

Fibre grade PP resins are mainly isotactic homopolymers. PP homopolymer when drawn or oriented gives a material with good tensile, stiffness and tear strength, and clarity due to the molecular alignment. Several important fibre technologies take advantage of the drawability of PP resins and are major consumers of PP. Low melt flow resins are used for monofilaments and slit film applications. Medium to high melt flow rate polypropylenes are used to produce continuous, fine denier filaments, extruded through spinnerets in melt spinning. Spunbonding usually requires a narrow-molecular-weight distribution and high MFR resins, typically of 30–80 MFR. Melt blowing is compatible with a variety of narrow-molecular-weight melt flow rate resins, typically in the range of 30–1500 MFR. [13] The choice of the grade has a major impact on the properties of the resultant nonwoven fabric, including fibre diameter, and so it is an important consideration when trying to make final products to a defined product specification.

Polyethylene (PE) resins are made by polymerising the ethylene monomer. It can also be copolymerised with other materials to modify or enhance certain properties. For example, the density of polyethylene can be manipulated by the type and amount of comonomer reacted with ethylene to make the polymer. This comonomer, in combination with the manufacturing process, affects the type, frequency, and length of branching that occurs in the molecule. This variation results in different types of polyethylenes [13]. There are three basic types of poly-ethylenes, namely:

1. HDPE resin. The term HDPE is an abbreviation for high density polyethylene. The typical density of this resin is $0.950 \, g/cm^3$ and higher.
2. LDPE resin. The term LDPE is an abbreviation for low density polyethylene. The typical density of this resin ranges from 0.910 to $0.925 \, g/cm^3$.
3. LLDPE resin. The term LLDPE is an abbreviation for linear low density polyethylene, which typically ranges from 0.915 to $0.930 \, g/cm^3$.

Fibre grade PE resins are mainly HDPE and LLDPE. Low melt flow rate HDPE resins are used for filament applications. Medium to high melt flow rate LLDPE resins are used to produce continuous fine denier filaments. Both spunbonding and meltblowing require medium and high melt flow rate PE resins. The typical range is 0.5–300 MFR.

The extrusion and spinning characteristics of all polyolefin resins are quite distinct. It is known that polypropylene resin is more difficult to extrude than polyethylene [16]. This is mainly due to the high shear sensitivity of polypropylene resin and to a limited extent, the higher melting temperature. Generally speaking, the output for polypropylene from a given size of extruder is lower than that of polyethylene and has a greater tendency to surge [16]. On the other hand, both polypropylene and polyethylene resins are relatively easy to spin into fine denier filaments provided the resins have a narrow-molecular-weight distribution and an appropriate MFR.

Polyolefin technology has grown faster than any other polymer technology with respect to nonwovens. To understand new developments, we must first understand the catalysts in general. In polyolefin manufacturing, the monomers are reacted using a catalyst. All catalysts have reactive sites enabling them to perform their function, which is linking individual molecules to form a polymeric chain. Conventional catalysts have many reactive sites located randomly on the surface of the catalyst. This produces different and varying polymer structures. The new catalyst system, which is known as a single-site catalyst, also has many reactive sites, but all sites are identical. This in turn gives identical polymers and minimises the variability of polymers [15,17,18].

The most commonly used single-site catalyst to manufacture polyolefin resins is metallocene. Companies around the world have been switching to production of metallocene-based polyolefin resins [18]. The metallocene-based resins offer various attributes, as described below:

- Control of the molecular structure of polyolefins.
- The virtual elimination of nontargeted molecular weight species in resins.
- Incorporation of comonomers and termonomers with greater precision.
- Greater control of MWD and comonomer incorporation than other types of catalysts.
- Only a small catalyst residue is left in the finished product.

The fibre grade metallocene-based polyolefin resins used in spunbonding and meltblowing offer the following advantages over conventional resins: [13]

- Finer denier fibres than that possible with conventional resins.
- Optimum bonding temperatures in spunbonding are lower because of the lower melting point.
- Fabric strengths are comparable.
- Excellent spinning continuity.
- Can be spun at higher draw force.
- Substantial reduction in volatile deposits.
- A broader MFR range, especially for meltblowing.

6.3.3 Polyesters

Polyester is used in a number of commercial spunbond products, and offers certain advantages over polypropylene, including potential for higher thermal stability, although it is slightly more expensive. Applications for polyethylene terephthalate (PET) spunbonds include for example, hot water filter media, moulded carpet backings, primary and secondary backings for floorings, air filter carriers and reinforcements for bitumen membranes. Unlike polypropylene, polyester scrap is not readily recycled in spunbond manufacturing without additional polymer processing. The tensile strength, modulus, and heat stability of polyester fabrics are superior to those of polypropylene. Polyester fabrics are easily printed with conventional nonaqueous processes. PET is the most commonly used type of polyester for spunbonding and meltblowing processes. Fabrics produced from PET have higher tensile strength, modulus, and heat stability compared to those of PP, but PET is more expensive to process, since it requires efficient drying, higher temperature for melting and extrusion, and is more difficult to process than PP [4].

PLA is a biodegradable and biobased aliphatic polyester derived from renewable sources such as corn sugar, potato, and sugar cane [19]. PLA decomposes under composting conditions and it has potential uses in medical, biodegradable packaging materials, hygiene, and other nonwoven applications [20]. The University of Tennessee Knoxville started spunbond- and meltblown-based PLA research in 1993. He and Wang of Kimberly-Clark Worldwide, Inc. treated PLA with alcohol to modulate molecular weight, viscosity, and provide a higher melt flow index, making the polymer more suitable for the meltblown process [21]. Spunbonded PLA nonwovens have relatively good environmental resistance and are suitable for medical supplies and wipes [20]. Spunbonded needle punched PLA nonwovens are soft, antiwrinkle, durable, with a silk-like feel for skin contact applications [20]. Biax-Fibre Film developed a meltblown production line to produce high strength PLA nonwovens [20]. Nonwoven fabrics made from PLA have higher absorption capacity and higher selectivity in oil/water separation compared to PET [22]. Because of the attractive combination of properties, PLA has the potential to replace petroleum-based PET in the nonwoven sector. However, some properties need to be improved including melting temperature, hydrolytic stability against laundering as well as dyeing, and finishing capability [23].

6.3.4 Polyamides

Spunbond fabrics are made from both nylon-6, and nylon-6,6. Nylon production is highly energy intensive and therefore more expensive than polyester or polypropylene. Nylon-6,6 spunbond fabrics are produced with basis weights as low as $10 \text{g}/\text{m}^2$ and with excellent cover, dyeability, resilience, and strength. Unlike olefins and polyester fabrics, nonwovens made from nylon more readily absorb water by means of hydrogen bonding between the amide group and water molecules. Spunbond polyamide nonwovens are mainly used in durable applications for filtration, home furnishings and other applications, where the relatively high-temperature resistance (compared to PP) is important. Cerex is one example of spunbond nylon 6.6. fabric that has been produced for many years.

6.3.5 Polyurethanes

Following development work by various groups, spunbond fabrics made from TPU was commercially developed in Japan. Unique properties are claimed for this product, which are well suited to applications in apparel, hygiene products and other products requiring good stretch, recovery, and fit. Polyurethane has elastomeric properties and nonwovens produced from polyurethane can be used in single-use products such as diapers, masks and medical tape, as well as elastic filling materials [4]. Various polymer suppliers produce TPUs and while problems can be experienced in processing due to static electricity and other practical problems, the use of these polymers has gained increasing interest for both meltblown and spunbond fabrics.

For example, spunbond fabrics containing core-sheath bicomponent filaments with an elastomeric TPU core and a very thin sheath of a polyolefin were commercialised by ADC, a joint venture between BBA Fiberweb and Dow. Resulting fabrics possessed a soft handle combined with excellent stretch-recovery behaviour. The potential for nonaqueous dyeing, finishing and printing of such fabrics could be improved by selection of different polymers for the sheath, such as PET.

6.3.6 Rayon

Many types of regenerated cellulosic materials or rayons (including viscose and cuprammonium rayon) have been successfully processed into usable spunbond webs using wet-spinning technology. One long-established example is Asahi Kasei's Bemberg cupro spunbond fabric. The main advantage of rayon is that it provides good drape properties and softness to web. Rayon is hydrophilic by nature, and it can provide unique performance in fabrics when combined with a polyester that is hydrophobic. Spunbond rayon nonwovens are used in medical, cosmetic and industrial applications.

6.3.7 Polyetherimide

Polyetherimide (PEI) is a high-temperature engineering thermoplastic polymer. It was first developed by the General Electric Company (now produced by SABIC) in 1982, under the trade name Ultem resin. This polymer has a very high glass transition temperature and outstanding chemical and thermal stability, high strength to weight ratio, and excellent mechanical and electrical properties. The Ultem resin has good processability characteristics and can be applied in the aerospace, automotive, and insulation industries [24]. Researchers have explored this polymer for producing nonwovens by meltblown technology. Bhat et al. [24] discussed the possibility of manufacturing high-temperature, high chemical resistance, and high-pressure operating filter media from Ultem resin using meltblowing. It was observed that PEI has very good potential as a filter medium, with excellent chemical resistance and mechanical properties, and can be used at high temperatures.

6.4 Spunbond fabric production

The primary factor in the production of spunbond fabrics is the control of four simultaneous, integrated operations: (i) filament extrusion, (ii) drawing, (iii) laydown, and (iv) bonding. The first three operations are directly adapted from conventional man-made filament extrusion, and constitute the fibre spinning and web formation phase of the process, while the last operation is the web consolidation or bonding phase of the process, hence the generic term spunbond [5].

In general, spunbonding consists of several elements including a polymer feed, an extruder, a metering pump, a die assembly, filament spinning, drawing and deposition system, collecting belt, bonding zone and a winding unit [25]. Fig. 6.1 shows a schematic diagram of Hill's open spunbond process. It is believed that the term spunbond was coined by DuPont in the early 1960s. All spunbond manufacturing processes have two aspects in common: [26] they all begin with a polymer resin and end with a finished fabric; all spunbond fabrics are made on an integrated and continuous production line.

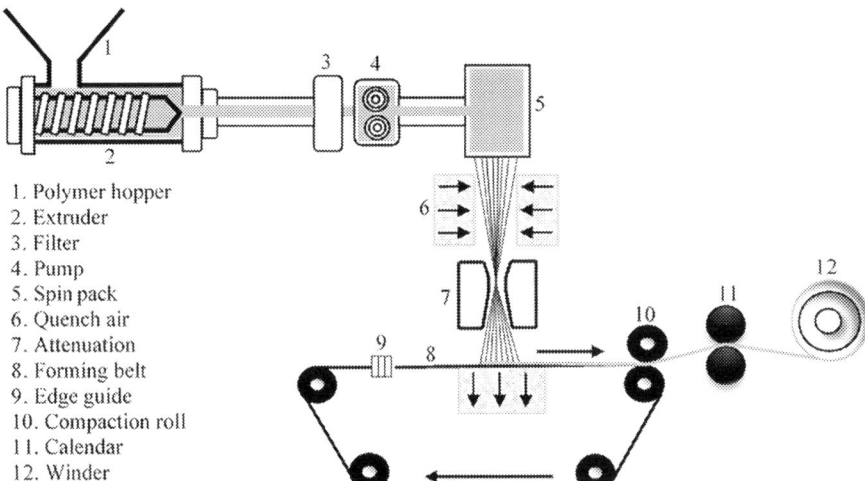

1. Polymer hopper
2. Extruder
3. Filter
4. Pump
5. Spin pack
6. Quench air
7. Attenuation
8. Forming belt
9. Edge guide
10. Compaction roll
11. Calendar
12. Winder

FIG. 6.1 Schematic diagram of the spunbond process. *Based on references N.V. Fedorova, Investigation of the Utility of Islands-in-the-Sea Bicomponent fiber Technology in the Spunbond Process. PhD dissertation, North Carolina State University, 2007. https://repository.lib.ncsu.edu/bitstream/handle/1840.16/5145/etd.pdf?sequence=1; G. Bhat, S.R. Malkan, J. Appl. Polym. Sci. 83(3) (2002) 572–585; Anonymus: Meltblown Process, 2020. http://simonyan-company.com/en/products/complete-plants-for-the-artificial-and-synthetic-fibers-industry/meltblown-process/.*

6.4.1 Extrusion spinning

One of three generic extrusion spinning techniques (melt, dry and wet) is employed in any spunbonding process. In spunbonding, each is directly adapted from conventional filament fibre spinning. Melt spinning is by far the most commonly used. A brief summary of the three approaches is given below: [27]

1. Melt spinning involves melting a thermoplastic fibre-forming polymer and its extrusion into air or an alternative gas, where cooling and solidification of the filament is accomplished (Fig. 6.2). This is sometimes referred to as direct spinning. Polyolefins (principally PP and PE), polyester, and polyamide are amongst the most common thermoplastic polymers used in meltspun spunbond fabrics. Additionally, by modifying the spinneret and polymer feed system, bicomponent (Bico) or conjugate filaments are produced, which are composed of different polymers arranged in various configurations in the cross-section. With certain bicomponent filaments it is possible to produce fabrics containing submicron filaments in the fabric by either, chemical dissolution of one of the polymers, or by longitudinal splitting of the filament components using a suitable source of mechanical energy.

2. Dry spinning involves continuous extrusion of a solution of the fibre-forming polymer into a heated chamber to remove the solvent, leaving the solid filament, as in the manufacture of cellulose acetate.

3. Wet spinning involves continuous extrusion of a solution of the fibre-forming polymer into a liquid coagulating medium, where the polymer is regenerated, as in the manufacture of viscose rayon or cuprammonium rayon. Calcium alginate spunbonds as well as those based on regenerated cellulose have also been produced using this approach.

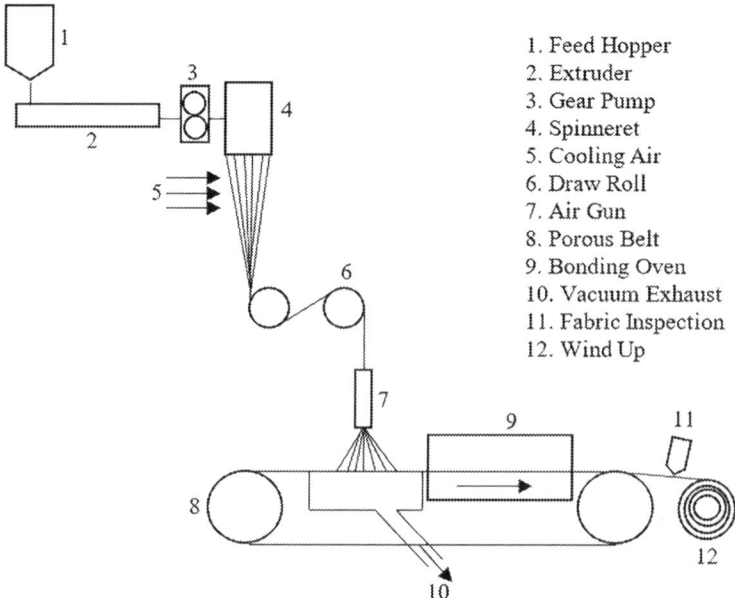

1. Feed Hopper
2. Extruder
3. Gear Pump
4. Spinneret
5. Cooling Air
6. Draw Roll
7. Air Gun
8. Porous Belt
9. Bonding Oven
10. Vacuum Exhaust
11. Fabric Inspection
12. Wind Up

FIG. 6.2   Schematic of a melt spinning spunbonding process. *Based on reference G. Bhat, S.R. Malkan, J. Appl. Polym. Sci. 83(3) (2002) 572–585.*

All of the above spinning techniques can be used to make spunbond fabrics. However, the melt spinning technique is by far the most widely used, partly because of its simplicity and attractive economics. In addition, it should be noted that at least from a technical viewpoint, most other filament extrusion processes can be adapted to form spunbond fabrics, including for example, gel spinning and centrifugal spinning. Detailed discussions of the different techniques are available in the published literature [28].

6.4.2 Basic stages of spunbond fabric production

In its simplest form, a spunbond line consists of the following elements: an extruder for forming filaments; a metering pump; a die assembly; a filament spinning, drawing and deposition system; a belt for collecting the filaments; a bonding zone; and a winding unit. Fig. 6.3 shows a flow diagram of the spunbonding process based on melt spinning. The stages involved in producing a spunbond fabric using melt spinning are now discussed in more detail.

6.4.2.1 Polymer melting

The polymer pellets or granules are fed into the extruder hopper. A gravity feed supplies pellets to the screw, which rotates within the heated barrel. The pellets are conveyed forward along the hot walls of the barrel between the flights of the screw [29]. As the polymer moves along the barrel, it melts due to the heat and friction of viscous flow and the mechanical action between the screw and barrel. The screw is divided into the feed, transition, and metering zones. The feed zone preheats the polymer pellets in a deep screw channel and conveys them

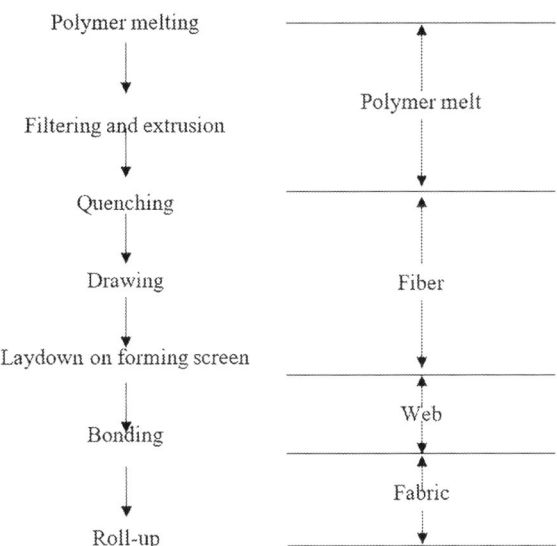

FIG. 6.3   Basic stages in the production of spunbond fabrics.

to the transition zone. The transition zone has a decreasing depth channel in order to compress and homogenise the melting plastic. The melted polymer is discharged to the metering zone, which serves to generate maximum pressure for pumping the molten polymer. The pressure of the molten polymer is highest at this point and is controlled by the breaker plate with a screen pack placed near the screw discharge. The screen pack and breaker plate also filter out dirt and unmelted polymer lumps [30]. The pressurised molten polymer is then conveyed to the metering pump.

6.4.2.2 *Metering of the melt*

A positive displacement volume metering device is used for uniform melt delivery to the die assembly. It ensures the consistent flow of clean polymer mix under process variations in viscosity, pressure, and temperature. The metering pump also provides polymer metering and the required process pressure. The metering pump typically has two intermeshing, counter-rotating, toothed gears. The positive displacement is accomplished by filling each gear tooth with polymer on the suction side of the pump and carrying the polymer around to the pump discharge. The molten polymer from the gear pump goes to the feed distribution system to provide uniform flow to the die nosepiece in the die assembly (or fibre-forming assembly) [1].

6.4.2.3 *Die block assembly*

The die assembly is one of the most important elements of the spunbond process. The die assembly has two distinct components: the polymer feed distribution section and the spinneret.

6.4.2.4 Polymer feed distribution

The feed distribution in a spunbonding die is more critical than in a film or sheeting die for two reasons. First, the spunbonding die usually has no mechanical adjustments to compensate for variations in polymer flow across the die width. Second, the process is often operated at a temperature range where thermal breakdown of polymers proceeds rapidly. The feed distribution is usually designed in such a way that the polymer distribution is less dependent on the shear sensitivity of the polymer. This feature allows the processing of widely different polymeric materials using just one distribution system. The feed distribution balances both the flow and the residence time across the width of the die. There are basically two types of feed distribution that are employed in the spunbonding die, the T-type (tapered and untapered) and the coat-hanger type. An in-depth mathematical and design description of each type of feed distribution is given by Mastubara [31–35]. The T-type feed distribution is widely used because it gives both even polymer flow and even residence time across the full width of the die.

6.4.2.5 Spinneret

From the feed distribution channel, the polymer melt goes directly to the spinneret. The spinneret is one of the components of the die assembly. The web uniformity partially hinges on the design and fabrication of the spinneret; therefore, the spinneret in the spunbonding process requires very close tolerances, which has continued to make their fabrication very costly. A spinneret is made from a single block of metal having several thousand drilled orifices or holes. The orifices or holes are bored by mechanical drilling or electric discharge machining (EDM) in a certain pattern. The spinnerets are usually circular or rectangular in shape. In commercial spunbonding processes, the objective is usually to produce a wide web (of up to about 5 m), and therefore, many spinnerets are placed side by side to generate sufficient fibres across the width [36]. The grouping of spinnerets is often called a block or bank. In commercial production lines, two or more blocks are used in tandem in order to increase the coverage of the filaments.

6.4.2.6 Filament spinning, drawing, and deposition

The proper integration of filament spinning, drawing, and deposition is critical in the spunbonding process. The main collective function is to solidify, draw, and entangle the extruded filaments from the spinneret and deposit them onto an air-permeable conveyor belt or collector.

6.4.2.7 Drawing

Filament drawing follows spinning. In conventional extrusion spinning, drawing is achieved using one or more set of draw rollers. While roller drawing can certainly be used in spunbonding, and some systems still operate using this approach, specially designed aerodynamic devices such as a Venturi tube are most commonly adopted in modern systems.

6.4.2.8 Deposition

Filament deposition to form the final web follows the drawing step. Filament deposition is also frequently achieved with the aid of a specially designed aerodynamic device referred to as a fanning or entangler unit. The fanning unit is intended to cross or translate adjacent filaments to increase cross-directional web integrity.

6.5 Spunbond production systems

Many filament spinning, drawing, and deposition systems have been patented and commercialised. Some of the basic principles involved as proposed by Hartman [37], are shown in Fig. 6.4 and are explained below.

- Route 1: Fig. 6.4A is a system using longitudinal spinnerets, with air slots on both sides of the spinneret for the expulsion of the drawing gas '1' (primary air). The room air (secondary air) '2' is carried along and after laydown of the filaments, the air is removed by suction '3'. This process is well suited for tacky polymers, such as linear polyurethanes. The web is truly 'spunbonded', that is, the continuous filaments after web collection bond themselves (self-bond) at their crossover points due to their inherent tackiness. Crystallisation, which then sets in, subsequently eliminates the stickiness of the filaments after the bonding step.
- Route 2: Fig. 6.4B shows how a higher draw ratio can be achieved, which results in increased molecular orientation of the filaments. Filaments are drawn with several air or gas streams using drawing conduits. The air is removed by suction '4' after web formation. This process has advantages in preparing lightweight spunbond webs containing fine filaments with a textile-like appearance and handle.

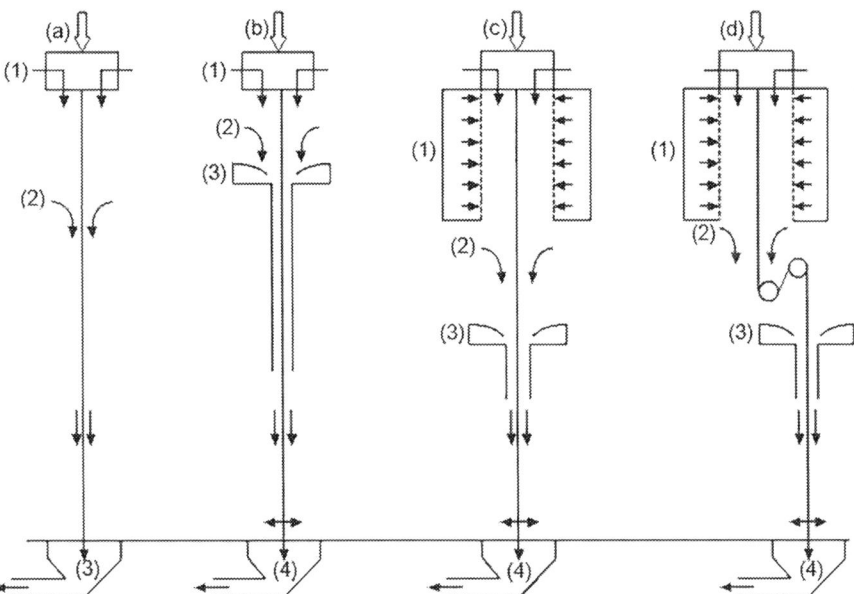

FIG. 6.4 Schematic of filament spinning, drawing and deposition systems. *Based on reference L. Hartmann, Text. Manuf. 101 (1974) 26–30.*

- Route 3: Fig. 6.4C operates with regular cooling ducts '1' and drawing jets '3'. The drawing and cooling arrangements can be operated to give very high spinning speeds with the result that highly oriented filaments are produced. The room air '2', of controlled temperature and moisture content, can be entrained to control the development of filament properties. The air is removed by suction '4' after web formation.
- Route 4: Fig. 6.4D has a mechanical drawing step '2' between the spinneret and laydown zones. This route is similar to conventional spinning and is especially useful for polymers, which in regular air drawing do not give optimum filament molecular orientation. Again, air streams are used for cooling '1' and for laydown '3' and '4'. Webs with high strength and low elongation can be made using this particular route.

A large number of different spunbond processes can be classified according to one of these four routes with appropriate modifications or refinements. There are many examples of process improvements in the patent literature relating to drawing and web deposition. Some of the most successful spinning, drawing, and deposition spunbond systems merit a brief discussion.

6.5.1 Docan system

Developed by Lurgi Kohle and Mineraloltechnik GmbH, Germany in 1970, many nonwoven companies have licensed this process from the Lurgi Corporation for commercial production. The Docan route is based on melt spinning, where the melt is forced by spin pumps through special spinnerets having a large number of holes [1]. By suitable choice of extrusion and spinning conditions, the desired filament denier is obtained. The blow ducts located below individual spinnerets continuously cool the filaments with conditioned air. The force required for filament drawing and orientation is produced by an aerodynamic system. Each continuous filament bundle is picked up by a draw-off jet operated at high pressure and is passed through a guide tube to a separator which affects separation and fanning of the filaments [38]. Finally, the filament fan leaving the separators is deposited as a notionally random web on a moving mesh belt. Suction below the sieve belt assists the filament laydown.

6.5.2 Reicofil system

This route was developed by Reifenhauser GmbH, Germany and can be purchased as a turn-key installation. Commercially, many nonwoven companies use the Reicofil system of which there are now multiple generations, the most recent one being Reicofil 5. This route is based on melt spinning and the melt is forced by spin pumps through spinnerets having a large number of holes [1]. Primary blow ducts, located below the spinneret block, continuously cool the filaments with conditioned air. The secondary blow ducts, located below the primary blow ducts, continuously supply the auxiliary air at room temperature. A ventilator operating across the width of the machine, generates underpressure and sucks the filaments together with mixed air down from the spinnerets and cooling chambers. The continuous filaments are sucked through a Venturi (high-velocity low-pressure zone) to a distributing chamber, where fanning and entangling of the drawn filaments take place. Finally, the entangled filaments are deposited on a moving suctioned mesh belt to form a web.

Filament orientation in the web is influenced by turbulence in the air stream, which generally serves to increase randomisation. However, a small bias in the machine direction is usual due to some directionality imparted by the speed of the conveyor belt [1].

6.5.3 Lutravil system

Developed by The Carl Freudenberg Company, Germany in 1965, this spunbonding process is proprietary and is not available for commercial licensing. The melted polymer is forced by spin pumps through the spinnerets. The primary blow ducts, located below the spinneret block, continuously cool the filaments with conditioned air. The secondary blow ducts, located below the primary blow ducts, continuously supply controlled room temperature air. The filaments are passed through a device where high-pressure tertiary air is introduced to draw and orient the filaments. Finally, the filaments are deposited as a web on a moving mesh belt.

6.5.4 Ason/Neumag system

This process uses a slot design to generate higher filament speed. During the 1990s, various attempts were made by spunbond machine manufacturers to produce lighter-weight spunbond fabrics, particularly for hygiene applications. To achieve this, there is a need to produce webs containing finer filaments and consequently there has been much discussion about the prospect of spunbond fabrics replacing meltblowns in certain product applications. The Ason-Neumag process is claimed to be quite flexible with several benefits such as the capability to process a multitude of polymers at high spinning speeds. Another benefit is a smaller filament diameter. Finer (subdenier) filaments allow greater fabric coverage and weight uniformity, improved opacity and a softer but frequently stronger fabric (due to the increased filament surface area). The elevated spinning speed also allows for a stronger filament due to higher molecular orientation as well as, in some cases, increased fabric elasticity.

6.6 Bonding techniques

The conveyor belt carries the unbonded web to the bonding zone. There are three basic bonding options in a spunbond process: thermal, chemical/adhesive, and mechanical. The choice of technique is dictated mainly by the web weight and the intended application of the fabric. Occasionally, a combination of two or more bonding techniques is employed.

6.6.1 Thermal bonding

Thermal bonding is the most widely used method in spunbonding. Bonding is achieved by fusing thermoplastic filaments in the web. The fusion is achieved by the direct action of heat and pressure via calender rollers (contact bonding) or an oven (through-air bonding). The degree of fusion determines many mechanical and physical fabric properties including fabric handle.

6.6.1.1 Point and area bonding

Point bonding is the most common thermal bonding route in spunbond fabric production and involves fusing filaments in multiple, small, discrete, and closely spaced areas in the web using both temperature and pressure to achieve a thermal bond. Heated calender rollers in which one or more of the rollers are embossed with a point-pattern are used to achieve this. Since point bonding can be accomplished with as little as 10% bonding area (90% unbonded area), such fabrics are considerably softer and more textile-like in handle than area bonded webs [36]. Notionally, area-bonding uses all available bond sites in the web. Using a thermal calender, heat and pressure is applied to the web across the entire structure, causing fusion between a larger proportions of filaments than in point bonding. In area-bonded webs, not every contact point is necessarily bonded, since not every contact point is capable of forming a bond [39]. Calendered webs of this type are stiffer and more paper-like in appearance than a point bonded web.

6.6.1.2 Ultrasonic bonding

Ultrasonic bonding is similar to thermal point bonding, except that heating of the web is achieved by converting mechanical energy applied during the bonding process. To achieve this, frequencies higher than 18,000 Hz are used. In one example, 50–60 Hz of electrical power is converted to about 20,000 Hz of electrical energy, which is conducted to an electro-mechanical converter that transforms high frequency electrical oscillations into mechanical vibrations [40]. A waveguide assembly that is made up of a booster and a horn transfers the mechanical vibrations to the web, normally as the web passes over an embossed roller surface. The local heat generated in the web causes bonds or weld points at these locations. The three critical parameters in ultrasonic bonding are pressure (on the web), amplitude and time.

6.6.1.3 Thermal bonding approaches

In addition to thermal fusion of the filaments in the web to achieve bonding, various additional methods of thermally bonding for spunbonds can be used, as listed below:

- Combining filaments of the same generic polymer type having different softening points, for example, Typar is bonded by utilising undrawn PP filaments as a binder for spunlaid webs of PP, which produces a softening point difference of 3°C.
- Distribution of other thermoplastic materials in the form of powders through the spunlaid web, for example, copolyesters and copolyamides. With this technique, it is very difficult to achieve uniform bonding.
- As an alternative to homo-filaments, core-sheath (or core-skin) and side-by-side bicomponent filaments can be spunlaid. The two components normally have different melting points, so that during thermal bonding only one of the polymers softens, leaving the second intact. Clearly, in the case of core-sheath filaments, the sheath has the lower melting point. Core-sheath filaments are usually easier to manufacture than side-by-side. By varying the characteristics of the sheath polymer, the ratio of core-to-sheath polymers, and the proportion of the hetero-filaments in the web, a wide variety of webs can be engineered. Examples of bicomponent filament compositions include polyamide 66/polyamide 6, polyester/polyamide 6, polyester/low melting polyester, and PP/PE [41].

6.6.2 Chemical and solvent bonding

In chemical or adhesive bonding, a polymer latex or a polymer solution is deposited in and around the web structure and then cured thermally to achieve bonding. The bonding agent is usually sprayed onto the web or saturated into the web. In spray bonding, the bonding agent usually stays close to the surface, resulting in a web with little strength and high bulk. In saturation bonding, all the filaments are bonded in a continuous matrix, which tends to give high rigidity and stiffness to fabrics. Fabrics produced by applying adhesive all over the surface, become stiff and rigid and final products reflect the properties of the adhesive, rather than the filaments. Printing adhesive on parts of the web surface can be helpful in minimising this problem. In spunbonding, a form of solvent bonding or partial solvation bonding is also sometimes used. This method, which is applicable only to fabrics composed of polyamide, uses gaseous hydrogen chloride or alternative solvents. The gaseous HCl solvates the outer surface of the carrier filaments by disrupting hydrogen bonds and bonds the fibres together at all cross-over points by melding together the contacting solvated surfaces. The removal or neutralisation of the solvent is the final step in bond formation [42]. An example of this type of spunbonded product is Cerex (by Cerex Nonwovens). The effect of bonding on the properties of the fabric is the same for any type of bonding [43].

6.6.3 Mechanical bonding

The two most common mechanical bonding processes are needlepunching and hydroentangling. In needlepunching of spunlaid webs, filaments are mechanically entangled by repetitive penetration of barbed needles through the structure. Needlepunching is easily adapted to spunbonding and requires less precise control than thermal bonding. Compared to thermal bonding it also has the advantage of improving delamination strength, since segments of filaments lying in the surface of the web are to some extent migrated through-thickness. In addition, it is the only bonding method suitable for the production of heavy-weight spunbonded fabrics, for example up to $800 \, \text{g/m}^2$. It is, however, only suitable for the production of uniform fabrics over $100 \, \text{g/m}^2$, since needling tends to concentrate filaments in areas resulting in loss of visual uniformity at lower weights [36]. An example of a spunbonded needlepunched PET fabric is Marix (Unitika). Hydroentangling (also called spunlacing) is becoming increasingly important for manufacturing spunbond fabrics containing microfilaments. This is the basis of Freudenberg's Evolon process. High-pressure water jets during bonding, both mechanically split the bicomponent filaments in the web, and form a mechanically entangled microfilament network. Such fabrics have remarkable strength combined with attractive aesthetics following hydroentangling.

6.6.4 Winding

The spunbond fabric is usually wound on a cardboard core and processed further according to the end-use requirements. The combination of fibre entanglement and fibre-to-fibre bonding generally produces sufficient strength to enable the web to be readily handled without further treatment.

6.7 Operating variables in the spunbond process

The key variables can be divided into two categories: (i) material variables, and (ii) operational variables. By manipulating both, a variety of spunbond fabrics can be produced with the desired properties. Each of these variables plays a significant role in process economics and product reliability. Therefore, it is essential that each of these variables be precisely defined and understood in order to optimise the spunbond process.

6.7.1 Material variables

The key material variables include polymer type, molecular weight, MWD, polymer additives, polymer degradation, and polymer forms such as pellets or granules. Basically, any fibre-forming polymer with an acceptable melt viscosity at a suitable processing temperature and is capable of solidifying before landing on the collector screen is potentially compatible with spunbond processing. The polymer's molecular weight and MWD are important material variables. The spunbond process requires moderately high-molecular-weight and broad MWD resins to produce uniform webs. PP and PE polymers of low MFRs have been used successfully for the spunbond process. They range from about 12 to 70 MFR. Typically, but not exclusively, polypropylenes of about 20–25 MFR are used in Europe, and 35 MFR PP is used in the USA. A wide range of polymers have been spunbonded, as will be explained.

6.7.2 Operational variables

The operational variables can be classified as (i) on-line and (ii) off-line. Both variables directly affect the filament diameter, filament structure, web laydown, physical properties, and tactile properties of the web. By manipulating these variables, a variety of spunbond webs can be engineered. The on-line variables can be changed according to requirements during production and vary according to the spinning, drawing, and deposition system used. Essentially, polymer throughput, polymer/die temperature, quench air rate and temperature, take-up speed, and bonding conditions are the main variables [1].

- Polymer throughput rate and polymer/die temperature basically control the final filament diameter and to a certain extent the texture of the filaments.
- Quench air rate helps to control the draw-down and air drag. Quench air temperature controls the cooling of the filament and hence the development of microstructure.
- Take-up speed controls the final draw-down and filament deposition on the conveyor belt, which also influences orientation and therefore the anisotropy of the web structure.
- Bonding temperature and pressure (assuming calender bonding is used) influence the tensile properties of the final fabric.

Off-line processing variables can only be changed when the production line is not in operation, for example, spinneret hole size, spinneret-to-collector distance, and the bonding system, amongst others. Most of the off-line processing variables are fixed for a particular product line.

6.7.3 Process parameters influencing fibre formation and properties

The important process variables that have an impact on filament properties and subsequently on the structure and properties of the final fabric are listed in Table 6.3 [44]. The structure and properties of the filaments formed are determined by the dynamics of the threadline and the effect of air drag on spinline that are dependent on elongational deformation and crystallisation during solidification. Malkan et al. [45] studied the effect of several process variables such as melt temperature, quench air temperature, air pressure, Venturi gap setting and air suction using a 35 MFR PP polymer.

Filament linear density is one of the most important properties. In the results reported herein, filament diameters were determined by microscopic analysis of final fabrics. Polymer melt temperature has a slight effect on filament diameter, the value decreasing with an increase in the melt temperature. This is because at a higher melt temperature the viscosity is lower and it is easier to draw-down the filaments. While choosing the melt temperature, the flexibility may be limited as at lower temperatures, filament draw-down and diameter reduction become difficult and at higher temperatures, there is the possibility of polymer degradation leading to filament breaks and spot formation on the conveyor belt. The filament diameter increases with throughput, in spite of an increase in air suction to keep a balance of air-to-polymer ratio, because of a decrease in the cooling rate of the filaments and higher die swell resulting in lower effective draw-down. It has been observed that filament diameter values can be different depending on whether they are measured before or after bonding.

Usually, increasing the primary air temperature decreases the filament diameter. Such a trend was explained by Misra et al. in terms of the spinline stress based on their modelling studies [46,47]. The model consisted of a set of differential equations developed from the application of fundamental physical principles such as conservation of mass, energy, and momentum, together with polymer-specific information such as the apparent elongational viscosity, crystallisation kinetics, polymer density, etc. With an increase in cooling air temperature, the spinline cools much more slowly and the crystallisation occurs much farther from the spinneret. The increase in temperature also allows the viscosity of the polymer to remain lower allowing higher draw-down leading to lower fibre diameter. Unlike the experimental case where the temperature range studied was narrow (10–21°C), the model considered a

TABLE 6.3   Variables that determine properties of spunbond fabrics.

Filament variables	Filament arrangement	Bonding variables
Linear density	Filament separation	Binder nature
Tenacity	Fabric weight uniformity	Binder concentration
Elongation	Random vs directional	Binder distribution
Modulus		Self-bonding
Cross-section		
Crimp		
Morphology		

wider range of quench air temperature (120–150°C). The model predicted that with an increase in quench air temperature, there is also a decrease in final crystallinity and orientation as the draw-down takes place under a low spinline stress. Model predictions by Smith and Roberts showed similar trends [48]. Bhat et al. [49–54] conducted an experimental study to understand the effect of certain processing variables, where filaments before bonding and after bonding were carefully analysed. In that study, it was observed that with increase in primary air temperature in the small range of 10–25°C, the filament diameter increased with increase in quench air temperature (Fig. 6.5). These results contradicted what was reported earlier and the model predictions. One has to be careful in interpreting these sets of data as the model predictions considered a very wide range of temperatures, and in the earlier experimental investigations, fibre diameters were determined from the bonded fabric.

The filament diameter increase with an increase in quench air temperature is due to the fact that the low temperature is helpful in generating higher spinline stress that leads to reduction in fibre diameter. This complex phenomenon is a result of compensating effects of changes in the elongational viscosity and the spinline stress. The changes taking place during melt solidification in a spinline are quite complex involving rapid changes in temperature, viscosity, orientation, crystallinity, etc. The change in cooling conditions causes a shift in the drawing zone along the spinline. As the cooling temperature decreased, the orientation and crystallinity of the fibres increased with a decrease in fibre diameter, unlike the model predictions where the finer fibres were predicted to have lower birefringence and crystallinity. Smaller fibre diameter in combination with higher orientation indicates that the diameter reduction takes place under stress, and is not just due to lower melt viscosity. Another reason for possible differences with the model predictions could be that diameter predictions were based on on-line studies and there is possibly some relaxation of the fibres that leads to changes in the final fibre diameter.

To understand some of the intricate changes that take place in the spinline, Misra et al. [46,47] investigated the spinnability of nine different PP resins having a range of MFRs and MWDs, and the structure and properties of the filaments meltspun from them. It was observed that increasing molecular weight (with same MWD) led to higher spinline stress, faster filament draw-down, higher crystallisation rates, higher levels of crystallinity and

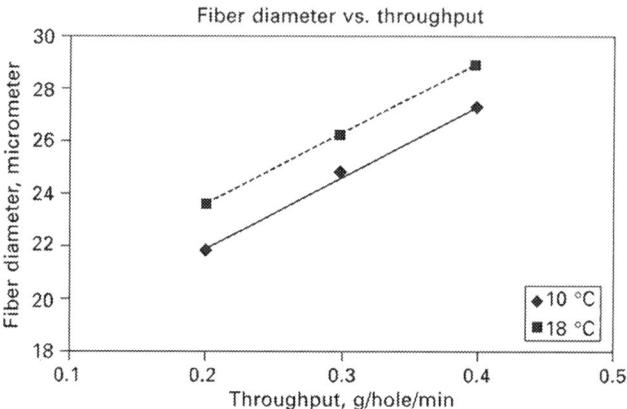

FIG. 6.5   Effect of throughput and quench air temperature on filament diameter.

orientation, higher tensile strength and lower breaking elongation for filaments spun under similar conditions. Breadth of MWD also had a great influence on the spinline behaviour and the structure and properties of the filaments. While broad MWD resulted in fibres with higher density and lower birefringence than the samples from narrow MWD, fibres spun from narrow MWD had higher tensile strength and lower elongation to break. Broader MWD polymer showed more elongation thinning and higher elongational viscosity, and a higher tendency to undergo stress induced crystallisation. These led to a shift in draw-down points closer to the spinneret although the birefringence rises earlier along the spinline, develops more slowly and reaches a lower final value for broader MWD.

For the PP copolymer investigated [55], changes in processing conditions showed a trend similar to that observed for the homopolymer. However, the actual values of crystallinity and birefringence were lower for the copolymer filaments even though the filament diameters were comparable to that of the homopolymer PP. Also, copolymer filaments had lower tensile strength and modulus, and higher breaking elongation.

An increase in the quench air pressure, which was accomplished by adding more auxiliary air resulted in a decrease in the filament diameter. Obviously, an increase in air pressure led to an increase in the spinline draw ratio of the filaments. The effect of a Venturi gap on the filament diameter did not show any trend. The Venturi gap has an effect on filament laydown and fabric properties rather than on the filament properties as such.

Air suction has an obvious effect on the filament structure as the air suction directly corresponds to take-up speed. The filament diameter decreases with increase in air suction speed. The effect of take-up speed on residual draw ratio of the spun filaments for PET and PA were also similar [56].

The relationship between filament diameter and birefringence is shown in Fig. 6.6. The general observation is that with a decrease in filament diameter there is an increase in birefringence values, indicating higher molecular orientation. Filaments with higher molecular orientation had higher crystallinity values as well. The tensile properties are strongly related to birefringence. With an increase in birefringence, the filaments had higher tensile strength

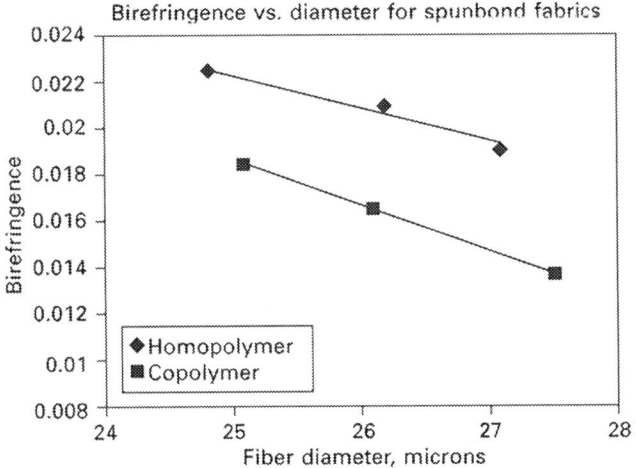

FIG. 6.6   Relationship between filament diameter and birefringence.

and lower breaking elongation, which is an expected result due to the higher molecular orientation.

Air drag has been shown to have a major role in determining the morphology of the filaments in a spunbonding process [57]. In the Ason-Neumag process, air drag in the spinline is manipulated by slot attenuation with low-pressure air [58]. With the introduction of the waveform, there is an increase in drag force that results in a rapid increase in birefringence values with spinning speed (Fig. 6.7).

The crystalline structures of the filaments vary to a great extent depending on the resin characteristics and the processing conditions. WAXD scans of filaments produced under different throughput temperatures (Fig. 6.8) clearly indicate that not only do crystallinity values change, the crystal structure might be different as well. The differences in morphology of the filaments can be clearly seen from the thermomechanical responses of filaments produced at different cooling air temperatures. The filaments produced at lower primary air temperatures, which had higher crystallinity and orientation, were more stable than those spun at

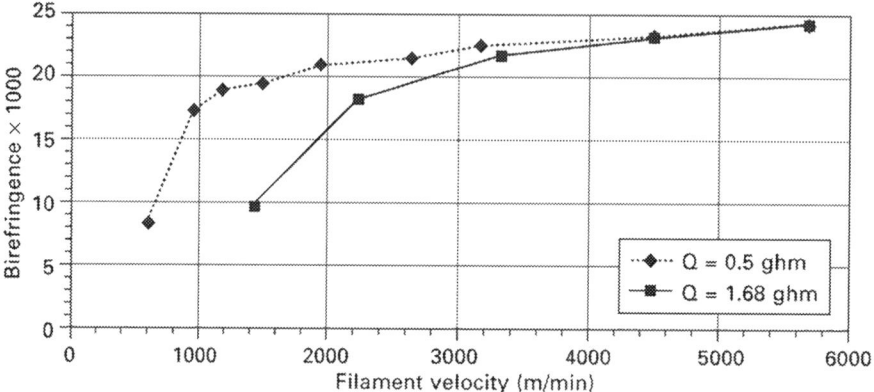

FIG. 6.7   Change in birefringence with filament velocity in the Ason process.

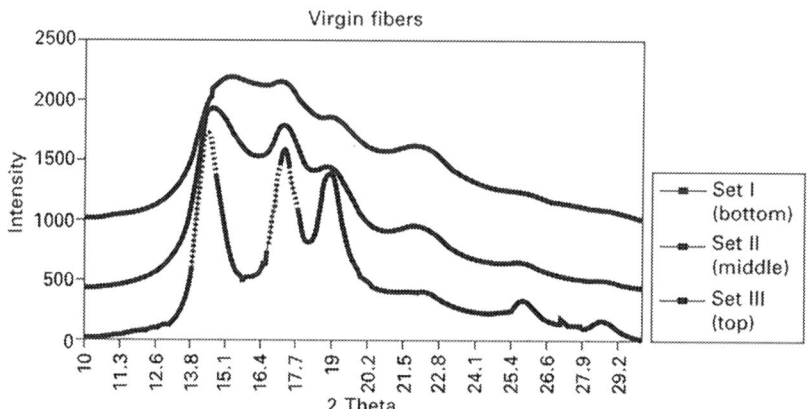

FIG. 6.8   WAXD scans of spunbond filaments produced under different conditions showing differences in crystalline structure.

higher temperatures. This deformation behaviour has a great significance for thermal bonding as will be explained in later sections.

In traditional filament extrusion spinning in the textile industry, some orientation is achieved by winding the filaments at a rate of approximately 3000 m/min to produce partially oriented yarns (POY). The POYs are mechanically drawn in a separate step to enhance their strength. In spunbond production, filament bundles are partially oriented by pneumatic acceleration speeds of up to about 6000 m/min. Such high speeds result in partial orientation and high rates of web formation, particularly for lightweight structures ($17 \, \text{g/m}^2$). For many applications, partial orientation sufficiently increases strength and decreases extensibility to give a functional fabric. However, some applications require filaments with very high tensile strength and a low degree of extension. For such applications, the filaments are drawn over heated rolls with a typical draw ratio of 3.5:1. The filaments are then pneumatically accelerated onto a moving belt or screen. This process is slower but gives stronger webs, which is particularly useful when fabrics are intended for durable applications such as in geosynthetics or construction.

The web is formed by the pneumatic deposition of the filament bundles onto a moving belt [36]. For the web to have maximum uniformity and cover, individual filaments must be separated before reaching the belt. This can be accomplished by inducing an electrostatic charge onto the bundle while under tension and before deposition. The charge may be induced triboelectrically or by applying a high voltage charge. The former is a result of rubbing the filaments against a ground, conductive surface. The electrostatic charge on the filaments must be at least $30,000 \, \text{esu/m}^2$. The belt is usually made of an electrically grounded conductive wire. Upon deposition, the belt discharges the filaments. This is a simple and reliable method. Webs produced by spinning linearly arranged filaments through a so-called slot die eliminate the need for such bundle separating devices. Mechanical or aerodynamic forces also separate filaments. One method utilises a rotating deflector plane to separate the filaments by depositing them in overlapping loops; suction holds the filament mass in place [36].

For some applications, the filaments are laid down randomly with respect to the direction of the laydown belt. In order to achieve a particular characteristic in the final fabric, the directionality of the splayed filament is controlled by traversing the filament bundles mechanically or aerodynamically as they move towards the collecting belt. In the aerodynamic method, alternating pulses of air are supplied on either side of the filaments as they emerge from the pneumatic jet. By proper arrangement of the spinneret blocks and the jets, laydown can be achieved predominantly in the desired direction. Highly ordered cross-lapped patterns can be generated by oscillating filament bundles.

If the laydown conveyor belt is moving and filaments are being rapidly traversed across this direction of motion, the filaments are being deposited in a zig-zag or sine wave pattern on the surface of the moving belt. The relationships between the collecting belt speed, period of traverse, and the width of filament curtain being traversed determine the appearance of the formed web. Fig. 6.9 shows the laydown for a process where the collecting belt travels a distance equal to the width of the filament curtain 'a' during one complete period of traverse across a belt width 'b'. If the belt speed is V_x and the traverse speed is V_t, the number of layers deposited, c, is calculated by,

$$c = \frac{a \times V_t}{b \times V_X} \tag{6.1}$$

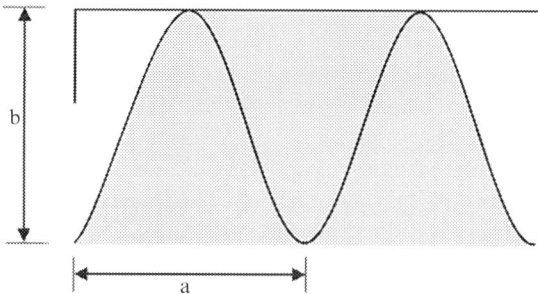

FIG. 6.9 Web laydown pattern. *Based on reference R.L. Smorada, Encyclopedia of Polymer Science and Enginnering, 10 (1985) 227–253.*

If the traverse speed is twice the belt speed and if 'a' and 'b' are equal, a double coverage occurs over all areas of the belt. Hearle et al. [59–61] investigated the filament laydown pattern by simulated spunbonding studies. It was observed that when a filament is fed perpendicularly onto a moving belt, the laid-down form is determined by filament properties such as linear density, bending rigidity, and torsional rigidity, the height of the feed point and the feed-to-belt ratio. When a filament is being fed onto a moving belt, it will be laid in a modified cycloidal form, with a shape depending on the ratio between feed and belt speeds. With higher throughputs, the filament diameters are larger and that leads to cycloids of larger diameter.

The Venturi gap in the Reicofil machine has a notable effect on filament laydown as the change in gap alters the air velocities and air profiles considerably. It was observed that for any basis weight, smaller Venturi gaps resulted in a less uniform web than one with larger gaps [45]. With small Venturi gaps, there is higher cabin pressure that imparts higher oscillations and instabilities to the filament stream, which results in a nonuniform web structure. With an increase in the Venturi gap, the uniformity of the web increases considerably.

6.8 Structure and properties of spunbond fabrics

Spunbond fabrics are produced by an integrated process of spinning, attenuation, deposition, bonding, and winding to form rolls in a single machine. The fabrics are up to 5.2 m wide and are usually not less than 3.0 m in width for acceptable productivity. Filament linear density ranges from 0.8 to 50 dtex (0.07–45 denier), although a range of 1.5–20 dtex (1.36–18 denier) is most common. A combination of thickness, filament denier, and the number of filaments per unit area determines the fabric basis weight, which ranges from 10 to 800 g/m^2. Most typically, fabric basis weight is from 15 to 180 g/m^2. Like all nonwoven fabrics, the properties of a spunbond fabric reflect both the polymer composition and the geometry of the fabric structure.

6.8.1 Composition

The method of manufacture determines the fabric geometry, whereas the polymer determines the intrinsic properties. Properties such as filament density, temperature resistance, chemical and light stability, ease of coloration, and surface energies, amongst others, are a function of the base polymer. Although any filament-forming polymer can be used in the spunbonding process, most spunbond fabrics are based on isotactic PP and polyester. Small quantities are made from nylon-66, and increasing amounts from HDPE. LLDPE is also used as a base polymer because it gives a softer fabric.

6.8.2 Characteristics and physical properties

Most spunbonding processes can yield a fabric with near planar-isotropic properties owing to the random laydown of the filaments, although they are rarely entirely isotropic. Unlike woven fabrics, such spunbond fabrics are nondirectional and can be cut and used without concern for stretching in the bias direction or unravelling at the edges. Anisotropic properties are obtained by controlling the orientation of the filaments during the preparation of the web. Commercially, in the majority of the cases, spunbond webs are anisotropic with preferred orientation in the machine direction. This is because the filaments are deposited on a high-speed conveyor. Anisotropy is determined by both filament diameter and the filament to belt speed ratio, the latter having a greater impact. All processing conditions that affect the diameter have an effect on laydown and thus the directionality of key properties. Uniformity of the fabric is normally measured in terms of basis weight variation and the spatial distribution of filaments over small areas. Image analysis has been shown to be useful for determining uniformity at various levels [62]. This particular technique, since it uses transmitted light, is suitable only for lightweight webs. Other advantages of image analysis are its ability to determine filament diameter, diameter variation and filament orientation by automated techniques [63].

The method of bonding greatly affects the thickness of the web and other characteristics. Spunbond webs bonded by calendering are thinner than needlepunched webs since calendering compresses the structure, whereas needlepunching moves a small proportion of filaments from the XY plane of the fabric into the z-direction (thickness). The tensile properties of the spunbonded fabric are dependent on the filament properties, web laydown, bonding, the composition of the final fabric and structure [39]. Because of a slight preferential orientation of filaments in the machine direction (MD), higher tensile strength is normally observed in the MD compared to the cross direction (CD) with a higher tear strength in the CD. The air permeability increases if filament diameter increases due to processing conditions. This increase in diameter results in lower fabric density after calendering, which is responsible for an observed increase in air permeability. Fabric handle, which depends on fabric rigidity and modulus is also a function of filament diameter, as well as bonding conditions. Fabrics composed of fine filaments are usually softer and more flexible compared to those containing higher diameters. Similarly, bonding at a lower temperature and a lower basis weight tends to result in increased softness.

It is clear that the structure and properties of spunbond fabrics can be varied by several means. It is possible to engineer the fabric properties by appropriate selection of polymer and processing conditions. This gives a great amount of flexibility. However, the effect of certain processing conditions on specific properties is still not clear because of the complexities involved. Additional research in some of these areas is on-going in an attempt to expand spunbond markets.

In summary, some of the main characteristics and properties of a spunbond fabric are: [45]

- Near random fibrous structure, with near planar-isotropic properties.
- Generally, the web is white, with a high opacity per unit area.

- Most spunbond fabrics are layered or have a shingled structure, the number of layers increases with increasing basis weight.
- Basis weight ranges between 5 and $800 \, g/m^2$ and is typically $10–200 \, g/m^2$.
- Filament diameter ranges between 1 and $50 \, \mu m$, but the preferred range is between 10 and $35 \, \mu m$.
- Web thickness ranges between 0.1 and $4.0 \, mm$ and is typically $0.2–1.5 \, mm$.
- High strength-to-weight ratios compared to other nonwovens, woven, and knitted structures
- High tear strength.
- Good fray and crease resistance.
- High in-plane shear resistance
- Low drape.

6.8.3 Models for the prediction of fabric properties

Modelling approaches have been successful in predicting some of the properties of spunbond fabrics. The stress–strain responses of certain spunbond fabrics can be predicted using the stress–strain behaviour of constituent fibres, the fibre orientation angle distribution, the Poisson ratio of the fabric and shear strain [64,65]. For Cerex spunbond fabrics, it has been shown that Cox's fiberweb model can be used to predict the tensile response [66]. Several theoretical models have been proposed to predict nonwoven fabric properties, with certain assumptions, that can be used or adapted for spunbonded filament networks. Examples include:

- Backer and Petterson [67]
  - The filaments are assumed to be straight and oriented in the machine direction.
  - Filament properties and orientation are assumed to be uniform from point to point in the fabric.
- Hearle et al. [68,69]
  - The model accounts for the local fibre curvature (curl).
  - The fibre orientation distribution, fibre stress–strain relationships, and the fabric's Poisson ratio must be determined in advance.
- Komori and Makishima [70]
  - Estimation of fibre orientation and length.
  - Assumed that the fibres are straight-line segments of the same length and are uniformly suspended in a unit volume of the assembly.
- Britton et al. [71]
  - Demonstrated the feasibility of computer simulation of nonwovens.
  - The model is not based on real fabrics and is designed for mathematical convenience.
- Grindstaff and Hansen [72]
  - Stress–strain curve simulation of point-bonded fabrics.
  - Fibre orientation is not considered.
- Mi and Batra [73]
  - A model to predict the stress–strain behaviour of certain point-bonded geometries.
  - Incorporated fibre stress–strain properties and the bond geometry into the model.

- Kim and Pourdeyhimi [74]
  - Image simulation and data acquisition.
  - Prediction of stress–strain curves from fibre stress–strain properties, network orientation, and bond geometry.
  - Simulated fibres are represented as straight lines.

6.9 Spunbond fabric applications

Spunbond fabrics find applications in a variety of end-uses (Fig. 6.10). In the early 1970s, spunbonds were predominantly used for durable product applications, such as carpet backing, furniture, bedding, and geosynthetics. By 1980, disposable, single-use applications accounted for an increasingly large proportion, primarily because of the acceptance of lighter-weight spunbond PP fabrics in coverstock for diapers and incontinence products [36]. Today, the main markets for spunbonds span both single-use and durable products, including for example, hygiene continence management products, medical devices and protective clothing, automotive backings and reinforcements, house wrap, building and roofing materials, geosynthetics, home furnishings, air and liquid filter media, crop cover and other uses in agriculture, as well as an array of packaging products.

6.9.1 Automotive

One of the major uses of spunbonds in automobiles is as a backing for tufted floorcoverings. Spunbonds are also used as reinforcing components in trim parts, trunk liners, interior door panels, and seat covers. Spunbonds account for around half of global

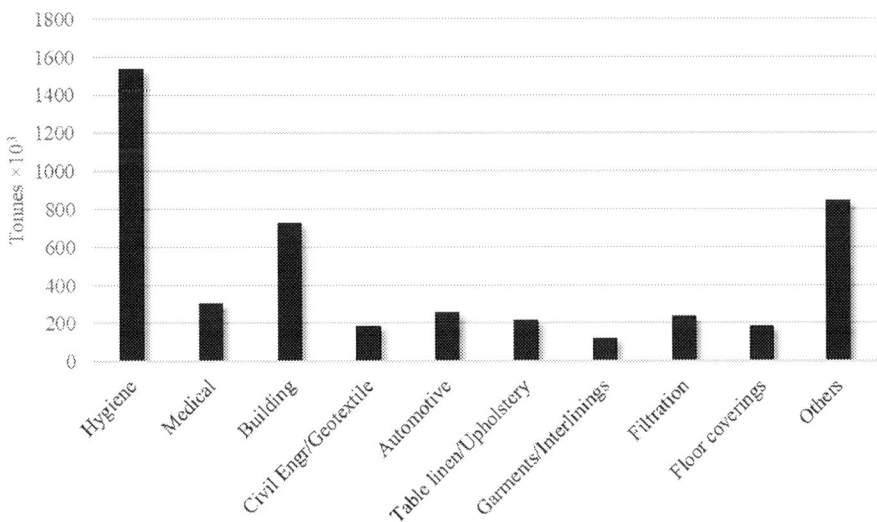

FIG. 6.10   Spunlaid nonwovens market in 2016. *Based on reference Anonymous: Smithers Pira, 2016.*

nonwoven consumption in the automotive sector. It is expected that the use of spunbond in automotive sectors will increase to around 9.5% by 2021 [75].

6.9.2 Civil engineering

Civil engineering remains a large market segment for spunbonds accounting for over 25% of the total sales. Geosynthetic applications include erosion control, revetment protection, railway bed stabilisation, canal and reservoir lining protection, and highway and airfield black top cracking prevention. In these applications, the high tear strength, UV resistance, bio-resistance, chemical resistance, resistance to other long-term environmental exposure, physical stability, high strength/cost ratio, and the engineering potential of spunbond fabric structures are attractive to end users [41]. Another important area of application is in roofing, including PET-based substrates for bitumen backing. Spunbonds are extensively used in house wrap and roof underlay fabrics. These materials, which can be combined with membranes or meltblowns, prevent outside air and rain from penetrating into buildings while allowing humid air to escape from the inside to the outside [75].

6.9.3 Hygiene and medical

At present, the largest commercial market for spunbond fabrics in terms of area is in the hygiene sector. Baby and adult incontinence diapers, as well as feminine hygiene products account for the largest volumes of over 1.5 million tonnes in 2016 [75]. Over the last few decades, growth of hygiene products containing spunbonds, particularly in continence care and management products has increased dramatically. This is mainly because of the attractive economics of polyolefin-based spunbonds, combined with their thermoplasticity (facilitating conversion of final products) and their intrinsically hydrophobic nature, minimising wetback during use, aiding user comfort [36]. For similar reasons, sanitary napkins and to a limited extent tampons are further applications. In the medical sector, traditional textile materials have been increasingly replaced by spunbonds, which due to their filamentous structure resist linting. This helps to minimise risk of contamination of biological tissues and fluid, and is valuable for barrier fabrics such as single-use protective clothing used in operating room gowns, masks, shoe covers and sterilisable packaging [36].

6.9.4 Packaging

Spunbond fabrics are widely used as packaging materials in place of paper products and where plastic films are not satisfactory. Important advantages of spunbonds in this market are the high resistance to tearing, retention of mechanical properties when wet, opacity and the ability to control barrier properties. Examples include metal-core wrap, sterile medical packaging, high-performance envelopes, and stationery products.

6.9.5 Filtration

Spunbonds are extensively used in filter media, either as the primary filter fabric, or as the reinforcing component, acting as a mechanical support to meltblowns, nanofibre webs, or membranes. A multitude of applications exist across liquid, air, and gas filtration. Spunbonds are used in coffee filters, edible and hot oil filters, milk and water filters, pool and spa filters, oil and fuel automotive filters, blood filters, and in tea bags. They are also found in engine and cabin air filters of vehicles and in domestic and industrial heating, ventilation, air conditioning, and vacuum cleaning. In these applications, the filament diameter, pore-size distribution, permeability and pressure drop, resiliency, strength, and uniformity are amongst the key parameters for end users [75].

6.10 Meltblown fabric production

Microfibres (also referred to as superfine or meltblown fibres) are less than about 10 μm in diameter, and typically 2–5 μm. Such fibres are found in nature, for example in spider silk and pineapple leaf fibres [76]. Man-made microfibres are produced using a variety of polymers and production techniques. Submicron glass fibres in 'glass wool' are a prime example. There are various methods of forming microfibres including direct extrusion spinning, bicomponent spinning (followed by splitting or dissolution of some components), spray spinning, electrostatic spinning, and centrifugal spinning. Commercially, the most important method is meltblowing, which involves the introduction of dissolved or molten polymers into high-velocity streams of air or gas, which rapidly convert the liquid into microfibres [77].

Similar to spunbond, meltblown is also produced directly from polymers. The meltblowing process falls under the general classification of spunlaid technology and is defined as follows: [78] Meltblowing is a process in which, usually, a molten thermoplastic fibre-forming polymer is extruded through a linear die containing several hundred small orifices. The number of orifices ranges from 10 to 20 orifices/cm, but can be up to ∼100/cm in multirow systems such as that found for example, in Biax Fiberfim's Spunblown technology. Convergent streams of hot air (exiting from the top and bottom sides of the die nosepiece or totally enclosing each die hole, according to the machine design) rapidly attenuate the extruded molten polymer streams to form extremely fine diameter fibres (normally averages in the range 1–5 μm). The attenuated polymer streams are blown by high-velocity hot air streams onto a collector (conveyor), thus forming a fine fibred, self-bonded nonwoven meltblown web (Figs 6.11 and 6.12). As in spunbonding, production throughput is linked to the number of rows and orifices per cm in meltblowing, and varies from about 10–100 kg/m/h (10–20 orifices/cm) up to 500 kg/m/h (with up to 104 orifices/cm).

The fibres in meltblown webs are held together by a combination of entanglement and cohesive sticking. If the fibres are still tacky on laydown then self-bonding is enhanced, and a subsequent bonding step, which normally involves thermal point bonding, might not be required. Because the fibres are drawn to their final diameters while still in the semimolten state, there is no downstream method of drawing the fibres before they are deposited onto the collector, and hence the webs exhibit low to moderate strength, compared to spunbonds. Examples of products containing meltblown fabrics, include for example, oil sorbent mattings, wipes, surgical gowns, surgical face masks and respirators, liquid and air filtration fabrics, lithium battery separators, clothing insulation and feminine hygiene products [78].

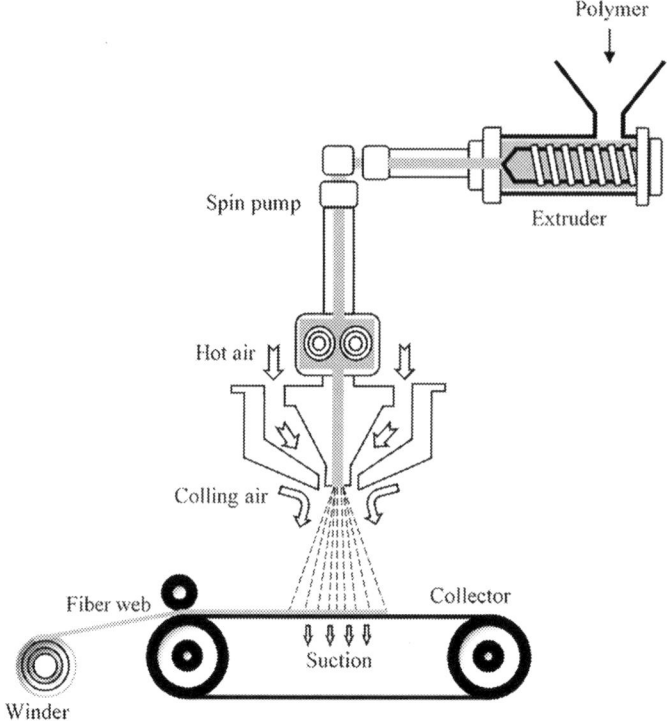

FIG. 6.11 Schematic of a meltblowing process. *Based on reference G. Bhat, S.R. Malkan, J. Appl. Polym. Sci. 83(3) (2002) 572–585; Anonymus: Meltblown Process, 2020. http://simonyan-company.com/en/products/complete-plants-for-the-artificial-and-synthetic-fibers-industry/meltblown-process/.*

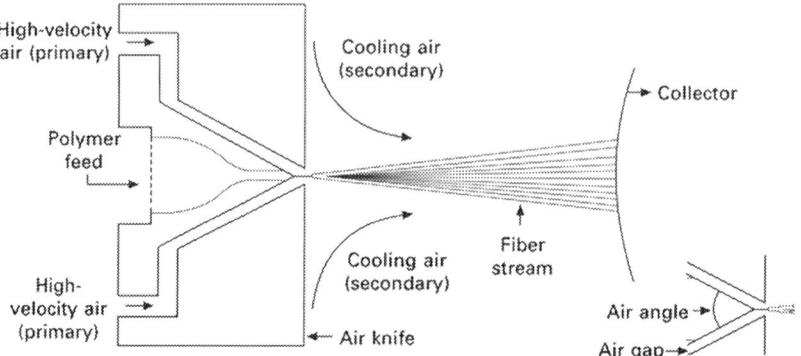

FIG. 6.12 Schematic showing primary and secondary air flow and web formation.

6.10.1 Meltblowing process technology

Despite extensive research and development in meltblowing, in the early stages of industrial development, there was a paucity of published research studies, mainly because of the secretive and competitive nature of the industry. However, numerous academic studies focused on understanding fundamental mechanisms, as well as a large body of patented literature reporting process enhancements and improvements are now available. The majority of the published papers are experimental in nature, with the exception of a few analytical studies. They can be classified into five main categories as follows: parametric, characterisation, polymer processing, electrostatic charging, and theoretical modelling.

The meltblowing process is intuitively simple, but scientifically, it is very complex, and therefore, there is considerable amount of literature on parametric studies of the process. These give significant insight into the fundamentals of the meltblowing process. The properties of meltblown fabrics are greatly influenced by the process technology used to make them. Some of the important process parameters are melt temperature, polymer mass flow rate (throughput), die geometry, airflow rate and air temperature, die-to-collector-distance (DCD) and collector speed. Fibre properties including cross-sectional shape and morphology, mean diameter and distribution as well as the presence of defects, web geometry and mechanical properties can be controlled by these parameters.

6.10.1.1 Effect of resin melt flow index and polydispersity

Jones [79], reported the effects of resin MFR and polydispersity on the mechanical properties of meltblown webs. In general, it was found that the strength peaked at an MFR of 300, and decreased as the MFR increased significantly above that. The fabric strength was believed to be a combination of individual fibre strength and good thermal bonding and/or fibre entanglements. The study also reported that the change in polydispersity of the polymers did not appreciably affect the mechanical properties. Only a slight downward trend of the fabric strength with increased degrees of polydispersity was observed. The elongation decreased with increased MWD. The decrease in elongation was attributed to the larger diameter fibres and fewer fibre entanglements, resulting from higher die swell in the broad MWD resins. Finally, the effects of varying extruder temperatures and throughput rate on the strength were readily evident. Decreasing the extruder temperature increased tenacity and bursting strength and increasing the throughput rate decreased the bursting strength. Malkan [78] has also reported the same trend.

Straeffer and Goswami [80] reported the effects of MFR and polydispersity index on the mechanical properties of meltblown webs. Three different mMFR resins with varying polydispersity indices were meltblown. The air flow rate used for the study was considerably lower than the one used in normal meltblowing conditions. The study concluded that 1100 MFR resin degraded more than the 300 MFR resin. This result was rather surprising. Malkan [81] previously reported the opposite trend. Usually, lower MFR resins degrade more than higher MFR resins. The study also showed that the fibre diameter distribution appeared to be narrower as the air velocity was increased, and there was no discernible difference in the crystallinity of the fibres.

6.10.1.2 Effects of pigments

Eaton et al. [82] studied the effects of pigments on the physical properties of meltblown nonwovens. The objective was to evaluate the feasibility of processing pigmented resins into meltblown webs. Different MFR pigmented PP resins were meltblown under different processing conditions. The physical properties of pigmented webs were compared with the properties of unpigmented webs. The study found that meltblown webs of uniform coloration were readily obtainable by the mass pigmentation of PP resins. It was believed that the pigments acted as nucleating agents in the high MFR resins that were meltblown. The average fibre diameter decreased slightly with increased MFR for both pigmented and unpigmented webs. The apparent colours of the webs became lighter with increased resin MFR. This effect was attributed to the increased ratio of fibre surface area to the amount of pigment resulting from the finer fibres produced with the higher MFR resins. The photomicrographs showed no notable differences between the pigmented and unpigmented webs in terms of fibre and web structure. The pigmented webs generally exhibited higher stiffness than the unpigmented webs.

6.10.1.3 Effects of machine settings on web structure, filament and shot formation

Wente and coworkers [9] first established the feasibility of manufacturing submicron fibres from a variety of thermoplastic polymeric materials. The main objective of their study was to design an apparatus that would produce submicron fibres. It was thought that a nonwoven material composed entirely of submicron fibres would be very useful for aerosol filtration, as well as for dielectric insulation. The early work of Wente describes the various aspects of the meltblown process, such as processing parameters, single fibre properties, and equipment design. It was found that since the molten polymer issues from the die nose tip directly into the confluence of the air streams, the greatest amount of attenuation occurs at the point of exit. Thus, the orifice size has little importance provided it is large enough to pass the melt without plugging. Malkan [83,84] also showed that the orifice size has very little effect on the final fibre diameter. The ability of a polymer to attenuate to a fine fibre is dependent on its melting point, viscosity-temperature characteristics, and surface tension. These findings appear to imply a streamline attenuation behaviour such as in air drag melt spinning. But according to Milligan and Haynes [85], the attenuation of fibres in meltblowing is primarily due to the shape they assume during their flight from the die face to the collector screen, otherwise known as 'form drag'. It was also found that a steeper air angle gives a higher degree of fibre dispersion and random orientation in the web during laydown. A smaller air angle yields a greater number of parallel fibres, greater attenuation, and reduces the probability of fibre breakage.

Buntin and Lohkamp [86], described the development of the meltblowing process for converting PP into low-cost fine fibrous webs. Two process design concepts of using multiple dies and a circular die were also discussed. The study found that by using multiple dies, high production rates can be achieved without increasing polymer throughput rates above the optimum. Multiple dies also produced more uniform webs with fewer fabric defects. The main disadvantage of the multiple dies operation was that for heavy basis weight, the bonding between layers was found to be unsatisfactory. In a circular die, a die with the orifices in a circle was used. The die was operated in such a way that the conical fibre blast from the orifices was focused roughly on a point where the fibres were formed into a continuous cylindrical roving. The technique was studied to make fine fibred cigarette filters. The study also reported that during the normal operation of the meltblowing process, fibre breakage occurred continuously and that fibre lengths were on the order of a few inches. The fibre breakage was associated with the incidence of 'shot' (they defined shot as a particle of polymer, considerably larger than the fibres, which is formed by the elastic 'snap back' of the fibre ends upon breaking). However, the current understanding of the meltblowing process is that the fibres in the final meltblown webs are mostly continuous in nature; therefore, the origin and nature of shot remain debatable. Below a certain melt temperature, a given thermoplastic polymer forms a large granule of nonfibrous material called 'shot'. High air pressure generally yields uniform and shot-free fibres but, at the same time, interferes with the separation of fibre and air because of the excess volume of fibres.

Early in the studies of the process, it was known that meltblown webs derive some strength from the mechanical entanglement and frictional forces existing in the fine network of microfibres in the web. Tsai et al. [87] studied the effects of water spraying near the die exit during meltblowing web production. The study showed that water spraying improved the mechanical properties of the web, such as tensile, bursting, and tear strength. Water spraying also reduces shot formation during meltblowing. The use of water spray did not affect the fibre diameter, air permeability, and filtration efficiency of the webs. Water spraying at higher processing and air temperatures was also suggested as a means of producing fine fibred webs with reduced shot intensity.

Wadsworth and Jones [88] demonstrated the effects of die-to-collector distances, polymer throughput rates, and die orifice diameters on meltblown fabric properties. The study found that machine direction (MD) tenacity decreased with increased die-to-collector distance whereas the cross direction (CD) tenacity and bursting strength remained relatively constant. Fibre diameters increased slightly with increased die-to-collector distance. The MD tenacity of the fabric increased with increased polymer throughput rates at corresponding collecting distances. On the other hand, the CD tenacity changed little. The study also showed that bursting strength decreased with increased throughput rates. The webs produced with a smaller orifice die were found to have slightly larger fibre diameters than webs produced with a standard orifice die. It was hypothesised that greater die swell could have occurred with the small orifice die, resulting in a net increase in filament diameter compared to the standard orifice die. This finding was in contrast to findings by Wente [89] and Malkan [78,83,84]. Wente and Malkan reported that the orifice size had very little effect on the final fibre diameter.

Khan [90] studied the effects of die geometry and process variables on fibre diameter, and shot formation of meltblown PP webs. The study included consideration of the die orifice L/D ratio, nose piece angle, air gap, nosetip setback, polymer throughput rate, air throughput rate, die-to-collector distance and the resin MFR. The statistical analysis showed that the L/D ratio, air gap, MFR, and polymer throughput rate have a significant effect on the average fibre diameter. Based on statistical interaction analysis, it was shown that all the variables had a significant effect on shot and fly generation. It was shown that by maintaining a constant air-to-polymer ratio, a web with known fibre diameter could be obtained. There was no conclusive evidence that the setback had a significant effect on the average fibre diameter. However, the shot intensity increased slightly with an increase in setback. Shot also notably increased with increases in the polymer throughput rate and orifice diameter. Wente [89] and Malkan [78,83,84] reported a similar trend. Part of the same study revealed that an increase in orifice size from 150 to 200 μm increased the resultant average fibre diameter. This increase was attributed to higher shear viscosity for the 200-μm orifice size. This finding was contradicted by the findings of Malkan [78], but agreed with the findings of Wadsworth and Jones [88]. The 60°C die tip resulted in slightly smaller diameter fibres than from the 90°C die tip. However, web properties such as air permeability, filtration efficiency, and bursting strength were higher for the 90°C die tip, compared to the 60°C tip.

Lee and Wadsworth [91] studied the relationships amongst the processing conditions, structure, and filtration efficiency of PP meltblown webs. The structure of the meltblown webs was varied using different die-to-collector distances and smooth roll thermal calendering. The study found that air permeability, and mean and maximum pore sizes increased with increased die-to-collector distance. The filtration efficiencies for water aerosols containing 0.5 and 0.8 mm latex spheres and for bacteria (*Staphylococcus aureus*) were found to be dependent on mean pore diameter and the parameter of basis weight divided by the square of the average fibre diameter. The calendering process decreased the web thickness, maximum pore diameter and air permeability, regardless of the original values.

Choi et al. [92] studied the tensile properties of meltblown webs. A method to obtain the tenacity and Young's modulus of meltblown webs without direct measurement of web thickness was proposed and tested using several web samples with different basis weights. The web tenacity and Young's modulus, which were, of course, normalised for weight, were found to be nearly independent of basis weight and gauge length. However, these properties were greatly affected by the processing conditions. Tenacity and Young's modulus decreased with an increase in the die temperature, air pressure at the die and die-to-collector distance. The study also compared the strength of single meltblown fibres with that of the web, as well as values obtained from high-speed melt spun filaments prepared from the same resins. Single fibre strength tended to be intermediate between the strength of the web and that of the high-speed melt spun filaments. The low strength of the single fibre was attributed to low molecular orientation, irregular diameter profiles along the length of the fibres, and the existence of voids in the fibres.

Malkan [78,83,84] studied the process–structure–property relationships in meltblowing for different PP resins. This extensive study considered various aspects, including process conditions, resin variables and the effects on web and single fibre properties. It was hypothesised that the reason for an increase in fibre diameter with increasing polymer throughput, was because of the increased die swell and the change in air-to-polymer ratio at higher polymer throughput rates. A 'processing window' was established to optimise the processing parameters to ensure good web quality. The DSC and X-ray diffraction results indicated that the observed double endothermic peaks might have been due to the presence of different crystal forms, namely, a-form (monoclinic) and b-form (hexagonal). It was hypothesised that higher polymer throughput rates could cause a change in the morphological or crystalline structure of meltblown PP fibres. It has been shown by Bodaghi [93] that water-quenched PP fibres have a para-crystalline structure, and that air-quenched fibres have a regular monoclinic crystal structure. However, there have been arguments against the possibility of different morphologies (folded chain crystals or extended chain crystals) since double endotherms were produced at an isothermal crystallisation temperature as low as 130°C. This temperature appears to avoid the possibility of extended chain crystal formation.

Milligan et al. [94] reported the use of a stream of unheated cross-flow air to make finer meltblown fibres. The authors claim that in addition to finer fibre diameters, the variation in fibre diameter was smaller with the use of cross-flow air. The motivation for this research was the earlier finding by Milligan and Haynes [85] that the fibre form drags, resulting from large amplitude flapping of the molten polymer jet, as it gets blown away from the die, can play a significant role in contributing to the total air drag force necessary for fibre attenuation during meltblowing. The study pointed out the possibility of substituting unheated cross-flow air for a portion of the primary hot air, resulting in energy savings. Haynes [94] studied the effect of air jet exit velocity of a single hole orifice meltblowing die on average fibre diameters. It was observed that initially, fibre diameter decreases with an increase in the air jet exit velocity, but after a certain level, the fibre diameter does not change as the air jet exit velocity increases. This is referred to as the critical air jet velocity. Haynes [94] observed that for 0.4 g/min polymer throughput, the critical air jet exit velocity is ~220 m/s. After this critical velocity, the exit air jet velocity increases up to 380 m/s, but average fibre diameter remains almost the same at around 5.5 μm.

According to Haynes [95], the smaller air gap (higher air velocity) gave a standard deviation of 5.53 μm, while with the larger gap (lower air velocity), the standard deviation was ~3.4 μm. It appears that the larger air gap (lower air velocity) produced a narrower fibre diameter distribution. This indicates that the velocity of primary air in itself does not affect the average fibre diameter significantly in the range of velocities discussed here (high end of the possible velocity scale). A careful examination of average-fibre diameters produced at low air velocities indicated a notable effect of air velocity on average-fibre diameter, even at constant SCFM. Further evidence to this observation is presented elsewhere by Haynes [95]. The fibre diameter decreased when the velocity was increased by increasing the mass flow rate of air at a constant air gap. This result implies that the primary controlling variable of fibre diameter is the mass flow rate of air, and therefore, increasing the velocity alone, without increasing the mass flow rate of primary air, is not of much significance with respect to fibre diameter. This finding was in contrast to popular belief that a smaller air gap gave finer fibres at a constant primary air flow rate.

Warner et al. [96] observed spherulitic crystalline morphology in meltblown fibres, and that fibres of >10 μm have essentially no molecular orientation. This implies that such fibres do not contribute to meltblown web strength. It should be noted that the mass of a fibre per unit length is proportional to the square of its diameter. Therefore, the mass diameter distribution of the web is arguably more appropriate than the number diameter distribution as a means to calculate the fibre strength realisation in a meltblown web.

6.10.1.4 Formation of ultrafine meltblown microfibres

Wadsworth and Muschelewicz [97] reported the results of a study designed to produce extremely fine meltblown fibres using 35, 300 and 700 MFR resins. The study established the optimised meltblown processing conditions to produce fine fibred webs, with less than 2 μm fibre diameters. The study found that the small orifice die (with orifice diameter in the range of 200–300 μm or 0.2–0.3 mm) resulted in statistically smaller mean fibre diameters than with dies of standard orifice size (approximately 400 μm). However, the actual difference was found to be minimal, considering the fact that the cross-sectional areas of the small die holes differed by a factor greater than two. The study also reported that increased airflow rates decreased mean fibre diameters. The air gap settings did not have any noticeable effects, except that much larger diameters were obtained with the standard die tip and the smaller air gap setting with 300 and 700 MFR resins. Notably higher bursting strength values were achieved with the small die holes, indicating increased fibre strength with greater drawdown. Meltblown webs containing very small diameter fibres, of <1 μm (down to 250 nm) can be produced from low viscosity melts, i.e., when using high MFR resins (e.g. 1500–1800 MFR), combined with high airflow velocities. Generally speaking, higher MFRs are needed to cost-effectively make meltblown filaments of 1 μm or below, which is also an important consideration in the manufacture of products such as respiratory masks and protective clothing intended to act as barrier materials.

6.10.2 Minimisation of energy consumption

Milligan [98] analysed several design concepts to minimise energy costs in meltblowing. The pressure losses associated with the air pipework and the air heater were two of the principal sources of energy consumption. Several design rules have been presented to minimise the cost of air utilisation based on Darcy's pressure drop relation:

$$\Delta P/L = \rho.f.V^2/2D \qquad (6.2)$$

Where,
ΔP = the pressure drop.
L = the length of the pipe.
D = inner diameter of the pipe.
P = air density.
F = pipe friction factor (dependent on D and pipe roughness).
V = air velocity in the pipe.
The design rules are:

1. The piping between the air compressor and the die assembly should be as short in length and as large in diameter as possible.
2. The air heater should be as close as possible to the die assembly.
3. The number of pipe fittings should be minimised, and all piping should be well insulated.

The study also highlighted the importance of correctly sizing the compressor in a meltblown pilot line. The suggested design rules minimise energy cost, specifically by reducing the pressure losses associated with the air piping and efficient location of air heaters.

Milligan, Wadsworth, and Cheng [99] investigated the energy requirements for the meltblowing of different polymers. The studied polymers were PP, LLDPE, nylon (PA), and polyester (PET). The energy requirements were reported in kW-HR per kg of polymer. It was apparent from the reported data that the energy cost per unit mass of the product greatly depended on the air flow rate, air temperature, polymer throughput rate, and polymer molecular weight. The results also showed that a large fraction (greater than 85% for all the materials investigated) of the energy required was associated with the utilisation of hot air streams. The study also found that the difference in total energy consumed and the actual energy required at the die can be attributed to improper compressor size, compressor cooling, and heat losses from the die and piping. The study suggested that substantially lower energy consumption is possible if a meltblowing line is carefully designed and operated with the objective of minimising energy consumption.

6.10.3 Other processes—Nanoval

Various other processes similar to classical meltblowing have been developed, but with differences in terms of how the polymer streams are attenuated or interact with the airflow.

One example is Nanoval technology (Nanoval GmBH), where Laval nozzles are deployed in the spinning beam to mix the airflow and liquid polymer differently to meltblowing, creating multiple separate polymer streams. Like meltblowing, continuous filaments are collected on the conveyor with mean diameters of about three microns and below. In addition to a large variety of thermoplastic polymers, Nanoval technology has also been harnessed to process nonthermoplastic lyocell spinning solutions, enabling the production of cellulosic nonwoven webs.

6.11 Meltblown web characterisation techniques

This section briefly reviews some of the characterisation techniques used to study meltblown web properties and related phenomena. Mostly, meltblown webs are characterised in terms of stress–strain properties, filtration efficiency, and air permeability. Many of these properties can be characterised using standard ASTM or INDA methods. However, researchers have developed other techniques to characterise web properties, such as fibre diameter, mean pore size analysis, filtration efficiency, and on-line fibre diameter measurements through light scattering and alternative techniques.

Tsai [100] proposed a mathematical relationship to characterise key web properties using an air flow technique. This mathematical relationship uses air permeability data and determines approximate fibre size and pore size. Naqwi et al. [101] developed an interferometric optical technique, referred to as an adaptive phase/Doppler velocimeter (APV) for in situ sizing of spherical and cylindrical objects as applied to spunbond and meltblown fibres. It has been suggested that this technique could be used to monitor changes in processing conditions during the meltblowing process, specifically fibre diameter. Bhat [102] used sonic velocity measurements to characterise meltblown webs. The sonic velocity can be used as an indication of the overall arrangement of structural elements in the fabric. The results showed a good correlation between the measured sonic velocity values and meltblown fabric mechanical properties.

Wallen et al. [103] investigated the use of small angle light scattering to study transient single fibre diameter and to monitor the fibre attenuation process as a function of distance from the die during the meltblowing process. The fibre diameters were determined from the total intensity of the scattered light. The study concluded that the meltblown process is not a steady state below a certain timescale and that the fibre attenuation process is not constant with respect to time or distance from the die. Bodaghi [93] has described many meltblown microfibre characterisation techniques. He found that water-quenched PP fibres have a para-crystalline crystal structure, whereas air-quenched fibres exhibit a regular monoclinic crystal structure.

6.11.1 Theoretical studies and modelling

Narasimhan and Shambaugh [104] attempted to model the meltblowing process based on a single die hole rig and a circular air slot surrounding the meltblown spinneret nozzle. However, the most commonly used meltblown die geometry consists of a row of spinnerets with sheets of hot air exiting from the top and bottom sides of the die. Shambaugh continued this study [105] and applied macroscopic energy balance and dimensional analysis concepts to meltblowing. These two concepts were analysed using different die geometries. Three operating regions were identified in the melt blowing process according to the extent of the air flow rate as follows:

- Region I has a low gas velocity similar to a commercial melt spinning operation in so far as the fibres are continuous.
- Region II is unstable as the gas velocity is increased. In this region, filaments can break up into fibre segments and undesirable lumps.
- Region III occurs at a very high air velocity and involves excessive fibre breakage.

The meltblown process predominately uses a low airflow rate (Region I) and is most energy efficient in this region. A monodisperse fibre distribution is claimed to require less energy to produce than a polydisperse fibre distribution. The dominant dimensionless groups in the meltblowing process are claimed to be the gas Reynolds number, the polymer Reynolds number, fibre attenuation, and the ratio of the polymer viscosity to the gas viscosity.

Milligan and Haynes [106] studied the air drag on monofilament fibres. The aim was to study the air drag by simulating actual meltblowing conditions. The experimental set-up closely simulated a commercial meltblowing operation except that the high-velocity air was at room temperature and one end of the monofilament was secured to a tensiometer. The air drag was studied as a function of fibre length, upstream stagnation pressure, air injection angle, and gravity orientation. Four series of experiments were conducted as follows:

1. Determination of air drag for a fibre of constant length over a range of air stagnation pressures.
2. Determination of air drag for a range of fibre lengths at a constant value of stagnation pressure.
3. Determination of air drag for a fibre of known length and stagnation pressure using different injection angles (15°, 30°, and 45°).
4. Determination of air drag by changing the die orientation, with respect to gravity, using a 30° air injection angle with different fibre length and stagnation pressures.

The study concluded the following:

- The drag increased with fibre length when all other parameters were constant. The drag peaked at a length of 2.5 cm. This was due to large amplitude flapping of the fibre.
- The drag increased linearly with stagnation pressure for stagnation pressures up to 207 kPa.
- The drag increased with decreased air injection angle for any particular stagnation pressure. This finding was in basic agreement with the fundamental momentum consideration.
- Orientation of the die with respect to gravity using an injection angle of 30° (45° above the horizontal to 90° below the horizontal) showed no measurable difference in drag. It was concluded that the viscous and pressure forces on the filament far exceed the gravitational force for the flow conditions investigated.

Uyttendaele and Shaumbagh [107] have reported various analytical studies involving mathematical modelling of meltblowing. Majumdar and Shaumbagh [108] have calculated the air drag on fine filaments in the meltblowing process using a wide range of filament diameters, gas velocities (primary air velocity) and Reynolds number.

6.12 Characteristics and properties of meltblown fabrics

It will be apparent that meltblown fabric properties can be tuned, depending on end-use requirements by adjusting polymer selection and grade, process variables, bonding conditions and finishing processes. Some of the main characteristics and properties of meltblown webs are: [109]

- Random fibre orientation, and a near-planar isotropic structure.
- Low to moderate web strength compared to spunbonds.
- Generally, the web is highly opaque (high cover factor).
- Meltblown webs derive their strength from mechanical entanglement and frictional forces.
- Most meltblown webs have a layered or shingled structure, the number of layers increases with increasing basis weight.
- Fibre diameter ranges between 0.5 and 30 μm, but the typical range is 2–7 μm.
- Basis weight ranges between 8 and 350 g/m^2, typically 20–200 g/m^2.
- Microfibres provide high surface areas for good insulation, filtration and barrier performance.
- The filaments have a smooth surface morphology and are mostly circular in cross-section.
- The filaments vary in diameter along their length as well as throughout the web.
- Close examination of approximately 800 photomicrographs revealed no 'fibre-ends' except a few near areas where 'shot' is present. Therefore, meltblown fabrics consist mainly of continuous filaments.
- Filaments are characterised by thermal branching, i.e., bifurcation along their length. The exact cause of thermal branching is debatable, but according to Malkan [78] branching occurs when propagating fibres collide with other propagating fibres, which in turn, strip off portions of polymer streams as fine branches (filaments). Hermans [110] pointed out that when the velocity of the molten jet relative to air jet increases, portions are stripped off as filaments. Bresee and Wadsworth [111] stated that fibre splitting (branching) occurs when extrudate is stressed in complex ways in flight towards the collector.

6.13 Meltblown fabric applications

Meltblown fabrics are used in some of the same applications as spunbond fabrics, and often they are combined. Key markets are in hygiene, medical, industrial, filtration, oil sorbents, insulation and wipes.

6.13.1 Filter media

The original development work on meltblowing was focused on the production of microfibres which could be used in high-performance filtration products. Therefore, the filtration market segment remains the largest single market for meltblown webs, representing about 30% of the total [112]. The future growth is projected to be quite strong. Generally, meltblown webs are used for the more critical filtration applications where the superior filtration performance of the fine fibre network can be exploited. Applications for meltblown fabrics include:

- Room air filter and recirculation.
- Precious metal filtration and recovery.
- Food and beverage filtration.
- Water and liquid filtration (including blood and body fluids).
- N95/N99/FFP2/FFP3 respirator and surgical masks.
- Coalescence filters (for example in the removal of water from diesel fuel).

The inherently large fibre surface area in meltblown fabrics associated with the small fibre diameters present, combined with the ability to modulate surface chemistry and electrostatic behaviour, results in some valuable properties [113,114]. These include the ability to achieve high filter efficiencies when dealing with small and submicron particles in air and liquid fluid flows, including dry particles and aerosols, as well as in liquid–liquid separation applications such as automotive filter media for the coalescence of water droplets from oil or fuel. Removal of water droplets is essential to prevent damage to vehicle engines, particularly given the increasing availability of biofuels. Meltblown fabrics have also attracted heightened attention during 2020 because of their importance in the manufacture of critical PPE products such as single-use respirators and surgical masks, which have to meet strict performance requirements in terms of aerosol or bacterial filtration efficiency, and breathing (airflow) resistance, amongst other criteria. For single-use respirators (respiratory masks), e.g., FFP3 (in Europe), electrostatically charged meltblown fibres within a certain diameter range are needed to meet prescribed filtration efficiency and airflow resistance values set out in the relevant test standards.

6.13.2 Hygiene and medical

Meltblown fabrics are extensively used in hygiene and medical products. The former includes feminine sanitary napkins, diaper topsheets and other components, as well as in disposable adult incontinence products. In addition to single-use respirator and surgical masks, the major markets for meltblown nonwovens in the medical sector are single-use gowns, drapes, and sterilisation wraps. Depending on the level of protection required, meltblown fabrics (M) for making gowns are normally delivered in the form of layered or composite nonwoven structures with spunbond fabrics (S). Note that increasing the basis weight, and particularly the weight of the meltblown layer, increases the degree of barrier performance. Typical constructions include:

- Four or five-layer SMMS or SMMMS: 35–70 g/m^2.
- Four or five-layer SMMS or SMMMS: $\geq$35 g/m^2 with additional liquid repellent compounds incorporated either into the resin prior to extrusion, or applied to the finished fabric as a chemical treatment to improve liquid repellency.
- Trilaminate of ca. 60 g/m^2 consisting of spunbond (S) or SMS outer layers, with a film membrane laminated in the middle to maximise resistance to liquid ingress.

6.13.3 Insulation

Meltblown webs provide good insulation because of the large surface area, which creates significant drag forces on air convection currents passing through the fabric. The trapping of still air as a means of providing high thermal insulation is a concept that has been exploited successfully in Thinsulate (3M) for outdoor sports and leisure clothing. Ando [114] has also presented applications for meltblown webs in thermal insulating media.

6.13.4 Absorption

Meltblown webs are widely used in non-aqueous liquid/oil absorption, where the hydrophobic and oleophilic nature of polyolefin-based meltblowns can be exploited. This market is growing due to growing governmental regulations concerning spillages, contamination of ground water, and environmental cleanliness. There are many meltblown products in the market that require liquid absorption such as industrial wipes, oil sorbent pads, and booms amongst others. White [115] has also shown the effective use of meltblown webs in food fat absorption, and with appropriate chemical treatment, e.g., following the application of surfactants, absorption of aqueous liquids is possible for some household and hygiene applications.

6.13.5 Miscellaneous applications

There are many other growing applications for meltblown fabrics, including but not limited to, disposable industrial apparel, substrates for synthetic leather, liner fabrics for data storage devices and elastomeric nonwovens.

6.14 Bicomponent spunlaid production processes

6.14.1 Bicomponent filaments

Bicomponents (or bicos) consist of at least two polymers of different physical and chemical properties within a single filament, and extruded from the same spinneret. The production process for bicomponent filaments is similar to that of single filaments for spunbonds or meltblowns, but with modification of the spin pack arrangement. Here, two separately extruded polymer streams pass through the filters, spin beam, spin pack, and then conjugate in the spinneret. After extruding from the spinneret, the bicomponent filaments are quenched, attenuated, deposited onto the conveyor belt, and then bonded (usually by thermal means or hydroentangled if the splitting is desired) before finally being wound on a package (Fig. 6.13) [25]. It is possible to produce microfilaments by this process.

Bicomponent filaments can take several forms. Some of the most common types are sheath-core, side-by-side, tipped, segmented pie and island in the sea (IS). In sheath-core filaments (Fig. 6.14A), a higher melting temperature polymer core is surrounded by a low melting polymer sheath. Common sheath-core combinations are PE-PP, PE-PET, and PA-PET. These types of bicomponents are largely produced in nonwoven industry to aid thermal bonding [116].

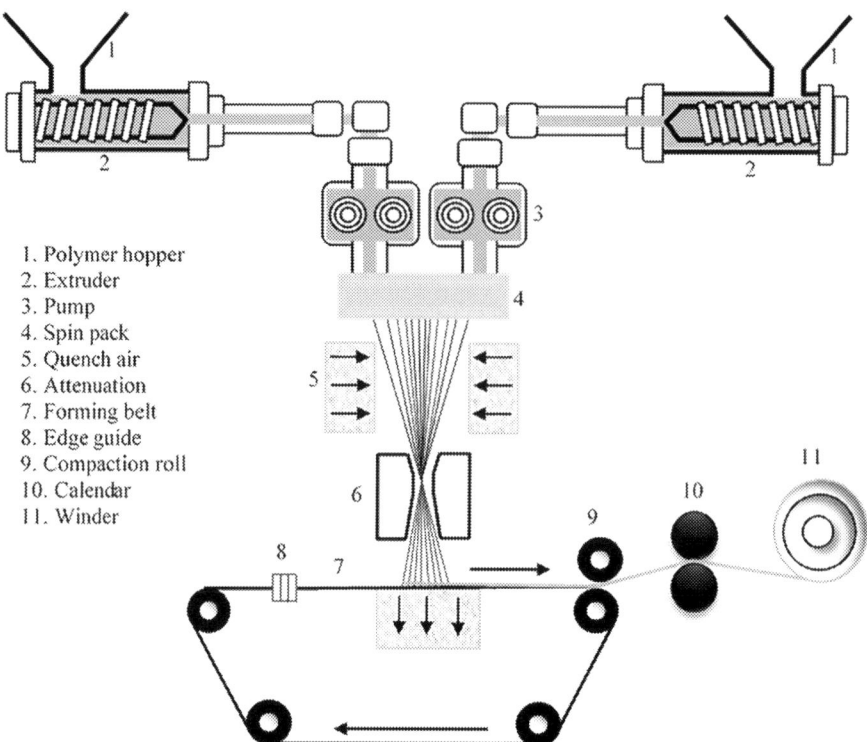

1. Polymer hopper
2. Extruder
3. Pump
4. Spin pack
5. Quench air
6. Attenuation
7. Forming belt
8. Edge guide
9. Compaction roll
10. Calendar
11. Winder

FIG. 6.13 Schematic diagram of the bicomponent spunbond process *N.V. Fedorova, Investigation of the Utility of Islands-in-the-Sea Bicomponent fiber Technology in the Spunbond Process. PhD dissertation, North Carolina State University, 2007. https://repository.lib.ncsu.edu/bitstream/handle/1840.16/5145/etd.pdf?sequence=1; Anonymus: Meltblown Process, 2020. http://simonyan-company.com/en/products/complete-plants-for-the-artificial-and-synthetic-fibers-industry/meltblown-process/; J. Liu, et al., RSC Adv. 7(69) (2017) 43879–43887.*

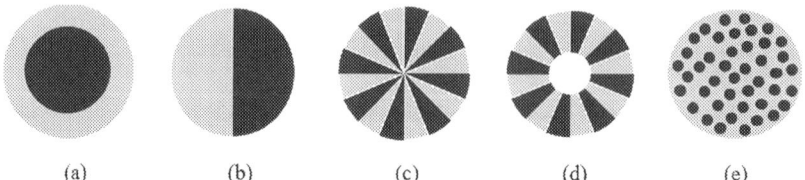

(a)   (b)   (c)   (d)   (e)

FIG. 6.14 Different bicomponent fibre (A) Sheath and core (B) side-by-side (C) segmented pie (D) hollow segmented pie (E) island-in-the-sea.

Side-by-side filaments (Fig. 6.14B) provide self-bulking functionality, where the two polymers within a filament possess differential shrinkage behaviour. In tipped filaments, one polymer is placed on the tip of a trilobal or delta cross-section of another polymer.

Spunlaid nonwovens containing microfilaments are normally produced by either splitting segmented pie bicomponent filaments, or by dissolving, or melting out one of the two polymer components in island-in-the-sea bicomponents, after the web or final fabric is formed.

In segmented pie bicomponents, the filament consists of multiple segments, either arranged in wedge-like segments (usually in filaments with a round cross-section) or in stripes (usually in filaments with a tape-like, or rectangular filament cross-section). They are normally made of two different polymers, which alternate between consecutive segments and can be split apart in downstream processing (Fig. 6.14C) [116]. It is possible to produce spunbonded micro- and nanofilaments of reproducible dimensions using the segmented pie method. In general, filaments with a diameter 0.1–5 μm are produced using splittable technologies [117].

In spunbonds consisting of segmented pie bicomponent filaments, mechanical splitting is normally accomplished by hydroentangling after the spunlaid web is formed. Splitting efficiency can be assisted when the filaments have a hollow core (hollow segmented pie filament) (Fig. 6.14D). Various production lines are in operation that integrate segmented pie bicomponent filament deposition with hydroentangling to produce microfilament spunbonded fabrics. For example, Evolon fabrics produced by Freudenberg are made of hollow and solid segmented bicomponent filaments that are split into microfibres during hydroentanglement of the spunlaid web. Originally, Evolon fabrics were based on 16 segmented pie bicomponents, but subsequent variants used 32 segmented pies to provide a smooth, uniform surface ideal for printing applications. The majority of such fabrics are based on PET/PA6.6 (polyamide 6.6) in a ratio of 65/35. Additionally, bicomponent fibres with an eccentric sheath-core arrangement are used to develop crimp in spunlaid fabrics by differential thermal shrinkage of the two polymer components. For ease of mechanical splitting, the two polymer components should have very little adhesion to one another. Hence selection of appropriate polymers is a key factor for ease of splitting and the generation of good quality microfilaments in the fabric. Spunbond fabrics produced in such a way containing microfilaments, yield high strength, opaque fabrics, which after appropriate finishing processes, have tactile properties similar to suede and leather products [118]. They have been used in durable wipes, medical dressings, shoe linings and for some durable sports goods as well as automotive headliners [119]. Several types of spunbonded bicomponent fabrics are available in the market.

In the second approach, known as island-in-the-sea (Fig. 6.14E), multiple microfilament (islands) of one polymer are dispersed in the matrix (sea) of another polymer. In island-in-the-sea bicomponents, the number of islands may be 16, 24, 32 or even >1000. As the number of islands increases, the diameter of the resultant filaments, following removal or disruption of the sea matrix, decreases. For island-in-the-sea bicomponents with 16, 37, and 1000 islands, the resultant fineness of the extracted filaments can be approximately 0.15, 0.05, and 0.001–0.01 denier (or 1, 0.3, and 0.1–0.5 μm) respectively [117]. Hills Inc. has produced diameters as small as 300 nm in this way, based on 1120 islands of 50% PP, PET, or nylon 6 with 50% of the filament comprising a sacrificial water-soluble PVOH matrix. Separation of bicomponent filaments by wet treatment after the fabric has been formed to dissolve out the matrix component with an appropriate solvent obviously means that there is no need to extrude very fine filaments in the first place. Generating the microfilaments after the web or fabric is formed from island-in-the-sea filaments can be accomplished not just by dissolving out the matrix, but also by fracturing it [25]. Hydroentangling is one way to do this, where the high-pressure waterjets used to bond the fabric, also induces physical disruption of the sea, releasing the embedded microfilaments.

Researchers have made detailed studies of fracturing the cross-section of island-in-the-sea bicomponent filaments by hydroentangling following spunlaid web formation, to yield high surface area and high strength fabrics [120]. Anantharamaiah et al. [120] observed that the fractured islands-in-the-sea and sheath-core filaments have remarkably high surface area because of the tiny dimensions of the islands, and of course, the fracturing approach has the added advantage of not requiring the matrix polymer to be dissolved out in a separate processing step.

6.14.2 Nanofibre layers

The properties of spunbond or meltblown fabrics can be significantly modified by combining them with nanofibre webs made by other processes. These are sometimes designated as SNS or SN fabrics, i.e., the nanofibre layer (N) is supported by one or more spunbond (S) layers. Nanofibre layers increase the fibrous surface area, which is useful in terms of engineering high efficiency filter media, tissue scaffolds for regenerative medicine, as well as water vapour breathable membranes. Nanofibre layers also provide an opportunity to further improve barrier properties, as well as sound absorption performance. There are several techniques by which nanofiber layers are produced, including electrospinning (see Section 6.16.2) and centrifugal spinning (Section 6.16.3). At present, electrospinning is the most widely used method of producing nanofibre layers, and web deposition is very often directly on top of a preformed spunbond. Lee and Obendorf used electrospinning technology to produce nanofibre layer fabrics, exhibiting good barrier properties at microscopic scale, to enhance resistance to liquid penetration in fabrics intended for medical personnel and chemical workers [121].

6.14.3 Composite nonwoven structures

Commercially, spunbond (S) and meltblown (M) web layers are combined in-line, in one integrated web-forming and bonding production line, to make composite fabric structures, such as the classical SMS structure. The SMS concept was first introduced and patented by the Kimberly-Clark Corporation, leading to the production of SMS composites for barrier application under the trade name of KLEENGUARD [122]. SMS and related composite nonwoven fabrics are now produced globally by numerous roll good manufacturers, and turnkey installations have been available for years.

SMS fabrics are mostly made of polyolefins such as PP and are therefore inherently hydrophobic, such that they act as excellent barriers, preventing liquid leakage and penetration of fine particles. Barrier and other performance characteristics can also be modulated by the addition of appropriate additives or fillers in the masterbatch prior to processing. The benefits of combining spunbond and meltblown webs include:

- Barrier to liquid permeation especially of bodily fluids.
- Increase in the cover of the base spunbond web.
- Barrier to penetration of solid particulates.
- Ability to combine high particle capture/filter efficiency with low flow resistance.
- Good tensile, tear and bursting strength.

There is a variety of structural combinations available. In addition to single-layer spunbonds, up to four beam spunmelt machines can be employed to make S, SS, SSS, and SSSS. Combinations with meltblowns include SM, SMS, SMMS, SMMMS, SMMMMS, SSMS, SSMMS, SSMMMS, SSMMMMS, SSMMSS, SSMMMSS, and SMMMMSS, depending on the required final product properties. The most popular combinations are SMS and SM composite fabrics. Broadly speaking, given it consists of finer filaments compared to the S layer, the M layer is responsible for controlling the liquid barrier, fluid flow resistance and particle capture properties of the fabric structure. Meanwhile, the S layer provides dimensional stability, abrasion resistance and can be used to modulate surface wetting behaviour and fabric appearance, e.g., through the use of masterbatch additives, surface coatings, mass colouration/dope dyeing, or printing.

The provision of multiple S or M beams in a production line means high linear output can be achieved while still ensuring the necessary S and M layer weights are laid down, and also provides flexibility in terms of the polymer grade and operating conditions, filament dimensions and other properties present in each layer. Continuous production lines such as this, mean that all the webs are brought together and bonded once, most commonly via thermal point bonding. This means the resulting barrier fabric remains highly porous, retaining most of its fibrous structure and can be made in very light basis weights if required. However, it also requires significant capital investment because of the multiple spinning beams required.

In an alternative approach, some manufacturers and converters buy rolls of preformed spunbond and meltblown fabrics and then laminate them together in a much less capitally-intensive production system. However, this does not yield SMS fabrics with an identical structure to those produced in a continuous multibeam production line, because the fabrics are already bonded (to allow their winding and unwinding) before they are laminated together, and this also places limits on their minimum basis weight.

Spunbonds and meltblown are also combined with other nonwoven processes to make a variety of co-formed or multilayer composite nonwoven fabrics. Composite nonwovens consisting of a spunbond layer (S), an airlaid wood pulp layer (P) and a spunbond (S) layer (or SPS) is an example of how fabric properties can be further developed by process integration. Such SPS fabrics enable for example a wet wipe substrate to be produced that is capable of holding large volumes of liquid, but will not disintegrate or lint during use. Similarly, combinations of spunbonds with wetlaid pulp and carded webs are also possible. In other applications, preformed films, including elastomerics are combined with spunbond and meltblown fabric components at the web formation stage.

6.14.4 Coform and similar technologies

Coform processes generally refer to methods of combining particles or short fibres with filaments during production, such that the resulting web and final fabric contains mixed materials. The Coform process operated by Kimberly Clark produces meltblown webs mixed with wood pulp to act as a liquid absorbent. During the process, the wood pulp in sheet form is fibrised and the separated pulp is injected from one side onto the still tacky meltblown filaments as they travel from the die to the collector. In this way, the wood pulp adheres to the filaments as they cool. The resulting meltblown fabrics can therefore consist of approximately 60%–70% wood pulp, yielding a remarkably thin liquid absorptive fabric, reinforced by a network of meltblown filaments. After the web is laid down, a preformed spunbond or meltblown fabric or film is thermally laminated to at least one side to form a dimensionally stable, liquid absorbent composite. By changing the number and composition of the various layers, a variety of products can be produced. These are used in liquid containment applications in the hygiene sector. Applications include wipes for domestic, hospital and nursing homes, incontinence, birthing, and nursing pads and fenestration areas on drapes.

Other integrated processes for producing commixed structures in a single step have been developed. They differ in their specific process arrangements and operational settings, but fundamentally, all involve combining extruded filaments with pulp, staple fibre or particles such as SAP. They include combining spunbond and/or meltblown with up to 90% fluff pulp for single-use hygiene applications, as well as multilayer structures combining spunmelt and airlaid to make wipes (C-Form and Phantom Technology, TKW Materials/P&G). An intended benefit of C-Form technology is that by combining spunbond or meltblown with fluff pulp a simpler process is provided for producing single-use wipe substrates compared with carding and hydroentangling, and of course, the energy costs of drying the fabric are avoided. Additionally, in the automotive sector, commixed nonwoven fabrics that combine meltblown filaments with PET staple fibre are manufactured to produce sound absorption materials (e.g. Neptune technology, Solar Nonwovens).

6.15 Mechanics of the spunbond and meltblown processes

Spunbonding and meltblowing incorporate many engineering concepts, some of which are discussed in this section.

6.15.1 Dynamics of melt spinning process

Spunbonding and meltblowing are based on multifilament fibre spinning and are an extension of conventional fibre spinning processes. To understand the theoretical framework of these processes, it is necessary to understand the dynamics of melt spinning. The basic equations involved in the melt spinning process, based on a single-filament model provide a useful starting point. The equations for multifilament spinning, related to spunbonding and meltblowing processes are quite complex. It is beyond the scope of this book to cover multifilament spinning but further reading is available [123,124]. The dynamics of melt spinning is based on mass balance, force balance, and energy balance principles. The balance equations are now considered.

Based on the principle of conservation of mass, the following mass balance relation can be written as: [125–127]

$$W = \rho A V \tag{6.3}$$

The above equation for a cylindrically shaped filament becomes:

$$W = \rho \pi (D/2)^2 V \tag{6.4}$$

The overall force balance for filament spinning can be written as follows: [28, 125–127]

$$F_{rheo} = F_0 + F_{inert} + F_{drag} - F_{grav} - F_{surf} \tag{6.5}$$

Where,

F_{rheo} = the rheological force in the filament.

F_0 = the rheological force at the beginning of the spinline.

F_{inert} = the inertial force as the filament is accelerated.

F_{drag} = the drag force caused by the filament moving through a stationary fluid.

F_{grav} = the gravitational force on the filament.

F_{surf} = the surface tension force at the filament–air interface and is generally considered negligible compared to the magnitude of the other forces.

The individual forces may be expressed as follows: [28,125–127]

$$F_{rheo} = W (V - V_0)$$

$$F_{drag} = \pi D \sigma_f dz$$

$$F_{grav} = \rho g \left(\pi D^2 / 4 \right) dz$$

$$F_{surf} = (\pi \sigma)/2 \left(D_0 - D \right) dz \qquad (6.6)$$

Where,

W = the mass throughput rate (related to mass balance).

V_0 = the average polymer velocity in the die.

D_0 = the die diameter.

σ_f = the shear stress at the fibre–air interface due to aerodynamic drag.

σ = the surface tension of polymer melt with respect to air.

z = the distance along the filament away from the die.

Neglecting the radial variations and the force due to surface tension, the gradient of axial tension along the spinline can be written using Eqs 6.3 and 6.4 as:

$$dF/dz = WdV/dz + 1/2\rho_a C_d V^2 \pi D - Wg/V \qquad (6.7)$$

Where,

D = the filament diameter.

V = the filament velocity.

g = the gravitational acceleration constant.

C_d = the drag coefficient and is defined using the equation for the shear stress at the fibre–air interface due to aerodynamic drag.

$$\sigma_f = 1/2\rho_a C_d V^2 \qquad (6.8)$$

The rheological force in the spinline is related to the axial spinline stress σ_{11} or σ_{zz} as follows:

$$\sigma_{11} = F_{rheo}/\left(\pi D^2 / 4 \right)$$

$$\sigma_{zz} = \eta(T)dV/dz \qquad (6.9)$$

Where, η is the viscosity at temperature T.

One of the principal unknowns in the force balance equation is the air drag force, which becomes increasingly important at high spinning speeds. Numerous theoretical investigations have focused on identifying the nature of the drag coefficient C_d, in the drag force equation. To evaluate C_d for melt spinning, Matsui [128] has formulated a simple expression for C based on air Reynolds number using turbulent theory:

$$C_d = KR_e^{-n} \tag{6.10}$$

Where R_e is the air Reynolds number and is given by:

$$R_e = (\rho_a VD/\mu_a)$$

$$K = 0.37 \text{ and } n = 0.61$$

The energy balance is required to determine the filament temperature as a function of distance from the spinneret. Heat transfer from the melt spinline to the ambient medium involves several mechanisms; radiation, free convection, and forced convection. The effect of radiation is strongly dependent on the temperature of the filament. In polymer melt spinning, the radiation contribution is usually negligible compared to the convective heat transfer. Heat released due to crystallisation can be neglected for slow crystallising polymers, but should be considered for fast crystallising polymers. Neglecting radial temperature variations and including heat of crystallisation, the differential energy balance equation can be written as:

$$dT/dz = -\pi Dh(T - T_a)/WC_p + \Delta H_f/C_p * dX/dz \tag{6.11}$$

Where,

T = the filament temperature.

T_a = the ambient air temperature.

ΔH_f = the heat of fusion.

X = the crystalline fraction.

h = the heat transfer coefficient.

C_p = the resin heat capacity.

Usually, the temperature of the filament during melt spinning is determined experimentally as a function of the distance from the spinneret, using an infra-red sensor. Then Eq. (6.11) is used to calculate the heat transfer coefficient as a function of the distance from the spinneret.

6.15.2 Deposition ratio

In spunbonding and meltblowing, the deposition of filaments on the conveyer belt is an important variable. The manner in which the filaments are laid down, dictates the web geometry in terms of the orientation distribution and uniformity and hence some key fabric properties.

It is difficult to quantify this step mechanistically, but a 'deposition ratio' (Dr) can be defined (see Eq. 6.12). By manipulating this ratio, the CD strength of the fabric can be altered. Higher ratios give higher CD strength and lower ratios give lower CD strength. However, the ratio has little effect on the longitudinal or MD strength of the web.

$$D_r = V_f/V_b \tag{6.12}$$

Where,

V_f = the filament speed.

V_b = the conveyor speed.

The filament speed can be calculated using the mass balance equation based on the initial and final diameter of the filament. The conveyor speed is given by:

$$V_b = m/(G \times W_c) \tag{6.13}$$

Where,
m = the polymer mass throughput rate.
G = the desired web weight.
Wc = the width of the conveyor belt.

6.15.3 Polymer residence time in the extruder

The polymer residence time in an extruder can be calculated as follows:

$$t = (\rho_{melt} \times v)/W \tag{6.14}$$

Where,
t = residence time in minutes.
ρ_{melt} = polymer melt density in g/cm^3.
v = the screw volume in cm^3.
W = the polymer throughput rate in g/min.

6.15.4 Determination of the airflow rate

In meltblowing and spunbonding, airflow plays an important role in filament formation. It is necessary to know the air mass flow rate, as well as the velocity of air in the system. The determination of airflow rate and velocity is complicated by the effect of temperature and pressure and also by the fact that the flow is likely to be unsteady or subject to disturbances. Therefore, measurement conditions, instruments, and methods should be specific and reliable [129]. In practice, the air mass flow rate is usually determined using an orifice plate or Venturi meter. The air velocity is usually measured using an anemometer or a pitot tube. The measurement procedures are established by the American Society of Mechanical Engineers (ASME) [130]. More detailed descriptions of air measurement techniques are available [131,132].

The following describes the determination of the air mass flow rate using orifice plates in a pipe. The orifice plate is usually flat and consists of a circular hole in the centre. Its most common use is as a flow quantity measuring device. The determination of the airflow rate using an orifice is based on the use of Torricelli's theorem and Bernoulli's equation [133].

The air mass flow rate using an orifice plate can be calculated as follows: [134]

$$W_h = 359 CFd^2 \sqrt{Yh_w} \tag{6.15}$$

Where,
W_h = the air mass flow rate in lbs./h (multiply by 1.26×10^{-4} for kg/s).
C = the discharge coefficient.
d = the diameter of the orifice in inches (multiply by 0.0254 for m).
h_w = the differential pressure in H_2O (multiply by 249.06 for Pa).
Y = the weight of air in lbs./ft^3 (multiply by 16.018 for kg/m^3) entering the orifice.

F = the velocity of approach factor and is expressed as:

$$F = \frac{1}{\sqrt{1 - \beta^4}} ; \beta = \frac{d}{D} \qquad (6.16)$$

Where, d is the orifice plate diameter and D is the pipe diameter.

The orifice diameter, d, is 1.335 in. (0.034 m) and the inside pipe diameter, D, is 2.9 in. (0.074 m). This gives a diameter ratio of 0.46. Evaluation of Eq. (6.16) gives a velocity of approach factor, F = 1.023. Thus, Eq. (6.15) becomes:

$$W_h = 654.64C\sqrt{Yh_w} \qquad (6.17)$$

The specific weight of the air entering the orifice, U, can be determined from the ideal gas equation of state:

$$Y = \frac{P}{RT} \qquad (6.18)$$

Where, P is the absolute pressure in lb./ft^2 (multiply by 4.88 for kg/m^2), T is the absolute temperature in ^{0}R, and R is the gas constant of air, 53.34 ft lb./lbm-^{0}R (multiply by 3.407×10^3 for kg.mol.^{0}R). The absolute pressure upstream of the orifice is given by:

$$P_1 = P_2 + \rho g h_w + P_\infty \qquad (6.19)$$

Where,

P_1 = the pressure upstream of orifice in psia (multiply by 6894 for Pa).

P_2 = the pressure downstream of orifice in psig (multiply by 6894 for Pa).

ρg = the specific weight of H$_2$O in 0.0361 lb./in^3 (multiply by 27×10^3 for kg/m^3).

P_∞ = the ambient pressure in psia (multiply by 6894 for Pa).

The coefficient of discharge, C, is a function of B and the Reynolds number, R_e, based on the inside pipe diameter. The expression for the Reynolds number is given by:

$$R_e = \frac{YVD}{\mu} \qquad (6.20)$$

Where,

V = the air velocity in ft./s (multiply by 0.30480 for m/s).

μ = the dynamic viscosity in lb./ft-s.

D = the inside diameter of pipe in ft. (multiply by 0.30480 for m).

The viscosity of air is a function of the temperature. The following expression, derived from the Sutherland equation is useful to determine the viscosity in terms of the absolute temperature:

$$\mu = 1.3183 \times 10^{-6} \frac{T^{1.5}}{(T + 200)} \text{ in} \frac{\text{lb}}{\text{ft} - \text{s}} \qquad (6.21)$$

Eq. (6.20) gives the air velocity in the pipe:

$$V = \frac{W_h}{Y\frac{\pi D^2}{4}} \frac{1}{300} \text{ in} \frac{\text{ft}}{\text{s}} \qquad (6.22)$$

The flow rate, Eq. (6.15), can be converted to standard cubic feet per minute (SCFM) using a standard density, evaluated at P = 14.7 psia and T = 68 °F (20°C). The standard density is = 0.07516 lb./ft^3 (multiply by 16.018 for kg/m^3). The volumetric flow rate in SCFM (multiply by 4.72×10^{-4} for m^3/s), is given by:

$$Q = \frac{W_h}{60 Y_s} \text{ in SCFM} \tag{6.23}$$

6.16 Alternative extrusion processes for nonwoven production

As mentioned in Section 6.4.1, other extrusion processes not based on melt spinning can, and have been developed for the manufacture of spunlaid nonwovens. Examples involving the processing of cellulose into spunlaid webs, include the Bemberg Cupro process, and the more recently launched Lenzing Web Technology. Historically, wet spinning, dry spinning, dry-jet wet spinning and flash spinning have all been harnessed to make spunlaid nonwovens, including fabrics composed of naturally-derived polymers. Although still a very small proportion of the market, those based on wet spinning in particular are noteworthy because of their ability to process regenerated polysaccharide materials such as cellulose, starch and alginate to produce nonwovens made from natural polymers. All involve continuous filament extrusion, web deposition and bonding taking place in one integrated process. Developments also include dry spinning cellulosic polymer solutions in a hybrid meltblowing process, to produce small diameter, blown filaments that solidify as the solvent evaporates. This is akin to meltblowing without the need for a polymer melt, which provides scope for the use of a wider range of naturally-derived polymer feedstocks, in place of those based on petrochemicals. Historically, key challenges relating to process economics in such spunlaid systems based on wet and dry spinning, are the relatively low concentration of polymer in solution that is normally involved, and the resultant necessity to ensure effective recovery and recycling of solvents.

6.16.1 Solution flash spinning

Flash spinning is an alternative technique for the conversion of fibre-forming polymers into spunlaid webs using a dry spinning technique. The process, which is proprietary and is not available for commercial licensing, was developed by DuPont USA in the 1960s. Under DuPont's Tyvek brand, the technology has been progressively refined and advanced over subsequent years.

Traditionally, flash spinning requires a spinning agent that: (a) only forms a solution with the polymer at high pressure and not below its boiling point; (b) enables a two phase dispersion to be created with the polymer as the solution pressure is decreased and, (iii) flashes off (vaporises) when fed to an area of much lower pressure. Many spinning agents have been evaluated, one of the oldest being trichlorofluoromethane. In the flash spinning process, a polymer, typically PE, is blended with a solvent (e.g. methylene chloride) under high temperature (about 25°C or more above the boiling point of the solvent) and under high pressure. The blended solution is then released under controlled conditions through a spinneret to produce what is effectively an explosive reaction in which the solvent flashes off to produce a three-dimensional network of thin, continuous interconnected ribbons many of which are less than 4 mm thick. The fibrous elements are usually termed film-fibrils or plexifilaments [135]. Sometimes dissolved inert gas, for example CO_2, is used to increase the degree of fibrillation [41].

The plexifilaments are electrostatically charged by a corona field to aid their separation prior to their collection on a moving, electrically grounded conveyor after which the web is consolidated by thermal calender bonding. Area bonding is typically used for applications such as postal envelopes, housewrap membranes for construction, sterile packaging and for filters whereas point bonding can be used for apparel applications. The individual plexifilaments have high molecular orientation, leading to high strength. Tyvek fabrics (DuPont) are produced using this method. Flash spinning is the most complex and difficult method of manufacturing spunbond fabrics because of the need to spin a heated and pressurised solution under precise conditions [36]. Other organisations have also developed flash spun processing technology and products, including Asahi Chemical Industry in Japan, under the LUXER brand for polyolefins.

6.16.2 Electrospinning systems

Fibres having diameters in the nanometre range (<500 nm) are becoming increasingly important. The unique characteristics of nanofibre fabrics concern their comparatively high porosity and pore volume, high-moisture vapour transport, small fibre diameter, high surface area, high absorbency, and the ability to possess large numbers of chemically functional groups. Nanofibre fabrics are already applied to improve electrolytes that heighten the efficiency and lifespan of rechargeable batteries, and for making artificial skin and highly efficient hazard suits that filter germs. Potential applications for nanofibres include filtration products, barrier fabrics, and biomedical devices. The application of electrical technology to produce nonwovens based on nanofibres is opening up many applications in the protective clothing areas, as well as other value-added products such as battery separators, specialty wipes and high-performance filters. Bhat and Lee summarised important developments, and the state of nanotechnology in the fibres and nonwovens areas [136].

In the electrospinning (ES) process, the tensile force is generated by the interaction of an applied electric field with the electrical charge carried by the jet rather than the spindles and reels used in the conventional spinning process. According to the literature, higher electrical fields can be attained in the high vacuum of space and electrospinning could be the most economical way to create filaments to carry high-tensile loads [137,138].

As shown in Fig. 6.15, there are four regions in the electrospinning process. The jet emerges from the charged surface in the base region, and then electrical forces accelerate the polymer liquid and stretch the jet. The diameter of the jet decreases and the length increases in a way that keeps the amount of mass per unit time passing any point on the axis in the jet region the same. In the splaying region, the radial forces from the electrical charges become larger than the cohesive forces, and the single jet divides into many charged jets with equal diameters. As the jet progresses from the base towards the collector, the forces from the external electric field accelerate and stretch the jet. Stretching and evaporation of the solvent molecules cause the jet

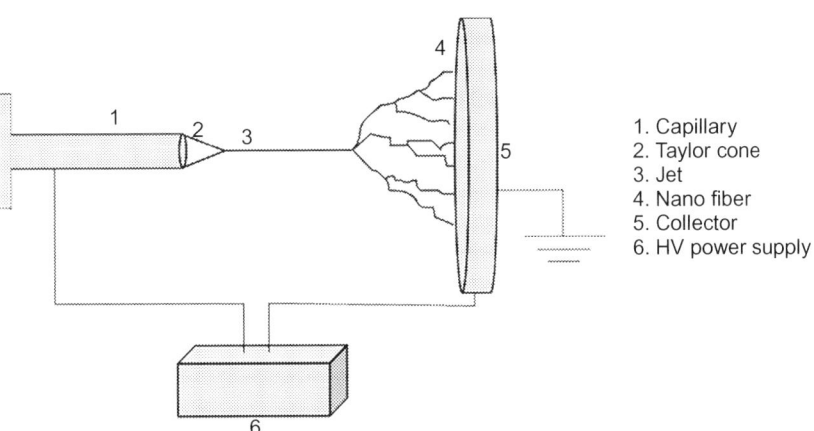

1. Capillary
2. Taylor cone
3. Jet
4. Nano fiber
5. Collector
6. HV power supply

FIG. 6.15   Basic setup of ES process.

diameter to become smaller. This whole process produces a large number of small electrically charged fibres moving towards the collector. The web that is collected consists of fibres having a wide range of diameters.

One of the disadvantages of nanofibres and their webs is that their mechanical properties such as tensile and tear strength are often very poor, unless very thick webs are produced. The best way to take advantage of the high surface area of nanofibres and to increase the durability and strength of products is to form composite structures with either spunbond or meltblown fabrics [139,140]. Typical spunbond webs have 15- to 20-μm-diameter filaments. It has been shown that by incorporating less than 10% of nanofibres, a very marked improvement in barrier properties can be accomplished. Composite spunbond–nanofibre–spunbond (SNS) fabrics are therefore of increasing industrial significance.

6.16.3 Centrifugal spinning systems

Many spunbond manufacturing companies have investigated making spunbond fabrics based on centrifugal spinning. In this system, a fibre-forming polymer is pumped into a die having a plurality of spinnerets about its periphery. The die is rotated at a predetermined adjustable speed, whereby the liquid, e.g., a polymer solution or polymer melt, is expelled from the die so as to form highly attenuated liquid streams, with solid fibres forming as a result of solvent evaporation or cooling (respectively). The fibres may be used to produce a continuous web, fibrous tow, and yarn through appropriate collection and take-up systems. Although there have been serious efforts to commercialise centrifugal spinning for nonwoven web production, and many pilot lines have been sold, commercial utilisation is still limited at present.

6.17 Recent developments

A selection of some key developments, both in equipment and associated technologies that have helped advance spunbond and meltblown processes are briefly discussed.

6.17.1 Multiple beams

By using multiple beams, two or more spunbond or meltblown web layers can be combined, and different types of composite nonwoven structures can be produced. Commercially, in spunbond production, four to five beams are quite common, and this allows better web uniformity and higher production speeds for lightweight fabrics. Reifenhauser's Reicofil technology centre in Troisdorf, Germany, has demonstrated combining up to seven-beams in-line (SSMMMSS) for hygiene and medical products, enabling very high production outputs [141].

6.17.2 Reicofil RF5 technology

The Reicofil 5 (RF5) system for the production of spunbond, meltblown and composite SMS nonwovens claims remarkable production speeds, combined with digitally-connected, smart factory functionality.

6.17.2.1 Production speeds

Throughputs of up to 270 kg/h/m width are claimed for spunbond production, with conveyor belt speeds up to 1200 m/min, while meltblown throughputs are up to 70 kg/h/m width. This is equivalent to a 30% increase over the previous system. To produce more uniform, light-weight webs at high speeds, the new system is also capable of producing spunbond filament diameters of less than 1 denier.

6.17.2.2 Machine intelligence

One of the most important developments, which is increasingly a feature of the latest nonwovens production equipment, is digitalisation linked to machine learning, i.e., artificial intelligence (AI). This goes beyond the deployment of sensors, diagnostics and automation, but also enables connectivity between individual manufacturing systems. By harvesting data from distributed sensors and the use of cloud-based machine learning, spunlaid processes now have the capacity to determine their own status, predict how different process settings will affect web/fabric properties, and even how customers will perceive the quality of the product. It also assists in scheduling maintenance of machine parts before being damaged [141]. Essentially, the approach constitutes a move towards smart connected factories, with the capacity for separate processes to detect anomalies and solve production tasks more independently [141,142].

6.17.2.3 Contamination

The RF5 system is claimed to eliminate 90% more hard piece contamination from fabrics, in comparison to the previous RF4 system [142].

6.17.2.4 Filament diameter

The RF5 system claims the ability to make filaments 20% smaller in diameter [141].

6.17.2.5 3D fabrics

By laying down filaments on customised conveyors with occluded and nonoccluded regions, localised variations in the air permeability of the belt can be created, enabling three-dimensional textural effects to be produced in the web during laydown. This is particularly useful to differentiate the visual appearance of topsheets, in for example baby diapers or feminine hygiene products, where line or dot patterns can be created. There is also the potential to modify the fluid handling behaviour of fabrics using this approach too.

6.17.3 Oerlikon digitisation

Oerlikon Manmade Fibres announced its fully digitising technologies for the conversion of polymers into filaments, yarns, and nonwovens in 2018 at the ITMA ASIA + CITME exhibition held in China. This relates to a fully networked production process that can automatically control products from supply chain to dispatch [141].

6.17.4 Tyvek expansion

DuPont recently announced its biggest investment of $400 million to expand the production capacity of Tyvek in Luxembourg. The new plant is expected to start operation from 2021. In 2019, Tyvek announced its sustainability goal, with increased efforts to reduce energy consumption and lower greenhouse gas emissions and waste.

6.17.5 Hartge

Hartge recently developed a spunbond process of reduced complexity and number of components. A versatile system occupying 70% less space, consuming 50% less energy, and the capability to run production within 20 min is claimed. An impingement spinning process that employs highly turbulent mixing is at the heart of the process [143].

6.17.6 Meltblown nanofibre production

Polymeric nanofibres have enormous specific surface area. Webs produced from nanofibres have large surface-to-volume ratio and have microscale and nanoscale pores, that are relevant to the performance of filter media, catalysis, super absorbents, scaffolds for tissue engineering and wound dressings, energy storage and electronic applications [144].

According to the National Science Foundation, nano relates to sizes less than 100 nm, whereas in the textile sector, fibres of less than 1 µm in diameter are generally considered to be nanofibres [144]. Recently, several researchers have developed new approaches for nanofibre production by melt blowing. Zuo et al. [145] successfully produced meltblown nanofibres of poly(ethylene-*co*-chlorotrifluoroethylene) with an average fibre diameter of 70 nm. Although there is potential for a high production rate, environmental concerns relate to the need for removal of the organic solvent, trifluoroacetic acid as part of the process. To address this, a water extractable melt blowing process was developed.

Developments in bicomponent fibre spinning are also relevant because of the potential to adapt them for bicomponent meltblowing. Tran et al. [146] produced PLA nanofibres by melt spinning with an average fibre diameter of 60 nm where PLA islands were separated after dissolving out the PVOH matrix with water. In melt spinning, polymer temperature should be high enough so that the viscosity of the polymer is sufficiently low to aid extrusion, but in the case of water-soluble polymers, this is not possible given the risk of polymer degradation. To minimise thermal degradation of PVOH and other water-soluble polymers, researchers have developed approaches based on plasticisation and blending [147]. Wang et al. [147] blended sulphopolyester (a thermally stable water-dispersible polymer) with PBT, and successfully produced nanofibres with an average diameter of 66 nm using a water extractable melt blown process. This new approach is also suitable for the production of multilayer, meltblown nanofibre and microfibre fabrics. It has been clearly shown that in modular melt blowing processes, it is possible to produce submicron fibres with good filter quality factors with the capability to be adopted by the industry [148,149].

6.17.7 Chemical recycling in PET spunlaid manufacture

Chemical recycling of waste PET can be integrated with spunlaid nonwoven manufacture to enable continuous manufacture of rPET nonwovens. This involves both melt filtration under vacuum of the recycled polymer feedstock to remove impurities, and the facility to increase the intrinsic viscosity (IV) of the PET recyclate prior to its introduction to the meltstream for extrusion. The ability to modulate the IV in this way enables excellent filament mechanical properties to be obtained from a lower grade PET feedstock, and can be considered a form of upcycling. Oerlikon Nonwoven and BB Engineering's VacuFil system is an example of this approach.

6.18 Concluding remarks

Spunbond, meltblown and related nonwovens production continues to grow due to the inherent cost-effectiveness of such direct polymer-fabric integrated processes, combined with the performance attributes of the resulting cost-effective fabrics. Composite multilayered fabrics containing multiple layers of spunbond, or spunbonds and meltblown continue to be an important part of the hygiene and protective fabrics markets, both of which are growing. The pandemic of H1N1 in 2019 and the current Coronavirus or COVID19 pandemic in 2019/20 onwards, has rapidly created additional very high demand for meltblown fabrics of the correct specification to meet requisite filter efficiencies combined with breathability for personal protective equipment (PPE), e.g., respirator masks and related PPE. As the demand for such products has grown, more companies are involved in meltblown nonwoven production and the total worldwide capacity is increasing. Globally, the enormous consumption of single-use PP-based PPE is also likely to renew concern about the disposal and environmental impacts of such products, and stimulate concerted efforts to find alternative polymer systems that can be cost-effectively deployed, without compromising performance. In addition to the academic literature, readers should continue to monitor trade periodicals to keep up with the fast changing developments in the industry, which is likely to include increased utilisation of recycled or repolymerised plastics, a gradual transition to nonpetroleum-based polymer materials in meltblowing and spunbond production, further development of smart, digitally connected manufacturing processes, and a greater variety of nonwoven product qualities in the future.

References

[1] S.R. Malkan, L.C. Wadsworth, International Nonwovens Bulletin, Fall 1992 and Winter 1993.
[2] R.G. Hill, Book of Papers, Fiber Producer Conference, Greenville, South Carolina, April 23–25, 3A1–4, 1990.
[3] L. Bradhe, Textile Asia 10 (4) (1973) 32–37.
[4] H. Lim, J. Text. Appar. Technol. Manag. 6 (3) (2010) 1–13. 2010.
[5] E.A. Vaughn, Nonwovens World Fact Book 1991, Miller Freeman Publications, Inc, San Francisco, California, USA, 1990.
[6] J.G. McCulloch, Int. Nonwovens J. 8 (1) (1999) 124–134.
[7] M.T. Morman, T.W. Odorzynski, G.L. Zehner, U.S. Patent No. 6,465,073, 2002.
[8] J.A. Degroot, G.J. Claasen, S. Bensason, M. Demirors, T. Gudmundsson, J.C. Brodil, U.S. Patent Application No. 10/131,114, 2018.
[9] V.A. Wente, et al., Manufacture of Superfine Organic Fibers, United Department of Commerce, 1954. Office of Technical Services Report No. PB111437, NRL 4364, April 15.
[10] M. Luo, V.A. Roscelli, S. Camarena, A.N. Neogi, J.S. Selby, U.S. Patent No. 6,773,648, 2004.
[11] M.A. Hassan, B.Y. Yeom, A. Wilkie, B. Pourdeyhimi, S.A. Khan, J. Membr. Sci. 427 (2013) 336–344.
[12] Z. Wang, X. Liu, C.W. Macosko, F.S. Bates, Polymer 101 (2016) 269–273.
[13] S.R. Malkan, Book of Papers, Hi-Per Fab 96 Conference, Singapore, April 24–26, 1996.
[14] A. Stahl, et al., 5th Annual TANDEC Conference, October 31–November 2, The University of Tennessee, Knoxville, TN, USA, 1995.
[15] A. Montgana, J. Floyd, Book of Papers, MetCon'93, May 26–28, Houston, TX, Paper No. 171, 1993.
[16] C.Y. Cheng, TAPPI Nonwovens Conference, February 14–16, Orlando, Florida, USA, 1994, pp. 39–49.
[17] Anonymous, Metallocene Catalyst Initiate New Era in Polymer Synthesis, Chemical & Engineering News, 15–20, September 11, 1995.
[18] Anonymous, Metallocenes, Modern Plastics, 1996, pp. 56–57.
[19] E. Castro-Aguirre, F. Iñiguez-Franco, H. Samsudin, X. Fang, R. Auras, Adv. Drug Deliv. Rev. 107 (2016) 333–366.
[20] S. Feng, X.N. Jiao, Adv. Mat. Res. 332 (2011) 1239–1242.
[21] A. He, J.H. Wang, U.S. Patent No. 9,091,004, 2015.
[22] J. Shi, L. Zhang, P. Xiao, Y. Huang, P. Chen, X. Wang, T. Chen, ACS Sustain. Chem. Eng. 6 (2) (2018) 2445–2452.
[23] C. Woodings, TechnicalTextile.Net-A fiber2Fashion Venture, 2000. https://www.technical. textile.net/articles/cropbased-polymers-for-nonwovens-3571.
[24] G. Bhat, V. Kandagor, D. Prather, R. Bhave, Int. J. Mater. Text. Eng. 11 (7) (2017) 482–486.
[25] N.V. Fedorova, Investigation of the Utility of Islands-in-the-Sea Bicomponent fiber Technology in the Spunbond Process, PhD dissertation, North Carolina State University, 2007. https://repository.lib.ncsu.edu/bitstream/handle/1840.16/5145/etd.pdf?sequence=1.
[26] H.L. Wooten, Book of Papers, Fiber Producer Conference, Greenville, South Carolina, April 24–26, 3A/17 - 3A/23, 1990.
[27] Anonymous, Man-Made Fiber Dictionary, Celanese Corporation, New York, 1978.
[28] A. Ziabicki, Fundamentals of fiber Formation: The Science of fiber Spinning and Drawing, Wiley, New York, 1976.
[29] N.G. McCrum, Principles of Polymer Engineering, Oxford University Press, New York, 1988, pp. 264–265.
[30] F. Rodriguez, Principles of Polymer Engineering, McGraw Hill, New York, 1982, pp. 330–331.
[31] Y. Mastubara, US Patent No. 4,285,655, 1981.
[32] Y. Mastubara, Polym. Eng. Sci. 19 (3) (1979) 169–172.
[33] Y. Mastubara, Polym. Eng. Sci. 20 (3) (1980) 212–214.

[34] Y. Mastubara, Polym. Eng. Sci. 20 (11) (1980) 716–719.

[35] Y. Mastubara, Polym. Eng. Sci. 23 (1983) 17–19.

[36] R.L. Smorada, in: H.F. Mark, J.I. Kroschwitz (Ed.), Encyclopedia of polymer science and engineering, Wiley, New York, vol. 10, 1985, pp. 227–253.

[37] L. Hartmann, Text. Manuf. 101 (1974) 26–30.

[38] Anonymous, Nonwovens Reports International, vol. 135, July 1982, pp. 7–10.

[39] G. Bhat, S.R. Malkan, J. Appl. Polym. Sci. 83 (3) (2002) 572–585.

[40] Z. Mao, B.C. Goswami, Book of Papers, INDA-TEC 99, Cary, NC, 1999.

[41] K. Othmer, Encyclopedia of Chemical Technology, Wiley, Hoboken, NJ, 2007.

[42] A.G. Hoyle, TAPPI J. 72 (1989) 109–112.

[43] R. Dent, in: J.W.S. Hearle, M.S. Burnip (Eds.), Nonwovens '71, The Textile Trade Press, Manchester, England, 1971, pp. 155–169.

[44] C.W. Ericson, J.F. Baxter, Text. Res. J. 43 (1973) 371–378.

[45] S.R. Malkan, L.C. Wadsworth, C. Davey, Int. Nonwovens J. 6 (2) (1994) 42–70.

[46] S. Misra, J.E. Spruiell, G.C. Richeson, INDA J. Nonwovens Res. 5 (3) (1993) 13–19.

[47] S. Misra, F.M. Lu, J.E. Spruiell, G.C. Richeson, J. Appl. Polym. Sci. 56 (1995) 1761–1779.

[48] C. Smith, W.W. Roberts Jr., Int. Nonwovens J. 6 (1) (1994) 31–41.

[49] G.S. Bhat, D. Zhang, S.R. Malkan, L.C. Wadsworth, Proceedings of the Fourth Annual TANDEC Conference, Knoxville, TN Nov. 14–16, 1994.

[50] D. Zhang, G.S. Bhat, S.R. Malkan, L.C. Wadsworth, Proceedings of the 1996 TANDEC Conference, Knoxville, TN, Nov. 1996.

[51] D. Zhang, PhD Dissertation, University of Tennessee, Knoxville, TN, 1996.

[52] G.S. Bhat, D. Zhang, S.R. Malkan, L.C. Wadsworth, Proceedings of the Joint Conference on Fibers & Yarns, Textile Institute, Manchester, UK, December, 1996.

[53] D. Zhang, G.S. Bhat, S.R. Malkan, L.C. Wadsworth, J. Thermal Anal. 49 (1997) 161–167.

[54] D. Zhang, G.S. Bhat, S.R. Malkan, L.C. Wadsworth, Text. Res. J. 68 (1) (1998) 27–35.

[55] G.S. Bhat, Proceedings of the Clemson University Polypropylene Conference, Clemson, SC August 23–24, 1995.

[56] R. Beyreuther, H.J. Malcome, Melliand Textilnerichte, 4, 287–290 and E133–135, 1993.

[57] C.H. Chen, J.L. White, J.E. Spruiell, B.C. Goswami, Text. Res. J. 53 (1) (1983) 44–51.

[58] F. Lu, Proceedings of the 1997 TANDEC Conference, Knoxville, TN, Nov. 1997.

[59] J.W.S. Hearle, M.A.I. Sultan, S. Govender, J. Text. Inst. 67 (11) (1976) 373–376.

[60] J.W.S. Hearle, M.A.I. Sultan, S. Govender, J. Text. Inst. 67 (11) (1976) 377–381.

[61] J.W.S. Hearle, M.A.I. Sultan, S. Govender, J. Text. Inst. 67 (11) (1976) 382–386.

[62] X.C. Huang, R.R. Bresee, INDA J. Nonwovens Res. 5 (3) (1993) 28–38.

[63] X.C. Huang, R.R. Bresee, INDA J. Nonwovens Res. 6 (4) (1994) 53–59.

[64] S. Bais-Singh, B.C. Goswami, J. Text. Inst. 86 (2) (1995) 271–287.

[65] S. Bais-Singh, S.B. Biggers Jr., B.C. Goswami, Text. Res. J. 68 (5) (1998) 327–342.

[66] P.P. Shang, PhD Dissertation, North Carolina State University, Raleigh, NC, USA, 1995.

[67] S. Backer, D.R. Patterson, Text. Res. J. 30 (1960) 704–711.

[68] J.W.S. Hearle, P.J. Stevenson, Text. Res. J. 38 (1968) 343–351.

[69] J.W.S. Hearle, P.J. Stevenson, Text. Res. J. 34 (1964) 181–191.

[70] T. Komori, K. Makishima, Text. Res. J. 47 (1977) 13–17.

[71] P. Britton, A.J. Simpson, Text. Res. J. 53 (1983) 363–368.

[72] T.H. Grindstaff, S.M. Hansen, Text. Res. J. 56 (6) (1986) 383.

[73] T.F. Gilmore, Z. Mi, S.K. Batra, Proceedings of the TAPPI Conference, May 1993.

[74] H.S. Kim, B. Pourdeyhimi, J. Text. Appar. Technol. Manag. 1 (4) (2001) 1–7.

[75] Anonymous, Smithers Pira, Global spunlaid nonwoven applications and trends, 2016, pp. 1–8. https://www.smitherspira.com.

[76] M. Ahmed, Polypropylene Fiber Science and Technology, Elsevier Scientific Publishing Company, New York, 1982.

[77] S.R. Malkan, L.C. Wadsworth, Int. Nonwovens Bull. 2 (1991) 46–52.

[78] S.R. Malkan, PhD Dissertation, The University of Tennessee, Knoxville, TN USA, May 1990.

[79] A.M. Jones, Book of Papers, Fourth International Conference on Polypropylene Fibers and Textiles, Nottingham, England, September 23–25, 47, 1987, pp. 1–10.

[80] G. Straeffer, B.C. Goswami, Book of Papers, INDA-TEC, Baltimore, MD, USA, June 5–8, 1990, pp. 385–419.

[81] S.R. Malkan, et al., Nonwovens—An Advanced Tutorial, TAPPI Press, Atlanta, GA, USA, 1989, pp. 101–129.

[82] G.M. Eaton, The Effects of Pigments on the Physical Properties of Melt Blown Nonwovens, Book of Papers, INDA-TEC 87, Hilton Head, SC, USA, May 18–21, 1997, pp. 1–11.

[83] S.R. Malkan, L.C. Wadsworth, INDA J. Nonwovens Res. 3 (2) (1991) 21–34.

[84] S.R. Malkan, L.C. Wadsworth, INDA J. Nonwovens Res. 3 (1991) 3.

[85] M.W. Milligan, B.D. Haynes, Am. Soc. Mech. Eng. 54 (1987) 47–50.

[86] R.R. Buntin, D.T. Lohkamp, Tappi 56 (4) (1973) 74–77.

[87] P.P. Tsai, L.C. Wadsworth, G. Richeson, Annual technical conference-Society of Plastics Engineers Inc. Danbury, CT, USA, 1995, pp. 1770–1770.

[88] L.C. Wadsworth, A.M. Jones, Book of Papers, INDA-TEC, Philadelphia, Pennsylvania, June 2–6, 1986, pp. 312–320.

[89] V.A. Wente, Ind. Eng. Chem. 48 (8) (1956) 1342–1346.

[90] A.Y.A. Khan, PhD Dissertation, The University of Tennessee, Knoxville, TN, USA, 1993.

[91] Y. Lee, L.C. Wadsworth, Polym. Eng. Sci. 30 (22) (1990) 1413–1419.

[92] K.J. Choi, et al., Polym. Eng. Sci. 28 (2) (1988) 81–89.

[93] H. Bodaghi, Book of Papers, INDA-TEC, Philadelphia, Pennsylvania, USA, May 30–June 2, 1989, pp. 535–571.

[94] M.W. Milligan, L. Fumin, Book of Papers, Second Annual TANDEC Conference, The University of Tennessee, Knoxville, TN, October 13–15, 1992.

[95] B.D. Haynes, PhD Dissertation, The University of Tennessee, Knoxville, TN, 1991.

[96] S.B. Warner, et al., INDA J. Nonwovens Res. 2 (2) (1990) 33–40.

[97] L.C. Wadsworth, A.O. Muschelewicz, Book of Papers, 4th International Conference on Polypropylene Fibers and Textiles, Nottingham, England, September 23–25, 47.11–47.20, 1987.

[98] M.W. Milligan, Nonwovens—An Advanced Tutorial, Tappi Press, Atlanta, GA, USA, 1989.

[99] M.W. Milligan, et al., Book of Papers, INDA-TEC, Philadelphia, PA, USA, May 30–June 2, 573–583, 1989.

[100] P.P. Tsai, Int. Nonwovens J. 2 (1999) 1–8. https://journals.sagepub.com/doi/pdf/10.1177/1558925099OS-800216.

[101] A. Naqwi, et al., 5th Annual TANDEC Conference, Knoxville, TN, USA, October 31–November 2, 1995.

[102] G. Bhat, INDA J. Nonwovens Res. 4 (3) (1992) 26–28.

[103] J. Wallen, et al., Int. Nonwovens J. 7 (3) (1995) 49–50.

[104] K.M. Narasimahan, R.L. Shambaugh, Book of Papers, INDA_TEC, Hilton Head, SC, USA, May 18–21, 1987, pp. 189–205.

[105] R.L. Shambaugh, Ind. Eng. Chem. Res. 27 (12) (1988) 2363–2372.

[106] M.W. Milligan, B.D. Haynes, Am. Soc. Mech. Eng. 54 (1989) 47–50.

[107] M.A.J. Uyttendaele, R.L. Shambaugh, AICHE J. 36 (2) (1990) 175–186. No. 2.

[108] B. Majumdar, R.L. Shambaugh, J. Rheol. 34 (4) (1990) 591–601.

[109] S.R. Malkan, L.C. Wadsworth, Polymer-laid systems, in: A. Turbak (Ed.), Nonwovens: Theory, Process, Performance, and Testing, Tappi Press, Atlanta, 1993, pp. 171–192.

[110] J.J. Hermans, Flow Properties of Disperse Systems, North-Holland Publication Corp, Amsterdam, 1953.

[111] R.R. Bresee, L.C. Wadsworth, Book of Papers, Exxon Melt Blown Seminar, Baytown, Texas, September 15-16, 1988.

[112] Anonymous, Meltblown Technology Today, Miller Freeman Publications, San Francisco, 1989.

[113] W.J.G. McCulloch, Book of Papers, Fiber Producer Conference, Greenville, South Carolina, USA, April 23–25, 1990.

[114] K. Ando, Book of Papers, International Symposium on Fiber Science & Technology, Hakone, Japan, August 20–24, 1985, p. 238.

[115] L.R. White, TAPPI J. 71 (2) (1988) 79–81.

[116] A.E. Wilkie, Int. Nonwovens J. 8 (1) (1999) 135–140.

[117] J. Hagewood, Int. Fiber J. 13 (5) (1998) 47–48.

[118] K.K. Cheng, Int. Fiber J. 13 (1998) 40.

[119] B. Pourdeyhimi, Nonwovens: Current Trends & Opportunities, Text. World (2019). https://www.textileworld.com/textile-world/features/2019/02/nonwovens-current-trends-opportunities/.

[120] N. Anantharamaiah, S. Verenich, B. Pourdeyhimi, J. Eng. Fibers Fabr. 3 (3) (2008) 1–9.

[121] S. Lee, S.K. Obendorf, Fibers Polym. 8 (5) (2007) 501–506.

[122] S. Gaddam, Masters Theses, University of Tennessee—Knoxville, 2002.

[123] H. Ishitara, et al., Int. Polym. Process. 4 (2) (1989) 91.

[124] A. Dutta, Polym. Eng. Sci. 27 (14) (1987) 1050.

[125] R. Patel, PhD Dissertation, The University of Tennessee, Knoxville, TN, 1991.

[126] S. Kase, T. Matsuo, J. Polym. Sci. A 3 (1965) 2541.

[127] S. Kase, T. Matsuo, J. Polym. Sci. A 11 (1967) 251.

[128] M. Matsui, Trans. Soc. Rheol. 20 (1976) 465.

[129] A. Barber, Pneumatic Handbook, Elsevier Advanced Technology, Oxford, UK, 1997.

[130] H.S. Bean, Fluid Meters their Theory and Application, ASME, New York, 1971.

[131] Anonymous, Flow Measurements, ASME, New York, 2004.

[132] E. Owner, R.C. Pankhurst, The Measurement of Air Flow, Pergamon Press, New York, 1977.

[133] T.B. Hardison, Fluid Mechanics for Technicians, Reston Publishing, Reston, VA, 1977.

[134] M.W. Milligan, F. Lu, R.R. Buntin, L.C. Wadsworth, J. Appl. Polym. Sci. 44(2), (1992) 279–288.

[135] H. Blades, J.R. White, US Patent No. 3,081,519, 1962.

[136] Y. Lee, G. Bhat, Recent Advances in Electrospun Nanofibers, 2002. *https://www.research*gate.net/publication/289270140_Recent_Advances_in_Electrospun_Nanofibers.

[137] D.H. Reneker, I. Chun, Nanotechnology 7 (3) (1996) 216–223.

[138] D.H. Reneker, I. Chun, D. Ertley, U. S. Patent No. 6,382,526, 2002.

[139] J. Doshi, M.H. Mainz, G.S. Bhat, Proceedings of the Tenth TANDEC Nonwoven Conference, Knoxville, TN, 2000.

[140] T. Grafe, K. Graham, Int. Nonwovens J. 1 (2003) 51–55.

[141] H. Davis, Adding intelligence direct from the polymer, Sustain. Nonwovens (2019). https://www.nonwovensnews.com/features/14448-adding-intelligence-direct-from-the-polymer.

[142] Reicofil., Setting the new standards for spunbond, meltblown and composite nonwovens, Reicofil Technol. (2019). https://www.reicofil.com/system/uploads/attachment/file/58f8ab2e59d9e623a3d0fa43/REI-Reicofil-Broschuere-Low.pdf.

[143] Anonymous., New Innovation on Spunbond Process, Nonwoven Technical Textiles Technology, 2016. http://www.nonwoventechnology.com/en/new-innovation-on-spunbondprocess/.

[144] G.S. Bhat, J. Nanomater. Mol. Nanotechnol. 5 (1) (2016) 23–24.

[145] F. Zuo, D.H. Tan, Z. Wang, S. Jeung, C.W. Macosko, F.S. Bates, ACS Macro Lett. 2 (4) (2013) 301–305.

[146] N.H. An Tran, H. Brünig, C. Hinüber, G. Heinrich, Macromol. Mater. Eng. 299 (2) (2014) 219–227.

[147] Z. Wang, X. Liu, C.W. Macosko, F.S. Bates, Polymer 101 (2014) 269–273.

[148] R. Uppal, G. Bhat, C. Eash, K. Akato, Fibers Polym. 14 (4) (2013) 660–668.

[149] W. Han, X. Wang, G.S. Bhat, J. Nanomater. Mol. Nanotechnol. 2 (3) (2013) 1–5.

7

Nanofibre and submicron fibre web formation

E. Stojanovska[a], S.J. Russell[b], and A. Kilic[a]

[a]TEMAG Labs, Istanbul Technical University, Istanbul, Turkey [b]University of Leeds, Leeds, United Kingdom

7.1 Introduction

Nonwoven fabrics have always provided an important vehicle for innovation and continue to enable improved products to be developed, as well as the creation of entirely new ones. For many decades, a variety of nonwovens have been manufactured from small-diameter microfibres (<1 dtex) for reasons as diverse as to mimic the properties of leather and suede, or to modulate fabric softness, prevent infection or filter submicron particles, to name just a few. As fibre science and manufacturing processes for making even smaller diameter fibres have advanced, the range of nonwoven materials containing submicron fibres has progressively grown, and developments in nanotechnology continue to have significant impact.

Nanofibres can be considered as one branch of the nanomaterials tree. They combine a high aspect ratio and are of nanoscale diameter. Although the general definition of a nanomaterial covers a size range between 10 and 100 nm, it is common for fibres with diameters up to 500 nm to be referred to as nanofibres, while those in the range from 500 to 1000 nm are usually described as submicron, i.e., <1 μm [1].

The ability to make nonwovens continuously from nanofibres with diameters that may be 100,000 times smaller than a human hair is the result of progressive developments in polymer, spinning and nonwoven production techniques. Uniform, coherent webs of nanofibres can now be manufactured continuously in one integrated process, and in contrast to most other nonwoven processes, an additional bonding step is not always required. Nanofibre webs are typically very thin, being referred to by some as 1D structures, although in reality, they are assembled from multiple layers of fibres. The webs have high surface area and porosity (Fig. 7.1), making them suitable for a wide range of applications. The importance of fibre diameter on the magnitude of surface area is given with Eq. (7.1).

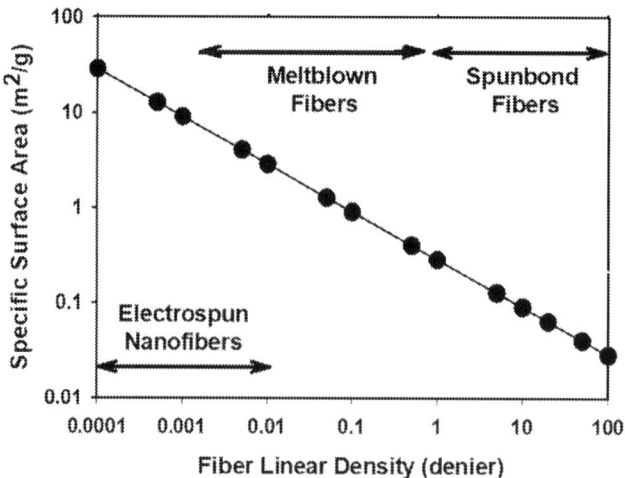

FIG. 7.1   Effect of fibre diameter on surface area.

$$A = \frac{\text{Denier}}{900,000 \cdot \rho} \qquad\qquad (7.1)$$

where A is the fibre cross-sectional area, and ρ is fibre density.

Given the important markets that exist for nonwoven fabrics with a high surface area, this chapter discusses the basic properties and recent advances in nonwovens containing nano- and submicron fibres. It is divided in to three parts, where first, we provide an insight into the diverse technologies used to produce such fibres, focusing mainly on methods of industrial manufacture. Then, we discuss some of the distinctive properties that nano and submicron fibres can provide when produced from different materials and in various architectures. Finally, the potential and real-world applications of these fibrous structures are reviewed across three broad application areas: energy, filtration, and healthcare.

7.2 Nanofibre and submicron fibre production techniques

Industrially, production of submicron and nanofibres for nonwovens can be achieved by either bottom-up or top-down approaches. Top down refers to processes that remove material from a larger bulk structure until the required small-diameter fibres are produced, whereas the bottom-up approach involves creating the required fibre dimensions from much smaller components, such as atoms or molecules. Whereas numerous methods of making nanofibre and submicron fibres have been described in the scientific literature, one of the most valuable aspects of any nonwoven production technique is its ability to be industrially scaled. Realistically, this means processes need to be technically feasible at scale (in terms of linear delivery speed and/or operating width), cost-effective and resulting products should have differentiated properties that meet unmet market needs. Similar to the production of

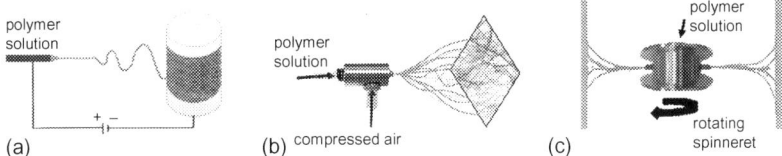

FIG. 7.2   Schematic representation of different nanofibre production methods: (A) electrospinning, (B) air or solution blowing (courtesy of Areka Nanofibres), and (C) centrifugal spinning. Note that in all cases the polymer solution can be replaced by a polymer melt.

microfibres, nano and submicron fibres for nonwovens are produced via a range of solution (wet), dry or melt spinning methods.

By careful control of polymer rheology, die geometry and feed rate, mainstream processes such as meltblowing can be adapted to produce submicron fibre webs and similarly, other extrusion techniques such as nanoval can be configured to produce fibres approaching submicron dimensions. However, as illustrated in Fig. 7.2, three general spinning technologies have emerged for one-step formation of nanofibre and submicron nonwoven fibre webs: electrospinning, solution blowing, and centrifugal spinning [1,2]. Note that hybrid systems such as electroblowing are also in available. Each method uses specific forces to attenuate the liquid polymer solution or melt into either submicron or nanofibrous dimensions, depending on operating conditions. As an integral part of these processes, attenuated liquid streams are converted into solid fibres that are deposited onto a collecting surface whereupon the final web structure is formed. The collecting surface is typically a conveyor belt or roller to enable continuous production. Because the resulting webs are generally thin and easily deformed, they are often deposited on top of a preformed fabric, perforated film or scrim, like a coating, or sandwiched between two or more other fabric layers to provide reinforcement. Of course, this also provides a valuable means of building composite nonwovens and controlling final fabric properties. Thicker, self-supporting webs can be wound up in to rolls without a supporting substrate.

Each of the three main nanofibre spinning techniques is used for industrial production of nonwovens, and turnkey equipment is commercially available. However, each has pros and cons at an industrial scale. While large scale electrospinning equipment is available in widths up to 1.6 m, operating with up to eight consecutive spinning electrode units over a common conveyor belt, their overall maximum production rates are still limited. Additionally, the high-voltage demand (up to 140 kV) influences cost effectiveness. Air blowing systems increase production rates significantly, but compressed air systems still consume large amounts of energy. Like electrospinning, a continuous air blowing process is readily facilitated by multiple-nozzle or multihead production lines. Centrifugal spinning is arguably capable of the highest production rates [2,3], and relatively low energy is needed to run the spinneret.

7.2.1   Electrospinning

Electrospinning, as the most widely used method for nanofibre web production, uses electrical force for attenuation of the polymer jets [1]. The principle is based on the interaction of a charged fluid, such as a polymer solution or melt, with a strong electrical field (typically in the range, 10–80 kV DC). The induced electrical charge in the viscoelastic fluid and the strong electric field forms a structure called a Taylor cone at the nozzle tip. When the electrical forces overcome the surface tension and viscosity of the fluid, the jet is ejected from the tip of the nozzle. Bending instabilities in the fluid jet results in whipping and progressive stretching of the jet to produce what become very small-diameter fibres. The solvent evaporates during the motion of the fluid jet towards the collector, leaving a solid nonwoven fibre web on the collector. The web weight is normally in the range $0.03–1.0 g/m^2$. Higher web weights and line speeds normally require multiple electrospinning electrodes operating in sequence over the conveyor system.

A simple electrospinning apparatus (Fig. 7.2A) consists of a high-voltage power supply, a grounded collector, and spinneret components that produce the polymer jets [4]. This set-up is very suitable for lab-scale production of nanofibres from a wide variety of organic materials. However, when large scale production rates are needed, more versatile electrospinning configurations operating with a larger number of nozzles (jets) or the use of nozzle-less electrodes are required. In free surface electrospinning, a moving electrode, which is typically a roller or metallic wire is continuously immersed or coated with polymer solution/melt to increase production rates [5]. This enables a very large number of Taylor cones to be formed, producing multiple jets without the use of nozzles or conventional spinnerets. In all cases, controlling the uniformity of the process due to the repulsion forces between adjacent jets is challenging. Various modified electrospinning set-ups have been also proposed with the purpose of increasing the flow rate during production and to reduce reliance on electrical forces [6]. Such methods tend to rely on a combination of electrospinning and other techniques, such as air blowing or centrifugal spinning, where the presence of compressed air or centrifugal forces, has the effect of reducing the demand for high voltage to increase production rates. Further details on the diversity of multijet electrospinning set-ups available are given by SalehHudin et al. [5]. While different electrospinning variants, including nozzle and collector modifications, that have been developed over the years are discussed by Wang et al. [7] and illustrated in Fig. 7.3.

7.2.2 Solution blowing

In the solution blowing system, compressed air generates forces that convert the molten/dissolved polymer into small-diameter fibres [2]. Here, the polymer solution or melt is extruded through the nozzle into a stream of highly pressurised air (Fig. 7.2B). The drag forces of the air on the surface of the polymer elongate and solidify the polymer jet. Here, nozzle geometry plays and important role in the fibre formation step. It should enable the air to form a uniform cone at the tip of the nozzle so that the polymer melt/solution can be split into jets [8]. The air/polymer boundary should remain in a conical shape throughout the whole process, so that the polymer will not be interrupted by turbulence, which will otherwise result in beads and/or droplets being formed. Additionally, process parameters, such as air pressure, air temperature, feeding rate, and nozzle–collector distance, should be controlled in order to obtain nano and submicron fibres within the target diameter range.

An alternative method that represents a synergy of both electrospinning and solution/melt blowing is electroblowing [9–11]. The combination of air-flow and an electric field facilitates

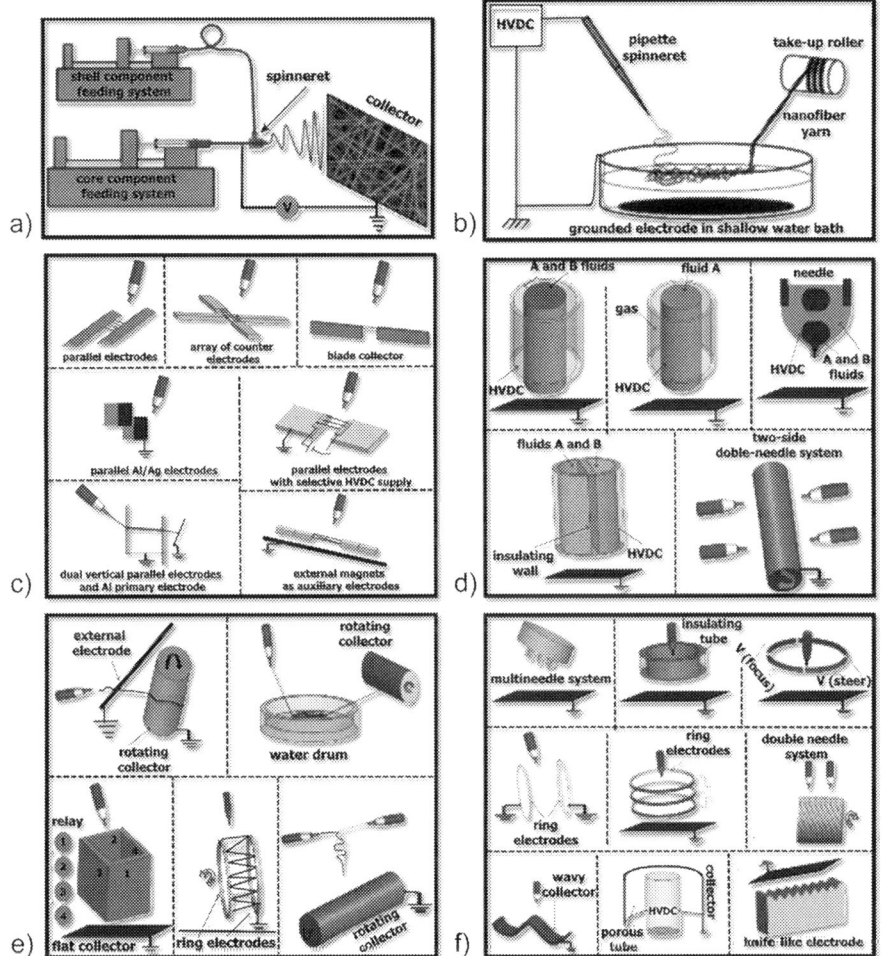

FIG. 7.3 Schematic representation of modified electrospinning systems: (A) Coaxial electrospinning system; (B) wet-electrospinning system; (C) electrospinning set-up with various collector configurations with counter electrodes; (D) multinuzzle electrospinning systems; electrospinning systems with (E) various auxiliary and (F) ring electrodes. *Reproduced and edited from C. Wang, et al., Fabrication of electrospun polymer nanofibers with diverse morphologies, Molecules 24(5) (2019), https://doi.org/10.3390/molecules24050834, with permission from MDPI, Copyright © 2019.*

jet initiation and continuity, improves the fibre deposition rate and the directional control of the fibre. Moreover, it reduces the need for high voltage and high air pressure (applied when electrospinning and solution blowing are used alone, respectively), enabling a reduction in the production cost of spinning large amounts of fibre. While the air-flow increases the evaporation rate of the solvent, the electric field helps the polymer jets to attenuate before they completely solidify. Fibres produced via this technique typically have diameters in the range between those obtained by electrospinning and solution blowing. Further modifications, such as air temperature and/or humidity control, lead to improved processability of different polymers and production of fibres with narrower diameter distributions [9].

7.2.3 Centrifugal spinning

Centrifugal (also called force- or rotary-jet) spinning uses centrifugal force to form the fibre. The spinning system is equipped with a spinneret filled with a polymer solution or melt (Fig. 7.2C) [2]. A multinozzle spinneret rotates in order to produce the centrifugal force necessary to overcome the surface tension of the polymer fluid to form a jet. The polymer jet further elongates into a fibre, guided by the centrifugal and frictional drag forces between the jet surface and the air. In addition to the angular velocity and the spinneret–collector distance, the type of nozzle should also be taken into account in determining the morphology of the final fibres. In this regard, various spinneret designs have been studied. A nozzle-less system has also been developed where the polymer fluid is spread onto a rotating disc to form a film, which then splits into fingers and develops into fibre jets [12,13]. Here, a balance between forces is achieved at lower rotational speeds, which is convenient for highly viscous polymer fluids [12]. In the case of nozzle systems [14], the size and orientation of the opening through which the polymer is ejected to form a jet plays an important role in the initial jet trajectory and the final fibre morphology [14]. Additionally, use of nozzles improves control over fibre diameter, allows narrow diameter distributions to be achieved and reduces the tendency for bead formation [13]. A modified centrifugal system where the spinneret is immersed in a precipitation bath has also emerged as a promising method for production of high performance submicron fibres [15].

7.2.4 Bottom-up processes for controlled nanofibre production

While electrospinning, solution blowing, and centrifugal spinning enable continuous production of nonwovens in relatively large volumes, the fibre diameter ranges that can be achieved vary between methods, and a wide distribution of fibre diameters and morphologies is typically produced. This can lead to quality control and consistency issues during production. Although not all are compatible with continuous high speed production of nonwovens, other techniques are available that allow nanofibres to be produced with well controlled dimensional properties and diameters below 100 nm [2]. Such methods can be generally divided into the following groups:

(a) Template synthesis [2] – nanofibrous structures are formed inside or on the surface of a mould that enables production of nanofibrous structures with well controlled structure and morphology.
(b) Phase inversion [2] – when one phase (usually the liquid solvent) is exchanged with (or converted into) another phase under specific conditions (usually freeze-drying) and is most suitable for production of nanofibrous 3D networks.
(c) Lithography or drawing techniques [16] – where tiny lines of polymer with well controlled interfibre distance are drawn on a given flat or otherwise shaped surface.
(d) Chemical vapour deposition (CVD) [17] or growth of fibres (or other nanomaterials) on a substrate as a result of catalytic reaction of a transition metal catalyst and gas molecules which results in the formation of short lengths of inorganic nanofibres with diameters below 500 nm.

Despite the potential to obtain nanofibres with very small diameters, less than 100 nm, together with well controlled diameter distributions, fibre morphology and web porosity, the major drawback of such techniques is their scalability [6]. One exception is the vapour deposition technique, which has been commercially applied for the production of short inorganic nanofibres, nanotubes, nanowires, graphene, and other 1D and 2D structures [18].

7.2.5 Top-down processes based on fibrillation, splitting, and dissolution

While electrospinning, solution blowing, and centrifugal spinning directly produce submicron and nanoscale fibres, other approaches can be harnessed, that effectively involve controlled disintegration of a larger bulk structure. Such 'top-down' manufacturing of nonwoven fabrics containing submicron fibres can be accomplished by exploiting the propensity for certain fibres with a fibrillar internal structure, to undergo longitudinal fibrillation, splitting or separation, so that much finer embedded fibrous components can be released from the 'parent' fibre. Typically, the approach is to make webs from island-in-the-sea (IS) or segmented pie bicomponent fibres, either as staple fibres, so that they are compatible with drylaid, wetlaid, and airlaid web formation platforms, or as filaments produced by spunmelt processes, usually spunbond. Then, after the web is made, they are fibrillated or split, separating out the embedded microfibres during mechanical bonding, typically by means of hydroentangling, or dissolved out after the web has been bonded, typically by use of solvents. Another approach is exemplified by Eastman Cyphrex technology, where short cut segmented bicomponent fibres were separated into microfibres with a ribbon-like cross-section (minimum thickness of ~2.5 μm) in the aqueous slurry used to feed wetlaid web formation.

(a) Splitting and fibrillation

Industrially, segmented pie bicomponent fibres and filaments can be split after the web is made by subjecting it to mechanical forces, yielding submicron fibres in the final bonded fabric. For example, spunlaid manufacture of splittable bicomponent filaments followed by hydroentangling at relatively low specific energy is the basis of Freudenberg's Evolon technology. By adjusting the aspect ratio of the filament, adhesion between the segments and the addition of a hollow core, ease of splitting can be controlled. While splitting of segmented bicomponent fibres is most usually associated with hydroentangling, the mechanical forces generated in staple fibre carding, needlepunching and ultrasonic bonding can also be harnessed.

Although it is sometimes thought of as a disadvantage in fibre processing, it is worth noting that a variety of natural and man-made fibres including cotton, flax, silk, lyocell (Tencel), and even p-aramids (e.g. kevlar) can be fibrillated during nonwoven processing when subjected to sufficiently high mechanical forces or kinetic energy, most notably in processes such as hydroentangling. This results in the partial release, as well as entanglement of microfibrils from the fibres and affects bulk properties of the fabric such as opacity, permeability, tensile, and tear strength. Similarly, in nonwoven fabric manufacture, notably following spunlaid web formation, island in the sea (IS) bicomponent filaments can be fractured during hydroentangling to release finer filaments embedded within the cross-section. By appropriate

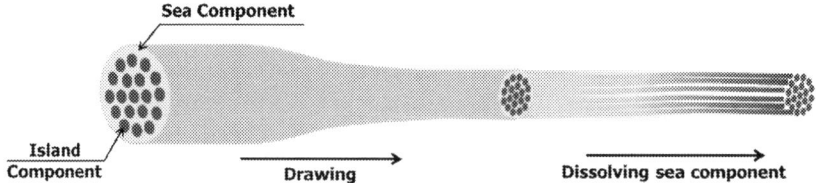

FIG. 7.4   Approach to submicron fibre production by dissolution.

selection of the constituent polymers, the island to sea (I/S) ratio and specific energy, it is possible to produce high strength fabrics, with residual filament diameters of 0.5 μm and below, dependent on the number of islands in the original filament cross-section.

(b) Dissolution

Dissolving out the sea component in IS bicomponents containing two or more polymers is a well-established approach to submicron fibre diameters of relatively uniform diameter and has been used for decades in the production of highly durable, artificial leather and suede. Of course, a key issue here from environmental and cost perspectives, is the choice of solvent chemistry, and the fact that a large proportion of the original polymer material in the fibre is removed as part of final fabric production. The basic principle of the dissolution route to submicron fibre dimensions in the production of nonwoven fabrics is illustrated in Fig. 7.4. First a bicomponent IS fibre is formed and drawn into the requisite dimensions for nonwoven processing, and subsequently, the sea component is dissolved out. Usually dissolution occurs after the nonwoven fabric has been bonded. This strategy forms the basis of the Alcantara and Ultrasuede manufacturing process. For example, after drylaid staple fibre web formation and needlepunching, the polymer forming the sea (continuous phase) is dissolved out of the substrate, releasing the embedded submicron island fibres (discrete phase). Strong, dimensionally stable, aesthetically pleasing, high density fabrics are produced following subsequent fabric finishing, which do not fray, have excellent tear resistance and are highly opaque.

7.3 Materials and structures

This section together with those that follow focuses on nonwoven materials produced by the three main methods of nanofibre production, based on electrospinning, air (solution) blowing, and centrifugal spinning. For simplicity, in the following discussion, reference to nanofibres also applies to submicron fibrous structures.

The unique properties of nanofibres are mainly due to the wide range of materials that nanofibres can be made from, and the fibre different diameter ranges and morphologies that can be achieved. Given that they are, almost always, produced from polymer solution (or melt), it is safe to say that the bulk properties of nanofibre nonwovens are largely governed by polymer selection. Monolithic polymeric nanofibres are the simplest structures that can be produced, but not always the best choice for a given application. Therefore, commixtures or polymer blends obtained by in situ grafting or cross-linking of at least two polymers are an alternative for improved functionality [19,20]. Polymer nanofibres also serve as precursors for the formation of inorganic nanofibrous structures. A general approach for obtaining such fibres is in situ doping of polymer solutions with organic/inorganic salts followed by thermal treatment, in order to burn out the polymer [21,22]. In a similar way, careful selection of a polymer type that will not be completely burnt off during heat treatment enables their conversion into carbon nanofibres (CNF) [2,23]. Thermal treatment of polymeric webs under controlled environmental conditions, heating rate and temperature is the most widely used approach for this purpose. Here also, in situ doping of the spinning solution, or posttreatment can be applied in order to obtain doped or composite CNFs with specific functionalities [4,24,25].

In situ functionalisation or doping with inorganic or organic particles enables formation of composite structures in which the polymer serves as matrix, while the particles are the reinforcement that imparts specific functionality to the nanofibre. This type of composite fibre is called a matrix-type structure [20,26,27]. Controlling the polymer/particle ratio in the spinning solution, as well as the viscosity are the most important parameters determining whether smooth, or fibrous structures with rough surfaces are produced with either protruding or aggregated particles [28,29].

Functionalisation of already spun fibres is also possible. In this case, the fibres act as a substrate that holds specific functionalities on their surface. This can be achieved by (1) hydro/solvothermal treatment – when fibrous webs are firstly dipped into a chemical solution and then heat treated under controlled atmospheric, thermal and temporal conditions; by (2) electrodeposition – suitable when the fibres are electrically conductive (e.g. when conductive polymers or carbon nanofibres are present); or by (3) dip coating of fibrous webs in a solution containing inorganic or organic agents that impart specific properties. The first two methods are usually used to decorate fibres by 'growing' inorganic structures on their surface to improve the (re)activity of the fibres towards specific targeted compounds [2]. On the other hand, with the latter method, functional materials are impregnated on the surface of the fibres in the web [26,30].

It is also feasible to entrap particles between the nanofibres by introducing them into the area between the spinneret and collector during spinning, or indeed incorporating them directly into the spinning solution or melt, provided their dimensions are sufficiently small to allow polymer streams to be properly attenuated. When nanoparticles are included, preparatory steps such as sonication of the spinning solution or melt may be required to inhibit agglomeration inside the resulting fibres. Numerous insoluble materials can be incorporated such as pigments, magnetic nanoparticles, ceramics, and antimicrobials.

Regarding the structures of nanofibre that can be produced, the basic variant is a solid, and usually monolithic (Fig. 7.5A) with a smooth surface [24,31,32]. As mentioned earlier, more complex composite fibre structures such as matrix-type (Fig. 7.5B) or surface functionalised (Fig. 7.5C) fibres with either smooth of rough surfaces can also be made [20,26,27,29,33–37]. Porous nanofibres can also be manufactured (Fig. 7.5D) [24,25,29,31], by (1) a phase inversion method [38]; or by (2) sacrificial template synthesis, in the case of CNFs, when inorganic particles or a second polymer is used to form the pores and then is later removed during heat treatment and/or by a subsequent leaching process [29]. Hollow (Fig. 7.5E) [24] and core-shell nanofibre structures (Fig. 7.5F) [20,26,32,35] are another group of architectures that enhance the porosity and functionality of resulting products. In both cases, coaxial spinning of two different solutions through the same nozzle is the starting requirement for manufacture.

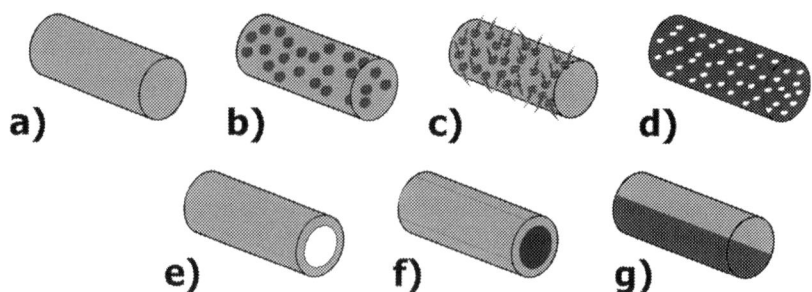

FIG. 7.5 Nanofibre architectures (A) solid, (B) matrix-type, (C) surface decorated, (D) porous, (E) hollow, (F) core-shell (coaxial bicomponent), (G) side-by-side bicomponent.

Addition of more than one coaxial channel enables formation of multicomponent fibres with more than one hole or core (island) present in the structure [27,39]. Extrusion of two or more spinning solutions/melts side by side from one nozzle is also a route to bicomponent fibres, or multicomponents based on other spinning arrangements (Fig. 7.5G) [27,33].

Besides the architecture of the fibres produced, the structure of the web that these fibres are formed in to can also be tailored to meet the requirements of final products. In this regard, multicomponent layered or sandwich structures can be easily made by layering various single nanofibrous webs on the top of each one (Fig. 7.6) [19]. Single layer, multicomponent fibrous webs can be produced using multiple nozzle spinning systems [40]. Here, each nozzle is fed with a different spinning solution or melt and integrated into a single web. The nozzles are positioned so that polymer jets are close enough to form homogenous fibrous mat on the collector. It is worth mentioning, that when tailoring such different fibrous architectures, the directional alignment (orientation) of the fibres within the web and its bulk density can play crucial role in many applications [34,37,41]. Therefore, the types of collector and collection speed are also parameters to be taken into account. 3D nanofibrous networks, aerogels or

FIG. 7.6 Schematic of sequential electrospinning process for multicomponent layered fibrous structures. *Reproduced and edited from C. Wang, et al., Fabrication of electrospun polymer nanofibers with diverse morphologies, Molecules 24(5) (2019), https://doi.org/10.3390/molecules24050834, with permission from MDPI, Copyright © 2019.*

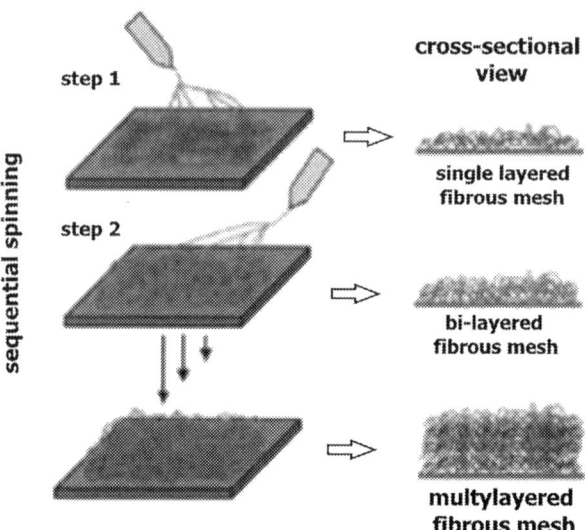

sponges can also be produced [25,41–43]. For the production of bulkier structures, methods other than electrospinning are preferred, because fibre deposition onto collectors that are not flat is more straightforward [42].

In all cases, polymer selection is determined by the properties required in the final product and the application for which the web will be used. Many biobased and synthetic polymers have been used in the production of nanofibre nonwovens, reflecting the diversity of applications that have been or are being developed. Cellulose, lignin, chitosan, collagen, gelatin, silk, alginate, hyaluronic acid, polyacrylonitrile (PAN), polyvinylpyrrolidone (PVP) polystyrene (PS), polyvinyl alcohol (PVA), polyethylene glycol (PEO), polyvinylidene fluoride (PVDF), polyamide (PA), polycaprolactone (PCL), poly(lactic-*co*-glycolic acid), and polylactic acid (PLA) are just a few amongst many. Metals, semimetals or metalloids, metal oxides, metal organic frameworks, ligands, bioactive molecules are general groups of inorganic and organic materials that have been delivered in nanofibrous webs, either as particles or precursors, to form polymer/particle composite or inorganic structures. Since the list is long and diverse, more detail on the specific materials used for nanofibre formation linked to application area is available elsewhere (Khulbe et al. [38], Zhang et al. [30], and Rasouli et al. [36]). By way of example, Table 7.1 provides an insight into the general groups of materials that have been used to make nanofibre materials across the three main application areas of filtration, energy, and healthcare.

TABLE 7.1   Nanofibre material compositions and application areas.

Materials	Application area	References
Polymer (monolithic and/or blended bio and synthetic)	Filtration and Separation, Energy, Energy Conversion, Biomedical	[19,20,22,25,32,34–36,38,44–47]
Conducting polymers	Energy Storage and Conversion	[22,24,48]
Inorganic (metal oxide, metal sulphide, nitrile, metal organic frameworks, spinel and olivine structured materials, bimetallic alloys)	Filtration and Separation, Energy, Energy Storage and Conversion	[4,21,22,24,38,44,49]
Carbon nanofibres (monolithic, doped, composite)	Filtration and Separation, Energy, Energy Storage and Conversion	[4,24,25,29,38,48,49]
Polymer/organic particles composites (chelating ligands, bioactive molecules)	Filtration and Separation, Biomedical (Drug Delivery, Wound Dressing)	[20,26,28,33,35,44,50,51]
Polymer/inorganic particles composites (metal oxides, metal organic frameworks, minerals)	Filtration and Separation, Energy Storage, Biomedical (Bone Scaffolds, Musculoskeletal Tissues, Wound Dressing)	[4,26,28,31,41,44]

7.4 Applications of nanofibrous webs and nonwovens

Nanofibrous nonwovens have been in commercial production for many decades, with filter media being amongst the earliest industrial markets to be established. In addition to air and liquid filtration, applications in biosensors, energy, electronics, and biomedical products are of growing importance [52]. As manufacturing turnkey capacity has increased, opportunities in consumer goods have also been developing from cosmetic face masks, to water and windproof breathable membranes used in outdoor clothing. Sound absorption, military protective clothing and composites are amongst other important applications. From 2018 to 2025, annual growth in nanofibre markets of 25% is expected, equivalent to a growth in value of US\$ 1.26 billion in 2017 to US\$ 7.24 billion in 2025 [53].

Given the particular importance of the filtration, energy, and biomedical markets, a brief overview of recent advances in nanofibrous materials for these sectors is included. Numerous review articles are available covering detailed developments relating to materials, properties, and performance for each of these application areas. Table 7.2 lists some of the relevant articles published in the period from 2015 to 2019, which will be beneficial for further reading.

7.4.1 Filtration and separation

With increasing concerns about environmental pollution and stricter emission regulations affecting industry, progressive development continues to target high filter efficiencies combined with low flow resistance (pressure drop), as well as longer operating lifetimes. As an industry with approximately 85 billion US\$ in annual revenues, the filtration and separation industry has five main market segments [63], with domestic, municipal and industrial waste-water management applications being the largest, followed by filtration and separation elements used in industrial processes. These applications include chemical, oil and gas, microelectronics, fluid power, pulp and paper and power generation. Pollution control and HVAC (heating ventilation and air conditioning) systems in residential, industrial, and healthcare buildings, as well as vehicles and transport systems are important part of the overall market. There are also growing applications in pharmaceutical and biological processing, medical devices, biomedical diagnostics and testing and in the food and beverage industry.

Globally, addressing clean air and water pollution requirements in terms of human health, ecosystems and food crop production, as well as the needs of the world's growing cities and built environment will continue to pose major challenges [64]. Various nanofibrous nonwovens are being directed to tackle associated filtration and separation needs, such as fabrics containing inorganic nanoparticles, carbon absorbents, and replacements for membranes [65]. As an alternative to cast, thin film membranes, nanofibre webs made of ultrafine fibres exhibit high filtration efficiency and when chemically functionalised, adsorption capacity as a result of their high surface-to-volume ratio and porosity [20,38]. Moreover, their porosity, pore size, and distribution can be easily tailored during production, by controlling fibre diameter and fibre alignment in the web [19]. Nanofibrous webs with cross-ply alignment have been shown to provide high filtration efficiency when used as air filters [19,20,44]. Functionalised nanofibrous membranes made of blended polymers or doped with inorganic

TABLE 7.2 Recent review articles on nanofibrous membranes for various application areas.

Application area	Topic	References
Filtration and Separation	Nanofibres, their functionalisation, characterisation, and application as membranes for metal ion removal	[20]
Filtration and Separation	Materials and properties of electrospun nanofibre membranes and their application and performances in various membrane distillation systems	[28]
Filtration and Separation	Materials, structures, and performances of nanofibrous membranes for wastewater treatment, chiral separation, and desalination	[38]
Filtration and Separation	Nanofibrous membranes for micro-, nano-, and ultra-filtration	[54]
Filtration and Separation	Materials and structures for nanofibrous membranes for oil–water separation	[25]
Filtration and Separation	Materials and structures for nanofibrous membranes for air filtration	[44]
Filtration and Separation	Production methods and properties of nanofibrous membranes for air filtration	[19]
Environmental and Energy	Ion exchange nanofibrous membranes for separation/purification purposes and as polymer electrolytes for fuel cells	[30]
Energy Storage	Ceramic nanofibrous electrolytes for all solid-state Li-ion batteries	[21]
Electronics	Nanofibres for pressure and gas sensors, photodetectors, conductive wires, and smart nanogenerators	[55]
Energy Storage	Carbon nanofibres and metal oxide, metal sulphide, nitride and flexible conducting polymer nanofibres and their composites as supercapacitor electrodes	[24]
Energy Storage	Graphene embedded nanofibres as supercapacitor and Li-ion battery electrodes	[48]
Energy Storage	Carbon-based nanofibrous structures for wearable electronics	[56]
Energy Storage and other	Carbon nanofibres – production methods and applications in energy storage and functional nanocomposites	[23]
Energy Storage	Nanofibrous anodes and cathodes for Na-ion batteries	[29]
Energy	Aligned nanofibres for fuel cell applications	[45]
Energy	Nanofibrous structures as proton exchange membranes for fuel cells	[46]
Energy	Nanofibrous structures as electrocatalysts for oxygen reduction reaction in fuel cells	[49]
Energy	Nanofibrous membranes with high ionic conductivity as polymer electrolyte fuel cell membranes	[57]
Energy Conversion	Nanofibrous materials and structures for solar energy conversion	[22]

Continued

TABLE 7.2   Recent review articles on nanofibrous membranes for various application areas—cont'd

Application area	Topic	References
Biomedical	Nanofibres for tissue engineering, bio-sensors, drug delivery systems, wound dressing, air, blood, and water purification	[36]
Biomedical	Electrospun nanofibres for drug delivery systems	[58]
Biomedical	Modified electrospinning techniques for production of nanofibres and their application in drug delivery systems	[27]
Biomedical	Nanofibres for immediate, prolonged, and biphasic drug release	[34]
Biomedical	Nanofibre shish kebab structure and its potential biomimetic application	[59]
Biomedical	Nanofibres – production and modification methods and their application for bone tissue engineering	[31]
Biomedical	Nanofibres for skin regeneration – materials, functionalisation methods, and performances	[26]
Biomedical	Nanofibres for bioinstructive scaffolds	[60]
Biomedical	Nanofibrous wound dressings for treating diabetic foot ulcer	[32]
Biomedical	Nanofibrous membranes for musculoskeletal regeneration	[41]
Biomedical	Nanofibrous membranes as wound dressings and drug delivery systems	[33]
Biomedical	Nanofibrous membranes for cuff repair	[43]
Biomedical	Nanofibrous membranes for cartilage tissue repair	[61]
Biomedical	Bio-polymer based nanofibres for targeted drug delivery systems	[35]
Biomedical	Nanofibres for in brain tumour therapy – mechanism and properties	[51]
Biomedical	In vivo activity of drug-loaded nanofibres for local anticancer therapy	[50]
Biomedical	Behaviour of nanofibre systems in ex vivo expansion of cord blood–derived haematopoietic stem cells	[47]
Biomedical	Nanofibres for mesenchymal stem cell–based bone engineering	[37]
Food and Packaging	Nanofibres based on proteins and carbohydrates for food processing and packaging	[62]

particles [38] can also provide improved adsorption, antibacterial, chemical, or physical functionality. Such nanofibre membranes are relevant to wastewater treatment [38,54], distillation [20,28], air filtration, purification, and conditioning [19,44].

7.4.2 Energy applications

1D nanofibrous materials provide one of the most promising types of electrode structure for energy applications. Nanofibre electrodes enable shorter ion diffusion lengths and improved surface area or higher electrical conductivity, such as in the case of carbon nanofibres

[66]. A further advantage is the possibility to use them as free standing electrodes or as membranes that separate the electrodes.

Tailored nanofibrous structures with optimised compositions are almost ideal candidates for battery or supercapacitor electrodes [4,24,29,48]. Such structures can be based on carbon, metal oxides, sulphides, nitrides, conductive polymers and even graphene-based nanofibre composites. Each of these different materials provide specific features that contribute to either, enhanced electron conductivity, electrochemical activity, and/or stability [4,24,29,48]. Appropriate dispersion of different particulate active materials or the amount of precursors are key elements to be taken into account when optimising the production conditions for composite nanofibrous electrodes [24,48]. Surface modified nanofibres with selected electrochemically active substances improve the capacity and cycle life of batteries. The flexibility of carbon nanofibres and especially that of polymer nanofibres embedded with electrochemically active and conductive components, further amplifies their application into the area of flexible energy storage devices and smart wearable electronics [24,56].

Nanofibrous structures are also promising candidates for separator/electrolyte membranes in all-solid state batteries. Flexibility, mechanical strength, porosity, and ionic conductivity are just a few of the advantages that polymer and composite polymer/ceramic nanofibrous membranes can provide when separating the positive and negative electrodes in a battery [67,68]. Moreover, nanofibrous structures contribute to reduction of the calcination temperature, and thus lower the cost of production in ceramic solid electrolytes for all solid-state batteries [21]. Nanofibre webs also find applications as ion-exchange membranes due to their high surface area, fully interconnected porous network, ion-absorption, and exchange capacity. Additionally, the opportunity to tailor the alignment of the fibres in the web, enables controlled and improved proton conductivity through the fibrous network [45,57]. They can also be embedded into an ion-conductive polymer matrix to form a porous membrane with enhanced ion-conductivity [45]. All these features, makes nanofibrous membranes suitable for fuel cell applications [30,46]. Furthermore, core-shell, porous, hollow composite or surface modified nanofibres are good candidates for electrodes intended for solar energy conversion, sensors, nanowires, and flexible electronic devices [22,55].

7.4.3 Biomedical and healthcare

In addition to their high surface area and porosity, the chemical and biological properties that can be engineered into nanofibrous structures, makes them suitable candidates for a variety of biomedical applications, many of which are in clinical development, including:

(a) Tissues for artificial skin [26,33],
(b) Bioactive scaffolds, for soft and hard tissues, including applications in heart valves, musculoskeletal, corneal and neural tissue repair and regeneration [31,41,43,60],
(c) Drug delivery systems [27,34,58],
(d) Materials for pharmaceutical and cosmetics industry.

In Europe, electrospun webs are commercially manufactured for medical devices according to ISO 13485 quality management requirements. Established applications for electrospun products include well-plates containing fibrous scaffolds that are used as R&D laboratory consumables and other fabrics are used in advanced wound healing products. Many more are under clinical evaluation, or are still in product development. For applications in wound healing, electrospun materials of various compositions, have been demonstrated with the ability to act as barriers to pathogenic organisms, to provide a moist would healing environment, to promote cell proliferation, to deliver bioactive agents and to reduce scar formation or provide regenerative function [33]. Industrial manufacturing of electrospun products for medical devices comes with challenges, because of the inherent variability in terms of fibre diameter, morphology and other dimensional and physical properties.

The process of electrospinning lends itself well to the incorporation of functional chemistry or nanoparticles within fibres or the covalent attachment of chemical compounds to the fibre surfaces, which transform bulk performance. Such functionalised nanofibres made of biocompatible synthetic and naturally-derived polymers have been widely investigated in the last decade for medical and healthcare applications [26]. This includes nanofibre scaffolds or occlusive membranes capable of directing cell response for musculoskeletal, vascular, immunological, and nervous system tissues [31,41,43,60].

Biopolymers in combination with calcium phosphate-based minerals are attractive materials for the development of new bone scaffolds. Accordingly, nanofibres made from natural or synthetic biopolymers combined with calcium phosphate materials during the spinning process or afterwards, have shown promising results for bone tissue regeneration [31]. The dimensional properties and hierarchical structure of electrospun scaffolds are also important in determining performance as a scaffold structure. For example, three-dimensional cell penetration and nutrient exchange can be highly challenging in electrospun structures because of the small pore size and high packing density. In relation to hard tissue regeneration, nanofibres can also be decorated with lamellar crystals orthogonal to the fibre axis in a quasiperiodic pattern. Such structures are referred to as nanofibre 'shish-kebab'. [59]. The 3D structure has a porous base, because of the presence of nanofibres, and the surface functionality promotes mineralisation, leading to promising substrates for bone scaffold applications. In addition to nonwoven webs, electrospun materials have also been formed into yarns, which of course then enables processing into conventional textile structures such as braids, woven or knitted constructs. Braids and woven fabrics produced from nanofibrous yarns have shown promising results in relation to musculoskeletal tissue repair and regeneration [43].

A further important application of nanofibrous webs is the potential for directed drug delivery, combined with controlled release. Rapid drug delivery (burst release) can be achieved by manufacturing nanofibres from water-soluble polymers or hydrogels with the drug carried within the amorphous regions [34]. In core-shell nanofibres, modulation of fibre diameter and combining polymers with different degradation behaviour enables greater control of the drug release kinetics to be achieved [34,58]. Such fibres are considered promising candidates for antiinflammatory, DNA, anticancer, antimicrobial, antiviral, transdermal, implantable, fast-dissolving, and other drug delivery systems [35,47,50,58].

7.5 Industrial manufacturing capacity

The academic literature is replete with references to real-world applications for nanofibre fabrics, but not all have reached commercial translation in terms of production. Globally, there has been progressive growth in the industrial manufacturing infrastructure to service existing and emerging markets. Freudenberg was one of the earliest adopters of nanofibre nonwoven production, with others such as Donaldson being major producers for many years. Table 7.3 gives a nonexhaustive list of commercial producers, application areas, and production technologies.

TABLE 7.3 Examples of commercial nanofibre producers, application areas, and production methods.

Company	Trademark/product	Application area	Nanofibre production method
Donaldson	Ultra-Web, Fineweb	Filtration	Electrospinning
Optimum Filtration Company	ProTura	Filtration	Electrospinning
Mann+Hummel	Micrograde NF	Filtration	Electrospinning
HiFyber	HiFyber	Filtration	Electrospinning
Liquidity Corporation	Naked Filter	Filtration	Electrospinning
SPUR Nanotechnologies	SpurTex	Filtration	Electrospinning
Sorbent JSC	Petryanov filter	Filtration	Electrospinning
Shijiazhuang Chentai Filter Paper Company	/	Filtration	/
Ahlstrom-Munksjö	NanoPulse	Filtration	Electrospinning
NASK	Smart Masks	Facemask	Electrospinning
N2Cell	Breath	Facemask	Electrospinning
M-TechX	Magic Fibre	Absorbent Materials	Air blowing (melt)
DuPont	/	Energy Storage (separator)	Electro-blowing
Ortho Rebirth	Rebossis	Biomedical	Electrospinning
Nanofibre Solutions	NanoAligned, NanoECM	Biomedical	Electrospinning
Electrospinning Company	Electrospinning Co	Biomedical	Electrospinning
Nicast	Avflo	Biomedical	Electrospinning
Leonardino Srl	SKE Research Equipment	Biomedical	Electrospinning
DiPole Materials	BioPapers	Biomedical	Electrospinning
SNS Nano	Nanosan	Biomedical	Electrospinning
Zeus	Bioweb	Biomedical	Electrospinning

Continued

TABLE 7.3   Examples of commercial nanofibre producers, application areas, and production methods—cont'd

Company	Trademark/product	Application area	Nanofibre production method
NXTNANO	nPel, nMax, nFlux, nTex	Filtration, Apparel	Spinning (High Yield Production Rate)
Nano4fibres Group	BreaSafe, Riftelen N15, NnF Cream, NnF Mbrane	Filtration, Inorganic and polymeric nanofibres	Centrifugal spinning
eSpin Technologies	nWeb, Cytoweb	Filtration, Biomedical, Apparel	Electrospinning
Verdex Technologies	Verdex	Filtration, Biomedical, Technical Textiles	Electrospinning, Air blowing (melt)
Revolution Fibres	Seta, ActivLayr, Phonix, XantuLayr, NanoDream	Filtration, Biomedical, Acoustic insulation, Composite reinforcements, Textile	Electrospinning
Ftene	Finetex EnE, Nexture	Filtration, Technical Textile, Biomedical and Healthcare, Electronics	Electrospinning
eSpin Technologies	Carbon nanofibre	Various	Electrospinning
ACS Material	Carbon nanofibre	Energy storage, Composite filler, Biomedical	CVD
Carbon Nano-material Technology	Graphite nanofibre	Composite filler, Energy storage, Electronics	CVD
Catalytic Materials LLC	Graphite nanofibre	Catalyst Support, Energy storage	CVD
ANF Technology	NAFEN – Alumina nanofibre	Catalyst, Energy, Electronics, Aerospace, Automotive	/
Nano Technology	Ceramic nanofibre	Various	/
Applied Sciences	Carbon nanofibre, Nanomat	Electronics, Composite reinforcement, Energy	CVD
Nanostructured & Amorphous Materials	Carbon nanofibre	Various	CVD
Grupo Antolin	GRAnPH – Carbon nanofibre	Various	CVD
US-Nano	Carbon nanofibre	Various	CVD

7.6 Future developments

The combination of nanoscale or submicron fibre dimensions, high surface area to volume ratio, high porosity and the wide spectrum of compatible polymer materials that are compatible with the manufacturing platforms has fuelled a steady growth in demand for nonwovens across a diverse range of application areas.

Electrospinning, air blowing, centrifugal spinning, and related hybrid technologies have emerged as the core methods underpinning nanofibre nonwoven manufacturing. While academic literature emphasises the importance of electrospinning, not least because of its utility in small-scale laboratory research, performance limitations persist in high speed industrial production [21]. Therefore, development of industrial scale nanofibre nonwoven process technologies continues to be focused on increasing productivity, quality control and reduction of variability and improving the integration of nanofibre web formation with other nonwoven processes, to enable continuous, cost-effective production of composite nonwovens.

Production of nonwovens containing nano or submicron fibres produced by current techniques still predominantly relies on polymeric feedstocks, although posttreatment of the fibres is capable of delivering inorganic materials. In terms of fabric structure, both the morphology of individual fibres and the architecture of the fibrous assembly itself can be influenced to a large extent by controlling the properties of the spinning solution or melt or process conditions. Hence, it is possible to make nano and submicron fibre assemblies from one feedstock material but deliver it in a variety of different structural configurations. Whereas, fibre spinning technologies are the basis for bulk production of nanofibrous and submicron fibre nonwovens, other bottom-up manufacturing methods meet the requirements for producing well-organised and precisely sized and structured nanofibrous structures. Techniques that control the formation of each fibre individually and enable production of micro-sized products with extremely strict dimensions and order of elements, are essential for some applications in areas such as electronics.

References

[1] R. Seeram, L. Teik-cheng, F. Kazutoshi, An Introduction to Electrospinning and Nanofibers, World Scientific, 2005.

[2] E. Stojanovska, et al., A review on non-electro nanofibre spinning techniques, RSC Adv. 6 (87) (2016) 83783–83801, https://doi.org/10.1039/C6RA16986D.

[3] Y. Lu, et al., Centrifugal spinning: a novel approach to fabricate porous carbon fibers as binder-free electrodes for electric double-layer capacitors, J. Power Sources 273 (2015) 502–510, https://doi.org/10.1016/j.jpowsour.2014.09.130.

[4] E.S. Pampal, E. Stojanovska, B. Simon, A. Kilic, A review of nanofibrous structures in lithium ion batteries, J. Power Sources 300 (2015) 199–215, https://doi.org/10.1016/j.jpowsour.2015.09.059.

[5] H.S. SalehHudin, E.N. Mohamad, W.N.L. Mahadi, A.M. Afifi, Multiple-jet electrospinning methods for nanofiber processing: a review, Mater. Manuf. Process. 33 (5) (2018) 479–498, https://doi.org/10.1080/10426914.2017.1388523.

[6] R. Nayak, R. Padhye, I.L. Kyratzis, Y.B. Truong, L. Arnold, Recent advances in nanofibre fabrication techniques, Text. Res. J. 82 (2) (2012) 129–147, https://doi.org/10.1177/0040517511424524.

[7] C. Wang, et al., Fabrication of electrospun polymer nanofibers with diverse morphologies, Molecules 24 (5) (2019), https://doi.org/10.3390/molecules24050834.

[8] E.S. Medeiros, G.M. Glenn, A.P. Klamczynski, W.J. Orts, L.H.C. Mattoso, Solution blow spinning: a new method to produce micro- and nanofibers from polymer solutions, J. Appl. Polym. Sci. 113 (4) (2009) 2322–2330, https://doi.org/10.1002/app.30275.

[9] M. Pokorny, V. Rassushin, L. Wolfova, V. Velebny, Increased production of nanofibrous materials by electroblowing from blends of hyaluronic acid and polyethylene oxide, Polym. Eng. Sci. 56 (8) (2016) 932–938, https://doi.org/10.1002/pen.24322.

[10] A. Balogh, et al., Electroblowing and electrospinning of fibrous diclofenac sodium-cyclodextrin complex-based reconstitution injection, J. Drug Delivery Sci. Technol. 26 (2015) 28–34, https://doi.org/10.1016/j.jddst.2015.02.003.

[11] I.C. Um, D. Fang, B.S. Hsiao, A. Okamoto, B. Chu, Electro-spinning and electro-blowing of hyaluronic acid, Biomacromolecules 5 (4) (2004) 1428–1436, https://doi.org/10.1021/bm034539b.

[12] H. Xu, H. Chen, X. Li, C. Liu, B. Yang, A comparative study of jet formation in nozzle- and nozzle-less centrifugal spinning systems, J. Polym. Sci. B Polym. Phys. 52 (23) (2014) 1547–1559, https://doi.org/10.1002/polb.23596.

[13] R.T. Weitz, L. Harnau, S. Rauschenbach, M. Burghard, K. Kern, Polymer nanofibers via nozzle-free centrifugal spinning, Nano Lett. 8 (4) (2008) 1187–1191, https://doi.org/10.1021/nl080124q.

[14] S. Padron, A. Fuentes, D. Caruntu, K. Lozano, Experimental study of nanofiber production through forcespinning, J. Appl. Phys. 113 (2) (2013), https://doi.org/10.1063/1.4769886, 024318.

[15] G.M. Gonzalez, et al., Production of synthetic, para-aramid and biopolymer nanofibers by immersion rotary jet-spinning, Macromol. Mater. Eng. 302 (1) (2017) 1600365, https://doi.org/10.1002/mame.201600365.

[16] H. Lee, I.S. Kim, Nanofibers: emerging progress on fabrication using mechanical force and recent applications, Polym. Rev. 58 (4) (2018) 688–716, https://doi.org/10.1080/15583724.2018.1495650.

[17] M. Endo, et al., Vapor-grown carbon fibers (VGCFs): basic properties and their battery applications, Carbon 39 (9) (2001) 1287–1297, https://doi.org/10.1016/S0008-6223(00)00295-5.

[18] G.G. Tibbetts, Vapor-grown carbon fibers: status and prospects, Carbon 27 (5) (1989) 745–747, https://doi.org/10.1016/0008-6223(89)90208-X.

[19] Y. Akgul, Y. Polat, E. Canbay, A. Demir, A. Kilic, 20—Nanofibrous composite air filters, in: M. Jawaid, M.M. Khan (Eds.), Polymer-Based Nanocomposites for Energy and Environmental Applications, Woodhead Publishing, 2018, pp. 553–567.

[20] O. Pereao, C. Bode-Aluko, K. Laatikainen, A. Nechaev, L. Petrik, Morphology, modification and characterisation of electrospun polymer nanofiber adsorbent material used in metal ion removal, J. Polym. Environ. 27 (9) (2019) 1843–1860, https://doi.org/10.1007/s10924-019-01497-w.

[21] A. La Monaca, A. Paolella, A. Guerfi, F. Rosei, K. Zaghib, Electrospun ceramic nanofibers as 1D solid electrolytes for lithium batteries, Electrochem. Commun. 104 (2019), https://doi.org/10.1016/j.elecom.2019.106483, 106483.

[22] D. Joly, J.-W. Jung, I.-D. Kim, R. Demadrille, Electrospun materials for solar energy conversion: innovations and trends, J. Mater. Chem. C 4 (43) (2016) 10173–10197, https://doi.org/10.1039/C6TC00702C.

[23] J.C. Ruiz-Cornejo, D. Sebastián, M.J. Lázaro, Synthesis and applications of carbon nanofibers: a review, Rev. Chem. Eng. 36 (4) (2020) 493–511, https://doi.org/10.1515/revce-2018-0021.

[24] X. Lu, C. Wang, F. Favier, N. Pinna, Electrospun nanomaterials for supercapacitor electrodes: designed architectures and electrochemical performance, Adv. Energy Mater. 7 (2) (2017) 1601301, https://doi.org/10.1002/aenm.201601301.

[25] J. Ge, Q. Fu, J. Yu, B. Ding, Chapter 13—Electrospun nanofibers for oil–water separation, in: B. Ding, X. Wang, J. Yu (Eds.), Electrospinning: Nanofabrication and Applications, William Andrew Publishing, 2019, pp. 391–417.

[26] S.P. Miguel, et al., Electrospun polymeric nanofibres as wound dressings: a review, Colloids Surf. B: Biointerfaces 169 (2018) 60–71, https://doi.org/10.1016/j.colsurfb.2018.05.011.

[27] M. Liu, et al., Recent advances in electrospun for drug delivery purpose, J. Drug Target. 27 (3) (2019) 270–282, https://doi.org/10.1080/1061186X.2018.1481413.

[28] C.-Y. Pan, et al., Electrospun nanofibrous membranes in membrane distillation: recent developments and future perspectives, Sep. Purif. Technol. 221 (2019) 44–63, https://doi.org/10.1016/j.seppur.2019.03.080.

[29] E. Stojanovska, F.N. Buyuknalcaci, M.D. Calisir, E.S. Pampal, A. Kilic, 12—Nanofibrous composites for sodium-ion batteries, in: M. Jawaid, M.M. Khan (Eds.), Polymer-Based Nanocomposites for Energy and Environmental Applications, Woodhead Publishing, 2018, pp. 333–360.

[30] S. Zhang, A. Tanioka, H. Matsumoto, Nanofibers as novel platform for high-functional ion exchangers, J. Chem. Technol. Biotechnol. 93 (10) (2018) 2791–2803, https://doi.org/10.1002/jctb.5685.

[31] S. Chahal, A. Kumar, F.S.J. Hussian, Development of biomimetic electrospun polymeric biomaterials for bone tissue engineering. A review, J. Biomater. Sci. Polym. Ed. 30 (14) (2019) 1308–1355, https://doi.org/10.1080/09205063.2019.1630699.

[32] Y. Liu, S. Zhou, Y. Gao, Y. Zhai, Electrospun nanofibers as a wound dressing for treating diabetic foot ulcer, Asian J. Pharm. Sci. 14 (2) (2019) 130–143, https://doi.org/10.1016/j.ajps.2018.04.004.

[33] S.P. Miguel, et al., An overview of electrospun membranes loaded with bioactive molecules for improving the wound healing process, Eur. J. Pharm. Biopharm. 139 (2019) 1–22, https://doi.org/10.1016/j.ejpb.2019.03.010.

[34] S. Kajdič, O. Planinšek, M. Gašperlin, P. Kocbek, Electrospun nanofibers for customized drug-delivery systems, J. Drug Delivery Sci. Technol. 51 (2019) 672–681, https://doi.org/10.1016/j.jddst.2019.03.038.

[35] S. Fahimirad, F. Ajalloueian, Naturally-derived electrospun wound dressings for target delivery of bio-active agents, Int. J. Pharm. 566 (2019) 307–328, https://doi.org/10.1016/j.ijpharm.2019.05.053.

[36] R. Rasouli, A. Barhoum, M. Bechelany, A. Dufresne, Nanofibers for biomedical and healthcare applications, Macromol. Biosci. 19 (2) (2019) 1800256, https://doi.org/10.1002/mabi.201800256.

[37] S. Wang, et al., Design of electrospun nanofibrous mats for osteogenic differentiation of mesenchymal stem cells, Nanomed.: Nanotechnol. Biol. Med. 14 (7) (2018) 2505–2520, https://doi.org/10.1016/j.nano.2016.12.024.

[38] K.C. Khulbe, T. Matsuura, Art to use electrospun nanofbers/nanofber based membrane in waste water treatment, chiral separation and desalination, J. Membr. Sci. Res. 5 (2) (2019) 100–125, https://doi.org/10.22079/jmsr.2018.87918.1197.

[39] F. Li, Y. Zhao, Y. Song, Core-shell nanofibers: nano channel and capsule by coaxial electrospinning, Nanofibers (2010), https://doi.org/10.5772/8166.

[40] K. Ulubayram, S. Calamak, R. Shahbazi, I. Eroglu, Nanofibers based antibacterial drug design, delivery and applications, Curr. Pharm. Des. 21 (15) (2015) 1930–1943, https://doi.org/10.2174/1381612821666150302151804.

[41] S. Sankar, C.S. Sharma, S.N. Rath, S. Ramakrishna, Electrospun nanofibres to mimic natural hierarchical structure of tissues: application in musculoskeletal regeneration, J. Tissue Eng. Regen. Med. 12 (1) (2018) e604–e619, https://doi.org/10.1002/term.2335.

[42] H. Wang, et al., Ultralight, scalable, and high-temperature–resilient ceramic nanofiber sponges, Sci. Adv. 3 (6) (2017), https://doi.org/10.1126/sciadv.1603170, e1603170.

[43] N. Saveh-Shemshaki, L.S. Nair, C.T. Laurencin, Nanofiber-based matrices for rotator cuff regenerative engineering, Acta Biomater. 94 (2019) 64–81, https://doi.org/10.1016/j.actbio.2019.05.041.

[44] S. Zhang, et al., Chapter 12—Electrospun nanofibers for air filtration, in: B. Ding, X. Wang, J. Yu (Eds.), Electrospinning: Nanofabrication and Applications, William Andrew Publishing, 2019, pp. 365–389.

[45] P. Kallem, N. Yanar, H. Choi, Nanofiber-based proton exchange membranes: development of aligned electrospun nanofibers for polymer electrolyte fuel cell applications, ACS Sustain. Chem. Eng. 7 (2) (2019) 1808–1825, https://doi.org/10.1021/acssuschemeng.8b03601.

[46] R. Sood, S. Cavaliere, D.J. Jones, J. Rozière, Electrospun nanofibre composite polymer electrolyte fuel cell and electrolysis membranes, Nano Energy 26 (2016) 729–745, https://doi.org/10.1016/j.nanoen.2016.06.027.

[47] M.S.V. Ferreira, S.H. Mousavi, Nanofiber technology in the ex vivo expansion of cord blood-derived hematopoietic stem cells, Nanomed.: Nanotechnol. Biol. Med. 14 (5) (2018) 1707–1718, https://doi.org/10.1016/j.nano.2018.04.017.

[48] K. Javed, M. Oolo, N. Savest, A. Krumme, A review on graphene-based electrospun conductive nanofibers, supercapacitors, anodes, and cathodes for lithium-ion batteries, Crit. Rev. Solid State Mater. Sci. 44 (5) (2019) 427–443, https://doi.org/10.1080/10408436.2018.1492367.

[49] M. Rauf, J.-W. Wang, P. Zhang, W. Iqbal, J. Qu, Y. Li, Non-precious nanostructured materials by electrospinning and their applications for oxygen reduction in polymer electrolyte membrane fuel cells, J. Power Sources 408 (2018) 17–27, https://doi.org/10.1016/j.jpowsour.2018.10.074.

[50] L. Poláková, J. Širc, R. Hobzová, A.-I. Cocârță, E. Heřmánková, Electrospun nanofibers for local anticancer therapy: review of in vivo activity, Int. J. Pharm. 558 (2019) 268–283, https://doi.org/10.1016/j.ijpharm.2018.12.059.

[51] M. Norouzi, Recent advances in brain tumor therapy: application of electrospun nanofibers, Drug Discov. Today 23 (4) (2018) 912–919, https://doi.org/10.1016/j.drudis.2018.02.007.

[52] Technavio, Global Nanofiber Market 2019–2023, Technavio, May 2019. [Online]. Available from: https://www.technavio.com/report/global-nanofiber-market-industry-analysis.

[53] The Insight Partners, Nanofiber Market to 2025—Global Analysis and Forecasts by Material and Application, Jan 2019. www.researchandmarkets.com. [Online]. Available from: https://www.researchandmarkets.com/reports/4747994/nanofiber-market-to-2025-global-analysis-and.

[54] C. Cheng, X. Li, X. Yu, M. Wang, X. Wang, Chapter 14—Electrospun nanofibers for water treatment, in: B. Ding, X. Wang, J. Yu (Eds.), Electrospinning: Nanofabrication and Applications, William Andrew Publishing, 2019, pp. 419–453.

[55] W. Han, et al., Fabrication of nanofibrous sensors by electrospinning, Sci. China Technol. Sci. 62 (6) (2019) 886–894, https://doi.org/10.1007/s11431-018-9405-5.

[56] S.C. Dhanabalan, B. Dhanabalan, X. Chen, J.S. Ponraj, H. Zhang, Hybrid carbon nanostructured fibers: stepping stone for intelligent textile-based electronics, Nanoscale 11 (7) (2019) 3046–3101, https://doi.org/10.1039/C8NR07554A.

[57] M. Tanaka, Development of ion conductive nanofibers for polymer electrolyte fuel cells, Polym. J. 48 (1) (2016) 51–58, https://doi.org/10.1038/pj.2015.76.

[58] B. Ghafoor, A. Aleem, M. Najabat Ali, M. Mir, Review of the fabrication techniques and applications of polymeric electrospun nanofibers for drug delivery systems, J. Drug Delivery Sci. Technol. 48 (2018) 82–87, https://doi.org/10.1016/j.jddst.2018.09.005.

[59] A.C. Attia, T. Yu, S.E. Gleeson, M. Petrovic, C.Y. Li, M. Marcolongo, A review of nanofiber shish kebabs and their potential in creating effective biomimetic bone scaffolds, Regen. Eng. Transl. Med. 4 (3) (2018) 107–119, https://doi.org/10.1007/s40883-018-0053-3.

[60] D.T. Bowers, J.L. Brown, Nanofibers as bioinstructive scaffolds capable of modulating differentiation through mechanosensitive pathways for regenerative engineering, Regen. Eng. Transl. Med. 5 (1) (2019) 22–29, https://doi.org/10.1007/s40883-018-0076-9.

[61] Y. Liu, L. Liu, Z. Wang, G. Zheng, Q. Chen, E. Luo, Application of electrospinning strategy on cartilage tissue engineering, Curr. Stem Cell Res. Ther. 13 (7) (2018) 526–532, https://doi.org/10.2174/1574888X13666180628163515.

[62] T.S.M. Kumar, K.S. Kumar, N. Rajini, S. Siengchin, N. Ayrilmis, A.V. Rajulu, A comprehensive review of electrospun nanofibers: food and packaging perspective, Compos. B Eng. 175 (2019) 107074, https://doi.org/10.1016/j.compositesb.2019.107074.

[63] T.H. Ramsey, Filtration & Separation Industry – The American Filtration and Separations Society (AFS), 2017, [Online]. Available from: https://www.afssociety.org/afs-launches-pov/.

[64] H. Ritchie, M. Roser, Air Pollution, Our World in Data, Apr. 2017.

[65] M.M. Khin, A.S. Nair, V.J. Babu, R. Murugan, S. Ramakrishna, A review on nanomaterials for environmental remediation, Energy Environ. Sci. 5 (8) (2012) 8075–8109, https://doi.org/10.1039/C2EE21818F.

[66] H.-G. Wang, S. Yuan, D.-L. Ma, X.-B. Zhang, J.-M. Yan, Electrospun materials for lithium and sodium rechargeable batteries: from structure evolution to electrochemical performance, Energy Environ. Sci. 8 (6) (2015) 1660–1681, https://doi.org/10.1039/C4EE03912B.

[67] S.-S. Choi, Y.S. Lee, C.W. Joo, S.G. Lee, J.K. Park, K.-S. Han, Electrospun PVDF nanofiber web as polymer electrolyte or separator, Electrochim. Acta 50 (2) (2004) 339–343, https://doi.org/10.1016/j.electacta.2004.03.057.

[68] G. Cheruvally, et al., Electrospun polymer membrane activated with room temperature ionic liquid: novel polymer electrolytes for lithium batteries, J. Power Sources 172 (2) (2007) 863–869, https://doi.org/10.1016/j.jpowsour.2007.07.057.

CHAPTER

8

Mechanical bonding

S.C. Anand[a], D. Brunnschweiler[†], G. Swarbrick[b],
and S.J. Russell[c]

[a]University of Bolton, Bolton, United Kingdom [b]Groz-Beckert UK Ltd., Wigston, United Kingdom
[c]University of Leeds, Leeds, United Kingdom

8.1 Introduction to stitch bonding

Stitch-bonded fabrics are those in which fibres, yarns, fibres and yarns, or fibres and a ground fabric are held together by subsequent stitching, or knitting-in of additional yarns [1]. The Textile Institute defines stitch-bonded or 'sew-knit' fabric as a multicomponent fabric, one component of which is a series of interlooped stitches running along the fabric length [2]. The other components may be a fibre web, yarns, or preformed fabric. Some stitch-bonded fabrics can fall outside industry definitions of nonwovens, but are discussed here for completeness.

A number of different stitch bonding systems have been developed and commercially employed since the end of the Second World War; including Ara machines (Czech Republic); Mali machines (Germany); Liba machines (Germany), and A Ch V-Sh VP machines (Russia). Within each of these machine families, a modular concept was developed and utilised so that a variety of product combinations could be produced using one basic stitching head or one headstock, with the possibility of feeding different substrates or combination of substrates. Commercially, a wide spectrum of products is produced using stitch bonding including apparel, household, and technical fabrics. This is because different combinations of substrates, raw materials, or even preformed fabrics can be combined during stitch bonding to modify or improve the characteristics of the final product.

Currently, the full range of stitch bonding equipment is produced by only one manufacturer, with another well-known company offering a multiaxial stitch bonding system for

[☆]Professor David Brunnschweiler OBE sadly passed away on the 29 April 2014 and he is very greatly missed. This chapter is dedicated to his memory.

[†]Deceased.

which they are universally recognised. Historically, a Russian company, which ceased trading in the 1970s, produced a limited number of VP machines that saw operation mainly within Eastern Europe. In 1964, the production of Arachne machines started in Czechoslovakia and during the 1970s, Arachne offered a full range of stitch bonding machinery including the Arabeva, Arachne, Aranit, Arutex, Araloop, and Bicolor Araloop systems. This type of machine was extensively operated throughout Europe and many other countries until the early to mid-1990s and Arachne finally discontinued production of new equipment. Liba Maschinenfabrik GmbH, Germany, manufactures Multiaxial stitch bonding equipment and in the past has also manufactured bespoke stitch bonding equipment exclusively for specific companies, including Cosmopolitan Textiles Ltd., UK. In what was then the German Democratic Republic, the Institute for Textile Machines, in Karl-Marx-Stadt (now Chemnitz), developed a 'sew-knit' machine based on the patents of Heinrich Mauersberger. The principle of sewing was used for producing bonded nonwoven fabrics and took advantage of the high sewing speeds that could be obtained with suitably designed mechanisms [3]. Malimo Maschinenbau GmbH, Chemnitz, Germany, was acquired by Karl Mayer Textilmaschinenfabrik GmbH, Obertshausen, Germany, in 1992. Subsequently, Karl Mayer has redesigned these stitch bonding systems and further commercial developments have been introduced.

8.2 The Maliwatt and Malivlies stitch bonding systems

8.2.1 Maliwatt

The Maliwatt system consists of the following components:

1. Stitch bonding unit with drive system for the working elements.
2. Web feed system.
3. Yarn feeding and monitoring systems.
4. Take-up of stitch-bonded fabric and fabric storage or plaiting.
5. Cutting and tearing unit, and a machine control and drive system.

The main elements of a Maliwatt stitching head are illustrated in Fig. 8.1 and a close-up view of the stitching action is given in Fig. 8.2. The horizontal compound needle and closing wire system, which operates in conjunction with the knock-over sinker and the supporting rail, penetrates through the substrate which is normally a cross-laid web. The sheet of stitching yarn, which is inserted via the guides into the open hooks of the compound needles, forms stitches that penetrate the web. Pillar stitch and tricot stitch are possible on the basic version with one guide bar (cam shogging). On two-guide-bar machines, by shogging the first or second guide bar via pattern discs, all the basic two-guide-bar structures can be produced up to a repeat length of 16 courses. By adjusting the compound needle and closing wire system, it is possible to incorporate fibres from the web into the stitches at the same time, thereby preventing the stitches from running from the end knitted last.

With tricot lapping, a parallel warp yarn system can be placed into the web and subsequently incorporated into the stitch-bonded fabric. Retaining pins, together with the supporting rail, prevent the web from moving during penetration by the needles.

The knock-over sinkers, which are arranged on the opposite side, allow the stitches to be knocked over while the web is being held back. The distance between the retaining pins and

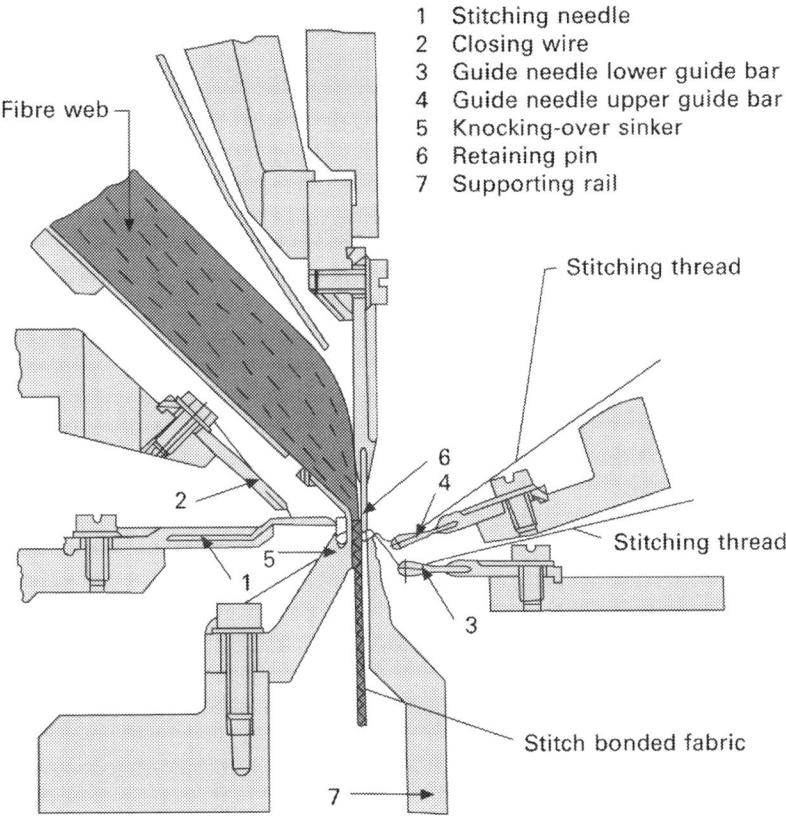

1	Stitching needle
2	Closing wire
3	Guide needle lower guide bar
4	Guide needle upper guide bar
5	Knocking-over sinker
6	Retaining pin
7	Supporting rail

FIG. 8.1   Main elements of a Maliwatt stitch bonding machine.

the knock-over sinker can be adapted to suit the thickness of the web, depending on the lift of the compound needles. A web can be fed either discontinuously in roll form, or continuously to the stitching head. In the latter case, the web-forming system consisting of a card and cross-lapper is linked directly to the stitching head. It is also possible to reinforce the fabric using spunlaid or other types of nonwoven fabric.

8.2.1.1 The Intor system

The Intor system was developed and licensed by the SVUT Textile Research Institute, Liberec, Czech Republic and integrates the web-forming and stitch bonding system to achieve a considerable reduction in space and investment costs. The fibres are supplied either as slivers in up to 32 cans or in the form of two sliver laps, each 265 mm wide with a maximum diameter of 600 mm. The fibre is fed to an airlaid machine operating with a cylinder speed of between 2400 and 4200 revs min^{-1} to produce a web. The web is transferred to a horizontal web-laying unit that feeds the stitch bonding machine at a width of 500 mm and at a maximum speed of 45 m min^{-1}. The width of the web can be varied between 1400 and 2650 mm. An additional web-nipping mechanism in front of the stitch bonding point aids transport of the web right up to the knitting point of the machine. Two versions of the Intor

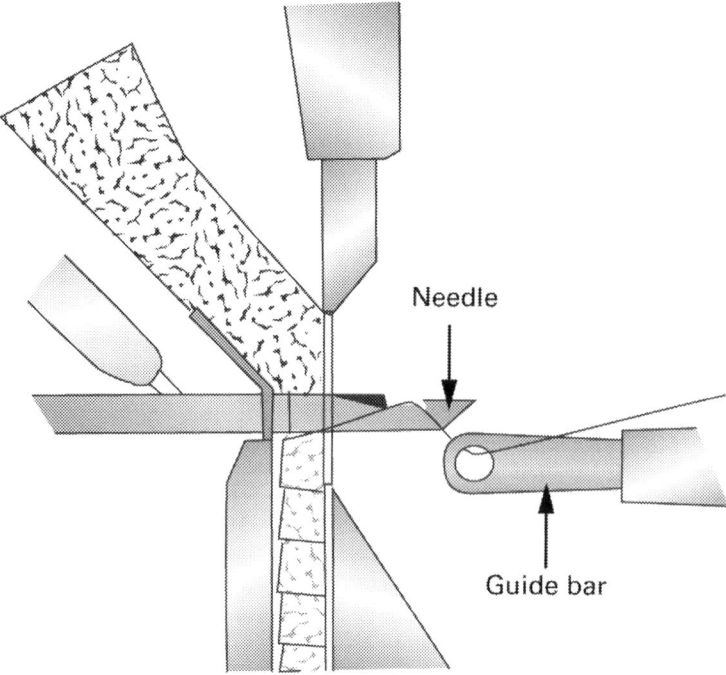

FIG. 8.2   Stitching action of Maliwatt stitch bonding machine.

system are available depending on fibre length, the Intor PN is suitable for short to medium staple length fibres and Intor L is designed for long staple fibres, such as jute and other bast or leaf fibres.

8.2.1.2 Fabric structure and applications

Polyester filament is mainly used as the stitching yarn, but other polymers are also utilised depending on the end product. The fabric structure is influenced by the use of one or two guide bars, and by changing the stitch set-up of the guide bars. The web component in the fabric varies between 80% and 95%, area densities range from 15 to 3000 g m^{-2} and the fabric thickness from between 0.5 and 20 mm. Fig. 8.3 shows the structure of a single-guide-bar Maliwatt fabric. The main applications of Maliwatt fabrics are soft furnishings, upholstery

FIG. 8.3   Structure of a single-guide-bar Maliwatt fabric.

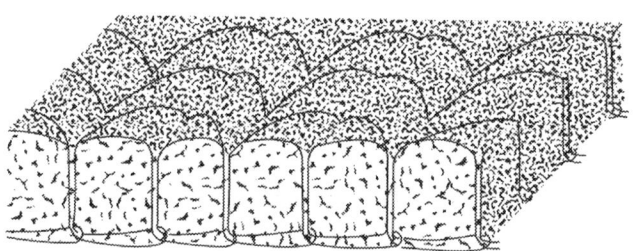

fabrics for mattresses and camping chairs, blankets, transportation cloth, cleaning cloths, secondary carpet backing, lining fabrics, interlining for shoes and apparel, adhesive tapes (e.g. those used for harnessing electric cables in automobiles), velcro-type fasteners, fabrics for hygiene and sanitary purposes, laminating and subupholstery fabrics, insulating materials, coating substrate, geotextiles, filter fabrics, composites, and flame-retardant fabrics.

8.2.2 Malivlies

The main working elements of the Malivlies system and their relative positions are shown in Fig. 8.4. A view of the stitching action is illustrated in Fig. 8.5 and the structure of a typical Malivlies fabric (through the courses) is shown in Fig. 8.6. Laying-in sinkers prevent the web from moving during penetration. As the compound needles move back to their knock-over position, fibres lying across the front of the web are hooked by the open hooks of the needles, held within the needle hook by the closing wires and pulled through the web thickness. As these fibres are pulled through the stitches formed by fibres on the previous course, which are still hanging on the needle stems, the newly formed stitches are pulled through the existing stitches, which are cast-off the closed hooks of the needles. A loop structure that resembles the technical face of a warp-knitted fabric is produced on the side facing the knock-over sinkers. The laying-in sinker, which is drawn back opposite the

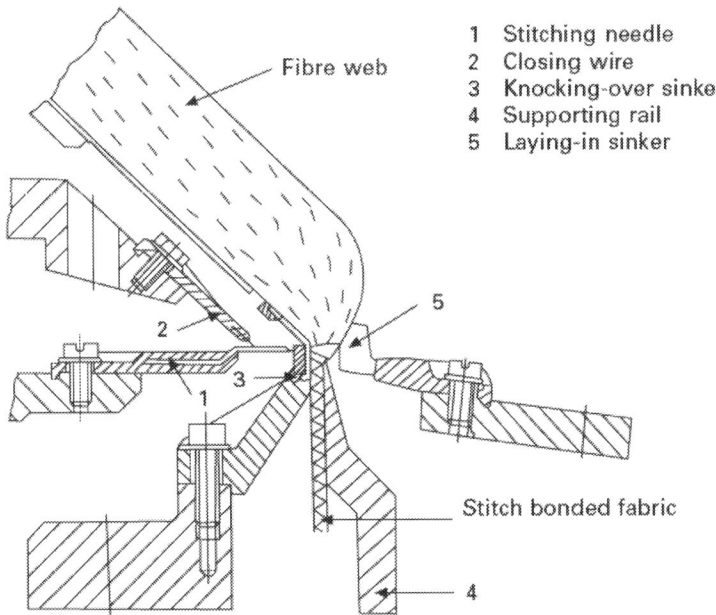

FIG. 8.4   Main elements of a Malivlies stitch bonding machine.

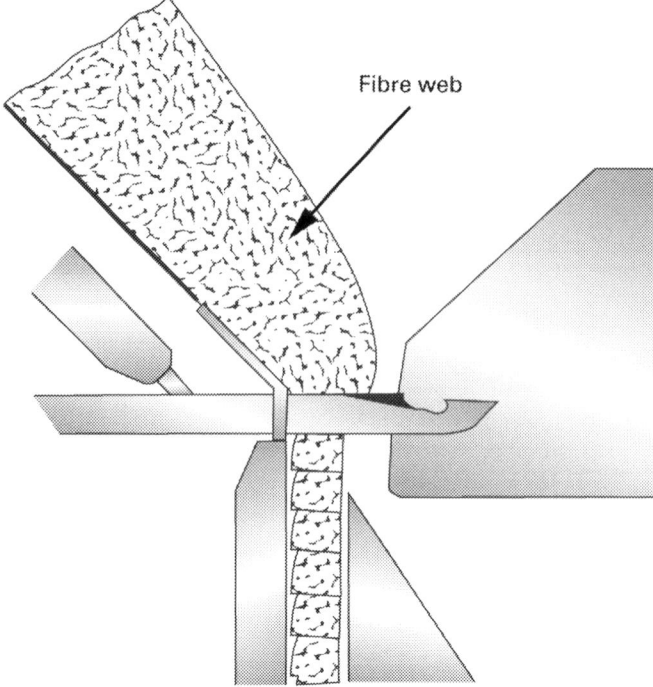

Fibre web

FIG. 8.5   Stitching action of Malivlies stitch bonding machine.

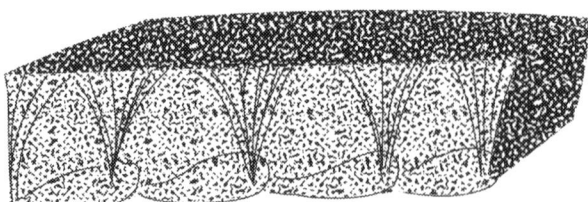

FIG. 8.6   Structure of Malivlies fabric.

supporting rail, enables the fibres to be grasped firmly by the compound needles due to a build-up of the web.

The technical specifications of Malivlies machines are identical to those of the Maliwatt. Malivlies fabrics composed entirely of fibres and no filaments are mechanically recyclable. The main fibre types used are polyester, polypropylene, viscose, and reclaimed fibres and the fabrics range from 120 to $1200 \, \mathrm{g \, m^{-2}}$. The main applications are internal lining of cars (headliners, rear, and side linings), wall coverings and furnishing felts, packaging, insulating, absorbing and polishing cloths, filter fabrics and geotextiles, coating substrates, laminating and bonding, medical, hygiene and sanitary products, and secondary carpet backing.

8.3 The Malimo stitch bonding system

Malimo stitch bonding machines comprise the following components:

1. Stitch bonding head with gearboxes to drive the working elements (stitch bonding elements):
   * Weft-yarn layer formation and feeding device.
   * Warp-yarn let-off motion.
   * Stitching-yarn let-off motion (both with yarn monitoring devices).
   * Fabric take-down motion.
   * Machine control and drive system.
2. Warp-beam(s) let-off frame/package creel for stitching and warp threads.
3. Package creel for weft threads.
4. Fabric storage/winding or cutting mechanism.
5. Foot pedals at the front and back of the stitch bonding head.

Fig. 8.7 shows the relative positions of the different elements of the Malimo stitch bonding unit. The compound needles pierce the yarn layers (warp and weft yarns), webs, backing fabrics, films, paper or any other sheet material that may be inserted. The guides place the stitching yarns into the open hooks of the compound needles. The previously formed stitches slide up the compound needle shanks and closing wires. The needles start to withdraw to their knock-over position, the hooks of the compound needles with the new stitching yarn overlaps are now closed by the closing wires so that the old stitches can slide off over the top of the needles. The old stitches are knocked over and the new loops are drawn through them to complete the new course. Guides also shog to place the stitching yarns in the correct position for the next machine cycle, which is a new course. This shogging movement is called the underlap movement. Based on the Malimo platform, different versions and auxiliary

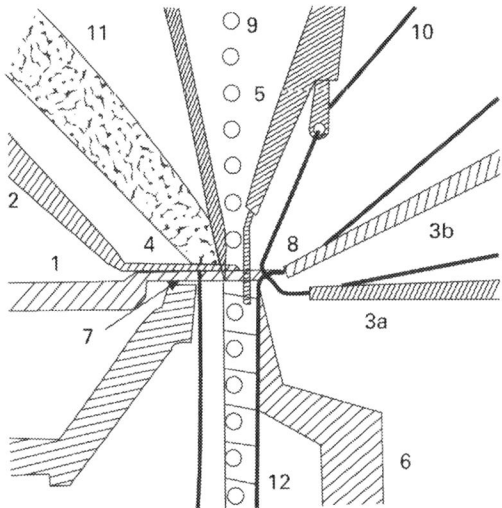

1 = Compound needle
2 = Closing wire
3a = Guide, 1st guide bar
3b = Guide, 2nd guide bar
4 = Knocking-over sinker
5 = Retaining pin
6 = Backing rail
7 = Old loop
8 = New overlaps
9 = Weft threads
10 = Warp threads
11 = Fibre web
12 = Malimo fabric

FIG. 8.7 Main elements of a Malimo stitch bonding machine.

devices have been developed to enable complex fabric structures to be produced, particularly for the production of technical textiles.

These developments enable [4]:

- Cross weft insertion (a system is shown in Fig. 8.8).
- Noncontinuous parallel weft insertion and continuous parallel weft insertion.
- Multiaxial constructions.
- Glass fabric manufacture.

Karl Mayer also developed a Malimo system with an electronically controlled warp yarn racking device. The warp yarns are fed right up to the stitch bonding point via guide tubes, which are attached to individual yarn guide blocks. These blocks of different widths are easily moved as they are mounted on a guide rail parallel to the stitch bonding point and can be connected up to a toothed belt driven by a servomotor. However, they can also be fixed or connected up via spacers or springs. The servomotor is controlled by a single-axle positioning module and allows the blocks to move over the whole working width of the machine. The repeat length depends on the storage capacity of the computer and is currently 27,000 courses. As there are two racking lines for the warp yarn feed, it is possible to cross over the warp yarns during racking. As two stepper motors are used and the yarn guide blocks can be

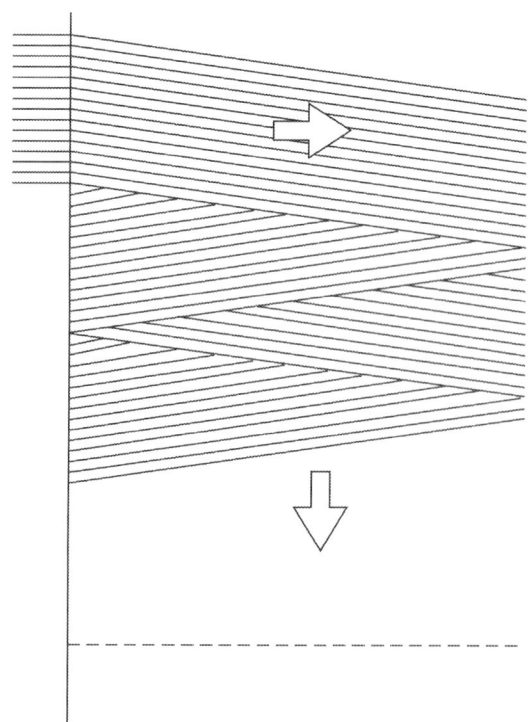

FIG. 8.8   Cross weft insertion on Malimo machine.

connected to both sides of the toothed belt loop, a total of four different basic movements, two of which are symmetrical, are possible [5].

8.3.1 Cross weft insertion

Weft yarns can be fed from a stationary package creel located at the side of the Malimo stitch bonding system via yarn guide elements to the centre of the machine and into the weft yarn-laying device. The end of the weft yarn is held by a clamp, which is stationary in the running direction. When the weft yarn-laying device has reached the other side of the machine, the weft yarn is again clamped. By lowering the stationary clamps on both sides simultaneously, the yarns are transferred to a yarn transport device and cut between the stationary clamp and the transport device on the side where the weft is deposited. On the weft insertion side, the clamps are blow-cleaned. With each movement of the weft carriage, 18 weft yarns are drawn off and inserted into the clamps. The Malimo P1 has a maximum working width of 1800 mm and the gauge ranges from 7 to 18 needles/25 mm. One or two guide bars are used, and a warp yarn system and a backing substrate are introduced if required.

8.3.2 Parallel weft insertion

In a typical example of this system, 16 weft yarns are transported 0.5 in. apart and simultaneously inserted into the hooks of the moving hook chain and a stationary auxiliary device. After the weft yarn is inserted and secured on one side, the weft carriage travels to the other side. Once the weft yarns reach the opposite insertion position, the pulled-out weft yarn loop is transferred to the transport chain by moving the control hook bar, which is activated by a double-stroke magnet, to the first side thus giving a parallel weft yarn layer. The resulting yarn loops are cut off and removed by a suction device. A warp yarn or backing fabric feed mechanism may be fitted and parallel weft insertion system are available.

8.3.3 Manufacture of glass composite preforms

A system for the production of preforms for composites was developed by Karl Mayer. Typical end-uses are:

- Structural sections and panels for use in the construction industry and machine-building.
- Boat building, surfboards, skis, the vehicle industry.
- Pipes, poles, and container making.
- Mouldings (press moulding).
- Coating, pipe repairs.
- Sports hall floors.
- Rotors for wind-driven power generators.
- Components for aviation and the aerospace industry.
- Composites.

A chopping device located behind the stitch bonding machine enables continuous glass rovings to be cut into defined lengths and deposited in a random arrangement onto a continuous conveyor belt situated underneath. The cut length of the glass fibres is normally 25, 50, or 100 mm and the resulting web of chopped glass is continuously fed to the stitch bonding point. A weft-laying device can also place glass filament yarns onto the web before the stitching process and warp yarns can also be inserted. The products may be combined with an additional nonwoven web or other substrate resulting in a highly complex integrated fabric structure. The nominal width of these machines is 1600 or 2400 mm, with machine gauges ranging from 7 to 18 needles/25 mm. In some cases, it is possible to feed the warp yarn at a coarser gauge [6]. Fig. 8.9 illustrates the structure of a biaxial glass fibre fabric and Fig. 8.10 gives a plan view of the technical back and a cross-section through the weft (courses) of a biaxial plus chopped glass fibre web fabric. Both structures are popular substrates for PVC coating and as reinforced mats for composite products. It is common to use high tenacity polyester filament yarn as stitching yarn in such materials.

8.3.4 Multiaxial weft insertion

Multiaxial stitch bonding systems enable a diverse variety of yarn types and linear densities to be processed in many fabric constructional variations. Warp glass filaments up to 2400

FIG. 8.9   Structure of biaxial fabric.

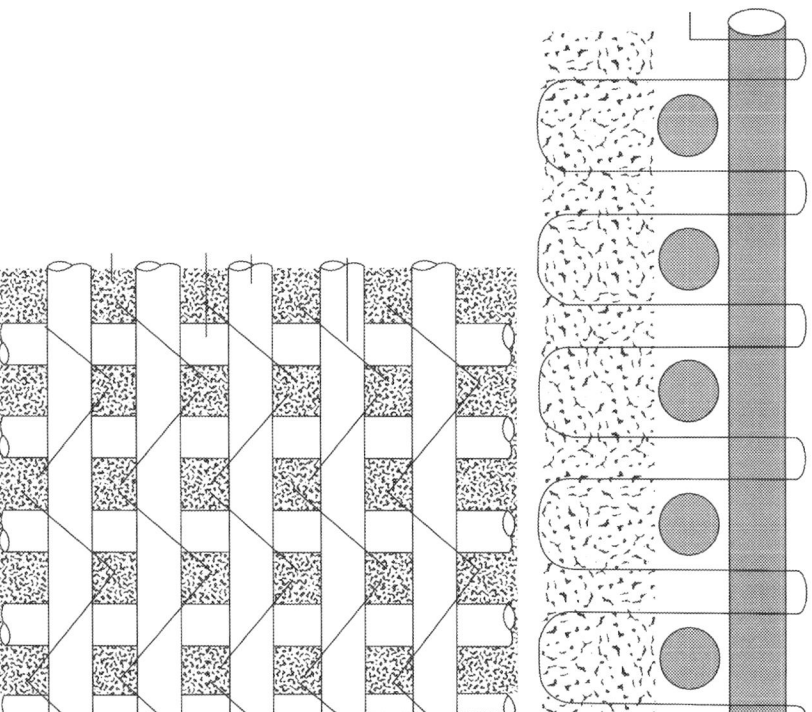

FIG. 8.10 Structure of biaxial fabric with glass mat: plan view (technical back) and cross-section through courses.

tex, weft glass filaments up to 1200 tex, and diagonal glass filaments up to 800 tex may be combined in fabrics with an area density from 300 to 3000 g m^{-2}.

Composite materials with a high degree of structural complexity are produced. In multiaxial stitch-bonded structures, it is important to optimise the number, angle, and linear densities of each individual yarn layer in the composite as well as the machine gauge, the specification of the stitching yarn, and the geometry of the fabric. The above parameters are of course dictated by the specific application of the composite. Fabric structure can be based on [7]: one warp yarn layer (0°); two weft yarn layers of ∼90°; two weft yarn layers of ∼ −45°; two weft yarn layers of ∼90°; two weft yarn layers of ∼+45°; and one stitching yarn layer or one chopped glass web layer. Diagonal layers orientated at 45° can therefore be introduced. In the Malimo weft yarn-laying technique the yarn-laying angle varies from 1° to 5° from the angles stated above, which is claimed to enable:

- Improved load-bearing yarn distribution.
- Improved wettability and resin impregnation in the finished product due to the lower packing density.
- Improved drape.
- Reduced potential for delamination in the composite material.

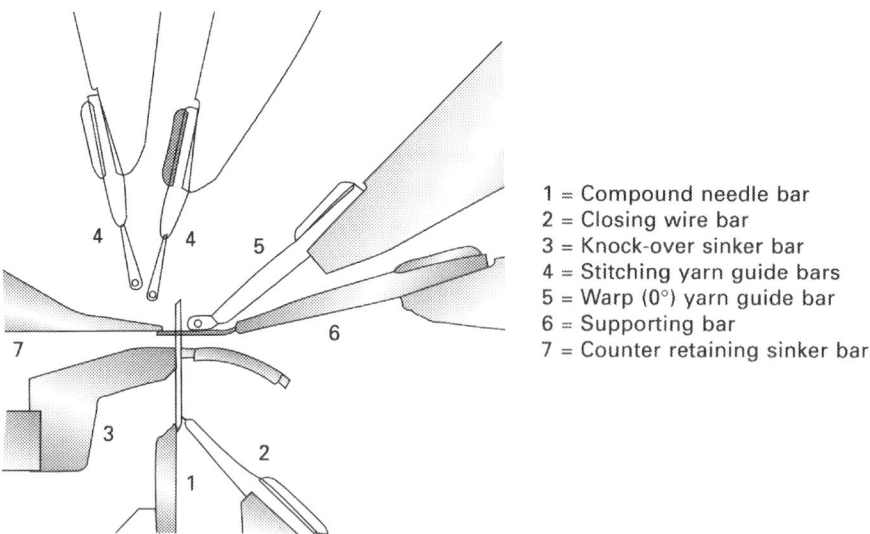

1 = Compound needle bar
2 = Closing wire bar
3 = Knock-over sinker bar
4 = Stitching yarn guide bars
5 = Warp (0°) yarn guide bar
6 = Supporting bar
7 = Counter retaining sinker bar

FIG. 8.11   Main elements of Malimo multiaxial stitch bonding machine.

The maximum working width is normally 1600 mm, which can be reduced if required, and machine gauges range from 3.5 to 18 needles/25 mm. The stitch bonding point of a Malimo multiaxial machine is illustrated in Fig. 8.11.

This consists of a pointed head compound needle bar and closing wire bar, knock-over sinker bar, supporting bar, counter retaining sinker bar, stitching yarn guide bars, and filler (warp 0°) thread guide bars. Multiaxial multiply fabrics are used to reinforce different matrices. The combination of multidirectional fibre layers and matrices has proved capable of absorbing and distributing extraordinarily high strain forces in use. The thread angles may be varied from 30° to 60° for diagonal yarns (+45° and −45° yarns). The major attributes of multiaxial reinforcement for both flexible and rigid composites are dimensional stability in any direction (high shear strength in the bias or diagonal directions), isotropic mechanical properties, reduced delamination tendency, noncrimped and parallel yarn sheets, low specific area density, adjustable stiffness between extremely stiff and high extensibility, noncorrosive, nonmagnetic, resistant to chemicals, high mechanical load resistance, and, above all, high resistance to crack propagation.

Major applications for multiaxial structures are for inflatable structures such as airships and boats, flexible roofing membranes, rotor blades for wind power stations, moulded parts for automotive application, aircraft and ship building, and equipment for sports and leisure activities such as skis, snowboards, surfboards, and boats. Most materials are compatible with the multiaxial system, including fibreglass, aramid, carbon, high-tenacity polyester, polyamide, polyethylene, and polypropylene. Thermosetting and thermoplastic matrix materials are used and even pressure setting matrix materials such as concrete and cement have been successfully utilised. Fig. 8.12 shows stacking of the different yarn layers, including a nonwoven web or preformed substrate. All five or more layers are stitched together with the stitching yarn, which is commonly high-tenacity polyester filament. The structure of a multiaxial fabric is shown in Fig. 8.13.

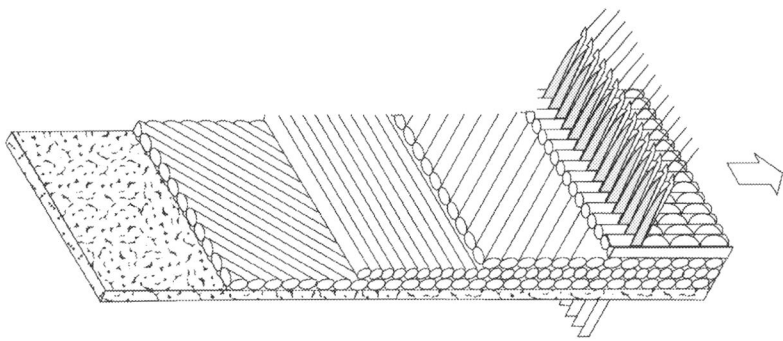

FIG. 8.12 Different layers stacked and stitched together.

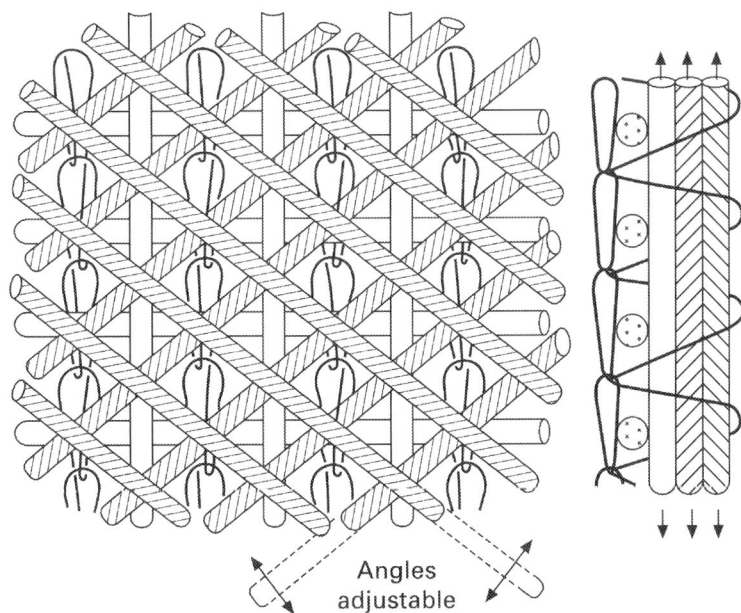

Angles
adjustable

FIG. 8.13 Structure of multiaxial stitch-bonded fabric.

8.3.5 The Schusspol technique

This modification of the Malimo system produces fabrics for floor coverings, upholstery, and furnishings. The machine uses two guide bars, stitching yarn, pile yarn, and a weft yarn. The pile, usually between 5 and 11 mm in height, appears only on the face of the fabric and is bound firmly into the ground fabric. This is achieved by lapping the pile yarn and the stitching yarn system in a special manner. A pile sinker is also utilised to create the pile on one face of the fabric.

8.3.6 The Malifol technique

Films, rather than yarns, are used for the warp and weft threads, while conventional yarns are used for the stitching threads. The warp film in open width is fed into the stitching head directly from the warp beam unwinding frame, while the weft films are slit on the film web-forming mechanism and, as on the Malimo machine, they are suspended in the hook needles of the weft carrier chains by the weft yarn-laying device. These stitch-bonded fabrics have a low area density with good fabric cover and are dimensionally stable. Their polymer composition is normally chosen to enable UV, rot and insect resistance to be achieved since the main areas of application are in coating substrates, packaging, insulation, secondary carpet backing, filtration media, and geotextiles.

8.4 Malipol

Malipol stitch bonding systems have the following main elements:

- Pile yarn feed.
- Ground fabric.
- Stitch bonding head.
- Fabric take-down and batching.

The main stitching elements are shown in Fig. 8.14. The compound needles penetrate the ground fabric, and the stitching or pile yarn is overlapped in the needle hook. The pile yarn is also laid on top of the pile sinker at the same time so that a tricot movement, i.e., 1–0/1–2, is used to create the pile and knit the yarn into the ground structure. The machine is available in gauges of 10, 12, and 14 (needles/25mm), pile sinker heights from 1 to 11mm and stitch lengths of 1 to 3mm can be achieved by using the change gears. The machine speed usually

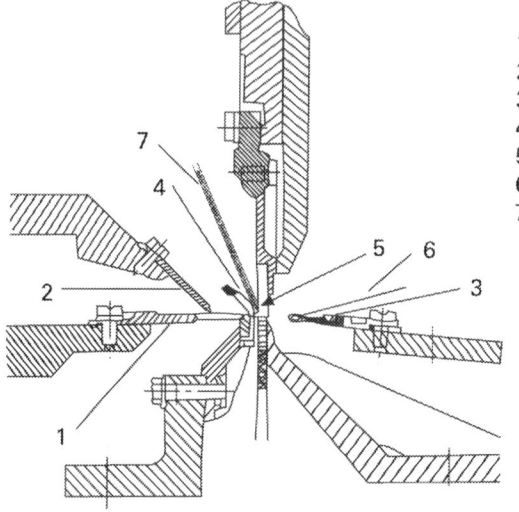

1 = Compound needle bar
2 = Closing wire bar
3 = Guide bar for pile yarn
4 = Knock-over sinker bar
5 = Pile sinker bar
6 = Pile yarn
7 = Ground fabric

FIG. 8.14   Main elements of Malipol stitch bonding machine.

ranges from 900 to 1300 courses/min. The pile yarn is fed either from a creel of the single end or magazine type, or from a warp beam arrangement.

The choice of feed system is governed by the quality of the product and the product versatility required. The optimum pile yarn linear density is influenced by the machine gauge and ranges from 140 tex for a 10-gauge to 50 tex for a 14-gauge machine. Any substrate through which the compound needles can penetrate may be used as the ground fabric, as long as it will remain intact. Twill and satin woven fabrics are the most suitable, although loose plain weave fabrics are compatible. Alternatives include stitch-bonded fabrics, foams, knitted fabrics, and films. Woven cotton or viscose fabrics of $100\text{–}200\,\mathrm{g\,m^{-2}}$ are the most common for blankets and upholstery fabrics, and polyamide or polyester filament fabrics of $50\text{–}100\,\mathrm{g\,m^{-2}}$ are favoured for lining plush, soft-toy plush, and imitation fur. The major end-uses for Malipol fabrics include blankets (raised on one or both sides), beachwear and leisurewear, babywear, bathroom sets, bath robes and gowns, upholstery fabrics, imitation furs, soft-toy plush, and floor coverings.

8.5 Voltex

Voltex fabrics are high pile or high plush fabrics based on two principal preformed elements, a ground fabric and a web, which are continuously introduced. No stitching yarn or yarn preparation, such as winding, or warping are required. A cross-section through the main stitching parts of a Voltex system is shown in Fig. 8.15. A continuous Voltex system consists of a web-forming line coupled to the stitch bonding unit. Typical working widths are 1700 and 2500 mm and machine gauge ranges from 7, 10, 12, 14 (needles per 25 mm). Pile sinker heights vary from 1 to 23 mm (depending upon the lift of the compound needles) and stitch lengths from 0.55 to 5.0 mm. Machine speed depends on the stitch length, pile sinker height, and the line speed of the web formation unit and is adjustable from 500 to 1500 $\mathrm{r\,min^{-1}}$. Voltex fabrics have found applications in lining fabrics, imitation furs, soft-toy plush, shoe uppers and shoe lining, floor coverings, and upholstery fabrics.

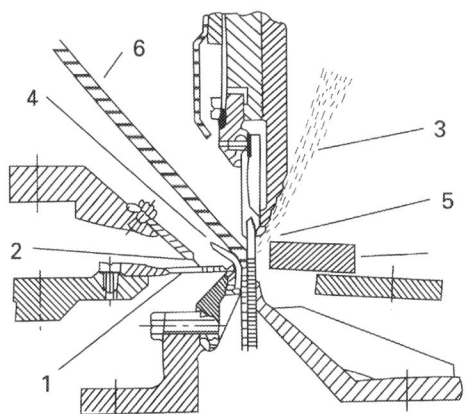

1 = Compound needle bar
2 = Closing wire bar
3 = Fibre web
4 = Knock-over sinker bar
5 = Pile sinker bar
6 = Ground fabric

FIG. 8.15   Main elements of Voltex stitch bonding machine.

8.6 Kunit

Kunit and Multiknit are two important developments of the Malimo system launched in 1991 and 1993, respectively. Both were developed with the aim of producing fabrics directly from fibres without the need for yarn. In the Kunit process, fibres are fed to the stitch bonding head in the form of either a thin web or a batt. Voluminous, three-dimensional pile fabrics can be produced that have a distinctive folded pile; the pile can have a variable thickness and density if required. A compound needle having a round head is used. It uses a brushing bar in conjunction with the stitch-forming elements, whose oscillating path may be varied between 6 and 51 mm by exchanging the cams. The setting governs the height of the pile fold.

The elements of a Kunit machine and their relative positions are shown in Fig. 8.16. The flat, oscillating brush compacts the lightweight web, whose fibres are mainly orientated in the machine direction, so that the fibres are pressed into the needle hooks and formed into stitches. The fibres that are not knitted are arranged as cross-orientated pile folds. Parallel-laid webs are particularly suitable composed of fibres of 40–120 mm and 1.7–3.3 dtex. The brush oscillation setting (6–51 mm brush stroke) is influenced by the fibre length in the incoming web. For short fibres (<60 mm) an 8 mm maximum oscillating cam stroke is recommended while for long fibres (>60 mm) a 34 mm maximum oscillating cam stroke is used. The web area density is normally $20–80 \, g \, m^{-2}$ and stitch lengths between 0.55 and 5.0 mm.

Machine speed ranges between 500 and 1200 $r \, min^{-1}$. The delivery speed of the web-forming unit and the oscillation of the brushing device affect the machine speed. Kunit systems have been produced in gauges from 3.5 to 22 (needles per 25 mm) and working widths of 1700–2800 mm. Final fabric area density ranges from 90 to $700 \, g \, m^{-2}$. Fig. 8.17 illustrates the

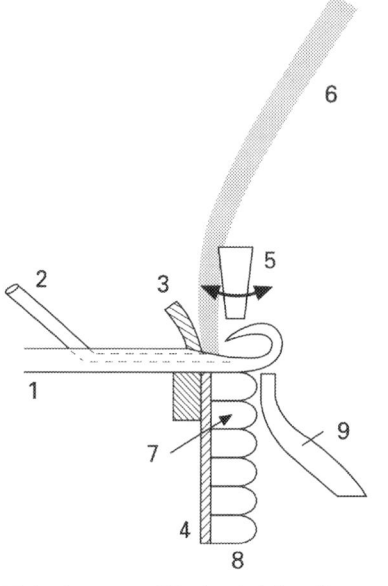

1 = Stitching needle bar
2 = Closing wire bar
3 = Knocking-over sinker bar
4 = Backing rail
5 = Oscillating brushing device
6 = Fibrous web
7 = Fibre stitches
8 = Kunit fabric
9 = Support rail

FIG. 8.16   Main elements of Kunit stitch bonding machine.

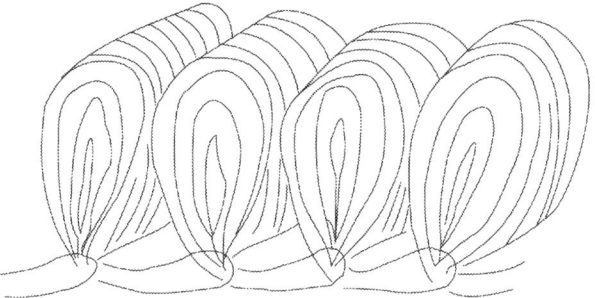

FIG. 8.17   Structure of Kunit fabric.

loop structure of a Kunit fabric. Kunit fabrics have been used in linings, soft-toy fabrics, filtration media (particularly depth filters), covering materials for polishing discs, coating substrates, and as upholstery materials for car interiors. Kunit fabrics are sometimes finished and the processes used depend on the intended end-use and the fibre composition. No finishing is required for many applications or when the fabrics are to undergo further processing on the Multiknit machine. Finishing processes applicable to Kunit materials are back-coating, raising, polishing and shearing, tumbling to achieve surface effects, coating, and laminating with other substrates.

8.7 Multiknit stitch bonding systems

In the most basic form of Multiknit, both sides of the fabric are formed into a dense knitted construction by intermeshing pile fibres on the surface of the incoming Kunit fabric. The fibres in the pile folds are stitch bonded to produce a double-sided, three-dimensional, nonwoven fabric with an inner pile structure connecting the two faces. The pile surfaces in two separate Kunit fabrics can be joined in this way to make one integrated multilayer structure. Other structures, such as fabrics, webs, or even fibres and powders, can be incorporated within the base web and covered by stitch bonding the multilayer material to produce a composite material. A Multiknit line typically includes a card that continuously supplies a Kunit machine that stitch bonds the web on one side only and in the second stage, the stitch-bonded fabric is fed continuously to the Multiknit machine where it is stitch bonded on the other side. The stitch bonding heads of Kunit and Multiknit machines are similar, but in the case of the Multiknit system, a pointed head needle is employed, and different adjustable components are used for stitch formation.

A retaining or sinker bar is also used when further processing a Kunit fabric. The stitch bonding parts of a Multiknit machine are shown in Fig. 8.18, which, if required, can also be combined with a warp yarn feed system, fabric feed system, or a scattering device, enabling a wide range of multilayer structures to be produced with a stitch-bonded structure on both faces of the fabric. Fabric area densities currently range from around 120–800 g m^{-2} for single-layer Multiknit nonwovens and from about 150–1500 g m^{-2} for multilayer constructions.

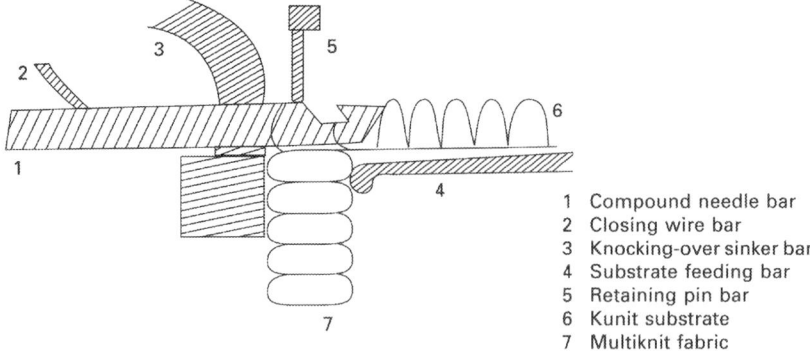

1   Compound needle bar
2   Closing wire bar
3   Knocking-over sinker bar
4   Substrate feeding bar
5   Retaining pin bar
6   Kunit substrate
7   Multiknit fabric

FIG. 8.18   Main elements of Multiknit stitch bonding machine.

On upgraded Kunit and Multiknit models, maximum working widths of 1.7, 2.9, and 4.15 m are supplied, and the width can be reduced steplessly. A maximum machine speed, up to 1800 r min^{-1}, is attainable depending on the type of fibre being processed and the required thickness of the fabric. The final area density of the single-layer fabric is between 120 and 800 g m^{-2} and the fabric thickness, varies between 2 and 11 mm. These fabrics are also finished, if required, by processes including heat-setting, chemical and thermal bonding, coating, and lamination. The structure of a single-layer Multiknit fabric is illustrated in Fig. 8.19.

Multiknit fabrics have excellent compressibility, low area density and low bulk density, excellent heat, noise and vibration insulation, excellent mouldability with a smooth and uniform surface on both sides, and are weldable if the composition is predominantly thermoplastic. Multiknit fabrics have been used in upholstered furniture parts, seating for automobiles and other furniture as a PU foam replacement material. Other applications include filter fabrics, insulation materials, garment interlinings, and nondelaminating mouldable composite materials.

Over 90 million cars were produced worldwide in 2019 with each car using between ca. 10 and 12 m^2 of seating fabric. Traditionally, the seating fabric is a laminate consisting of a

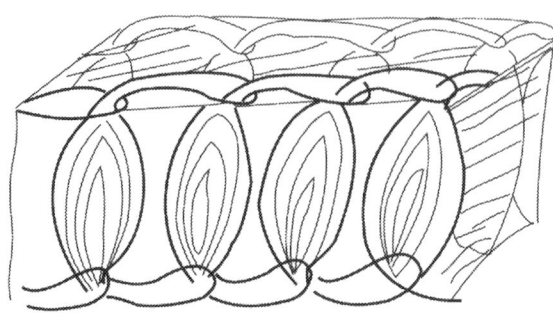

FIG. 8.19   Structure of Multiknit fabric.

face fabric composed of either air textured or false twisted polyester yarn, flame laminated to PU foam, and a warp knitted scrim, in such a manner that the foam is sandwiched between the polyester face fabric and polyester warp knitted scrim. The PU foam has excellent elastic compression recovery and imparts softness and bulk to the seating fabric, which in turn makes the seat comfortable. The foam, however, is negatively associated with emissions from additives, including flame-retardant chemicals and full life-cycle problems relating to after-use disposal and recycling. Other problems are that during the flame lamination process, singed substances can be responsible for fogging. Foam seating materials also exhibit poor air permeability, moisture and water vapour permeability, inhibiting moisture and temperature regulation. Caliweb is the registered trademark of products manufactured by the Kalitherm-Technique. Caliweb is used for mechanically and thermally bonded nonwovens as well as for lamination of nonwoven composites with fabrics such as those used for car seating. In this system, Kunit or Multiknit fabrics that have been thermally bonded are utilised to produce a foam substitute.

Calibration is preferably carried out using flat-bed laminating systems. The nonwoven foam substitute requires a proportion of low melt or bicomponent fibres (sheath-core type) to effect stabilisation in thermal bonding. If produced from 100% polyester, these materials can be completely recyclable and have high air permeability, moisture and water vapour permeability, and good ageing behaviour. They have low emissions and do not cause fogging. This is only one example of the successful utilisation of three-dimensional nonwoven fabrics made by the Kunit and Multiknit techniques. These lightweight, low-density, high-volume, 100% nonwoven fabrics with structured and smooth surfaces either on one face (Kunit) or on both faces (Multiknit) are successfully used as heat, noise, vibration, and sound insulation materials, as dust filters, as adhesive tapes, as shoe linings, as antidecubitus mattings and as medical, hygiene, and sanitary materials.

8.8 Developments in stitch bonding

8.8.1 Maliwatt and Malivlies machines and products

Since 1993, Maliwatt and Malivlies stitch bonding machines have been progressively redesigned and upgraded to set new standards in quality and performance. Systems are up to 6150 mm working width, which can be decreased as required. The yarn let-off motion for the two guide bars is via an electronically controlled let-off system EBA, which is standard in many types of high-speed warp knitting machines. Both guide bars can also be controlled by pattern discs and can have stitch set-up or structure for stitching repeating up to 16 courses. The speed of Maliwatt and Malivlies machines can be varied between 1500 and 2200 r min^{-1}, depending upon a number of factors associated with product design and specifications. The machine gauge can vary from 3.5 to 22 needles per 25 mm. The increased performance and higher fabric quality are achieved by:

- A low vibration machine frame.
- Precise gauge accuracy of the stitch bonding elements by automatic bar heating facility, which features narrow temperature tolerances for heating all the bars with additional monitoring function.

- Electronic beam control EBA for yarn let-off mechanism.
- Pattern disc control of stitching yarn guide bars up to 16 courses repeat.
- Slider crank mechanism for compound needle and closing wire bars (running in oil bath).
- Reduced weight, high-strength components for all the moving elements.
- Up to 40% increase in productivity with up to 520 linear mh^{-1} for Maliwatt and 590 linear mh^{-1} for Malivlies extra wide 6150 mm machines.

Maliwatt G and Maliwatt C are used for specific technical textile products. Maliwatt G is a special machine for processing chopped glass mats. Randomly laid glass fibres either 50 or 100 mm long in a mat form are stitched with polyester filaments or glass yarns for textile reinforced composite materials. Maliwatt C is suitable for applications where several substrates or materials such as webs, yarns, fabrics, films, textile waste materials, powdered, or granular materials arranged in layers one on top of the other need to be bonded mechanically. Ten years ago, the majority of Maliwatt and Malivlies fabrics were marketed for household goods, such as bed ticking, curtains and curtain lining, bedspreads, wall coverings, garden furniture, transportation and packing blankets, etc. Currently, they are finding applications in technical textiles, such as adhesive tapes, roof lining, rear and side linings in cars, coating substrates, filter fabrics, geotextiles, and healthcare and hygiene products.

8.8.2 Biaxial stitch bonding

Karl Mayer has redesigned and improved the performance of Malimo biaxial stitch bonding machines. The new biaxial M/NM stitch bonding machines bear a closer resemblance to their RS2(3)MSUS Raschel machines. As with the RS2(3)MSUS Raschel units, Biaxial M/NM stitch bonding machines are intended to produce textile-reinforced structures for both flexible and rigid or load-bearing composite materials. High-performance yarns, such as fibre glass, carbon, aramid, high-tenacity, and high-modulus thermoplastic polymer filaments, may be processed to produce biaxial all yarn or biaxial yarn plus any nonwoven web, or a fabric, or any other substrate to produce the correct composite fabric for either saturation coating, lamination, or for combining with a suitable matrix to produce a complex composite material. A number of different versions have been designed:

- Biaxial system with parallel weft insertion that is in line with the stitch courses with or without the glass chopper mat attachment (Biaxial M or M/Ch).
- Biaxial system with parallel weft insertion that is not in line with the stitch courses with or without the glass chopper mat attachment (Biaxial NM or NM/Ch).

- Biaxial system with a cross weft (from 1° to 5°) insertion that is not in line with the stitch courses with or without the glass chopper mat attachment.

All versions have the possibility of feeding different base materials. The use of the established parallel weft insertion system in line with the stitch courses, or MSUS principle, is a distinctive feature of these machines. This system is already used in some warp knitting machines and is capable of handling a wide variety of yarn types and linear densities. The other major distinction between biaxial M and biaxial NM models is that in the former case, the stitching elements are positioned in a similar fashion to all other types of stitch bonding machine. In the case of model NM machines, the stitch bonding elements are positioned in a similar fashion to Raschel machines, with the compound needle bar and closing wire bar being positioned in a vertical direction. The other important features that influence the machine efficiency and fabric quality are:

- Electronically controlled warp beam let-off motion EBA or as an alternative braked warp beams with yarn feeding facility.
- Single-end monitoring of knitting yarns electronically enhanced by either drop wires or by laser stop.
- Use of the patented 'complementary weft insertion device' featuring almost constant weft yarn draw-off speeds (i.e. without any acceleration peaks).
- Chromium-plated yarn guiding and feeding rollers.

8.8.3 Multiaxial stitch bonding

Liba Maschinenfabrik GmbH, Germany, developed a commercial multiaxial warp knitting machine based on parallel weft insertion not in line with the courses known as the Copcentra Multiaxial. Karl Mayer's first multiaxial warp knitting machine with diagonal weft insertion in line with the courses was called the RS2-DS 'Carousel machine' and its first multiaxial machine model 14,016 based on the principle of crosswise weft insertion not in line with the courses followed. The latest Karl Mayer multiaxial stitch bonding machines have the following technical specifications:

1. Working widths: max. 1525, 2550, and 3300 mm; min. 1025, 2000, and 2600 mm. Working widths can be adjusted in steps of 25 mm.
2. Machine gauges: 3.5–14 (needles per 25 mm).
3. Yarn feeding devices for stitching yarns: electronically controlled EBA system.
4. Yarn stop motion: Protechna Laserstop (monitors single ends).
5. Speed: up to 1400 r min^{-1}.
6. Production: up to 4.4 m min^{-1}.
7. Number of guide bars: 1 or 2 for stitching yarn and 1 for warp yarn.
8. Number of weft insertion devices: up to 4, depending upon requirements.
9. Additional substrates possible: possibility of feeding in chopped glass strands above or below the weft layers. Other type of substrates can also be used.
10. Bobbin creel.

The industries of aerospace, aviation and space travel, shipbuilding, and high-performance automotive have always been important sectors but in energy generation, wind turbine sites have become established as a growing high value market and the number of multiaxial machines operating worldwide has increased [8]. Carbon fibre reinforced plastics (CRP) incorporating carbon tow multiaxial structures as reinforcement have made it possible to reduce the weight of composite materials considerably, they have a long life, are heat and fire resistant and are resistant to corrosion and chemicals. Multiaxial structures can absorb and distribute exceptionally large forces because of the ability to orientate yarn layers in different predetermined directions ($0°/90°/+45°/-45°$). A system of stitching yarns fixes these layers in position. This parallel orientation in which the yarns are free of crimp, permits optimum utilisation of the yarn strength in every stress direction, which is an advantage compared to woven fabrics. These preform structures have improved interlaminar shear strength, increased impact resistance and strength, dimensional stability in all directions, uniform elongation behaviour, and enable rapid wetting of the resin in composite manufacture. Additionally, the risk of delamination is minimised by using a system of stitching yarns in the Z-direction.

8.9 Introduction to needlepunching

The process of needlepunching, also known as needle felting, was originally developed to produce mechanically bonded nonwoven fabrics from fibres that could not be felted like wool. Fig. 8.20 illustrates the basic principle of a simple needleloom. The fibres are mechanically entangled to produce a fabric by reciprocating barbed needles through a moving web or batt of fibres in a needleloom.

Fig. 8.21 illustrates the action of the barbed needle, which is fundamental to the needling process. The barbed needles are fitted in a needleboard, which oscillates vertically between two fixed plates containing the moving batt, each plate being drilled with corresponding holes through which the needles move. A feed system introduces the batt between the lower bed plate and the upper stripper plate by nip rollers or aprons, while a nip roller system draws the consolidated structure away from the needling zone. As the batt moves through the loom, some fibre segments are progressively reoriented and tensioned, as a result of their interaction with the needle barbs and a coherent fabric structure is formed.

Needling therefore aims to increase the frictional resistance between fibres to generate greater network strength. This is why needling is not suited to the bonding of very short fibres

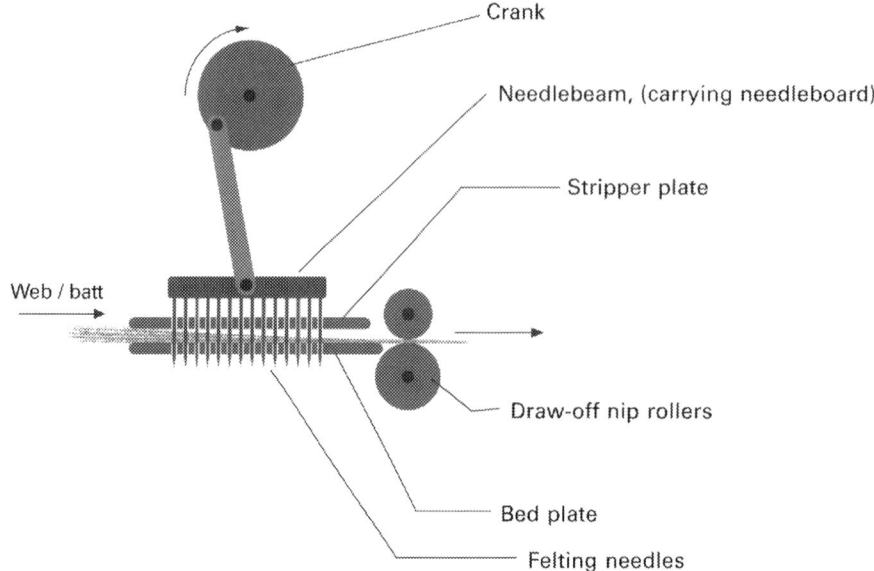

FIG. 8.20   Operation of a simple needlepunching machine.

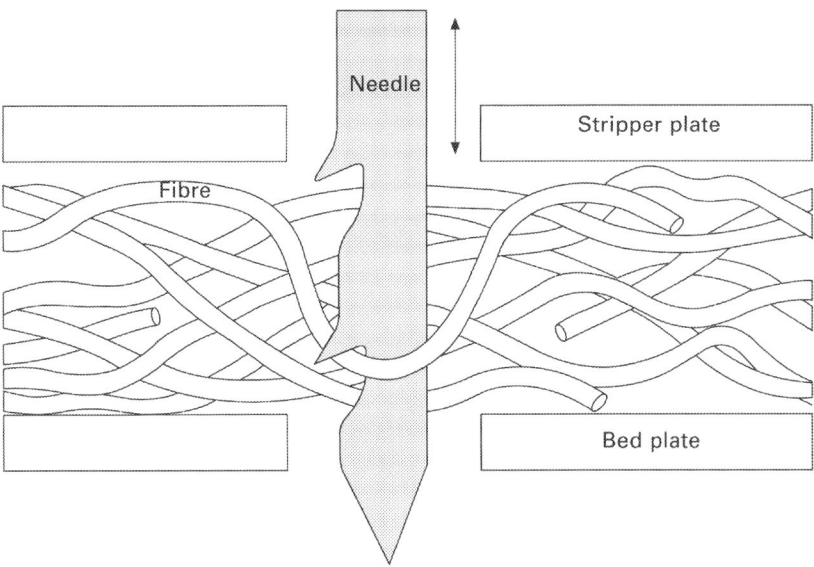

FIG. 8.21 Action of a barbed needle.

or wood pulp. Fibres are tensioned along their length during transport in the barbs, and this can lead to fibre breakage if the elongation at break is low. This is particularly important when levels of fibre entanglement and frictional resistance between fibres in the batt are already very high. Also, if the fibres are brittle, breakage is highly likely as the penetration depth and the punching rate of the needle increases.

For any significant bonding to occur, fibres need to be capable of being collected and transported by the moving barbs. Collection and transport require bending of fibres around the very small radii of the barbs, which is challenging if a fibre is very stiff (high bending modulus) or brittle and liable to breakage. Additionally, the number of fibres that can be collected by each barb or the 'carrying capacity' depends on fibre diameter. Another important factor is fibre friction (usually influenced by the selection of fibre finish), not just in terms of its contribution to frictional resistance and therefore the degree of bonding, but also in terms of the needling forces generated and the rate of needle wear during the process.

Originally, needlepunched fabrics were made from fibres such as jute, coir, hair, waste fibre, and shredded rags to produce carpet underlay, mattress padding, insulation, and rough blankets, the manufacture being relatively crude and dusty. While many of these products are still made using needlepunching, and especially since the availability of synthetic fibres, the process has evolved into a clean, high-speed manufacturing method of nonwoven production.

8.9.1 Batt formation

Prior to needling, several different methods of batt forming are of relevance, the commonest being one or more cards feeding to a cross-lapper (cross-folder) to form a cross-lapped batt of the required area density and width. High-speed profiling cross-lappers lay the carded web with a minimum of distortion to give the required laydown angle and therefore the fibre orientation in the cross-direction of the batt. Parallel-laid batts can also be formed by superimposing webs from several cards and width expanders or spreaders give a degree of cross-orientation, if required. More isotropic batts without a pronounced layered cross-section are produced by long-fibre air-laying, but needlepunching is not compatible with lightweight air-laid webs containing very short fibres. Garnett machines are still used in some sectors to make batts where coarse fibres or textile waste is mechanically recycled, usually in tandem with a cross-lapper. Some heavyweight spunlaid fabrics composed of continuous filaments are also needlepunched for geosynthetics and other durable product applications.

8.9.2 Drafting

Drafting to reduce the fabric area density and to modify the MD/CD ratio can take place (i) on the batt before needling, (ii) during needling or, (iii) after preneedling between individual needlelooms. The drafting process essentially involves tensioning the batt or preneedled fabric, inducing controlled fibre slippage. Heavy webs from a slow running cross-lapper may have their linear speed increased and their weight per unit area reduced prior to needling using a batt drafting machine employing a series of drafting zones between three or more roller nips which successively elongate the structure. Some draft inevitably occurs during the first needling operation as the batt is drawn through the bonding system. Such draft is uncontrolled and must be minimised by considering the advance per loom cycle and the needle penetration.

Drafting after preneedling can provide better control of fibre reorientation and provides a means of adjusting the MD/CD ratio for strength in cross-laid structures. This is a common strategy in, for example, the manufacture of geosynthetics, where a relatively low MD/CD ratio may be required. While some production lines take advantage of the needle loom as a drafting unit, great care is needed to avoid the introduction of short-term irregularities. In drafting after preneedling, the fabric is wrapped successively around the upper and lower peripheries of a series of rollers to increase the tension and create elongation. The progressively increasing surface speed and high friction coverings of the drafting rollers are designed to provide a controlled and adjustable draft. The calculated draft between the entry and exit nip rollers must be adjusted to take account of roller slip and the elastic recovery of the fabric when the tension is reduced.

8.9.3 Basics of needleloom operation

The needleloom consists of a heavy, substantial frame carrying the fixed bed plate and stripper plate between which the batt passes and the vertically reciprocating needleboard carrying the needles, which generates significant vibrational forces at high speed, which the frame must absorb. The needleboard is driven with simple harmonic motion, the method of suspension varying according to the machinery maker. The batt, which is normally of very low density, must be supported by aprons or rollers as it is carried into the gap between the bed plate and the stripper plate. Two batt compression aprons, which converge towards the entry point, are commonly used to positively feed the batt into the gap between the stripper and bed plates. This helps to prevent differential slippage of fibres in the exterior and interior portions of the batt. The gap between the stripper and bed plates is also adjusted to control the batt compression during needling and may be wedge-set.

Clearly, batt compression is particularly important in preneedling where the consolidation in terms of thickness is largest. After needling, the fabric is transported away from the needling zone by take-up rollers, the movement of which may be intermittent or continuous depending on the design and age of the machine. The object of an intermittent take-up, which was the original method, is to have the batt stationary, while the needles are penetrating the web to minimise needle breakage and to avoid marking of the fabric. This is desirable for weak batts at low speed, but for synthetic fibres and high needling speeds, continuous take-up systems predominate.

8.10 Needle design and selection

The design of the felting needle, its thickness (gauge), length, cross-sectional shape, and the number, projection, spacing, and dimensions of the barbs have an important effect on the needlepunching process and the properties of the final fabric. Batt weight per unit area, fibre type, and fibre dimensions are essential considerations when selecting needles for various applications. Generally, coarser needles with larger barbs are used for coarse fibres and vice versa. Fibre breakage needs to be minimised and fibre lubrication can be helpful to reduce friction in the process. Synthetic polymer fibres, particularly PET and PP, exhibit good strength and abrasion resistance as well as consistent fibre properties enabling high needling speeds and fabric production rates to be achieved.

Fig. 8.22 shows a typical felting needle with a view of the working blade and barbed apices. The shank locates in the hole in the needleboard and the crank is clamped between the upper surface of the needleboard and the needlebeam, thus holding it firmly and vertically aligned in the loom. Traditionally, the cross-section of the felting needle is triangular and carries a total of nine barbs, three per apex. The dimensions of the barb, their relative spacing and proximity to the point are varied depending on the application and machine operation.

8.10.1 Needle reduction

An outline of two types of needle used by the industry is shown in Fig. 8.23. The single reduction needle has two sections, the shank and the blade. The shank normally has a diameter of 15-wire gauge (1.83 mm) although others are available. The function of this part is to hold the needle in the needleboard. The blade is the working part of the needle. It is triangular in cross-section and can be made in a variety of diameters, from 43 gauge (the finest) up to 17 gauge (the coarsest), depending upon the fibre type and fibre dimensions to be processed. The barbs are formed or pressed into the apices of the triangular blade. The double reduction needle also has a middle or intermediate section, which is a transition stage between the different diameters of the shank and the blade. It is round in cross-section and is usually 18 gauge (1.21 mm) in diameter. Other diameters of 17 or 16 gauge are sometimes used. The single reduction needle is much stiffer than the double reduction type and is usually made only for coarse gauge needles. It is used for punching stiff fibres including some ceramic materials,

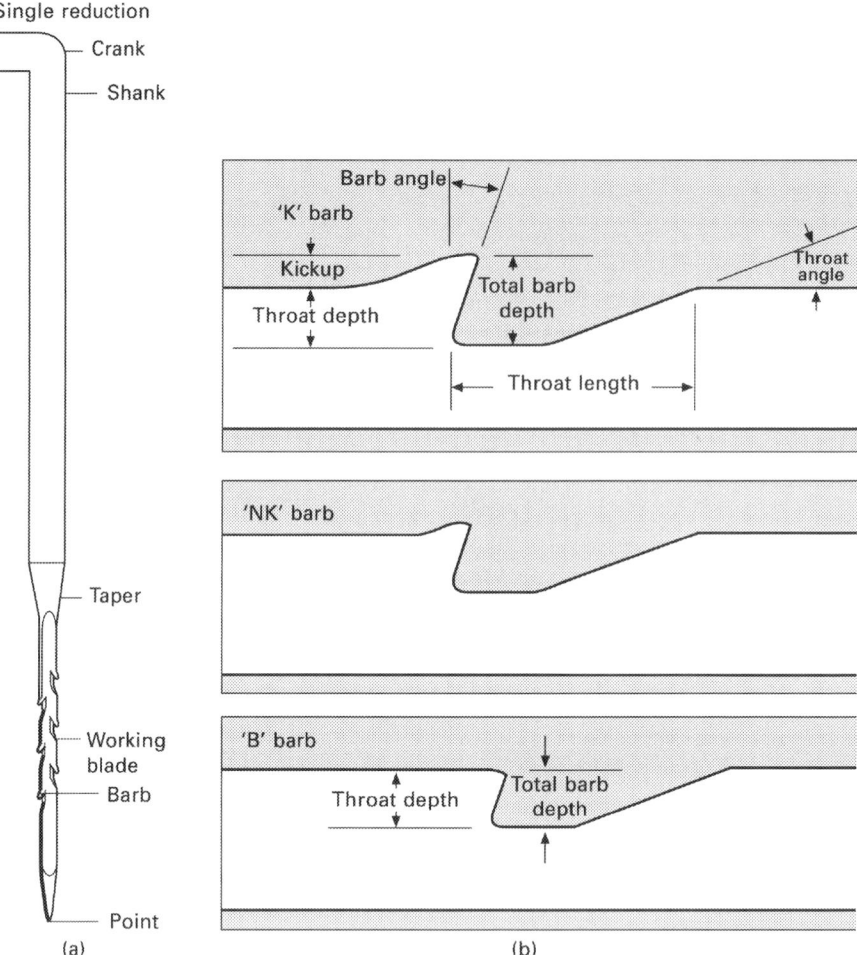

FIG. 8.22   Typical felting needle (A) and close-up of different barb designs (B).

waste fibre blends, and shoddy where the needle penetration forces are high. Needlepunching of spunlaid webs can be associated with high needle forces.

8.10.2 Barb spacing

Barb spacing must be considered in association with needle penetration depth. Common barb spacings together with the corresponding needle cross-section are given in Fig. 8.24; other spacings are made for specialist applications. Regular barb (RB) spacing is perhaps the most widely used. These needles have nine barbs evenly spaced on a blade of about 30 mm in length. It has been found to balance the need for effective fibre entanglement while minimising fibre breakage and damage. This type of barb spacing is commonly used for preneedling and light tacking applications. In the majority of cases, the penetration depth

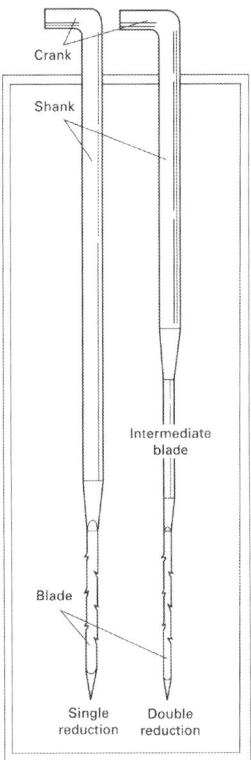

Crank

Shank

Intermediate
blade

Blade

Single
reduction

Double
reduction

FIG. 8.23 Single and double reduction needles.

with this needle is in the range 8–17 mm. The lower penetration depths are normally associated with finish needling, where a uniform surface appearance is required.

The upper three barbs on the needle therefore contribute very little to fibre entanglement. To increase entanglement, needles with a shorter blade carrying only six barbs may be employed. The shorter blade is also stiffer and less prone to breakage. The close barb (CB) spacing is frequently described as being more 'aggressive' in its action because it transports a larger number of fibres per needle stroke at low penetration depths. At a fixed punch density, fibres are intensively entangled but frequently at the expense of an uneven fabric surface. The close barb spacing is intended to enable low penetration depths during final needling, or where a high level of entanglement is required at high loom output speeds as, for example, in the production of needled spunbond fabrics. The dwell time of the needle in the fabric is minimised allowing high line speeds.

8.10.3 Barb dimensions and shape

The traditional barb shape is known as the conventional barb or cut barb. It is the simplest type and the easiest to form on the blade. It is made by the action of a chisel-like tool applied to the apices of the blade. The action of the chisel tool on the needle blade raises sharp edges on

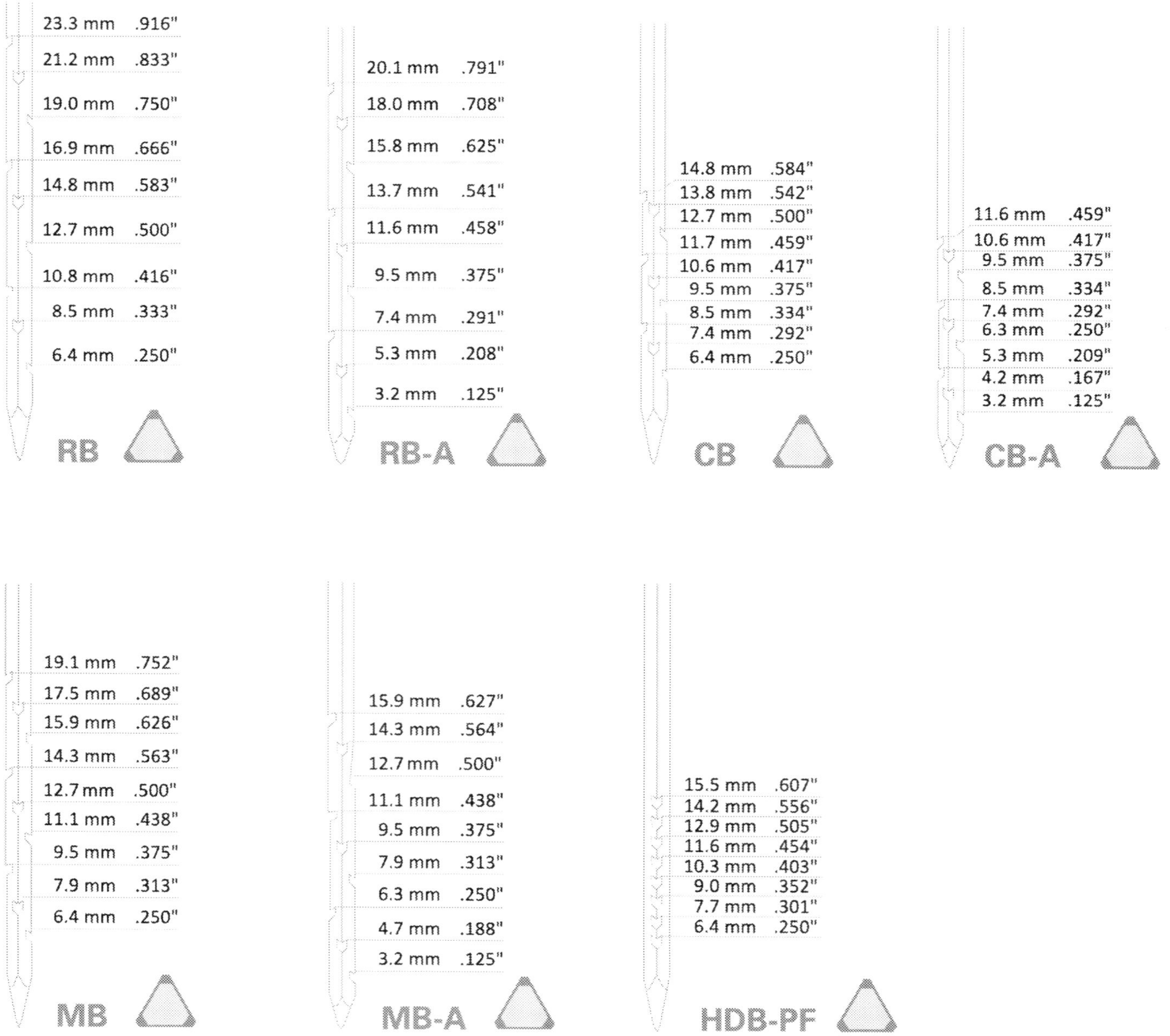

RB

23.3 mm	.916"
21.2 mm	.833"
19.0 mm	.750"
16.9 mm	.666"
14.8 mm	.583"
12.7 mm	.500"
10.8 mm	.416"
8.5 mm	.333"
6.4 mm	.250"

RB-A

20.1 mm	.791"
18.0 mm	.708"
15.8 mm	.625"
13.7 mm	.541"
11.6 mm	.458"
9.5 mm	.375"
7.4 mm	.291"
5.3 mm	.208"
3.2 mm	.125"

CB

14.8 mm	.584"
13.8 mm	.542"
12.7 mm	.500"
11.7 mm	.459"
10.6 mm	.417"
9.5 mm	.375"
8.5 mm	.334"
7.4 mm	.292"
6.4 mm	.250"

CB-A

11.6 mm	.459"
10.6 mm	.417"
9.5 mm	.375"
8.5 mm	.334"
7.4 mm	.292"
6.3 mm	.250"
5.3 mm	.209"
4.2 mm	.167"
3.2 mm	.125"

MB

19.1 mm	.752"
17.5 mm	.689"
15.9 mm	.626"
14.3 mm	.563"
12.7 mm	.500"
11.1 mm	.438"
9.5 mm	.375"
7.9 mm	.313"
6.4 mm	.250"

MB-A

15.9 mm	.627"
14.3 mm	.564"
12.7 mm	.500"
11.1 mm	.438"
9.5 mm	.375"
7.9 mm	.313"
6.3 mm	.250"
4.7 mm	.188"
3.2 mm	.125"

HDB-PF

15.5 mm	.607"
14.2 mm	.556"
12.9 mm	.505"
11.6 mm	.454"
10.3 mm	.403"
9.0 mm	.352"
7.7 mm	.301"
6.4 mm	.250"

FIG. 8.24 Examples of barb spacing specifications.

the barb. These edges lead to fibre breakage during needlepunching and can also damage the reinforcing scrim if one is incorporated into the fabric. The kick-up and overall barb depth are varied in order to vary the fibre carrying capacity of the barb. While the conventional barb is still used and is widely available, more efficient barb shapes have been developed, which enable the fibres to more effectively engage and transport fibres while minimising fibre damage. These are commonly known as formed barbs or die-pressed barbs and their shapes are more rounded than the conventional barb. The barb side walls are smoother, minimising fibre damage. Moreover, since these barb shapes include a wrap-round along the side of the barb, they engage and hold fibres more efficiently. Longer barb life, increased fabric strength, and smoother fabric surfaces are the claimed benefits of these types of barb compared to the conventional barb.

The terms used to define barb shape and dimensions and the commonly encountered types of kick-up (K=high, NK=low, and B=no kick-up) are given in Fig. 8.25. Formed barbs with high kick-up and no kick-up barbs are illustrated in Fig. 8.26. Open barbed needles where the barb is let into the point find application in the needling of ceramic fibres. The barb depth relative to the fibre diameter is particularly important because it affects the capacity of the needle to collect and transfer fibres into the vertical plane. The theoretical number of fibres that may be collected in the barbs of a needle can be calculated:

$$n_f = \frac{2b_d}{d_f} \times n_b \tag{8.1}$$

where, b_d is the barb depth, d_f is the fibre diameter, n_b is the number of acting barbs on the needle, and n_f is the number of fibres collected by the barbs.

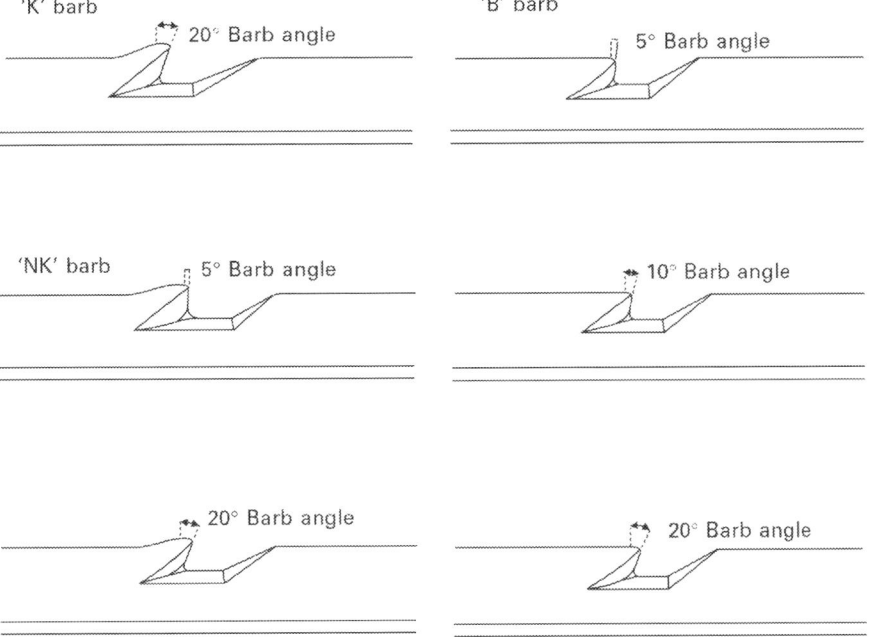

FIG. 8.25   Advanced barb shapes. *Courtesy: Foster Needle.*

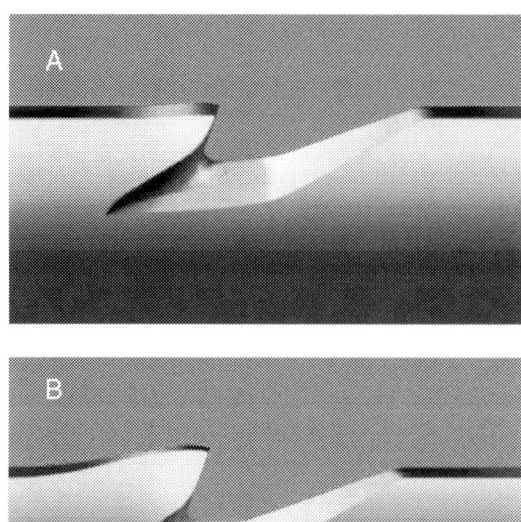

FIG. 8.26 Low kick-up (A), and high kick-up (B), formed barbed needle showing wrap-around section.

Clearly, the depth of penetration, number of barbs per needle, and the fibre to metal friction, among other factors will also influence the actual number of fibres that a needle is able to collect during its vertical travel. It should also be noted that the lead barb on a needle normally carries the largest number of fibres and successive barbs carry progressively fewer. A rough guide for the needle gauge, and therefore the barb dimensions that are used for polyester, polypropylene, nylon, and acrylic fibres is shown in Table 8.1.

It is useful to understand how the barb influences key fabric properties and dimensions. To increase fabric thickness, one strategy is to select smaller and fewer barbs per needle to decrease the barb angle and to increase the barb spacing by selecting the regular barb spacing. To

TABLE 8.1 Needle selection for different fibre fineness.

Fibre linear density (denier)	Needle blade gauge (SWG)
0.5–1.5	42
1.5–6	38–40
6–10	38
10–18	36–34
18–30	36–32
30+	30–coarser

decrease the fabric thickness, the opposite approach may be adopted, which includes the selection of a closer or high-density barb spacing. Permeability is important in geosynthetic fabrics, filtration media, and papermakers' felts among others and may be increased by selecting a coarser gauge blade and using larger barbs with perhaps a higher kick-up. To maximise the surface smoothness of the fabric, finer gauge needles are normally selected usually with the regular barb spacing and a zero kick-up barb. Triangular blades and blades having barbs on only one or two apices are also thought to be beneficial in promoting a good surface finish.

8.10.4 Types of needle and needle selection

A wide variety of needle types and design configurations are used in the needlepunching industry and there are no well-defined rules about which type of needle should be used in particular applications. Needle selection depends upon the desired fabric characteristics, and fibre linear density is a major deciding factor as well as fibre type and needleloom type. Fig. 8.27 shows a variety of commonly selected needle blade cross-section types.

8.10.4.1 Triangular blade needles

The blade cross-section of classical needles is triangular, with one of more barbs formed on each apex. Typically, on standard needles, there are three barbs per apex.

8.10.4.2 Cross STAR needles

The blade of the Cross STAR needle of Groz-Beckert has four apices and has a cruciform cross-section. The needle is therefore intended to carry an increased number of fibres per penetration as compared to the standard triangular needle, which can increase fabric strength. The additional fibre carrying capacity is also intended to increase the production rate by enabling higher line speed. The cross-sectional shape stiffens the needle, which is intended to

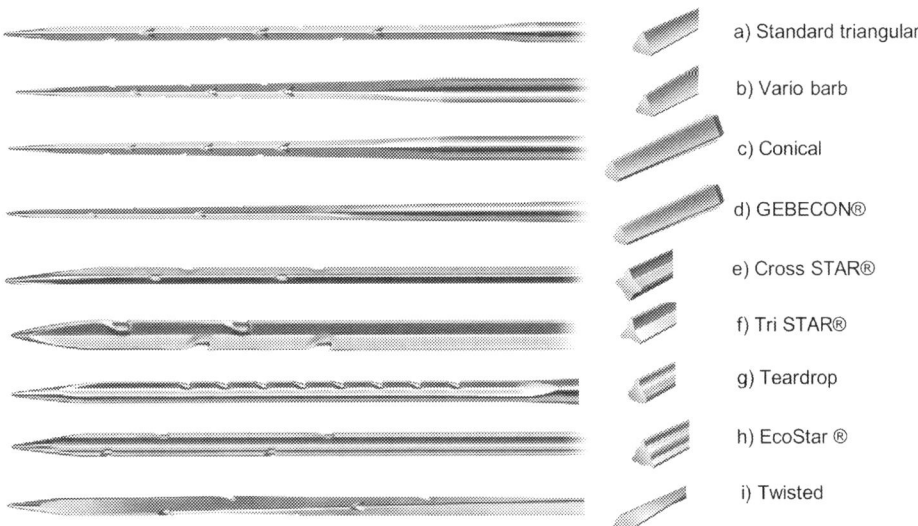

a) Standard triangular

b) Vario barb

c) Conical

d) GEBECON®

e) Cross STAR®

f) Tri STAR®

g) Teardrop

h) EcoStar ®

i) Twisted

FIG. 8.27 Examples of needle types. *Courtesy: Groz-Beckert.*

reduce needle breakage. The star blade needle is used when fabric strength has to be maximised in short, high-speed lines or where a smooth surface is required in the finished fabric at low penetration depths in finish needling.

8.10.4.3 Fork needles

Fork needles are used for structuring and patterning preneedled fabrics. They have no barbs but rather a forked opening at the end of the needle that is capable of transporting large numbers of fibre as the needle penetrates. The throat of the needle can be varied to adjust carrying capacity. When the fork is orientated in line with the linear direction of the fabric, a velour surface structure is produced. When the fork opening is orientated perpendicular to the linear direction of the fabric, a ribbed or rib-cord structure is produced (Fig. 8.28). Coarser gauge fork needles are used only in conjunction with lamella bed plates.

The rib frequency in the finished fabric is a function of the spatial arrangement of the forked needles across the width of the loom. The spacing can be uniformly periodic or more complex depending on patterning requirements. The height of the ribs is a function of the needle penetration and is limited by the fibre length in the batt and fibre mechanical properties, particularly fibre elongation. There is significant interfibre friction during this process,

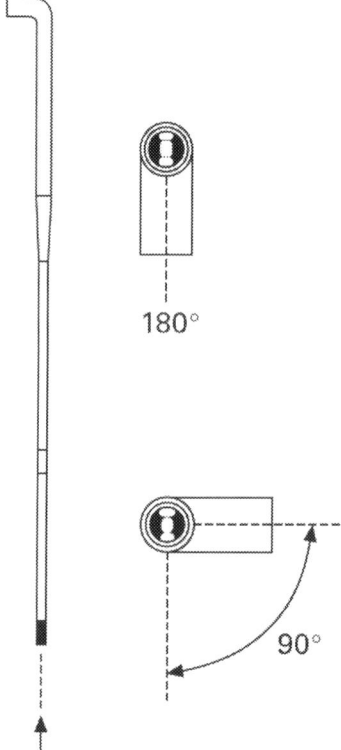

FIG. 8.28   Fork needle orientation.

which generates heat and necessitates fibre lubrication to minimise the possibility of thermal damage to thermoplastic fibres. Fine gauge fork needles of 38–42 gauge are also utilised in random velour needlelooms, which have a moving brush bed rather than the normal bed plate.

8.10.4.4 *Crown needles*

Crown needles are designed to be used in conjunction with random velour 'structuring' needlelooms to introduce fibre loops that protrude from the surface of the fabric. A crown needle has only three barbs or openings, one on each apex (Fig. 8.29). These barbs are equally spaced from the point, normally at a distance of 3.2 mm. Like forked needles, the barbs on crown needles are intended to carry large numbers of fibres to the reverse side of the fabric for the purpose of structuring the surface of a preneedled fabric.

8.11 Penetration depth and other factors affecting needle use

The vertical distance through which the needles penetrate the batt during needlepunching has a direct influence on fabric properties. The needle penetration depth is defined as the distance between the upper surface of the bed plate and the tip of the needle when the needles are located at bottom dead centre. Since the amplitude of the reciprocating needles is constant, the penetration depth is normally adjusted by raising or lowering the bed plate. On all needlelooms, there is some means of mechanically achieving this, the settings being indicated by a measuring index on the loom frame, or by an electronic sensor. The penetration depth is important because it determines the number of barbs entering the batt on each stroke, and hence the level of fibre entanglement and bonding that can be achieved. The barb spacing

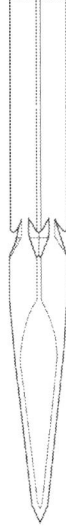

FIG. 8.29   Crown needle.

of the needle is therefore an important consideration when changing the penetration depth. The penetration depth also influences the linear speed or advance per stroke of the needleloom. If the penetration depth is large, the advance per stroke on modern continuous take-off machines must be small to avoid the possibility of needle breakage and drafting of the fabric. To produce a strong fabric while minimising draft and therefore needle penetration depth, needles with regular barb spacing may be selected with a shorter distance from tip to first barb, for example, 3.2 mm instead of 6.4 mm.

8.11.1 Punch density

The punch density defines the number of needle penetrations per unit area (punches/cm^2) and directly affects fabric properties and dimensions. The effects on fabric thickness, density, and mechanical properties are particularly important. Punch density is a function of the fabric throughput speed, the stroke frequency (punches/min) of the loom, and the number of needles per unit width of the needle board. The punch density may be calculated as follows:

$$P_d = \frac{n_n}{A} \qquad (8.2)$$

where, P_d is the punch density (punches/cm^2), A is the advance per stroke (or the output per loom cycle) (cm), n_n is the number of needles per cm width of the needleboard and,

$$A = \frac{P}{S_f} \qquad (8.3)$$

where, P=fabric production speed (cm/min) and S_f=punch (stroke) frequency (punches/min). Therefore, when the needle board density is constant and for a given stroke frequency, the punch density is determined by adjusting the fabric throughput speed (continuous) or indexing distance (intermittent). To obtain high punch density in a needled fabric without compromising linear processing speed, usually requires two or more needlelooms. Depending upon the scale of production, these sequential needling steps may take place as separate operations, or more usually they form part of a continuous production line, sometimes with a fabric drafting unit fitted between consecutive needlelooms to minimise the anisotropy of the resultant fabric, particularly in respect to tensile properties.

8.11.2 Barb wear

One of the principal quality control issues in needlepunching concerns needle wear, specifically barb wear. As needles engage fibres during needling, fibre to metal frictional forces are generated, which leads to progressive barb wear. The rate of barb wear can be high in heavily entangled fabrics, and when needlepunching fibres such as silica, glass, stainless steel, carbon, and aramids, frequent needle changes are required to avoid batch to batch variations in fabric properties. Ultimately, barb wear affects the shape and surface features of the barb and changes the needle's capacity to hold and release fibres over time. Consequently, barb wear affects the bonding and generation of strength in the fabric, as well as its appearance, directly impacting the quality of the needlepunched product. It is therefore necessary to understand the rate of wear and to take proper action to prevent the gross deterioration of fabric quality.

The rate of barb wear is governed by many factors, including the fibre type, lubrication (fibre finish) and metal to fibre friction, the original barb shape and kick-up, the needlepunch density, depth of penetration, and fabric density. The general pattern of progressive barb wear due to needlepunching is illustrated in Fig. 8.30. The back wall and the kick-up of the barb are the first sections to show signs of wear. Gradually, the back wall is worn away from its normal angle of 17°–20°, to near 0°. As the angle is reduced, the fibre-carrying capacity of the barb is also reduced and without some intervention, the quality of the fabric will deteriorate to unacceptable levels.

At angles less than zero degrees, the barb contributes little fibre entanglement and is reflected in fabric tensile properties. The greatest wear tends to take place on the barb closest to the tip of the needle and decreases as the barb distance moves further away from the tip. The first few rows of needles also tend to wear more quickly than those further back and therefore, it is important to renew the needles at periodic intervals. Another strategy that is used to maximise needle life and maintain acceptable fabric properties is to increase the

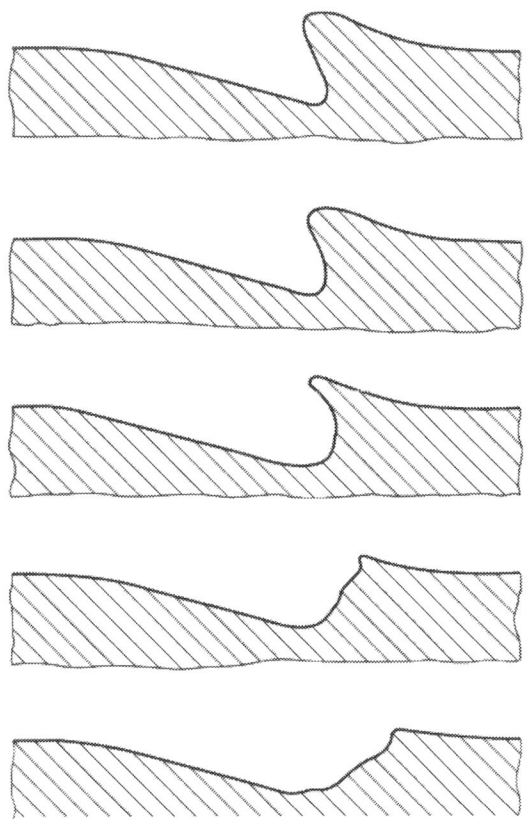

FIG. 8.30   Barb wear.

penetration depth thereby introducing more of the less worn barbs into the needling zone. However, an increase in penetration depth to introduce new barbs can only be accomplished if the needleloom settings and fabric properties allow and therefore it is not always a practical solution. For example, if the line speed is high, the increase in needle penetration would increase the dwell time leading to greater drafting and needle breakage.

8.11.3 Needle rotation

It is normally not recommended to replace all needles in the board at the same time but rather to carry out a partial replacement in which needles are replaced in sections. This is because the properties of a fabric made with worn, low-efficiency needles are markedly different from those produced with new needles. It is better practice to change only sections of the needleboard at any one time. To illustrate this, let us assume the lifetime of the needle is 30 million punches. The needleboard can be divided into three sections: the first section can be changed after 10 million punches, the second section changed after a further 10 million punches, and the third section changed after 10 million punches. In this way, over the total needle lifetime, new needles are periodically introduced or rotated through the board. This procedure has been founded to maximise the uniformity of fabric properties with respect to time.

In a production situation, it is not always easy to calculate the number of punches since the last needle change, particularly when many different fibres and batches are processed on the same needleloom. More often, other parameters are monitored and used as a guide such as:

- Linear production (number of metres of fabric produced).
- A significant change in a fabric characteristic is detected: this method is usually effective when the fabric is made to a narrow specification and parameters are regularly measured such as a decrease in tensile strength or an increase in thickness. Clearly, such an approach lends itself to the use of statistical control techniques.
- Total weight of fibre processed by the machine.
- Running time of the machine.

The physical wear rate can be found only by direct inspection of the needle, which of course requires the machine to be stopped. Periodically, needles should be taken from the board after a predetermined interval and checked for wear. Using a systematic approach, it is possible to correlate the observed barb wear to quantitative data obtained for corresponding fabric properties or dimensions, principally thickness.

8.11.4 Needleboard changeovers

To save loom downtime, it is usual to carry spare needleboards so that broken needles can be changed or worn needles removed in sections. Fig. 8.31 shows one approach, whereby the needleboard is clamped to the needlebeam and held in place by pneumatic bags, enabling boards to be changed in less than 3 min. Conventionally, needles are changed manually in a labour-intensive process, in which the operator pushes out old needles using a special tool and a mallet. Various attempts to automate this process have been devised using mechanical

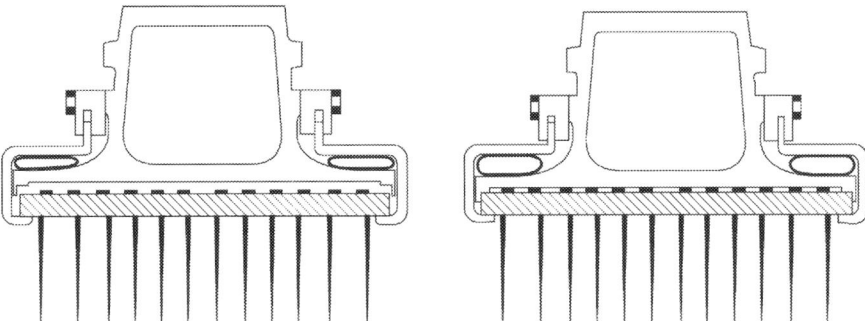

FIG. 8.31   Clamping of needleboards to permit rapid changeovers.

means but none has found widespread acceptance because these systems do not entirely remove the need for labour and their flexibility is limited.

8.11.5 Needle arrangement

A further quality control issue in needlepunching, particularly using boards with needles arranged in rows, is the presence of needle marks or tracks in the fabric. Longitudinal, lateral, and diagonal tracks may be produced in needlepunched fabrics due to the position and pitch of needles in the needleboard and the advance per stroke. New needleboard patterns have been introduced to eliminate interference patterns. Tracking in the MD is normally the result of broken or bent needles. The widthwise variations tend to be associated with the needle pitch and corresponding advance per stroke. In practice, tracking can be difficult to predict because of drafting or lateral contraction of the fabric during needlepunching. Sometimes, the patterns emerge following a change in the needleloom settings. Computer simulations are used by machine makers to visualise the position of needle penetrations in the fabric for a given combination of conditions and this can be used to optimise the position of needles in needleboards to avoid the introduction of such tracking defects.

8.11.6 Scrim reinforcement of fabrics

For some needling applications, the modulus and dimensional stability of the fabric can be usefully increased by the incorporation of a reinforcing scrim, which usually consists of a preformed woven fabric, grid, or net structure. Scrims can be incorporated inside or often beneath the batt prior to needling. Scrims are available in many different aperture sizes up to about 20 mm × 20 mm. Alternatively, reinforcing yarns can be combined with the cross-laid batt in the machine direction prior to needling. Scrim or yarn reinforced needle felts are particularly common in the production of filter media and papermakers' felts, as well as historically, in the manufacture of needlepunched blankets. Other applications include geosynthetics, roofing fabrics, floor coverings, and some woundcare products, where dimensional stability during use is critical.

Minimisation of mechanical damage to the scrim by the needle points and barbs is necessary to prevent a reduction in fabric strength. Specifically, the approach angle of the needle barbs with respect to the yarns in the scrim has to be taken into consideration. Ball point needles with single apex barbs are produced for paper felt production, where scrim damage is of critical concern. The stress–strain properties of scrim and yarn reinforced needlepunched fabrics show two peaks, one corresponding to the failure of the reinforcing scrim and the other to the surrounding needlepunched fabric. Scrim damage caused by the needles can seriously affect tensile properties if precautions are not implemented. Dilo Group's HyperTex system is an example of a standalone production unit, which produces a scrim, feeds it between two preneedled fabrics and then integrates them all into a composite assembly by needlepunching at output spends up to $40\,\mathrm{m\,min^{-1}}$.

8.12 Needlepunching technology

Needlepunching machines can be classified as single-board, multiboard, structuring, or speciality and their use varies depending on application. Single-board machines are either down stroking or up stroking and have one needle beam. Multiboard machines can be arranged in the following combinations:

- Double boards (down stroking).
- Double boards (up stroking).
- Twin board (two boards up stroking and down stroking in the same vertical plane).
- Tandem or twin boards (up and down stroking in alternation in two sequential needlepunching zones).
- Four boards or quadpunch (up and down stroking for simultaneous double-sided needlepunching with two sets of up- and down-stroking boards, each set arranged in the same vertical plane). This arrangement may be referred to as a double-twin board arrangement.

While there are numerous possible needleloom installations depending on the intended application for flat needlepunching, there are two very common needleloom sequences found in industry. A common arrangement is a down-stroking preneedling loom, followed by either an up- and down-stroking loom, or a four-board up- and down-stroking machine. A variety of different needlepunching arrangements is illustrated in Fig. 8.32, all of which require more than one needleloom operating in sequence. Examples of related applications for each of the different arrangements are listed as follows:

(a) Filter media.
(b) Synthetic leather—preneedling is in the order of 50–80 punches $\mathrm{cm^{-2}}$ followed by flat needling of 1000–3000 punches $\mathrm{cm^{-2}}$.
(c) Underlay composed of recycled fibres.
(d) Floor coverings—preneedling is in the order of 75–150 punches $\mathrm{cm^{-2}}$ and finish needling is around 150–300 punches $\mathrm{cm^{-2}}$.
(e) Automotive headliners.
(f) Blankets.

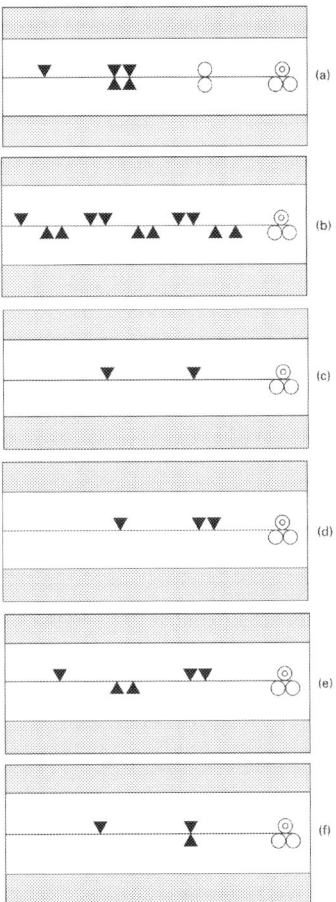

FIG. 8.32 Examples of needlepunching machine sequences.

In general, 'flat' needling of preneedled fabrics is intended only for bonding, whereas 'structuring' is designed to introduce three-dimensional visual effects such as ribs, velour-effects, and geometric patterns.

8.12.1 Preneedling or tacking

Initially, the batt is preconsolidated by a needleloom, and a low needleboard density of between 1000 and 6000 per linear metre is not unusual. The preneedling loom (or preneedler) normally punches from one side only and the aim is to gently consolidate the batt and to introduce some fibre entanglement thereby decreasing batt thickness prior to full consolidation or finish needling. Some preneedlers are fitted with two, rather than a single down-stroking board. The punch density of the preneedled fabric depends upon the desired product characteristics but 10–75 punches cm^{-2} is not unusual. Punching speeds range from less than

500 rpm up to 1500 rpm. Preneedlers normally have some means of compressing the batt as it is fed between the bed and stripper plates. A pair of driven batt compression aprons mounted just in front of the needling zone is commonly utilised. These aprons converge to enable progressive compression and to minimise slippage of the outer layers of the batt relative to the inner layers. Examples of different batt feeding methods in preneedling are illustrated in Fig. 8.33, (a) by apron, (b) by roller, and (c) by locating the needle beam within a driven perforated cylinder, which forms the stripper and bed plates. The latter arrangement is referred to as a rotary tacker.

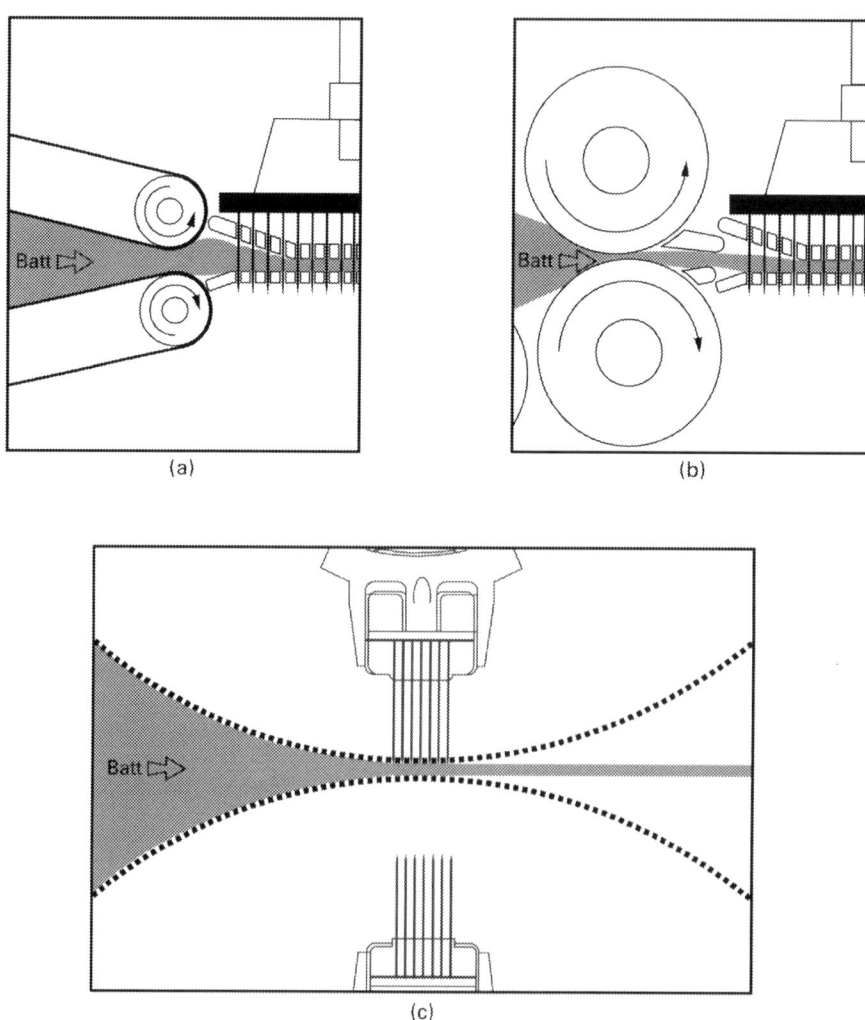

FIG. 8.33   Methods of batt feeding in preneedling; (A) apron; (B) roller; (C) perforated cylinder. *Adapted from: ANDRITZ Asselin-Thibeau.*

Needlelooms are frequently operated with the stripper plate angled down towards the output side of the needling zone to assist in batt compression during bonding. This together with a wide bed plate to stripper plate gap setting helps to accommodate the batt. Since the bed plate to stripper plate gap can be large, 90 mm ($3^1/_2$ in.) needles, or sometimes longer are required. The penetration depth or stroke of a preneedling loom is usually larger than for finishing looms since there is a greater batt thickness to consolidate.

To produce a uniform fabric with a smooth surface finish, a needle with a regular barb spacing is normally selected. The wide barb spacing along the blade of such needles ensures small groups of fibre are engaged by the barb through the cross-section of the batt without the need for a large penetration depth. Reducing the spacing between barbs in preneedling can lead to the reorientation of a larger number of fibres, producing a cratered surface in the fabric. Such crater marks in the surface are difficult to mask in subsequent needling. The consolidation of the fabric with a closely spaced barbed needle is however more rapid as compared with a regular spaced barbed needle.

It should be noted that not all needlepunching applications require two-stage needling, and in some cases, a preneedling loom alone is used. For example, in the manufacture of wound dressing fabrics composed of fibres such as calcium alginate, a preneedling loom is preferred as a means of producing low-density, highly absorbent structures. Similarly, in the production of ceramic fibre fabrics for high-temperature insulation and in the waste-recycling business making capillary mattings and similar products, low needle punch densities help to minimise fibre and needle breakage. In these applications, high degrees of fabric consolidation are not necessarily needed. Additionally, it is not uncommon to find a preneedling loom as a preliminary bonding operation in a thermal or chemical bonding installation, and preneedling prior to hydroentangling can also significantly reduce overall energy consumption in the manufacture of heavyweight fabrics.

8.12.2 Flat finish needling

The purpose of flat needling is to achieve a high degree of fibre entanglement and to increase fabric strength while producing a smooth surface. One or more looms may be installed in line, punching sequentially from one side then the other, from both sides simultaneously or alternately. Such continuous multiloom production lines are found, for example, in the manufacture of automotive headliners, synthetic leather, and geosynthetic fabrics. The needle density is considerably higher than used in preneedling, with up to 32,000 needles per metre, being employed for final consolidation and finishing. The cumulative punch density in finish needling ranges from about 100 punches cm^{-2} to >1000 punches cm^{-2} depending on the required density, fabric weight, fibre composition, and physical properties of the final product. It is desirable to punch from each side as this normally promotes a stronger, more uniformly consolidated fabric.

Modern finishing looms run at high punch frequencies of 1000–3000 rpm and consequently, they tend to operate with a shorter needle penetration depth than preneedling needlelooms. Needle penetration depths are lower than in preneedling and therefore, shorter 76 mm (3 in.) needles are commonly selected. These short blade needles are stiffer than the 90 mm ($3^1/_2$ in.) type and are suited to low penetration depths and high needling densities.

Since preneedled fabrics are thinner than the original batt introduced to the preneedling loom, the bedplate to stripper plate gap setting is kept small and is only marginally greater than the preneedled fabric thickness. In finishing looms, it is usual to use fine gauge needles carrying small barbs. Although the number of fibres that can be carried in each of these small barbs is comparatively low, higher needle densities (punches per cm^2) compensate. This approach also gives good fabric strength and produces an even fabric surface, free of the perforation patterns that large needles can introduce.

One of the more versatile looms used in the finishing process is sometimes referred to as a double punch or quadpunch loom. This loom has four needleboards, two punching from the top and two punching from below. A needling line incorporating one or more quadpunch looms is more compact than a line consisting of two or more looms punching from only the top or below. Each needleboard holds up to 8000 needles per metre of the working width. Needling from opposite sides gives greater fabric strength than needling from only one side. Both the upper and lower fabric surfaces have the same appearance using this approach. Modern four-board looms run at high speed and are arranged so that the penetration depth of each needleboard can be independently adjusted. Different needle specifications can be fitted in the upper and lower needleboards on the in-feed side to the upper and lower boards on the draw-off side, if necessary, to achieve desired felt characteristics. For example, if the in-feed needleboards are fitted with close barb spaced needles and the draw-off boards are fitted with regular barb spaced needles, a well consolidated fabric with a smooth surface will result.

Some needlepunching factories prefer to have their finishing looms off-line with only the preneedling looms being incorporated into the carding and cross-lapping line. This is common in the manufacture of filtration fabrics. The batt is preneedled and carefully wound onto rolls or 'A' frames and they are then transported to a finishing loom or line of finishing looms. Many filter fabrics are made up of multiple layers of preneedled fabric which are assembled and needled together in one or more passes through a finishing loom. A woven reinforcing fabric or scrim is normally incorporated during this process to increase dimensional stability of the product.

Examples of high speed needlepunching systems for different applications are shown in Fig. 8.34A–C. Needling is used in some cases to bond heavyweight spunlaid webs containing continuous filaments. For needlepunching such webs, high line speeds are required, and either high capacity looms or multiple individual looms operating in sequence are needed to balance production. Fig. 8.34C is a three-loom line suitable for spunlaids, operating with two down-punch and one up-punch machine, where up to 35 m min^{-1} is claimed at a loom speed of 2000 r.p.m., and more than 60 m min^{-1} at 3000 r.p.m.

Other developments include the Cyclopunch, Hyperlacing, and Variopunch systems of Dilo. The Dilo Cyclopunch system operates with four needle boards and up to 20,000 needles m^{-1} working width. It is intended to enable 'micropunching' using needles with only a single small barb, to minimise fibre transport and resultant pillar dimensions. This is useful for needling lightweight fabrics containing fine fibres. The Cyclopunch system also has a circular beam movement, enabling high linear speeds. More than one Cyclopunch unit can be coupled to make a 'Hyperlacing' production line to enable highly bonded lightweight fabrics to be made at high speed. This approach has been claimed to offer an alternative to hydroentangling. The Dilo Variopunch VPX 2020 system is

(A)

(B)

(C)

FIG. 8.34 Examples of high speed needlepunching systems: (A) Multiple needleloom production line, e.g., for synthetic leather. (B) DI-LOOM double needleloom, incorporating four needleboards. (C) Multiple needleloom production line. *Courtesy: (A and B) DILO Group and (C) AUTEFA Solutions.*

principally intended to improve surface uniformity by compensating for the unwanted streaks and patterns caused by the normal situation of needles being fixed in a single position within the board. Adaptive needling modules are employed, the position of which can be readily changed to maximise the uniformity of the bonded fabric. Visual uniformity in the finished fabric is particularly important in applications such as automotive headliners.

8.12.3 Elliptical needlepunching

The use of an elliptical needle path enables a large advance per stroke and therefore, very high line speeds (Fig. 8.35) in both preneedling and finish needling applications. In this system, the needles move with the fabric during needle penetration and therefore the bed and stripper plates have slotted holes to allow for the needle motion. The elliptical motion is

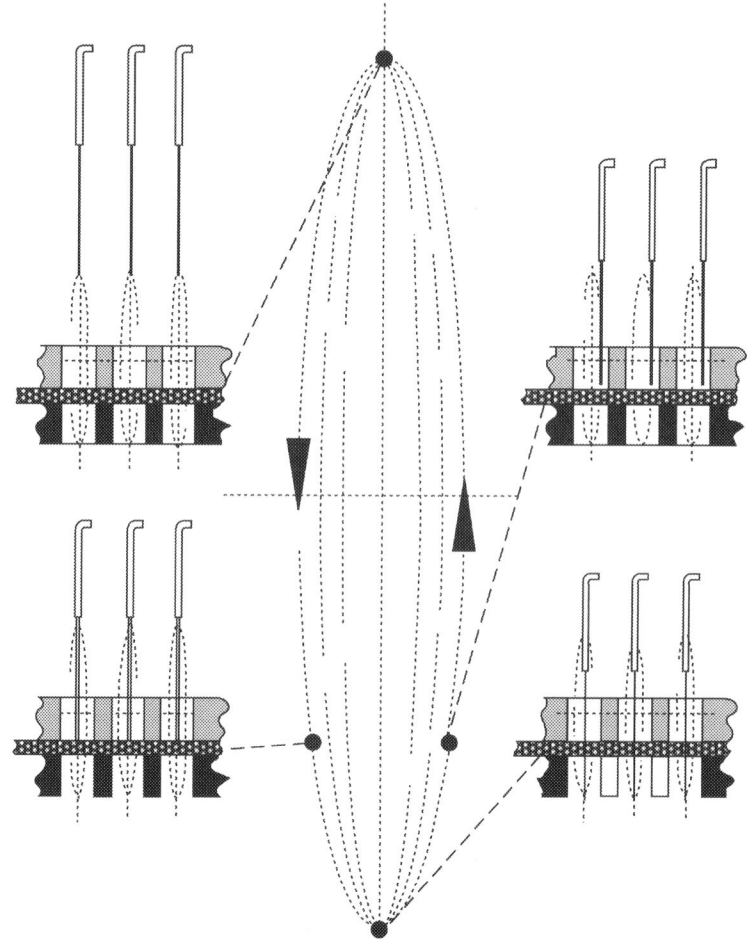

FIG. 8.35 Principle of elliptical needlepunching. *Courtesy: Dilo Group.*

claimed to reduce drafting, providing the needle penetration depth is low, and to give a uniform surface finish. Synthetic leather production is one of the markets for which it is intended as well as in the needlepunching of spunlaid webs and the production of paper machine felts. Elliptical needling can also be incorporated in high-speed structuring looms to produce rib, diagonal, diamond, and hobnail patterns.

A number of different technologies are available, including the original Hyperpunch system (Dilo) and the Gamma system of Tectex.

8.12.4 Inclined angle (oblique) needlepunching

An early example of needling from both sides of the batt at an inclined angle was the Chatham fibrewoven process for blankets, where the needles were angled at 20° to 30° to the plane of the batt. In the 1960s, this was a very sophisticated machine in concept and engineering design, producing blankets with good strength and dimensional stability but output speeds were low by modern standards. Another example was the Fehrer H1 system, which employed a curved bed and stripper plate with a corresponding needleboard. The changing curvature varied the angle of needle penetration as the fabric passed through the needling zone, giving fibre pillars of different angular inclinations within the cross-section. A claimed effect was an improvement in the isotropy of the fabric with respect to tensile properties.

8.12.5 Structuring needlelooms

Surface textured needlepunched fabrics with a pile surface, or those with three-dimensional structural patterns are produced using structuring needlelooms. Such fabrics are commonly used for a variety of applications including floor coverings and automotive interiors. Structuring looms work with a preneedled fabric and the process involves two main changes to the strategy used for bonding. The first is the replacement of the bed plate with either lamella plates (to produce ribs, loop pile, and other effects) or a brush-bed (to produce random velour effects). The second is the use of structuring needles, which are different to those used for bonding (Figs 8.28 and 8.29).

8.12.5.1 Rib fabrics

Ribbed and looped pile fabrics are commonly produced using forked needles that are designed to transport groups of fibres between fixed lamella plates (or strips) that serve as the bedplate (Figs 8.36 and 8.37). The fork acts like one large barb collecting multiple fibre segments as it penetrates between the lamella plates, pushing them towards the reverse side of the fabric to produce a structural projection in the form of a rib or pile. Patterning is introduced by varying the position of needles in the board and by controlling the advance per stroke. By lifting and lowering the lamella table, the height of the rib or pile is adjusted. In certain systems, the needleboard position is raised or lowered to enable the rib or pile height to be adjusted. Other important parameters are the fork dimensions, needleboard density and punch density. Such structuring looms are typically down-punch machines producing rib or pile surface structures depending on the orientation of the needle fork relative to the incoming fabric.

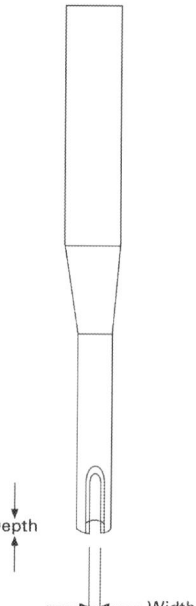

FIG. 8.36   Forked needle.

8.12.5.2 *Velour fabrics*

Classical fibrous pile or velour fabrics can be produced using coarse gauge fork needles orientated in the correct direction and operating in conjunction with a lamella bed plate. By contrast, in the production of random velour fabrics, a continuous, moving brush bed conveyor is employed in place of the lamella bed plate, to produce a high-density velour finish. Fine gauge fine fork needles and crown needles, sometimes in combination, are commonly used to manufacture random velour fabrics. Such structuring needles are required to ensure a uniform surface pile structure is formed. The pile surface on the fabric is formed in the brush bed conveyor during needling. The design of the brush, particularly its density, brush filament diameter and stiffness, height and uniformity, influence the appearance and structure of pile surface produced on the fabric. Needling directly into the brush bed conveyor inevitably leads to its progressive damage and wear, which over time can lead to visible quality problems such as variations in pile height, shade, and surface uniformity.

In double random velour systems, more than one needling head is positioned over a common brush bed conveyor to give a high pile density, and sometimes to introduce coloured effects by means of yarns or other material. Such yarns are tacked into the fabric structure after being introduced from an overhead creel. An example of a random velour structuring machine is shown in Fig. 8.38. Use of two needleboards allows the introduction of a backing fabric before the second needling zone, which gives a close pile structure and added stability to the finished product. Rib and velour fabrics with large repeat patterns with a patterned surround or border on all sides can also be made instead of using two machines in tandem.

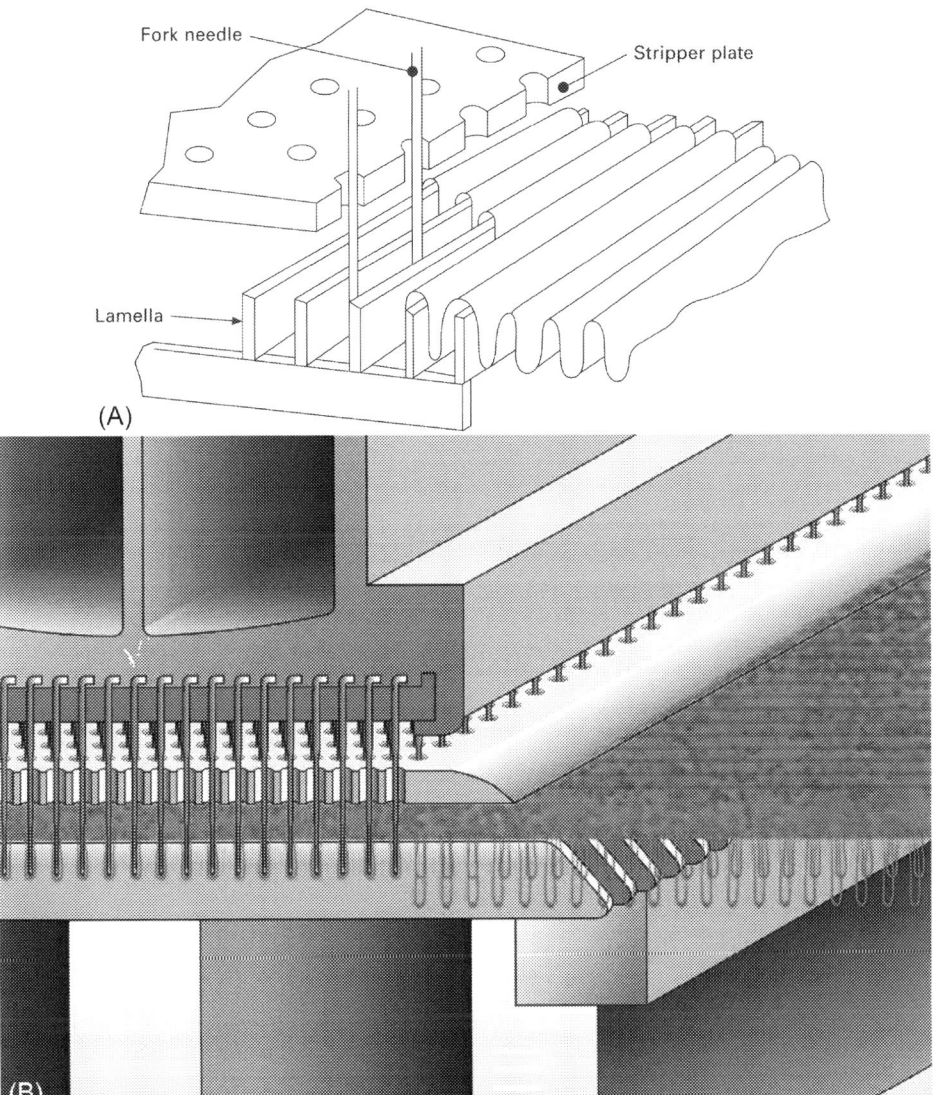

FIG. 8.37 Structuring using lamella plates (A) main components; (B) the DI-LOOP system. *Courtesy: Dilo Group.*

This is particularly useful in the manufacture of needled carpets and can be produced by two independent needle zones that are electronically synchronised to maximise pattern flexibility at high speed. Simulated oriental carpet patterns are claimed to be possible using such systems. In other systems, register control is provided between two looms when complex relief patterns need to be produced.

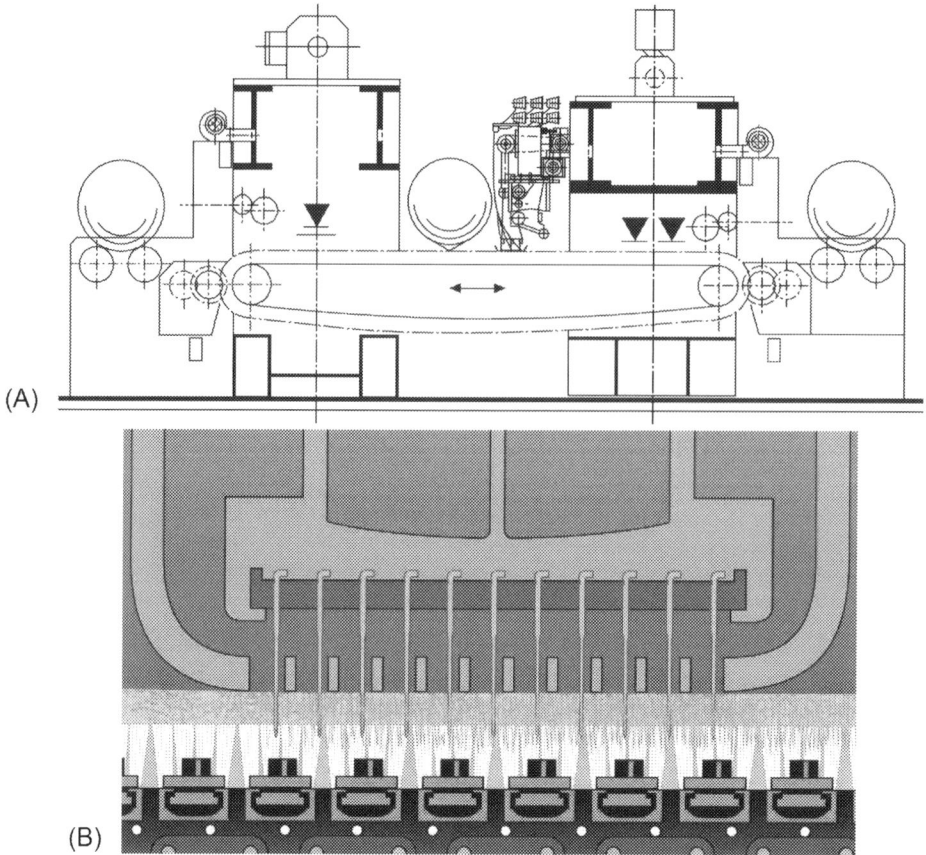

FIG. 8.38   (A) A velour structuring machine and (B) Needling into a brush bed conveyor to produce random velour fabric using the DI-LOUR system. *Courtesy: Dilo Group.*

8.12.6 Specialist needlelooms

8.12.6.1 Continuous belts

Needlepunched fabrics are produced in the form of a long continuous belt for use in the manufacture of paper. These needlepunched papermakers' felts are used in the pressing and drying stages of the papermaking process and have very large working widths of more than 12 m. The felts form a wide endless belt consisting of layers of preneedled or carded web needled into a special monofilament scrim. The quality control requirements are extremely high since structural imperfections in the belt affect the quality of the sheet that is made in subsequent papermaking.

8.12.6.2 Tubular fabrics

Tubular needlepunched fabrics can be produced using a specially adapted needleloom design developed by Dilo. These tubes normally have an inside diameter of 25–500 mm. In some cases, tubes with diameters of only 5 mm may be produced. The Rontex S 2000 loom employs two needling units acting from opposite sides with different angles of needle penetration. A continuous needlepunched spiral is produced which may be layered with the tube wall having different types of fibre.

8.12.6.3 Three-dimensional spacer fabrics

Needlepunched spacer fabrics can be manufactured in which two discrete needled layers are connected by fibrous pillars. Large channel-like voids within the fabric cross-section are created, which if required, can be simultaneously filled with other materials as part of the process. The Laroche Napco 3D web linker exemplifies the process (Fig. 8.39).

The needlepunching machine is fed with two fibrous webs (A and B) between two stripper plates and one or two spacer tables that consist of static bars or tubes. As the webs pass through the machine, the barbed needles drive fibres from one web into the other creating interconnecting fibre bridges or pillars. The voids between the spacer bars can be used to introduce functional components such as wires or cables. Spacer tubes allow the insertion of powders, foams, and other materials as the fabric is being made.

8.12.6.4 Yarn and fabric punching

The concept of yarn punching was developed by Fehrer (now AUTEFA Solutions) and involved needlepunching coarse yarns to increase their tensile strength. This was intended for carpet, mop and effect yarns as well as for friction-spun and open-end yarns requiring sheath-

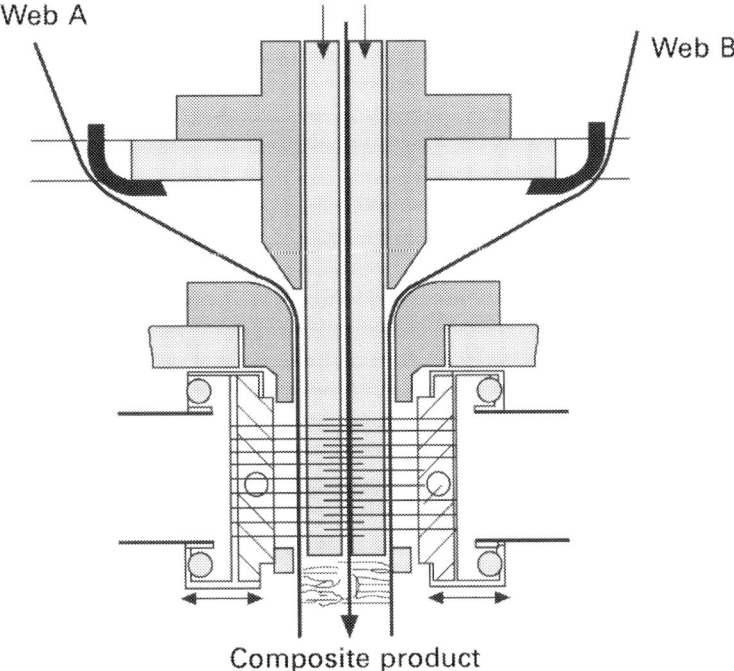

FIG. 8.39   The Laroche Napco 3D web linker.

core stabilisation. The yarn is fed along a narrow channel through which the needles pass in a reciprocating motion. Needlepunching has also found commercial application as a finishing process for woven coating substrates.

8.13 Needlepunched fabric structure and strength

There is a considerable body of published research on the complex subject of needled fabric structure and properties. Needling induces fibre migration and the formation of 'pillars' or 'pegs' of fibres in the cross-section due to the action of the barbed needles. These pillars are readily observed by means of tracer fibres and optical microscopy, or by microcomputed tomography. Normally, only segments of fibre are reoriented by the action of the needle barbs, and it would be wrong to imagine that most fibres are oriented in the z-plane or thickness of the fabric after the process. In practice, depending on the needling intensity, a minority of fibres are incorporated into these pillars along most of their length, while the majority remain predominantly oriented in the x–y fabric plane. In a down-stroking needling process, the first two or three barbs engage the largest number of fibres in the upper surface of the batt, which has the effect of tying-in these fibres to the lower surface, thereby providing cohesion and a reduction in fabric thickness.

Initially, as pillars are created, the fabric strength increases, but after peaking the fabric strength subsequently decreases as fibres are broken and the fabric begins to perforate. It should be remembered that the presence of the structural pillars of fibre in the cross-section depends both on fibre and process factors. Few pillars are formed and therefore fabric strength increases little unless fibres can deform and extend while remaining in contact with the barbs. The number of fibres in each pillar, their frequency and interconnection, are a function of barb dimensions, needle punch density, the depth of penetration, and the advance per stroke, among other factors. It is therefore possible to engineer the structure of a needlepunched fabric to a large extent by considering these aspects.

Fibre composition, length, diameter, fibre tensile properties, fabric density, and thickness are particularly important with respect to fabric properties. Fabric mechanical properties most frequently measured are tensile, tear, bursting, and puncture resistance. In the early days, comparisons were always made with the strength of woven fabrics, but since man-made fibres became available in multiple lengths, diameter, and crimp, tailored to the needlepunching process, the properties of needlepunched fabrics are considered in their own right. It is the versatility and attractive balance of physical properties and cost that has made needlepunched fabrics so well suited to a variety of different applications.

8.14 Applications of needlepunched fabrics

The major applications of needlepunched fabrics include geosynthetics for civil engineering, agriculture, and landscaping; automotive interior and boot lining fabrics; filter media, synthetic leather, contract floor coverings, and glass fibre composites. The full range of applications is extensive and extends into many niche product areas including, but not limited to medical wound dressings, wipes, composite breather felts, capillary mattings for horticulture, fire barriers, and ballistic-impact-resistant fabrics. Some of the product applications are briefly described below but this list is not exhaustive.

8.14.1 Geosynthetics

Needled fabrics are used in civil engineering applications that require deformability, high tensile and bursting strength as well as controlled permeability and weight, for road reinforcement, subsoil stabilisation, pond liners, hazardous waste containment protection liners, and drainage. Staple fibre geotextiles are typically composed of polypropylene, polyester, or polyamide. Where spunlaid webs are needlepunched, they are produced at high linear speed with a relatively low degree of needling. The Foster star blade needle (now Groz-Beckert) having four barbed edges has found use in such applications where high strength geotextiles made from both staple fibre and spunlaid webs are produced. Needlepunched geosynthetic clay liners are installed for low water permeability in landfills, canals, ponds, and pollution-prevention barriers in highway and airfield construction. Rolls of bentonite clay are sandwiched between two fabrics, one or both of which are nonwoven. Needling the composite locks the bentonite clay in place. The clay is the critical component as it provides extremely low hydraulic conductivity. In needling, the abrasive nature of the clay means that manufacturing conditions and needle design must be controlled to minimise needle wear. Conventional needlepunched spunlaid fabrics for geosynthetic applications are made from about 120 to $340\,\mathrm{g\,m^{-2}}$. Needlepunched spunlaid fabrics are utilised in the laying of asphalt as stress-absorbing-membrane interlayers; fabrics dipped in bitumen help to improve the adhesion between layers of asphalt.

8.14.2 Filter media

Fabric porosity and permeability are properties relevant to the filtration of gases and liquids, and depth filters are particularly suited to needlepunched fabrics because of their substantial thickness. Woven scrim reinforcement is needed in industrial bag house applications, to provide dimensional stability. For general filtration applications, PET, PA, and PP are common, but for high-temperature or corrosive environments other high-performance organic and inorganic fibres are needlepunched to make chemically or thermally stable filter fabrics including glass, silica, basalt, aramid, PTFE alone or in blends, polyimide, and stainless steel, among others. Electret filters are also needlepunched based on drylaid blends of electrostatically charged staple fibres selected for their relative position in the triboelectric series. The surface of needlepunched fabrics may be coated, singed, or calendered to adjust the surface structure and therefore both the cleaning and filtration efficiencies of the media. The fabric porosity may also be graduated through the fabric cross-section by adjusting needle penetration and needlepunching density, which influences filtration efficiency in use. Both roll products and tubular needlepunched fabrics are used as filtration media.

8.14.3 Synthetic leather

Synthetic or artificial leather fabrics made of microfibres imitate the natural product with a densely entangled fibre construction that is impregnated with a polyurethane (PU) resin, frequently in the form of an emulsion, to give a smooth surface free from needle marking and with a high surface abrasion resistance. Typical production lines may have as many as eight needlelooms in sequence from preneedling to finish needling. The density is progressively increased in successive needlepunching stages. To avoid marking of the fabric and to minimise needle force, fine gauge needles with small barbs and small tip to first barb distance are selected. Sometimes, single barb needles combined with high needlepunching densities are employed to produce highly dense fabrics with a uniform surface. Only small needle penetrations are required in the process using such needles. To further increase the density of the fabric, a proportion of high-shrink thermoplastic fibre may be included in the blend, which after heating induces contraction of the fabric.

For over 50 years, carded, cross-lapped, and needlepunched batts of island-in-the-sea bicomponent staple fibres have been used as a basis to make synthetic leather and related suede products. Once the fabric is made, the sea component is dissolved out, releasing the very fine island fibres. Toray's highly tactile Ultrasuede fabrics, originally launched in 1970, are produced in Japan from PET or PA island-in-the-sea staple fibres, which after drylaid formation and needling to form a fabric (which can be scrim-reinforced), undergo PU impregnation and solvent treatment, to yield a soft microfibre pile. Similar technology was launched by Kuraray in 1964, under the Clarino brand.

In 2016, Toray's Ultrasuede PX was launched containing a proportion of plant-based PET obtained by polymerising with ethylene glycol sourced from waste molasses. Ultrasuede BX followed in 2019, containing ca. 30% plant-based PET and PU made from castor oil–derived polyol. Alcantara, produced in Italy, originates from the same basic technology as Ultrasuede, but the two products have evolved separately in terms of development and market positioning. Related monolayer nonwoven fabrics produced by Toray known as Ecsaine and GS Felt, also contain PET island-in-the-sea fibres impregnated with PU elastomer. Progressive efforts are being made to reduce the dependency on organic solvents during manufacturing. For example, Tirrenina from Kuraray is a low PU variant of the Clarino product, made with a water-soluble sea component in the fibre.

Splittable bicomponent fibres have also been developed capable of separating during needlepunching in a similar manner to hydroentanglement. Traditionally, needlepunched synthetic leather is utilised by the fashion industry in clothing and footwear, with premium, highly aesthetic fabrics such as Alcantara being used at the luxury end. Nonapparel applications of needlepunched synthetic leather include, luggage, sports goods, automotive interior trim such as seat and steering wheel coverings, upholstery and wall coverings. Some are also marketed as components for audiovisual and office equipment, cameras and PCs, as well as for semiconductors.

8.14.4 Waddings and paddings

Based on fibre consumption, waddings and paddings are one of the largest single application areas for needlepunching and the fabrics are incorporated into mattresses and furniture as insulator pads (in contact with the sprung unit) as well as comfort layers to provide support, carpet underlay, sound, and heat insulation for automobiles and other industrial uses. Fibre selection ranges from mechanically recycled natural and synthetic fibres, usually obtained from pulled waste clothing, jute, sisal, coir, and cotton as well as virgin synthetic fibres, particularly PET, PP, and acrylic.

8.14.5 Floor coverings

Flat floor coverings generally consist of a face layer, a scrim, and a bottom layer and are produced by preneedling, flat needling, and, in many cases, structuring. Needling from both sides tends to increase the wear resistance of the face layer. As described previously, structuring involves needlelooms with lamella bed plates in which forked needles are used, or random velour needlelooms with brush conveyor belts in which fine gauge fork needles or special crown needles are used. Needlepunched floor coverings are commonly produced from spun-dyed PP and blends of PP and PA, which is blended prior to carding, cross-lapping, and needlepunching. In some low traffic applications, where softness is needed, PET is sometimes selected. Different fibre linear densities are frequently blended to adjust the durability and compression recovery properties of the floor covering without compromising aesthetics. Fibre linear densities are about 12–20 denier and fabric weights are in the order of 300–800 g m^{-2}. A small proportion of very coarse fibres is commonly added to a base blend of 17 dtex fibres to promote durability.

8.14.6 Automotive fabrics

End-uses for needled fabrics consist of decorative interior trim including headliners, door trims, seatbacks, boot liners, load floors, and package trays. In the USA, the interior trim was traditionally composed of spun-dyed PP, whereas PET is important elsewhere in the world. Fibre linear densities range from 15 to 18 denier in the USA, where the abrasion resistance specifications are high, to as low as 6 denier in the Far East and Japan. Other products include sound dampers, underfelts, padding, performance gaskets, seals, filters, and shields. Extensive use is made of random velour fabrics in automotive applications particularly in small-to-medium-sized cars and structured needlepunched fabrics are encountered in more expensive interiors. In moulded floor fabrics, up to about 65% of the total fibre content is visible as surface pile. The quality control issues are particularly stringent and a major consideration in interior trim and headliners is colour consistency and matching between batches, even in solid shades. There is also growth in the composites field, and blends of wood fibres with synthetic fibres are needled prior to resin impregnation and the formation of rigid panels. Glass fibre composites are made in a similar manner and there is increasing acceptance of natural fibres including hemp, flax, sisal, and other bast fibres in automotive composites. Composites are now fitted around the firewall, dashboard, speaker, engine bay, and parcel shelf structures. Another developing area is the application of thermoformed carbon fibre composites, made from drylaid and needlepunched blends of recycled carbon and thermoplastic fibre.

8.14.7 Insulation

Both thermal and acoustic insulation fabrics are made by needlepunching. In high-temperature applications, ceramic and other inorganic blown or spun fibres such as those based on alumina, silica, glass, and basalt are produced in batt form and then needlepunched to make finished fabrics of up to 75 mm thick. To aid processing, they are sometimes mixed with sacrificial organic fibres or binders, which degrade away in situ following installation, leaving behind only the high-temperature–resistant material. Avoiding damage to brittle ceramic and glass fibres during needling is challenging but is necessary to achieve acceptable fabric strength and minimise short fibre and dust formation. Such high-temperature resistant thermal insulation is used in the automotive and aerospace sectors, as well as for insulating pipework, machinery and components operating at hundreds of degrees Celsius. Many of these applications are subject to strict regulation. Low-temperature insulation fabrics for both thermal and acoustic applications are also made by needling drylaid batts made of synthetic fibres, often in blends, as well as recycled fibre extracted from clothing and low-grade wool and cotton.

8.14.8 Blankets

This was one of the earliest applications for needling, and while high-quality natural and synthetic fibres are still used occasionally, the process is commonly found where cheap blankets made from blends of regenerated fibres are needed, including emergency and disposable blankets. It is common for reinforcing yarns or scrims to be incorporated into such needled fabrics to improve dimensional stability.

8.14.9 Wipes

Needling is used for some heavy-duty household and industrial wipes and polishing cloths. Use of spun dyed viscose staple blended with a thermoplastic bonding fibre is popular and many other fibres including lyocell have been needled for personal care and industrial wipe applications. In premoistened wipes, needlepunched fabrics are competing in some sectors because it is possible to store larger volumes of lotion in the structure compared to a hydroentangled wipe. One application is in postoperative wipes containing antibacterial soap.

8.14.10 Roofing

Needlepunched spunlaid fabrics composed of PET find applications as bitumen-coated roofing felts because they have good puncture and tear resistance. Glass scrim reinforcement may be introduced during needling to improve dimensional stability.

8.15 Introduction to hydroentanglement

Hydroentangling, spunlacing, hydraulic entanglement, and water jet needling are synonymous terms describing the process of bonding fibres (or filaments) in a web by means of high-velocity water jets. The interaction of the energised water with fibres in the web and the support (conveyor) surface increases fibre entanglement and induces fibre displacement and rearrangement. In addition to mechanical bonding, structural patterns, apertures, and three-dimensional emboss effects can be produced, if required, by the selection of appropriate support surfaces. If required, hydroentanglement also provides a convenient method of mechanically combining two or more webs to produce multilayer fabrics.

The early work on the process of hydroentanglement has been principally attributed to Chicopee (division of Johnson and Johnson, now part of Berry Global Inc) and DuPont in the USA. The respective technical contributions of these companies are described in a series of detailed patents filed from the 1950s to the early 1970s [9–11].Originally, the utilisation of relatively low-pressure water jets (<150 psi) and porous conveyor surfaces to locally rearrange fibres in a web, principally for the purpose of aperturing and the production of patterned fabrics, was developed. Chemical bonding at low binder applications was required to fully stabilise the fabric, and Keybak-apertured fabrics were commercialised based on this technology.

In the 1960s, efforts by DuPont to increase the level of fibre entanglement and to maximise bonding led to the utilisation of higher water pressures and additional process developments. An early motivation was the production of substrates for making artificial leather (Corfam). The progression towards higher water pressures eventually gave way to the production of both fully bonded flat and apertured hydroentangled fabrics without the need for a secondary bonding process. Commercial hydroentangled products, notably wipes, appeared during the early 1970s under the DuPont trade name of Sontara (now produced by Jacob Holm). The process technology and commercial manufacture of hydroentangled fabrics was restricted to a handful of companies until the mid-1970s. Other companies began to develop their own proprietary systems and in the 1980s, new turnkey hydroentanglement machinery became available for purchase (or licensing).

This combined with increased market demand led to a proliferation of hydroentangling installations worldwide, particularly in Europe, Asia, and the USA. Machinery installations have grown strongly, particularly in China and in the USA, as suppliers have sought to move geographically closer to their overseas markets [13]. During the late 1990s and early 2000s, significant growth in the installed hydroentanglement capacity was fuelled by developments in the consumer products sector, particularly wipes [12]. Diversification of the wipes sector, including development of water dispersible (flushable) products based on wetlaid and short fibre airlaid platforms, as well as the growth of durable speciality fabrics has led to further growth. Gradually, hydroentangling systems have been coupled with most web formation systems to enable a greater variety of products to be made. Notable developments have taken place from 2010 to 2020 in hydroentangling spunlaid webs, particularly for producing fabrics from microfilaments aimed at long-life, durable nonwovens. Advances in process technology have also been directed at reducing energy consumption during production, as well as the launch of new turnkey systems for specialist markets, such as wetlaid-hydroentangling for dispersible wipes. Despite potential overcapacity in certain sectors, global hydroentangled output has continued to grow.

8.15.1 The principles of hydroentanglement

Hydroentangling can be coupled with carded, carded and crosslapped, airlaid, wetlaid, or spunlaid web formation systems, depending on the intended application of the fabric. Of

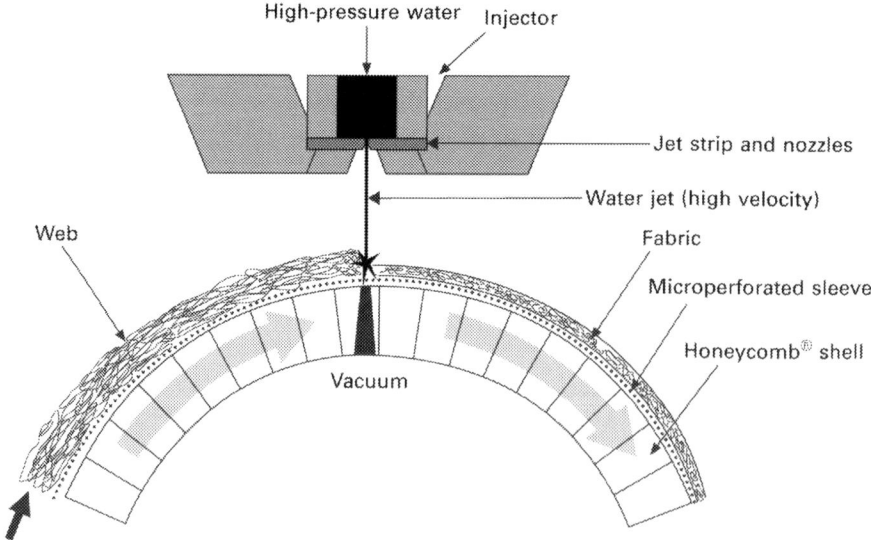

High-pressure water   Injector

Jet strip and nozzles

Water jet (high velocity)

Web

Fabric

Microperforated sleeve

Honeycomb® shell

Vacuum

FIG. 8.40   Basic elements of hydroentanglement. *Courtesy: ANDRITZ Perfojet.*

these, carding is the most common. The basic elements of a hydroentangling system are illustrated in Fig. 8.40.

A curtain of multiple small diameter, high-velocity water jets is produced by pumping filtered water at high pressure through small cone-capillary-shaped nozzles in a jet strip clamped into an injector (manifold). These high-velocity jets are directed at the web supported on a moving conveyor ('wire' or 'support'), which may be of flat bed or cylindrical configuration. Fibre entanglement resulting from the applied kinetic energy is introduced by the combined effects of the incident jets and the turbulent water generated within the web, which intertwines neighbouring fibres. The conveyor belt or cylinder sleeve being permeable enables most of the deenergised water to be drawn into the vacuum box for recycling and reuse. Some of the remaining process water continues with the web, some drains from the side of the support surface, and some is atomised depending on the water pressure. Normally, multiple injectors running at progressively increasing water pressures are arranged in sequence to produce a fully bonded fabric with the requisite physical properties and appearance.

8.15.2 Specific energy

Hydroentanglement relies on the transfer of kinetic energy from the water jets to the web and the constituent fibres to introduce mechanical bonding. The degree of bonding, fabric properties, and economic efficiency of the process are influenced by the energy introduced to the web, which can be calculated. It is normally expressed as the specific energy consumed by a unit mass of fibres in the web, K ($J\,kg^{-1}$), and depends on the flow rate, water pressure, and residence time of fibres under the jets as follows [14].

$$K = \frac{1}{b\,m\,V_b} \times \frac{C_d\,\pi\,\sqrt{2}}{4\,\sqrt{\rho_w}} \times \sum_{i=1}^{M} n_i\,l_i\,D_i^2\,p_i^{3/2} \tag{8.4}$$

where,

b = width of the web (m)

m = area density of the web (kg m^{-2})

V_b = conveyor belt velocity (m s^{-1})

p_i = water jet pressure at the i_{th} injector (Pa)

C_d = nozzle discharge coefficient

n_i = number of jets in the i_{th} injector (per m)

l_i = width of i_{th} injector (m)

D_i = diameter of jet nozzles in the i_{th} injector (m)

M = number of injectors

ρ_w = water density (kg m^{-3}).

For capillary cone down nozzle configurations, the value of C_d is usually about 0.60–0.66. The pressure–volumetric flow curve normally corresponds to a root function, the precise shape of which depends on the nozzle diameter, the number of nozzles per width, and the nozzle geometry [15]. Unfortunately, not all the energy applied to the web is utilised in directly entangling the fibres. The energy consumption required to produce a serviceable fabric also depends on the physical and mechanical properties of the constituent fibres in the web, the fibre orientation, web thickness and density (which influences the position of fibres relative to the incident jets and their mobility), the porosity and design of the conveyor surface (which influences the probability of turbulent flows in the web that are thought to increase entanglement), and the water content of the web (standing water reduces energy transfer to the fibres). Other potential sources of energy consumption include fibre-to-fibre frictional forces, fluid drag forces, and web compression. The drag forces are associated with the development of fibre entanglement.

Fundamentally, the level of fibre entanglement depends on the quality and stability of the jets and the mechanical and physical properties of the fibres in the web. Particularly important is ability of fibres to wet out, deform, and entangle in response to the applied mechanical forces. To maximise economic efficiency, acceptable fabric strength has to be obtained at the lowest possible energy consumption and highest achievable delivery speed. These competing requirements are difficult to balance and partly depend on raw materials and process conditions.

Because the economics of hydroentanglement depend on energy consumption, the selection of an appropriate jet pressure profile that minimises the specific energy consumption while obtaining satisfactory bonding is important to ensure economic efficiency. The specific energy coefficient (SE_c) is the ratio of energy consumption to the fabric tensile strength [16] and is a useful means of assessing the efficiency of hydroentanglement in terms of the development of fabric strength.

$$SE_c = \frac{K}{T_s} \tag{8.5}$$

where, K = the total energy consumption (sum of all injectors) and,

$$T_s = \frac{T_{s(MD)} + T_{s(CD)}}{2} \times \frac{100}{m} \qquad (8.6)$$

$T_{s(MD)}$ = tensile strength of fabric in the machine direction, $T_{s(CD)}$ = tensile strength in the cross direction, and m = fabric weight per unit area.

8.15.3 Jet impact force

The jet impact force affects the consolidation, thickness, and entanglement of the web during hydroentanglement and depends on the water pressure p and the jet impact area, which is influenced by the jet diameter d. The actual diameter is affected by jet constriction.

$$F \propto p \frac{\pi}{4} d^4 \qquad (8.7)$$

Therefore, at a fixed pressure, the impact force will increase if the nozzle diameter increases [15] providing a potentially more effective means of increasing fibre entanglement by involving more fibres (fibre segments) in the entanglement. The ratio of the impact force F to conveyor speed V_b influences the structure of the fabric particularly the transverse fibre orientation, local density variations, and fabric thickness. In practice, increasing the F/V_b ratio tends to encourage fibres to stick to the conveyor surface, and in extreme conditions, to web perforation. For composite hydroentangled fabrics, consisting of more than one web layer, the ratio can be adjusted to control the degree of interlayer bonding. At high pressure, physical modifications such as longitudinal splitting of segmented pie bicomponent fibres, and fibrillation of fibres such as lyocell can be observed.

8.16 Mechanism of hydroentanglement and fabric structure

There is an incomplete picture of the mechanism of fibre entanglement during hydroentanglement, which reflects the difficulties of observing, as well as simulating, the complex dynamic interactions of fibres with water jets. Understanding is based on analyses of fabric microstructure and theoretical considerations of the probable jet–fibre interactions as well as the interaction of the fibre with the wires and holes in the support conveyor. In hydroentangled fabrics, the fibres or more specifically fibre segments are tangled, intertwined, and interlaced with others.

The rearrangement of fibres is influenced by the interaction of fibres with the water as well as with the support surface. As the incident water jet penetrates the web towards the support surface, some fibre segments are deflected downwards or displaced sideways and entanglements are produced by eddies present within the fluid medium [17]. A method for determining the clustering of transverse fibre segments by analysis of photomicrographs based on the ratio of areas has been described [18]. The number of fibres impacted by the water jet depends on their spatial arrangement, which is governed by the web structure and by the dimensions and frequency of the issuing water jets. Particularly in lightweight fabrics, it has been suggested that high energy fluid flow reflected from the support surface contributes to fibre entanglement [17]. The complex nature of the turbulent effects created in the web near the support surface and the drag forces on fibres, as well as the turbulence created between adjacent water jets, continue to be studied. The importance of turbulent effects and the influence of water jet dispersal on fibre entanglement and fabric structure were acknowledged in the early days of the technology [19].

The potential for effective energy transfer depends on the degree of dispersal of the water jet and the energy dissipation, web density and thickness, jet pressure, and the open area of the support surface. Assuming excess water can be quickly drained from the web, energy transfer is maximised when the support surface is solid, and Suzuki [20] introduced this concept for lightweight fabrics. Rearrangement of fibre segments occurs in both the planar and transverse directions during hydroentanglement. However, it is important to realise that while some deflected fibre ends and looped segments may be observed in the fabric cross-section, hydroentangled fabrics are structurally quite different to needlepunched fabrics and do not have the pronounced periodic fibre pillars associated with the latter. Most fibres are in-plane.

The relatively few transversely orientated fibres present in hydroentangled fabrics are subject to marked variation in terms of relative orientation, periodicity, and depth. This variation is particularly noticeable in the early stages of hydroentanglement at low water pressure. The proportion of fibre segments reoriented in-plane and in the transverse direction is dependent on process conditions such as the water pressure and the ratio of impact force to web speed, as well as web geometry and density. For a given pressure, the number of fibre segments oriented in the transverse direction is inversely related to line speed.

While locally, fibres are rearranged by the jets in the impact zones, there is conflicting evidence about the effect of hydroentanglement on the global fibre orientation distribution of the fabric [21–23]. If the initial web is random or isotropic in terms of fibre orientation and physical properties, this may change during hydroentanglement. The MD/CD ratio of tensile strength in hydroentangled fabrics produced from cross-laid webs can sometimes approach unity following hydroentanglement [14].

Early work reported an increase in the MD/CD ratio with pressure and a combing effect of the jets has been suggested [21], which aligns segments of fibres because of drag forces. However, increasing the jet pressure will also increase the force needed to remove the fabric from the conveyor belt by peeling. The effect of this applied force cannot be ignored. Where transfer is from flat-belt systems, peeling of the fabric from one conveyor to the next can be expected to increase the MD orientation [22] and the tension involved may be expected to increase with jet pressure. Where tension on the web can be minimised, as in rotary systems, changes in fibre orientation can be very small or undetectable.

A characteristic feature of hydroentangled fabrics is jet marking. Jet marks are continuous indented parallel tracks running in the MD of the fabric, the position of which corresponds to the jet impact spacing. Their visibility is minimised by:

– Ensuring successive jets do not impact the same area.
– Increasing the number of injectors in series or use of multirow jet strips.
– Use of smaller diameter jets in the final injectors.
– Reciprocating an injector from side to side.

Jet marking is more pronounced if a high pressure is utilised early in the process or when prewetting is incomplete prior to hydroentanglement. Additional textural features in the fabric are associated with the structure of the support conveyor, which are transferred to the fabric during the process.

8.16.1 The degree of bonding

Initially, as the water pressure is incrementally increased in small steps from zero, no significant increase in fabric strength may be observed, depending on fibre properties until a threshold pressure is reached. After this point, additional small increments in pressure lead to large increases in tensile strength; as the pressure increases, the tensile strength of the resulting fabric increases to a maximum fabric strength before levelling.

The initial rate of increase of the strength depends on fibre and process-related parameters. Fibre fineness and wet modulus are particularly important. Depending on conditions and fibre type, as the pressure continues to increase, the breaking load may eventually decrease due to fibre damage. The maximum fabric strength (MFS) obtained by increasing the water pressure varies depending on several factors. A critical pressure can be identified for a given web of fibres [14] that results in the highest strength and modulus. The fabric strength can be developed with a few injectors operating at relatively high pressure or by an appropriate pressure profile. The choice of approach influences the energy consumption of the process. Some of the considerations affecting the MFS are as follows:

- Fibre type. Strong fibres normally have a high MFS, but the energy expended in reaching that maximum may be unacceptably high from an economic viewpoint. With high modulus fibres, the increased water pressure required can lead to fibre damage, which prevents the theoretical MFS from being reached in practice.
- The pressure profile and the ratio of specific energies applied to the face and back. A lower MFS may be obtained if the fabric is hydroentangled from one side only as compared to both sides. An alternating treatment of the web is preferred.
- The web weight. Normally, a higher MFS is obtained if web weight is increased.

Fig. 8.41 exemplifies the nature of initial increases in fabric strength obtained as energy begins to increase for different fibre types. It illustrates that different energy levels are required to initiate increases in fabric strength depending on fibre type.

The impact force rather than specific energy is also important factor [24] since fabrics produced at similar specific energies using different process parameters do not always yield the same fabric strength [25]. Fabrics produced at the same specific energy, but with a different combination of jet pressures, a different number of injectors or a different face and back treatment of the web, often yield different tensile properties.

It should also be remembered that the aim is not always to produce the highest possible fabric strength, since this may compromise other desirable properties such as the bulk density and absorbency. In practice, the specific energy required to satisfactorily bond fabrics ranges from about 0.1–$0.85\,\mathrm{kWh\,kg^{-1}}$ depending on web weight [15] and there is an approximately linear relation between the energy consumption and web weight.

The entanglement completeness and the entanglement frequency can be used to express the degree of bonding in hydroentangled fabrics based on average values from tensile strip tests conducted in the MD and CD. The entanglement completeness is a measure of the proportion of fibres that break (rather than slip out of the fabric) based on fabric breaking loads obtained at different gauge lengths and strip widths, whereas the entanglement frequency is a measure of the bond point frequency along individual fibre lengths in the bonded fabric [10].

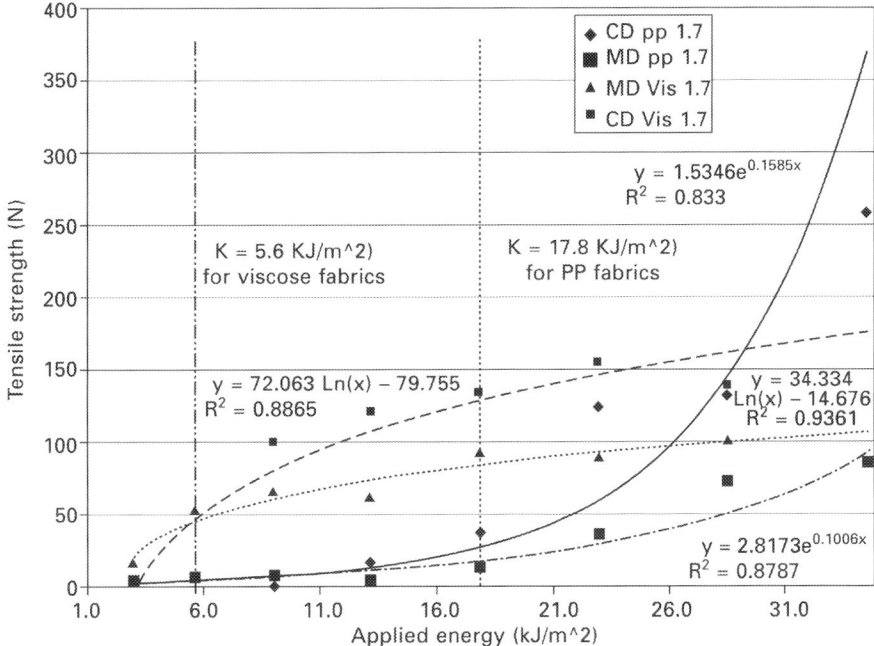

FIG. 8.41 Influence of applied energy on the initial increases in tensile strength of hydroentangled fabrics (1.7 dtex PP and 1.7 dtex viscose rayon in the machine (MD) and cross-directions (CD)).

High entanglement frequencies are associated with improved pilling resistance and fabric surface stability. Durable fabrics can be expected to have an entanglement frequency of 7.9 cm^{-1} and an entanglement completeness of at least 0.5 [10].

Another approach is to estimate the theoretical hydroentanglement intensity [26,27] introduced by the jets. This is based on a simplified model of part of the bonding mechanism, which considers the number of fibres impacted, N, in a unit area of the web by the incident water jets, and the resulting bending of these fibres according to their mechanical properties, y. This approach does not consider the additional complex interactions between the water and the fibres at the support surface, or the intertwining of adjacent fibres. The deflection depth of a fibre segment in the z-direction, y, can be calculated based on the dynamic impact force of a water jet on a fibre segment [26],

$$y = \frac{\sqrt{2}\pi g}{12} \frac{\rho_f}{Ed_f^2} \frac{\rho_w^{0.5} C_d p^{0.5} D^2 d_x^3}{mv_b} \times \left(1 + \sqrt{1 + \frac{48}{\pi^2 g^2 \rho_w} \frac{Ed_f^2}{\rho_f} \frac{mv_b^2 K_2}{nC_d^2 p D^4 d_x^3}} \right) \qquad (8.8)$$

where,

ρw: water density (1000 kg m^{-3})
ρf: fibre density (kg m^{-3})
C_d: water flow discharge coefficient
D: orifice diameter (m)

d_f: fibre diameter (m)

d_x: diameter of a water jet at the web surface (m)

E: Young's modulus of the fibre (Nm^{-2})

K_e: kinetic energy of the water jets consumed in bending the fibres per unit area of the web (Jm^{-2})

m: area density of the web ($kg\,m^{-2}$)

n: number of jets in the jet strip (jets m^{-1})

p: hydrostatic pressure drop (Nm^{-2})

v_b: conveyor belt velocity (ms^{-1})

y: deflection depth of a fibre due to a dynamic impact (m).

If it is assumed that the coefficient of jet constriction is equal to one, then the applied energy can be written as follows [26].

$$K_e = \frac{1.11 n C_d^3 D^2 p^{1.5}}{v_b \rho_w^{0.5}} \; (J/m^2) \tag{8.9}$$

The number of impacted fibre segments N is proportional to the total impact area of the water jets on the web, and the deflection depth of each impacted fibre segment depends both on fibre properties and the applied energy.

$$N = \frac{m A_w}{\frac{\pi}{4} d_f^2 d_x \rho_f} \frac{1}{v_b} = \frac{4mn}{\pi d_f^2 \rho_f} \; (\text{Number of impacted fiber segments/m}^2) \tag{8.10}$$

where, A_w is the impact area of a single water jet on the web per unit time (m^2), N is the number of fibre segments deformed in a unit area of the web during jet impact (fibres m^{-2}), H is the hydroentanglement intensity (m) [26].

$$H = yN = \frac{\sqrt{2}g \, n\rho_w^{0.5} C_d p^{0.5} D^2 d_x^3}{3E d_f^4} \frac{1}{v_b} \times \left(1 + \sqrt{1 + \frac{48}{\pi^2 g^2 \rho_w} \frac{E d_f^2}{\rho_f} \frac{m v_b^2 K_2}{n C_d^2 p D^4 d_x^3}}\right) \tag{8.11}$$

The influence of fibre stiffness on the hydroentanglement intensity can be obtained from,

$$H \approx 2.43 * 10^2 \left(\frac{1}{R^{0.5} \rho_f d_f}\right) \left(\frac{m}{v_b}\right)^{0.5} \left(\frac{n C_d^{1.5} p^{0.75} D d_x^{1.5}}{\rho_w^{0.25}}\right) \text{when}$$

$$\sqrt{\frac{1.03 * 10^{-6} R}{d_f^2} \frac{m v_b^2 K_2}{n \rho C_d^2 D^4 d_x^3 p}} \gg 1 \tag{8.12}$$

where, R is the flexural rigidity of a fibre with a circular cross-section (Nm^{-2}).

8.17 Fibre selection for hydroentanglement

Virtually, all polymeric fibres of a wide range of dimensions are compatible with hydroentanglement, providing they can be first formed into a web at commercially acceptable production speeds. However, the process efficiency, fabric properties, and economics of hydroentanglement vary depending on fibre selection. It is important in hydroentanglement to maximise bonding while minimising the energy. Fibre properties have a significant influence on the degree of bonding that can be achieved for a given energy consumption.

8.17.1 Fibre stiffness

Manufacture of coherent hydroentangled fabrics with minimum energy consumption requires flexible, deformable fibres that can be readily intertwined and entangled. Fibre flexural rigidity depends on diameter, Young's modulus, and cross-sectional shape and density. Depending on the fibre type, some of these properties are strongly moisture dependent, particularly the modulus, which is particularly pertinent in hydroentanglement. Fibre flexural rigidity values can influence hydroentanglement efficiency, but the fibre bending deformations in the process are believed to be three, rather than two-dimensional in nature and the torsional rigidity of the fibre should also be considered [17]. Viscose rayon has a low wet modulus, and this partly explains the ease with which this fibre can be hydroentangled. At low pressure and energy, heavily entangled fabrics are produced from viscose rayon, whereas under the same conditions, the entanglement is significantly lower in fabrics containing, for example, PP.

While it is possible to hydroentangle high modulus fibres such as glass and carbon fibre, the strength realisation tends to be poor because of the limited fibre entanglement that is introduced. Attempts to increase the entanglement by increasing the water pressure can lead to fibre damage and some fibres can be broken rather than entangled, even at relatively low pressure. Fine fibres are therefore more flexible than coarse fibres of the same composition, they hydroentangle more intensively for a given energy consumption and produce stronger fabrics. The accompanying increase in fibre surface area and the increased number of fibre intersections as the fibre diameter decreases contributes to increased fabric strength. For a given polymer type, fibre cross-sectional shape also affects bending and entanglement. Triangular-shaped fibres require more energy than do round fibres, and elliptical and flat fibre cross-sections hydroentangle quite efficiently.

8.17.2 Wettability

Effective hydroentanglement requires uniform and rapid prewetting of the web. The choice of spin finish is particularly important to ensure proper wetting-out of hydrophobic fibres such as PP and PET and to minimise foaming during the process as the finish is removed in the wastewater. The antistatic component of the finish can be particularly associated with foaming, but formulations have been developed that minimise the problems. In one example, a finish containing one monoester of glycerol and a fatty acid having from 6 to 14 carbon atoms is applied as an aqueous dispersion [28]. The durability of hydrophilic finishes tends to be low, such that they are effectively removed during hydroentanglement. This can necessitate reapplication of hydrophilic agents to synthetic fibre fabrics intended for absorbent products following hydroentanglement. Durable hydrophilisation has been demonstrated by some fibre manufacturers [29] and PP fibres specifically for hydroentanglement are produced [30]. Treatments such as those involving plasma can also be used to modify the surface chemistry of synthetic polymers, with hydrophilic groups that modulate the wetting behaviour.

8.17.3 Fibre dimensions

Hydroentangled fabrics can be made from webs containing fibres, continuous filaments, or a combination or both, but the majority are made from staple fibres. For hydroentangling, fibre linear density normally, but not exclusively, ranges from 1.1 to 3.3 dtex.

Because hydroentangling relies on frictional forces to generate fabric strength, very short fibres of 1–4 mm, including wood pulp cannot be entangled to any significant degree, and need blending with longer fibres. However, the consolidation resulting from flattening of the web during hydroentangling can often produce a degree of strengthening, due to increased normal forces acting on each fibre-to-fibre contact point. To bond wetlaid or airlaid webs containing large proportions of wood pulp normally requires inclusion of short cut man-made fibres of 5–15 mm in the blend. For hydroentangling of carded webs, common linear density and fibre length combinations are 1.7 dtex/38 mm, 3.3 dtex/50 mm, and 3.3 dtex/60 mm. The fibre length and slenderness ratio are particularly important in determining compatibility with the process as well as with preceding web forming.

Fine fibres have a high specific surface area for a given fabric weight leading to good fabric strength. There is also a direct relation between fibre diameter and bending stiffness, which influences entanglement efficiency. For good fabric formation, fibres must be flexible and capable of deformation and deflection in the web. While short cut fibres above 4–5 mm hydroentangle quite efficiently, there is a positive relationship between fibre length and fabric strength within a narrow length range. For carded-hydroentangled fabrics when fibre fineness is constant, fabric strength increases with fibre length up to a maximum length of about 50–60 mm [16]. For viscose rayon, fabric strength has been shown to increase with fibre lengths up to 51 mm [17]. Generally, hydroentangling of continuous filaments in spunlaid webs requires relatively high water pressure, to overcome the frictional resistance and enable slippage and entanglement.

While crimp promotes cohesion in carded webs prior to bonding, it also influences the strength of fabrics produced by low to medium pressure hydroentanglement systems [16]. As a consequence of the high resistance to compression, helically crimped fibres require higher energy inputs to obtain fabrics with acceptable strength. The effect of crimp is particularly noticeable in the hydroentanglement of fine wool, which is resilient and has a high crimp frequency such that relatively high, water pressure is required to produce a coherent, abrasion-resistant fabric.

8.17.4 Fibre types

Commercially, wood pulp, staple fibre viscose rayon, lyocell, and PET are most widely used in hydroentanglement as well as blends thereof. Industrially, hydroentangled fabrics are also made from a range of other fibre types including cotton, flax, PP, PLA, as well as bicomponent fibres and aramids. Partly because of the growth of hydroentangled wipes, hygiene, and medical products, the consumption of viscose rayon and lyocell has increased. The high wet strength of lyocell is particularly valued in absorbent hydroentangled wipes, as well as the excellent pattern definition it provides in apertured fabrics. The wet strength and volume of viscose rayon fabrics is commonly increased by blending with PET or PP. Wood pulp is widely utilised in hydroentangled fabrics as a low-cost absorbent for wipes, surgical gowns and drapes but only in blends or multilayer structures. In airlaid or wetlaid production, wood pulp is blended with either short cut viscose, lyocell, or PET and hydroentangled. The short cut fibre provides a reinforcing matrix for the pulp in the fabric. Assuming appropriate water filtration is available as part of the hydroentanglement system, webs composed of linters and noils, as well as bleached cotton are hydroentangled to produce absorbent products including wet wipes, medical gauze, and cosmetic pads. The fibres are usually 7–25 mm and 1.2–1.8 denier [31] and a large proportion of cotton wax can be removed from unbleached fibre during hydroentanglement at comparatively low energy (below 1 MJ kg^{-1}), which increases the hydrophilicity. The strength of hydroentangled cotton fabrics tends to increase as the fibre fineness decreases but the fabrics have a harsher handle. One limitation of bleached cotton is that it can be difficult to card, which affects the production rate of a hydroentanglement line.

8.17.4.1 High-temperature fibres

High-temperature–resistant hydroentangled fabrics have been produced for years from high-performance fibres, largely pioneered by DuPont. Meta-aramid hydroentangled fabrics are found in aerospace and protective clothing markets [32,33] together with para-aramids for high-temperature protection and blends of meta and para-aramids are utilised in thermal barrier, heat-shielding materials. Hydroentangled melamine and aramid fibres have been developed for fire-retardant clothing [34] and hydroentangled Basofil fabrics have also been produced. Provided the impact forces are low to minimise fibre damage, inorganic fibres such as glass and silica can be hydroentangled to produce pre-pregs for composites. The hydroentanglement of ceramic fibres has been discussed [35]. Fibres with a high metal oxide composition including silica (SiO_2) are liable to break during low-pressure hydroentanglement and fabrics produced from such fibres tend to have low integral strength because of the limited fibre entanglement that can be introduced.

8.17.4.2 Splittable bicomponent fibres and filaments

The formation of microfibres in situ by splitting segmented pie bicomponent fibres or continuous filaments within webs during hydroentanglement is well established. This is different to separating island-in-the-sea bicomponents using solvents following nonwoven production, which is the traditional method of synthetic leather production. Splitting during hydroentanglement takes place at the same time as bonding and relies on a low interfacial adhesion between the two polymers in the bicomponent, as well as the impact and shear forces delivered by the incident water jets (Fig. 8.42). High-performance, durable wipes for applications such as optical glass cleaning, synthetic leather coating substrates, and upholstery fabrics are produced in this way. In the drylaid route, carded and cross-lapped batts containing segmented-pie bicomponent fibres are hydroentangled to produce microfibres by splitting. The spunlaid route obviously relies on hydroentangling webs comprising segmented bicomponent filaments.

FIG. 8.42   Splitting of bicomponent fibre as a result of hydroentangling.

Subsequently, fabric density can be further increased by inducing thermal shrinkage of additional thermoplastic fibres blended or mixed with the splittable bicomponent fibres in the web. The elimination of conventional leather production steps such as shrinking, splitting, and grinding leads to a significant saving in raw materials. Synthetic leather fabrics have excellent strength and durability and depending on polymer composition can be dyed and finished to develop attractive softness and handle characteristics.

The specific cross-sectional configuration of splittable bicomponent fibres and filaments affects splitting efficiency and fabric properties. The splitting efficiency of segmented pie bicomponents improves if there is a hollow core. Rectangular-striped cross-sections also increase splitting efficiency because of the aspect ratio. The impact force generated by the jet is a function of the pressure and jet diameter, and these must be selected to obtain uniform splitting throughout the fabric. At a fixed pressure, web weight and number of injectors, the splitting efficiency increases with jet diameter. Partial separation of some splittable bicomponent fibres can occur during high-speed carding, and this can limit the production rate of both the card and hydroentanglement process. The critical pressure at which splitting is induced ranges from 50–100 bar and varies depending on the stored strain in the individual fibres and their geometric position within the web cross-section. To achieve high splitting efficiency, water pressures up to 250–400 bar can be necessary partly because in heavyweight webs, the degree of splitting varies through the cross-section. It is necessary to ensure, at least initially, that fibres are properly entangled through the fabric cross-section before splitting is complete. For these reasons, the water pressure in the first few injectors must not induce excessive splitting, but rather should entangle the fibres to introduce satisfactory bonding through the cross-section.

The fineness of the fibres or filaments after splitting depends on the number of segments in the bicomponent cross-section. Typically, after splitting, the fineness ranges from about 0.05 to 0.3 denier, depending on the cross-sectional configuration of the bicomponent. Experimentally, island-in-the-sea bicomponent filaments have been produced with as many as 600–1120 individual PP filaments embedded in a soluble PVA matrix within the cross-section to make exceptionally fine microfilaments, but their release normally relies on washing out rather than hydroentangling the structure [36]. Hydroentangled nanofibre fabrics produced from splittable bicomponents is also technically feasible, and hydroentanglement of nanofibre filaments has been demonstrated [37].

8.17.4.3 *Fracturing bicomponent filaments*

In addition to splitting segmented bicomponent filaments during hydroentanglement of spunlaid webs, an alternative route involves fracturing island-in-the-sea bicomponents to release microfilaments. During hydroentangling, the sea (or 'matrix') is mechanically fractured to release the embedded island components, and clearly modulation of polymer combinations, cross-sectional structure, and the number of island components, provides significant opportunity for engineering fabric properties.

8.17.4.4 *Fibrillating fibres*

Longitudinal splitting of microfibrillar fibres is possible at high impact forces, which can be exploited to engineer the physical properties of fabrics. Such fibrillation during hydroentanglement can be observed in natural cellulosics including cotton and bast fibres as well as lyocell, polynosics, polyacrylonitrile, and even para-aramids if the water pressure is sufficiently high. Fibrillation usually occurs by splitting of the fibre from the outside in. The fibrils are partially exposed and project from the surface of the parent fibre, such that they are readily entangled with neighbouring fibrils. The water pressure at which fibrillation begins varies depending on the fibre type and specification, and for many commercial fabrics, hydroentanglement conditions are purposely selected to avoid the onset of fibrillation. The increase in surface area associated with mass fibrillation modifies the physical, optical, and mechanical properties of the fabric. There is a decrease in fabric permeability, which is useful in the design of filter media, and an increase in the fabric opacity. Following mechanical finishing, a microfibre pile can be created on the fabric surface, transforming the aesthetics. If finishing is not carried out, fibrillation tends to lead to a paper-like surface, which can give the fabric a harsh handle. Lyocell has attracted interest partly because of its propensity to fibrillate at high pressure, which can be harnessed to both increase entanglement and modify key fabric properties such as tensile strength, permeability, and liquid transport. Applications for hydroentangled fibrillated lyocell fabrics include filtration and wipes.

8.18 Process layouts in hydroentanglement

The market for hydroentangled fabrics has driven significant growth in the installed manufacturing capacity across Europe, the USA, and Asia. Turnkey hydroentanglement machinery suppliers include ANDRITZ Perfojet who supply the Jetlace and Spunjet range of systems, Trützschler Nonwovens (Fleissner AquaJet system), and AUTEFA Solutions (V-Jet FUTURA). These companies have been responsible for the supply of many new installations, since the 1980s, in the case of Perfojet, and since the late 1990s, in the case of Fleissner AquaJet technology. Hydroentangling machinery is also supplied by SICAM, as well as machine builders in China, and historically installations have been constructed by companies such as CEL in the UK, Mitsubishi Engineering in Japan, among others.

Hydroentanglement has become increasingly cost-effective and accessible to manufacturing companies globally. Over the last 15 years, flexible, high production turnkey systems have been developed suitable for a range of markets, as well as for specific products, e.g., wetlaid coupled with hydroentangling (the 'wet on wet' process) for dispersible (flushable) wipes. Turnkey installations are available coupled with drylaid, wetlaid, or spunlaid web formation systems. Technically, all generic web types can be hydroentangled including combinations of different web types, although carding is still the most common. Higher productivity and process efficiency continue to be major drivers for development and production rates up to $1100 \, \text{kg} \, \text{h}^{-1} \, \text{m}^{-1}$ have been reached. In the production of lightweight fabrics of $20–120 \, \text{g} \, \text{m}^{-2}$ for wipes, hygiene, and medical applications, line speeds of $200 \, \text{m} \, \text{min}^{-1}$ to over $250 \, \text{m} \, \text{min}^{-1}$ are common, but with newer systems, production speeds of $300–400 \, \text{m} \, \text{min}^{-1}$ are possible with 99% water recycling. For heavier weight fabrics up to ca. $400 \, \text{g} \, \text{m}^{-2}$, carding, cross-lapping, and hydroentangling lines produce automotive, filtration, artificial leather, and coating substrate fabrics in widths up to about 6–7 m, but inclusion of a cross-lapper limits line speeds to about $100 \, \text{m} \, \text{min}^{-1}$. Coupling hydroentangling with spunlaid web formation has enabled maximum operating speeds to reach ca. $1000 \, \text{m} \, \text{min}^{-1}$, with widths up to 7 m. However, not all hydroentangled fabrics are produced at such high speeds. In the production of cosmetic pads composed of cotton, for example, machine widths are generally less than 3.6 m, and line speeds can be as low as $30\text{-}60 \, \text{m} \, \text{min}^{-1}$. Given the high capital investment costs of hydroentangling installations, about 15 years ago smaller systems were launched to provide a more cost-effective entry point, e.g., the Fleissner LeanJet system [38] and MiniJet (now Trützschler Nonwovens), which when paired with a short staple flat card, provides a versatile R&D installation. In addition to contemporary hydroentangling systems, older installations remain in operation. To remain competitive, many have been extensively modified and updated since their original installation.

The water pressure, number of injectors, and machine width influence the costs of the process. Consequently, minimising these while obtaining satisfactory fabric properties, particularly strength, can be advantageous providing the versatility of the line is not lost. The investment costs of an installation have been claimed to approximately double for every 100 bar increase in pressure [39] taking into account the costs of the water pump, electric motor, inverter, and high-pressure pipework. The electrical power consumption of an injector is linked to the pressure used. Based on a 3.5 m operating width, approximately 155 kWh is consumed by an injector at 200 bar, 544 kWh at 400 bar, and 1088 kWh at 600 bar [40]. Over about 150 bar, the costs increase further with machine width. Total energy consumption for the hydroentangling part of a production line is about $1\text{-}2.5 \, \text{kWh} \, \text{kg}^{-1}$.

Commercially, there is no universal layout; the machine configuration depends on:

- Raw material properties and the method of web formation used to supply the machine.
- Fabric weight.

- Cost, particularly with respect to energy consumption.
- The need to introduce other components such as webs, reinforcing scrims, or preformed fabrics to manufacture multilayer structures.
- The required patterning or aperturing capabilities.
- Whether or not additional bonding will take place, e.g., thermal, or chemical bonding.
- The intended product application and the degree of machine versatility required.

Injectors may be arranged over the top of a flat conveyor or around the circumference of a rotary drum or cylinder. In practice, both arrangements are frequently incorporated as modules in sequence. Examples of integrated machine systems are shown in Fig. 8.43. Rotary drums are favoured if, as is usual, a dual-sided treatment of the web is required, although historically this has also been accomplished in flat belt systems using an overhead transfer and turn-back arrangement fitted between modules [41]. Dewatering, fabric threading, floorspace requirements, and peeling of the fabric from the conveyor surface are improved with the rotary system. In the flat belt system, fibre ends can partially penetrate the conveyor and when dragged over the suction slots during processing this inhibits web removal.

The precursor web fed to the hydroentanglement machine largely determines the isotropy and quality of the final fabric, and in a production line web formation is often the rate-determining step. Web weights can range from 15 to $400\,\mathrm{g\,m^{-2}}$ but commercially, most are 40–$150\,\mathrm{g\,m^{-2}}$ depending on fibre fineness.

8.18.1 Carding and other drylaid-hydroentanglement installations

Commercially, web formation by carding predominates in hydroentangled fabric production. Common applications are wipes, cosmetic pads, and woundcare fabrics. Lightweight webs are usually prepared by straight-through carding and the fibre length ranges from 25 to 60 mm. Production lines making heavier weight fabrics, particularly those for multiple use, or durable product applications are fed with carded and cross-lapped batts. Cross-lappers in a hydroentangling line normally need to be capable of high speeds, e.g., up to $190\,\mathrm{m\,min^{-1}}$, to maximise the overall production efficiency. Traditionally, in the manufacture of cotton make-up removal pads, the webs from multiple cards are combined to produce web weights up to 200–$250\,\mathrm{g\,m^{-2}}$.

Parallel-laid carded webs tend to require higher energy to obtain adequate CD strength than do cross-laid webs. In straight-through carding systems, one or more double doffer cards are used to supply web weights from about 20 to $80\,\mathrm{g\,m^{-2}}$. Where a card and profiling cross-lapper are utilised, basis weights range from 80 to $400\,\mathrm{g\,m^{-2}}$. Web spreading using rollers is sometimes used to increase the width and to modify fibre orientation prior to hydroentanglement. The design of high production nonwoven cards for producing light-weight, isotropic webs that can balance the high delivery speeds in hydroentanglement continues to be a challenge. High-speed cards capable of $400\,\mathrm{kg\,h^{-1}\,m^{-1}}$ (depending on fibre specifications) allow web geometry to be modified by scrambling or randomising to minimise the MD/CD ratio. However, scrambling rollers decrease the linear output speed of the carded web.

Card widths extend to 5.1 m for producing hydroentangled fabric although trim widths may be lower at about 4.6 m [42]. A line designed for the production of 'straight-through',

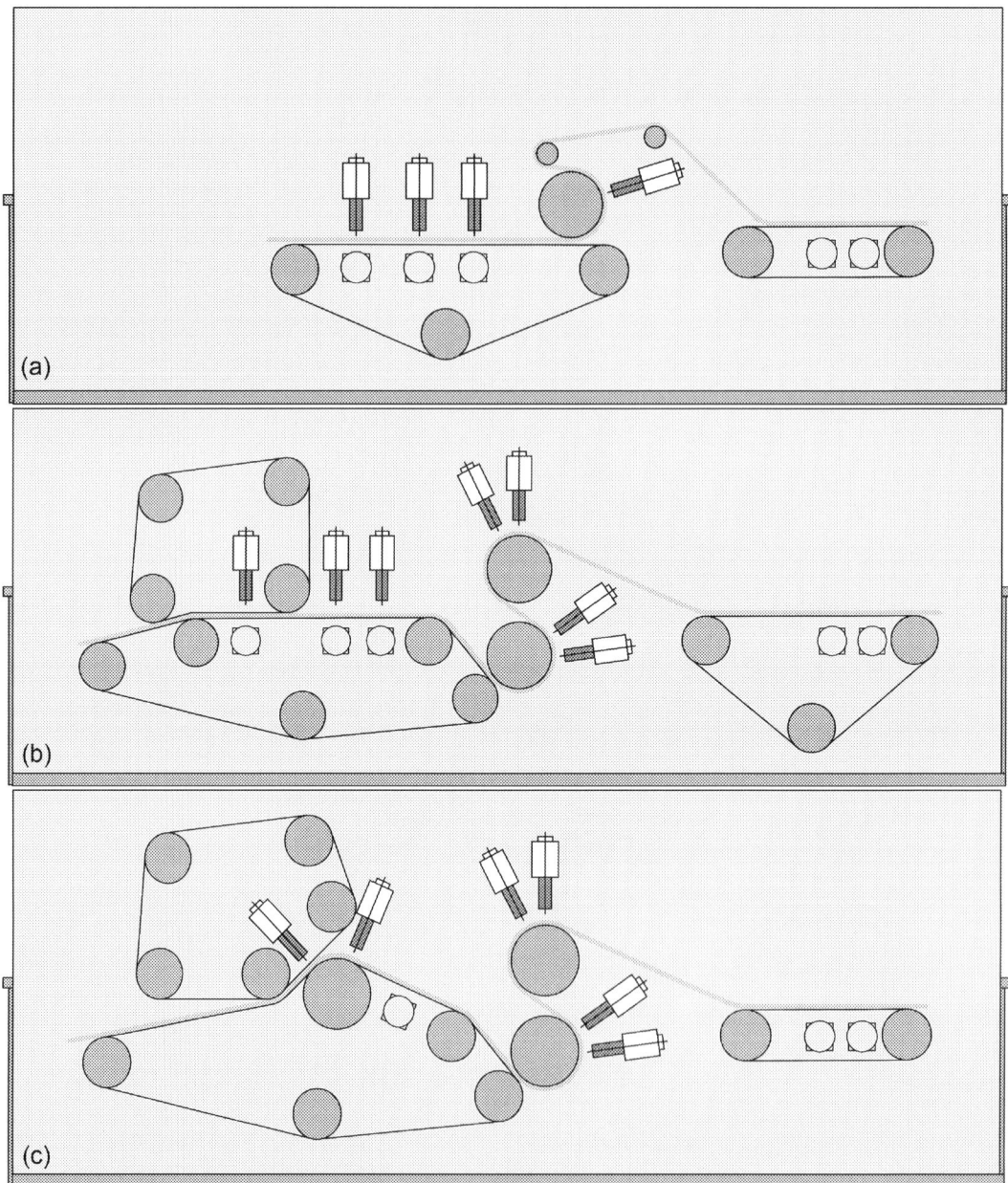

FIG. 8.43 Hydroentanglement machine configurations: Aquajet systems (A) one hydroentangling drum; (B) two hydroentangling drums, and (C) three hydroentangling drums. *Courtesy: Trützschler Nonwovens.*

or 'direct' carded, low to medium weight fabrics of ca. 20–120 g m^{-2} at delivery speeds up to 400 m min^{-1} might consist of the following components: two in-line worker-stripper nonwoven cards and a twin cylinder hydroentanglement module operating with multiple injectors, followed by hydroentanglement on a flat belt using two injectors, and through-air drying and winding. If higher weights up to 400 g m^{-2} are required with good isotropy, carding and cross-lapping are typically employed for web formation.

Some lines operate with twin cards, and in this case one of the cards usually operates with a cross-lapper to increase versatility. Following the formation of a batt, batt drafting enables modification of the MD/CD ratio to increase isotropy. Four cylinders are employed on the hydroentangling system, with alternating injectors before the final flat belt section, which in this example is fitted with one injector. Through-air double drum drying precedes winding up.

Hybrid card-airlay systems have also been introduced for producing webs with very low MD/CD ratios. Nevertheless, regardless of the isotropy of the initial web, the resulting hydroentangled fabric can have a higher MD/CD ratio. Rather than using wetlaid web formation, short fibre airlaid webs consisting of wood pulp blended with short cut regenerated cellulose fibres of typically 5–12 mm can also be hydroentangled to make water dispersible (flushable) wipe substrates, as well as other products. Another example of an airlay-hydroentangling system, is offered by SICAM to produce cotton pads in the range 140–240 g m^{-2} at 10–25 m min^{-1} over operating widths of 1.6–2.5 m. Drying of hydroentangled fabrics commonly makes use of a through-air, single-drum dryer before wind-up.

Preformed tissue, other nonwoven webs, reinforcing scrims, and filaments can be incorporated with carded webs prior to hydroentanglement.

8.18.2 Carding and preformed tissue hydroentanglement installations

To produce composite nonwovens consisting of carded webs and tissue layers, the tissue can be introduced onto the prebonded carded web between the first and second hydroentanglement modules. Carding is followed by prewetting, and the web is then hydroentangled using twin-cylinder modules fitted with alternating injectors. The carded web and tissue are combined on a flat belt hydroentanglement module using two injectors to integrate the two components. Through-air drum drying and winding then follows.

8.18.3 Carding and pulp hydroentanglement installations

Two-layer carded-pulp (CP), as well as three-layer carded-pulp-carded (CPC) hydroentangled fabrics can be produced by combining carded and either wetlaid, or airlaid webs. When based on airlaid pulp, three-layer fabrics are sometimes referred to as, carded-airlaid pulp-carded (CAC). CP and related fabrics are particularly suitable for wet wipe applications including premoistened baby and body wipes, where the ability to make substrates from 100% cellulosic fibres is attractive. To produce CP hydroentangled fabrics, pulp is either airlaid or wetlaid and combined with a carded web, which is usually composed of regenerated cellulose fibres such as viscose. Historically, blends have also incorporated PET and PP when using short fibre airlaying. Preformed tissue can be combined with the carded web as an alternative to in-line airlaid or wetlaid web formation, reducing the capital cost of the production line. To make hydroentangled CPC fabrics, a second carded web is laid on top of the CP layer prior to prewetting. Twin-cylinder hydroentanglement follows with four or more injectors, followed by optional hydro-patterning of the fabric. Dewatering, through-air drum drying and winding up completes the installation. CP hydroentangled fabrics can also be produced by bypassing the second card, but CPC products help to prevent dusting of the pulp from the wipe during use. Examples include the ANDRITZ neXline wetlace CP and CPC systems, which couple wet-laying, carding, and hydroentangling. Operating widths are up to 4.8 m, with production capacities of ca. 25,000 t per annum.

8.18.4 Wetlaid-hydroentanglement installations

Wetlaid-hydroentangling installations have been around for years, as part of lines making mostly lightweight fabrics for a variety of single-use applications. However, in recent years, there has been substantial growth in the installed capacity mainly driven by developments in the demand for wipe substrates. The 'wet-on-wet' process, which couples wetlaid web formation and hydroentangling, is particularly relevant (but not exclusively) for the manufacture of INDA/EDANA guideline (GD) compliant dispersible (flushable) wet wipe substrates. Turnkey installations include the Voith–Trützschler Nonwovens Aquajet wetlaid-spunlace (WLS) system, and the ANDRITZ Wetlace system. An inclined wire wetlay former allows webs to be produced containing a high proportion of wood pulp, blended with short cut fibre, usually composed of regenerated cellulosics. This provides a route to a 100% cellulose-based substrate, free of chemical binders. In modern systems, the white water from the wetlaid former is filtered along with the water from the hydroentangling section, so that it can be recycled as part of the process. In addition to applications in hygiene, the wetlaid-hydroentangled process also facilitates production of fabrics made of short inorganic, metal coated, or other high-performance fibres for long-life, high-value market applications.

8.18.5 Spunbond-hydroentanglement installations

Hydroentangling spunlaid webs is particularly valuable for the purpose of manufacturing fabrics suitable for multiple and long-life applications. Although high water pressures of 300–400 bar are potentially required to adequately bond spunlaid fabrics, one of the attractions is the opportunity to produce at high linear delivery speeds up to >600 m min^{-1} in widths up to 7 m. Since spunlaid-hydroentangled fabrics are free of thermo-fused regions introduced by calender bonding, the fabrics have comparatively high bulk and good tactile characteristics as well as high tear strength and a comparatively low flexural rigidity. The risk of thermal degradation of the polymer during bonding is also obviated. In one example of a spunlaid-hydroentanglement installation, a web of 10–400 g m^{-2} is compressed and hydroentangled on a multidrum hydroentanglement installation using five injectors. The fabric is then hydroentangled on a flat belt module fitted with at least one injector. Through-air drum drying and winding up completes the line.

Using a spunlaid platform followed by hydroentanglement, low to medium weight fabrics, particularly in the hygiene, filtration, and geosynthetic sectors have been targeted because, for example, the polymer type, the additives, and the linear density of the filaments can be changed to suit the application as required, prior to hydroentanglement. Examples of spunlaid-hydroentangled fabrics include Berry Global's Spinlace fabrics produced from 0.5 to 3 denier continuous filaments, RKW's HyJet hydroentangled spunbond nonwovens and Freudenberg's Evolon fabrics produced from splittable bicomponent filaments. The Spunjet system of ANDRITZ Perfojet is an example of a turnkey spunlaid web formation and hydroentanglement installation suitable for either bonding, or bonding combined with splitting of bicomponent filaments. Typical applications include geosynthetics (Spunjet Bond) and hot gas filtration (Spunjet Splittable). Owing to their continuous filament rather than staple fibre composition, spunlaid-hydroentangled fabrics tend to have a flat surface free of projecting fibre ends, making them excellent coating substrates. Hydroentangled fabrics made from microfilaments can also be coated, e.g., with TPU for applications in the automotive and furniture industries.

8.18.6 Combination bonding

In practice, thermal or chemical bonding is sometimes used after hydroentanglement to produce the final fabric. Traditionally, apertured fabrics produced at low pressure have been chemically bonded to adequately stabilise the fabric. In household wipes applications, approximately 7% to 27% of the final fabric weight can be binder, with the higher proportions being applied to viscose rayon fabrics, to enhance wet strength. In viscose-PET blends, the binder content can be reduced. Rotary screen printing is one way to apply the binder and pigment to the hydroentangled fabric. This also provides a facility to visually differentiate the product, as well as to modulate the surface roughness, to aid wiping efficiency. Fabrics can be produced using one to three low-pressure injectors operating at pressures below 80 bar. Thermal or chemical bonding then follows to develop the required fabric properties. For chemical bonding, it is common to apply a low acrylic binder add-on of 2% to 5% owf by foam padding. This increases the modulus, tensile strength, and abrasion resistance of the fabric while minimising associated increases in stiffness [43] and such fabrics may be flat or apertured. The hydroentanglement, followed by chemical bonding route, is used commercially to produce linings and interlinings and some wipes products. Chemically bonded hydroentangled fabrics overcome the delamination problems associated with traditional carded-chemically bonded fabrics.

Hydroentanglement followed by thermal bonding provides a potential means to minimise energy costs and to permit high production rates since both the water pressure utilised in hydroentanglement and the temperature and pressure applied in thermal bonding are lower than would be selected if one method alone was employed. Resulting fabrics also tend to be softer than those produced by thermal bonding alone since extensive thermo-fusion of the thermoplastic fibres in the fabric is not required to develop acceptable fabric properties. Thermal bonding is also used in some nonwoven manufacturing lines to fuse bicomponent fibres in the fabric, following an initial hydroentangling step.

In some cases, hydroentangling is utilised after, rather than before thermal bonding, for example, to reduce the stiffness of fabrics, as well as introduce additional texture, with can aid aesthetics. In addition to being the primary bonding process by which some spunlaid nonwovens are made, hydroentangling can also be applied to thermally bonded spunbond fabrics to further entangle continuous filaments in the structure, enabling production of highly durable and dyeable nonwovens [44].

8.19 Hydroentanglement process technology

8.19.1 Web/batt compaction and prewetting

Prewetting evacuates air from the web or batt prior to hydroentanglement in order to:

- Prevent uncontrolled disturbance of the fibre arrangement to minimise changes in the MD/CD ratio of the web prior to bonding.
- Minimise jet marking when the web is impacted by the main jets.
- Enable the web to pass between the first injector and the support surface.
- Lightly adhere the web to the conveyor to prevent slippage.

Prewetting must be uniform to minimise variations in the degree of bonding introduced by subsequent injectors. In practice, various methods of prewetting have been employed.

A low-pressure injector can be used, but jet marking can be a common problem with such an approach, and it reduces the efficiency of the hydroentanglement line. Similarly, spraying systems introduce imperfections in the web if not precisely controlled. If properly designed, weir systems, which apply a continuous curtain of water across the web, can be effective. Compression of the web by rollers in a water bath has also been utilised. Modern approaches integrate mechanical compaction of the web and prewetting using at least one low-pressure injector. The web is sandwiched by either two permeable belts, a permeable belt and a roller or two permeable rollers depending on the machinery supplier. Using such arrangements, it is possible to use higher water pressure without destroying the web and to minimise unwanted drafting of the web. In Fig. 8.44, the web is mechanically compressed between

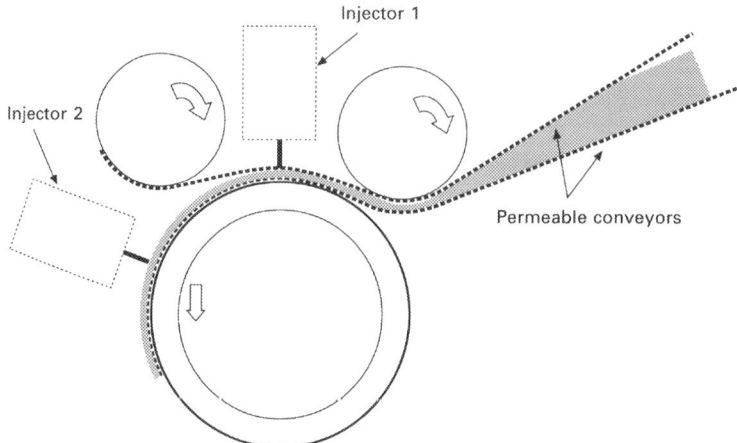

FIG. 8.44 Mechanical web compaction and prewetting system. *Courtesy: Fleissner GmbH, now Trützschler Nonwovens.*

two permeable, converging conveyor belts that transport the sandwiched web towards a roller and a low-pressure injector.

In the case of, for example, cotton pad production, where compression needs to be minimised to maximise the porosity of the final fabric, a drum and belt system may be selected. Once at the nip point, low-pressure water jets from a single injector are directed towards the web through one of the permeable conveyor surfaces. This introduces some fibre entanglement depending on the water pressure. The excess water is removed by means of a suction slot and this must be effective to avoid a decrease in fabric strength. Interestingly, although low water pressure (<30 bar) is normally used in a prewetting injector, large increases in fabric strength can be obtained by operating the prewetting injector at high pressure. In practice, this can enable a reduction in the number of subsequent high-pressure injectors needed to produce a serviceable fabric. The ISOjet system developed by ANDRITZ Perfojet was introduced to decrease the MD/CD ratio in hydroentangled fabrics to about 1.2:1 to 4:1 depending on the line speed [45].

8.19.2 Support surface

The design of the support surface, i.e., the conveyor, or 'wire' on which the web is carried during hydroentangling influences the final fabric strength, fabric structure, visual appearance, and energy consumption.

The conveyor surface may be a permeable, continuously woven mesh of metal or polymeric construction, a solid metal roll, or a sleeve, which is normally perforated. The latter enables precise control of the open area and surface structure as well as allowing dewatering and may be engraved to transfer patterns or apertures during hydroentanglement. To promote bonding and the formation of dense fabrics by maximising energy transfer into the web, mesh structures with a small open area of 15% to 25% are preferred, whereas meshes with larger open areas produce permeable fabrics with lower tensile strength. During operation, the interstices in the conveyor belt can fill with fibre debris and contamination, negatively affecting drainage and causing the web to stick. When woven belts are utilised it has been claimed that energised water hitting the surface is randomly reflected and is not directed back into the web to increase fibre entanglement [40]. Consequently, because of the energy absorption by the mesh belt, more injectors are required to compensate for the low energy transfer efficiency. In early Perfojet systems, injectors were operated in conjunction with opposed deflector plates that were designed to reflect energised water back into the web creating turbulence and more intensive bonding. Utilisation of up to 40% of the initial water jet energy has been claimed using this approach [46].

Historically, impermeable drum and conveyor surfaces have been found to increase the energy transfer to the web and produce complex turbulent effects on the support surface that are believed to account for increased fibre entanglement. In the original Unicharm system, drums of 50–300 mm diameter were arranged in series operating in combination with at least one injector [47]. Lightweight webs were bonded at relatively low energy by injectors positioned on one side of the web surface. The maximum web weight in these systems was limited to 15–100 g m^{-2}, although 20–60 g m^{-2} was preferred. In addition to defects in the fabric, impermeable drums tend to flood, which limits the maximum flow rate of each injector. Water drains along the rollers and at the edges of the web by gravity for collection in a drip tray under the machine. To avoid flooding in all hydroentanglement systems, the flow rate introduced by the injectors must be balanced by the water removal rate from the web and support surface. Perforated rollers operating with an internal suction slot assist with dewatering. The perforations are organised in regular patterns, and their size and the open area are sufficiently large to permit efficient water drainage at high flow rates. The solid surface also provides potential for good energy transfer and therefore an increase in the final fabric strength and energy efficiency.

To avoid the introduction of unwanted shadow marking and striping of fabrics produced with perforated rollers having a zig-zag arrangement of perforations, sleeves having a random distribution of smaller microperforated holes (normally 250–300 µm) have been introduced with an open area of 3% to 12% to maximise water drainage [48]. Commercial systems have consisted of a thin serigraphic nickel sleeve mounted on a metal honeycomb support cylinder, which has an open area of about 95% to maximise drainage [40]. Spacers extend beyond the honeycomb cells to maximise the open area immediately below the sleeve to minimise flooding. Such microperforated sleeves increase the energy efficiency of hydroentanglement by increasing the fibre entanglement for a given energy input. As the open area of the support surface increases, the probability of energised water dissipating from the system is higher and therefore lower fibre entanglement is produced. Using microperforated sleeves, high fabric strength can be obtained using relatively low pressure, which is energy efficient. It has been claimed that if the open area of the sleeve is increased from 8% to 15%, the number of injectors must be doubled to obtain the same fabric tensile strength [40].

Wire meshes with a pronounced knuckle height or drum surfaces with raised projections are used to produce apertured fabrics with a gauze-like or three-dimensional appearance [49]. The water jets displace fibre segments from the surface of these projections producing apertures, the shape, size, and frequency of which are directly affected by the three-dimensional geometry of the support surface. The methods used to produce the drum surface or sleeve, that are readily linked to computer-aided design packages, include serigraphy, which relies on electrolytic deposition of nickel or other metals to build up the required surface and laser engraving, which provide extensive design opportunities for customisation of hydroentangled fabrics by emboss patterning.

8.19.3 Injector operation

Injectors (or manifolds) are constructed of steel and are robustly designed to withstand high water pressures. They can be built to widths of 6.6–7 m, but there are many more machines in operation of 1.6–2.5 m wide. While commercially, hydroentanglement systems capable of water pressures up to 600–1000 bar, have been designed most commercial installations operate at lower pressures up to about 250 bar. This is partly to minimise the energy consumption, but also to maximise the life of the jet strips, which wear out quickly at high pressures. To minimise energy consumption, it is desirable to use the minimum possible water pressure needed to obtain acceptable mechanical and physical fabric properties. Continuous improvements in the geometry, design, and construction of injectors and jet strips as well as the use of microperforated support surfaces have led to improved energy transfer from the jets to the web.

Consequently, fabric tensile properties can be achieved today at lower pressure than was possible in the past; however, pressures of up to about 300 bar are still required to fully bond heavyweight webs of, for example, 400 g m^{-2} [50] and webs composed of splittable bicomponent fibres, where high splitting efficiency is required also require high-pressure inputs [40]. Separate high-pressure pumps serve each injector, which enables independent adjustment of pressure. This is more energy efficient than using one pump connected to multiple injectors since throttling losses are avoided. It also allows greater flexibility in processing. Piston pumps are normally preferred over centrifugal pumps and pump pressure is regulated by variable-frequency-controlled AC motors.

The design of injectors has been subject to extensive modelling and simulation work by the machinery manufacturers to improve energy efficiency and to ensure uniform flow conditions across wide operating widths. Computational fluid dynamics (CFD) modelling has proved invaluable in this area. Following fundamental studies, a significant increase in energy efficiency has been achieved by replacing drilled injectors with the fine slot type, and as result it is claimed that webs of 300 g m^{-2} that previously required 300 bar for entanglement now require only 180 bar [40].

Drilled and slot-type injectors have been described [40,51]. The drilled type consists of a main body with an upper and lower chamber that is capable of withstanding high pressure. In the upper part, there is a cylindrical chamber into which the high-pressure water is fed. Inside the chamber is a cartridge, which may consist of a perforated cylinder lined with a metal sleeve that acts not only as a filter but also as a water distributor. The high-pressure water fed into the chamber then passes through cylindrical holes arranged at intervals across the width of the injector and these holes are between 4 and 10 mm in diameter and 3–5 mm apart. These holes, which can be conical at the outlet, feed the high-pressure water to the lower part of the chamber where the water flows towards the jet strip nozzles. At pressures >50 bar, the geometry of such injectors can lead to turbulence in the lower chamber, specifically in the regions between the consecutive hole outlets in the lower chamber resulting in energy losses. This can lead to heterogeneous bonding and variations in fabric density and appearance. An alternative high-pressure injector design [40,51] consists of a cylindrical feed chamber inside which the high-pressure water flows through a filter and then enters a distribution region, which transports the water towards the jet strip nozzles. Inside the feed chamber there is a cartridge consisting of a perforated cylinder lined with a filter system. The pressurised water is fed down to the jet strip via a narrow rectangular slot that extends the full width of the injector. The ANDRITZ Perfojet neXjet injector is fitted with a cartridge inside the injector immediately before the jet strip to aid water filtration and has a seal-holding system enabling ease of injector maintenance.

Energy savings continue to be an important motivation for product development in injector design. To achieve higher fabric tensile strengths at the same water, pressure and energy consumption is therefore attractive. This is one of the claimed benefits of the V-jet injector (AUTEFA Solutions), which is designed to reduce the operating distance between the jet strip nozzles and the bottom of the injector from the currently used 15–25 mm to as little as 0.5 mm (Fig. 8.45). This still allows operating widths of 3.6 m and pressures up to 340 bar.

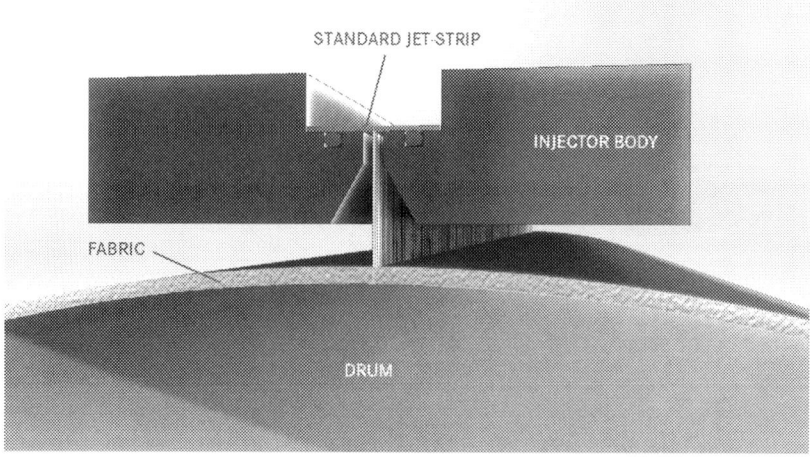

A

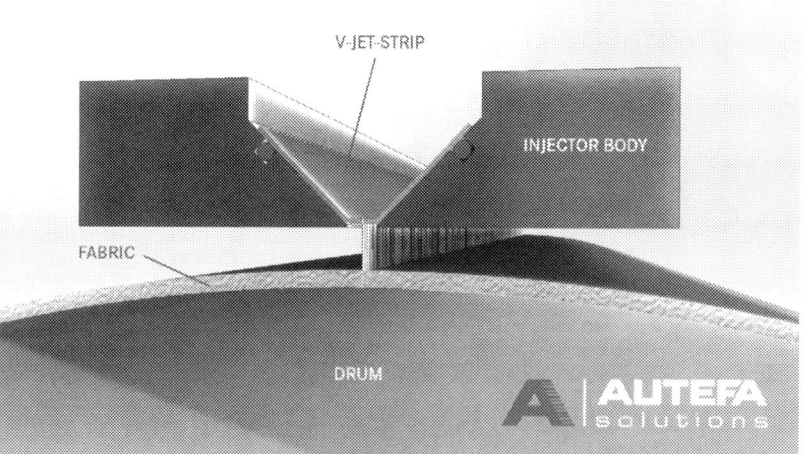

B

FIG. 8.45   Standard (A) and V-Jet (B) injector configurations. Courtesy: AUTEFA Solutions.

Commercially, the surface uniformity of hydroentangled fabrics can be important, and to address this, reciprocating injectors have been introduced to minimise jet marks. High-capacity injector systems have also been developed that incorporate either two (duplex) or three (triplex) jet strips in a single injector [31].

8.19.4 Arrangement of the injectors

The number of injectors fitted to commercial hydroentanglement installations varies, but typically 5–8 injectors are required to provide fabrics with adequate bonding and visual uniformity assuming no additional bonding processes precede or follow. Some installations operate with only 2–4 injectors and then follow with chemical or thermal bonding to complete fabric production. In contrast, machines with more than 10 injectors have been constructed commercially that are capable of very high line speeds and jet strip changes on the run. The number of injectors and the maximum operating pressure depends partly on the line speed and the degree of versatility that is required by the roll goods manufacturer. It is possible to hydroentangle at hundreds of metres per minute ($>300\,\mathrm{m\,min^{-1}}$) provided the production can be balanced by the web formation system and sufficient energy can be transferred to the web by the injectors to produce a fabric with satisfactory properties. Although increasing the number of available injectors increases the versatility and potential line speed of the installation, not all the injectors are necessarily employed to increase the degree of bonding.

The final injectors may be set up to improve the visual uniformity of the fabric; relatively low pressure and fine nozzles are employed for this purpose, or to introduce embossed patterns or apertures. It is possible to produce high-strength fabrics using only a few injectors operating at high pressure. While this approach minimises production costs and simplifies the process, it can lead to quality problems such as pronounced jet marking in the fabric. However, the large increase in strength after the first injector gives rise to a lower risk of drafting as the web is transferred from the cylinder to the next stage.

The development of the original German Norafin process helped to establish that an alternating face and back treatment of the web by successive injectors leads to the largest increases in fabric strength. This alternating treatment is particularly important for heavyweight fabrics of 200–600 g m^{-2} to avoid problems of delamination. In most modern hydroentanglement systems, alternating groups of injectors (1–4 in each group) arranged in succession direct jets onto the face and reverse sides of the web in sequence. The jet pressure profile describes the position and operating pressure of each successive injector with respect to the web. The pressure is usually profiled from the entry to the exit of the machine and the smallest pressure is usually encountered at the beginning. The pressure profile affects the specific energy ratio [52], which is the ratio of the specific energy applied to the face side K_f to the total applied specific energy K_t:

$$\text{Specific energy ratio} = \frac{K_f}{K_t} \tag{8.13}$$

Even if the specific energy ratio applied to a web is the same, the resulting fabric properties can be quite different depending on the pressure profile that is used. One example is the bending rigidity, which tends to vary face and back in the fabric. An example of the variation in density arising from different pressure profiles in a nonapertured fabric is illustrated in Fig. 8.46.

8.19.5 Jet strips and nozzles

The water, which should be uniformly distributed inside the lower section of each injector, is forced through nozzles made in a thin, metal jet strip clamped to the injector, usually by

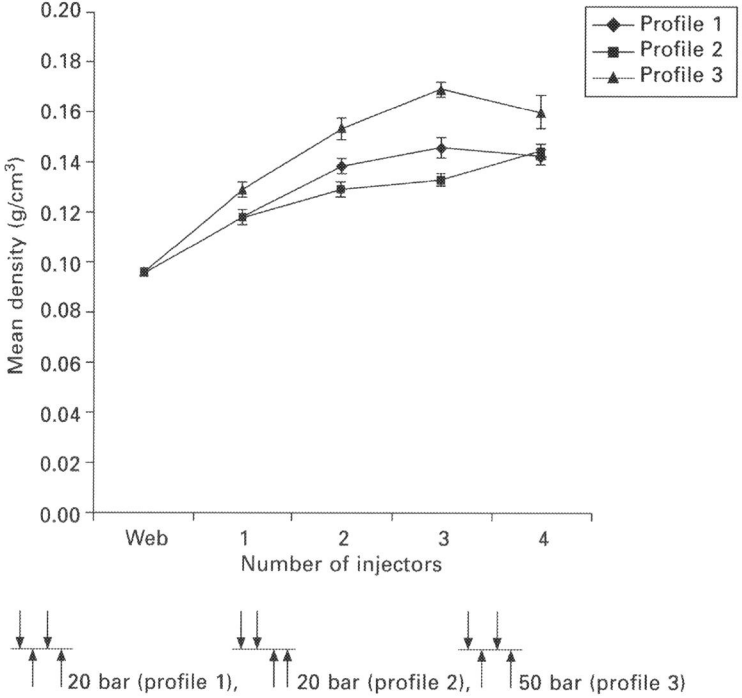

FIG. 8.46   Example of change in fabric density with the number of injectors for different jet pressure profiles (nonapertured fabric).

hydraulic means or a self-sealing mechanism based on the water pressure in the injector. The jet strip is typically 0.6–1 mm in thickness, 12–25 mm wide, and has between 1 and 4 rows of nozzles [53]. The jet velocity in a high-pressure system is about 100–350 m s^{-1} and issues from nozzles with a diameter of between 60 and 150 μm (although some can be up to 300 μm). The spatial frequency of the nozzles is from 40–120 per 25 mm [15]. The nozzles used in hydroentanglement normally have a capillary section with straight sides that connects to a cone section.

Conventional capillary cone nozzles are formed by punching the strip. The ratio of the length of the capillary section to the inlet nozzle diameter is normally about 1. Lower aspect ratios can be used to promote a constricted jet. To obtain stable, high-velocity columnar jets, nozzles are usually operated in the 'cone-down' rather than the 'cone-up' position. The capillary portion of the nozzle, which forms the inlet side of the jet, therefore influences the jet diameter. The energy efficiency of the process is largely dependent on the formation of a constricted jet, which remains intact between the nozzle and the web. Break-up or dispersal of the jet once it emerges from the nozzle results in poor energy transfer to the web and reduces the overall efficiency of the hydroentanglement process.

For cone-up nozzles, there tends to be greater jet instability and the water pressure more strongly influences the discharge coefficient [54]. Operation in the cone-down position reduces the discharge coefficient and increases the velocity coefficient with a uniform jet. It also assists in the prevention of cavitation, which tends to break up the jet. The nozzle geometry, particularly around the capillary inlet, is one of the factors influencing jet break-up. The nozzle aspect ratio is also important. The flow characteristics of jets emerging from nozzles and the effect of nozzle geometry on jet stability have been extensively studied by means of computer simulations as well as experimental observations [55–57].

Industrially, one of the limitations in hydroentanglement is the life of the jet strip, which extends to about 4000 h at best or only 100 h depending on operating conditions, particularly water pressure and water composition [58]. Nozzle damage due to cavitation, abrasion or chemical degradation alters nozzle geometry and resultant jet formation may be affected giving rise to instability of the jet and fabric quality problems such as variations in fabric density and texture, as well as reduced energy transfer efficiency. At high pressures, such as 400 bar, stainless steel jet strips can deteriorate quickly within hours. One solution is to use nozzle inserts where a very hard material surrounds the orifice and stainless steel or a hard coating comprises the rest of the strip [59]. The hardness of conventional stainless steel jet strips is about 250 shore but to increase wear resistance, which is essential in high pressure systems, modern strips constructed from new alloys have a hardness of 1200 shore or more [40].

In practice, the nozzle inlet diameter, and the number of nozzles/m in the jet strips fitted to each consecutive injector head varies. For example, in the first few injectors, strips with relatively large nozzles (120–150 μm) are fitted to maximise impact force and fibre entanglement. Jet strips in the final injectors commonly have finer nozzles (80–100 μm), which reduce the appearance of jet marks and produce a smoother fabric surface. In practice, hydroentanglement of webs of $>200\,\mathrm{g\,m^{-2}}$ may limit the nozzle diameter to about 100 μm to minimise flooding. Jet strips are produced by several companies including Groz-Beckert, Ceccato, Nippon Nozzle, and Enka Tecnica.

8.19.6 Dewatering

Suction is used to aid removal of excess water from the support surface during hydroentanglement to prevent flooding. Flat belt systems are particularly prone to flooding. Excess water not removed by suction is allowed to drain below the machine, and this is particularly effective when cylindrical support surfaces are used. Flooding leads to energy losses that can cause reduced fabric strength and interference with the bonding process. Flooding also produces defects in the fabric. Approximately, 100–1000 mm head of water is needed, 500 mm head of water is preferable. Dewatering is further improved by mangling the fabric prior to drying by means of squeeze rollers. Roberto rolls constructed from fibre can draw out water from the fabric as it passes through.

8.19.7 The water circuit and filtration

The filtration system is a major cost in a hydroentanglement installation and water quality affects process efficiency. Some of the main problems associated with filtration systems are blocked jet strips leading to jet marking and variations in fabric uniformity, the high cost of frequent filter bag replacement, bacterial growth, the potential loss of sand and damage to machinery, excessive discharge of water in the backwashing of filters, and the requirement to replenish sand filters. In terms of water quality, a neutral pH and a low content of metallic ions, for example, calcium is required. Depending on machine size, the quantity of water in the circuit is about 40–100 m^3 h^{-1} and the trend is to reduce the circulating water to improve process efficiency. For a 3.5 m wide machine producing fabric for wet wipes, the quantity of circulating water has been estimated to be about 100 m^3 h^{-1} [40].

Industrially, a large proportion of the wastewater produced in hydroentanglement is recycled and recirculated to the main high-pressure pumps. Fibre finish, fibre debris and other impurities present in the wastewater must be removed by the installed filtration system. Whether or not this is practicable depends on the design of the filtration system and the volume of impurities removed from the fibre during the process. Self-cleaning filters are employed to minimise cost and for cotton or pulp, flotation and sand filters are selected. Progressive developments in filtration systems have made it possible to process a greater variety of fibre types although traditionally, more sophisticated filtration systems are required for cellulosics such as cotton, pulp, and viscose rayon compared to synthetics such as PET. Chemical mixing and flocculation, dissolved air flotation units, and sand filters are employed. Sand filtration systems enable the removal of suspended solids and a reduction in fibre finish in the water circuit.

The choice of filtration system largely governs the versatility of the hydroentanglement line in terms of the compatible fibre types as well as cost. In a system processing cellulose pulp, a closed loop arrangement consists of a flotation unit that sends water to a sand filter operating with a back-washing recovery system. The back-washed water is returned to the flotation unit, while the remaining water from the sand filter is UV sterilised and sent to a bag filtration system prior to being returned to the beginning of the process [60]. For cotton, filtration systems have been designed to treat the waste water directly in sand filters without the need to use an initial flotation unit [61]. A two-stage process can be adopted where back-washed water from the sand filter is sent to a flotation unit and additional sand filter before being returned to the circuit.

8.19.8 Drying

Immediately after the fabric is hydroentangled, a proportion of the water held interstitially within the hydroentangled fabric is mechanically removed by suction, which for synthetic fibres is quite effective in reducing the moisture content to well below 100%. One advantage is lower overall drying costs. For cellulosics and other hydrophilic fibres, the water content is much higher even after mechanical extraction, which places a high demand on the subsequent drying process. Drum drying, can drying and through-air flat conveyor drying are all found in operation although through-air drum drying is the most common solution. More recently, a dryer combining belts and a drum has been introduced (AUTEFA Solutions Square Drum Dryer) to increase drying efficiency, by increasing the overall length over which the fabric is dried, while maintaining a small footprint.

8.19.9 Aperturing and patterning effects

Apertured fabrics are produced by hydroentangling on support surfaces with small projections, e.g., raised cross-over points or 'knuckles' in wire mesh around which fibres are directed and entangled. In addition to woven wire support surfaces, the projecting points can be formed on metal support sleeves or even printed onto flat surfaces too. Depending on the geometry of these projections, particularly the wall angle relative to the conveyor, fibre segments on top of the projections are displaced to adjacent regions. Because of this, the local density in the regions adjacent to the apertures is significantly higher than the global fabric density. Aperturing can also occur on support surfaces with a high open area. The support surface therefore influences the periodic structure and texture of the fabric as well as the geometry and spacing of the apertures. Analysis of the forces involved in rearranging fibres in the web into a bonded structure concluded that the work is only about 1% of the input energy [24].

Normally, aperturing takes place after the web has been preentangled to maximise pattern definition, assuming fibre segments are still mobile. Aperture definition tends to improve as fibre length decreases because fewer fibre segments bridge between the apertures that are formed during fabric formation.

Three-dimensional emboss-type patterns, ribs, logos, and surface pile effects are introduced by hydroentangling a preentangled web on a support surface containing recesses into which fibres are pushed. Support surfaces can be designed so that embossed effects and apertures can be produced in the fabric. Early in the development of hydroentanglement, embossing geometric patterns was identified as a potential way of producing textile-like fabrics having the appearance of woven fabrics [11]. Further developments have led to the adoption of more complex support surface patterns produced by CAD and laser engraving on to which webs are hydroentangled. Both aperturing and hydro-emboss patterning provide a convenient means of visually differentiating hydroentangled fabrics, as well as influencing their local liquid transport properties.

Water consumption can sometimes be higher in the production of apertured and patterned fabrics since large-bore nozzles of 120–150 μm are preferred to obtain good definition. A high flow rate and impact force are therefore very important in producing high-quality apertured structures. Jet strips with three rows of holes are used by some manufacturers to increase the water flow rate to the fabric. Apertured and patterned lightweight fabrics are particularly important in single-use products for the medical (e.g. replacement gauze dressings) and hygiene sectors (e.g. wipes and coverstock).

Apex technology was introduced by PGI (now Berry Global) for producing complex, embossed patterns in hydroentangled fabrics in the range 50–400 g m^{-2}. These early Miratec fabrics and the derivatives, Mirastretch and Miraguard, exhibited good elastic recovery and barrier properties, respectively. Some of the complex structural patterns introduced in Miratec fabrics resembled the appearance of woven and knitted textiles. Fabrics are produced by hydroentangling webs on laser imaged three-dimensionally patterned support surfaces to enable transfer of these patterns into the fabric. To improve image definition in the fabric, the web tension should be low and there should be no differential in the imaging surface and web speeds. After hydroentanglement, polymeric binders are added to stabilise the fabric or to introduce elastic properties and finishing processes such as compressive shrinkage (e.g. compaction by Sanforising) are undertaken to further adjust the softness and drape of the fabrics.

In variations to the process, blends containing fusible binder fibres avoid the need for chemical bonding after hydroentanglement, and scrims can be introduced to increase durability and pattern definition in the fabric [62]. The complex patterns enabled the cleaning performance of wipes and dusters to be improved, for example, recessed pockets may be formed in fabrics to improve the collection of low-viscosity contaminants present on skin [63]. Original fabric applications have included food service towels, automotive interiors, window dressings, upholstery fabrics, pillow covers, wall coverings, bedspreads, and apparel fabrics. Spunlaid web formation combined with Apex technology, forms part of Berry Global's Spinlace capability, and can produce durable fabrics that are compatible with traditional finishing processes such as padding, spraying, and dipping [64].

8.19.10 Microfilament fabrics based on bicomponents [65,66]

The Evolon family of hydroentangled fabrics were an outcome of Freudenberg's Omega project, which aimed to develop technology combining the benefits of staple fibre (carded) nonwovens such as high softness, drape, bulkiness, and resilience with the benefits of spunbond fabrics, particularly the high MD and CD tensile strength. To mimic the characteristics of staple fibre nonwovens using a spunlaid platform producing continuous filaments, two bicomponent extrusion technologies can be exploited. The first relies on the differential shrinkage and self-crimping that can take place in an asymmetric or eccentric sheath-core bicomponent filament via quenching, drawing, and heat treatment.

The resulting crimp in the filaments increases the bulk of the fabric. The characteristic softness and textile-like handle of Evolon fabrics principally relies on the use of segmented-pie splittable bicomponent technology. A spunlaid web was produced from splittable bicomponent filaments (originally 16 segmented pie, PET/PA6.6 in a weight ratio of 65%/35%) of about 2 dtex, which is hydroentangled at up to 400 bar to split the filament cross-section into multiple microfilaments. These microfilaments have a linear density of 0.09–0.13 dtex. More recently, this has developed into fabrics based on a 32 segmented pie structure.

Splitting efficiency, which can reach 97% or more, is maximised using hollow-core segmented-pie filaments and is influenced by the nozzle diameter in the spinneret, quenching, the stretching rate and, of course, the water pressure. To maximise fabric softness, the fabrics are mechanically finished using processes such as tumbling. Owing to the combination of fabric softness, high filament surface area, strength, and abrasion resistance in Evolon fabrics, applications include some clothing markets in sports and activity wear (e.g. hiking, skiing, cycling) where fabric weights are in the range 100–220 g m^{-2}. Other applications include workwear, automotive (e.g. interior trim, carpets for sound absorption), shoe components (e.g. linings, heel grips, PU coating substrates), luggage, and home furnishings (e.g. bed linen). Evolon fabrics can be jet dyed and finished and are said to withstand multiple domestic washes. Additional potential applications include cleanroom wipes.

Another example is Madaline fabrics produced by Mogul, which contain a binary mixture of trilobal microfilaments produced by hydroentangling splittable filaments and PET/PA bicomponent filaments in a spunlaid web. The fabrics are manufactured in the range of 40–200 g m^{-2} and have excellent abrasion and tear resistance, permeability, and softness. Again, the intended applications are mostly in longer-life, rather than single use products, where the combination of aesthetics and durability is attractive.

8.19.11 Multilayer composite hydroentangled fabrics

Webs (as well as prebonded nonwoven fabrics) can be simultaneously bonded and combined during hydroentanglement to produce flexible multilayer, composite fabrics. Where spunbond and meltblown webs are combined in this way, the approach can be viewed as the mechanically bonded analogue of thermally bonded SM or SMS composites. Hydroentangled multilayer fabrics are found in both single-use and durable applications although it is in absorbent wipes products that most of the development has taken place. Developed twin-layer hydroentangled fabrics include:

- Spunbond with airlaid pulp (SP).
- Spunbond with carded web (SC).
- Spunbond with wetlaid (pulp, glass, or other short fibre papers).

Three or more layer hydroentangled fabrics include:

- Spunbond-pulp-spunbond (SPS)—where the pulp may be in the form of a preformed roll or deposited directly by means of a short fibre airlaying system.
- Carded-pulp-carded (CPC), or CAC (carded-airlaid pulp-carded).
- Carded-pulp-spunbond (CPS).
- Carded-spunbond-carded (CSC).
- Carded-net-carded (CNC).

Scrims or net structures can be combined with carded webs during hydroentanglement to modify tensile properties, particularly in lightweight fabrics, and is generally referred to as scrim reinforcement. Hydroentangling fibres or pulp with elastomeric foam [67], filaments, or perforated films have all been explored as a means of increasing fabric elasticity. Among other end-uses, CPC fabrics are intended for single-use wipes, CPS in incontinence products such as nappies and both CNC and CSC in industrial wipes. SPS composites consisting of a spunbond-airlaid pulp-spunbond sandwich are intended for wet wipes as well as for absorbent medical applications. The economics of SPS fabrics are attractive because of the low cost of pulp. The pulp acts as an absorbent core and the spunlaid layers provide abrasion resistance and structural reinforcement when the pulp is wet. The spunlaid fabrics also resist linting, dusting, and pilling in use. To reduce the fabric density and to increase the softness, the filaments in the two spunlaid layers can be crimped during production.

The pulp in CPC, CPS, and SPS fabrics is either airlaid directly onto the web or is introduced as preformed cellulosic tissue (paper) before hydroentanglement. As an alternative to pulp, airlaid cotton linters have also been used. The carded web components normally consist of PET, PP, viscose, or cotton. Composite hydroentangled fabrics are expected to play an increasingly important role in the future of hygiene and medical products and have the advantage that no thermal lamination or bonding is required. Opportunities in heavier weight long-life applications in, for example, roofing fabrics composed of PET, geosynthetics, automotive interior trim, and filter media, using both drylaid and spunlaid webs are also evolving.

8.20 Applications of hydroentangled fabrics

While hydroentangled fabrics have become heavily associated with wipes, their use spans a much greater variety of applications, both in single-use and longer-life articles. Diverse products include protective clothing, e.g., medical gowns, synthetic leather, filtration media, wound dressings, composites, and garment linings. The utilisation of hydroentanglement to mechanically join rather than thermally laminate webs or fabrics together has also created product development opportunities. There has been rapid growth in global hydroentanglement capacity with strong demand for higher productivity with wider and faster machinery. Price pressure in some consumer markets and increasing competition is intensifying demand for product differentiation, and has stimulated interest in new market opportunities. Some high value, niche markets for hydroentangled fabrics do not necessarily require the high production capacities of many existing hydroentanglement installations.

8.20.1 Wipes

Hydroentangled fabrics for wipes have been produced since the 1970s. The soft, strong, flexible and in most cases, absorbent characteristics of the fabrics combined with increasingly attractive economics and a textile-like handle have brought hydroentanglement to the fore in this sector. One of the earliest applications was as replacements for woven gauze in products such as laparotomy and X-ray detectable sponges [68]. The wipes industry is now remarkably diverse encompassing baby wipes, personal care, including facial cleansing and make-up removal, food service, industrial/technical, medical, and household cleaning products. Baby wipes are one of the biggest single markets for hydroentangled fabrics, but there has been significant growth in personal care wipes markets, including hydroentangled fabrics for dispersible premoistened toilet tissue and feminine hygiene wipes. Industrial wipes include those for the aerospace, automotive, optical, and electronics industries as well as other speciality products, and include multiuse or washable products made from splittable bicomponents. Such hydroentangled fabrics made from microfilaments also find applications as sports towels and facial cleaning cloths.

Over the years, airlaid–thermal-bonded wipes have been increasingly replaced by hydroentangled fabrics because of their softer handle, good strength, and low thickness. Hydroentangled fabrics are also important in the premoistened or wet wipes market, alongside airlaid and thermal-bonded products. Some examples of the composition of commercial hydroentangled wipes are shown in Table 8.2 [12]. Hydroentangled floor wipe compositions are traditionally, 70%/30% or 65%/35% viscose-PET in fabric weights of $90–100\,\mathrm{g\,m^{-2}}$. In floor wipes, a binder is applied, which increases the durability of the fabric in the wet state; it also provides a means of adding pigments. For dusters, compositions include 75%/25% and 91%/9% PET/PP blends and fabrics with embossed 3D patterns or mock pile surfaces are produced to improve dust pick-up. Fabric weights range from 30 to $65\,\mathrm{g\,m^{-2}}$ and scrim reinforcement may be used to increase the dimensional stability. Among the wipes for cleanroom applications, lyocell and PET blends are produced as well as 100% PET hydroentangled wipes [69].

TABLE 8.2 Examples of dry wipe products.

Product	Fabric weight	Composition
Baby wipe	$50 \, \mathrm{g \, m^{-2}}$ or $55 \, \mathrm{g \, m^{-2}}$	70% viscose rayon 30% polyester
Baby wipe	$55 \, \mathrm{g \, m^{-2}}$	50% viscose rayon 50% wood pulp
Food service wipe	$68–80 \, \mathrm{g \, m^{-2}}$	80% viscose rayon 20% polyester
Swiffer type dry wipe	$68 \, \mathrm{g \, m^{-2}}$	Carded polyester + polypropylene scrim + carded polyester composite

Hydroentangled cotton fabrics for gauzes, cosmetics (e.g. make-up removal), and certain premoistened wipes products are produced in fabric weights ranging from 30 to $250 \, \mathrm{g \, m^{-2}}$. Cotton pads for make-up removal are traditionally produced by bonding a drylaid web using one of three approaches: surface impregnation with a binder, thermal bonding with bicomponent fibres and, embossing (under compressive force) with engraved pattern rollers. To enable the manufacture of 100% cotton products with acceptable surface abrasion, cotton pads are also produced using hydroentanglement. Hydroentangled cotton pads and wipes are produced from 30 to $300 \, \mathrm{g \, m^{-2}}$ with apertured or smooth surfaces [73]. Low-pressure hydroentanglement (up to about 40 bar) allows bonding to be concentrated at the surface thereby maximising fabric porosity. Therefore, whereas the fabric surfaces are well entangled to minimise linting in use, the core is lightly entangled to maximise the absorbent capacity and softness. Localising the entanglement to the surfaces of a fabric is a strategy used in the formation of other wipes products [70].

The ability to differentiate and personalise wipes products is important for manufacturers to avoid the potential for commoditisation. Emboss-patterning and aperturing in hydroentangled fabrics provide a means of adding value. To underpin the expanding technical requirements of the wipes sector and to enable further diversification, production of composite hydroentangled fabrics, in which different web types are combined during hydroentanglement is a growth area. Such composites enable significant improvements in wiping performance, dimensional stability, absorption, and soil cleaning according to the composition of the layers that are combined. Composite hydroentangled substrates for wipes are mostly based on combinations of carded + airlaid wood pulp, carded + scrim, and spunlaid + wood pulp combinations. Carded staple fibre webs and airlaid wood pulp blends are combined to produce wipes with good absorbency in baby wipes, bodycare, food service, and industrial cleaning.

It has been a challenge to produce fabrics with the required softness and drape that do not lint or produce dust using wood pulp. To achieve this, composite hydroentangled fabrics composed of $40–45 \, \mathrm{g \, m^{-2}}$ wood pulp sandwiched between two carded PET webs of $8–10 \, \mathrm{g \, m^{-2}}$ were developed. In twin-layer composites, the carded web weight is increased to $25 \, \mathrm{g \, m^{-2}}$ [40]. To increase durability for applications such as industrial wipes, wood pulp is hydroentangled into a spunlaid web. Scrim-reinforced carded-hydroentangled fabrics are utilised in domestic wipes such as P&G's Swiffer.

Another market where there has been rapid development of manufacturing capacity is water dispersible (flushable) wet wipe substrates. Continued technical development of flushable wipes substrates has been essential in response to the introduction of rigorous, industry-led guidelines (GD4) for determining the dispersibility of nonwoven fabrics, led by INDA/EDANA. More recently, regulation has emerged, notably the single-use Plastics Directive 2019/904 (SUP directive) in the EU that is impacting the technical specifications of premoistened, wet wipe products.

The technical development of flushable wipe substrates for wet wipe applications has been a long, progressive process that is still on going. Historically, some of the early substrates relied on ion-sensitive cationic polymers applied as binders to webs, or lightly bonded hydroentangled fabrics. In another approach, a hydroentangled fabric consisting of three laminated layers with a pulp core was produced with some areas left unbonded to promote disintegration of the structure in the sewer system [71]. Another early example is Hydraspun 784, a dispersible wet wipe substrate composed of a latex binder-free hydroentangled fabric [72]. Composed of a synthetic and natural fibre blend the fabric achieved a flush index (tube test-first break) of seven to eight turns depending on its weight (55 or 65 g m^{-2}). More recently, water dispersible premoistened wipes made entirely of 100% cellulosic, biodegradable materials have been a major development focus, not least because of the SUP directive in the EU. One example is the latest GD4 compliant Hydraspun family of wipe substrates (Royal, Essential, Flow, and Plus) from Suominen, based on cellulosic fibre content. Generally, for dispersible wet wipe applications, hydroentangled fabrics based mainly on wetlaid webs, containing for example, 70% to 80% wood pulp blended with 30% to 20% short cut cellulosic fibres, e.g., lyocell of 5–12 mm are particularly important. Partly for economic reasons, products containing higher proportions of wood pulp, blended with short cut regenerated cellulosic fibres have been a particular focus of development. In general, one of the major challenges in manufacturing flushable wet wipe substrates is achieving sufficiently high wet strength to facilitate the conversion process, and physical integrity during use, without compromising rapid dispersibility in the toilet.

8.20.2 Washable domestic fabrics

Hydroentangled cotton fabrics for semi-durable bedsheets, napkins, and tablecloths have been produced that can be washed up to 10 times before disposal. Since they are cotton based, such fabrics can be dyed or printed. Impregnating hydroentangled fabrics with 0.2% to 1.0% owf of polyamide-amine-epichlorohydrin resin [74] is claimed to enable repeated washing of hydroentangled cotton fabrics. Durable hydroentangled fabrics intended for repeated laundering have also been developed by stitching the fabrics over the top, by applying a binder or by incorporating thermally fusible binder fibres in the fabric [75]. Such fabrics can also be dyed and finished. A further method of increasing the durability of fabrics for washing without the need for chemical or thermo-fusible binders is to hydroentangle at high water pressure, but the energy consumption is high.

8.20.3 Hygiene

For reasons of aesthetics, function as well as increasingly, to enable hygiene fabrics to be produced entirely from cellulosic materials, such as cotton or lyocell, a variety of applications for hydroentangled substrates exist in baby diapers, continence management, and femcare products. This is despite the higher cost relative to spunbonds. In femcare, the main application is in textile-like topsheets, as well as acquisition and distribution layers (ADLs) to aid fluid management and control wetback. Hydroentangled fabrics are also used as components in high stretch diaper ears, where the low modulus but high strength of the fabric can be used to reduce the weight of the elastic film component, while decreasing the risk of damage to the ear due to the tape hook. In the personal care market, hydroentangled fabrics are also used to manufacture wax strip substrates for hair removal.

8.20.4 Masks for facial skincare

The application of cosmetic and skin conditioning treatments via impregnated hydroentangled fabrics made into facial and eye mask products has become an important market. The softness, conformability, and fluid handling properties of such fabrics containing regenerated cellulosics are particularly valued by users. Fabrics are cut to shape and once applied to the skin, help to modulate the temporal release and delivery of cosmetic formulations to the skin. Such products are based on wetlaid-hydroentangled and drylaid-hydroentangled platforms. Blends of viscose-polyester, lyocell, or 100% cotton, among others are manufactured. Spunlaid-hydroentangled substrates containing microfilaments have also been developed capable of multiple use.

8.20.5 High-temperature protective clothing

Hydroentangled aramids, including blends of meta- and para-aramids, are well established as protective liners and moisture barrier substrates in fire-fighting garments. In addition to the high-temperature resistance derived from the polymer, the fabrics are favoured because of their softness, drape, and light weight. Fabrics composed of 50% Basofil/50% aramid are utilised in thermal protection lining fabrics that form components in fire-fighting garments and such linings are commonly quilted to a woven aramid face fabric. A protective liner fabric composed of hydroentangled FR cotton and Basofil is produced to improve comfort in high-temperature clothing. Meta-aramid fibres are commonly found in thermal protection liners and 50% Lenzing FR/50% aramid blend hydroentangled fabrics have also been developed for fire-fighting jackets to give improved resistance to flame break-open. In addition to fire-fighting garments, hydroentangled fire-blocking fabrics for upholstery and mattresses are produced from Type Z11 Kevlar, and Type E92/Kevlar blends are produced as fire blockers for aircraft seats and as thermal liners in protective clothing [76].

8.20.6 Artificial leather

The production of synthetic leather and coating substrates using hydroentanglement is well established. Hydroentangled fabrics are widely used as backings for PU-coated synthetic leather substrates for durable applications such as seat coverings, bags, belts, sports goods, and numerous other products. In addition to the traditional method of dissolving out part of the matrix in island-in-the-sea bicomponent fibres using a suitable solvent, synthetic leather articles are also produced by hydroentangling splittable bicomponent fibres composed typically of 6, 16, or 36 segments in the cross-section. Such fabrics find applications as specialist wipes, particularly for cleaning glass.

8.20.7 Surgical fabrics

Hydroentangled fabrics have long been used to manufacture surgical gowns, scrub suits, sheets, and drapes for their excellent comfort and softness, particularly when compared to SMS and two-ply laminate products. They are frequently chemically treated to improve re-pellency of low surface tension fluids. In surgical gowns, fluid barrier performance (governed in Europe by EN 13795–2011) is a key requirement for infection control, and composite spunmelt nonwovens containing multiple meltblown layers, films, or coatings are favoured over hydroentangled fabrics where there is a need for high-barrier protection. Medical protection garments and warming gowns are well-established applications as well as composite hydroentangled fabrics composed of wood pulp hydroentangled onto a PET layer. Single-use hot-water–soluble PVA hydroentangled fabrics for surgical scrub suits, gowns, and drapes have also been developed.

8.20.8 Medical gauze and wound dressings

Traditionally, yarn-based wound contact layers in dressings have heavily relied on gauzes composed of cotton, or cotton blended with up to 50% viscose rayon. As an alternative, apertured hydroentangled fabrics have led to significant cost savings in this market partly because fewer layers are required for exudate management. In practice, hydroentangled gauze fabrics composed of, for example, 70% viscose rayon and 30% polyester provide high absorbency and low linting properties. Resistance to linting is particularly important to re-duce potential for contamination of the wound during dressing changes. A variety of nonapertured hydroentangled fabrics are produced for use as low adherent primary dressings.

8.20.9 Linings and clothing

Undyed, dyed, or printed hydroentangled fabrics find use in shoe and clothing linings. Hydroentangled fabrics composed of merino wool have also been developed [77] as linings for performance outerwear products such as outdoor clothing. Sports shirts and vests have also been produced from hydroentangled fabrics.

8.20.10 Filtration

Hydroentangled fabrics intended for air filtration, particularly bag house applications are manufactured as well as pleated hydroentangled media for liquid filtration. Hydroentangled PET media [78] as well as those composed of PPS, aramid, or PTFE have been developed for industrial dust filtration. One example is Hydrolox (Bondex) filter media, which combines both needling and hydroentanglement to enable a better balance between filtration efficiency and air permeability than is possible with membrane laminated felts. As is common in needlepunched fabrics for filtration, scrim-reinforced hydroentangled fabrics are manufactured to improve dimensional stability. In another example, Hycofil is a pleated textile or metal scrim-reinforced hydroentangled fabric for flue gas filtration composed of polyimide or aramid fibres [79].

8.20.11 Automotive

Hydroentangled fabrics composed of 100% PET have been developed for backing materials for one-step injection moulding of interior trim automotive components replacing heavy weight needlepunched fabrics. This process is used by manufacturers to produce automotive interior trim. The fabric is laminated to a knitted facing before the injection moulding process and the use of hydroentangled fabrics is claimed to improve acoustics in the car. The Zetajet technology of Tenowo is one example of hydroentangling being used to produce fabrics for vehicle interiors [80]. Experimentally, hydroentangled flax and hemp fabrics up to $1500\,\mathrm{g\,m}^{-2}$ have also been produced for use in automotive composites.

8.20.12 Other applications

The applications for hydroentangled fabrics are developing rapidly, particularly where long-life, multiple use functionality is required. In the composites sector, hydroentangled PET is used as a carrier web in pultrusion products, as well as for filament wound pipe production. Similar fabrics are also useful as surface veilings for fibreglass-reinforced plastics. Some hydroentangled geosynthetic and construction materials, as well as microfibre filter fabrics are commercially produced based on spunlaid web formation platforms. The high strength to weight ratio of hydroentangled fabrics means there is potential to reduce fabric weight and obtain raw material savings in the roofing sector. Experimental thermal insulation fabrics, in which nanoporous silica gels are introduced within cavities in the cross-section of hydroentangled fabrics, have been developed using Hydrospace technology [81]. Such fabrics have also been designed for the storage and controlled release of actives, and by appropriate control of the fabric permeability, delivery can be confined mainly to one side of the fabric. Outside the nonwoven sector, hydroentanglement technology has been developed to enhance the appearance and physical properties of conventional textile fabrics in a process known as hydroenhancement to control the permeability of airbag and filtration fabrics and to improve seam strength and fabric softness. Attachment of secondary carpet backings by hydroentangling webs onto the back of the carpet has also been developed as a means of simplifying the traditional production process.

Acknowledgments

The authors are indebted to a number of people for their valuable contributions, advice, and support during the preparation of this chapter. They are extremely grateful to Mr Matthew Yeabsley, of Karl Mayer Textile Machinery Ltd., UK, for providing the technical literature, and above all for supporting a visit to Karl Mayer Malimo company at Chemnitz, Germany. His kindness and friendship are always appreciated. The authors would also like to thank Mr Alexander Battel, Dr Holger Erth, Mr Axel Wintermeyer, and Mr Daniel Standt. Finally, we are also grateful to Dr Monica Seegar and Mr Wolfgang Schilde of Sächsisches Textil Forschungs Institute. V., Chemnitz, Germany, for their kindness and friendship during visits to their Institute. A great deal of information has been used here directly from the Karl Mayer and machinery manufacturer literature that could not be individually referenced. The authors also wish to thank Groz-Beckert, Dilo Group, ANDRITZ Perfojet, Trützschler Nonwovens, and AUTEFA Solutions for their assistance and valuable contributions to this chapter.

References

[1] P.J. Cotterill, Production and properties of stitch bonded fabrics, Text. Prog. 7 (2) (1975) 101. The Textile Institute.
[2] Textile Terms and Definitions, eleventh ed., The Textile Institute, 2002, p. 333.
[3] R. Krcma, Nonwoven Textiles, Textile Trade Press, Manchester, 1967, p. 156.
[4] Kettenwirk Praxis, 1993 (1) E22.
[5] Kettenwirk Praxis, 1994 (2) E5.
[6] S.C. Anand, 'Karl Mayer warp knitting equipment at I T M A'99, Asian Text. J. 9 (1999) 49.
[7] Kettenwirk Praxis, 1994 (2) E8.
[8] Kettenwirk Praxis, 2002 (2) 25.
[9] F.J. Evans, US3,485,706, 1969.
[10] W.W. Bunting, F.J. Evans, D.E. Hook, US3,508,308, 1970.
[11] F.J. Evans, C. Shambelan, US3,498,874, 1970.
[12] P. Coppin, The future of Spunlacing, Nonwovens World (2001) 60–67. January 2002.
[13] http://www.nonwovens-industry.com/articles/2005/03/feature2.php.
[14] E. Ghassemieh, M. Acar, H.K. Versteeg, Improvement of the efficiency of energy transfer in the hydroentanglement process, J. Compos. Sci. Technol. 61 (12) (2001) 1681–1694 (ISSN 0266–3538).
[15] A. Watzl, New Concepts for Fiber Production and Spunlace Technology for Microdenier Bicomponent Split Fibers—From Polymer to Final Product, INDEX 99 Congress, Manufacturing Session 1, Geneva, April 1999, pp. 1–13.
[16] W. Moschler, Some results in the fields of hydroentanglement of fibrous webs and their thermal after-treatment, in: International Nonwovens Symposium, EDANA, 1995, pp. 1–23.
[17] D. Bertram, Cellulosic fibers in hydroentanglement, INDA J. 5 (2) (1993) 34–41.
[18] W.W. Bunting, F.J. Evans, D.E. Hook, US3,493,462, 1970.
[19] J.A. Guerin, H.T. Jeandron, US3,214,819, 1965.
[20] M. Suzuki, T. Kobayshi, S. Imai, US4,718,152, 1998.
[21] W. Moschler, A. Meyer, M. Brodtka, Influences of fibre and process on the properties of spunlaced fabrics, ITB Nonwovens Ind. Text. 2 (1995) 26–31.
[22] H. Zheng, A.M. Seyam, D. Shiffler, The impact of input energy on the performance of hydroentangled nonwoven fabrics, Int. Nonwovens J. 12 (2) (2003) 34–44.
[23] B. Pourdeyhimi, A. Minton, Structure-process-property relationships in hydroentangled nonwovens, part 1: preliminary experimental observations, Int. Nonwovens J. 13 (4) (2004) 15–21.
[24] A.M. Seyam, D. Shiffler, An examination of the hydroentangling process variables, Int. Nonwovens J. 14 (1) (2005) 25–33.
[25] A. Pourmohammadi, S.J. Russell, S. Hoeffele, Effect of water jet pressure profile and initial web geometry on the physical properties of composite hydroentangled fabrics, Text. Res. J. 73 (6) (2003) 503–508.
[26] N. Mao, S.J. Russell, A framework for determining the bonding intensity in hydroentangled fabrics, J. Compos. Sci. Technol. 66 (1) (2006) 80–91.
[27] N. Mao, S.J. Russell, Towards a quantification of the structural consolidation in hydroentanglement and its influence on the permeability of fabrics, in: Nonwovens Research Academy Proceedings, EDANA, 2005.
[28] R. Mathis, US6,190,736, 2001.
[29] U. Yoshiharu, K. Wakesaka, JP2002146630, 2002.
[30] FiberVisions Hy-Entangle WA. www.fibervisions.com.
[31] https://www.truetzschler-nonwovens.de/en/.
[32] Forsten, H.H., New Sontara Spunlaced aramid structures, Nonwovens Symposium, 1985, p. 251.
[33] www.aramid.com.
[34] K.D. Kelly, T.A. Hill, F. Lapierre, S. DeLuca, S.D. DeLeon, US6,764,971, 2001.
[35] J.J. Rogers, J.L. Erickson, S.M. Sanocki, US5,380,580, 1995.
[36] www.hillsinc.net/nanofiber.shtml.
[37] J. Zucker, WO2004092471, 2004.
[38] http://www.fleissner.de/ne_25112005_e.htm.

[39] F. Noelle, Spunlace: improvements which enhance production efficiencies and reduce operating costs, Tech. Text. 44 (2001) 100–101.

[40] K. Völker, Advancements in spunlacing technology, Nonwovens World 4/5 (2002) 97–103.

[41] J.R. Starr, Water jet entangled nonwovens expanding rapidly, Nonwovens World 3 (2) (1988) 62–68.

[42] Anon, Nonwovens report international, 2005, p. 52. issue 6, December.

[43] A. Shahani, D.A. Shiffler, Foamed latex bonding of spunlace fabrics to improve physical properties, Int. Nonwovens J. 8 (2) (1999) 41–48.

[44] M. Putnam, R. Ferencz, M. Storzer, J. Weng, PCT:WO 02/05578 A1, July 18, 2002.

[45] www.andritz.com.

[46] A. Vuillaume, The Perfojet entanglement process, Nonwovens World 2 (1) (1987) 81–84.

[47] M. Suzuki, T. Kobayashi, GB2114173, 1983.

[48] F. Noelle, US5,768,756, 1998.

[49] F. Kalwaites, US3,033,721, 1962.

[50] D.T. Ward, Spunlace line offers energy economy and versatility, ITB Nonwovens Ind. Text. 4 (1997) 38–42.

[51] L. Schmit, B. Roche, US6,474,571, 2002.

[52] T.F. Gilmore, N.B. Timble, G.P. Morton, Hydroentangled nonwovens made from unbleached cotton, TAPPI J. 80 (3) (1997) 179–183.

[53] www.nippon-nz.com.

[54] E. Ghassemieh, H.K. Versteeg, M. Acar, Effect of nozzle geometry on the flow characteristics of hydroentangling jets, Text. Res. J. 73 (5) (2003) 444–450.

[55] V.T. Tafreshi, B. Pourdeyhimi, Simulating the flow dynamics in hydroentangling nozzles: effect of cone angle and nozzle aspect ratio, Text. Res. J. 73 (8) (2003) 700–704.

[56] V.T. Tafreshi, B. Pourdeyhimi, R. Holmes, D. Shiffler, Simulating and characterising water flows inside hydroentangling orifices, Text. Res. J. 73 (3) (2003) 256–262.

[57] A. Begenir, V.T. Tafreshi, B. Pourdeyhimi, Effect of nozzle geometry on hydroentangling water jets: experimental observations, Text. Res. J. 74 (2) (2004) 178–184.

[58] A. Fechter, U. Münstermann, A. Watzl, Latest developments in hydroentanglement, Chem. Fibers Int. 60 (2000) 587–588.

[59] G. Fleissner, DE10059058, 2002.

[60] www.idrosistem.com.

[61] www.nonwovens-industry.com/articles/2004/08/feature2.php.

[62] S.K. Black, S. Deleon, EP1434904, 2004.

[63] K.S. Chang, S.D. Edward, EP1454000, 2004.

[64] https://www.berryglobal.com/en/product/product-item/spunlace-wiping-substrate-13473760.

[65] D. Groitzsch, Ultrafine microfiber spunbond for hygiene and medical application, in: NT New Textiles, EDANA Symposium, 2000.

[66] R. Groten, G. Riboulet, The Evolon Project, vol. 41, TUT, 2001, pp. 27–28.

[67] F. Zlatkus, US6074966, 2000.

[68] G. Mansfield, H_2O tricks, Textile World (2004) 28–31.

[69] www.contecinc.com.

[70] P. Barge, N. Carter, US2004068849, 2004.

[71] M.C. Ngai, EP1354093, 2003.

[72] Ahlstrom, Fiber Composites Hydraspun 784 Dispersible Wet Wipe Leaflet, 2005.

[73] A. Watzl, J. Eisenacher, Spunlace Process for Cotton Pads and Other Products, ITB, Nonwovens & Industrial Textiles, 2000, pp. 16–18.

[74] A. Vuillaume, J.C. Lacazale, US5393304, 1995.

[75] M.J. Putnam, H. Hartgrove, USP6,669,799, 2003.

[76] https://www.jacob-holm.com/us/en-us/our-markets/high-performance-materials/aramid-fabrics/sontara-kevlar-sl-type-z-11/.

[77] http://www.technical-textiles.net/archive/htm/att_20020601.475879.htm.

[78] C.R. Pearce, S. DeLeon, WO2004073834, 2004.

[79] E. Schmalz, Hycofil: Spunlace Scrim Supported Nonwoven, vol. 65, TUT, 2005, pp. 27–30.

[80] https://www.tenowo.com/en/markets/automotive/interior/.

[81] S. Hoeffele, S.J. Russell, D.B. Brook, Lightweight nonwoven thermal protection fabrics containing nanostructured materials, Int. Nonwovens J. 14 (4) (2005) 10–16.

Chemical bonding

R.A. Chapman[a], M. Molinari[b], S. Rana[c], and P. Goswami[c]

[a]3 The Wardens, Kenilworth, United Kingdom [b]Department of Chemical Sciences, School of Applied Sciences, University of Huddersfield, Queensgate, Huddersfield, United Kingdom [c]Technical Textiles Research Centre, University of Huddersfield, Huddersfield, United Kingdom

9.1 Introduction

Textile Terms and Definitions [1] defines a binder as an adhesive material used to hold together the fibres in a nonwoven structure. The word 'binder' describes the function of a composition in the final product. The terms 'binder', 'binding agent', 'binder composition', 'binder system', 'nonwoven binder', and 'chemical binder' are used in the literature to describe the polymer, polymer and carrier, part-formulation or total formulation used in chemical bonding – the meaning shifts according to the context. A binder not only 'holds the fibres together' but also affects the final properties of the nonwoven fabric including its strength (both tensile and compressive), stiffness, softness, waterproofness, breathability, and other functional properties. The choice of binder also influences the capability of the fibrous assembly to be recycled or biochemically degraded at the end of its useful life. Chemical bonding remains popular because of the large range of adhesive binders available, the durability of the products, and the broad variety of final properties that can be engineered in the fabrics.

While in the early days of development, natural binders such as starch and rubber were used, synthetic polymers now dominate the industry. Mostly in response to the needs for more environmentally sustainable materials, concerns about free formaldehyde and ease of end-of-life disposal, there is a resurgence of interest in biodegradable binders derived from agricultural sources for particular applications. These include polysaccharides, pectins, oils, and casein amongst others. Binders are also applied to nonwoven fabrics that are already prebonded to provide additional functionality, since the binder can be mixed with active components or solids such as flame retardants and functional finishes, ceramics, and metals. For example, in the manufacture of wipe products, pigments and chemical binders are commonly printed onto the surface of hydroentangled fabrics to increase the wet strength, to control wet pick-up, and to improve the visual appearance of the product. It is increasingly common to use such 'combination bonding' procedures in which several different bonding methods are used in succession. Thermally bonded airlaid fabrics, needlepunched, and hydroentangled fabrics are frequently subject to secondary chemical bonding to modify fabric properties or appearance.

Binder polymers can be dissolved in a solvent including water, but most commonly they are in the form of a dispersion or emulsion. The most important binders are latices of emulsion polymers, which are generally referred to as latex binders. These are fine dispersions of specific polymers in water. They are applied in a number of different ways to nonwoven substrates and because their viscosity is close to that of water, they can easily penetrate thick or dense nonwoven structures by simple immersion. After application of the binder by, for example, immersion, they are dried to evaporate the solvent. Typically, the binder forms an adhesive film across or between fibre intersections and fibre bonding is obtained. Binders create a network of interlocked fibres, which can be throughout the fabric structure or in selected areas depending on the required end-use.

The distribution of the binder in the fabric structure and its properties can be affected by the use of coagulating and crosslinking agents as well as the application method utilised. In chemically bonded fabrics, the concentration of binder on the surfaces and in the interior may not be uniform and this affects fabric stiffness, handle, and the probability of delamination in some cases. The concentration of binder may be graduated in the fabric cross-section, for example, it may decrease from the fabric surface towards the middle due to migration of the binder towards the surfaces during drying. Alternatively, as in some foam bonding operations, for example, the application of the binder may be purposely designed to be concentrated differently throughout the fabric cross-section. The binder system wets the fabric and following drying and/or crosslinking, forms a bonded structure. Although homopolymer emulsions can be utilised, copolymers or blends with fillers are common. Copolymers provide some tailoring of the main homopolymer properties, for example, to enable increased softness, and fillers help to reduce cost and provide additional useful properties such as improved thermal resistance, abrasion resistance, flame retardancy, water repellency, or antistatic properties. Generally, fillers are an economical way of achieving such properties in contrast to changing the fibre composition.

Commercially, binder systems are applied at levels between about 5% and 150% on the dry weight of fabric. A 5% binder addition is often sufficient to bond fibres at the surface. Addition levels as high as 150% are sometimes used to make stiff reinforcement components such as those found in shoes.

9.2 Chemical binder polymers

9.2.1 Introduction

Various binder polymers are used including vinyl polymers and copolymers, acrylic ester polymers and copolymers, rubber and synthetic rubber, and natural binders, principally starches and other polysaccharides. These are usually applied as aqueous dispersions but can be supplied as polymer solutions provided they have sufficiently low viscosity to allow penetration into the web [2]. Table 9.1 gives the main types of binder in use and their chemical structures are provided in Table 9.2.

TABLE 9.1 Summary of the main binder types.

Vinyl based

Acrylic esters and copolymers

Polyurethane and copolymers

Elastomers including silicone

Thermosetting resins: epoxy, polyester, urea formaldehyde, melamine, alkyd

Natural binders: starches, natural rubber, regenerated proteins

TABLE 9.2 Chemical structures of commonly used chemical binders [3–6].

Chemical binders	Chemical structure
Ethylene	$CH_2{=}CH_2$
Vinyl	$CH_2{=}CH{-}O{-}\overset{\overset{O}{\|\|}}{C}{-}CH_3$ (Acetate) $CH_2{=}CH{-}Cl$ (Chloride) $CH_2{=}\overset{\overset{Cl}{\|}}{C}{-}Cl$ (Vinylidene chloride)
Acrylonitrile	$H_2C{=}\underset{H}{C}{-}CN$
Acrylics	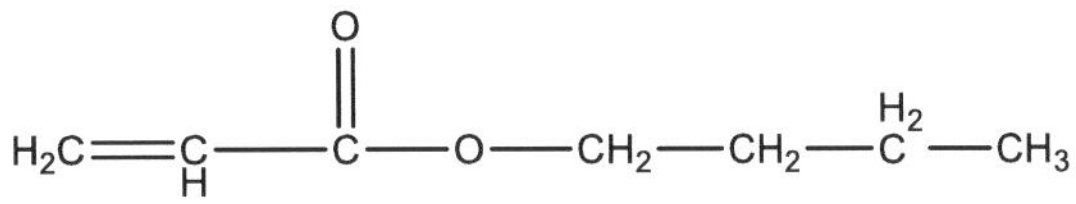 $H_2C{=}\underset{H}{C}{-}\overset{\overset{O}{\|\|}}{C}{-}O{-}CH_3$ (Methyl acrylate) $H_2C{=}\underset{H}{C}{-}\overset{\overset{O}{\|\|}}{C}{-}O{-}CH_2{-}CH_3$ (Ethyl acrylate $H_2C{=}\underset{H}{C}{-}\overset{\overset{O}{\|\|}}{C}{-}O{-}CH_2{-}CH_2{-}\overset{\overset{H_2}{}}{C}{-}CH_3$ (n-butyl acrylate)

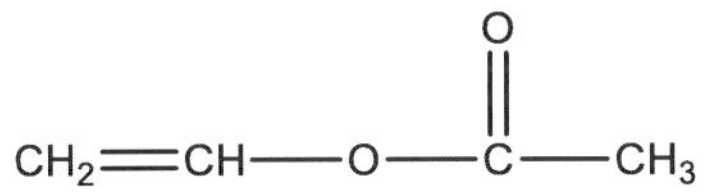

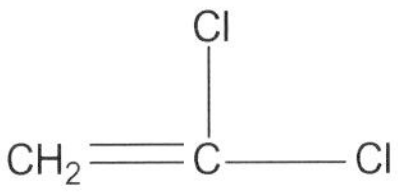

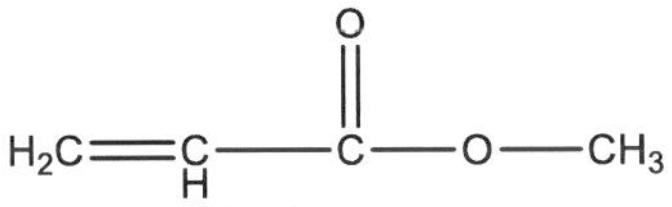

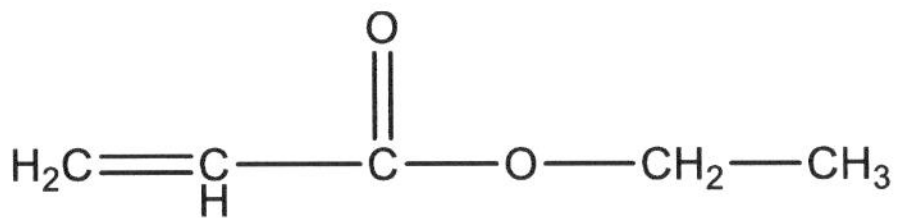

Continued

TABLE 9.2   Chemical structures of commonly used chemical binders [3–6]—cont'd

Chemical binders	Chemical structure

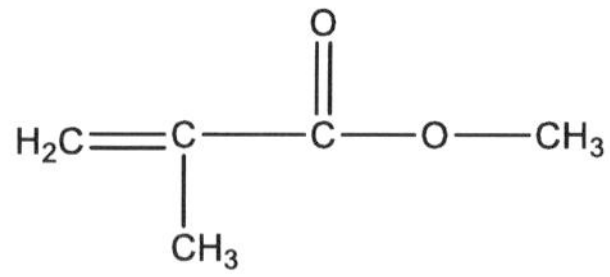

$H_2C{=}C$ — C — O — CH_3, with $=O$ above the C and CH_3 below the first C

(Methyl methacrylate)

$H_2C{=}C$ (with H below) — C — OH, with $=O$ above

(Acrylic acid)

$H_2C{=}C$ — C — OH, with $=O$ above and CH_3 below

(Methacrylic acid)

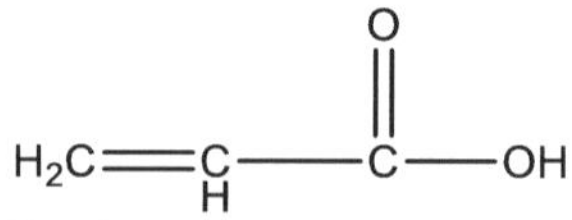

$H_2C{=}C$ (with H below) — C — NH_2, with $=O$ above

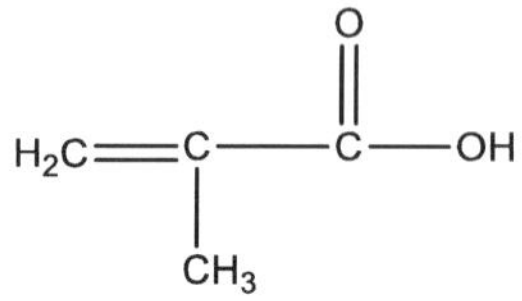

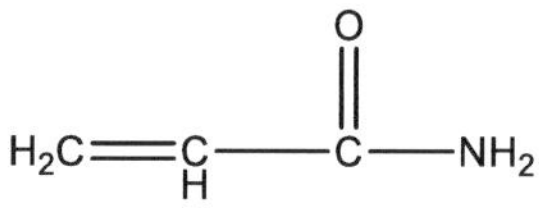

(Acryl amide)

Butadiene polymers

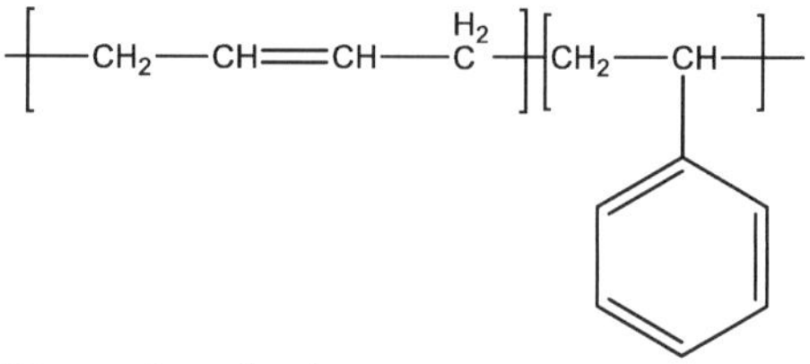

$-[CH_2-CH{=}CH-C^{H_2}]-[CH_2-CH]-$ (with phenyl ring on CH)

(Styrene butadiene)

$-[CH_2-C{=}CH-C^{H_2}]-$, with Cl below C

(Chloroprene)

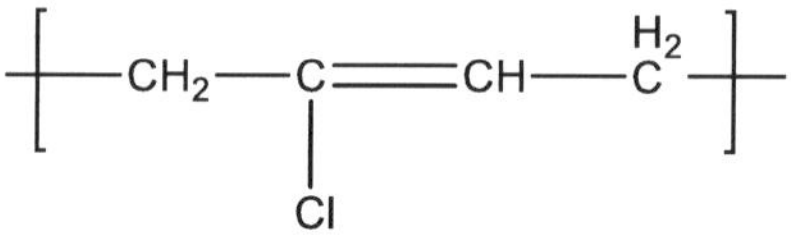

$-[CH_2-C(H){=}CH-C^{H_2}]-[CH-C^{H_2}]-$, with CN below CH

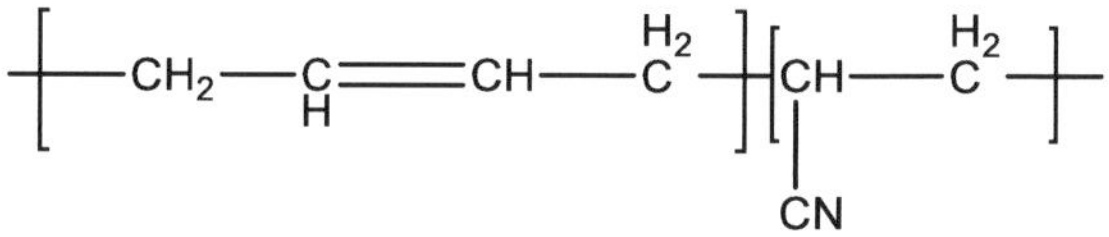

(Nitrile rubber)

TABLE 9.2   Chemical structures of commonly used chemical binders [3–6]—cont'd

Chemical binders	Chemical structure
Natural binders	(Starch) (Natural rubber) (Poly lactic Acid)
Thermosetting resins	(Urea formaldehyde) (Melamine formaldehyde) (Epoxy) (Polyester)

Acrylic thermoset resins have also been developed based on low molecular weight polyacids (polyacrylic acid) and an accelerant (sodium hypophosphite). These are intended to be formaldehyde resin alternatives and applications include glass fibre insulation [7]. Commercially, latex polymers are the most commonly encountered binder systems because of the wide variety available, their versatility, ease of application, and cost effectiveness.

9.2.2 Latex polymers [8,9]

9.2.2.1 Emulsion polymerisation

An emulsion polymer is a colloidal dispersion of discrete polymer particles with a typical particle diameter of 0.01–1.0 μm in a medium such as water. Common polymers used are acrylates, styrene–butadiene copolymers, acrylonitrile–butadiene copolymers, and ethylene vinyl acetate (EVA). A latex polymer is prepared by the controlled addition of several components either in a batch or a continuous monomer addition process. The components are water, monomers (the polymer building blocks), initiator (to start the polymerisation process), surfactant (to stabilise the emulsion particles as they form by preventing coalescence), and chain transfer agent (to control the final polymer molecular weight). The role of each component will be discussed later.

9.2.2.2 Process of latex formation

The process starts with a distribution of monomer droplets in water, stabilised by emulsifiers that have accumulated at the interface to the water phase. Emulsifier molecules have hydrophobic and hydrophilic parts. In Fig. 9.1, the line represents the hydrophobic part of the molecule and the dot represents the hydrophilic part. If the concentration of the emulsifier is above a critical value, a spheroidal collection of them form. This is called a micelle and contains about a hundred emulsifier molecules. The hydrophilic parts project into the water producing a hydrophobic interior. The hydrophobic interior is able to accommodate other hydrophobic substances, for example, monomer molecules. The initiator decomposes to form water-soluble free radicals.

Nearly all of the monomers are present in the form of monomer droplets, but there is a very small proportion that is dissolved in the water. When a free radical encounters monomer molecules dissolved in the water, it reacts successively with several to form a short polymer chain. This short chain, called an oligomer radical, is no longer soluble in water. It precipitates and is stabilised by the emulsifier, which accumulates at the newly formed interface. This is now a latex particle. Provided that there is enough emulsifier available, more oligomer radicals can be stabilised and grow into latex particles. However, if there is insufficient emulsifier, the insoluble oligomer radicals aggregate, presenting a smaller surface that requires less emulsifier to be stabilised. The result is that fewer but larger particles form. In addition to this process, we have to consider the emulsifier micelles. Monomer molecules diffuse into these. If an oligomer radical meets an emulsifier micelle, which contains monomer, the monomers polymerise and form another latex particle. This can occur only if the concentration of emulsifier is high enough (above the 'critical micelle concentration').

Finally, it is possible for a growing oligomer radical to meet a monomer droplet and initiate polymerisation, forming a latex particle. In this case, the latex particle would be large – about

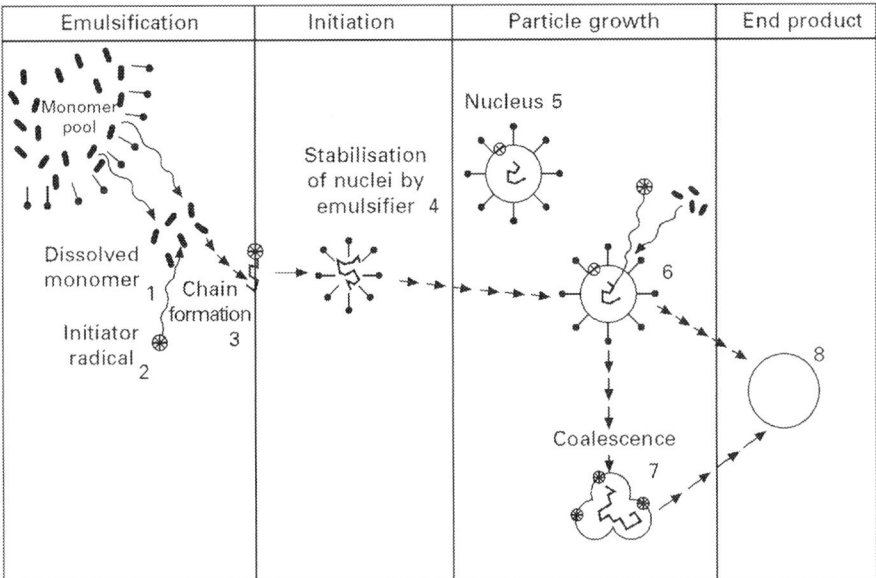

| Emulsification | Initiation | Particle growth | End product |

FIG. 9.1 Particle formation during emulsion polymerisation. Note: The emulsifier molecules are represented by a *line* and a *dot. Adapted from Polymer Latex 1 refers to* How do aqueous emulsions form? *by Polymer Latex GmbH & Co. KG.* *0/8000136/360e.*

the size of the original monomer droplet. While possible, it is a rare occurrence. Now the formation of latex particles is completed and growth starts. There is a flow of monomer from the water and the monomer droplets to the latex particles where polymerisation occurs. The latex particle grows larger and rounder and it can contain hundreds or thousands of closely packed molecules in one particle. If there is a shortage of emulsifier, then the growing particles do not grow as above but coalesce. As propagation proceeds, more particles are added in layers to form a larger latex particle.

9.2.2.3 Binder components

(a) Polymers

The monomers are selected to form the basic building blocks of the binder. The selection of monomers is determined by cost and the final fabric properties required. Polymers are often characterised as 'hard' or 'soft' depending on their glass transition temperature, T_g. The binder T_g influences fabric handle and the perception of softness in use. Indicative glass transition temperatures for typical homopolymers used for making binders vary slightly depending on the source. The values in Table 9.3 are approximate values.

A calculated estimate of the T_g of any copolymer can be obtained using the Fox equation [10]:

$$1/T_g = W_x/T_{gx} + W_y/T_{gy} \tag{9.1}$$

TABLE 9.3   Glass transition temperatures of homopolymers produced from the listed monomers.

Monomer	T_g (°C)
Soft	
Ethylene	−120
2-Ethylhexyl acrylate	−85
Butadiene	−78
n-Butyl acrylate	−52
Ethyl acrylate	−22
Hard	
Methyl acrylate	+9
Vinyl acetate	+30
Vinyl chloride	+80
Methyl methacrylate	+105
Styrene	+105
Acrylonitrile	+130

where T_{gx} and T_{gy} = glass transition temperatures of polymers x and y respectively and W_x and W_y = weight fraction of polymers x and y respectively, and $W_x + W_y = 1$.

In addition to affecting the handle and bending stiffness of the final product, the choice of monomers affects the hydrophilic or hydrophobic properties of the fabric. This directly reflects the hydrophilicity of the monomers used in the assembly of the binder polymer. For example, butyl acrylate is relatively hydrophobic and vinyl acetate is relatively hydrophilic. Clearly, the wet stability of the binder is a consideration in some applications such as disposable wipes and incontinence products. It is also important in the design of single-use, water-dispersible wipes, where solubilisation of the binder may be required. Binder extensibility is inversely related with the T_g and is also influenced by the molecular weight.

(b) Surfactants

Surfactants perform several functions in emulsion polymerisation of which the most important is providing latex stability both during and after polymerisation. The surfactants used are either anionic, cationic, or nonionic. In emulsion polymerisation, anionic and nonionic types are normally used [11]. Typical anionic surfactants are sodium lauryl sulphate or sodium lauryl ether sulphate. The molecule contains both polar (hydrophilic) and nonpolar (hydrophobic) groups. The surfactant works by stabilising latex particles using electrostatic repulsion forces to prevent particle attraction. Nonionic surfactants, for example, ethoxylated lauryl alcohol, are used to improve the mechanical and freeze–thaw stability of a latex. They work by steric hindrance. The choice of surfactant affects the charge on the emulsion, the particle size, surface tension (which affects the wetting behaviour of the binder on the fibre), fibre adhesion, film formation, and emulsion stability [12]. The wetting behaviour is particularly important to ensure the binder is properly distributed over the fibre surfaces in the web or fabric.

(c) Initiators

The initiator, which is commonly ammonium persulphate, decomposes on heating to form free radicals that start the polymerisation process.

(d) Chain transfer agents

Sometimes it is desirable to limit the molecular weight of the polymer by introducing a chain transfer agent such as dodecyl mercaptan. The growing polymer radical combines with the chain transfer agent to stop chain growth. A short chain radical also forms from the chain transfer agent, which reacts with a monomer molecule to form a new polymer radical that starts to grow.

(e) Buffers

A buffer is used to control the pH during the polymerisation process. Some monomers may hydrolyse if the pH is not controlled.

(f) Other additives

Sometimes alkali (e.g. sodium hydroxide) is added to the latex to increase the pH and improve its stability.

9.2.3 Latex polymer binder systems

The main systems are based on vinyl, acrylate (also called acrylic), and butadiene polymers. Choice depends on cost, stiffness, binder hardness and softness (which influences fabric handle), toughness, water and solvent resistance, as well as ageing properties.

9.2.3.1 *Vinyl polymers*

Vinyl monomers contain carbon–carbon double bonds and form polymers of the type – $[CH_2–CR·CR']_n–$. A range of vinyl polymers are available, of which acrylates are a subdivision as shown in Fig. 9.2. Examples of vinyl polymers include polyvinyl acetate, polystyrene, and polyvinyl chloride. Vinyl homopolymers such as vinyl chloride and vinyl acetate are hard and have strong adhesion to a wide range of fibres. Because of their hardness, they are often plasticised with internal or external plasticisers such as phthalates. Ethylene is not used as a homopolymer in binders but in a copolymer such as EVA or ethylene vinyl chloride and provides flexibility.

9.2.3.2 *Vinyl acetate*

Vinyl acetate binder polymers have a T_g of around 30°C and are quite hard and tough. The hardness can be reduced using acrylates or ethylene as comonomers. The polymers are hydrophilic and tend to yellow on heating. Self-crosslinking versions provide improved

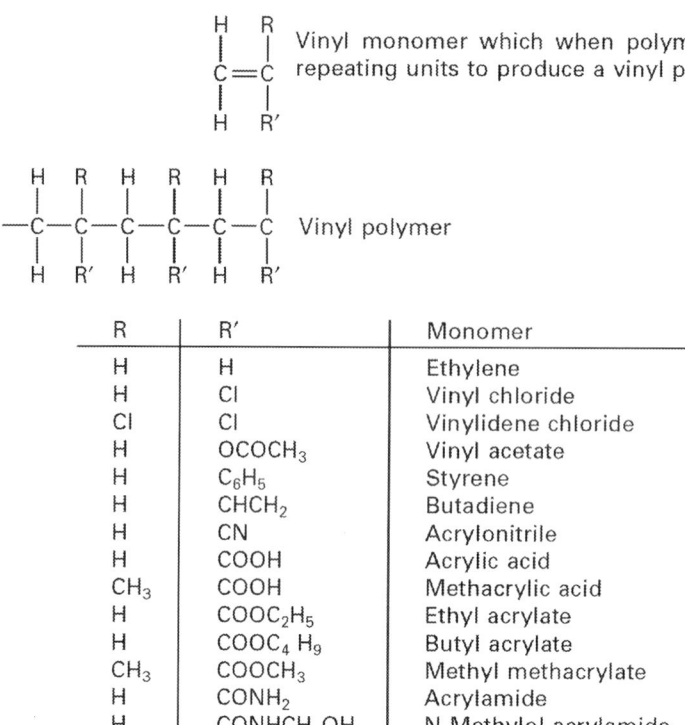

R	R'	Monomer
H	H	Ethylene
H	Cl	Vinyl chloride
Cl	Cl	Vinylidene chloride
H	$OCOCH_3$	Vinyl acetate
H	C_6H_5	Styrene
H	$CHCH_2$	Butadiene
H	CN	Acrylonitrile
H	COOH	Acrylic acid
CH_3	COOH	Methacrylic acid
H	$COOC_2H_5$	Ethyl acrylate
H	$COOC_4H_9$	Butyl acrylate
CH_3	$COOCH_3$	Methyl methacrylate
H	$CONH_2$	Acrylamide
H	$CONHCH_2OH$	N-Methylol acrylamide

FIG. 9.2   Vinyl monomers. *Adapted from R. Pangrazi, Chemical binders for nonwovens – a primer, INDA J. Nonwovens Res. 4(2) Spring (1992) 33–36.*

stability to water. They are relatively cheap. Vinyl chloride is sometimes included to enable nonwovens to be bonded by dielectric heating because the polymer has a comparatively low softening temperature.

9.2.3.3 *Vinyl chloride*

Vinyl chloride is a hard polymer ($T_g \sim +80°C$) and is therefore unsuitable for many non-woven products. Copolymerising with the softer acrylic monomers reduces the hardness of the homopolymer improving its utility for nonwoven applications. These polymers are often used because of their inherent flame retardancy arising from the chlorine content. They are also thermoplastic and can be welded using dielectric heating but like vinyl acetate, they tend to yellow on heating.

9.2.3.4 *Ethylene vinyl chloride*

These binder polymers can be considered similar to vinyl chloride polymers but with the ethylene monomer acting as an internal plasticiser to provide greater polymer ductility. This class of binders has a slightly broader range of stiffness than vinyl chloride but without the need for an external plasticiser. The vinyl chloride monomer also provides some attractive flame retardant properties and can be welded using dielectric heating. They bond well to synthetic fibres and provide good abrasion and acid resistance.

9.2.3.5 Ethylene vinyl acetate

EVA polymers can be made with a wide range of softness properties. They tend to be cheaper than acrylics and have good adhesion to many synthetic fibres. They are less resistant to solvents than acrylics but provide a good combination of high wet strength, excellent absorbency, durability, and softness. They are often used in disposable hygiene products such as wipes and are increasingly used in the bonding of short-fibre pulp airlaid fabrics used in disposable hygiene fabrics.

9.2.3.6 Vinyl acetate acrylate

These are mostly based on butyl acrylate. They can be regarded as being a compromise between vinyl acetate and acrylics, both in performance and cost. The vinyl acetate monomer is generally cheaper than the acrylics. The acrylic monomer decreases the sensitivity to moisture and solvents.

9.2.3.7 Acrylonitrile

The homopolymer is not used as a binder by itself but when the acrylonitrile monomer is used to make nitrile rubber, it provides excellent resistance to solvents, oil, and moisture.

9.2.3.8 Styrene

Styrene monomers are hard and provide stiffness and hydrophobicity. Polystyrene homopolymer is hard and brittle at room temperature and does not easily form a film. For this reason, it is difficult to use this as an effective binder.

9.2.3.9 Acrylate polymers

Polyacrylates (commonly referred to as acrylics) are a type of vinyl polymer. The most important are copolymers of acrylic acid derivatives, especially acrylic acid and methacrylic acid esters. They are made from acrylate monomers, which are esters containing vinyl groups. There are more than 30 monomers used. Their hardness and solvent resistance decreases with increasing chain length of the alcohol moiety. Polymethacrylates have higher film hardness than polyacrylates [13]. Crosslinking improves their resistance to washing at the boil and dry cleaning but they tend to be more expensive than other binders. Common examples are ethyl acrylate and butyl acrylate. To increase stiffness, these may be copolymerised with methacrylate, methylmethacrylate, or styrene. For increased hydrophilicity, methyl acrylate monomers are used. Specifically, for increased hydrophobicity, 2-ethylhexyl acrylate or styrene can be used [14]. Styrenated acrylics are hydrophobic, tough binders, which are relatively cheap. They are used where high wet strength is needed but at some detriment of UV and solvent resistance. A typical particle size in an acrylate binder formulation is around 200 nm.

9.2.3.10 Butadiene polymers

Polymers based on butadiene CH_2=$CH \cdot CH$=CH_2 usually have relatively high elasticity and toughness and have been used since the early years of the nonwovens industry. They include natural rubber latex (polyisoprene), polychloroprene, styrene–butadiene rubber (SBR), and nitrile–butadiene rubber (NBR).

9.2.3.11 Natural rubber latex

This was one of the earliest binders used in the manufacture of nonwoven fabrics and was superseded by styrene–butadiene and nitrile–butadiene rubbers. After drying, the temperature is increased to initiate polymer crosslinking (vulcanisation). It provides an excellent soft handle and high elasticity.

9.2.3.12 Chloroprene

Polychloroprene binders are unusual in that they crystallise, causing an increase in stiffness. Their resistance to organic solvents and oils is not quite as high as NBR copolymers but they are exceptionally acid resistant. Their resistance to weathering is better than NBR and SBR but their discoloration is greater and they are used for some nonwoven shoe materials.

9.2.3.13 Styrene–butadiene rubber

SBR binder polymers are tough, flexible, and have excellent solvent resistance. Their stiffness and hardness increase with the level of styrene. They are cheaper than acrylates and nitrile rubbers (but less elastic than the latter). Crosslinking gives them excellent water resistance.

9.2.3.14 Nitrile rubber

These are butadiene–acrylonitrile copolymers. Increasing the level of acrylonitrile in such rubbers increases the hardness. Compared to other polymers used as binders, they have low thermoplasticity and so can be sueded (or subject to intensive mechanical abrasion) without melting the fibres in the fabric. They also have high abrasion resistance and are often used to make synthetic leather.

9.2.4 Other polymer binders

9.2.4.1 Polyurethane

Polyurethane (PU)-based binders have been favoured for many years in the manufacture of synthetic leather nonwoven fabrics as well as products where good extensibility is required. They are applied from solvent or are produced as aqueous dispersions. Today, most PU binders are actually polyester–polyurethane copolymers. They tend to have excellent adhesion and film-forming properties. Film structure is controlled by the pH of the aqueous dispersion, and acid coagulation can enable the formation of microporous films used in breathable membranes and coatings. In the case of solvent systems, the PU is first dissolved in dimethylformamide (DMF) and after the fabric has been impregnated, the DMF is displaced and the PU is coagulated. During drying, a dense and porous structure is formed

[15]. PU polymers provide soft, elastic binders and films with comparatively good resistance to hydrolysis and light fastness. Water-based dispersions are increasingly favoured over solvent-based PUs but they tend to have lower wet stability. PU is frequently used to produce low-cost hydrophilic breathable coatings on fabrics. Traditionally, poly(ethylene oxide) (PEO) is used to increase the inherent hydrophilicity of these PU materials.

9.2.4.2 Phenolics

Phenolic binders are occasionally used for full saturation bonding of fabrics to make durable filter fabrics and for fabrics requiring high abrasion resistance that are operated at high temperature, for example, in clutch and brake pads.

9.2.4.3 Epoxy resins

Waterborne epoxy resins are used for bonding nonwovens when high chemical resistance, stability at high temperatures, or electrical insulation properties are required. Epoxy resins are particularly important in the field of fibre-reinforced composites. Obviously, the chemical and mechanical properties of the constituent fibres need to be carefully selected to achieve the correct blend of overall properties [16].

9.2.5 Bio-based binders

Historically, the nonwovens industry has used binders based on natural rubber, carbohydrates, e.g. starches, as well as proteins obtained from renewable bio-based sources, but usage progressively reduced as synthetic alternatives were developed. Although chemical binders derived from petroleum sources are now most common, increasing concerns about environmental sustainability, not least the end-of-life disposal of plastics, has renewed interest in bio-based chemical binders from renewable materials. Examples of carbohydrate binders are dextrin and maltodextrin [17,18]. Chemical binders based on aqueous emulsions containing modified biopolymers and natural plant compounds, e.g. OC-Biobinder (OrganoClick) have also been commercially developed for use in nonwovens, which are both biodegradable and compostable.

9.2.5.1 Soy protein isolate

Amongst bio-based proteins, soy protein isolate (SPI) has been used for chemical bonding of nonwoven fabrics. The chemical structure of SPI is provided in Fig. 9.3. It is already in use as an adhesive in a range of industrial applications. SPI has many useful properties such as low cost, biodegradability, high hydrophilicity, and no formaldehyde release, making it an attractive alternative to petroleum-based chemical binders [19]. Kumar et al. used SPI as a binder for viscose-based nonwoven fabrics and compared their properties with those bonded with a commercial acrylic binder [20]. The binders were applied to a prebonded hydroentangled fabric using a foam application method, as schematically shown in Fig. 9.4 and both types of bonded fabrics showed similar mechanical, thermal, and moisture absorption properties [20].

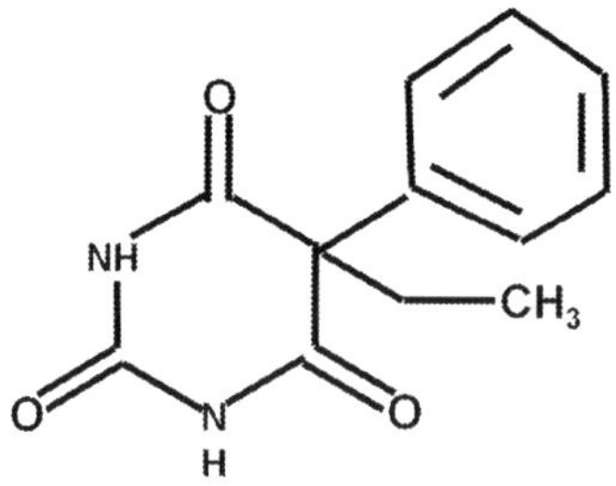

FIG. 9.3   Chemical structure of SPI binders (https://www.chemsrc.com/en/baike/1198690.html).

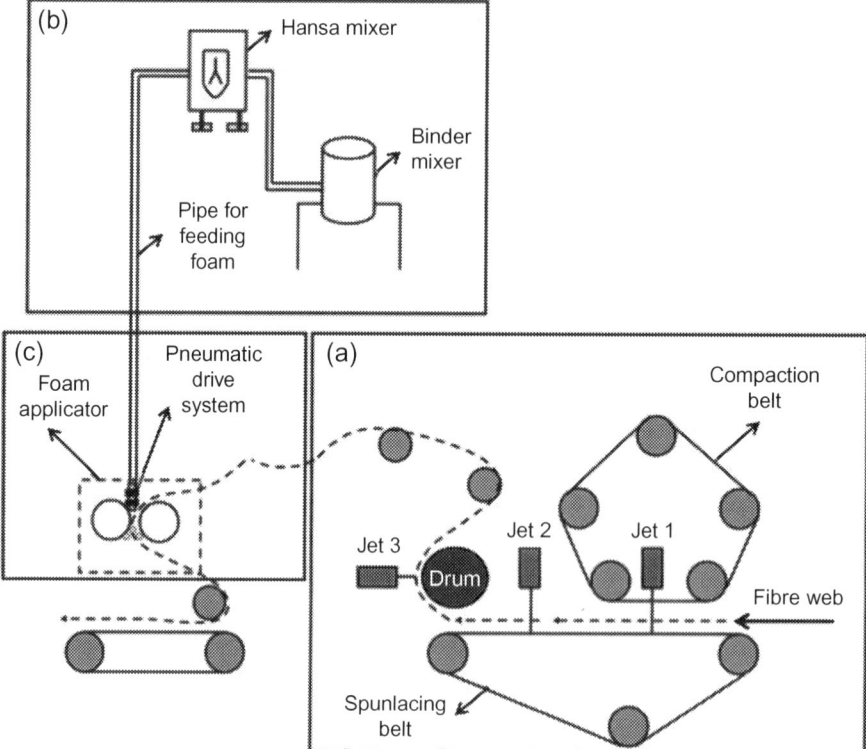

FIG. 9.4   Application of SPI to viscose nonwoven fabrics: (A) hydroentanglement, (B) foam generating machine, (C) foam applicator for chemical bonding [20].

9.2.5.2 Nanocellulose binders

Recently, nanocellulose has been found to be an effective binder in different applications including paper-coating processes and nonwoven fabrics, particularly when made of cellulosic fibres [21]. Nanocellulose is extracted from renewable cellulosic materials in different forms such as nanofibrillated cellulose (NFC) and nanocrystalline cellulose (CNC) or is produced using bacteria in the form of bacterial cellulose (BC) [22,23]. Nanocellulose has been applied in a wide range of industries due to its remarkable physical, mechanical, and chemical properties [22,23]. A schematic of CNC extraction from cellulose after removing the amorphous regions through acid hydrolysis is presented in Fig. 9.5, along with the morphologies of different forms of nanocellulose. The presence of nanocellulose has shown an excellent bonding with a smoother coating on papers and paperboards, such that it may be possible to reduce the amount of latex binder that is normally required [21].

Earlier studies demonstrated the successful use of nanocellulose to improve binding of different materials such as clay, pigments, calcium carbonate, starch, and graphite to paper surfaces [17]. The capability of nanocellulose to act as an effective binder is mainly attributed to the formation of hydrogen bonds between nanocellulose and cellulose-based fibrous substrates [21], as schematically illustrated in Fig. 9.6, and its excellent mechanical properties, as listed in Table 9.4.

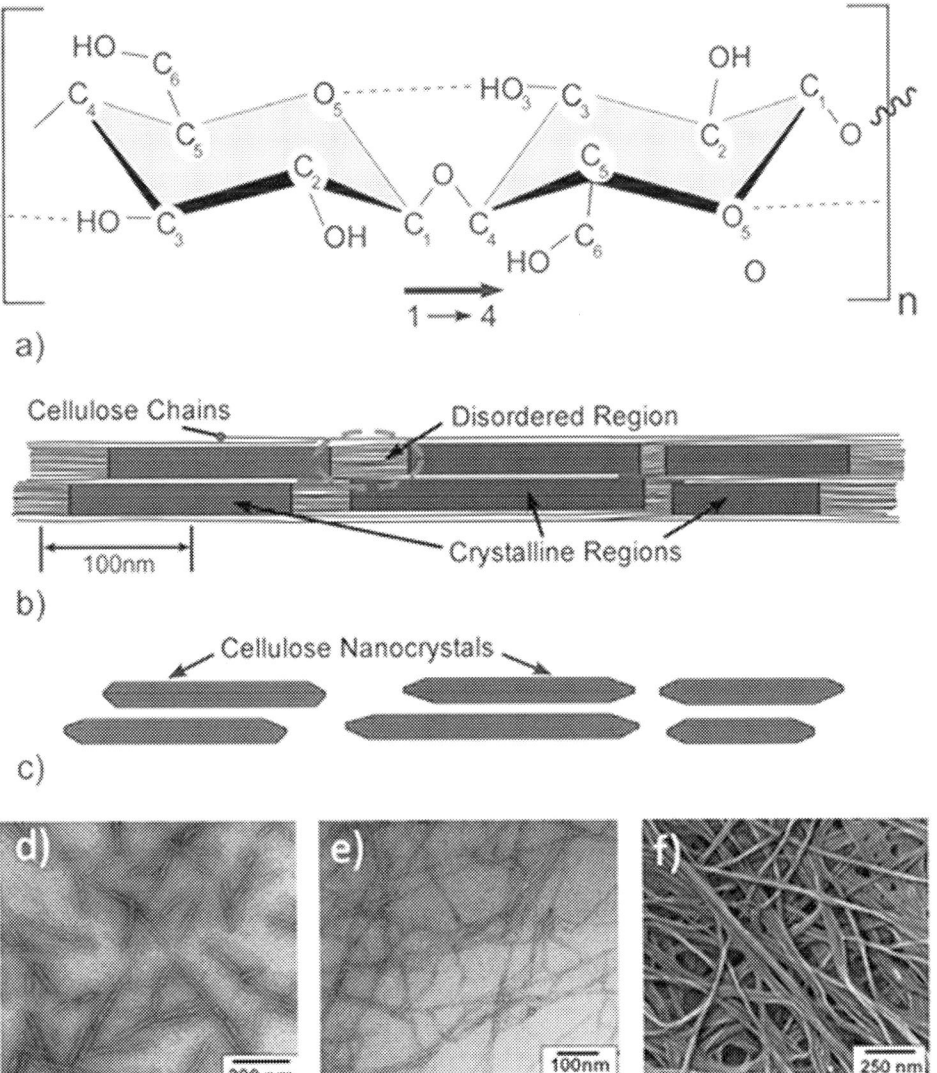

FIG. 9.5 Schematic of (A) single cellulose chain repeat unit, (B) cellulose microfibrils containing crystalline and amorphous regions and (C) CNC and TEM image of CNC (D), NFC (E) and SEM image of BC (F) [23–25].

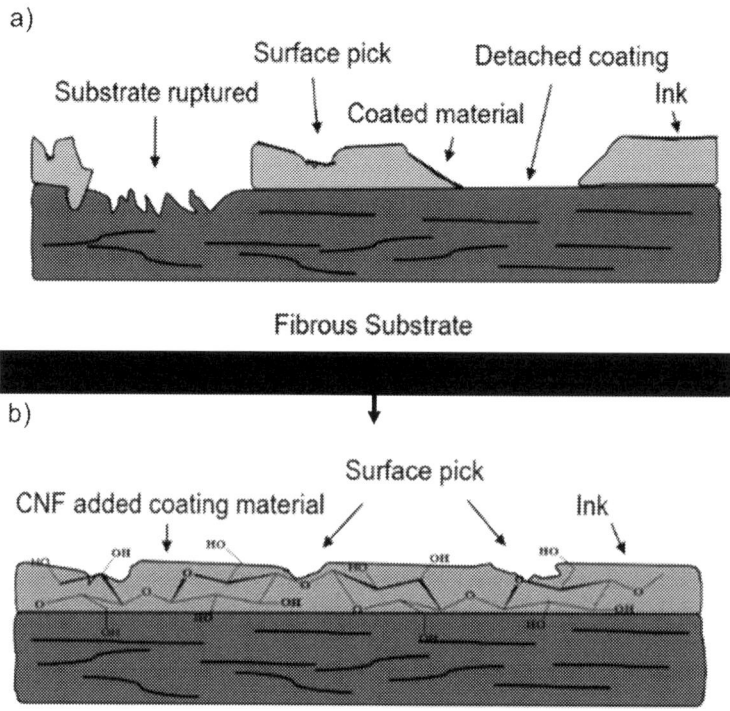

FIG. 9.6 Schematic showing effect of NFC in improving bonding of coating: coating without NFC (A) and with NFC (B) [21].

TABLE 9.4 Comparison of properties of nanocellulose with other high strength materials [22,23].

Material	Density (g/cm^3)	CTE (10^{-6}/K) axial	Tensile strength (GPa) axial	Elastic modulus (GPa)	
				Axial	Transverse
Crystalline cellulose	1.6	0.1	7.5	120–220	11–57
Kevlar-49 fibre	1.4	2	3.5	124–130	2.5
Clay nanoplatelets	–	–	–	170	–
Carbon nanotubes	–	–	11–63	270–950	0.8–30
Boron nanowhiskers	–	6	2–8	250–360	–

TABLE 9.5 Different types interactions between nanocellulose and cellulose substrate [21].

Adhesion mechanism	Bond length	Bond strength (kJ/mol)
Diffusion	<2000 μm	—
Physical entanglement	0.01–1000 μm	
Van der Waals	0.5–1 nm	8.4–21
Acid–base interaction	0.1–0.4 nm	—
Hydrogen bond	0.235–0.27 nm	4.2–188
Covalent bonding	0.15–0.45 nm	147–628

Different types of interactions which can occur between nanocellulose and cellulose-based substrates are listed in Table 9.5. It is possible to note that the type of interactions ranges from molecular (e.g. hydrogen bonding and Van der Waals forces) to micron levels (e.g. physical entanglements due to mechanical fibrillation) leading to a strong adhesion between nanocellulose-based coating and cellulose substrates [17].

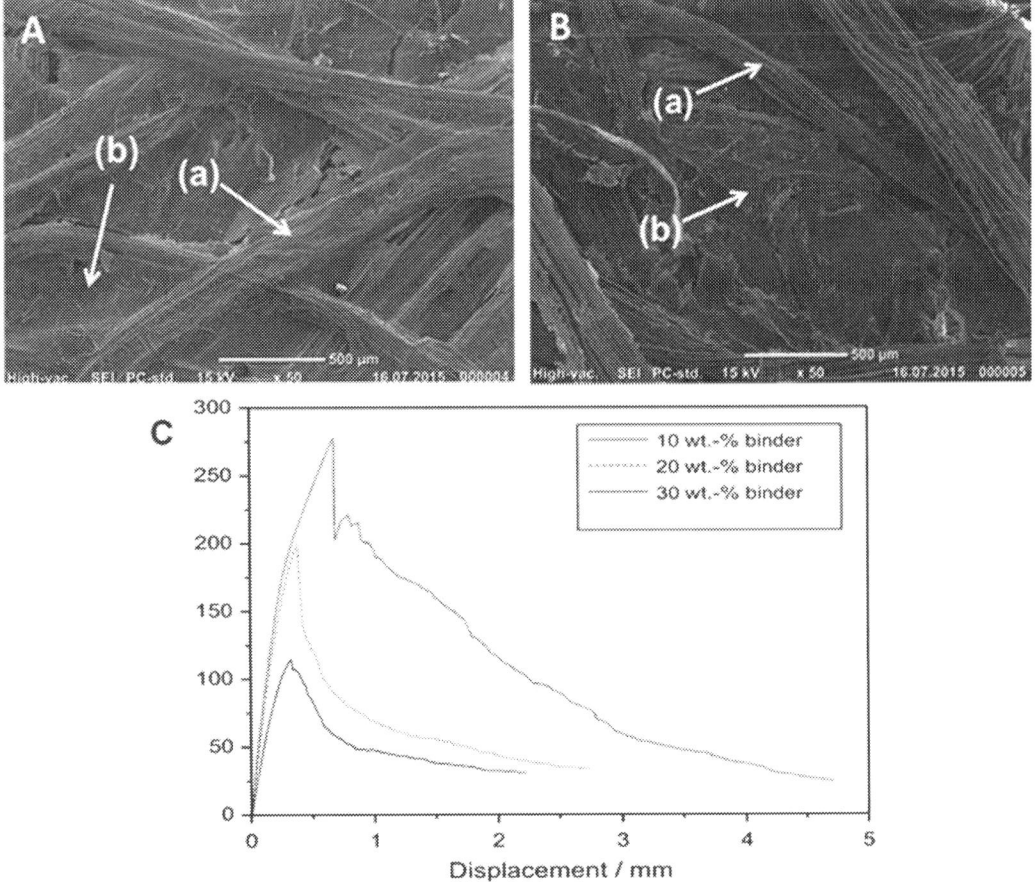

FIG. 9.7 Nonwoven fabrics containing 10 wt% BC (A), 10 wt% NFC (B) and load-elongation curves at different binder percentage (C) (the *arrows* indicate: (a) Flax fibres, (b) BC or NFC [26].

Nanocellulose has also been successfully used as a binder in flax-based nonwoven fabrics [26]. Two types of nanocellulose were explored, namely NFC and BC, and flax fabrics were soaked with 10%, 20%, and 30% suspensions using a single-step and a layer-by-layer wet-laying process. Both NFC and BC were found to be excellent binders and resulted in lower porosity and higher packing as compared to nonwoven fabrics prepared using cellulose pulp binders. The morphology of NFC- and BC-bonded nonwoven fabrics and the effect of nanocellulose percentage content on the mechanical properties are shown in Fig. 9.7 [26]. It is clear that both the strength and elongation of the bonded fabrics improved significantly with nanocellulose content, indicating the effective binding capability of the material.

9.2.5.3 Poly (lactic acid) binders

Poly (lactic acid) or PLA is a biodegradable, thermoplastic polymer synthesised from plant-based sources such as corn, wheat, grain, beet, etc. [23]. Development of PLA biodegradable polymer emulsions for chemical bonding, as well as those based on other renewable materials is continuing. Of course, nonwoven fabrics can already be produced with PLA fibres using drylaid formation followed by hydroentangling or needling, spunbond with thermal bonding, spunbond with needle punching, or meltblown formation [27]. PLA fibres possess good mechanical and physical properties along with biodegradability and therefore, attractive for nonwoven applications. A comparison of PLA fibre properties with polyester (PET) is presented in Table 9.6. Also, for thermal bonding, PLA fibre is already available as a thermoplastic binder to produce nonwoven fabrics at a relatively low bonding temperature [27]. Such PLA binder fibres have also been successfully used to thermally bond cotton-based nonwoven fabrics [28]. However, the relatively high cost of PLA limits applications in some sectors of the nonwoven industry.

TABLE 9.6   Comparison of PLA fibre and PET fibre properties [27].

Properties	PLA	PET
Breaking strength (CN/dtex)	2.5–5	3.4–6.5
Breaking elongation (%)	35	15–40
Moisture regain (%)	0.4–0.6	0.2–0.4
Melting point (°C)	175	260
Young's modulus (GPa)	6–7	10–13
Dyeability	Disperse dye, 130°C	Disperse dye, 130°C
Flammability	Burning for 2 min after removal from the flame	Burning for 6 min after removal from the flame
Drapability	Good	Bad
Lustre	Low to especially high	Low to moderate
Crease resistance	Excellent	Good
Wicking function	Good	Good

9.2.6 Characteristic properties of latex polymer dispersions

Suppliers of chemical binders typically provide data sheets indicating the binder properties as shown in Table 9.7 [29].

9.2.7 Minimum film-forming temperature

The minimum film-forming temperature is the lowest temperature at which an emulsion polymer can form a continuous film. It is usually several degrees above the glass transition temperature (T_g). An emulsion polymer comprises about 50% by weight of polymer particles in water. As the water evaporates, the particles move closer together and become less mobile until they touch each other. They can be imagined as an agglomeration of spherical particles packed closely together in layers. At their closest packing, the level of solids is about 75%. As the water evaporates from the surface of the agglomerated spheres, it is replaced by water from lower layers. Very thin water layers form between the particles and effectively become tiny capillaries. The high capillary forces squeeze the water out, further compressing the particles together. If the polymer globules are too hard and dimensionally stable, a tightly packed heap of solid globules is produced (a powder). If the particles are soft enough, they deform under the capillary forces and become polyhedra. The remaining water is squeezed out and the polyhedra coalesce to form a film.

TABLE 9.7 Items listed in the specifications provided by manufacturers of chemical binders.

Binder composition/characteristics	Notes
Monomers	Many latices are in fact copolymer systems
Crosslinking	Whether or not the system is self-crosslinking or can be externally crosslinked
Solids content	Typically is 50% solids and can be between 30% and 60%
Average particle size	Ranges from 0.01 to 1 µm. The particle size and size distribution affects the properties of the binder and the ease of film formation
Residual monomer level	Some monomers can present a hazard to health
Ionic nature	Polymer dispersions are commonly anionic or nonionic
pH value	Is normally between 2 and 10
Viscosity	Varies between 50 and 50,000 Pa s
Glass transition temperature (T_g)	Used as an indicator of polymer hardness and stiffness
Minimum film-forming temperature	(See Section 9.2.7)
Nature of the film	For example, tacky or soft
Film mechanical properties	For example, elongation and tensile strength at break
Resistance to washing at the boil	Y/N
Resistance to dry cleaning	Y/N
Shelf life	6 months–5 years
Suitability for various methods of application	E.g. saturation, foam, spray, and print

The binder 'hardness' or 'softness' that affects the ease of film formation depends on how the polymer chains in the latex particles are packed together. If they have no side chains (branches), they can pack together closely and become relatively immobile, needing significant energy, for example, heat energy, to separate them. These structures are 'hard'. If the molecular chains have side branches, they cannot uniformly pack together and remain more mobile. These are 'soft' structures. This polymer chain mobility depends therefore on the polymer structure and the temperature (heat energy).

As the temperature is raised, the mobility of the molecules reaches a point called the minimum film-forming temperature. Above this temperature, the latex particles are able to merge to form a film – polymer chains on adjacent latex particles entwine and fuse the particles together. The ease of film formation can be enhanced by the use of plasticisers which facilitate the movement of the polymer molecules. Water can act as a plasticiser. If water-soluble molecular units, for example, acrylic acid or methacrylic acid, are incorporated into the latex particle, they act as plasticisers. Conversely, the mobility can be impeded by crosslinking the molecular chains. The crosslinks inhibit the deformation of the globules to form polyhedra and the ability of the molecules to interpenetrate one another at the polyhedra boundaries.

Although the ability to form a film is necessary for bonding using latices, not all latex binders used in a formulation need to be film forming. Sometimes a formulation will include two latices, one a high-styrene latex that will not form a film in the drying process, and the other, one that is capable of film formation. The combination of the two introduces the required high stiffness into the product.

9.2.8 Functionality of latex polymers

In addition to monomers that provide the backbone of the polymer and determine the key physical properties of the binder, other specific monomers are added to provide specific functionality. These are of particular importance to the processes of coagulation and crosslinking.

9.2.8.1 Coagulation and migration

As an impregnated nonwoven dries, the temperature difference through the cross-section of the fabric can cause the binder polymer to migrate to the higher temperature regions. This differential migration results in a nonuniform distribution of binder where the surfaces tend to have a higher concentration of binder than the core of the fabric, which is depleted. This can lead to problems such as fabric delamination but can be beneficial in some applications such as in the manufacture of synthetic leather. In this process, the impregnated nonwoven is split through its thickness, as is natural leather, and it is important that each 'split' component has similar properties. Thickeners have been used to inhibit migration but they reduce penetration during impregnation and slow the process down. Some polymers can be modified so that during drying, they coagulate and do not migrate. This is achieved by making them thermosensitive. When the binder reaches a particular temperature, the coagulation temperature, the latex particles coagulate on the fibres rather than migrating through the fabric. The ability of a binder system to be heat-sensitised depends on the particular monomers and the level and type of surfactant present. Nitrile rubbers and high-styrene SBR polymers can be heat-sensitised. Several heat-sensitising systems are known, for example, based on polyvinyl alkyl ethers, polypropylene glycols/polyacetals, divalent metal cations/amine, and organopolysiloxanes. Latices tend to be increasingly unstable as the pH is reduced and low pH aids heat sensitisation. The pH is adjusted with, for example, acetic acid or ammonia. Small amounts of some nonionic surfactant stabilisers are added. These have lower solubility in hot water than cold. They aid room temperature stability and become less stable as the temperature is raised, helping gelation [30]. Migration of the binder can also occur at the drying stage due to differential temperatures in the cross-section of the fabric.

Troesch and Hoffman [31] commented that the binder system migrates by capillary flow during the early stages of drying. The heat causes thermo-sensitised dispersions to form agglomerates whose diameters are larger than the capillaries; coagulation happens in a 'shock-like' manner. They point out that the difference between the wet-bulb temperature of the material in drying and the coagulation temperature of the binder is crucial. For complete prevention of binder migration, the coagulation temperature must be at least 5°C below the wet-bulb temperature which is typically 70–80°C. They describe agglomerate structure as being fine, coarse, or compact; the structure is a characteristic of the binder and is only slightly influenced by the type of coagulant. The agglomerate structure of the binder can also influence fabric mechanical properties.

9.2.8.2 Crosslinking

Crosslinking the binder polymer can increase stiffness and waterproofness of the bonded nonwoven by providing covalent bonds between polymer chains, which reduce their mobility. The crosslinking potential of a binder system can be classified as follows:

- Noncrosslinking
- Crosslinkable
- Self-crosslinking
- Thermosetting

The most well-known example of crosslinking is the vulcanisation of natural latex or butadiene polymers with sulphur, an accelerator, and zinc oxide. The process is complicated and the crosslinked product tends to discolour. Functional groups are introduced into the binder polymers to make self-crosslinking systems, which are initiated by heating. Alternatively, groups can be introduced which can react with a curing resin. Acrylic emulsions typically contain about 1%–3% of functional groups such as amine, epoxy, carboxyl, ketone, hydroxyl, and amide, associated with the copolymer backbone which react on heating to induce self-crosslinking.

9.2.8.3 Crosslinkable polymers

Functional monomers which contain hydroxyl or carboxyl groups can be introduced into the polymer. These can be crosslinked after impregnation using melamine formaldehyde or urea formaldehyde.

9.2.8.4 Self-crosslinking polymers

If *N*-methylol functional groups are introduced into the polymer, for example, as *N*-methylolacrylamide, they can react with themselves when the impregnated nonwoven is heated to form covalent bonds. The problem is that such emulsions contain free formaldehyde. This is present during the preparation, storage, and use of the binder. Formaldehyde is now recognised as presenting a risk to health. As a result, latex suppliers developed formaldehyde scavengers such as acetoacetamide [32] and are developing formaldehyde-free binders [33].

9.2.9 Formulated binder systems

The properties of binder systems are enhanced (or the cost reduced) by the addition of other materials. This is both necessary to facilitate processing and to enhance the properties of the bonded nonwoven, or to reduce cost. These additions are done just before application to the nonwoven batt, web, or fabric. Examples of such auxiliaries are listed in Table 9.8. A description of the factors affecting latex stability and rheology is given by Dodge [35]. When developing a new product, polymer suppliers can often provide initial formulations.

TABLE 9.8   Auxiliary materials used in formulated binder systems.

Fillers	Added to reduce tackiness, cost or to reinforce; e.g. calcium carbonate and china clay (5%–20% is added to reduce tackiness and 10%–40% for filling purposes)
	Functionality of the bonded nonwoven can be improved by other filler types such as carbon black
Flame retardants [34]	E.g. halogenated organics with antimony oxide, aluminium trihydrate, diammonium acid phosphate
Antistatic agents	E.g. sodium formate
Hydrophobic agents	Used to reduce wicking or water absorption, e.g. waxes, fluorocarbons, and silicones
Hydrophilic agents	E.g. additional surfactants (anionic and nonionic)
Thickeners	Increase the viscosity for some processes such as knife coating or to aid foam stabilisation, e.g. polyacrylate salts, methyl cellulose, or carboxymethyl cellulose
Pigments	
Optical brighteners	
Surfactants	To improve the stability (including foams), fibre wetting and penetration into the nonwoven
External crosslinking	To increase stiffness and water resistance, e.g. agents melamine formaldehyde
Catalysts	To aid crosslinking
Antifoaming agents	E.g. silicone emulsions
Dispersing agents	For added pigments or fillers, e.g. ammonium salts of acrylic polymers
Other latices	To provide additional properties, e.g. high stiffness from the use of two latices, one to film form and one

9.3 Mechanism of chemical bonding

9.3.1 Introduction

The physical properties of a bonded nonwoven, especially the strength, are determined by the fibre, the polymer, the additives, and the interaction between them, as well as their relative spatial arrangement, surface, and bulk properties.

The strength of the bonded nonwoven does not derive solely from the strength of the unbonded web and the accumulated strengths of the component fibres nor the dried binder composition, but from the interaction between them. Whereas it is normal to think of adhesives bonding together two substrates, in the chemical bonding of a nonwoven fabric, there is a range of potential bonding surfaces to consider. These include:

- *Binder polymer to fibre*. Different fibres will behave differently according to their surface properties. Essentially, the binder to fibre adhesion will vary.
- *Binder polymer to fibre finish*. It is unlikely that the surface of the fibre will be free of finish or contaminants. Many surface finishes act as wetting agents for binder formulations. Some fibres have silicone preparations deliberately applied to inhibit wetting. Hydrophilic fibre finishes are applied to hydrophobic fibres such as polypropylene for hygiene applications, including wipes to aid wetting out and processing efficiency in hydroentanglement. A further complication is that these finishes are rarely applied uniformly.
- *Binder polymer to added filler*. Fillers such as china clay are frequently added to binders.

In some chemical bonding situations, the weight of binder in relation to the fibre material being bonded is low. At high levels of binder to fibre ratio, for example 1:1, the system can be considered not as a binder adhering fibres together to form a network, but rather as a continuous or porous binder matrix, filled (or reinforced) with a fibrous network, and possibly an inorganic filler, such as calcium carbonate or china clay. We also need to consider the cohesive properties of the binder polymer itself. Its function is not only to adhere the fibres together but also to contribute to the performance of the finished product, for example, by providing toughness, stiffness, or elasticity.

9.3.2 Wetting

9.3.2.1 Science of adhesion

The adhesion between two materials, for example, fibres and the chemical binder is a complex phenomenon and a number of models have been used to explain the interaction. Some selected models that are relevant to nonwoven substrates are discussed below:

9.3.2.2 Mechanical adhesion model [36]

According to this model, the adhesion between the nonwoven substrate and the bonding agent occurs due to the pores and physical irregularities present in the substrate. The liquid bonding agent tends to penetrate into the pores of the nonwoven substrate due to the capillary action between adjacent fibre surfaces. The mechanical adhesion is not a molecular level phenomena and it only assists in the wetting and adsorption of the bonding agent by the substrate.

9.3.2.3 *Wetting model [36]*

The wetting of a nonwoven substrate by a chemical bonding agent can be described by the thermodynamic adsorption model. According to this, when a liquid droplet sits on a solid surface, it should take a configuration that will minimise the energy of the whole system. The configuration of the liquid droplet at the equilibrium state is presented in Fig. 9.8. This equilibrium will be controlled by a number of interactions such as Van der Waals forces and hydrogen bonds.

Here, γ_{SV} is the surface energy of the solid in contact with liquid vapour (the surface energy of the solid becomes γ_S in vacuum), γ_{LV} is the surface energy of the liquid in contact with its vapour, which is equal to γ_L (surface energy of the liquid), γ_{SL} is the interfacial solid–liquid energy and θ_{SL} is the contact angle of the liquid against the solid.

At equilibrium condition,

$$\gamma_{SV} = \gamma_{SL} + \gamma_{LV} \cos \theta_{SL} \tag{9.2}$$

The spreading pressure, $\boldsymbol{\Pi}$ can be expressed as,

$$\Pi = \gamma_S - \gamma_{SV} \tag{9.3}$$

Therefore, Eq. (9.2) becomes,

$$\gamma_S = \gamma_{SL} + \gamma_{LV} \cos \theta_{SL} + \Pi \tag{9.4}$$

The reversible energy of adhesion, W_{SL} of a liquid droplet sitting on a solid surface can be expressed by the following equation:

$$W_{SL} = \gamma_S + \gamma_L - \gamma_{SL} \tag{9.5}$$

The energy of cohesion, W_{LL} of the liquid is expressed by,

$$W_{LL} = 2\gamma_L \tag{9.6}$$

The solid surface is perfectly wetted by the liquid if the energy of adhesion $\geq$ the energy of cohesion of the liquid. This criterion can be further expressed in terms of the critical surface tension for wetting of a solid. According to this, the solid is perfectly wetted by the liquid if the critical surface tension of the solid is $\geq$ the surface tension of the liquid.

FIG. 9.8   Vectorial representation of liquid–solid interactions according to Young's model.

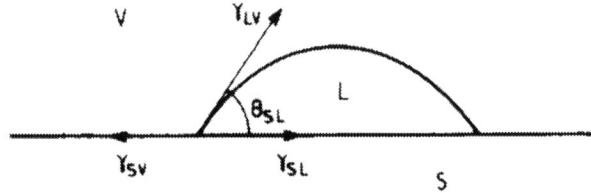

The wetting ability of the liquid is expressed by the spreading coefficient, S which is given by,

$$S = W_{SL} - W_{LL} \tag{9.7}$$

S represents the interfacial energy gained by the liquid–solid system due to the wetting process. Using Eqs. (9.4) and (9.7), S can be expressed as,

$$S = \gamma_S - \gamma_L - \gamma_{SL} \tag{9.8}$$

$$S = \Pi - \gamma_L(1 - \cos\theta_{SL}) \tag{9.9}$$

Therefore, the liquid will more spontaneously wet the solid surface for higher values of γ_S and Π and lower values of γ_L, γ_{SL}, and θ_{SL}.

9.3.2.4 Chemical adhesion model [36]

This model represents the adhesion between two materials due to covalent bond formation between them at the interface. As covalent bonds present much higher interaction energies ($\sim$60–700 kJ mol^{-1}) as compared to the energies present by the various physical interactions ($\sim$2–40 kJ mol^{-1}), adhesion due to chemical interactions is much stronger. One common example of chemical adhesion is the use of coupling agents such as silane to improve bonding of glass fibres with epoxy adhesives [37]. Silanes possess two terminal functional groups with different chemical reactivities, the one of which is generally an alkoxy group that reacts with the silanol functional groups of glass fibres and the other one is an amino or vinyl group capable of reacting with the functional groups of epoxy adhesives.

9.3.2.5 Diffusion adhesion model [36]

According to this model, adhesion can result from the interdiffusion of macromolecules of two polymers which remain in intimate contact with each other. A transition zone or 'interphase' is formed due to this interdiffusion phenomenon, as schematically presented in Fig. 9.9. The interdiffusion process follows the Fick's classical law:

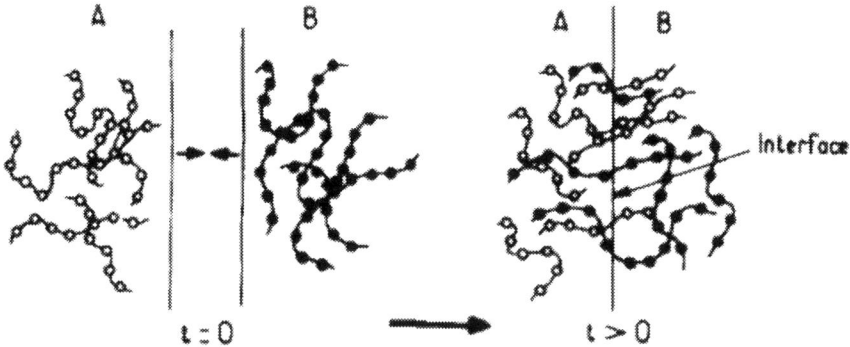

FIG. 9.9   Interdiffusion of macromolecules across the interface.

$$x \propto \exp\left(-\frac{E}{2RT}\right) t^{\frac{1}{2}} \tag{9.10}$$

where, x is the interpenetration depth, E is the diffusion activation energy, t is the contact time, R is the molar gas constant, and T is the temperature. For the interdiffusion to occur, the two polymers in contact must be compatible with each other and also, there should be sufficient mobility of the macromolecules induced by the increased temperature. The diffusion adhesion has been studied for the auto-adhesion of the thermoplastic polymers, when two identical polymers were kept in contact with each other. It can, therefore, play a significant role in the adhesion between a nonwoven fabric with a compatible polymeric binder.

Therefore, according to the above theory, for adhesion to occur, an adhesive first needs to wet the substrate, in this case, the fibre and the binder carrier needs to spread across the surface of the fibre. This requires the fibre surface to have a higher surface energy than the binder polymer. It is important to realise that the surface of, for example, a polyester fibre will usually have an applied finish, and possibly contaminants on it. This finish might not be continuous but present as 'islands'. The potential bond therefore might be between binder polymer and fibre polymer, binder polymer and finish or in some cases between binder polymer and contaminants. Some fibres are treated with silicone finishes to provide water repellence or increase fabric softness and such finishes are particularly difficult to bond with existing binders.

Based on the levels of binder that are most often applied, the bonded nonwoven is not a fibre-filled polymer matrix, in fact, there are large spaces between the fibres. The binder system, normally an aqueous dispersion, is free to move by capillary forces as the water evaporates. Therefore, it can bridge two fibres where they come close together or touch, thus a bonded network is created and the bonded fibres are able to contribute to the overall strength of the nonwoven. The contribution that the binder makes to the overall mechanical properties of the fabric depends on: (i) binder polymer cohesion properties (i.e. bonding to itself); (ii) binder polymer adhesion properties (i.e. bonding to fibre, finish, filler, etc.); and (iii) distribution of the binder and the volume of binder present in relation to the volume of fibre.

9.3.3 Binder polymer cohesion properties

The cohesion of a liquid is the attraction between its molecules that enables droplets and films to form. For a film to form, the polymer particles must coalesce. This happens as the carrier (e.g. water) evaporates. During evaporation, capillary forces between the emulsion particles causes them to squeeze together to form either a powder or a film. For good bonding, formation of a film is required. Smaller binder polymer particles will form a more effective film than larger particles [38].

Adhesion can be defined as the intermolecular forces that hold the touching surfaces of the fibre and binder polymer together. For good adhesion, the polymer particles and carrier, e.g. water, need to wet the fibre adequately. For this to happen, the binder carrier and the binder polymer need to have a lower surface energy than the fibre. Water, even at temperatures close to 100°C as in a dryer, is still above the surface energy of many fibres. Therefore, for wetting to occur, a wetting agent usually needs to be present. Polypropylene fibres have a low surface energy (about 23 mN/m) [39] and are difficult to bond. Corona and plasma treatments are sometimes used to change the chemical nature of the surface and improve wetting.

Polyester fibres have a higher surface energy of about 42 mN/m. Cellulose fibres not only have a higher surface energy than both polypropylene and polyester fibres, they are also relatively porous enabling liquid to penetrate and to present a higher surface area that gives better bonding. Commercially available fibres have surface chemical finishes present. These can be left over from the fibre manufacturing process or are deliberately applied to the fibre to facilitate wetting or to modify fibre friction and electrostatic charge generation in carding prior to bonding. They can be present as continuous or discontinuous thin layers. Wetting agents are often added to aid liquid spreading. Even if the surface energy of the binder system is lower than that of the fibre, wetting can be impaired if the viscosity of the binder is high. Factors that impair film formation such as crystallinity or high T_g can hinder adhesion by reducing polymer flow [40]. After the fibre surface has been wet, various interactions between the fibre and binder result in a bond. If crosslinking groups are present, further heating will increase the bonds between the binder polymer molecules and the cohesive strength of the film. Some attempts have been made to engineer binders to migrate to fibre crossover points and not to coat the fibres [41]. Rochery et al. [42] studied the interaction of fibres and binders in chemically bonded nonwovens and concluded that fibre to matrix adhesion depends on many phenomena including the fibre surface, the way the latex is processed, and the choice of latex reactants.

9.3.4 Distribution of the binder and the binder to fibre ratio

Assuming that the binder carrier wets the fibre, if there is a high density of fibre crossover points and sufficient binder, capillary forces can attract it to the crossover point to form a bond. If there are relatively few crossover points or a low level of binder, then the fibres will wet and coat with binder as the water evaporates and there is reduced opportunity for migration of the binder to a crossover point. If the binder system has been sprayed on, the spray droplets may not land near a fibre crossover point and consequently, no bonding will occur in that particular region. The physical structure of the bond points that develop therefore depends on the web structure, particularly fibre orientation and fabric density, the level of binder application, the binder flow properties, and the method of application. The resulting physical structure of the bond points and their relation with surrounding fibres directly influences fabric mechanical properties including fabric bending rigidity that is related to handle. Note that the binder can also migrate within the fibrous web during the drying process, as the water is gradually evaporated, affecting its spatial distribution.

Commonly, the role of the binder is simply to bond the fibres together to achieve an increase in network strength and the binder is not intended to play a major role in dictating other properties of the final product. Nonwoven fabrics are porous materials varying in porosity from about 50% to >99% for fabric such as high loft waddings and battings. A typical polyester needlepunched fabric of 300 g m^{-2} might be 2 mm thick. Typically, the fibre in such a structure will occupy only about 10% of the total volume or space in the fabric. As the binder takes up more and more space in the fibre network, i.e. the ratio of binder to fibre is increased, the role of the latex as a binder for the fibres becomes less important. Ultimately, as the binder content increases, the nonwoven effectively becomes a fibre-reinforced polymer composite, the properties of which are dependent on the relative proportion of fibre polymer and binder polymer present. The cohesive strength of the binder polymer then strongly influences the strength of the fabric. This situation arises in the saturation bonding of some shoe reinforcement materials.

As discussed, during drying, the binder system can migrate to the surfaces of the nonwoven, resulting in binder-rich surfaces and a binder-starved core. This can result in poor laminar fabric strength and is also important in the manufacture of synthetic leather where the impregnated nonwoven is split and is important that each 'split' has similar levels of binder. Additionally, the chemical binder concentration can vary through the fabric cross-section due to the method used to apply binder. Surface bonding and graduated binder content can be obtained depending on requirements.

For certain fibres, adhesion without modification of the surface by, for example, plasma or corona treatments, is not possible. If a high level of binder polymer is used, then the fibres in the web can be in effect surrounded or encapsulated by the binder polymer without effective surface adhesion between the two.

9.3.5 Modelling of chemically bonded nonwovens

Nonwovens have widened the application of fibrous structures across multiple market sectors; however, their complex physical behaviour is deeply related to their microstructural geometry, fibre orientation and porosity, as well as the constituent fibre type, fibre length distribution, diameter, crimp, etc. Designing nonwovens tailored to specific end-use requirements is a time-consuming and resource-expensive process. To optimise this process, modelling can be applied to help the experimental characterisation, particularly when it comes to linking the microstructure of the nonwovens to their mechanical behaviour and the properties of the component fibres. Computer simulations can reduce the number of trials in the search for new nonwoven prototypes. After knowing a target property, instead of generating many nonwoven materials in the laboratory and choosing the desired one afterwards, predictive modelling can provide suggestions for the selection of 'promising' candidates beforehand. Of course, laboratory synthesis and analysis will still be paramount in finding the best candidate material.

There are many different types of modelling techniques that can be implemented in order to gain insights into the nonwoven materials. From quantum mechanics to the macroscale passing through the atomistic scale, each modelling has its own benefits and drawbacks. True multiscale modelling that covers the different length and time scales is still in its infancy, as the different models can help describe specific aspects of the material, but it is not a trivial task to handle the incorporation of different computer simulations. However, realistic representations of experimental samples can now be obtained using micro-CT and other scanning methods, overcoming one of the major hurdles of modelling [43,44].

Not only is the experimental complexity difficult to fully account for in a computational model, there are also issues with the complete enumeration of the equations involved in the calculations. For example, most modelling implements so-called three-dimensional boundary conditions, where a small simulation cell is repeated in all directions, infinite times.

This has the advantage of treating a small portion of the sample, but it is likely that the omission of a random realistic microstructure within the small simulation cell can lead to inaccurate repetitive distributions, thus generating artefact in the calculation of the physical behaviour of the nonwoven.

At the macroscale, finite element methods have been widely used in the analysis of nonwovens [45,46]. The nonwoven material can be treated either as a continuous or as a discrete medium. In the former case, the modelling lacks information on the microstructure which is the main drawback when it comes to simulating deformation and damage initiation and propagation. However, when considering nonwovens discretely, fibres are modelled explicitly in a specific geometry that has to take account of the complexity of the structure, including the irregularities in the local microstructure of real fabrics. A large number of equations have to be solved simultaneously, which becomes prohibitive because of the randomness of the structures and the nonlinear material properties of single fibres in the nonwoven.

Although numerical methods provide a good level of approximation to solve large systems of equations associated with the structure and properties of the material, they provide little information on the arrangement of fibres and atomic-scale insights on the constituent building blocks. The smallest scale of modelling is the quantum scale where the reactivity can be fully accounted for, as quantum mechanics methods can provide changes in the electron configurations and thus in the bonding networks [47]. These methods will be the most useful when studying chemically bonded nonwovens by providing an accurate description of the bonding between polymer chains. Although the chemical reactions at the base of these processes are generally well known in chemistry, the computational costs of quantum mechanics methods hinder their full applicability to the study of nonwovens. Rather, classical atomistic methods and coarse-grained models are preferred, although their application to nonwovens is relatively restricted within the context of nonwovens industry. Depending on the research questions, there is modelling targeted at capturing the specific chemical details of the polymer and thus more appropriate when comes to generating quantitative predictions of the materials properties (e.g. classical atomistic molecular dynamics) and modelling targeted at reproducing generic structure and dynamics of the polymer material (coarse-grained models). Whereas the former would require quantitative inputs on the local geometries of the polymer, the latter accounts only for the topological properties of the polymer (e.g. connectivity). Modelling and simulation of polymers is widely used in the computational chemistry to unravel the structure–dynamics relations of single fibres [48], but is still quite restricted when it comes to the context of nonwovens, and generally is limited to the generation of molecular models [49].

9.4 Methods of binder application

The most common methods of applying a binder system to drylaid webs or prebonded nonwoven fabrics are saturation, foam, spray, and print bonding. Coating methods are also used. For wetlaid nonwovens, most of the same methods can be used but bonding is usually applied after partial drying. For printing, the web must be dry.

9.4.1 Saturation

Saturation bonding involves the complete immersion of the web or prebonded nonwoven either in a binder bath or by flooding the nonwoven as it enters the nip of a pair of rolls. The rate at which the binder is taken up depends on its permeability and ease of wetting. Nip rolls or vacuum slots remove excess binder and regulate the applied binder concentration. This method can provide high binder to fibre levels uniformly throughout the nonwoven.

Figs. 9.10 and 9.11 show the basic principle of applying a binder using padding. The nonwoven is guided through the saturation bath by rollers and then passes between a pair of nip rolls to squeeze out excess liquid. Clearly, this also compresses the substrate reducing its thickness. Sometimes three rolls are used to spread the binder more evenly and give greater penetration. In some systems, the nonwoven is pressed while it is in the bath using an immersed nip. This enables air to be removed and the nonwoven to wet faster giving more even distribution.

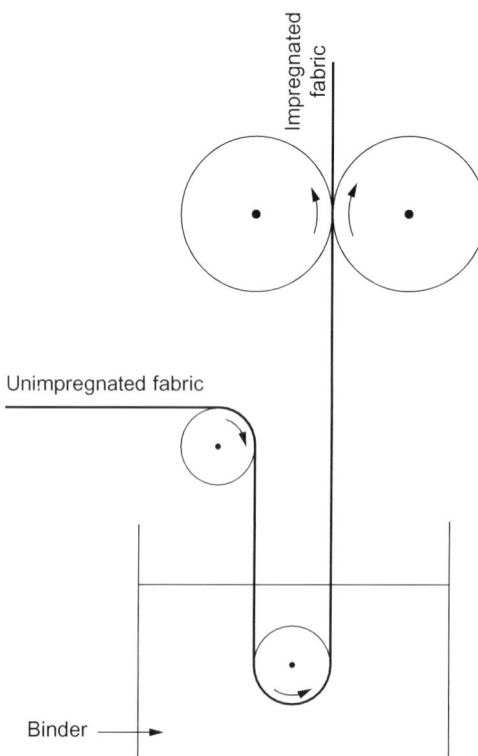

FIG. 9.10   Binder application using a padder (Configuration A).

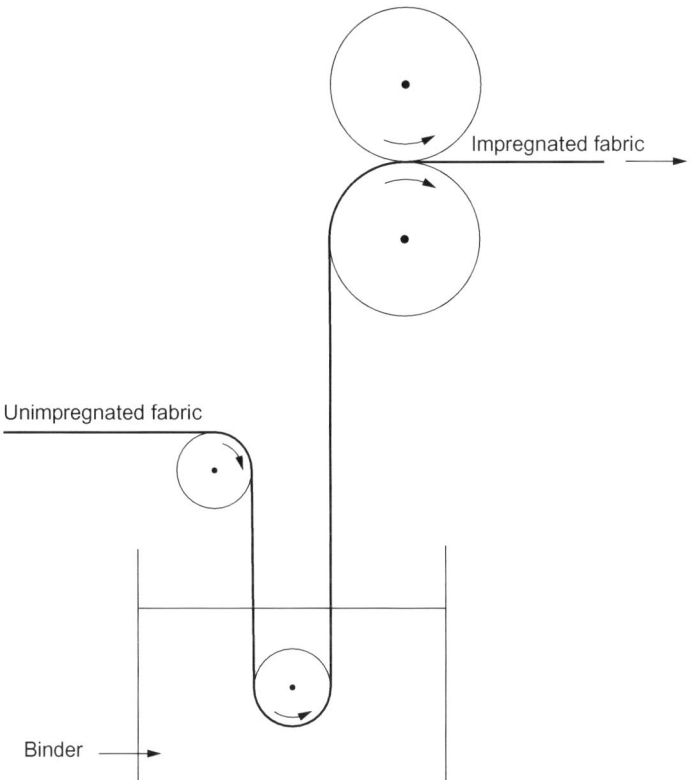

Impregnated fabric

Unimpregnated fabric

Binder

FIG. 9.11   Binder application using a padder (Configuration B).

The amount of binder taken up by the nonwoven depends on its weight per unit area, the length of time in the bath, the wettability of the fibres, and the nip pressure. The nip gap is usually set and maintained by applied pressure. In systems where no gap setting is required, only pressure, it is usual for one of the rolls to be rubber coated, the other usually being chrome steel. The trough and nip system is often called padding or 'dip and squeeze'. Padders are usually either vertical or horizontal (Figs. 9.10 and 9.11).

In a pad machine, the binder system is usually pumped around continuously and the level and concentration kept constant. Obviously, nonwovens need to be sufficiently strong to be self-supporting when passing through the trough. Sometimes they are prebonded by, for example, needlepunching or thermal bonding to confer sufficient strength. It is not essential to have a bath or trough. In the horizontal flooded nip system shown in Fig. 9.12, the nonwoven passes through a pool of binder held above the rolls. Advantages over the vat method include the use of less binder and easier cleaning. Disadvantages include the short wetting time, which means that the method is really suitable only for lightweight highly permeable nonwovens.

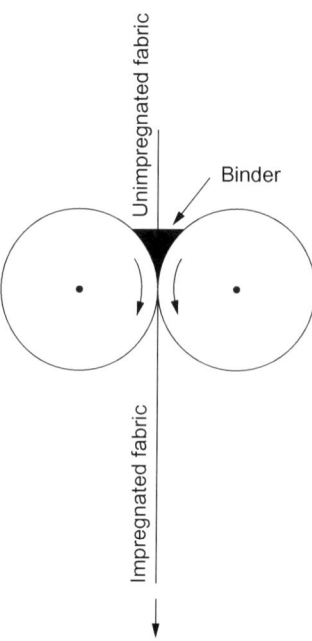

FIG. 9.12 Flooded nip binder application.

Other methods exist to saturate weak fabrics, for example, carrying the fabric through the vat between (i) perforated screens and (ii) a perforated screen and a perforated cylinder. Saturation is not a metered system. In practice, to ensure the level of binder pick-up is correct, a few metres of saturated nonwoven is run through the machine, a sample is cut out and dried and then the pick-up level is calculated. Adjustments to the nip setting are then made to adjust the pick-up until the required level is obtained. Many of the physical properties of a saturation-bonded fabric derive from the fact that all or most of the fibres are covered with a film of binder. This is particularly true of the handle and hydrophobicity or hydrophilicity, which will derive from the binder rather than the fibres [38].

9.4.2 Foam bonding

In foam bonding, air is used as well as water to dilute the binder system and as the means to carry the binder to the fibres. One advantage of diluting with air rather than water is that drying is faster and energy costs are reduced. Foam can be applied so as to remain at the surface or can be made to penetrate all the way through the fabric cross-section. Foam is generated mechanically and can be stabilised with a stabilising agent to prevent collapse during application. One or two reciprocating foam spreaders are commonly used to distribute the foam across the width of the fabric.

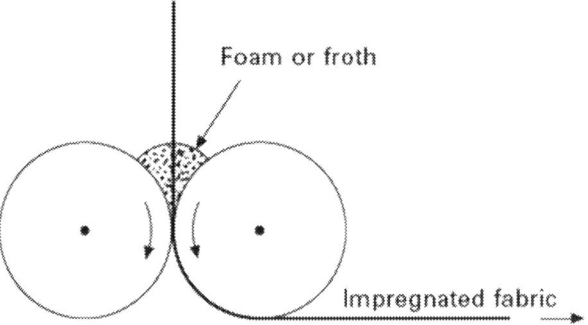

Foam or froth

Impregnated fabric

FIG. 9.13 Binder application in a Froth padder.

After foam application, the substrate is passed through a nip. The foam add-on and the degree of penetration are determined by the foam density or 'blow ratio' and the nip setting. To minimise the amount of energy used in drying, the solids content of the binder system must be high and the foam weight must be low. The ratio of these two is determined by the amount of foam applied and the rate of penetration. These in turn depend on the fibre type, surface structure of the nonwoven, fibre linear density, and the fabric weight per unit area.

The key advantages of foam bonding are more efficient drying and the ability to control fabric softness. It is possible for the foaming process to be done in such a way that the foam structure is maintained within the foam [50]. Disadvantages include the difficulty in achieving adequate foaming and in controlling the process to give a uniform binder distribution. Nonstabilised foams called 'froths' are sometimes used. The foam is applied in the flooded nip system or through a slot followed by a vacuum extractor (Fig. 9.13). It breaks down as it is applied and so is like saturation. As some of the 'carrier' is air, less drying is needed than for saturation [51]. Froth application can be thought of as an alternative method of saturation bonding. The properties and uses of the fabrics are identical [38].

9.4.3 Spray bonding

In spray bonding, binder systems are sprayed onto moving webs or prebonded nonwoven fabrics in fine droplet form. Spray bonding is used to make highly porous and bulky products such as high-loft waddings, insulation, filtration media, upholstery, absorbent, and sanitary product components as well as some industrial fabrics. This is possible because the substrate does not need to pass between nip rollers. The liquid is atomised by air pressure, hydraulic pressure, or centrifugal force and is applied to the upper surfaces of the nonwoven in fine droplet form through a system of nozzles, which can be statically mounted across the machine or traverse from one side to the other. It is important that the latex has adequate shear stability. The depth of penetration of the binder into the substrate depends on the wettability of the fibres, the permeability and the amount of binder.

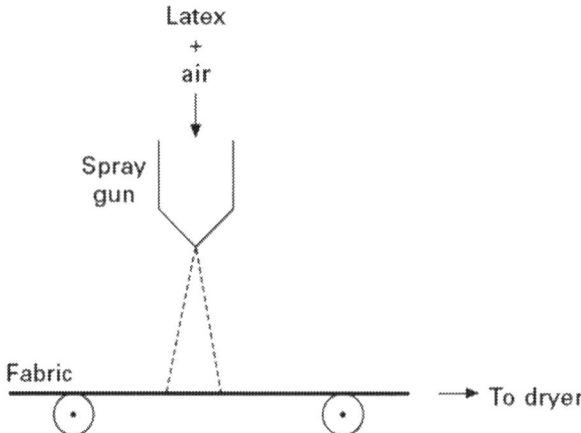

FIG. 9.14   Spray bonding.

If it is necessary to spray both sides of the substrate, an additional conveyor is used, which has a second spray system. Drying is required after each spray application. The levels of binder that can be applied are typically $10–30\,\mathrm{g\,m^{-2}}$. If crosslinking of the binder is required, the substrate passes through a third heater. A typical spray bonding system is illustrated in Fig. 9.14. The main advantage of spray bonding is that the substrate is not compressed and the original bulk and pore structure of the incoming web or fabric is maintained. Disadvantages include lack of control over the uniformity of binder level across the surface of the nonwoven, relatively poor binder penetration, high levels of overspray and waste, and the possible lack of shear stability of the binder.

9.4.4 Print bonding

Print bonding applies the binder only in predetermined areas as dictated by the pattern in the printing surface. The aim is to provide adequate tensile strength, but to leave areas free for water absorption and permeability. By limiting the binder coverage, the handle of the fabric is also comparatively soft. Typical applications are wipes and coverstock. In wipes, the fabric may be first hydroentangled. The design of the print influences softness, liquid transport, strength, and drape [12]. In deciding the shape of the bonding points, it is important to consider the geometry of the web in terms of fibre orientation to ensure adequate MD and CD strengths are obtained. The substrate is often prewetted to aid printing. The two most common methods of printing used are screen printing (rotary printing) and rotogravure printing. The binder is normally thickened to a paste. In screen printing (Fig. 9.15), the binder is forced through a rotating roll that is perforated in the desired pattern. The binder is forced into the substrate by the pressure of the roll and the squeegee inside the roll.

It is possible to impregnate both sides of the substrate with different binders by passing it between two screens rotating against each other. In a design by Stork Brabant, the two

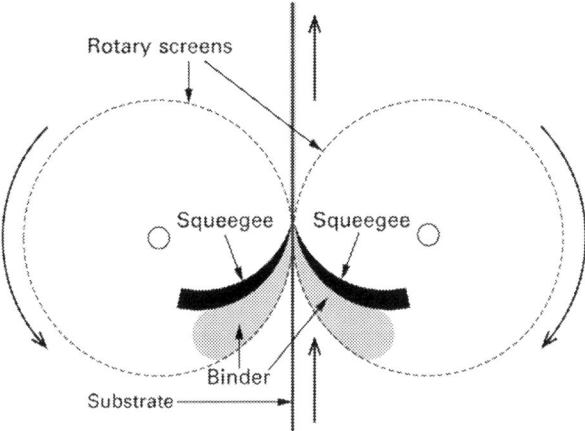

FIG. 9.15   Rotary screen bonding.

squeegees inside the two screens are placed so that each acts as the counter pressure roller for the other screen. In rotogravure or engraved roll printing (Fig. 9.16), the binder is picked up in the grooves of the roll. The level of binder add-on depends on the engraved area, depth, and level of binder solids. The excess binder is removed with a doctor blade. As the substrate passes the engraved roll, it is pressed against the surface by a counter roll, transferring the binder to the fabric. This method is suitable only for applying low levels of binder to the surface where a textile-like handle is needed. Applications include disposable clothing, coverstock, and wiping cloths especially those for washing up and dusting.

9.4.5 Coating or scraper bonding

Another technique for applying binder at the surface is by scraper or knife coating. A scraper knife is placed above the horizontal nonwoven. The thickened binder paste (or foam) is fed upstream of the knife and forms a 'rolling bank' on the moving nonwoven. There are several variations of this method depending on what type of conveyor is used and the surface of the nonwoven. These are knife over air, knife over blanket, and knife over roller. The degree of penetration of the binder system into the nonwoven depends on the nature of the counter surface under the nonwoven, the shape of the edge of the knife, its angle with respect to the nonwoven fabric, binder viscosity, fabric line speed, and fabric wettability.

In knife over air, the nonwoven is unsupported as it passes under the blade and so it is important that the nonwoven is able to withstand stretching. Although this method is often used for coating, it is rarely used for impregnation as only a low level of add-on is possible. After coating, the nonwoven passes through a nip and then to the dryer. Knife over blanket is used when an intermediate level of add-on is required. The nonwoven passes over a blanket, which passes around two rollers. The method is particularly suitable for nonuniform nonwoven substrates. Knife over roller is used for relatively high levels of add-on. The substrate

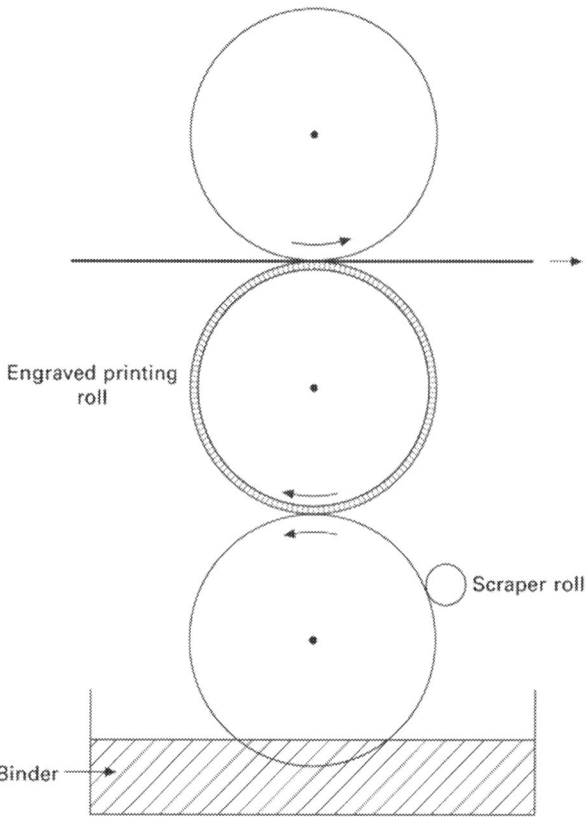

Engraved printing
roll

Scraper roll

Binder

FIG. 9.16   Bonding with an engraved roll.

passes in a 90° pass around a roller and beneath a doctor blade. The binder is applied at the entrance to the gap.

Although all of these methods are designed for coating, some binding in the depth of the nonwoven can be achieved by following the coating head with a nip to push the binder into the substrate. In reverse roll coating, the nonwoven passes between two rollers which rotate in the same direction. One applies the binder and the other provides counter-pressure. The binder add-on is determined by the gap between the rollers.

9.4.6 Solution and partial solution bonding

Solution bonding has been used for both drylaid and wetlaid webs, usually with water-soluble polymers. Traditionally, to make such binders water-resistant, they are cured with melamine or urea formaldehyde. Partial solution bonding (or solvent bonding) is still used in some specific applications. A latent solvent for the fibre is first applied and is then concentrated in order to partially solvate the fibre surfaces and enable them to be fused together at their crossover points [52]. These are sometimes known as spot-welds and the process is normally initiated at elevated temperature.

One of the oldest methods involved applying cyclic tetramethylene sulphone to acrylic fibres as they were fed into a carding machine followed by bonding at 115–160°C. Solvent bonding of diacetate and cotton fibres using cellulose solvents has also been demonstrated. In the case of cotton, for example, interfacial bonding between fibres can be induced by zinc chloride. By applying zinc chloride to decrystallise the cellulose, subsequent washing out of the chemical leads to recrystallisation and autogeneous bonding of the fibres at the cross-over points. A preferred method of applying solvents is by spraying of the web or batt prior to heating. In an alternative approach, Cerex nylon spunlaid fabrics, originally developed by Monsanto, are autogeneously bonded using HCl gas. Solvent-bonded products have been used for high loft waddings and in the case of solvent-bonded diacetate for cigarette filter tips.

9.4.7 Powder bonding [53]

In this process (schematically shown in Fig. 9.17), a thermoplastic polymer binder is applied to a nonwoven web. A sprinkler system, consisting of a groove dosing roller and a brush roller, is used to apply the binder onto the web. The binder powder is fed to the groove roller from the feed container and subsequently, brushed off from the grooves as the finely distributed powders are deposited onto the nonwoven web. Softening of the thermoplastic polymer and subsequent bonding can be achieved by passing the nonwoven web through hot air oven for less compacted nonwoven products or through heating rollers when dense or more compacted products are required. The solidification of thermoplastic polymer occurs when cooled leading to bonding of the web.

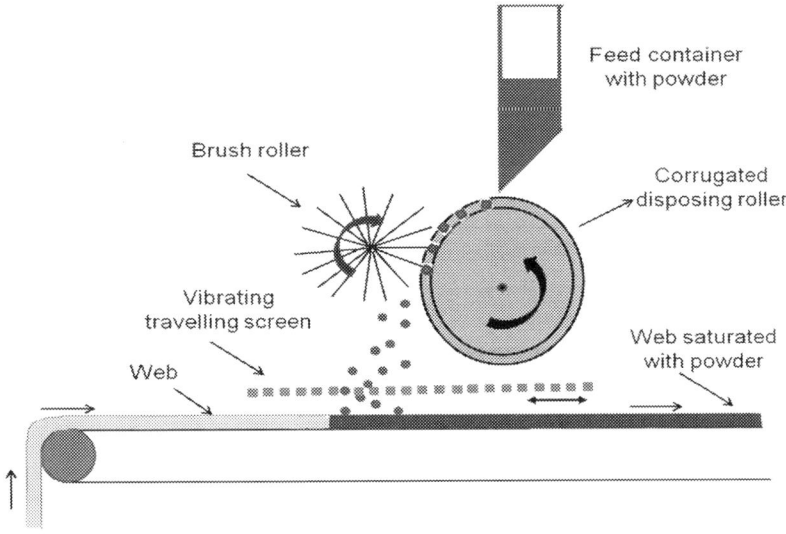

FIG. 9.17   Schematic of the powder bonding process in nonwoven manufacturing [53].

9.5 Drying

9.5.1 Introduction

After the binder system has been applied, the web or prebonded fabric is dried to evaporate the latex carrier (water) and allow the latex particles to bond the nonwoven. Crosslinking (external or internal) is usually carried out in the same dryer. During drying, film forming or coagulation takes place as well as evaporation of the water and crosslinking (if crosslinking groups are present in the binder formulation). There are several types of dryer available of which probably the most well-known are the drum dryer, flat belt dryer and stenter-based dryers. The selection of the type of dryer depends on:

- The type and amount of bonding agent.
- Weight per unit area, strength, density, and permeability of the wet nonwoven.
- Properties required in the finished product. For example, the method of drying can affect the surface finish and stiffness of the product.
- Production speed required.

The three most common drying methods used in chemical bonding rely on the following heat transfer mechanisms: convection, conduction, and radiation.

9.5.2 Convection drying

In convection drying, hot air is introduced to the nonwoven to heat and evaporate the water. If the nonwoven is sufficiently permeable, the hot air can be drawn through it and this is called 'through-air' drying. If the nonwoven is not permeable, hot air can be blown towards it from one or both sides. This is called 'air impingement' or 'nozzle aeration'. In a variant of this drying method, air is directed parallel to the surface of the nonwoven. The air can be heated directly, for example, via, heat exchangers or indirectly, for example, with gas.

9.5.2.1 Through-air dryers

In a through-air dryer, the hot air is sucked through the material, leading to a very effective heat and mass transfer. The nonwoven is guided over a perforated conveyor surface (usually a drum or flat belt conveyor) through which heated air passes. The most common arrangement is a perforated drum and large radial fan. Air is withdrawn from the inside of the drum, heated and returned to the drum surface, this producing suction which holds the nonwoven against the drum, preventing the formation of creases.

An arrangement of several drums in sequence is common. These can be arranged horizontally, or to save space, vertically. The use of several drums enables a temperature profile to be set through the dryer. For example, the first part of the arrangement might be for drying and the second part for crosslinking.

As the nonwoven passes from drum to drum, the air is able to penetrate from both sides. The nonwoven travels almost tension free through the dryer. The perforated drum is designed to maximise air throughput, for example some use a honeycomb structure to maximise the permeability and open area. By varying the fan speed, it is possible to adjust the drying capacity of the line according to the characteristics (e.g. air permeability and weight per unit area) of the nonwoven [12]. Fabric shrinkage can be achieved by overfeeding the non-woven onto the first drum. The flow of air is designed to control the temperature at the surface to within 1°C. The maximum operating temperature is typically 250°C. An alternative ar-rangement is a conveyor dryer. This provides continuous suction through the nonwoven, un-like an array of perforated drums [54,55].

9.5.2.2 Air impingement dryers

These dryers are also known as nozzle aeration dryers. For high density or low permeabil-ity nonwovens, including paper, air impingement dryers are used. Nozzles direct the air from one or both sides to speed up evaporation by applying turbulent airflows close to the non-woven surface. Moist air is swept away and recirculated, with some dry air being introduced. Figs. 9.18 and 9.19 show examples of single-belt and twin-belt dryers. The single-belt dryer is for drying and chemical bonding of lightweight nonwovens around $20\,g\,m^{-2}$ and less than 3 mm thick and the double-belt system is for thicker nonwovens.

The temperature of the air feeding the top and bottom can be controlled separately. A variation of the air impingement dryer is the flotation dryer (Fig. 9.20), which is often used

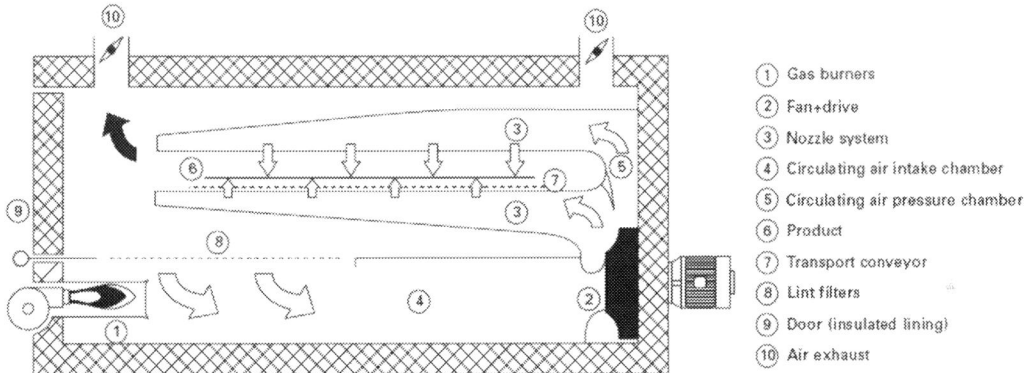

① Gas burners
② Fan+drive
③ Nozzle system
④ Circulating air intake chamber
⑤ Circulating air pressure chamber
⑥ Product
⑦ Transport conveyor
⑧ Lint filters
⑨ Door (insulated lining)
⑩ Air exhaust

FIG. 9.18 Single-belt dryer.

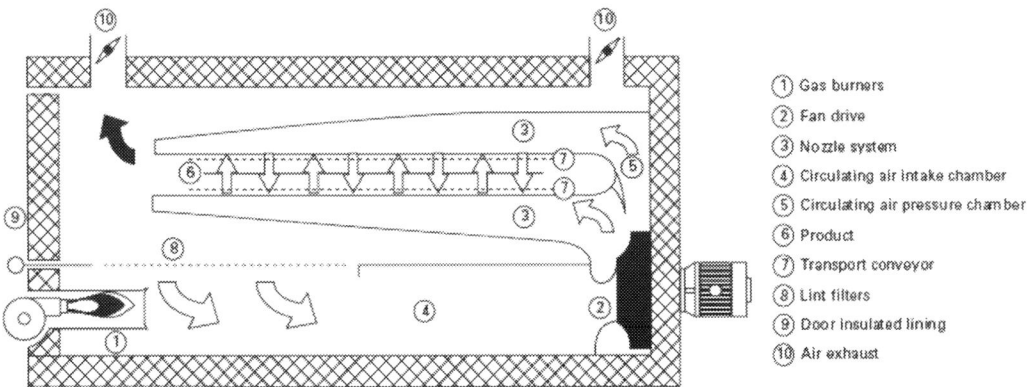

① Gas burners
② Fan drive
③ Nozzle system
④ Circulating air intake chamber
⑤ Circulating air pressure chamber
⑥ Product
⑦ Transport conveyor
⑧ Lint filters
⑨ Door insulated lining
⑩ Air exhaust

FIG. 9.19 Double-belt dryer.

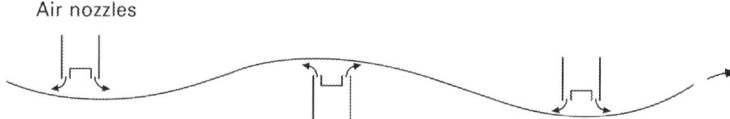

FIG. 9.20   Air-flotation dryer.

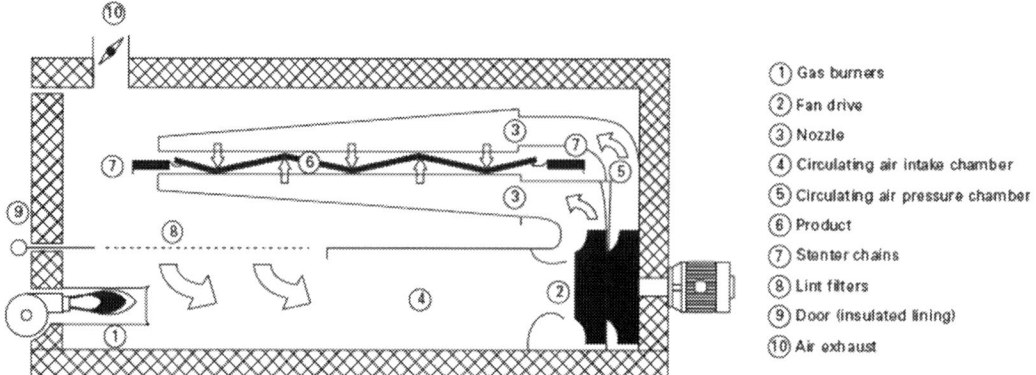

(1) Gas burners
(2) Fan drive
(3) Nozzle
(4) Circulating air intake chamber
(5) Circulating air pressure chamber
(6) Product
(7) Stenter chains
(8) Lint filters
(9) Door (insulated lining)
(10) Air exhaust

FIG. 9.21   Stenter dryer.

for delicate fabric structures and is widely used for paper products. The nonwoven floats through the space between alternate nozzle arrays and very high line speeds can be achieved as compared to other systems. Fig. 9.21 shows an example of a stenter dryer. The nonwoven is held at its edges by clips or pins on revolving stenter chains as it passes through a series of oven chambers. The stenter dryer can have two heater fans that can be separately controlled.

9.5.3 Conduction dryers

Conduction or contact dryers are sometimes used for thin, impermeable nonwovens because of their relatively low capital cost and high evaporative capacity. They are particularly used for nonwovens that have high steam permeability especially wetlaid webs. They usually comprise a line of revolving heated drums over which the nonwoven passes in alternate directions, giving a wrap angle that can be as high as 300°. The surface of the nonwoven adjacent to the drum heats up and water evaporates and moves through the thickness of the nonwoven heating and evaporating in successive layers. Light nonwovens are often carried on backing felts for support. The disadvantages of contact drying compared to through-air drying include a slower heat transfer rate and an increase in the thermal insulation of the nonwoven as it dries [51]. Sometimes a dryer is followed by a calender, hot or cold, to reduce the gauge (thickness) or smooth the surface of the impregnated nonwoven.

9.5.4 Infra-red dryers

Infra-red (IR) dryers work on the principle that water shows a marked absorption of IR energy, which rapidly converts into heat leading to evaporation. IR dryers require low capital investment but have high running costs. They are often used in front of other dryers to predry the surface. For example, they are used to prevent the first drum of a drum dryer being coated with binder and to coagulate the binder to prevent migration. They are also sometimes used after another dryer to complete crosslinking.

9.6 Applications of chemically bonded nonwovens

9.6.1 Introduction

Any new nonwoven fabric can be considered from the point of view of its architecture, i.e. its composition (fibre, binder, additives) and its structural geometry (dictated by the methods of web formation and bonding). Often more than one bonding process is used in combination. Literature covering nonwoven applications shows that although many of today's nonwoven products had been identified some years ago, the fabric architectures have been gradually changing as new ways of improving the manufacture of nonwovens have developed. Accordingly, in many major markets, dominant product architectures have not always emerged and competing manufacturers meet the same market need using different fabric structures. Examples of some of the existing nonwoven products, for which chemical bonding continues to be used are presented in Fig. 9.22 and will be discussed in the following sections.

9.6.2 Wipes

Chemically bonded wipes extend from lightweight, single use fabrics, including flushable (water-dispersible) products to strong, solvent-resistant, washable wipes. To ensure a high void volume for liquid absorption, the binder is usually applied by spray or print bonding. For personal care wipes, softness and high-water absorbency are needed and so absorbent fibres such as viscose rayon are often used (sometimes in blends with polyester to improve

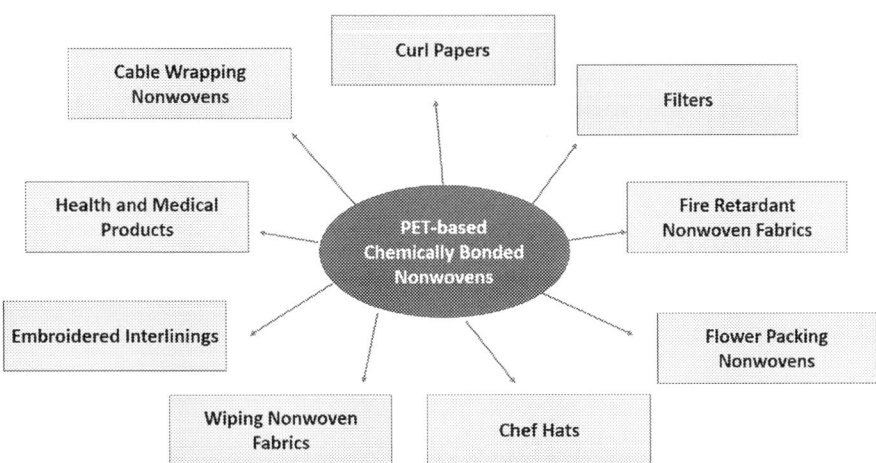

FIG. 9.22   Applications of PET-based chemically bonded nonwoven fabrics. *Source: Ningbo Ciheng Import & Export Co., Ltd.*

the fabric wet strength). In such cases, a soft binder will be selected, such as a nitrile rubber having a T_g less than $-18°C$. These are often carboxylated so that they can be crosslinked to increase their wet strength. Acrylics are also used. An example, for hard surface cleaning, e.g. floor wipes is print-coated apertured hydroentangled nonwoven fabrics, which provide softness, good wiping efficiency, absorbency, and adequate wet strength. Sometimes EVA is used for disposable wipes in place of acrylics, to reduce cost.

For industrial wipes, crosslinkable styrene–butadiene is often used as this has good solvent resistance and handle. EVA is also used because it is cheaper but it is not as resistant to solvents. When high resistance to solvents and oil is required, acrylonitrile is used. Flushable (dispersible) wipes are also of increasing importance in some markets, where, for example, soluble polymer binders allow rapid dispersion of the fibres during flushing, as the bond strength deteriorates in aqueous conditions. Various methods of producing flushable nonwovens rely on the appropriate choice of binder. A comprehensive list of methods to form flushable nonwovens has been given by Woodings [56]:

- Hydrogen-bonded cellulose.
- Hydrogen-bonded and hydroentangled cellulose.
- Man-made fibres bonded with water-soluble polymers, for example, starches, carboxymethyl celluloses, polyethylene oxides, polyvinyl alcohols, polyacrylates amongst others. This includes salt-/ion-sensitive water-soluble binders.
- Polyolefin fibres or films loaded with water-soluble polymers, for example, polyethylene oxide or derivatives engineered for better melt spinnability.
- Biodegradable polymers, for example, PLA, blended with water-soluble polymers to make fibres and fabrics.
- Fabrics made from, or pulp bonded with, water-soluble fibres.
- Fabrics bonded with crosslinked water-soluble polymers, for example, superabsorbents in fibre or powder form.
- Fabrics bonded with bicomponent fibres having a water-soluble or hydrolysable polymer sheath.
- Fabrics bonded with soft synthetic latices, which may be incompletely cured.
- Fabrics made from cellulose/synthetic blends bonded by heat where the cellulose/synthetic thermal bond is easily disrupted by cellulose swelling.
- Laminates with water soluble films or layers bonded with water-soluble adhesives.
- Thin films extruded onto flushable nonwovens, which are waterproof when the film side is wetted, but easily fragmented when both sides are wetted.

9.6.3 Interlinings

Garment interlinings require resistance to dry cleaning, washing, and yellowing, as well as a level of stiffness needed for the particular part of the garment. Acrylic binders are commonly used, for example, ethyl and butyl acrylates which are self-crosslinking to give good resistance to dry cleaning. Also, carboxylated butadiene methacrylates are used with a T_g of 20°C and 30% butadiene. A particular example is a self-crosslinking binder based on ethyl acrylate and methacrylate; the methacrylate increases the stiffness. This has a T_g of 49°C and gives a clear, colourless film. It is used for impregnating polyester nonwovens by spray or foaming. Wetlaid, carded, airlaid, hydroentangled, needled, and spunbonded nonwovens have all been used in interlinings.

9.6.4 Hygiene and medical products

Many incontinence products are chemically bonded. These are often based on carded or airlaid webs and require a high rate of absorption, high capacity, and some softness. Very soft grades of styrene–butadiene or self-crosslinking butyl acrylates are used. Some medical products require barrier properties. Nowadays this is provided by selection of an appropriate nonwoven fabric structure, for example, meltblowns and SMS composites or the insertion of continuous films. However, some products, for example, head and shoe covers, as well as some surgical drapes, are chemically bonded. Acrylic binders are often used as they are not degraded by sterilisation and are soft and hydrophobic. Another application is adhesive tape backings for medical applications, which are made by coating carded webs with acrylic binders.

9.6.5 Footwear

Impregnated nonwovens have been used for many years as a substitute for leather in footwear. All the leather parts of a shoe except the sole can be replaced by nonwovens; however, particular nonwoven technologies are favoured for particular components. The outer part of the shoe, the upper, is often made from microfibre nonwovens saturated with a coagulated polyurethane. These materials have excellent handle and abrasion resistance but are expensive.

The toe and heel of a shoe are often stiffened by the incorporation of a nonwoven stiffener material. These are usually needlepunched fabrics saturated with carboxylated styrene–butadiene. High binder to fibre levels are used to achieve high stiffness and adequate resilience. In shoemaking, the stiffeners are heated in the shoe in order to shape them. The T_g of the binder has to be chosen (i) sufficiently low so that it is possible to shape the stiffener in shoemaking without damaging the leather but (ii) sufficiently high so that the stiffener does not lose its shape in the shop window or storage. Polystyrene binders are sometimes used. Toe stiffeners made with this latex can be activated for reshaping using either heat or solvent. Plasticisers or other latices have to be added to bring the film-forming temperature to below 100°C.

The sole and upper are attached to the insole. The two technologies used to make a suitable nonwoven insole material are (i) the manufacture of cellulose board using styrene–butadiene as a binder and (ii) a polyester needlepunched fabric saturated with a self-crosslinking styrene–butadiene binder. The requirements of an insole are high perspiration absorption, stiffness, and the ability to stick the upper and sole onto it. These structures are relatively porous and provide the necessary perspiration absorption and some adhesion through mechanical bonding.

The lining at the back of a high heel shoe is designed to hold the shoe on by gripping the heel. Heel linings are often polyester needlefelts saturated with nitrile–butadiene rubber. This provides excellent abrasion resistance and softness. As these materials are often impregnated and dried before splitting into two, four, or more 'splits', the latex is designed to coagulate to prevent migration.

9.6.6 Automotive

Nonwovens are used in car carpets, tray coverings, luggage compartment linings, head-liners, and door coverings. Needlepunched nonwovens are favoured, which are subsequently secondary bonded by chemical or thermal processes. The most common binder is styrene–butadiene although some polyvinylacetate and acrylates are used. For moulded parts to be thermoformed, the add-on is up to 60%. Saturation bonding is being replaced by foam bonding [54,55]. Fibre-reinforced composites consisting of glass or increasingly biocomposites incorporating bast fibre nonwovens composed of flax, jute, or sisal alone or in blends are impregnated with polyester thermoset resins for use as automotive components. Door panels, parcel trays, and an increasing variety of additional low-stress bearing components are being made in this way using nonwoven reinforcements. Thermoplastic PP resins are also being utilised in the manufacture of automotive components using porous drylaid and needlepunched nonwovens as the fibre component.

9.6.7 Furniture

For the bonding of fibrefill for upholstery and bedding, self-crosslinking styrene–acrylic copolymers with a T_g of around 8°C are used. This gives a soft, only slightly extensible and tack-free film at room temperature. It has good resistance to a wide range of chemicals and good ageing properties. It can be applied by saturation, spray, or foam.

9.6.8 Examples of other applications

Additional chemical bonding applications include the following:

- Spray bonding of high loft airlaid batts with acrylic binders for insulation.
- Polyamide fibre abrasive pads for scrubbing ceramic products, automotive metal component finishing, hard floors, and other solid surfaces. Batts are spray bonded with phenolic binders premixed with abrasive particles of carefully selected size to prevent scratching.
- Cleaning cloths for the outside of aircraft consist of a needlepunched fabric sprayed on one side with a styrene–butadiene slurry containing hydropropyl methylcellulose and nitrile rubber particles [57].
- Prefilters can be made from high-loft airlaid nonwovens sprayed with a crosslinkable styrene–butadiene binder.
- Roofing membranes impregnated with bitumen can be based on needlepunched fabrics saturated with a relatively hard styrene–acrylate binder.
- Chemically bonded nonwovens are used for liquid and air filter media including HVAC filtration, ink mist collection, paper dust, and containment media. Nonwoven filtration products offer a number of advantages including excellent filtration efficiency, long product life cycle, reduced waste, and cost effectiveness.

References

[1] J.E. McIntyre, P.N. Daniels, Textile Terms and Definitions, tenth ed., The Textile Institute, Manchester, 1995.

[2] R. Pangrazi, Nonwoven bonding technologies: there's more than one way to bond a web, Nonwovens Industry (1992). October 32–34.

[3] G. Mehta, L.T. Drzal, A.K. Mohanty, M. Misra, Effect of fiber surface treatment on the properties of biocomposites from nonwoven industrial hemp fiber mats and unsaturated polyester resin, J. Appl. Polym. Sci. 99 (3) (2006) 1055–1068.

[4] R.I.C. Michie, R.H. Peters, W. Taylor, Nonwoven fabric studies: part I: properties of laboratory-made fabrics bonded with natural rubber, Text. Res. J. 33 (5) (1963) 325–329.

[5] N. Sherwood, Binders for nonwoven fabrics, Ind. Eng. Chem. 51 (8) (1959) 907–910.

[6] R.L. Adelman, G.G. Allen, H.K. Sinclair, Binders for nonwoven fabrics. Evaluation of binders based on copolymers of vinyl acetate, Ind. Eng. Chem. Prod. Res. Dev. 2 (2) (1963) 108–113.

[7] G. Clamen, R. Dobrowolski, Acrylic thermosets: a safe alternative to formaldehyde resins, Nonwovens World 13 (2) (2004) 96–102.

[8] Polymer Latex 1 refers to *How Do Aqueous Emulsions Form?* by Polymer Latex GmbH & Co. KG. 0/8000136/360e.

[9] Polymer Latex 2 refers to *How Are Films Produced?* by Polymer Latex GmbH & Co. KG. 0/8000136/366e.

[10] A.E. Wang, S.L. Watson and W.P. Miller, Fundamentals of binder chemistry, *J. Coat. Fabr.* 11, n.d. 208–225, April. (undated).

[11] W. Wilson White, Functionalised styrene-butadiene latexes for non-wovens, in: Nonwovens Binders: Additives, Chemistry and Use Seminar, 1985. TAPPI Notes, September 30–October 2.

[12] R.J. Pangrazi, Chemical bonding: spray, saturation, print, & foam application methods and uses, in: Inda-Tec 97 Book of Papers, 1997, pp. 6.0–6.5.

[13] Ullmann's Encyclopedia of Industrial Chemistry, Nonwoven Fabrics, vol. A17, 1996, Wiley-VCH.

[14] H.C. Morris, M. Mlynar, Chemical binders and adhesives for nonwoven fabrics, in: INDA-TEC Conference, 1995, pp. 123–136.

[15] J. Lunenschloss, W. Albrecht, Non-Woven Bonded Fabrics, John Wiley & Sons Inc, New York, 1985.

[16] K.L. Powell, Waterborne epoxy resins, in: Nonwovens Binders: Additives, Chemistry and Use Seminar, 1985. TAPPI Notes, September 30–October 2.

[17] C.M. Hawkins, J.M. Hernandez-Torres, L. Chen, Bio-Based Binders for Insulation Andnon-Woven Mats, Patent 20110086567A1, USA, 2011.

[18] H.W.G.V. Herwinjen, E. Pisanova, B. Sterfke, Renewable Binder for Nonwoven Materials, Patent 7,893,154 B2, USA, 2011.

[19] R. Kumar, D. Liu, L. Zhang, Advances in proteinous biomaterials, J. Biobased Mater. Bioenergy 2 (2008) 1–24.

[20] R. Kumar, D. Moyo, R.D. Anandjiwala, Viscose fabric bonded with soy protein isolate by foam application method, J. Ind. Text. 44 (6) (2015) 849–867.

[21] A.H. Tayeb, E. Amini, S. Ghasemi, M. Tajvidi, Cellulose nanomaterials—binding properties and applications: a review, Molecules 23 (10) (2018) 2684.

[22] S. Parveen, S. Rana, R. Fangueiro, Macro-and nanodimensional plant fiber reinforcements for cementitious composites, in: Sustainable and Nonconventional Construction Materials Using Inorganic Bonded Fiber Composites, Woodhead Publishing, 2017, pp. 343–382.

[23] R.J. Moon, A. Martini, J. Nairn, J. Simonsen, J. Youngblood, Cellulose nanomaterials review: structure, properties and nanocomposites, Chem. Soc. Rev. 40 (7) (2011) 3941–3994.

[24] T. Saito, S. Kimura, Y. Nishiyama, A. Isogai, Cellulose nanofibers prepared by TEMPO-mediated oxidation of native cellulose, Biomacromolecules 8 (2007) 2485–2491.

[25] S. Ifuku, M. Nogi, K. Abe, K. Handa, F. Nakatsubo, H. Yano, Surface modification of bacterial cellulose nanofibers for property enhancement of optically transparent composites: dependence on acetyl-group DS, Biomacromolecules 8 (2007) 1973–1978.

[26] M. Fortea-Verdejo, K.Y. Lee, T. Zimmermann, A. Bismarck, Upgrading flax nonwovens: nanocellulose as binder to produce rigid and robust flax fibre preforms, Compos. A: Appl. Sci. Manuf. 83 (2016) 63–71.

[27] S. Feng, X.N. Jiao, The application of PLA resin on nonwovens production, in: Advanced Materials Research, vol. 332, Trans Tech Publications, 2011, pp. 1239–1242.

[28] G. Bhatt, Nonwoven technology for cotton, in: S. Gordon, Y.L. Hsieh (Eds.), Cotton: Science and Technology, Woodhead Publishing, 2007, pp. 501–527.

[29] P.A. Mango, Binder quality—measurement, maintenance, and effect on the end user, in: Nonwovens Binders: Additives, Chemistry and Use Seminar, 1985. TAPPI Notes, September 30–October 2.

[30] Synthomer 1 refers to *Heat Sensitisable Binders for Non-Woven Fabrics*, Synthomer Limited, Harlow, Essex, NON2 7/99.

[31] J. Troesch, G. Hoffman, The effect of binder distribution and structure on the physical properties of nonwovens, in: Paper Synthetics Conference, TAPPI, 1975, pp. 25–35.

[32] B. North, D. Whitley, Polymer Emulsion, European Patent Office, Pat. No. 0438284, 1991.

[33] K.-H. Schumacher, R. Rupaner, A formaldehyde-free acrylic binder for constructing completely recyclable high performance nonwovens—reuse of fibers and binder, Vliesstoff Nonwoven Int. 11 (6–8) (1996) 181–182.

[34] E.D. Weil, Flame retardants for nonwovens, in: Nonwovens Binders: Additives, Chemistry and Use Seminar, 1985. TAPPI Notes, September 30–October 2 (undated).

[35] J.S. Dodge, Colloid chemistry fundamentals of latexes, in: Nonwovens Binders: Additives, Chemistry and Use Seminar, 1985. TAPPI Notes, September 30–October 2.

[36] G. Fourche, An overview of the basic aspects of polymer adhesion. Part I: fundamentals, Polym. Eng. Sci. 35 (12) (1995) 957–967.

[37] S. Parveen, S. Pichandi, P. Goswami, S. Rana, Novel glass fibre reinforced hierarchical composites with improved interfacial, mechanical and dynamic mechanical properties developed using cellulose microcrystals, Mater. Des. 188 (2019), 108448.

[38] A.R. Horrocks, S.C. Anand, Handbook of Technical Textiles, Woodhead, Cambridge, 2000.

[39] B.S. Gupta, H.S. Whang, Surface wetting and energy properties of cellulose acetate, polyester and polypropylene fibers, Int. Nonwovens J. 8 (1) (1999). Spring, 36–45.

[40] F.V. Di Stefano, Chemical bonding of air laid webs, Nonwovens Ind. 16 (6) (1985) 19–20. 22, 24.

[41] R.M. Blanch, A. Blanch, G. Borsinger, C.F. Lences, M.K. Seven, Binders Based on Alpha-Olefin/Carboxylic Acid/Polyamide Polymers and Their Ionomers, 2002. United States Patent Application 20020077011, June.

[42] M. Rochery, S. Fourdrin, M. Lewandowski, M. Ferreira, S. Bourbigot, Study of fiber/binder adhesion in chemically bonded non-wovens, in: 47th International SAMPE Symposium, May 12–16, 2002, pp. 1755–1766.

[43] F. Battocchio, M.P.F. Sutcliffe, Modelling fibre laydown and web uniformity in nonwoven fabric, Model. Simul. Mater. Sci. Eng. 25 (2017), 035006.

[44] S. Gramsch, D. Hietel, R. Wegener, Optimizing spunbond, meltblown, and airlay processes with FIDYST, Melliand Int. 21 (2) (2015) 115–117.

[45] K. Singha, et al., Computer simulations of textile non-woven structures, Front. Sci. 2 (2) (2012) 11–17.

[46] F. Farukh, et al., Nonwovens modelling: a review of finite-element strategies, J. Text. Inst. 107 (2) (2016) 225–232.

[47] A. Gooneie, et al., A review of multiscale computational methods in polymeric materials, Polymers 9 (2017) 16.

[48] T.E. Gartner III, A. Jayaraman, Modeling and simulations of polymers: a roadmap, Macromolecules 52 (2019) 755–786.

[49] C. Krausse, et al., Molecular modeling of amorphous, non-woven polymer networks, J. Mol. Model. 21 (2015) 263.

[50] J. Parsons, Chemical binder application technology, in: 30th Nonwoven Fabrics Symposium, Clemson University, 1999. 21–24 June.

[51] M.M. Mlynar, E.J. Sweeney, Processing aids for resin bonded nonwoven webs, in: Principles of Nonwovens, INDA Publication, 1993, pp. 249–257.

[52] A.G. Hoyle, Bonding as a nonwoven design tool, in: TAPPI Nonwovens Conference, 5–8 April, 1988, pp. 65–69.

[53] M. Patel, D. Bhrambhatt, Nonwoven Technology for Unconventional Fabric. https://textInfo.wordpress/2011/10/25/nonwoven-technology-for-unconventional-fabric/.

[54] A. Watzl, The modern concept of through drying for the nonwoven and paper industries, in: Nonwovens Conference, TAPPI Proceedings, 1989, pp. 87–104.

[55] A. Watzl, Production Lines for Nonwovens used in the Automobile Industry, n.d., Fleissner Publication, (undated).

[56] C. Woodings. http://www.nonwoven.co.uk/reports/flushability.htm.

[57] L.J. Mann, P.M. Winter, Cleaning Articles and Method of Making, United States Patent Application 20020173214, 2002.

CHAPTER
10

Thermal bonding

A. Pourmohammadi

Department of Mechanical Engineering, University of Payam Noor, Tehran, Iran

10.1 Introduction

Thermal bonding can be used to bond drylaid, wetlaid, and spunlaid (meltblown or spunbond) webs, as well as composite multilayered webs, provided thermoplastic polymer content is present. The basic concept of thermal bonding was introduced by Reed in 1942. He described a process in which a web consisting of thermoplastic and nonthermoplastic fibres was made, and then heated to the melting or softening temperature of the constituent thermoplastic fibres followed by cooling to solidify and bond the area. In the early development of thermal bonding, nonthermoplastic rayon fibres (base fibre component) were blended with plasticised cellulose acetate or vinyl chloride fibres (binder fibre component). A carded web consisting of a blend of such base and binder fibres when hot calendered and subsequently cooled, produced a thermally bonded fabric structure. The resulting thin, strong, and relatively dense fabric was more akin to a paper product than a textile material. Production costs for this fabric were very high, primarily because the available binder fibres were expensive. Applications were initially limited to products requiring a smooth surface, low porosity, high strength, and low thickness. Given the product limitations and the high cost of such binder fibres, nonwoven producers continued to prefer latex bonding using chemical binders.

The rising cost of energy and greater awareness of the environmental impact of latex bonding led to a change in direction. The energy consumption of various web bonding processes in Fig. 10.1 illustrates the relative savings associated with the thermal bonding process [1]. The high production rates possible with thermal bonding and the significant energy savings as compared to chemical bonding, due to the absence of significant water evaporation during bonding, makes the process economically attractive. In contrast to chemical bonding, the environmental impact of the process is also significantly reduced. The growing market demand for single use and durable nonwoven products spurred developments in new thermoplastic and thermoset materials in the form of powder, films, webs, hot-melt compounds as well as

Handbook of Nonwovens
https://doi.org/10.1016/B978-0-12-818912-2.00007-0

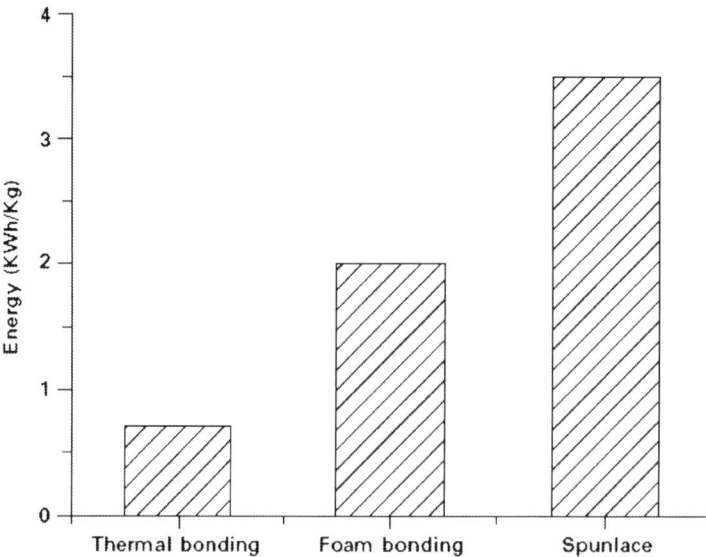

FIG. 10.1 Energy consumption of different bonding processes [1].

improved production methods such as point-bonding calenders, through-air bonding and belt bonders. This has greatly increased the diversity of products that can be manufactured by the thermal bonding process.

10.2 Principle of thermal bonding

Thermal bonding requires a thermoplastic polymer to be present in the form of a homofil fibre, bicomponent fibre, or powder. Thermoplastic films, preformed webs and hot-melt adhesives are also used. In practice, heat is applied until the thermoplastic component softens or melts. The polymer flows by surface tension and capillary action at fibre-to-fibre crossover points where bonding regions are formed. These bonding regions are fixed by subsequent cooling, and if a nonmelting base fibre is also present, e.g., a regenerated cellulosic, no chemical reaction takes place between the binder and the base fibre at the bonding sites.

When binders melt and flow around fibre crossover points, and into the surface crevices of fibres in the vicinity, an adhesive or mechanical bond is formed by subsequent cooling. Such adhesive bonds involve physiochemical bonding at the interface of two dissimilar materials. In the thermal bonding context, a mechanical bond is formed as a result of thermal shrinkage of the bonding material, which while in the liquid state encapsulates the fibre crossover points [2]. In contrast, if at the binder–fibre interface both components soften or melt, interdiffusion and interpenetration of the molecules across the interface can occur and the interface may disappear. This arises where compatible polymers are present with near comparable solubility parameters. Bonds formed in this way may be called cohesive bonds. Some of the main advantages of thermal bonding are as follows:

- Fabrics can be relatively soft and textile-like depending on blend composition and the overall bond area.
- Good economic efficiency compared to chemical bonding with lower thermal energy requirements and less expensive machinery.

  ···· High bulk products can be bonded uniformly throughout the web cross-section.
  ···· 100% recycling of fibre components is possible.
  ···· No latex binders are required, reducing environmental impacts.

10.3 Raw materials

Thermally bonded fabrics are produced both from webs made entirely from thermoplastic fibre materials and from blends containing fibres that are not intended to soften or flow on heating. The nonbinder component in blends may be referred to as the base fibre component and commercially, a variety of different types are used. The binder fibre component in blends normally ranges from 5% to 50% of the total weight of fibre, depending on the physical property requirements of the final product.

10.3.1 Base fibres types

The base fibre contributes to key physical, chemical, and mechanical properties of the fabric derived from the polymer from which it is composed. The choice of fibre depends on considerations such as the required wetting and liquid absorption behaviour, dyeing characteristics, flame resistance, tensile and attritional properties, hydrolytic resistance, and biodegradability, amongst many other properties. The commonly used base fibres include wood pulp and natural fibres (e.g. bast, vegetable, and protein fibres such as wool), regenerated cellulosic fibres, synthetic fibres (e.g. polyester, polypropylene, acrylic, nylon, aramid, and many others), mineral fibres (e.g. glass and silica) and metallic fibres.

10.3.2 Binder materials

Binder components are produced in many different forms including fibre or filament (homofil or bicomponent sheath/core or side-by-side melt-bonding fibres), powder, film, low melt webs, and hot melts. The physical form of the binder affects its distribution throughout the web, which has a significant impact on fabric properties. The amount of binder also plays an important role in determining the properties of the resultant nonwoven fabric. If the binder content is more than 50% of the total blend, the fabric behaves like a reinforced plastic. At a binder content of 10%, the fabric will be bulky and flexible and have a relatively low tensile strength.

To minimise energy costs, it is desirable that binder fibres have a high melting speed, a low-melting shrinkage, and a narrow melting point range. The most widely used thermoplastic binder polymers are given in Table 10.1. Decreasing the melting temperature of certain polymers, for instance PET, from 260°C to 135–190°C, requires the use of copolymers produced by

TABLE 10.1   Thermal transition points in common thermoplastic binder polymers.

Fibre type	Glass transition Temperature (°C)	Melting Temperature (°C)
Polyvinyl chloride (PVC)	81	200–215
Polyamide (PA)	50	210–230
Polyester (PET)	69	245–265
Polypropylene (PP)	−18	160–175
Polyethylene (PE) (low density)	−110	115

poly-condensation. The melting speed of these copolymers is very high; hence the thermal shrinkage is reasonably low.

When thermoplastic fibres or powders are used as binders, their melting temperature is significantly lower than the base fibres in the web, which helps to prevent thermal degradation. In low-melting temperature homopolymers, or copolymer binder fibres or powders, complete melting can occur, and the polymer becomes a fluid. If the viscosity of the molten polymer is sufficiently low, it flows along the surface of the base fibres and is collected at the fibre crossover points to form bonding points in the shape of beads by subsequent cooling. In webs composed of bicomponent fibres (of the sheath/core type), the sheath polymer does not need to completely melt but softens enough to form a bond. However, if it does melt and flow, the bonding mechanism becomes similar to that of homopolymer binder fibres. The advantage of bicomponent fibres is that every crossover can be potentially bonded and also since the physical structure of the core component is not degraded, thermal shrinkage is minimised, web structure remains essentially intact and fabric strength is usually higher. Binder fibres are selected by their suitability for the different thermal bonding processes, as well as considering the conditions under which the final fabric will be used.

10.3.3 Bicomponent binder fibres

Bicomponent or 'Bico' fibres and filaments, also referred to as conjugate fibres, particularly in Asia, are composed of at least two different polymer components. They have been commercially available for years; one of the earliest was a side-by-side fibre called Cantrece developed by DuPont in the mid-1960s followed by Monsanto's Monvel, which was a self-crimping bicomponent fibre used by the hosiery industry during the 1970s. Neither of these fibres was commercially successful because of complex and expensive manufacturing processes.

Later in 1986, commercially successful bicomponent spinning equipment was developed by Neumag, a producer of synthetic fibre machinery [3]. Use of bicomponent fibres accelerated dramatically in the early 1990s partly because of the need to uniformly bond the entire thickness of nonwoven fabrics, which in heavy weight structures could not be satisfactorily accomplished by chemical bonding. The market for bicomponent fibres has greatly developed, with significant early work being contributed by organisations in Japan and Korea [4]. Some of the various polymers used as components in bicomponent fibres are listed in Table 10.2. Examples of common polymer combinations in bicomponent binder fibres are:

- Polyester core (250°C melt point) and CoPolyester Sheath (melting points of 110–220°C).
- Polyester core (250°C melt point) and polyethylene sheath (130°C melting point).
- Polypropylene core (175°C melting point) and polyethylene sheath (130°C melting point).

Bicomponents are particularly important in drylaid and spunlaid manufacturing, as well as in some wetlaids for thermal bonding.

10.3.3.1 Bicomponent fibre classification

Bicomponent fibres or filaments are commonly classified by the structure of their cross-section as side-by-side, sheath–core, island in the sea or segmented pie. Of these, the side-by-side and sheath–core arrangements are most important for thermal bonding applications.

TABLE 10.2 Polymers used in manufacturing bicomponent fibres.

Polymer	Notes
PET and coPET	Melt temperatures range from 110°C to c. 250°C. Water soluble, alkali soluble, elastomeric and biodegradable coPETs available
PTT	Polytrimethylene terephthalate, e.g., Corterra
PETG	PET glycol
PBT	Polybutylene terephthalate
PEN	Polyethylene naphthalate
PP	Kromalon dyeable, Kromatex PP, SPP (syndiotactic PP)
	PE/PP copolymer
PLA	Polylactic – melting temperature ranges from 130°C to 170°C
HDPE	High density PE
LLDPE	Linear low density PE
PA (polyamide)	PA 6 (nylon 6), PA 6,6 (nylon 6,6), PA 11 (nylon 11) PA 12 (nylon 12), PEBAX copolyamide
PPS	Polyphenylene sulphide
PCL	Polycaprolactone
PS	Polystyrene
PVDF	Polyvinylidene fluoride
PVOH	Plasticised polyvinyl alcohol
TPU	Thermoplastic polyurethane
EVOH	Ethylene vinyl alcohol
PAN	Amlon PAN (polyacrylonitrile)

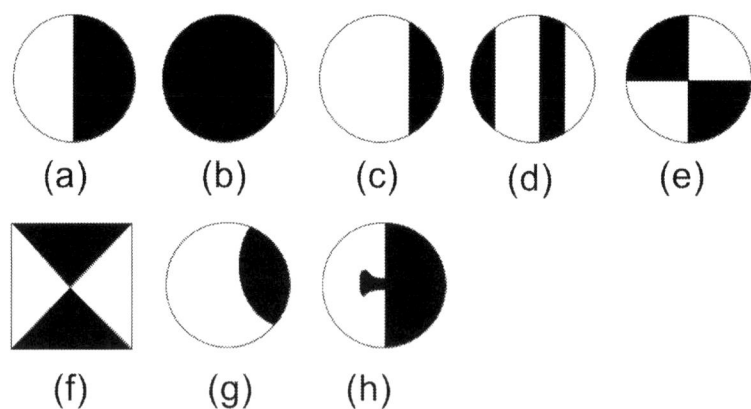

FIG. 10.2   Schematic diagram of side-by-side bicomponent fibre cross-sections.

10.3.3.2 Side-by-side (S/S)

Two components are arranged side by side and are divided along their length into two or more distinct regions (see Fig. 10.2). The components must have good adhesion otherwise two fibres of different composition will be produced. There are several ways of producing side-by-side bicomponent fibres described in the literature [5].

The geometrical configuration of side-by-side bicomponent fibres particularly asymmetry, makes it possible to achieve an additional three-dimensional crimp during thermal bonding by differential thermal shrinkage of the two components, for example. This latent crimp gives rise to increased bulk stability and a softer fabric handle. The characteristics of the crimp are determined by factors such as polymer properties, the weight ratio of the two polymers and the structure of the web, which can be varied according to the method of web formation. An increase in the crimp level from 15% to >30% and in the number of crimps/cm from 6.5 to >22 can be introduced using this differential thermal shrinkage approach.

10.3.3.3 Sheath–core

In sheath–core bicomponents, one of the components (the core) is fully surrounded by another component (the sheath). The arrangement of the core is either eccentric or concentric depending on the fabric properties required. If high fabric strength is required, the concentric form is selected, whereas if bulk is required, the eccentric type is employed [6]. Adhesion is not always essential for fibre integrity. A highly contoured interface between the sheath and core can provide the mechanical interlock that is desirable in the absence of good adhesion (see Fig. 10.3). One advantage of sheath–core fibres is the ability to produce a surface with the required lustre and handle characteristics, while the core dominates the tensile properties. The sheath–core structure also provides a means of controlling costs by engineering the relative proportions of the two polymer components. Traditionally, the ratio may be 50:50 or 30:70 but a variety of others are possible, including 10:90.

The first industrial exploitation of bicomponent fibres involved the use of Co-PET/PET or PE/PP fibres for hygiene applications as well as for high-loft waddings, wiping cloths, medical wipes and filter media. The difference in the sheath–core melting temperature in PE/PP is

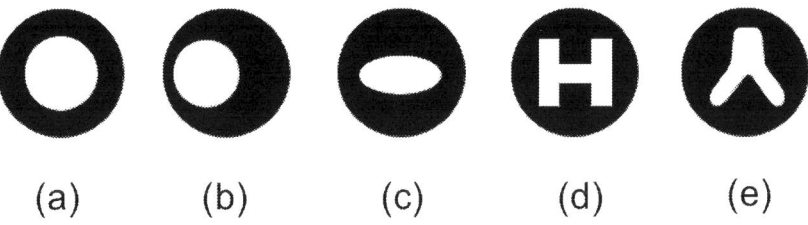

FIG. 10.3   Cross-sections of sheath–core bicomponent fibres.

about 40°C. In Co-PET/PET bicomponents, the sheath melts at 100–110°C while the core melts at 250–265°C. For drylaid applications using bicomponent fibres, blend ratios are generally in the range of 10%–50%, depending on the application and process parameters. A useful experimental guide is given in Table 10.3. Depending on the type of base fibres in the blend, Co-PET/PET bicomponent fibres can form strong primary bonds between themselves, and therefore, a framework structure in which the base fibres are embedded is produced. It is also possible to modify the fibres so that secondary bonds are formed between bicomponent fibres and the base fibres [7]. The marked difference between the melting temperatures of the PE sheath (125–135°C) and PET core (250–256°C) brings a number of advantages for PE/PET sheath–core bicomponent fibres in thermal bonding of nonwovens.

A wide variation in bonding temperature can be tolerated since the core component is largely unaffected by temperature variations that may inadvertently occur during the thermal bonding process. In contrast to other sheath–core combinations, the core remains stable to mechanical deformation even at high temperature after the sheath has melted, which facilitates the production of high-quality nonwovens [8]. By appropriate selection of polymer composition, polymer ratio and fibre cross-sectional geometry, it is possible to engineer bicomponent fibre structures for improved economic efficiency, cost-effectiveness and functionality.

10.3.4 Powder binders

Powdered polymeric binders can be applied to fibres during web or batt formation or following web formation and prebonding. A thermoplastic polymer with a low softening temperature is desirable that requires a short exposure to heat in order to melt and fuse the powder. For ease of operation, the thermoplastic powder should have a low melt viscosity

TABLE 10.3   A practical guide for producing nonwoven fabrics with different handle characteristics from Co-PET/PET bicomponent fibres.

Parameter	Nonwoven fabric handle		
	Soft	Medium	Harsh
Bicomponent fibre content (%)	10–20	15–30	>30
Bonding temperature (°C)	140–150	150–160	160–180
Fibre fineness (dtex)	1.7–3.3	3.3–6.7	>6.7

and the transition from melt to solid should occur over the shortest possible temperature range. Polymers such as polyethylene, low molecular weight polyamide and copolymers of vinyl chloride and vinyl acetate, are generally used. This method of thermal bonding is limited by difficulties in obtaining polymers with a suitable range of particle sizes to suit the base web. Obtaining a uniform powder distribution throughout the web, particularly through the thickness, is also problematic. Powder bonding is suited to lightweight webs where an open structure is required with a soft handle, or in the production of reinforced, moulded products. Applications include feminine hygiene, adult incontinence, medical and automotive products, wipes, computer disks, apparel and shoe composites [9].

10.4 Calender bonding

Thermal bonding relies on the application of heat energy to melt or soften one or more polymers in the web to achieve bonding. There are different methods of applying heat energy to the web and the heat transfer mechanism can take different forms: conduction, convection, and heat radiation. The widely used methods are discussed in this section.

Thermal calender bonding is a process in which a fibrous web containing thermoplastic material is passed continuously through a heated calender nip that is created by two rollers pressed against each other. Multinip calenders are also employed depending on the web weight, throughput speed and degree of bonding required. Both rollers are internally heated to a temperature that usually exceeds the melting point of the binder components in the web to ensure there is sufficient heat transfer to induce softening at the prevailing line speed. As the web passes between the calender nip, fibres are both heated and compressed. This causes the binder components of the web to become soft and tacky and induces polymer flow in and around the constituent fibres. Bonding sites are formed at fibre crossover or contact points. Cooling leads to solidification of the polymer and finalises bonding.

Calender bonding is mainly applicable to light and medium-weight webs because the fibres in a thick web insulate heat from the interior of the structure, leading to a temperature gradient and variation in the degree of bonding through the cross-section. To increase the efficiency of the process, the web can be preheated immediately prior to entering the calender nip, using for example, infrared heaters. Commercially, thermal calender bonding is used for light-weight webs of $7.5–30\,g/m^2$, including spunbonds and SMS structures intended for hygiene and medical applications, as well as medium-weight webs of $100\,g/m^2$ for interlining and filtration applications [10].

The degree of bonding that is achieved depends on temperature, pressure and the production or throughput speed, which determines the contact or dwell time between the rollers. The properties of the fabric are also influenced by the total bond area (expressed as %), which in turn depends on whether the calender roller surface is smooth or has a 'point pattern'. In practice, point bonding is achieved when the roller surface carries an engraved or 'emboss' pattern (bond area < 100%), or area bonding when the roller surface is smooth (bond area = 100%).

10.4.1 Area bonding

Two or more smooth rollers heat the entire surface of the web in the roller nip. Bonding at the crossover points between binder fibres or in blends, between the binder fibres and base fibres takes place. The applied nip pressure and extensive bond area can lead to fabrics that are thin and stiff with a paper-like handle and relatively low permeability. In area bonding, metal on metal or a combination of metal and/or elastic covered rollers are used. Elastic roller surfaces are composed of a deformable material such as urethane, silicon rubber, wool or cotton filling or nylon shells. More than two rollers are used to create multiple bonding zones for bonding heavy-weight webs. Generally, in the three-roller calender, the heated roller is in the middle, whereas in the four-roller configuration the heated rollers are on the top and bottom with the composition rollers in the middle. Note that in some applications, calender bonding is used in combination with other bonding processes. For example, in the manufacture of some high-loft waddings, through-air thermal bonded fabrics are passed between hot calender rollers at very low nip pressure to provide a flat, film-like surface on one or both of the fabric surfaces, without compressing the bulk structure.

10.4.2 Point bonding

Point bonding is based on a calender nip consisting of an engraved roller and a smooth roller. In some cases, both rollers are engraved. When the web enters the nip of the rollers, the temperature increases to a point at which softening and melting of the polymer causes fibre segments, which are held between the tips of the engraved lands and the smooth roller, to bind together (Fig. 10.4). Normally, both rollers are heated, but depending on the application, the bottom roller may not be heated. The degree of bonding in the fabric is heavily influenced by the design of the engraved point pattern on the calender roller, in terms of its geometry, land frequency, shape, and size, which influences the structure of the bond points. It is normal to have a bonded area in the final fabric of 10%–40%. This allows fabrics to be produced that remain soft and comfortable with unbonded, permeable regions between the bonded regions. In point-bonded fabrics, a further consideration is the fibre orientation in the web in relation to the geometric arrangement of the thermal point pattern and the shape of the bond points. This influences the resulting fabric properties, including the anisotropy.

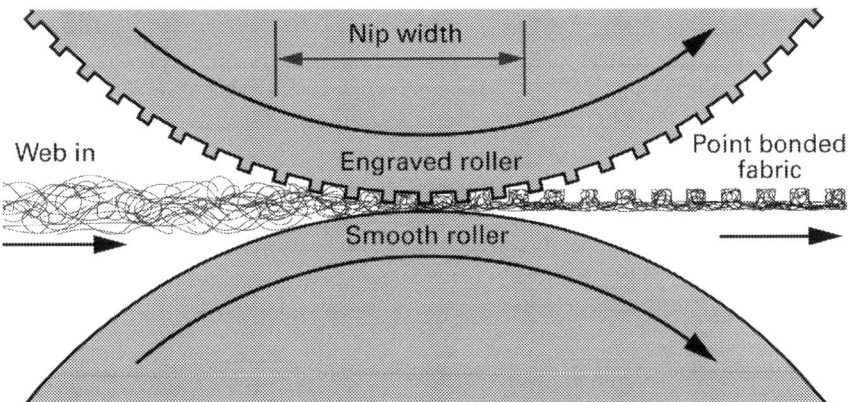

FIG. 10.4 Typical thermal point-bonding roller arrangement.

10.4.2.1 Novonette system

In point bonding, webs are bonded by passing between an engraved calender roller and a solid smooth roller. This can produce an embossed fabric with an impression on one side only, while the other side remains smooth. Although less common in the nonwovens industry, it is also possible for both calender rollers to be engraved with an identical pattern of raised and depressed areas, so that a raised area on one roller matches a raised area on the other roller to create a highly compressed area. However, two engraved rollers can also be designed in such a way that the raised area of one, registers with the depressed area of the other. The main problem in this case is that exact positioning of both rollers is essential to achieve proper pattern definition which is difficult and expensive. The Novonette pattern is a patented calender system developed by the Kendall Co. Two identical helically engraved steel rollers are used with lands and grooved areas. Owing to this helical pattern on each calender roller, a multiplicity of lands on one roller is constantly engaged with a multiplicity of lands on the other roller. The pressure distribution is therefore even, and there is no need for exact positioning of the roller.

The width of the lands and grooves can be varied, as well as the angle of the rollers, to alter the physical and aesthetic properties of the nonwoven fabric. As the web passes through the heated rollers under pressure, a repeating pattern is formed as shown in Fig. 10.5. Three defined areas are shown; (i) the dark section represents the area where the land of one roller crosses the land of the other roller and maximum pressure is applied to produce a high degree of bonding, (ii) the grey block represents the area where the land of one roller passes over a groove of another roller and (iii), the white block is where the groove of one roller passes over the groove of the second roller. Various factors influence the effect of the calender, associated with mechanical process conditions, and the web structure, see Table 10.4.

10.4.3 Effect of calender bonding process parameters

The main process variables influencing thermal calender bonding are roller temperature, the line (nip) pressure and residence time in the nip of the calender. Residence time is

FIG. 10.5   Pattern formed in a Novonette system.

TABLE 10.4 Factors influencing fabric properties during calendering.

Web composition and structure	Process variables
Fibre type and polymer specifications	Roller type: steel smooth, steel engraved, elastomer, or cotton covered
Fibre orientation	Temperature
Area density (fabric weight)	Nip pressure
Initial thickness	Contact time

determined by the production speed and roller diameters. Some of these parameters also interact with one another. For example, the nip width varies with the square root of pressure, which affects nip residence time.

10.4.3.1 *Bonding temperature*

Bonding temperature determines the structure of fibres at the bonding points. In areas where fibres are flattened but not completely melted to a film-like structure, fabrics with high strength are produced. Therefore, it is important that the temperature of the roller surfaces is adjusted in a way that the so-called sintering [11] of the fibre surfaces can be achieved while avoiding complete fibre melting and film formation. Fabric softness can also be maintained by keeping the fibres between bonding points at a temperature below the melting point. This may be accomplished by using different temperatures for the smooth and engraved rollers. In this way, fibre characteristics in the area between the raised points can be maintained.

The main method of heat transfer in calender bonding is conduction, and the effect of convection and radiation is limited [10]. The process of heat transfer through the thickness of a fabric can be explained using the finite plate model [10]. However, as the web passes through the nip, the density increases causing changes in the heat transfer coefficient, thermal conductivity, and web thickness. Web conductivity increases close to that of the solid polymer as a result of web compression and removal of air from the fabric. A relatively void-free bonding point may be obtained; in practice a 5% void content has been measured [12].

Particularly given the short time the web is present in the roller nip, the temperature in the centre of the web during bonding can be substantially lower than the surface. Consequently, heat transfer alone cannot always provide sufficient bonding through the entire cross-section and additional thermal inputs are required. A preheating stage is sometimes employed to improve bonding; however, this can increase the crystallinity of the web, necessitating higher bonding temperatures. In the case of heavier webs, delamination can occur if the centre is not properly bonded, and so the temperature should be instantaneously raised while the web is in the calender nip. Schwartz [13] suggested that this might be done by introducing a thin jet of steam into the web, coincident with its introduction into the nip. In practice, the choice of calender temperature involves balancing the requirements of fabric softness and flexibility with tensile strength.

Increasing bonding temperature up to a certain point increases the tensile strength of the fabric due to the formation of a well-developed bonding structure. SEM images of bond points produced at high temperature show a regular shape and a smooth fabric surface. However, further increases in temperature can reduce tensile strength, which may be attributed to the loss of fibre integrity and the formation of film-like spots, as well as the reduction in load

10. Thermal bonding

transfer from fibres to the bonding points [14,15]. Overbonding of this kind leads to 'popping' of the structure under tensile load as the fabric fails at the point-bond locations. Increasing the bonding temperature also leads to higher shear modulus and bending rigidity in calender bonded fabrics [16].

10.4.3.2 Calender nip pressure

The influence of nip pressure in calender bonding has been extensively investigated [17–19] and optimum values depend on the polymer composition of the web. For any web, an appropriate optimum pressure needs to be found to obtain maximum strength, while maintaining other properties, in particular softness and fabric handle. As the nip pressure increases, the fibres are deformed and the air that acts as an insulator is removed from the web. This maximises the overall fibre contact area with the rollers and consequently, increases the rate of heat transfer from the rollers to the web. Therefore, the nip pressure influences the heat transfer to the fibres, and through the web structure, as well as the melting point, flow and viscosity of the polymer. Thermal bonding calenders operate with line (nip) pressures between about 35 and 260 N/mm, but up to a maximum of about 150 N/mm is most common.

10.4.3.3 Influence of pressure on heat conduction

In calender bonding, the heat is transferred by conduction, and therefore, the thermal conductivity of the fibres in the web is important (see Table 10.5). There is a temperature gradient across the fibre cross-section as shown in Fig. 10.6, i.e., the highest temperature is at the surface in contact with the heated roller. The general equation used to explain heat transfer due to conduction is as follows,

$$Q = K \cdot A \cdot \frac{dT}{dX}$$ (10.1)

where

Q = Heat transferred per unit time.
K = Thermal conductivity.
A = Area (perpendicular to the direction of heat transfer).
$\frac{dT}{dX}$ = Temperature gradient.

If no pressure is applied to the web, the contact between the fibres and the heated calender would only be a thin line, which means that A would be zero and theoretically there would be no heat transfer. It is therefore necessary to have sufficient contact between the fibres and the

TABLE 10.5   Thermal conductivity of some polymers, steel and air.

Materials	Heat conductivity (w/m K)
Steel	48.0
Polyamide	0.22
Polyester	0.36
Air at high temperature 250°C	0.04

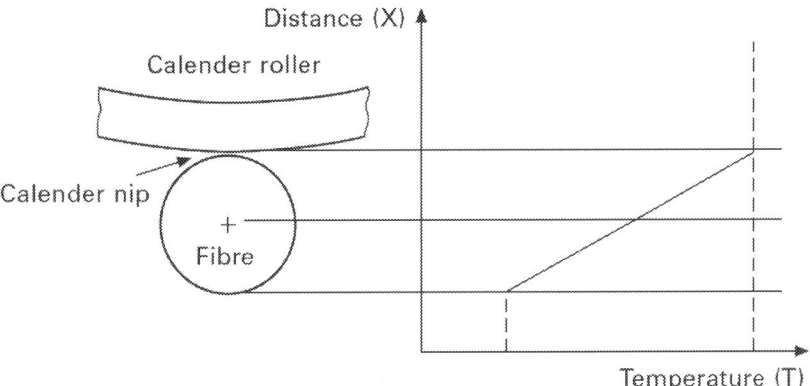

FIG. 10.6   Temperature gradient across the fibre cross-section.

rollers in the calender nip to achieve adequate heat transfer and bonding, because the nip pressure determines the level of roller–fibre contact.

10.4.3.4 *Heat of deformation*

The heat of deformation is generated by the high compressive forces applied in the nip of the calender rollers and this heat is sufficient to raise the polymer temperature by as much as 30–35°C [12] in the case of polypropylene fibres. This phenomenon is called deformation-induced heating (DIH) and can be estimated using an analysis method developed by Mayer et al. [20].

10.4.3.5 *Clapeyron effect*

Polymer molecules subjected to compression require more thermal energy to melt compared to when they are at atmospheric pressure; this concept is known as the Clapeyron effect. Wunderlich [21] estimated that pressure could increase the melting point temperature of polypropylene by 38 K/Kbar. Warner [12] reported a 10°C increase in melting temperature of PP fibres at the bond points. Pressure is therefore thought to limit the extent of melting under the lands and prevents sticking of the polymer to the rollers in the nip.

Polymer flow is also important in the formation of a proper thermal bond. The flow of polymer in the calender nip is mainly influenced by nip pressure, the available volume, polymer viscosity and residence time in the calender nip. Increasing the temperature and nip pressure increases the flow rate, but pressure also increases the polymer melting temperature. Polymer diffusion may also occur providing sufficient time is available. However, the polymer flow can be expected to be more critical to the formation of a strong bond than diffusion.

10.4.3.6 *Contact time*

The contact time of the web in the calender nip, is a function of the production speed and the roller diameters. Generally, the contact time is very short, in the order of milliseconds, which can be readily calculated from geometry (see Fig. 10.7). Using the equation in Fig. 10.7 and substituting the variables $D = 450$ mm, $T_{in} = 1.8$ mm, $T = 0.2$ mm, and $T_{out} = 0.75$ gives a contact area of 30 mm^2. If the production speed is in the range of 100–300 m/min, the

$$\text{Contact time} = \frac{\text{contact area}}{\text{speed}} = \frac{\sqrt{D/2} \left(\sqrt{T_{in} - T} + \sqrt{T_{out} - T} \right)}{V}$$

where
      D : roller diameter
      T_{in} : thickness of input web
      T : distance between two roller
      T_{out} : thickness of bonded web

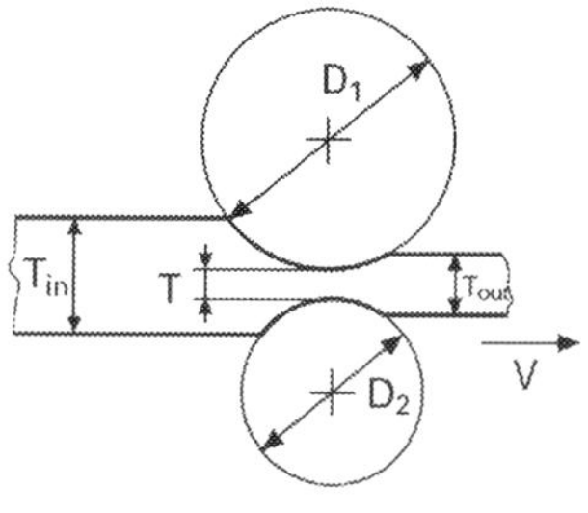

FIG. 10.7   Calculating the contact time of the web in the calender nip.

estimated contact time will be 6–18 ms. In practice, the contact time for medium-weight fabrics is from about 0.1 to 0.7 s (100–700 ms), and for light-weight fabrics such as coverstocks, about 0.001 s (1 ms).

Fig. 10.8 illustrates the relationship between the contact time and production speed at different nip pressures. This diagram is valid for lightweight fabrics. For thicker webs, a longer contact time is required. Lightweight fabrics are calendered at very short contact times compared to heavier web weights due to the latter's slower production speed and higher web thickness. When considering the impact of temperature, nip pressure or contact time on thermal calendering it is the temperature of the fibres themselves in the calender nip that is of particular importance. This temperature is not the same as the roller temperature because heat transfer controls the temperature gradient and depends on temperature, pressure, and time, all of which interact with each other in a complex manner.

10.4.3.7 *The effect of engraving*

The characteristics of calender bonded fabrics are influenced by the pattern and size of the engraved surface on the calender roller. Fig. 10.9 shows a typical engraved roller. Clearly, with a higher density of bond points, fabric strength tends to increase but fabric stiffness may be adversely affected.

10.4.3.8 *Design of the calender roller system*

At the nip, at least one of the calender rollers is made of steel. The exterior shell of the steel rollers for calender bonding is usually made of an alloy that can withstand the high temperatures and various stresses inherent in thermal bonding. In some circumstances, only one steel roller is employed with the other having a resilient roller surface made of either polyurethane, wool, or cotton. Such combinations are sometimes used to achieve high-density webs, particular surface patterns or increased dwell times since resilient roller surfaces provide a wider nip area under the applied pressure.

10.4.3.9 *Roller width*

A rule of thumb is that the width of the roller should be the width of the web plus 25 cm on each edge. This additional roller face is more critical as the webs become lighter and are affected by the heating system that is used. Thermal fluid systems require the least additional roller face [22].

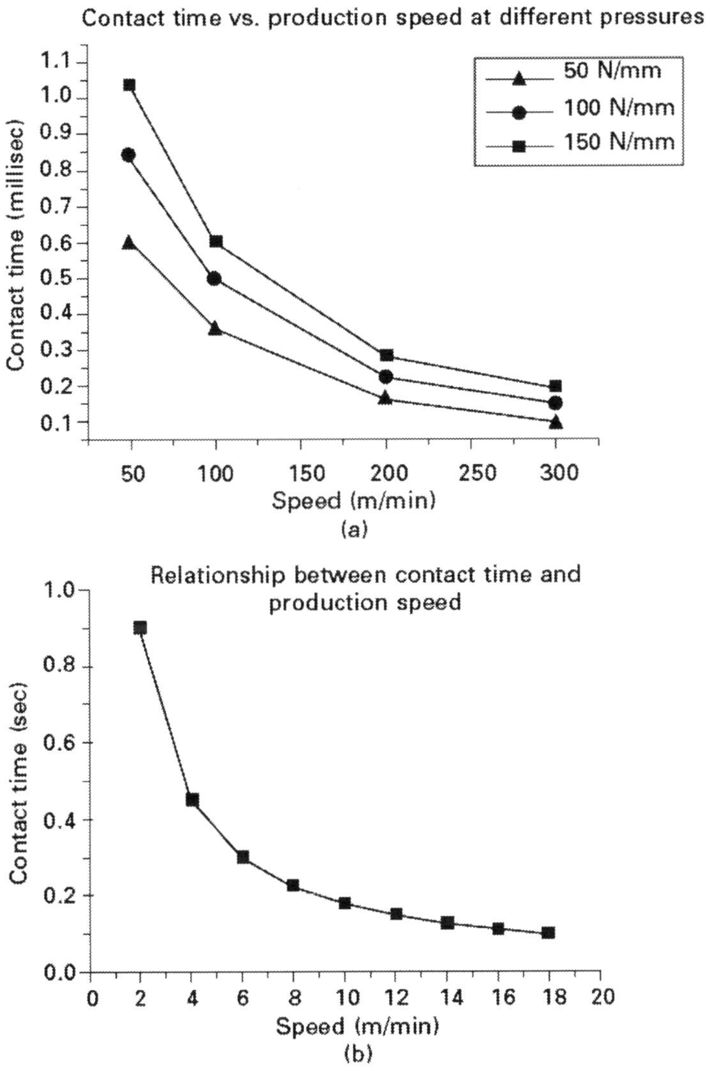

Contact time vs. production speed at different pressures

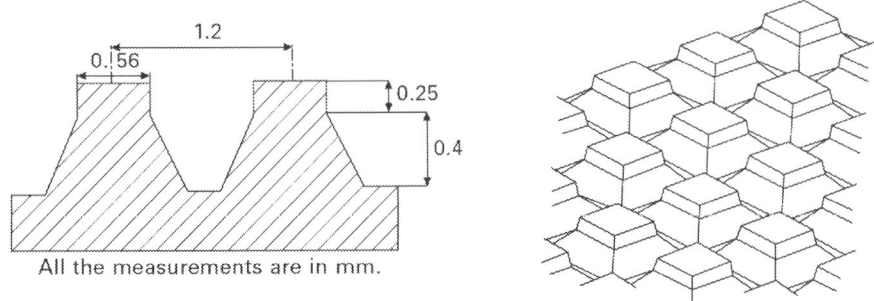

(a)

Relationship between contact time and production speed

(b)

FIG. 10.8 Relationship between contact time and production speed.

1.2

0.56

0.25

0.4

All the measurements are in mm.

FIG. 10.9 A typical engraved roller with 22% bonding area and 60 bonding points/cm^2.

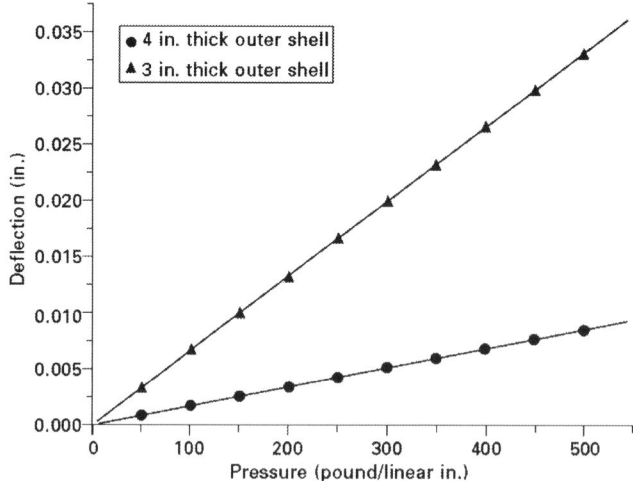

FIG. 10.10   Relationship between deflection and pressure [23].

10.4.3.10 Roller deflection/bending

Roller deflection across the width is a common problem in thermal calender bonding, because it can result in variations in contact pressure along the roller nip. The classic sign of this problem is heavily bonded edges with poorly bonded areas in the centre. Fig. 10.10 illustrates the relationship between deflection and pressure. Roller deflection mainly occurs in nipped rollers composed of steel and is less evident in nips formed between resilient rollers with a deformable surface. In cases where roller diameter is more than one-third of the roller length and pressures are less than 25 N/mm (140 lbs./in), no particular problem will be encountered with deflection.

As the roller is made wider without a corresponding increase in roller diameter, deflections in the rollers are caused by the application of pressure, resulting in variable pressure across the web. Various methods have been adapted to compensate for such deflection.

10.4.3.11 Crowning one or more rollers

If a roller is formed in a way that the diameter at the ends of the roller is less than the diameter at the centre of the roller (barrel shaped) it is referred to as being crowned, see Fig. 10.11. If the degree of crowning is properly chosen, the crown roller will compensate for the roller deflection present with the uncrowned roller. However, the method is relatively inflexible, since the degree of crowning compensates correctly for only a small range of loading pressures.

FIG. 10.11   Example of a barrel shape roller.

10.4.3.12 Skewing (cross-axis)

The longitudinal axes of the rollers are slightly skewed instead of being parallel. This wraps the rollers around each other under load, thus compensating for the roller deflection induced by the loading pressure. Pressure is applied to the roller outside the main roller bearings to reverse the deflection and thus compensate the deflection.

10.4.3.13 Floating

Floating or 'swimming' roller arrangements can be set up in which a rotating shell is fitted over a fixed core; this shell can be mechanically crowned in various zones across the width of the roller to compensate for deflection by controlling the internal pressure. The limitations of this approach relate to bearing life and the temperature that the oil on the rotating parts can withstand.

10.4.3.14 Surface finish

The steel roller may need to be plated or coated to improve wear, dent and abrasion resistance of the surface. This is achieved by chrome plating, nitrating the surface or flame hardening. Generally, a surface finish of 12–16 RMS (15–20 AA) is acceptable for most processes. A 1 RMS is the smoothest finish, while a 64 RMS is a rough machine finish.

10.4.4 Methods of heating calender rollers

Various methods have been devised to heat calender rollers and the commonly used systems are briefly explained. Modern thermal bonding calenders are usually made up to 5.8 m wide (sometimes up to 7 m) and are capable of maximum production speeds of 1300–1500 m/min. High speed calenders capable of over 1000 m/min are particularly relevant to thermal bonding of lightweight spunbond webs, and to achieve such speeds, twin calender systems can be employed. One example of a high-speed calender is the NeXcal twin pro (Andritz) system.

10.4.4.1 Hot water/liquid or steam

The electro-hot liquid roller is centre bored and sealed to produce an enclosed chamber. This chamber is partially filled with liquid. An electrical heating coil within the chamber heats the liquid producing pressurised steam. The steam condenses on the inner boring's surface and the heat is transferred into the steel and through the steel to the roller surface. Advantages are good temperature uniformity, an enclosed heating system (no external heating system is required), low investment and low maintenance cost. To reduce the steam pressure at high temperatures, liquids with a higher boiling point than water can be used. This system is limited in heat capacity and normally used for a maximum temperature up to 180°C.

10.4.4.2 Electrical or cal-rod system

Electrically heated rollers are heated internally and controlled by measuring the roller shell temperature near the surface, and so are much more easily and quickly controlled compared to the liquid heated systems. Also, they do not use any hot or pressurised liquids and high surface temperatures up to 420°C are possible. The drawbacks are high initial investment and maintenance cost. Electric conductive heated rollers have a dual shell design or a centre boring. An electric heating coil can be wound on the inner shell, and the top shell is pulled over the inner shell. In centre bored rolls, a heated coil is mounted in the inner boring. All remaining spaces are filled with a heat transfer medium. Thermal losses are limited to normal convection losses from the roller surfaces.

10.4.4.3 Gas heating system

This type of heating system normally uses a ribbon type burner in a bored steel roll. The maximum temperature is about 260°C. This system employs a live flame which could be a safety problem in many facilities. Also, the burner pipe inside the roll is difficult to manufacture over 4 m long and to remain straight at these temperatures.

10.4.4.4 Hot oil heating system

A straight turbulator is fitted to a bored steel roller. Hot oil is passed through the spiral and the bore using a closed-loop system. This is normally a good method for lower BTU transfer applications. This system can also be used with a multiple drilled roller. Small peripherical drillings are made close to the roller surface where the hot oil flows through the holes. Due to the small distance between the holes and roller surface, fast and uniform heat transfer to the roller surface can be achieved. With oil-heated systems, disposal of potentially toxic carbonised oil resulting from constant thermal cycling and high temperature is an ongoing challenge. However, new hydrocarbon oils have become available that handle higher operating temperatures without degrading.

10.4.5 Belt calendering for bonding

Belt calendering or belt bonding is a modified form of roller calendering with two main differences: higher residence time in the nip, combined with lower applied pressure. In roller calendering, the heating time is measured in milliseconds whereas in belt bonding, time in the nip is extended to 1–10 s. Pressure in a calender roll nip is in the range of 35–260 N/mm, while in belt bonding the pressure does not normally exceed 9 N/mm.

The belt bonder consists of a heated roller and a synthetic rubber blanket. The roller diameter ranges from 40 to 250 cm and is usually coated with PTFE to increase its life. The resilient, heat-resistant (up to a temperature of 250°C), silicone rubber blanket wraps around the heated roller, covering up to 90% of the roller surface. The nonwoven web is bonded by running it between the roller and the blanket when the heat and pressure are applied simultaneously. Pressure is applied by varying (i) the tension of the blanket against the heated roller and (ii) the pressure on the exit guide roller, see Fig. 10.12.

The resulting fabric is much less dense and papery compared to roller calendered products. This method also facilitates the use of binders with sharp melting point and flow properties, which present difficulties in roller calenders. It is also used in some cases to bond webs containing 100% sheath–core bicomponent fibres, where high pressure is not required to generate good interfibre bonding, and where excessive deformation of the fibres needs to be avoided. Both area bonding and point bonding can be achieved using different types of blanket. The embossing can be done in line using an embossing roller positioned after the web

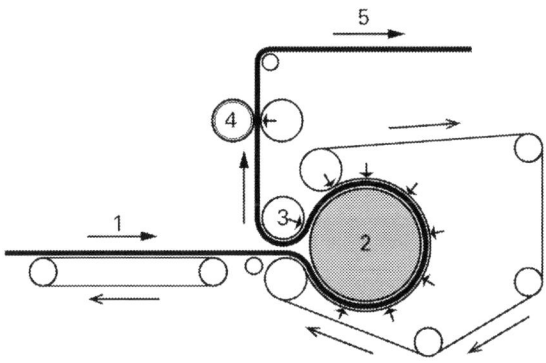

1. Unbonded nonwoven web
2. Heated drum
3. Pressure roller
4. Embossing unit
5. Thermal bonded
   nonwoven fabric

FIG. 10.12   Example of a belt bonder.

exits the belt section. A double drum belt bonder suitable for thick fabrics is also available. Working widths up to 6 m, and production speeds of up to 100 m/min are claimed for belt bonding.

10.5 Through-air and impingement bonding

The application of through-air (or thru-air) bonding is important for producing low density or high bulk thermally bonded nonwoven fabrics. Through-air bonding is used to make fabrics for durable as well as single use nonwoven products. In the latter, lightweight fabrics with good softness, drape, re-wet and high bulk can be produced compared with chemical bonding.

10.5.1 Principle of through-air bonding

Through-air bonding usually takes one of two different forms. These are the perforated drum or rotary system and the perforated conveyor or flatbed system. A schematic view of a typical rotary through-air bonding system is shown in Fig. 10.13. The main component in this system is a rotating air-permeable drum with a high open area onto which the web is transferred and supported by a travelling/carrying wire. The perforated drum is covered with a hood from where the heat is delivered; the hot air is drawn through the web cross-section by means of a suction fan.

Unlike other bonding techniques, through-air bonding almost exclusively uses bi- or tri-component staple fibres in the form of sheath–core or side-by-side configurations. The main polymers used are polyester, polypropylene, polyethylene or *co*-polyester [1,24]. In through-air bonding, web density and air permeability are critically important. Fig. 10.14 illustrates the relationship between web permeability, air speed and the required suction pressure. Referring to this figure, the higher the web density (and lower the permeability), the higher the pressure required to circulate the air through the fibres to achieve adequate bonding.

The key technological challenge in through-air bonding is to control the temperature and air flow. It is essential that the web is quickly heated to the melting temperature of the binder

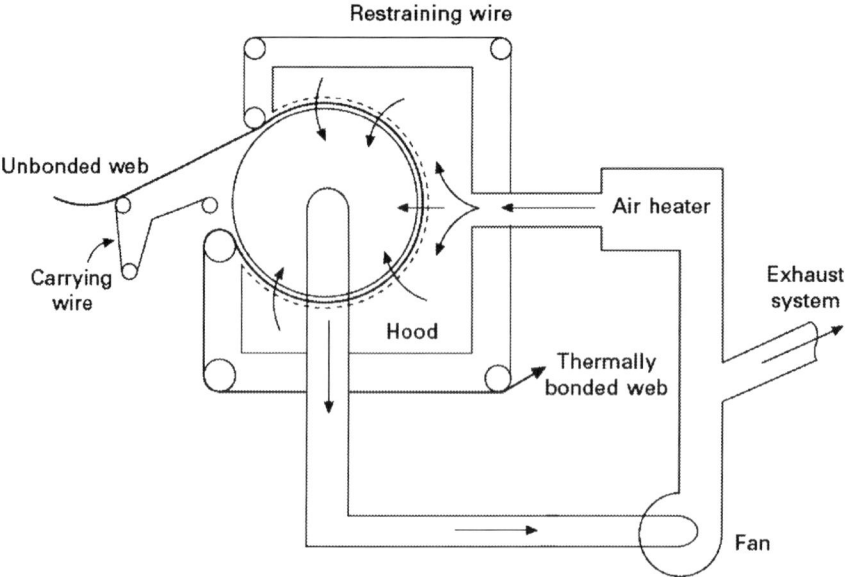

FIG. 10.13   Schematic diagram of a typical through-air bonding system.

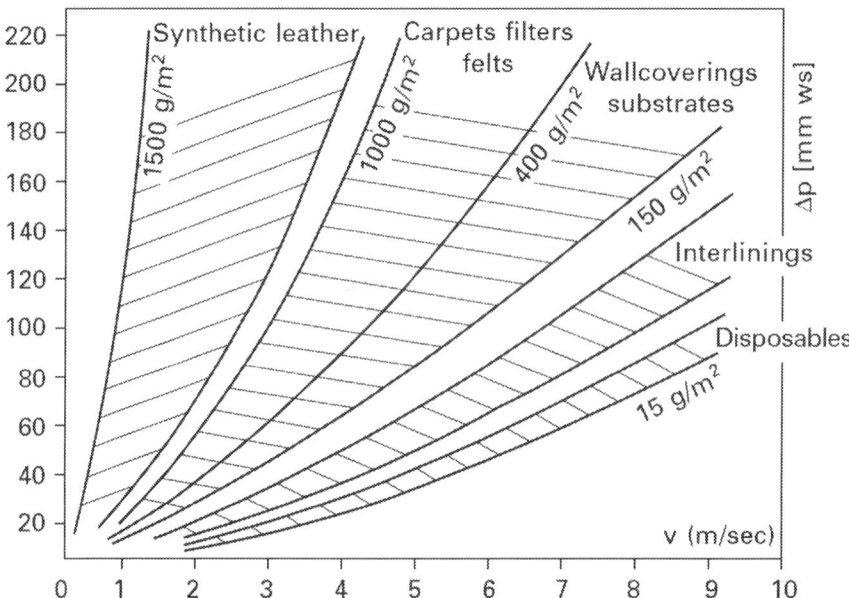

FIG. 10.14   Relationship between web air permeability, air speed, and required suction pressure [25].

fibres and then the air flow is reduced to avoid any undesirable change in web thickness. On the other hand, high air speed results in high fabric strength due to a reduction in web thickness and an increasing number of interfibre bonding points. Changes in web thickness also occur because of thermal shrinkage which can be minimised using bicomponent fibres since they are subject to smaller shrinkage. The web shrinkage may be controlled using a pressure belt or wire mesh which sandwiches the web on the perforated drum; the tensioned wire mesh tends to reduce the web shrinkage. This also improves the bonding by introducing a degree of web compression. The tension is released after the web is cooled down, see Fig. 10.15.

10.5.2 Perforated drum, through-air bonding

In this method, a web wraps around the circumference of a porous drum at an angle of 300 degrees. The remaining part of the drum is covered with a fixed shield positioned inside the drum. Heated air supplied to the area adjacent to the outside of the web is drawn through the entire width of the product by suction from a rotary fan. The perforated drum and fan are combined in one chamber which results in high flow efficiency. The open area depends on the shape and size/diameter of the perforation and can be up to 48%. When using a square perforation for special applications (e.g. sanitary products) it is possible to achieve open areas of up to 75%. Light-weight webs of $10\,g/m^2$, up to very heavy but permeable webs and felts of $3000\,g/m^2$ are also processed. The heating system is determined by the required temperature using steam, thermal oil, direct gas, hot water, or electric energy. Process speeds of 300 m/min are possible.

The drum system has several advantages compared with the belt method:

– Compact design (i.e. perforated drums, fans, and radiators are installed in an insulated housing).

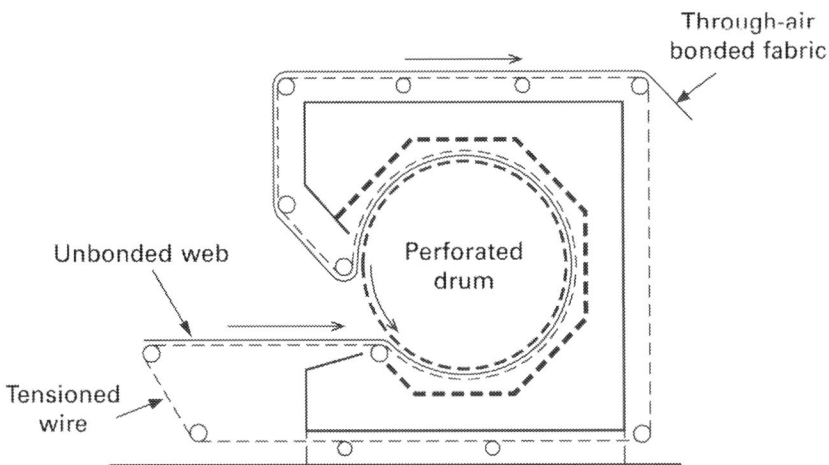

FIG. 10.15   Through-air bonding with restraining/tensioned wire mesh.

- Reduced energy consumption (no thermal losses by the conveying device, i.e., the drum remains inside the insulated chamber).
- Through-air bonding provides automatic heat recovery from the material, unless the line is combined with a calibrating unit.

Perforated drums are available in diameters from 1 to 3.5 m and working widths from 0.4 to 7 m or more. Generally, one-drum or two-drum units are in use; however, there are some multidrum lines. Fig. 10.16 shows a schematic diagram of a two-drum configuration arranged horizontally where a shield covers half the circumference of each of the cylinders.

In some through-air bonding systems, a pair of pressure rollers with adjustable roller gaps known as a calibrating unit is placed at the end of the line (see Fig. 10.16). This allows bulkier webs or batts to be produced to a prespecified thickness, as well as the production of denser webs having high tensile strength and a smooth surface. Perforated drums with a high open area of up to 96% are designed for bonding webs with a low air permeability and when high production speeds or high temperature and flow uniformities are required. Production speeds of more than 1000 m/min and working widths of up to 10 m are available [25].

10.5.3 Perforated conveyor–through-air bonding

In flat conveyor systems, the web is carried without the need for control by suction draught. This enables bonding of voluminous nonwoven fabrics as used, for example, in the production of airlaid waddings and other drylaid thermal insulation fabrics. Thickness changes depend on the degree of thermal shrinkage arising from unrestricted fibre shrinkage, and therefore, bicomponent fibres with low shrinkage properties are preferred. A uniform air flow and temperature distribution across the working width is essential to avoid irregular thermal shrinkage and bonding in the fabric. Generally, the perforated belt system is particularly suitable for bulky, low-density webs.

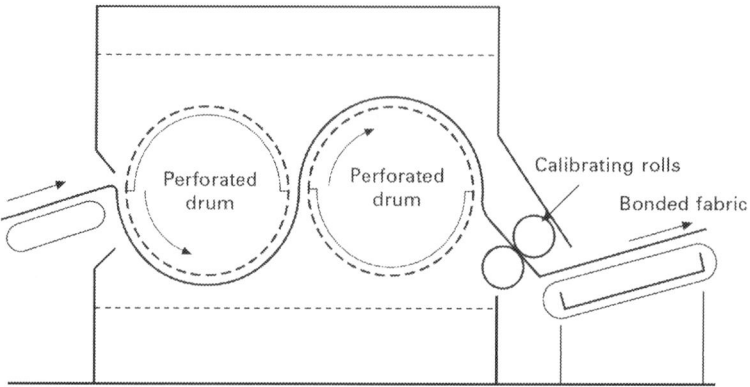

FIG. 10.16 Example of two-drum unit combined with calibrating rolls: perforated drum lines.

10.5.4 Impingement bonding (air jetting system)

Impingement systems are traditionally associated with the drying of paper products, but they can be adapted for thermal bonding of nonwovens. In such air jetting systems, hot air is blown onto the web from one or both sides by means of a nozzle system at a speed of up to approximately 40 m/s. The air flow approaches the web vertically from above and on contact it is deflected 90 degrees so that an air flow parallel to the web surface is formed, see Fig. 10.17.

In the case of double-sided air jetting the web is not pressed against the belt, but rather it floats on the bottom air flow and both sides of the web are bonded. Fibres within the web structure are less effectively heated by the hot air, and therefore, limited bonding occurs within the cross-section of the structure.

The difference between the through-air and air jetting approaches is given by the heat transfer coefficient:

$$\frac{\alpha_{\text{through air}}}{\alpha_{\text{air jetting}}} \approx 3:1$$

This technique is preferred for products where a pile is to be raised by adjusting the top and bottom air flows. It is mostly used in perforated belt systems. Fig. 10.18 illustrates the arrangement of typical one- and double-sided air jetting units.

10.6 Thermal radiation/infrared and ultrasonic bonding

Thermal radiation has been extensively used in textile finishing processes, in particular for drying. More recently, it has been utilised for bonding nonwoven structures containing thermoplastic fibres. Thermal radiation is the energy radiated/emitted by an object that is at a temperature above absolute zero ($-273°C$). This energy is transported by electromagnetic waves (or alternatively, photons). While the transfer of energy by conduction or convection requires the presence of a medium, radiation does not need any medium, and in fact transfer takes place more effectively in a vacuum. Thermal radiation encompasses a wide range of wavelengths from about 0.1 to 100 μm which includes a portion of the UV and all the visible (from about 0.35 to 0.75 μm) and infrared (IR) spectrum. The energy radiated per unit time and unit area by the ideal radiator (or blackbody) is proportional to the fourth power of the absolute temperature and is given by the Stefan-Boltzmann Law [26]:

$$E_{\text{b}} = \sigma T^4 \tag{10.2}$$

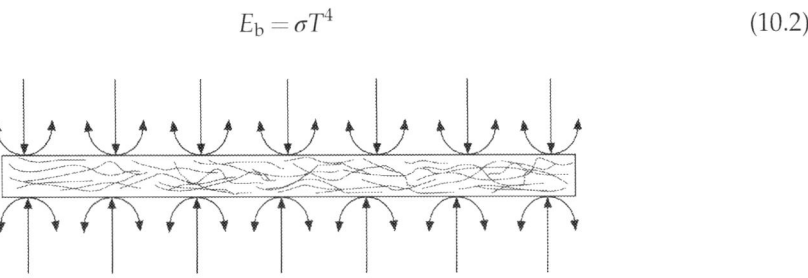

FIG. 10.17 Air jetting method of convection bonding.

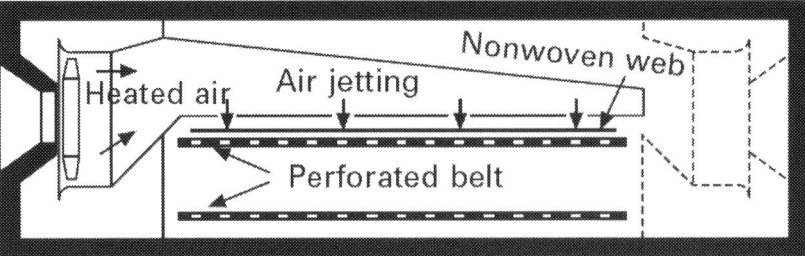

One-sided air jetting

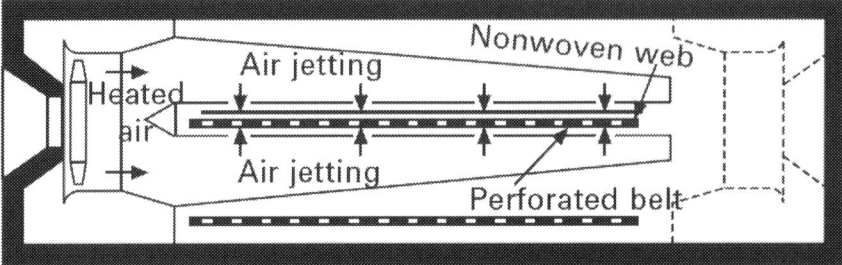

Double-sided air jetting

FIG. 10.18 Typical one-sided and double-sided air jetting systems [25].

where E_b is the heat flux (W/m^2) of a blackbody (ideal radiator), σ is the Stefan-Boltzmann constant (5.67×10^{-8} W/m^2 K^4) and T is the absolute temperature of the object (K). The heat flux emitted by a real surface is less than that of a black body and is expressed in the form:

$$E = \varepsilon \sigma T^4 \tag{10.3}$$

where ε is a radiative property of the surface termed emissivity with values in the range $0 \le \varepsilon \le 1$.

Radiation energy is reflected, absorbed and transmitted by the product. In heating, the aim is to increase the energy absorption by the material. Absorption characteristics of the materials are influenced by the product colour (darker colours tend to be more energy absorbent), surface finish (smooth products reflect more energy) and radiation wavelength. For example, water has a high absorption capacity in the wavelength ranging from 1.8 to 3.5 μm. Therefore, the wavelength ranges must be chosen to achieve maximum absorption by the web, for example, by changing the emitter temperature. The emission spectrum of the emitter cannot be varied as required; therefore, the radiation method is not flexible and has a limited application in thermal bonding.

Fig. 10.19 illustrates a basic arrangement for infrared bonding. The main advantage of this method is that the heat is transferred to the web without disturbing the structure since there is no contact with metal parts or a hot air stream. This allows for bonding of the web without the shrinkage or distortion of fibres associated with other methods of heating. It is also of particular use in powder bonding. The major drawback is the fact that the surface of the web is heated faster than the centre and uniform bonding through the cross-section of the web therefore presents practical difficulties.

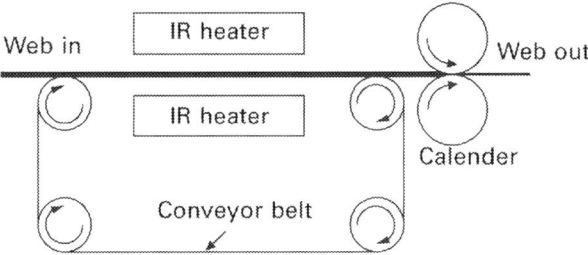

FIG. 10.19   Typical arrangement of infrared bonding process.

10.6.1 Ultrasonic bonding

The application of ultrasonics to thermally bond nonwoven webs, or to laminate layers or fabric is well known. In contrast to other methods of bonding, which use either hot air or hot surfaces, in ultrasonic bonding, thermal energy is conveyed by a mechanical hammering action (high frequency vibration) of the web surface. This involves a brief contact time with a limited pressure between the ultrasonic horn and the web carried out at an ultrasonic frequency of 20,000 cycles/second or more. Energy is therefore transferred to well defined, restricted areas in the web to induce thermal bonding as the mechanical energy applied to the fibres is converted into heat.

A schematic view of the basic ultrasonic bonding process is shown in Fig. 10.20. A fibrous web is compacted between an embossed patterned roller (referred to as an anvil) and an ultrasonic tool (horn). The horn is vibrated at high frequency in the range of 20–40 kHz; the lower limit is normally used for thermal bonding of webs. Materials used for manufacture of vibrating tools (horn) should have a high wear resistance; titanium and aluminium are widely used. The vibration of the horn imparts intense thermal energy to the web

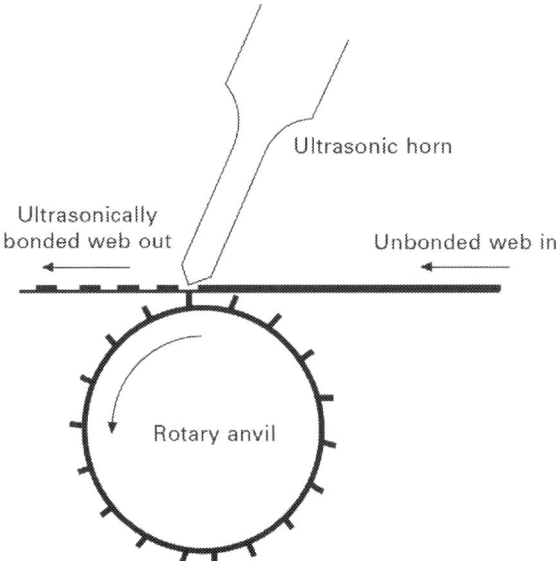

FIG. 10.20   A schematic ultrasonic bonding with raised bonding points.

immediately above the raised points on the patterned roller causing fibres to melt and undergo compression in the area above the raised points.

There is no need for a second binder component when synthetic fibres are used since these are self-bonding. However, to bond natural fibres, about 50% by web weight of binder fibres is usually required. An advantage of this method is that energy is imparted only to the points at which bonding takes place and no energy is transferred to the surrounding unbonded areas of the web. Localisation of bonding energy leads to fabrics with well-defined bond points and an excellent aesthetic appearance. The main problem is transferring uniform vibrational energy across the full width of wide fabrics. This requires many overlapping horns to be used which results in visible lines of overlap in the fabric. Also, ultrasonic horns must be carefully maintained, and the process is slow if reasonably large bond areas are to be obtained of >25%. Fabrics manufactured by ultrasonic bonding are soft, breathable, absorbent and strong. Ultrasonic bonding is also suitable for manufacturing patterned composites and laminates, such as quilts and outdoor jackets.

10.7 Thermally bonded fabric structure

One of the main attributes determining the application of a fabric is tensile strength. It is known that stronger fibres make stronger fabrics when all the other constructional factors are similar. However, in thermally bonded fabrics, it is possible to make a weak fabric using strong fibre if bonding conditions are not set appropriately. The failure mode of fabrics changes with bonding conditions. For example, tensile failure in a light-weight PET (20–30 g/m^2) calender-bonded fabric can be explained by three main mechanisms depending on manufacturing process conditions.

1. Failure of adhesion/cohesion between fibres within the bonding sites (bond point disintegration).
2. Failure of individual fibres, where fibre breakage occurs at the perimeter of bond points where they are attached to the bond points or somewhere along the side of the fibre length between bond points.
3. Fracture of bond points.

The occurrence of the above mechanisms depends on how the fabrics are made. Increasing the bonding temperature increases the likelihood of the failure mode changing from that of the first to the second mechanism. Failure by the third mechanism may occur at very high temperature and in very stiff fabrics [27].

10.7.1 Effect of fibre structure on properties of thermally bonded fabrics

The published research reviewed by Dharmadhikary [10] suggests that changes in fibre structure produced during point bonding impacts the properties of the fabric, and these changes require a comprehensive understanding of the structure and properties of individual polymer types. The role of the binder and base fibre structures is of particular importance. A study of PET area bonded fabrics with respect to the changes in binder morphology [28] suggested that crystalline binder fibres produce fabrics with a higher tenacity, modulus and breaking extension than amorphous binder fibres. In terms of fabric strength, these differences may be difficult to distinguish below a critical temperature and pressure.

Another study by Wei et al. [29] used commercial polypropylene fibres of varying draw ratios. Fibres with a lower molecular orientation compared with fibres with a higher orientation and microfibrillar structure, had a higher tensile strength and flexural rigidity. They suggested that this was due to the less orientated amorphous regions and lamellar crystal structure promoting fibre fusion, while the orientated microfibrillar structure was inhibiting fibre fusion during thermal bonding. Fabric shrinkage was also determined by the fibre morphology; they suggested that this resulted from molecular retraction in the amorphous regions. The highly orientated fibres also showed higher shrinkage, and this resulted in increased fabric thickness.

Point-bonded fabrics consist of a network of fibres, bonded in localised regions (bond points), by the application of heat and pressure. This results in partial melting of the crystals, which is essential for the formation of bond points. Although the point-bond process is simple, a clear understanding of the properties of point-bonded fabrics has yet to be established and there is only limited information about the ideal fibre structures required for point bonding and the changes that occur in fibre structure and properties during bonding. The morphology of the bond points and the bridging fibres is an important influence on properties. Bond strength influences fabric strength and this has been studied by Mi et al. [30]. Results from their model suggest that high strength bonds, (fabric failure caused by failure of the bridging fibres) led to the strongest fabrics. The failure of a fabric will be determined both by the nature and character of the bond points and the stress/strain relationship of bridging fibres.

During point bonding, depending upon the specific process variables that are employed and the bridging fibres, the properties associated with the bond point differ from those of the virgin fibre. Various workers have referred to this aspect, Warner [12] suggested that fibres break at the bond periphery because of the local thermo-mechanical polymer history and that the strength of point-bonded fabrics is therefore controlled by the bond periphery strength. Wei et al. [29] commented that the physical properties of thermally bonded fabrics are a result of the nature and quality of the bonding regions. The influences of pressure on polymer properties during point bonding are not well understood. Pressure might be expected to increase the melting point and the glass transition temperature, and therefore might influence the crystallisation rate. Crystal nucleation and growth is also influenced by pressure and could produce complex interactions.

10.8 Applications of thermally bonded fabrics

There are now a vast number of different end-uses, almost too numerous to mention, for thermally bonded nonwoven fabrics, across all sectors of the industry, from single use hygiene products to long life applications such as in building and construction. A major application is in the manufacture of light-weight fabrics for hygiene, such as spunbond PP coverstocks for sanitary and incontinence products of $7-30 \, g/m^2$, based on calender point-bonded fabrics. Other single use products include wipes produced from airlaid short fibres, which are through-air bonded to make fabrics in the 25–150 g/m^2 weight range, and through-air bonded carded wipe products of about 100–250 g/m^2. Through-air bonding in this application is preferred to maximise the bulk of the fabric. Thermally bonded wipe substrates have seen strong competition from carded-hydroentangled fabrics. Some wetlaid fabrics intended for tea-bag applications are through-air bonded. Calender bonding is routinely utilised in the manufacture of spunbond and SMS fabrics for numerous applications, including surgical gowns and drapes. Durable thermally bonded products produced from spunlaid webs can be either through-air bonded or calendered. Those using through-air bonding in an oven are frequently based on bicomponent filament webs. Sheath–core PET-PA6 filament spunbonds have applications in carpet backing whereas sheath–core PE-PP spunbonds have applications in geosynthetics and as air and water filtration media, horticultural products, and in clothing and footwear. Thermally bonded spunbond geosynthetic fabrics of 80–250 g/m^2, find uses in various civil engineering applications. Roofing felts or carriers of about 150–350 g/m^2 are an application for thermally bonded spunlaid fabrics whereas needlepunched fabrics are used as bitumen carriers in flat roofs. Thermally bonded spunlaid fabrics are installed in pitched roofs as bituminous underslating. Such fabrics are normally composed of PET partly to provide better heat stability during bitumen coating. Through-air bonded spunbond fabric of 150–200 g/m^2 is also produced for carpet backing applications.

Many drylaid filter fabrics are needlepunched, particularly those intended for high-temperature applications, but others are through-air bonded in weight ranges from 100 to 1000 g/m^2. A traditional thermal bonding application is in the manufacture of linings and interlinings ranging in weight from 25 to 150 g/m^2 using either calender or through-air bonding methods. In some applications, point bonding using a calender follows mechanical bonding to produce the final product. A variety of high loft insulation fabrics and waddings are produced by airlaid web formation and through-air bonding to ensure thickness and volume in the final product. Powder and thermo-dot bonded, fusible and nonfusible nonwoven fabrics for garment interlining applications, from polyester and EVA are also made. Shoe lining fabrics of about 150–200 g/m^2 containing bicomponent fibres in blends are thermally point bonded with a calender.

Drylaid, thermally bonded reinforcement fabrics, and substrates for electrical insulating materials, pressure sensitive tapes and filtration membranes are made from PET. Other applications include furniture and bedding components, horticultural and agricultural fabrics including crop cover and nonwovens used by the automotive industry. Spunlaid filtration fabrics stabilised by thermal bonding find applications in cabin filters in the automotive industry and are usually pleated to increase the surface area available for filtration.

References

[1] A. Watzl, Fusion bonding, thermobonding and heat-setting of nonwoven—theoretical fundamentals, practical experience, market trends, Melliand Engl. 10 (1994) E217.
[2] S.K. Batra, B. Pourdeyhimi, Thermal Bonding, Nonwovens Cooperative Research Center, North Carolina State University, Raleigh, NC, 2002.
[3] H. Rave, M. Schemken, A. Beck, State of the art of bicomponent staple fibre production, Chem. Fibers Int. 52 (2002) 52–58.
[4] . http://www.ifg.com/issue/june98/story3.html.

[5] R. Jeffries, Bicomponent Fibres, Merrow Publishing Co. Ltd, 1971. BP 1048370, NAP 66-12238, Shell International Research.

[6] B. Marcher, Tailor-made polypropylene and bicomponent fibres for the nonwovens industry, TAPPI J. 74 (12) (1991) 103–107.

[7] F. Thonnessen, J. Dahringer, Trevira bicomponent fibres for nonwovens, Chem. Fibers Int. 53 (12) (2003) 422.

[8] P. Raidt, Polyester/Polyethylene Bicomponent Fibres for Thermal Bonding of Nonwovens, Index 87 Congress, 1987.

[9] T.S. Hoag, From time-tested methods to recent innovations, bonding exhibits versatility, Nonwovens World 4 (1) (1989) 26.

[10] P.K. Dharmadhikary, T.F. Gilmore, H.A. Davis, S.K. Batra, Thermal bonding of nonwoven fabrics, Text. Prog. 26 (1995) 26.

[11] D.H. Muller, Improvement of thermalbonded nonwovens, Melliand Textilber. 70 (7) (1989) 499–502. E210.

[12] S.B. Warner, Thermal bonding of polypropylene fibres, Text. Res. J. 59 (3) (1989) 151–159.

[13] R.J. Schwartz, US Patent 4100319, 1978.

[14] R. Haoming, G.S. Bhat, Preparation and properties of cotton-ester nonwovens, Int. Nonwovens J. 12 (2) (2003) 55.

[15] W.K. Kwok, J.P. Crane, A. Gorrafa, Polyester staple for thermally bonded nonwovens, Nonwovens Ind. 19 (6) (1988) 30–33.

[16] H.S. Kim, B. Pourdeyhimi, P. Desai, A.S. Abhiraman, Anisotropy in the mechanical properties of thermally spot-bonded nonwovens: experimental observations, Text. Res. J. 71 (11) (2001) 965.

[17] D.H. Muller, How to improve the thermal bonding of heavy webs, INDA J. Nonwovens Res. 1 (1) (1989) 35–43.

[18] C.J. Shimalla, J.C. Whitwell, Thermomechanical behaviour of nonwovens, part i: responses to changes in processing and post-bonding variables, Text. Res. J. 46 (1976) 405–417.

[19] V. DeAngelis, T. DiGioacchino, P. Olivieri, Hot calendered polypropylene nonwoven fabrics, in: Proceedings of 2nd International Conference on Polypropylene Fibres and Textiles, Plastics and Rubber Institutes, University of York, England, 1979, pp. 52.1–52. 13.

[20] J.W. Mayer, R.N. Haward, J.N. Hay, Study of the thermal effects of necking of polymers with the use of an infrared camera, J. Polym. Sci. Polym. Phys. Ed. 18 (1980) 2169–2179.

[21] B. Wunderlich, Macromolecular Physics, vol. 3, Academic Press, New York, 1986.

[22] D.S. Gunter, Calender selection for nonwovens, Tappi J. 81 (1) (1998) 208.

[23] D.S. Gunter, Thermal bonding utilising calender, TAPPI J. 77 (6) (1994) 221.

[24] E.L. Wuagneux, Full of hot air, Nonwovens Ind. 30 (4) (1999) 52–56.

[25] A. Watzl, Instruction Manual of Fleissner Through Air Bonding Machinery, Internal Fleissner Machine Document, 2003.

[26] J.P. Holman, Heat Transfer, ninth ed., McGraw Hill, 2002.

[27] P.E. Gibson, R.L. McGill, Thermally bondable polyester fibre: the effect of calender temperature, TAPPI J. 70 (12) (1987) 82.

[28] S.R. Dantuluri, B.C. Goswami, T.L. Vigo, Thermally bonded polyester nonwovens: effect of fibre morphology, in: Proceedings of INDA Technical Symposium, 1987, pp. 263–270.

[29] K.V. Wei, T.L. Vigo, B.C. Goswami, Structure-property relationship of thermally bonded polypropylene nonwovens, J. Appl. Polym. Sci. 30 (1985) 1523–1524.

[30] Z.X. Mi, S.K. Batra, T.F. Gilmore, 'Computational Model for Mechanical Behaviour of Point-Bonded Web' First Annual Report, Nonwovens Cooperative Research Center, 1992.

CHAPTER

11

Finishing of nonwoven fabrics

M.J. Tipper and R. Ward

Nonwovens Innovation & Research Institute Ltd (NIRI), Leeds, United Kingdom

11.1 Introduction

It is possible to enhance the technical performance, appearance, and aesthetics of nonwoven fabrics after they have been produced, using a variety of finishing processes. Fabric properties can therefore be modified at a late stage after roll good formation, providing opportunities to diversify the product offering and customise performance according to end-user requirements. The majority of nonwovens find applications in products other than clothing outerwear, and consequently technical performance, rather than appearance, is usually the major driver in determining economic value.

The critical element in selecting appropriate nonwoven finishing processes is a correct definition of the required performance specification and an understanding of the unmet needs of end users. A common challenge in manufacturing is understanding the needs of the ultimate end user of the product, as well as responding to the requirements of the immediate customer in a supply chain [1]. It has been estimated that 70%–80% of new product development fails, not for lack of advanced technology, but because of a failure to understand end-user needs [2]. Shifting from producer-centred (technology push) to user-centred innovation can be an effective strategy to produce more commercially successful products, but there are many examples of users' not knowing what they need, until new products become available – a number of sustainability innovations fall into this category [3].

Fundamentally, the engineering of nonwovens depends on a careful consideration of their fundamental building blocks, including the (i) chemical composition, (ii) fibre/filament properties, (iii) fabric manufacturing conditions, and (iv) structural architecture. In general terms, chemical composition relates to the chemistry of the materials selected to make the fibre/filament, e.g., polymers, metals, ceramics, glasses, or metalloids, as well as any chemical additives, fillers, finishes, dyes, coatings, or other chemistry added during any part of the manufacturing process. Fibre/filament properties are of course heavily dependent not only on their chemical composition, but also on their selected dimensions, surface morphology, and cross-sectional shape. Fabric manufacturing conditions relate to the specific nonwoven processing routes and machine settings that are selected for web formation and bonding, as well as any later-stage processes. Finally, the multiscale structural architecture of the final fabric, which relates to its geometric and dimensional characteristics, is largely governed by the choices that are made around how the fabric is manufactured, including the selection of process conditions.

In this chapter, some of the major processes for nonwoven finishing and late-stage functionalisation are discussed, where the objective is to enhance final fabric performance and ensure fitness for purpose.

11.2 Finishing of nonwoven fabrics

The physical properties and performance of nonwoven fabrics is, of course, largely determined by the selected fibre/polymer composition, as well as by the manufacturing processes used to make the fabric. Once a manufacturer has committed to using particular raw materials and manufacturing processes in the production of a nonwoven fabric, the extent to which final fabric properties can be modified is significant, but is subject to limitations.

Investment in new nonwoven web forming and bonding lines is highly capital intensive, and therefore having the ability to modify fabric properties at a later-stage, after fabric production, through finishing, can be highly attractive, as a means to securing a competitive advantage. Finishing also provides an important way of differentiating products made from the same base fabric, in terms of technical functionality, appearance and aesthetics, as well as customising performance to the needs of different markets. In essence, this gives manufacturers greater agility when responding to new market opportunities and enables them to add value to the product offering, improving competitiveness.

Therefore, the advantages of nonwoven fabric finishing include:

- Reduced investment costs compared to the installation of new nonwoven roll good manufacturing equipment.
- Ability to modify or enhance fabric properties without changing or interrupting existing nonwoven production lines.
- Ability to modify or enhance fabric performance, tailored to different end-user requirements.
- Ability to modulate fabric properties without modifying the fibre composition of the fabric.

By conferring the required fabric performance during finishing, rather than relying entirely on fibre selection prior to nonwoven fabric manufacture, it is possible to substantially increase margins without compromising overall performance. One example of this strategy is the use of wetting agents on inherently hydrophobic PP spunbonds to enable rapid liquid inlet in diaper topsheets. Many different properties and performance attributes can be controlled by means of fabric finishing treatments, some examples of which, together with applications, are listed in Table 11.1.

Nonwoven finishing processes are diverse, encompassing dry- and wet-state treatments, as well as gaseous application processes. In dry finishing, the fabric is often subjected to

TABLE 11.1 Examples of fabric performance attributes enhanced through fabric finishing.

Performance attribute (examples)	Applications (examples)
Gas adsorption/permeability	Filtration, packaging, protective clothing
Oleophilicity/oleophobicity	Protective clothing, filtration
Chemical stability/resistance	Filtration, protective clothing
Hydrophilicity/hydrophobicity/wetting	Diapers, wipes, coalescing filters
Frictional behaviour	Brake pads, wipes
UV stability/resistance	Geosynthetics, crop cover
Antimicrobial behaviour	Wipes, protective clothing, filtration
Thermal stability	Fire protection, arc protection
Odour control/fragrant behaviour	Wet wipes, diapers, ostomy pouches
Abrasion resistance/resistance to linting	Floorcoverings, wipes, wound dressings
Bending rigidity	Pleated filter media, interlinings
Temperature regulation	Performance apparel, packaging
Tactile properties, including softness	Femcare, continence management products
Acoustic absorption and transmission	Automotive, building and construction
Thermal resistance and insulation	Clothing, building and construction
Flame retardancy	Mattresses and upholstery, protective clothing
Opacity/transparency	Food packaging, e.g. tea/coffee bags
Electrical conductivity/electrostatics	Protective clothing, filtration
Electromagnetic response/interference	Automotive/consumer electronics

entirely mechanical processes to modify fabric properties, whereas in wet finishing, liquid treatments are involved. Traditional textile dyeing and finishing processes are also employed in the production of some nonwoven products, including the ubiquitous techniques of impregnation, coating, calendering, and lamination.

11.2.1 Application of chemical finishes

Chemical finishes are topically applied or impregnated into nonwoven substrates to enhance performance. Clearly, the resulting modification in fabric properties is influenced by the chemistry and concentration of the chemical compounds involved, but also by the method by which they are applied. Not all chemical finishes are durable to wet treatment, washing or other agencies of wear, but this is not necessarily problematic, particularly when applied to single-use products. A large variety of chemical finishes are applied to nonwoven fabrics to enhance physical, chemical, and biological properties, whether they relate to bulk properties (e.g. tensile strength, flexural rigidity, or flammability) or surface properties (e.g. liquid wetting, electrostatic, or frictional behaviour). Some of the most commonly encountered chemical finishes are briefly reviewed.

11.2.1.1 Antistatic agents

Resistance to electrostatic charging is important for a variety of reasons, including to resist soiling in needlepunched floorcoverings, to minimise fouling in dust filter media and to ensure safety and comfort in protective clothing. The mechanism by which antistatic agents operate varies depending upon their chemistry. One group of antistatic compounds increase electrical conductivity through hydrophilic compounds applied to fibre surfaces, dissipating charge through the presence of moisture. Another group work by imparting a charge opposite to that normally generated, neutralising the build-up. Antistatic chemical finishes are available in durable, semidurable, and nondurable forms.

11.2.1.2 Antimicrobial finishes

In simple terms, microorganisms include a broad spectrum of bacteria, fungi, algae, protozoa, and viruses. Antimicrobial finishes are applied where protection is required from biological degradation, or where there is a risk of contamination, resulting from the growth of undesirable or pathogenic organisms. Strategies to combat microbial activity associated with for example, bacteria, include inhibiting adhesion and growth (bacteriostatic) and killing the organism (bactericidal). The latter relates to finishes capable of providing disinfectant or antiseptic function, but only certain chemistries have broad spectrum effects across all types of microorganisms.

Traditionally, the primary objective of antimicrobials has been to prevent physical degradation of cellulosic fabrics and the odour associated with microbial activity, e.g. mildew. Proprietary formulations, metallic compounds and even natural biopolymers such as chitosan, derived from crab shells have been developed for use as antimicrobials.

Common biocides in antiseptics and disinfectants that have been combined with nonwoven substrates for use in applications such as wipes and other cleaning products include: alcohols (most commonly ethanol or isopropanol), biguanides (e.g. chlorhexidine), halogen-releasing agents (e.g. iodine compounds), peroxygens (e.g. peracetic acid), and quaternary ammonium compounds or 'quats' (e.g. benzalkonium chloride). However, the selection of disinfectants has to be carefully considered in light of the intended mode of use of the final nonwoven product. For example, in nonwoven surface and infection control wipes as well as some woundcare products, the potential for skin-contact related irritation or sensitisation should be considered. Similarly, the potential for progressive degradation of polymeric surfaces to be cleaned is a further concern, particularly in healthcare settings.

Aqueous dispersions of chlorinated phenoxy-compounds containing pyrithione are effective against bacteria, algae, yeast, and fungi and can be applied to fabrics other than those composed of polypropylene. Nonionic AOX-free synthetic biocides such as isothiazolinones confer excellent fungiostatic and bacteriostatic properties but are restricted in the EU because of potential for skin irritation. More recently formulations containing monoatomic state silver and copper have been employed because of their inherent low toxicity and increasing prevalence of bacterial resistance to antimicrobials [4]. An example of a sliver containing finish is Ultra-Fresh Silpure FBR-5 (Thomson Research Associates Inc.), which is reported to be effective against Gram-positive *Staphylococcus aureus* [5]. Nonwovens with antimicrobial finishes find applications in insulation fabrics, sportswear, mattress ticking and bedding components, domestic furnishings, floor and wall coverings, hygiene, woundcare, and other healthcare products.

Antiallergenic nonwoven fabrics can be finished with antibacterial compounds designed to reduce the bacterial/fungal growth on which the dust mites feed. Avgol Nonwovens developed the beneFIT Control product for skin contact hygiene applications. The technology is reported to preserve natural skin flora whilst protecting against troublesome organisms [6]. Outside wipes applications, the importance of viricidal nonwoven fabrics for PPE such as face masks and products such as hospital curtains surged during 2020/21 as a result of the COVID-19 pandemic. Some antibacterial formulations are also effective against viruses, particularly those based on ionic metals. Viraloff (Polygiene), is one such formulation, claiming to reduce SARS-CoV-2 by 99% within 2h when tested to ISO18184:2019 [7]. Another antiviral formulation, HeiQ Viroblock NPJ03, combines silver and a vesicle technology. The technology uses nonphospholipid lipid vesicles to inactivate enveloped viruses by depleting the viral membrane, allowing the silver ions to attack the core of the virus [7,8].

11.2.1.3 Lubricants

Lubricants are applied to nonwoven fabrics for several reasons, most commonly to reduce fibre to fibre, and fibre to metal friction in subsequent processing and converting of fabrics into final products. Examples include high-speed sewing, e.g. for nonwoven interlinings, and for nonwoven primary carpet backings (normally spunbonds), where lubrication can improve compatibility with the tufting process.

11.2.1.4 Wetting agents

Wetting agents are topically applied to hydrophobic nonwoven fabrics, such as those made of PP, to enhance their liquid wetting and transport characteristics. These finishes may be temporary, durable, or semidurable depending on the final application. They are usually applied by lick coating, impregnation or spraying to either, or both faces of the fabric. Surfactant-based finishes act by reducing the solid–liquid interfacial tension as well as the gas–solid interfacial tension. The main application for these finishes is to reduce the liquid inlet time of nonwoven topsheets for single-use diaper and feminine hygiene products made from hydrophobic polymers. Such finishes are usually fugitive, i.e. they are partially washed off with each liquid insult during use of the product, resulting in an increase in topsheet hydrophobicity, which then inhibits rewet back through the topsheet. Cirrasol PP844 (Croda) is one example of a nonwoven fabric finish for diapers, adult incontinence and feminine hygiene products. Applied directly to the topsheet, it promotes rapid strikethrough of the liquid insult through the nonwoven into the absorbent core below [9]. ADLs (acquisition distribution layers) in baby diapers and continence management articles also require finishes to optimise liquid transport. In another example, HANSA PP 3 [10] promotes liquid transport to all areas of the absorbent hygiene product enhancing core utilisation. It is approved by several leading converters for this application [10].

11.2.1.5 Flame retardant finishes

In addition to the selection of inherently flame retardant fibres and additives prior to nonwoven production, flame retardant finishes applied to nonwoven fabrics provide opportunities to enhance performance. High temperature resistance and flame retardancy are important in numerous applications, including nonwoven fabrics used by the automotive industry inside and proximal to the engine bay and exhaust system, furniture and mattress components, insulation and fire protection materials for buildings and personal protective garments and workwear, such as those used by fire-fighters and the military. The purpose is normally either to prevent the onset or propagation of fire, or smoke and to protect people, products, or infrastructure in the event of fire.

Flame retardant finishes can be applied to reduce flame propagation, afterglow and charring and the suppression of toxic smoke emission. Flame retardance can be conferred either by physical or chemical means, as well as combinations of the two. In simple terms, physical approaches can involve the formation of a protective layer across the surface of the fabric, and cooling and dilution with inert substances, whilst the chemical approach can exploit reactions in the condensed or gas phase of combustion [11]. Flame retardant formulations are designed specifically for fibres and are commonly applied with binders or catalysts to promote durability.

Back coating of organic phosphorous salts is used in automotive fabrics to impart flame retardancy in combination with a polymer dispersion coating. In fabrics intended for use by the construction industry, inorganic mineral fillers act as flame retardants and can be padded or coated in conjunction with a polymer dispersion coating. Common finishes for cellulosic fabrics are based on nitrogen–phosphorous compounds, often applied with hygroscopic auxiliaries which help to suppress flammability. The Proban process (Solvay) utilises phosphorous organic compounds with an ammonia after cure to cross link the applied polymer in the core of the fabric. Problems with flame retardant finishes can include yellowing of the fabric, reduced tensile strength, and colour change. Environmental concerns about some traditional flame retardant chemistries have led to progressive development of alternative formulations. For example, Archroma's Pekoflam Hfc is a powder-based flame-retardant additive for coating applications that meets the Oeko-Tex Standard 100. This confers nonhalogenated fire protection for polyamide-based fabrics, based on organic phosphorous/nitrogen chemistries. Another example is Eco-flam (Devan Chemicals), which is a range of heavy metal free, flame retardants suitable for nonwoven applications. Bio-flam (Devan Chemicals) finishes are bio-based FR treatments derived from renewable sources and can be applied to cellulosic nonwovens.

11.2.1.6 Hydrophobic finishes

In some applications, nonwovens are required to act as barrier layers, preventing the penetration of aqueous liquids. This can be achieved by various means including application of hydrophobic compounds and polymers onto nonwoven substrates, by means such as coating, to create a continuous or microporous polymeric film layer, by liquid impregnation, foam impregnation or spraying. Traditionally, hydrophobic finishes have relied heavily on silicone and fluorocarbon-based formulations, applied as aqueous dispersions. Fluorocarbon finishes produce very low surface tension on the fabric surface, inhibiting not only aqueous wetting, but also oil, diesel, and petrol on glass and synthetic fibres. Durable water repellent (DWR) technology, developed by the textile industry has improved the wash resistance of waterproof finishes, which may be applied using high-pressure curing to significantly reduce water use [12]. Environmental and safety issues associated with PFOA (perfluorooctanoic acid), a known carcinogen associated with fluorocarbons, have accelerated the search for alternative chemistries. Whilst fluorocarbon producers have improved safety by transitioning from C8 to C6 carbon chain formulations, to reduce the risk of bio-accumulation, significant research and development work is focused on developing alternative chemistries. Today, even after many years of development, few alternative formulations offer the same level of performance as fluoropolymers, including in terms of oil and chemical resistance.

11.2.1.7 Softeners

The perception of softness is important in nonwovens for skin contact applications, including topsheets for feminine hygiene and diapers, as well as wipes, and is commonly imbued by the application of nondurable or semidurable topical finishes. As with all skin-contact applications, the chemical composition of the finish is formulated with consideration of biocompatibility and sensitisation. Hydrophilic softeners, often known as re-wetters, also increase the wettability of hydrophobic nonwoven fabrics, such as PP spunbonds. Emulsion and micro-emulsion-based silicone softeners have also been extensively applied as softening agents for use in nonwovens.

11.2.1.8 Stiffening agents

Most nonwovens are relatively flexible with little resistance to buckling forces, although they do not drape or conform on three-dimensional surfaces in a comparable manner to conventional textile materials. Whilst flexibility is valuable in consumer applications such as wipes, many other products require dimensional stability, stiffness, and high flexural rigidity.

Applications of stiffened nonwovens include footwear components, such as shoe counter and toe puffs, internal components of luggage such as bags and suitcases, interlinings, floorcoverings, and filters. Stiffening agents, adhesive binders, and particulate fillers can be applied to add weight and stiffness to nonwoven fabrics. Polymer dispersions are applied by padding, curtain coating, foam, spray, and knife coating. Polymers with high glass transition temperatures (t_g) are inherently stiff at room temperature, and polymer dispersions can therefore be selected to modulate the stiffening effect, as well as harnessing cross-linking chemistry. Micro-dispersed anionic polystyrene copolymer is applied to needlepunched carpet fabrics as a stiffener. Self-crosslinking anionic acrylate polymers are applied to nonwovens to improve dimensional stability and wash resistance. Glass fabrics and polyester spunbonds are sometimes finished with self-crosslinking anionic polystyrene acrylate dispersions, which have excellent thermal resistance and dimensional stability. Thermoplastic binders are important in the manufacture of nonwoven fabrics intended for moulding operations. At present, conventional binders are predominantly made of petroleum-based compounds, rather than renewable materials. OC-biobinder (Organoclick) is one example of a fibre binding system produced from modified biopolymers and natural plant compounds.

11.2.1.9 UV stabilisers

Protection from UV light is important for nonwovens made of certain polymers that are prone to degradation when exposed to the natural environment unless prestabilised by masterbatch additives before the fibres are extruded. Applications where UV stability is important include, for example, crop cover as well as other fabrics used in horticulture and agriculture, vehicle interiors and covers, artificial playing surfaces and geosynthetics, which can be left out on site prior to installation. Crop cover constructed from PP spunbonds without any UV stabilisation has the potential to suffer deterioration in mechanical properties in a matter of months, depending on the extent of UV exposure. Tensile, tear, and bursting strength are all adversely affected by associated changes in fibre molecular and supermolecular structure [13]. The level of protection required depends on the incident UV exposure, which varies depending on geographical location, and UV formulations have to be tailored accordingly. If UV stabilisers are not added to the polymer prior to extrusion, then protection may be applied through fabric coatings. Some formulations are based on UV absorbers. Hindered amine light stabilisers (HALS) do not absorb UV but enable a complex reaction that protects the polymer from chemical breakdown. For adhesive polymers, edges exposed to light can degrade with loss of adhesive properties, even if shielded between fabric layers.

11.3 Wet finishing methods

The majority of chemical finishes are applied to nonwovens in the form of a treatment liquor or foam (containing less liquid than a liquor), followed by drying and then, in some cases, curing. In the majority of cases, the liquid consists of an aqueous medium in the form of a solution, dispersion, or emulsion, depending on the nature of the treatment chemistry. Padding and coating techniques are the most commonly encountered methods of applying chemical finishes, and each encompasses a variety of processing methods.

11.3.1 Padding and wet impregnation

Padding involves impregnating fabrics with the treatment liquor or foam containing the chemical finish, as well as other additives or auxiliary chemicals. The treatment liquor or foam is forced through the fabric and into the internal void structure, using at least one set of squeeze rollers. Excess liquid is removed in the squeeze roller nip, and a predetermined pick-up level is thereby applied to the fabric. To prevent excess liquid accumulating at the fibre interstices, pick-up levels are usually targeted below the saturation point, which is economically attractive since less liquid is consumed in the process. Once the liquor has been applied, the challenge is to prevent significant migration of the chemical finish during the subsequent drying process, where the water is evaporated. Even if initially, the finish is uniformly distributed through the fabric cross-section after impregnation, drying can result in migration of the finish to the outer surfaces as a result of capillary effects. This is particularly noticeable in excessively wet or heavy-weight fabrics. Of course, in some cases it may be preferred to localise the distribution of the finish to one or both surfaces of the fabric, and in such cases, migration effects can be exploited, together with appropriate modifications to the viscosity of the treatment liquor.

Fabrics can be processed using wet-on-dry or wet-on-wet padding methods, depending on the state of the fabric prior to processing. In wet-on-dry padding, the dry fabric is saturated with the treatment liquor. In this case, limited time is available to exchange the air volume that is present within the voids of the fabric for the treatment liquor, meaning achieving a uniform treatment can be challenging. Wetting, de-aerating or de-foaming agents may therefore be required to improve treatment uniformity. In the wet-on-wet padding method, the fabric is prewetted, but the water needs to be exchanged for the treatment liquor to ensure the required addition level can be achieved.

Padding is a continuous process, and the residence time of the fabric in the nip of the pad rollers is relatively short. In addition to the treatment liquor formulation and roller speeds, the roller design and the choice of surface covering can have a significant impact on the uniformity of the treatment. Owing to deflection, as the roller width increases, maintaining a uniform pressure along the full length of the squeeze roller nip, and therefore, a uniform pressure on the fabric during padding can be challenging. This has been addressed by means such as cambered rollers to provide uniform squeezing pressure at the ends, and in modern padding systems, roller pressures can be readily controlled to provide uniform addition levels.

The required addition level for wet-on-dry padding can be calculated as follows:

$$\text{Addition level}\,(\%) = \frac{1000}{\text{liquor pick up}\,(\%)} = \text{g product/litre of pad solution} \qquad (11.1)$$

Therefore, for an addition level of 1.5%, at a pad liquor pick up of 80%:

$$1.5 \times \frac{1000}{800} = 18.8\,\text{g}\,\text{L}^{-1}\,\text{product required per litre of pad solution} \qquad (11.2)$$

Padding is normally followed by drying and curing of the treated nonwoven fabric, commonly by means of an oven, including through-air dryers (TADs), or sometimes, can dryers. Significant energy costs can be associated with such drying processes, because of the quantity of water that has to be evaporated, and if the weight of water carried over is too high, migration of the finish during drying can lead to treatment uniformity issues. Clearly, the amount of water carried over into drying for hygroscopic nonwoven substrates is usually much higher than for hydrophobic nonwoven substrates, where a large proportion of the interstitially-held water can be mechanically removed, e.g. through squeeze rollers. Foam-based, rather than liquor-based treatments are also attractive, as well as other methods aimed at reducing liquor ratios to enable finishing using more concentrated finish formulations. However, in low liquor ratio scenarios, the main challenge is ensuring uniformity of the finishing treatment.

11.3.2 Coating

Nonwovens are used as coating substrates for a variety of short and long-life nonwoven product applications. Coating normally takes the form of a continuous process, and is an effective means of enhancing the quality and functional performance, as well as increasing the value of nonwoven fabrics. Coating processes also form the basis for subsequent lamination of two or more layers of nonwoven fabric, as well as combinations with other sheet-like substrates such as films and foils. The applied coating formulation can be in various forms depending on its chemical composition, including for example, liquid dispersions and emulsions, plastisols, polymer solutions, foams, gels, molten polymers, or dry particles.

A variety of different coating processes have been developed to suit each form, followed by subsequent drying or curing steps (for aqueous coating formulations). Coating of nonwovens is carried out either to enable subsequent manufacturing processes to be carried out or to enhance fabric physical properties and performance. For example, to enable subsequent lamination, nonwovens can be coated with a molten adhesive thermoplastic polymer, or self-adhesive coating. Another example is in the production of multilayer products such as PVC floorcoverings, where a nonwoven carrier fabric based on a glass fibre nonwoven or a PET spunbond has to be laminated to a backing layer.

The extent to which the coating can be localised on one or more of the fabric surfaces depends on a number of factors, such as the viscosity of the liquid formulation during coating, film forming behaviour, surface energy of the coating and fibre surfaces, applied pressure during the coating process, the drying and curing conditions (if applicable) and the porosity of the nonwoven substrate. To achieve thick or heavy coatings, successive layering may be required, assuming there is sufficient interlayer adhesion. The chemical formulation, type of coating method, the required thickness and weight of the coating and number of coating layers can all be adjusted depending upon end-use requirements.

Direct coating is normally conducted with the nonwoven in open width and requires the fabric to withstand applied tension imposed during continuous processing. A number of different coating techniques are employed including lick or kiss roll coating, knife-over-roll coating, floating-knife coating, as well as combinations of different methods.

11.3.2.1 Kiss-roll coating

In kiss-roll coating, also known as lick roll coating or slop padding, the nonwoven is passed over a rotating roller, referred to as a kiss roll (or roller) (Fig. 11.1). The kiss roller is partially immersed in a 'float' containing the liquid coating formulation, such that the exposed roller surface is continuously covered by a thin liquid film that is transferred directly on to the surface of the nonwoven fabric. In the majority of cases, kiss-roll coating is used to apply thin coatings, using low viscosity liquid formulations. The pick-up and degree of penetration into the fabric is controlled by the liquid viscosity, roller speed, and roller direction. A scraper bar or alternatively Meyer bars may be employed to remove any excess and ensure the requisite final coating thickness is achieved. Meyer bars are helically wound wire metering rods, the groove profile in which dictates the residual pick-up and coating thickness. Some kiss-roll

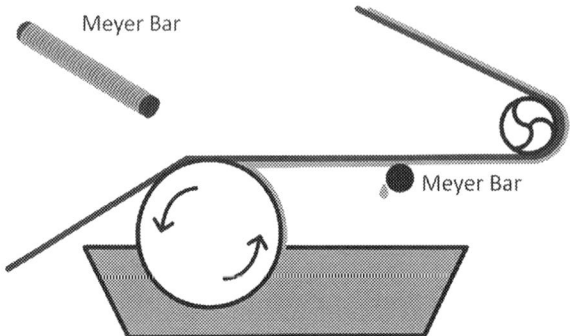

FIG. 11.1   Kiss-roll coating.

systems operate with a support roll above the lick roll, allowing pressure to be exerted on the fabric during pick-up. To enable coating of both sides of the fabric, knife-over-roller or floating knife coating with additional coating heads can be combined with kiss-roll coating systems. Kiss-roll coating is suited for coating weights in the range of ca. $25–500\,\mathrm{g\,m^{-2}}$.

11.3.2.2 Knife coating

Knife coating employs a height and angle adjustable knife or doctor blade, operating rather like a squeegee, to uniformly distribute the coating formulation across the width of the fabric and control the final coating thickness following application onto the fabric. Adjustment of the coating thickness principally relies on the gap setting between the fabric and the end of the knife, but the knife edge profile (sharp or rounded) and operating angle relative to the fabric can also be adjusted. Several knife designs are available depending upon the end-use application and generally, sharp edges leave a thin film layer, whereas rounded knives provide a heavier coating with slightly deeper penetration. Knife coating is most suited to high viscosity coating applications or foams. The knife is normally positioned directly over the top of the fabric and a support roller (knife-over-roll coating) (Fig. 11.2), or directly over the top of the fabric, carried between support rollers (floating knife coating) (Fig. 11.3). Knife-over-roll coating is applicable for coating weights of about $50–1200\,\mathrm{g\,m^{-2}}$ and $30–250\,\mathrm{g\,m^{-2}}$ for floating knife coating.

11.3.2.3 Reverse roll coating

Reverse roll coating employs an applicator roll that rotates in the opposite direction to the moving fabric (Fig. 11.4). It is a versatile process in terms of the range of coating weights and thickness that can be produced, combined with excellent coating uniformity. At least three rollers are arranged vertically, or in an s-shape, however, four roller configurations and pan fed configurations are also common. The coating formulation is applied to the nonwoven by an application roller as it passes around a support roller. The coating thickness presented by the application roller is controlled by the gap (nip volume) between the metering roller and the application roller. When operating with thixotropic-coating formulations, the viscosity can be affected by the shear forces created in the nip between the metering and application rollers.

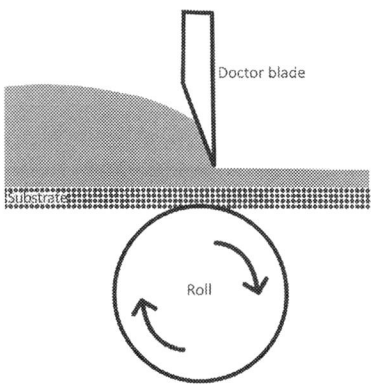

Doctor blade

Substrate

Roll

FIG. 11.2   Knife-over-roll coating.

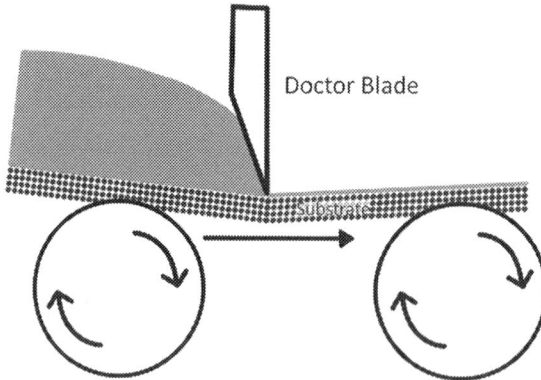

FIG. 11.3   Floating knife coating.

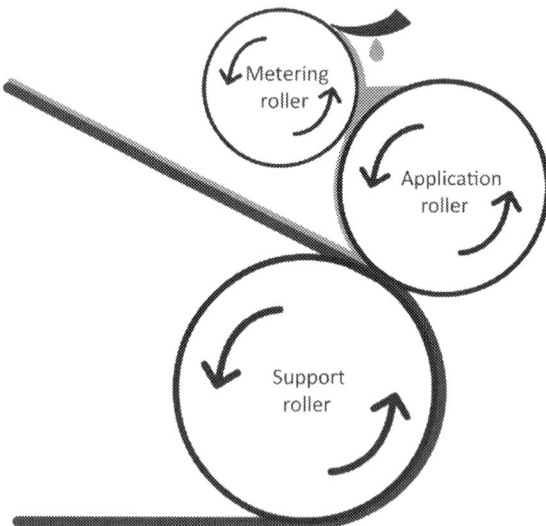

FIG. 11.4   Reverse roll coating.

11.3.2.4 *Magnetic roll coating*

Magnetic roll coating involves controlling the nip pressure between the support and applicator rolls using a magnetic loading system (Fig. 11.5). This is particularly valuable as a means to ensure the creation of uniform and thin coating weights ($10–300\,\mathrm{g\,m^{-2}}$), particularly over wide coating widths of 5 m or more. The coating formulation can be delivered in different ways, including directly by means of a transfer or kiss roll, or by means of single or double screen coating combined with magnet roll rods or magnet knife. The Zimmer Magnoroll/Magnoknife coating system is one example, that is often used for decorative patterning and printing, finishing and coating, paste dot applications, and adhesive applications due to its versatility. In addition to direct roll coating (Magnoroll), the coating formulation can also be delivered from within a perforated rotary screen roller, operating with a support

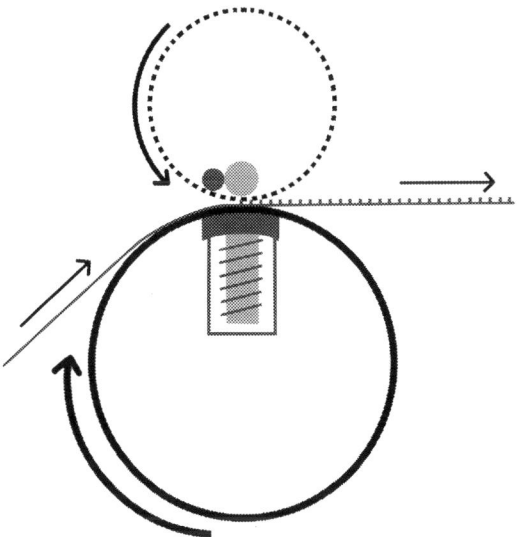

FIG. 11.5   Zimmer Magnoroll/Magnoknife. *Adapted from Zimmer.*

roller, to create a nip and a level of coating weight application, controlled by magnetic force. The coating weight is metered either with magnetically controlled roll rods or magnetically controlled knives (squeegees). The application level of the coating is controlled by magnetic force, which is applied to a profiled knife made from magnetisable steel. The magnet can be placed in different locations and is capable of applying coating weights of ca. 40–$60\,\mathrm{g\,m^{-2}}$.

Rotary screen coating is capable of applying coating formulations to discrete areas of the nonwoven by forcing the coating through a cylindrical perforated screen. This is also referred to as paste dot coating. The viscous liquid formulations, which can be thermoplastic are forced through the perforated screen using a squeegee bar situated within the screen. The frictional forces between the fabric and the rotary screen are minimal as the screen rotation is proportional to the fabric feed rate, which enables the formulation to be deposited onto the fabric and merges together to form a continuous coating. The add-on is controlled by the coating viscosity, and it is possible to apply 5–$500\,\mathrm{g\,m^{-2}}$. This process is suitable for fabrics such as nonwovens that have uneven surfaces where knives may cause uneven or streaky application. The rotary screen method can also be used to apply foamed coatings or for foam processing of fabric finishes instead of padding. Such coatings can be useful for controlling the frictional (e.g. antislip) and abrasive properties of nonwoven fabrics for applications in automotive fabrics, household floor wipes, and table coverings.

11.3.2.5 Gravure coating

Gravure coating enables discrete areas of the nonwoven to be coated rather than the entire surface, which is useful when increases in fabric stiffness or decreases in permeability, liquid transport and absorbency need to be minimised (Fig. 11.6). Various approaches are employed. In one embodiment, the nonwoven is fed between the nip of two rollers, one of which is engraved with a pattern and is partly submerged in a bath containing the coating formulation, and other acts as a pressure roller. The coating formulation is therefore applied to the fabric

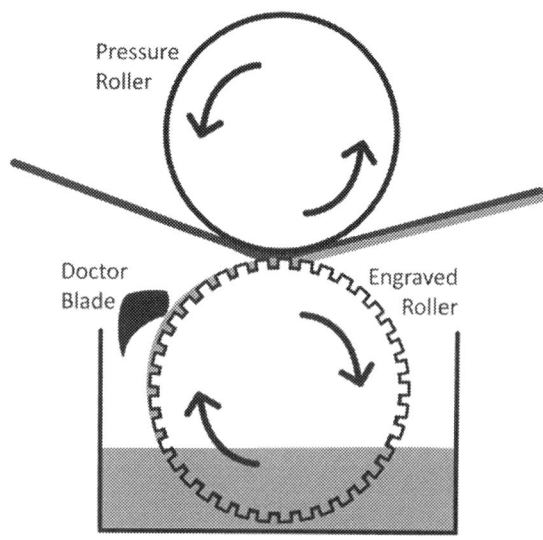

FIG. 11.6   Gravure coating.

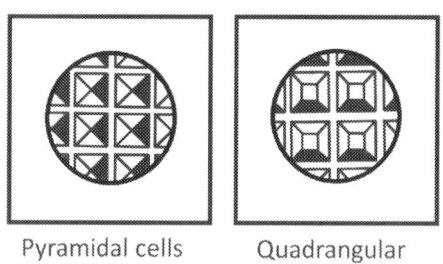

Pyramidal cells   Quadrangular

FIG. 11.7   Gravure roller profiles.

surface in discrete locations, determined by the design of the engraved pattern. Various gravure roller profiles are in use, including pyramidal and quadrangle, the choice of which governs the coating weights and final coating finish (Fig. 11.7). In some cases, an offset gravure is
used to transfer the coating onto an intermediate roller, before being deposited onto the fabric.

11.3.2.6 Slot die coating

A thin layer of liquid coating consisting of a dispersion, solution or an extruded polymer
film is applied directly to the nonwoven through a thin slot die across the width of the fabric.
The slot die has a high aspect ratio outlet, allowing the thickness and width of the liquid coating formulation to be controlled as it is applied of the nonwoven substrate. The resulting coating thickness depends on the feed rate and viscosity of the coating formulation through the
slot die, as well as the linear speed of the nonwoven substrate. Some coating systems are modular, allowing more than one type of coating method to be selected, depending on requirements, including slot die coating. One example of such an interchangeable, flexible coating
system is the Caviflex coating and laminating system illustrated in Fig. 11.8.

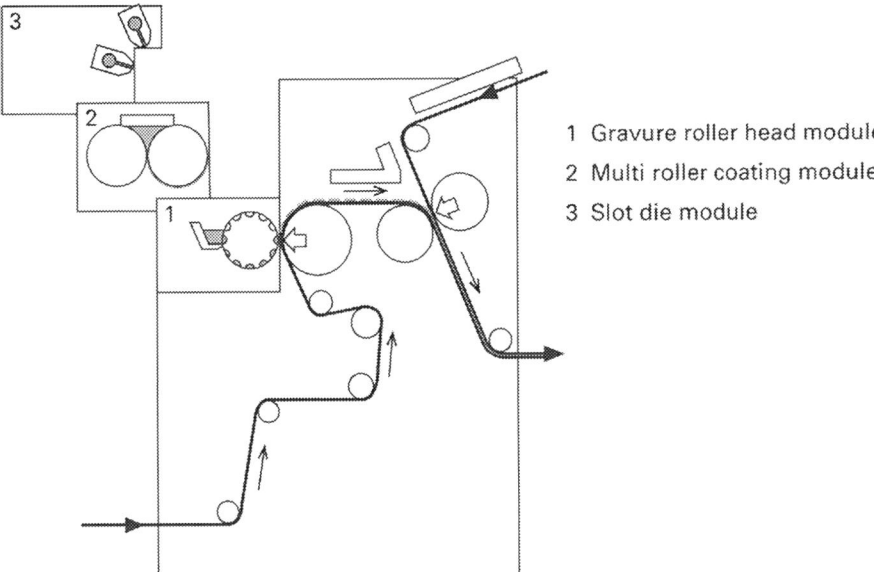

1 Gravure roller head module

2 Multi roller coating module

3 Slot die module

FIG. 11.8   Caviflex system with exchangeable coating modules. *Courtesy of Cavitec.*

11.3.2.7 Hot-melt coating

Hot-melt coating is particularly important for applying both thermoplastic and thermoset adhesives to nonwovens, including those based on copolyamides, polyesters, copolyesters, polyurethanes, coPVC, and ethylene vinyl acetate (EVA) polymers. Hot-melt coating is often followed by a lamination step. Commonly, the hot-melt coating is applied in discrete regions to the fabric, and rotary gravure, rotary screen printing, as well as slot die and spray coating can all be used to apply the adhesive. Unlike aqueous coating formulations, processing speeds are not limited by the need for drying capacity, instead cooling zones enable rapid solidification of the adhesives and for specific types of thermoset adhesive, gradual humidity-induced cross-linking can occur following coating. For thermoset hot-melt adhesives, application temperatures are typically in the region of 90–130°C and viscosities from 3000 to 35,000 cps, followed by crosslinking. For thermoplastic hot-melt adhesives, application temperatures range from about 150°C to 230°C and viscosities from 3000 to 150,000 cps. An example of hot-melt coating, combined with lamination is the Lacom system, which applies molten polymer to the nonwoven with pressure rollers enabling the production of the final laminate (Fig. 11.9).

11.3.2.8 Extrusion coating

Extrusion coating, involves coating the nonwoven with a liquid polymer melt composed of, for example, PP, LDPE, LLDPE, or EVA, by forcing it through a slot die at an angle to the nonwoven before calendering to spread the adhesive and ensure sufficient adhesion between the coating and the fabric. One or more sides of the fabric can be coated depending on the arrangement of dies in the coating system. Coating weights range from about 20 to

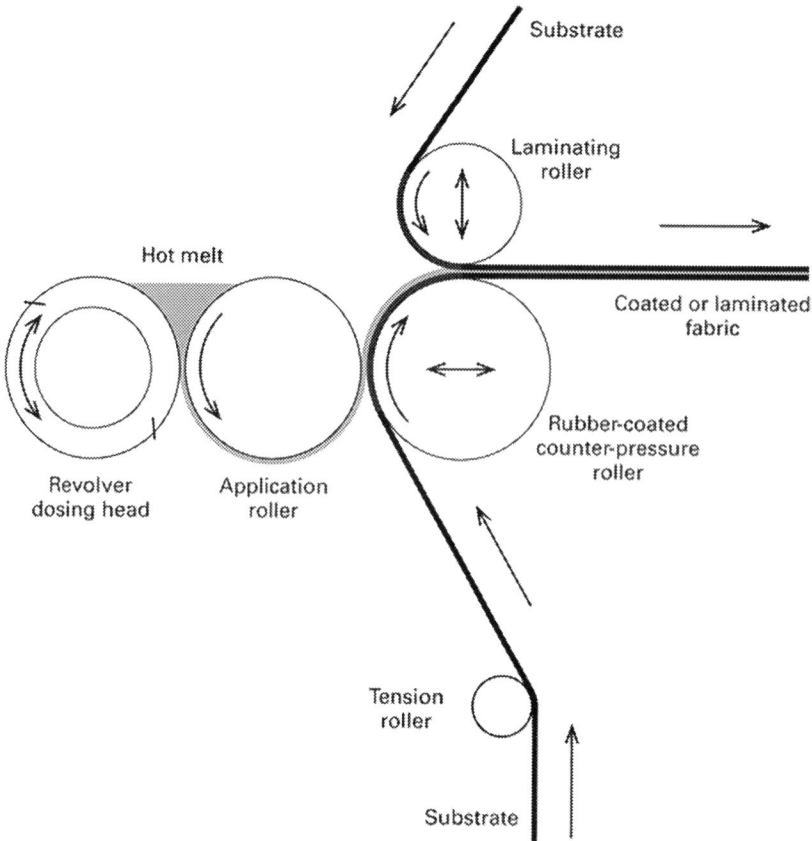

FIG. 11.9   Hot-melt coating and laminating. *Courtesy of Lacom.*

$500\,g\,m^{-2}$, depending on conditions, with operating widths up to 4m or more. Normally, a continuous film coating is produced that can either enable subsequent lamination or, production of fabrics with for example improved barrier performance in relation to water repellency, chemical resistance, or capability to be printed. In packaging applications, extrusion-coated LDPE provides a cost-effective water resistant barrier. In extrusion coating, the throughput speed of the fabric generally controls the thickness of the applied coating. Extrusion coating is also used as a precursor for extrusion lamination, where an additional layer, or layers of nonwoven, textile, film, or foil are combined with the coated nonwoven whilst the coating is hot. Note that in extrusion coating, the surfaces of the final product consist of the extruded polymer coatings, whereas in extrusion lamination the surfaces consist of the additional fabric, film, or foil layers.

11.3.2.9 *Transfer coating*

Transfer coating is often used to produce fabrics for waterproof protective clothing, upholstery, luggage, footwear components, automotive seat fabrics, and carpets due to its ability to provide a smooth film finish with uniform thickness. A coating is indirectly transferred onto a nonwoven fabric via a release carrier material, e.g. siliconised paper, initially, the polymer coating is applied to the release carrier material and then applied to the nonwoven at elevated temperature and pressure, or via solvent evaporation. Once the coating is applied, the release carrier material is peeled away. Coating quality is influenced by the uniformity of the non-woven fabric surface, its chemical composition and the adhesion between the nonwoven and coating formulation. The result is a thin, continuous film applied to the nonwoven surface.

11.3.2.10 Powder dot coating

Powder dot coating produces coated nonwoven substrates using dry thermoplastic particles consisting most commonly of co-PA, co-PET, PE, and PP applied to the nonwoven fabric surface. In a typical powder dot coating process, the fabric is pre-heated prior to being pressed against an engraved roller above which, is a hopper containing the thermoplastic particles. As the engraved roller rotates it delivers particles to localised areas of the fabric surface. The depth of the engraved regions and the pattern on the roller, controls the percentage coverage on the fabric and the coating weight per unit area. Subsequently, infrared heating normally follows to fuse the deposited particles and form the final coating.

Rotary screens can also be used to apply powders to the nonwoven in a similar manner. Applications include nonwoven interlinings, automotive seat padding and door interiors and shoe liners.

11.3.2.11 Scatter coating

Scatter coating involves depositing dry particles on to a moving nonwoven fabric supported on a conveyor belt. In addition to adhesives, particles can comprise granular additives such as activated carbon, inorganic fillers, superabsorbents, recycled particles, and other materials designed to enhance functionality. Typically, a hopper containing the particles is mounted over a rotating scatter roller covered in surface projections such as brush needles, which collect the particles as it rotates. A spring loaded doctor blade controls the weight of particles collected by the scatter roller. As the scatter roller continues to rotate, an oscillating or rotating brush unit removes the particles from the scatter roller, and they are deposited on the nonwoven fabric below. Coating weights of $3-2000\,\mathrm{g\,m^{-2}}$ are possible using modern equipment, but coating weight uniformity is obviously dependent on the nature of the particles used as the coating feedstock. Operating widths range from 0.7 to 5 m. Scatter coating is used for various applications, including some carpet backings and upholstery fabrics.

11.3.2.12 Double dot coating

Double dot coating is a related process in which a rotary screen printed fabric is made with paste-dots and then is immediately coated with powder using a scatter coating module. Infrared heating is then employed to produce the final coated fabric.

11.3.2.13 Foam coating

Foam coating is often used for apparel goods, floor coverings, wall coverings, black-out curtains and curtain linings and filter materials [14]. This coating method is suitable for high porosity nonwovens that cannot be direct coated as the rheological properties of the foam allow it to remain on the surface of the nonwoven rather than penetrate the internal pore structure of the fabric. The foam is dispersed uniformly over the nonwoven often using a knife, roller, or a rotary screen. The foam is then dried in a low moisture environment, typically followed by calendering. Foam coating enables low wet pick-up, reducing energy demands in drying, as well as reducing the volume of liquid effluent.

11.4 Lamination

To make final products, preformed nonwoven fabrics and other substrates, such as films, scrims, foils, foams, as well as conventional textile materials, are laminated together, typically by means of heat and pressure. Lamination is a valuable means of enhancing the performance of final nonwoven products in respect of, for example, liquid barrier, vapour permeability, surface abrasion, dimensional stability, heat transfer, and mechanical properties.

Thermoplastic or thermoset materials (dry lamination), or wet adhesive binders (wet lamination), can be employed to facilitate the lamination process and provide the interlaminar adhesion necessary to permanently join the layers and prevent delamination during use. Dry lamination uses thermoplastic and thermoset powders, films, scrims, and adhesives. The adhesive is applied either to the entire surface of the substrate, or in discrete patterns, localising the coverage and leaving unoccluded areas, free of adhesive.

In considering the required performance of the final laminated product, the composition and physical properties of the interlaminar layer/s needs to be as carefully considered as the laminated substrates themselves. For example, the thermal properties of the adhesive polymers used for lamination, particularly the glass transition temperature is important given its influence on hardness, toughness, softness, and flexibility at temperatures likely to be encountered during use.

Various lamination methods are found in the nonwovens industry, including but not limited to:

– Belt calendering.
– Lamination with infrared heating combined with calendering.
– Flatbed lamination.
– Extrusion coating and lamination.
– Hot-melt lamination.
– Flame lamination.

Generally in lamination, nonwoven substrates and other sheet substrates, such as films and foils, are unwound from separate rolls and joined following heating, using a series of nipped rollers under pressure.

In an example of dry lamination using a belt calendering installation, the first substrate is drawn from a feed roller and scatter coated with a particulate (powder) binder, composed of either a thermoplastic, e.g. TPU, co-PA, co-PET, PE, PP or EVA, or thermoset (e.g. epoxy, PET) polymer. The substrate is then preheated by infrared heating to soften the binder, before being combined with the second substrate, and then belt calendered to join the two layers. This involves sandwiching the layers to be laminated between the surface of a large rotating calender roller and a flexible belt. Lamination in the presence of a scatter coated binder can also be carried out by preheating both layers with infrared heaters before combining them in the nip of two or more calender rollers. This particular approach generally enables thicker laminates to be produced than is possible with belt calendering.

In flatbed lamination, heating and cooling during lamination can take place between two belts, such that the substrates do not need to pass around calender rollers. This can be particularly useful if the substrates are relatively stiff, although the process is not limited to such fabrics. There is also the capability to provide a gauge or gap setting during lamination, rather than a high-pressure roller nip. In flat bed laminating, nonwoven fabrics and thermoplastic films can also be combined before entering the lamination zone, which usually consists of belt, or plate type system. Within this zone, heat and pressure are usually applied for longer durations compared to other lamination systems that rely on the use of nip rollers. This longer residence time along with controlled temperature conditions can improve the quality and uniformity of the lamination. Additional advantages include versatility of operation, including the ability to operate piece-to-piece, or roll-to-roll, and a large variety of different substrates can be used.

In another example of dry lamination using calenders, heated calenders melt a film substrate and laminate by forcing the molten polymer into the nonwoven fabrics. Wind-up tension and cooling rates need to be considered to prevent sticking and deformation of the laminated substrates. Also, wrinkling is a common problem due to uncontrolled let off tension, and can be prevented by controlling the line speed, operating temperature and nip pressure. Fig. 11.10 illustrates a three stack calender system.

In extrusion lamination, a thermoplastic polymer is heated to a molten state and then extruded as a thin continuous sheet between the layers to be laminated, before the laminated structure is formed in the nip of two or more calender rollers. To increase the penetration of the thermoplastic polymer, increasing the pressure at the roller nip and the tension of the fabric can improve the adhesion between the layers. PE is one of the most commonly used polymers in extrusion lamination due to its low cost and ease of processing. Hot-melt lamination allows the molten adhesive to be applied in various patterns or sprayed onto the substrate prior to calendar-assisted lamination, so that it is present in the form of small dots, tracks, spirals, or even meltblown filaments. Hot-melt lamination facilitates high-speed processing and is relevant to many different nonwoven markets.

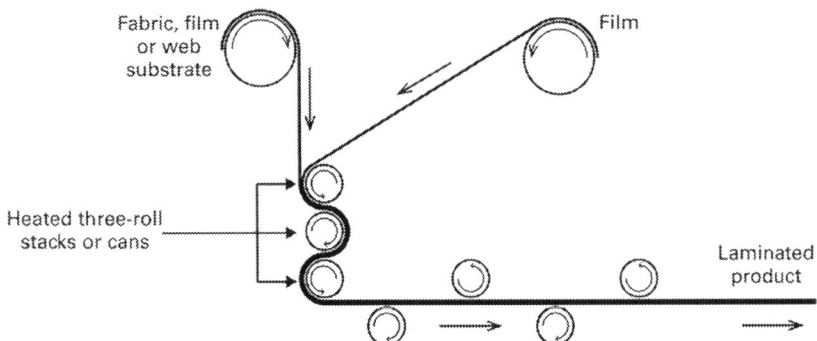

FIG. 11.10   Three-stack calender lamination system.

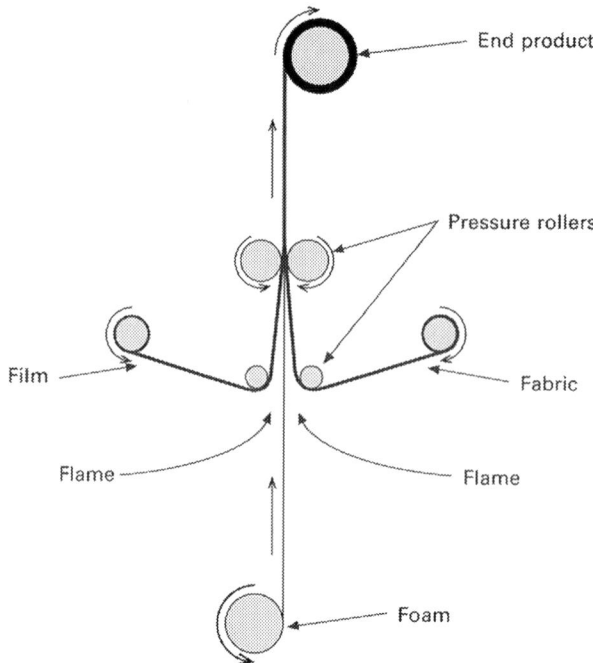

FIG. 11.11   Example of a flame lamination system.

Flame laminating is a traditional method of laminating nonwoven fabrics for automotive applications, including headliners, door panels, seat components, sun visors, head rests, carpets, and car boot liners. The surface of a foam substrate, e.g. comprising of open-cell polyester, polyether urethane, or cross-linked polyethylene is passed through an open flame, which partially melts the foam. One or both sides of the foam substrate can be exposed to an open flame, depending on the desired composite structure. Fig. 11.11, illustrates a 3-ply laminate produced by melting both surfaces of a foam layer, and then laminating a film and a nonwoven substrate, whilst the foam surface is still in a molten state.

Important process parameters affecting the resulting physical properties and bond strength of the laminate include the type of gas used to fuel the flame, the gas flow rate, or flame intensity, flame height, spread, and calender nip pressure.

11.5 Dry-particle impregnation

Traditionally, coating and lamination are considered to be the classical means of applying film forming compounds and other sheet-like materials onto nonwoven fabrics. However, it is also possible to impregnate small, dry particles into the structure of nonwovens, as well as onto the surface, after fabrics have been manufactured. Dry particles in this context can refer to powders, beads, capsules, and flakes. Examples of materials include but are not limited to, superabsorbents (SAPs), abrasives, intumescent particles, and odour adsorption materials such as zeolite and activated carbon, as well as other molecular sieves. Specifically, for enhancing the performance of nonwoven wipe substrates, a wider variety of materials can be considered as a means to add value after the nonwoven has been manufactured through dry impregnation of particles (Table 11.2). Applications for dry impregnation technology are developing for nonwovens used in the hygiene, personal care and cosmetics, filtration, building and automotive sectors.

TABLE 11.2 Particle formulations with potential for dry impregnation into preformed nonwoven wipe substrates.

Wipe application	Example additives
Flushables	Moisturisers (e.g. pantolactone, panthenol)
	Sodium chloride
Household	Super absorbent polymer (SAP)
	Antioxidants
	Moisturisers
	Surfactants
	Antimicrobials
	TiO_2
	Disinfectants
Industrial	Antigrease
	Adsorbents (e.g. activated carbon)
	SAP
	Disinfectants
	Surfactants
	Abrasives
Personal care	Antioxidants
	Vitamins E/C
	Herbal extracts
	Charcoal
	Preservatives (sodium benzoate, potassium sorbate)
	Moisturisers
	pH adjusters
Medical	Antimicrobials
	TiO_2
	Disinfectants
	Drug products

The processes developed for dry-particle impregnation differ from conventional methods of impregnation and coating because the fabric is not exposed to aqueous or molten liquids, heat or drying; compressive forces on the substrate are minimised; and there is also the possibility of impregnating admixtures of different types of particles. Therefore, the nonwoven retains its porosity and there is significant scope for modifying the specifications of the fabric. Furthermore, it is not always necessarily to apply adhesives, or covering layers, to retain the dry particles in the fabric, and potential line speeds of up to $600\,\mathrm{m\,min}^{-1}$ are possible, depending on the particle specifications and machine operating conditions.

An example of a dry-particle impregnation or 'inclusion' process is the D-Preg technology, and associated methods (T-Preg, S-Preg, and Y-Preg), developed by Fibroline (Fig. 11.12). Powders are first scatter coated onto the surface of a nonwoven carrier supported by a conveyor, and then injected into the interior of the fabric by means of a high voltage and low current alternating electric field. The process effectively energises the particles between two electrodes, enabling their transport into the internal pore structure of the nonwoven. Fibroline's T-Preg system is specifically designed for hygiene applications, S-Preg for extremely accurate impregnation control and Y-Preg Technology extends the process window to yarns, ribbons, and tapes. To minimise risk of particle migration following the dry impregnation process, thermal treatment can be applied to the fabric, if the thermoplastic fibres are present, or outer layers can be laminated onto the fabric surfaces.

As an alternative to impregnation using electric fields, ultrasonic energy can also be harnessed to rapidly inject loose particles into the interior of nonwoven fabrics, continuously whilst the fabric is carried on a conveyor.

By installing multiple scatter coating stations, and by appropriate selection of different particle compositions, it is possible to harness such dry impregnation methods as a basis to mass customise the performance of nonwoven products, as part of a continuous finishing or

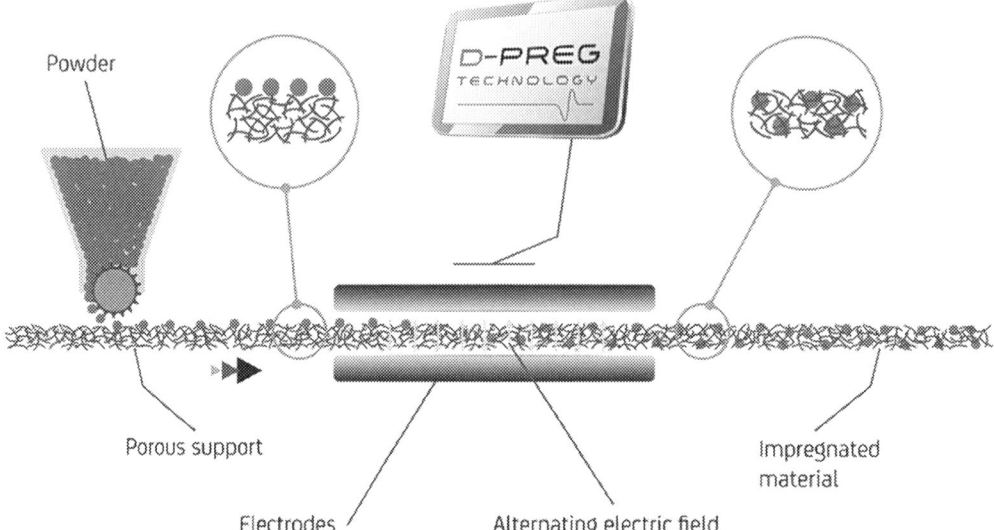

FIG. 11.12   Schematic of the D-Preg dry impregnation process. *Courtesy of Fibroline.*

converting process. Such an approach to mass customisation, is applicable to various high- and low-value nonwoven products, to reduce processing complexity and costs, whilst providing significant scope to enhance fabric performance, and add value at a late stage of production.

11.6 Grafting

Surface grafting enables the properties of nonwoven fabrics to be modified by introducing new chemical moieties, oligomers, and other polymers to the fibrous structure. Property advancements include but are not limited to increased liquid absorbency [15,16], water and oil repellency [17], flame resistance [18], mechanical stability [19], thermal stability, chemical resistance, elongation at break, water retention, antibacterial, and decreased surface resistivity.

Polymer synthesis enables excellent control of the bulk polymer properties; however for many applications, the surface interactions between the material and the environment can play a large role in determining the utility of the material. Therefore, grafting techniques such as chemical graft polymerisation, radiation-induced grafting, plasma-induced grafting, and light-induced grafting can be used to tailor the surface properties of polymers [20]. Grafting creates free radical sites within the macromolecules of the fibres in the substrate that can then be used as initiators for copolymerisation reactions. Parameters such as monomer concentration, treatment duration, catalyst type and concentration and radiation intensity can greatly affect the grafting efficiency and can be varied to control the grafting yield. These additional functional groups can be harnessed to modify the surface energy of the fibres, covalently bond additional molecules via linker mechanisms, and introduce specific surface chemistries.

Chemical graft polymerisation uses initiators to create radicals on the fibre surfaces that when reacted with monomers initiates copolymerisation reactions [20]. Chemical graft polymerisation is a well-known grafting technique; however, it has some major limitations due to the production of toxic by-products and difficulties in confining treatments to the surface of the material. Some specific examples of graft polymerisation are given in Table 11.3.

Radiation-induced grafting uses high-energy radiation to induce in situ formed free radicals on the surface of the material. Unlike chemical grafting, radiation-induced grafting does

TABLE 11.3 Examples of initiators used for graft polymerisation of textile surfaces [20].

Initiator	Reaction monomer	Purpose	References
Potassium permanganate	Acrylonitrile (AN)	Imparts high water absorbency on cotton	[15]
Ammonium peroxydisulfate	Benzyl methacrylate	Improves the tensile strength and reduces the elongation at break of wool fibres	[19]
Benzoyl peroxide	Perfluorooctyl-2 ethanol acrylic	Imparts water repellency, chemical/thermal stability and enhances adhesion on polyester fibres	[17]

not require chemical initiators and the grafting yield can be controlled by varying the irradiation dosages and exposure durations. High exposure can lead to a deterioration in the physical properties of the fabric, including photodegradation. Also, gamma-radiation is highly penetrable, so a high level of process control is required to avoid excessive disruption of chemical bonds.

Plasma-induced grafting uses activated species, such as ions, radicals, and metastable molecules to introduce activation sites on the surface of materials, which can be reacted and copolymerised with different monomers. These activation sites are introduced by exposing the substrate to plasma, which consists of partially ionised gas in the presence of electric, or both electric and magnetic fields. Steric hindrance can be an issue with plasma grafting due to instability of the frequency and distribution of activation sites. The presence of activation sites can be affected by the type of gas, gas volume, chamber temperature, chamber pressure, and intensity of the applied electrical energy. With the development of roll-to-roll plasma technology, plasma-induced technology is becoming increasingly popular for surface modification and grafting of fabrics. Plasma does not penetrate deep into the material; therefore, bulk properties are generally unaffected by the treatment and desirable properties can be maintained. Furthermore, the plasma treatment process does not generate toxic waste by-products. One example of plasma grafting is the linking of polysaccharides to PBT meltblown fabrics for the purpose of immunoadsorption in blood component filtration [21]. During filtration, an artificial antigen (polysaccharide) can be grafted to the PBT meltblown fibres by means of plasma-induced grafting to produce a filter medium capable of binding unwanted antibodies from donated blood plasma. Consequently, it is possible to produce universal plasma with the potential to be transfused into any patient, regardless of blood-type.

Light-induced grafting uses ultraviolet radiation to introduce radicals associated with the molecules of the fibres in the fabric. Similar to the other grafting techniques, the radicals are then used to initiate copolymerisation reactions with other molecules capable of conferring enhanced functional properties. Ultraviolet radiation is not as penetrative as other forms of radiation, e.g. gamma radiation; therefore, the radicals are positioned close the surface rather than uniformly distributed throughout the fibres [22].

11.7 Vapour deposition processes

Vapour deposition processes include vacuum-enabled techniques used to deposit a thin layer of functional material onto fibres within a nonwoven fabric. There are two main families of processes: chemical vapour deposition (CVD) and physical vapour deposition (PVD).

11.7.1 Chemical vapour deposition

The CVD process involves the formation of a thin solid coating onto a nearby heated substrate by the chemical reaction of gaseous reactants [23,24]. A schematic is given in Fig. 11.13. The atomistic deposition method can coat materials of high purity at the atomic or nanometre scale.

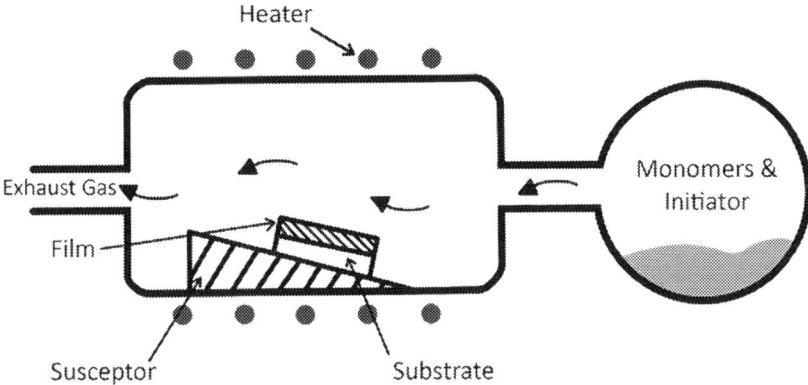

FIG. 11.13 Schematic of the chemical vapour deposition (CVD) process.

In one example of an application of this process, Conductive Composites (USA) developed a nickel coated nonwoven produced using CVD techniques [25]. The CVD-treated fabric was found to be as lightweight and conductive as carbon nanotube sheets, but could be produced much more economically. CVD-treated nonwovens have found applications in for example, fabrics requiring enhanced electrical conductivity, and electromagnetic shielding. The continuous surface coating enabled by the CVD process has been observed to both improve electromagnetic shielding, and be thinner and lighter compared to chemically bonded metal-coated fibre nonwoven fabrics [26].

11.7.2 Physical vapour deposition

Physical vapour deposition (PVD) processes involve the deposition of atomic or molecular species onto a substrate, such as a film or nonwoven. Within the PVD classification there are a number of different vacuum deposition methods which can be used to produce thin films and coatings on nonwovens; evaporation and sputtering being the most common processes.

PVD involves converting the coating material from a condensed phase to a vapour phase and then back to a thin film condensed phase onto the fibre surfaces of the substrate. Coating thicknesses range typically from a few nanometres to a few micrometres.

Deposition of aluminium atoms is possible to produce thin metalised layers on fabrics. Of course, this is a more sophisticated approach than lamination of aluminium foils or films onto nonwovens. General applications for aluminised nonwovens, but not necessarily conferred by vacuum deposition, include packaging, thermal protective clothing, sensors, wound care products, acoustic, and thermal insulation as well as breather membranes for thermal management in roof spaces. In vacuum deposition (vacuum evaporation), the coating material is thermally vapourised inside a vacuum chamber after which it condenses directly onto the nonwoven substrate with minimal collision with gas molecules between the source and substrate [27]. Within the vacuum chamber aluminium wire is fed onto heated evaporators, melting the aluminium and evaporating it into the chamber. The nonwoven substrate, supported on a chilled drum, passes the evaporation source at speeds of up to $1000\,\mathrm{m\,min^{-1}}$. Nonwovens can be difficult to process due to their high surface area, low dimensional stability, and hydrophilicity. Poor control of tension can result in low contact between the drum and the substrate, limiting cooling of the substrate, reducing the deposition rate and process speed [28].

11.8 Plasma treatment of nonwovens

Plasma treatment can be employed to modify the wetting characteristics, dye uptake, as well as other aspects of functionality in nonwoven fabrics. Plasma treatment utilises very small quantities of chemicals to functionalise the surface of nonwovens. Water and energy consumption are also very low compared to alternative wet treatment processes, with associated environmental benefits. Plasma comprises an admixture of ions, electrons, neutrons, photons, free radicals, meta-stable excited species, and molecular and polymeric fragments, generated by coupling electromagnetic power into a process gas [29].

Low temperature plasma technologies for nonwovens can be separated into two main types – low pressure and atmospheric plasma. Whilst low-pressure systems have traditionally found more commercial applications, more recently the latter is increasing in popularity. Atmospheric plasma systems negate the need for vacuum chambers and pumps making continuous production simpler and capital equipment cheaper [30]. Increasing the pressure from atmospheric towards partial vacuum increases the energy exchange between electrons and neutrals and offers better control over the process chemistry. Developments in atmospheric plasma technology have the potential to enable higher line speeds or increase coating thickness per unit time. Web speed in a roll-to-roll low-pressure vacuum plasma machine varies depending on the treatment. For activation processes, a typical web speed is between 5 and $50\,m\,min^{-1}$. For coating deposition processes, the speed is slower, usually between 0.5 and $10\,m\,min^{-1}$ [29].

Plasma treatment can be employed to enhance nonwovens for several purposes:

- *Fine cleaning of fibre surfaces*, e.g. before bonding or adhesion to other surfaces. The polymeric fibre surfaces within a nonwoven can be cleaned using pure gases such as oxygen and nitrogen. When plasma is used for cleaning it removes organic contamination at the microscopic level. Mechanistically, the chain length of contamination is reduced inducing volatility, which means it can be removed the plasma exhaust.
- *Activation of surfaces*, e.g. before printing, laminating, or coating. The fibre surfaces may be activated using plasma by creating functional groups on the outer polymer chains of the material. The activation increases the surface energy of the polymer improving the adhesion to coating formulations, adhesives, or printing inks.
- *Etching of surfaces*. Fibre surfaces may be microstructured using plasma. This usually involves applying direct plasma to remove material from the fibre surfaces. For polymeric materials, commonly used in nonwovens this process is slow, difficult to control and uneconomic for most common applications.
- *Coating of surfaces*, e.g. deposition of hydrophobic or hydrophilic layers. Plasma technologies may be used to deposit permanent coatings onto nonwovens with a typical thickness of 10–500 nm. Plasma polymerisation is a process in which reactive precursor gases are broken down into radicals that react on the substrate surface [31]. The functionality of the coating depends on the precursor gas, but hydrophilic, hydrophobic

(including superhydrophobic), oleophobic, scratch resistance, and barrier properties may be imparted [32].

11.8.1 Applications of plasma treated nonwovens

Although far from being ubiquitous, several industrial applications of plasma technologies have been established. These include but are not limited to:

- *Blood filter media*. A well-known application of plasma functionalisation is in the production of meltblown nonwoven blood filter media (e.g. for leukodepletion). Plasma treatment is typically performed roll-to-roll on meltblown nonwovens manufactured from polypropylene (PP) or polybutylene terephthalate (PBT), to improve fibre wetting behaviour and facilitate high efficiency blood component filtration.
- *Nonwoven battery separators for rechargeable batteries*. Hydrophilic functionalisation helps to improve electrolyte transfer in NiMHydride rechargeable batteries [32].
- *Respiratory face masks*. Hydro- and oleophobic coatings can improve filtration efficiency in electret meltblown filter media, challenged by oily particles (DOP). Oil repellency levels of 3 and 4 may be achieved, with reference to ISO 14419 [33].

Plasma treatment has been demonstrated to improve several aspects of filter performance, including increasing surface charge in electret filters and the permeability of reverse osmosis filters [34]. PP meltblown filter media, treated with methane and HMDSO plasma have been observed to achieve increased filtration efficiency against NaCl aerosol [35]. In GB2462159A, a plasma-coated, fibrous filter media are described using a hydrocarbon or fluorocarbon monomer (1H,1H,2H,2H-heptadecafluorodecylacrylate), to enable reduced filter cake formation and/or increased filtration efficiency [36]. Plasma technology can also be useful in some hygiene applications. Cold oxygen gas plasma treatment has been observed to significantly improve the dynamic water handling properties of PP nonwovens, providing potential for enhanced liquid transport in diapers, adult incontinence and feminine care products [37].

11.9 Molecular imprinting

Molecular imprinting is a technique whereby materials exhibiting molecular recognition or template polymerisation properties are produced. A molecularly imprinted polymer (MIP) is formed using processes for the preparation of synthetic polymers with specific binding sites for a target molecule, imbuing high selectivity performance. They are classified into two categories, covalent and noncovalent (alternative) molecular imprinting. MIP aims to biomimetically reproduce molecular recognition by biological receptors, such as antibodies or enzymes. The recognition sites on the polymer are able to specifically rebind the target molecule(s), whilst remaining inactive against closely related compounds.

Molecular recognition sites formed via electrospinning provide promising opportunities to develop nonwoven fabrics capable of binding target molecules. Applications for MIPs in nonwovens include the removal of uremic toxins to enhance kidney dialysis and the engineering of improved filter fabrics, biosensors [38], chemical sensors, catalysis and fibres for drug delivery [39].

11.10 Microencapsulation

Microencapsulation can be described as "a process by which individual particles of an active agent can be stored within a shell, surrounded or coated with a continuous film of polymeric material to produce particles in the micrometre to millimetre range, for protection and/or later release" [40]. There are a number of possible delivery systems including shell wall fracture and diffusion amongst others, which are activated by a stimulus to expose the core ingredient. In nonwovens, microcapsules are often used a delivery system for active compounds within personal care and cosmetic sectors; however, the technology is also relevant to industrial and pharmaceutical applications. Microcapsules can be incorporated within the fibres themselves, but it is more usual to apply them to nonwovens after roll good production, as part of the finishing process. The latter enables much higher microcapsule loadings, and therefore a more significant influence on bulk fabric properties, than is possible when they are incorporated within fibres. Table 11.4 illustrates a variety of functions/applications within different market sectors.

The most useful feature of microencapsulation is that the core materials, which can be either solid or liquid, are completely isolated from the external environment, provided the shell material and incorporation method are appropriately selected. The shell materials are formed using three main mechanisms: physical, chemical, and physiochemical, and these mechanisms are usually selected based upon the morphology of the core material. For the physical mechanism, the formation of the shell material depends upon the solid–liquid phase transition under heating or solubility reduction due to solvent evaporation. Monomers consisting of small molecules polymerise to form the polymer shell for the chemical mechanism and for the physiochemical release mechanisms, and the predissolved shell-forming materials precipitate from the solution following the variation of temperature, pH value, or electrolyte concentration, and gradually deposit on the surface of core material to form the shell [42].

TABLE 11.4   Examples of market applications for microencapsulated materials [41].

Personal care	Household care	Food industry	Pharmaceuticals
Fragrances	Fragrances	Probiotics	Drugs
Therapeutic/cosmetic actives	Ironing aids	Antioxidants	Enzymes
Antiperspirants	Antiwrinkle agents	Vitamins	Peptides
Cooling agents	Bleaches and bleach activators	Fish oil	DNA
Sunscreen actives	Enzymes for fabric care		
Whitening agents	Optical brighteners		
Pigments and colourants	Odour control agents		
Decorative liquid crystals			

Industrial sector	Printing industry	Textile industry	Agrochemicals
Catalysts	Ink	Phase change materials (PCMs)	Insecticides
Reagents		Antimicrobial agents	Fungicides
Adhesives		Fragrances	Herbicides
Flame retardants			

11.11 Splitting

Splitting is a mechanical separation technique used after the nonwoven has been manufactured to reduce its overall thickness. In some cases, it is only possible and economically viable to produce a thicker nonwoven than is ultimately desired, and then split it through the longitudinal plane to the required thickness. Examples include heavily bonded needlepunched fabrics, such as synthetic leather, heavy-weight chemically bonded and vertically-lapped nonwoven fabrics. Nonwoven fabrics are guided over a feed table and through feed rollers under carefully controlled tension before being exposed to a precisely adjusted rotating hoop, hot knife, or wire. The two or multiple nonwoven layers are then separated on to different winding beams by controlling the outfeed roller configurations.

11.12 Perforating

Perforation can increase the flexibility of dense nonwoven fabrics, such as synthetic leather substrates made from microfilaments. Nonwovens can be perforated to introduce multiple holes using heated needles or modified calender rollers. The vertical profile of the perforation is important and can be adjusted depending upon the type of materials used, the fabric structure and perforation depth requirement. Conical perforation profiles can be applied by using specially designed needles, which can also be harnessed to modify the drainage properties and wetback properties of nonwovens used in AHPs. Also, longer perforations or slits can be introduced into the fabrics using perforation techniques, and the length of the slit and the distance between the slits can be calculated to minimise the loss in fabric strength.

11.13 Microcreping

Microcreping is a mechanical treatment process for imparting fine corrugations and ridged morphologies in relatively thin nonwoven fabrics to significantly modify tensile modulus, extensibility, softness, drape, hand, bulk, absorbency, and create decorative effects. One application of microcreping is to modify the handle, bulk and mechanical properties of nonwovens used for wipes and hygiene products, including hydroentangled fabrics made from staple fibres. The fabrics are often processed dry; however, processing wet fabrics is also possible. Suitable fabrics for microcreping include but are not limited to wetlaid and drylaid fabrics, spunbonds, spray, and point bonded webs, as well as hydroentangled fabrics.

The fabric is fed through a converging cavity, which is bound by a drive roll and a series of spring steel blades. The fabric is compressed by overfeeding into the cavity in such a way that causes the fabric to buckle and create linear folds (Fig. 11.14). A retarder blade, which is etched with grooves, provides contours for the fabric to be compacted. The compaction is controlled by varying the rate of speed that the web is drawn away from the cavity by the take-up mechanism. Fine and coarse crepe configurations are possible with production speeds up to $300\,m\,min^{-1}$.

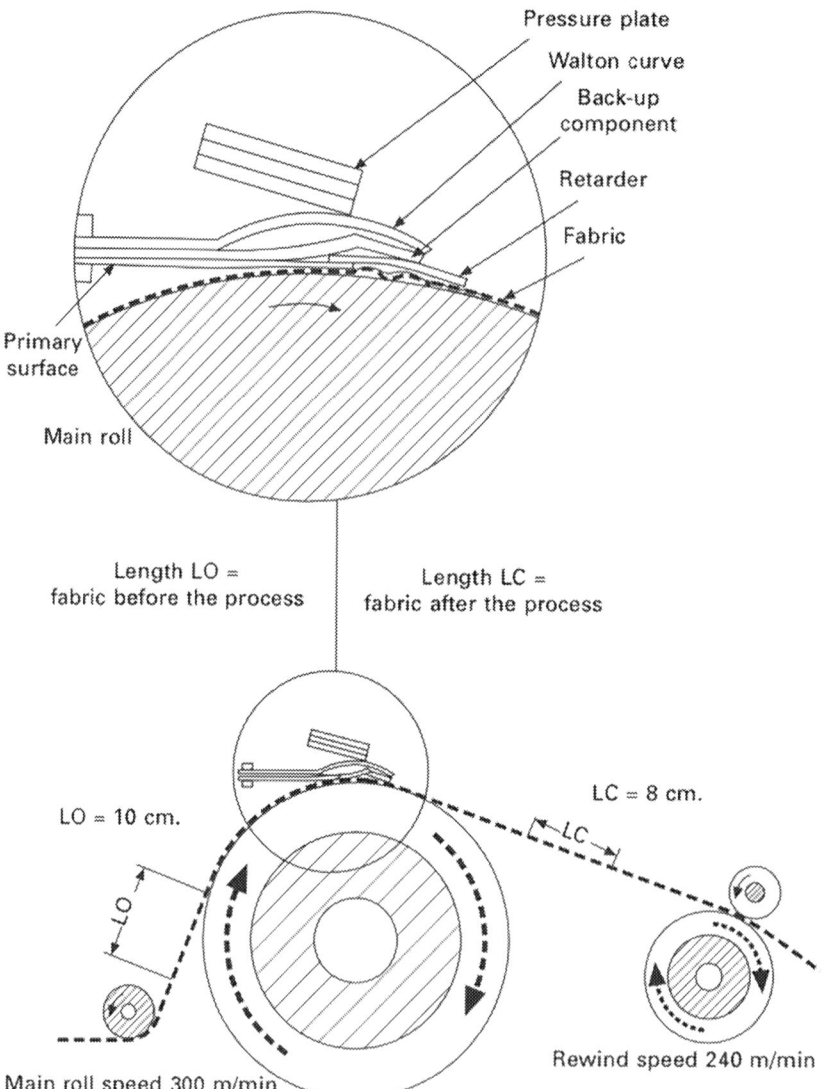

FIG. 11.14   Arrangement of the Micrex System. *Courtesy of Micrex Corporation.*

The tensile strength of microcreped fabrics can be affected by the processing temperature. For thermoplastic fabrics, increasing the temperature often decreases the softness and increases the elasticity of the fabrics. The wavelength and amplitude of the three-dimensional corrugated pattern produced by the process can be modified, and the degree of compaction can be calculated from the following formulae:

$$\text{Compaction } C\,(\%) = \frac{LO - LC}{LO} \times 100$$

$$\text{Stretch } S \, (\%) = \frac{LO - LC}{LC} \times 100$$

where LO is the original measured length and LC is the compacted measured length.

Micrex is widely associated with microcreping equipment, and available process configurations can incorporate a variety of different components, all of which influence the nature of the microcreped product, depending on the properties of the fabric and the required microcorrugated effects:

- Comb roll – For creping high basis weight substrates and to soften hydroentangled nonwovens.
- Rigid retarder – For softening wetlaid fabrics and creping paper.
- Bladeless – Designed to compact less structured materials and for softening lightweight nonwovens, bulk gauze, and a wide range of shrinking applications.
- Flat blade – Flexible configuration designed to enable flexible decorative styles, including plisse, crepe, and seersucker.
- Two roll cavity – Designed for minimising heat degradation at high production speeds and is often used for processing fabrics containing polymers with low melting points (e.g. PE and PP).

Microcreping is compatible with many types of fibre types, including cotton, viscose, wood pulp, wool, and synthetic fibres. The choice of fibres and blends is largely determined by the intended utility or function of the fabric. Several microcreping applications require heat-setting thermoplastic fibres. The thermal response characteristics of the fibres is critical, and microcrepers are capable of controlling thermal conditions by means of heating the main roller and by controlling frictionally generated heat during by the process.

11.14 Conventional wet finishes

In practice, the majority of nonwoven fabrics do not need to be washed (or scoured) prior to use, but in some cases it is required. A washing pretreatment step can facilitate other wet finishing techniques, particularly aqueous dyeing or chemical finishing, and is used to remove impurities without significantly modifying the chemical or physical properties of the fabric. Nonwoven fabrics are washed by immersion in aqueous solutions, usually containing a detergent. The detergent interacts with the contaminant, forming a separation layer, which reduces its adhesion to the fibres. The contaminant then forms into a globule and detaches from the fibres within the fabric. There are two main types of detergent in use: nonionic and synthetic. Nonionic detergents are widely used due to their fibre compatibility and stability in water with varying mineral concentrations. However, as they are effective degreasing agents, they can adversely affect the fabric softness, so anionics (synthetic detergents) are often selected as an alternative. Synthetic detergents are relatively inexpensive and easy to rinse, and in most cases encourage ionic charges at the fibre surfaces, which prevents re-deposition of the contaminant. Key factors affecting the washing efficiency include temperature, processing duration, mechanical agitation action, as well as detergency type and concentration.

11.15 Application of dyes and pigments

Unlike classical woven and knitted textile fabrics intended for clothing and household textiles, conventional dyeing processes are not necessary for most nonwoven product applications. When colouration of nonwoven fabrics is required, it relies on the application of either dyestuffs (soluble) or pigments (insoluble). Colouration of synthetic fibres for applications in for example, nonwoven floorcoverings, automotive headliners and interior trim is usually accomplished not by dyeing, but by mass pigmentation (dope dyeing) of the fibres. This results in excellent colour fastness, and where staple fibres are produced, enables different colours and shades to be produced by means of fibre blending. The disadvantage is that it requires manufacturers to commit to specific colours early on in the manufacturing process, rather than at a later-stage, after the fabric has been manufactured. Pigments can also be incorporated with adhesive binders used in chemical bonding, or can be added to coating or printing formulations applied to nonwoven fabrics after roll goods have been produced. The binding agent or resin fixes the pigment to the nonwoven during drying and thermal curing. The pigment particle size, particle dispersion uniformity within the resin, and resin application method can influence the uniformity of the shade.

Dyestuffs are applied to nonwovens using batch or continuous systems. In a short period, nonwovens typically dye a deeper shade than knitted or woven fabrics due to their high porosity and accessible fibre surface area, high permeability and absence of twisted yarns and yarn intersections. Nonwoven products that traditionally include a dyeing step, include high quality drylaid-thermally point bonded polyamide (PA) shoe linings. The type of nonwoven fabric and fibre composition, as well as the class of the dyestuff influence the quality and shade of the dyeing. Dyeing nonwovens containing a homogeneous fibre composition with a single dyestuff is relatively straightforward; however, the process can be complicated in blends, where the selected dye may not have affinity for all of the fibre components. Dyestuff fixation, levelness and shade reproducibility can be enhanced with chemical auxiliaries including, wetting and penetrating agents, antifrosting agents, blocking agents (if applicable in the case of blends), dispersing agents, pH adjusters and buffers, antifoaming agents, swelling and fixation chemicals.

11.15.1 Printing

Printing of nonwoven substrates as it relates to colouration, normally involves the application of a colourant, paste or ink to produce a design in a predetermined pattern. There are four main types of printing: direct, pigment, resist, and discharge printing. In direct printing, dyes in an admixture containing thickeners and auxiliaries are applied to the nonwoven and subsequently fixed by steaming. Residual dye and auxiliary chemistry left on the fibre surfaces are required to be washed off. During pigment printing, water insoluble pigments are dispersed within an adhesive binder system and applied to the fabric, followed by drying and/or curing to fix the pigment to the fibre surfaces. Resist printing involves printing a resist agent onto the nonwoven substrate, which prevents colour being applied to that area when subsequently overprinted. The principle of discharge printing involves printing predyed fabric with a reducing agent. The reducing agent decolourises the dye, leaving a white area where the print was applied against a darker background. For nonwoven applications, pigment printing is commonly employed as it is a relatively simple, economic process and is suitable for printing fabrics composed of multiple fibre types, including synthetics. As a mode of colouration, printing can reduce environmental impacts compared to dyeing, and pigment printing, for example, is claimed to save up to 80 L of water per printed metre [43].

Rotary and flat screen printing (rotary predominantly) remain the most important methods for printing high volumes of nonwoven fabrics, it has been suggested that greater than 60% of printed fabrics use these methods [44]. Screen printing techniques offer wide-ranging colour and design possibilities. The nonwoven fabric is fed continuously along the print table by a moving belt and passes either continuously (in the case of rotary printing) or intermittently (in the case of flat screen printing) below the print screen. The design may comprise a single colour whereby one screen is employed, or a series of screens may be used to build up the print design. The print medium is a paste comprised of pigment or dyestuff with chemical auxiliaries (thickeners, antifoams, adhesives, plasticisers). The paste is applied on the screen and colour applied using a squeegee, usually a blade, roller or magnetic rod, to push the paste through the screen in specific areas as determined by the print design.

Improvements to screen printing methods include reduced set up times, increased printing speeds, low tension fabric feed, precision systems for screen location, decreased changeover times between products, precision control of squeegee pressures and evenness and individual drive of printing positions for increased accuracy. Faults such as misfits, lint, missing colour and screen blockages can be automatically monitored for quality control purposes allowing waste reduction interventions to be made. Online print monitoring systems located after the last screen detect printing faults as they occur during the printing processes.

When pigment printing nonwovens, the admixture of chemicals used to apply and adhere the pigment to the substrate is important to ensure prolonged mechanical performance. Pigments are applied along with a suitable binder and auxiliary chemicals, which include thickeners (to improve the rheology of the paste), a cross linking catalyst for the binder and softeners. The pigment is applied together with the binder in the print-bonding process. After printing and drying, the fabric is baked to cure the binder and fix the pigment. For many nonwoven processes, washing is not always necessary, but pigment printed fabrics can suffer from reduced haptics, such as stiff handle in the printed areas. Concurrent with fixing the pigment to the fibre, binders have a secondary effect on the bonding of the nonwoven which can enhance fabric mechanical properties and dimensional stability.

Print bonding, in which a pigment is applied at the same time as the binder, is important in the manufacture of commercial and household wipes to increase wet strength, modify appearance and to control the wet pick up of the wipe in use. Recent developments have seen the manufacture of fully biodegradable and compostable J-Cloth fabrics compliant with the DIN EN 13432:2000-12 standard, whilst meeting the performance requirements for commercial kitchen use [45].

Both with dyestuffs and pigment printing, the viscosity of the print paste is important to ensure the required definition and clarity of the printed area. The viscosity is dictated by the thickener type and concentration and may include alginates, guar gum derivatives, or synthetic thickeners. The penetration of the paste into the fabric structure is important to prevent poor mechanical properties and resistance to abrasion, such as low crock (rub) fastness.

Moreover, the thickening agent must also be stable during any subsequent fixation treatments, such as steaming. Any breakdown of the paste system will result in flushing of the print and a reduction of print definition. Rheology modifiers such as thixotropic thickeners may also be used which shear under the mechanical action of the squeegee. Other additives to the print paste include wetting and levelling agents, fixing agents, antifoam, humectants, sequestrants, and antioxidants.

An alternative print process to screen printing is rotogravure printing. In this technique, tiny ink-capturing cells are laser engraved into a metal cylinder. Through a combination of the capillary action of the substrate and pressure applied to the engraved print cylinder the image is transferred to the nonwoven. This technique is suitable for long production runs of simple patterns as the engraved rollers are expensive to manufacture and change but offer high levels of repeatability and reliability. The patterns on baby diapers and feminine care products are traditionally printed using this method, although ink-jet printing is now also possible [46].

Discharge printing overcomes the issue of producing light colours on predominantly dark substrates. Generally, the fabric is predyed to the required dark shade and then overprinted with a paste containing a reducing agent to discharge some/all of the base dyed colour resulting in an area of lighter shade. If a different colour is required, the paste may also contain dyes stable to the discharging agent, so that as the base colour is discharged it is replaced by the stable applied dyestuff. Discharge techniques mean that light and bright shades can be produced on dark grounds with good print definition, which can be difficult to achieve if printing light and dark shades in intricate close proximity designs (print registration) using conventional techniques. Careful optimisation of the discharging procedure such as paste composition, dwell time and processing temperatures, is required as reducing agents can affect fibre (particularly mechanical) properties. In practice, discharge printing is rarely used on nonwovens due to the cost of both the dyeing and printing procedures.

Dye-sublimation or transfer printing techniques involve the sublimation of dyestuffs from a release paper preprinted with the design onto the nonwoven fabric. The release paper and fabric are brought together and passed around a heated drum or thermal press, the design being transferred by a combination of heat and pressure. Polyester fabrics are most suitable for printing with sublimable disperse dyestuffs, which are transferred at temperatures of around 200°C. It is important that the nonwoven fabrics are thermally stable at the application temperature, otherwise shrinkage may occur. The processing time will differ depending upon the molecular weight of the disperse dyestuffs as sublimation rate is affected. This type of printing is typically suitable for small print runs and is often a batch process although systems for ink-jet printing with sublimation inks is now a commercial possibility [47]. The main disadvantages of transfer printing is the limited penetration of dye into the nonwoven substrate, meaning the pattern is very one-sided and may be easily abraded [48].

Digital ink-jet printing is a nonimpact printing technique which enables intricate computer design patterns to be transferred onto nonwoven fabric substrates. Computer software is used to generate the print design with output to a wide bed printer of sufficient width to print directly onto fabric. Once considered slow and only suitable for pattern design work with shorter sampling times, inkjet printers are now capable of speeds of up to $90\,\mathrm{mm\,min^{-1}}$ making it competitive with rotary screen printing. In addition to speed, inkjet printing eliminates the fixed cost of roller or screen preparation and enables the rapid deployment of print information in digital format [49].

Digital printing is the fastest growing sector of the of textile printing sector, estimated to be growing at a rate of 25% per annum and to be worth €4.9 billion by 2023 [50,51]. Many recent printing system developments have focused on waterless digital textile print systems, reflecting the desire for greater sustainability from colouration systems. As well as reducing water and ink and chemical use, digital printing machines are also more energy efficient compared to conventional rotary screen printers, the former using 0.14 kw/m printed compared to 0.46 kw on average for screen printing. The costs of treatment plants for the effluent produced from screen printing is increasing capital investment for screen print producers as regulation increases, which encourages switching to digital systems. Conventional screen printing systems are liquid intense; unused colours, wash-off from the screens and waste water result in millions of litres of contaminated effluent each year [43].

Industrial ink-jet systems are supplied in multi or single-pass configurations. Multipass systems operate in a similar manner to home paper ink-jet printers, using a limited number of print heads which pass over the stationary fabric in the cross-direction before indexing the substrate to allow the next area to be printed. The capital cost is reduced by only needing a few print heads, but speed and print quality is compromised by the intermittent action of the printer. For high-speed continuous printing, single-pass printing systems have been designed. In single-pass printing multiple stationary printer heads are positioned across the complete width of the machine so the fabric can pass through the machine continuously, permitting increased printing speeds of $40\,m\,min^{-1}$ compared to $6\,m\,min^{-1}$ of multipass systems [52]. The print quality and sharpness are also improved as registration with previously printed areas across the width is not required as is the case with multipass systems.

Various systems are available, one example is the Drop on Demand (DOD) system which uses thermal or bubble-jet technology whereby the print head ejects a drop of dye at high temperature. The printing inks are designed specifically for the fabric composition to be printed and include reactive, acid and disperse dyes as well as pigments. Colourants should have low conductivity to be compatible with ink-jet systems; sodium chloride may be added to modify conductivity and ensure compatibility with continuous processing [53]. The high porosity and variable surface structure of many nonwoven fabrics necessitates their precoating or thermal calendering to provide a more suitable surface for ink-jet printing. Evolon hydroentangled nonwovens formed from splittable fibres are offered as suitable substrates for digital printing with latex inks [54].

Digital inkjet printing is finding new applications in 'smart' nonwovens and printed electronics. Electrically conductive polymer inks are deposited on to the nonwoven substrates to produce circuitry for multiple uses including wearable electronics, light-emitting diodes, temperature and motion sensors, heating elements, and bodily health monitoring [55,56]. Nonwovens have the potential to overcome limitations with woven substrates which tend to spread inks through interyarn interstices; important parameters affecting printability include fibre surface energy, fibre orientation, fibre diameter, fabric pore structure, and ink viscosity [57]. Liquid X Printed Metals (Pittsburgh, Pennsylvania) are using conformal inkjet printing on nonwoven fabrics with reactive metal inks for e-textile applications. The particle-free silver-based technology imparts conductivities of $<1\,\Omega/cm$ with high flexibility and resistance to straining and washing. Both polyester and polyamide nonwovens have been successfully printed using the Liquid X technology [58].

11.15.2 Applications of printed nonwovens

Nonwoven fabrics are printed for many applications and purposes, ranging from simple branding to full colouration/design. Several manufacturers, such as Autotech and Freudenberg-Vilene, are providing opportunities for personalisation through printed nonwoven automotive headliners. This enables OEM designers the ability to customise cabin interior designs whilst maintaining the stringent mechanical performance of the product [59]. Sales of nonwoven wallcoverings have more than doubled in the last 15 years, substituting traditional wallpapers. Wetlaid and spunbond nonwovens have been employed as digitally printed wallcoverings, offering improved breathability, tear, puncture, and abrasion resistance compared to traditional paper substrates [60]. Printed nonwovens are often employed in limited lifespan products such as those related to exhibitions, which includes displays, banners, and spunbond tote bags. Other applications for printed nonwovens include floorcoverings as well as tablecloths and suit-carriers.

11.16 Converting processes

After roll good formation, which may or may not be followed by finishing processes, nonwovens are converted into final products using a variety of highly specialised, automated, and semiautomated processes, prior to packaging and distribution. Converting lines mostly comprise a series of integrated, semi-, or fully automatic high-speed processes, which progressively convert the nonwoven roll good into individual finished products, which are then packaged, ready for distribution, as part of the process. A full overview of nonwoven converting process technology is beyond the scope of this book, but common processes that form part of nonwoven converting lines include unwinding, slitting, cutting, pressing, coating, impregnating, spraying, laminating, printing, folding, dosing and filling, injecting, sewing, welding, or heat sealing to make finished products.

Acknowledgements

The authors gratefully acknowledge the contribution of Mr. Steve Myers and Dr Idris Ahmed in assisting the preparation of the chapter based on the First Edition of the Handbook of Nonwovens. The contributions of the following organisations in the preparation of this chapter are also gratefully acknowledged: Fibroline (France), Micrex Corporation (United States), Cavitec (Switzerland), Lacom (Germany), and Zimmer (Austria).

References

[1] M.J. Tipper, Open innovation and a human-centred approach for the nonwovens industry, in: RISE 2019, INDA, Raleigh, 2019.

[2] E. von Hippel, An emerging hotbed of user-centered innovation, Harv. Bus. Rev. 85 (2) (2007) 27–28.

[3] R. Verganti, User-Centered Innovation Is Not Sustainable, Harvard Business Review Blog, 2010. 2017 http://blogs.hbr.org/cs/2010/03/user-centered_innovation_is_no.html.

[4] W. Sim, R.T. Barnard, M.A.T. Blaskovich, Z.M. Ziora, Antimicrobial silver in medicinal and consumer applications: a patent review of the past decade (2007–2017), Antibiotics 7 (2018) 1–15, https://doi.org/10.3390/antibiotics7040093.

[5] R. Erdem, S. Rajendran, Influence of silver loaded antibacterial agent on knitted and nonwoven fabrics and some fabric properties, J. Eng. Fibers Fabrics 11 (2016) 38–46, https://doi.org/10.1177/155892501601100107.

[6] Avgol to Unveil Next Generation Antimicrobial Solution at Anex, 2018. https://www.avgol.com/avgol-to-unveil-next-generation-antimicrobial-solution-at-anex-2018/16.

[7] Viraloff Textile Treatment Technology, 2020. https://polygiene.com/viraloff/.

[8] HeiQ Viroblock – Antiviral & Antibacterial Protection, 2020. https://heiq.com/technologies/heiq-viroblock/?no_popup=1.

[9] Croda, Cirrasol PP844 datasheet, 2020.

[10] Cht R. Beitlich GMBH, Brochure: Nonwoven – Multitalent in Versatile Applications, 2017, pp. 1–24.

[11] M.I. Kiron, An Overview of Flame Retardant Nonwoven Fabrics, 2020. https://www.technicaltextile.net/articles/an-overview-of-flame-retardant-nonwoven-fabrics-7125.

[12] G.S. Selwyn, Water-free finishing: combining DWR + Dye for sustainable manufacturing: a zero discharge technology, in: Textile Coating and Laminating, Berlin, 2019, 2019.

[13] A. Demšar, D.D. Žnidarčič, D.G. Svetec, Impact of UV radiation on the physical properties of polypropylene floating row covers, Afr. J. Biotechnol. 10 (2011) 7998–8006, https://doi.org/10.5897/ajb10.2538.

[14] K. Singha, A review on coating & lamination in textiles: processes and applications, Am. J. Polym. Sci. 2 (2012) 39–49, https://doi.org/10.5923/j.ajps.20120203.04.

[15] H.T. Deo, V.D. Gotmare, Acrylonitrile monomer grafting on gray cotton to impart high absorbency, J. Appl. Polym. Sci. 72 (7) (1999) 887–894.

[16] F. Hochart, R. De Jaeger, J. Levalois-Grutzmacher, Graft-polymerization of a hydrophobic monomer onto PAN textile by low-pressure plasma treatments, Surface Coat. Technol. 165 (2003) 201–210.

[17] M. Louati, A. Elachari, A. Ghenaim, C. Caze, Graft copolymerization of polyester fibres with a fluorine-containing monomer, Text. Res. J. 69 (5) (1999) 381–387.

[18] M.J. Tsafack, J. Levalois-Grutzmacher, Plasma-induced graft-polymerization of flame retardent monomers onto PAN fabrics, Surf. Coat. Technol. 200 (2006) 3503–3510.

[19] M. Tsukada, H. Shiozaki, G. Freddi, J.S. Crighton, Graft polymerization of benzyl methacrylate onto wool fibres, J. Appl. Polym. Sci. 64 (1997) 343–350.

[20] N. Abidi, Surface grafting of textiles, in: Q. Wei (Ed.), Surface Modification of Textiles, Woodhead Publishing Limited, s.l., 2009, pp. 91–107.

[21] M.J. Tipper, The holy grail of universal plasma, Filtr. Sep. 55 (6) (2018) 24–26. ISSN 0015-1882 https://doi.org/10.1016/S0015-1882(18)30376-8. https://www.sciencedirect.com/science/article/pii/S0015188218303768.

[22] R.M. Reinhardt, J.A. Harris, Ultraviolet radiation in treatments for imparting functional properties to cotton textiles, Text. Res. J. 50 (1980) 139–147.

[23] K.L. Choy, Chemical vapour deposition of coatings, Prog. Mater. Sci. 48 (2003) 57–170, https://doi.org/10.1016/S0079-6425(01)00009-3.

[24] C. Jones, M.L. Hitchman, Overview of chemical vapour deposition, in: Chemical Vapour Deposition: Precursors, Processes and Application, The Royal Society of Chemistry, 2009, pp. 1–36.

[25] Innovation in Textiles, Commercialization of CVD Coated Nonwovens, 2012. https://www.innovationintextiles.com/protective/commercialization-of-cvd-coated-nonwovens/.

[26] H. Burnette, Lightweight Metal CVD Coated Nonwovens Work Well, Cost Less, Wright-Patterson Air Force Base, 2012. https://www.wpafb.af.mil/News/Article-Display/Article/399506/lightweight-metal-cvd-coated-nonwovens-work-well-cost-less/.

[27] D.M. Mattox, Handbook of Physical Vapor Deposition (PVD) Processing, Elsevier, 2010, https://doi.org/10.1016/B978-0-8155-2037-5.00025-3.

[28] C.A. Bishop, Vacuum Deposition Onto Webs, Films, and Foils, second ed., Elservier, 2011, https://doi.org/10.1017/CBO9781107415324.004.

[29] R.L. Shishoo (Ed.), Plasma Technologies for Textiles, Woodhead Publishing Limited, 2007.

[30] R. Väänänen, P. Heikkilä, M. Tuominen, J. Kuusipalo, A. Harlin, Fast and efficient surface treatment for nonwoven materials by atmospheric pressure plasma, Autex Res. J. 10 (2010) 8–13.

[31] P. Lippens, Low-pressure cold plasma processing technology, in: Plasma Technologies for Textiles, Woodhead Publishing Ltd, 2007, pp. 64–78.

[32] E. Rogge, Environmental Benefits of Low Pressure Plasma Coatings, EDANA, Nonwovens Innovation Academy, Leeds, UK, 2015.

[33] E. Rogge, F. Legein, Environmentally friendly low pressure plasma nanocoatings for filtration and separation, Filtech 2013 – L16 – Functionalized Filter Media I, 2013, pp. 1–16.

[34] J. Kelly, W. Mecouch, W. Hooke, G. Maguire, Webinar – Plasma treatment to improve filter performance, International Technology Center Inc, INDA, 2011.

[35] W. Urbaniak-Domagała, H. Wrzosek, H. Szymanowski, K. Majchrzycka, A. Brochocka, Plasma modification of filter nonwovens used for the protection of respiratory tracts, Fibres Text. East. Eur. 83 (2010) 94–99.

[36] S. Coulson, S. Russell, M. Tipper, Plasma coated fibrous filtration media, 2010. GB2462159.

[37] Q. Wei, Q. Li, X. Wang, F. Huang, W. Gao, Dynamic water adsorption behaviour of plasma-treated polypropylene nonwovens, Polym. Test. 25 (2006) 717–722, https://doi.org/10.1016/j.polymertesting.2006.03.001.

[38] B. Ghorani, S.J. Russell, R.S. Blackburn, Approaches for the Assembly of Molecularly Imprinted Nonwoven Materials and Their Utilisation in Selected Target Recognition, Nonwovens Research Academy, 2010.

[39] G. Vasapollo, R. Del Sole, L. Mergola, M.R. Lazzoi, A. Scardino, S. Scorrano, G. Mele, Molecularly imprinted polymers: present and future prospective, Int. J. Mol. Sci. 12 (2011) 5908–5945, https://doi.org/10.3390/ijms12095908.

[40] F. Salaün, Microencapsulation technology for smart textile coatings, in: Active Coatings for Smart Textiles, Woodhead Publishing Series in Textiles, Woodhead Publishing, s.l., 2016, pp. 179–220.

[41] Z. Lidert, Microencapsulation: an overview of the technology landscape, in: Personal Care & Cosmetic Technology, Delivery System Handbook for Personal Care and Cosmetic Products, William Andrew Publishing, 2005, pp. 181–190. ISBN 9780815515043 https://doi.org/10.1016/B978-081551504-3.50013-4. https://www.sciencedirect.com/science/article/pii/B9780815515043500134.

[42] F. Fu, L. Hu, Temperature sensitive colour-changed composites, in: Advanced High Strength Natural Fibre Composites in Construction, Woodhead Publishing, 2017, pp. 405–423.

[43] D. McKeegan, ITMA. Waterless Digital Textile Printing Takes Centre Stage, 2019, FESPA Website https://www.fespa.com/en/news-media/features/itma-2019-waterless-digital-textile-printing-takes-centre-stage.

[44] A.R. Horrocks, Handbook of Technical Textiles, Woodhead Publishing, 2000.

[45] M. Malocho, J-Cloth biodegradable: the original and unique compostable food service wipe, in: International Nonwovens Symposium, 2019.

[46] V. Warner, H. Yang, Process and Apprataus for Printing Assembled Absorbent Articles With Custom Graphics, 2016. EP2890343B1.

[47] The Gill, HP Revolutionises Dye-Sublimation Printing With New HP Stitch Thermal Ink Technology, 2019. https://thegill.co.uk/advice/435-hp-revolutionises-dye-sublimation-printing-with-new-hp-stitch-thermal-ink-technology.

[48] J.R. Aspland, The coloration and finishing of nonwoven fabrics, in: Nonwoven Enhanc. Color. Finish, 2005.

[49] L.P. Chapman, Digital Printing Innovations, Textile World, 2010.

[50] CEMATEX, ITMA, Brochure: Innovating the World of Textile Printing, 2019.

[51] Smithers Pira, The future of digital textile printing to 2023, 2018.

[52] SPG prints, The ultimate guide to industrial digital textile printing, 2020.

[53] C.W.M. Yuen, C.W. Kan, A study of the properties of ink-jet printed cotton fabric following low-temperature plasma treatment, Color. Technol. 123 (2007) 96–100, https://doi.org/10.1111/j.1478-4408.2007.00068.x.

[54] Freudenberg Performance Materials, Evolon®, The Eco-Friendly and Innovative Printing Media for Wide-Format Advertising, 2020. https://evolon.freudenberg-pm.com/applications/Printing media.

[55] L. van Langenhove, Smart Textiles for Medicine ad Healthcare: Materials, Systems and Applications, Woodhead Publishing, Cambridge, 2007.

[56] Roepert, Smart textiles and wearable electronics – From Hype 2.0 to Industry 4.0, in: Dornbirn Glob. Fibre Congr, 2019.

[57] B. Karaguzel, Printing Conductive Inks on Nonwovens: Challenges and Opportunities, North Carolina State University, 2006.

[58] B. Babe, Conformal inkjet printing on nonwoven fabrics with reactive metal ink for e-textile applications, in: EDANA International Nonwovens Symposium, 2018.

[59] K. McIntyre, Nonwovens in Automotives, Nonwovens Industry, 2017.

[60] A. Wilson, Perfect Canvas, Nonwovens Industry, 2020.

CHAPTER

12

Characterisation, testing, and modelling of nonwoven fabrics

N. Mao[a], S.J. Russell[a], and B. Pourdeyhimi[b]

[a]University of Leeds, Leeds, United Kingdom [b]The Nonwovens Institute (NWI), North Carolina State University, Raleigh, NC, United States

12.1 Introduction: Characterisation of nonwoven fabrics

The physical, chemical, and mechanical properties of nonwovens governing their suitability for use depend on both the fabric's composition and structure. The composition in this context refers to fibre properties as well as any chemical binders, fillers, or finishes present on or between the fibres in the fabric. The structure relates to the architecture and dimensional properties of the fabric, including the way in which the fibres are arranged. This chapter focuses on how nonwoven materials are tested and characterised and considers the influences of fabric structure on the physical and mechanical properties of fabrics. Additionally, some of the models for describing the relationship between nonwoven fabric structure and important physical properties are introduced.

Nonwoven fabrics comprise a wide range of engineered fibrous assemblies that are responsible for the functional performance of a diverse range of products and the definition has continued to evolve over the years. A nonwoven is defined in both ISO 9092:1988 and BS EN 290925:1992 as *a manufactured sheet, web or batt of directionally or randomly orientated fibres, bonded by friction, and/or cohesion and/or adhesion, excluding paper and products which are woven, knitted, tufted, stitch-bonded incorporating binding yarns or filaments, or felted by wet-milling, whether or not additionally needled the fibres may be of natural or man-made origin. They may be staple or continuous filaments or be formed in situ* [1]. There is also a difference between wetlaid nonwovens and paper. Specifically, 'a material shall be regarded as a nonwoven if (a) more than 50% by mass of its fibrous content is made up of fibres (excluding chemically digested vegetable fibres) with a length to diameter ratio greater than 300; or, if the conditions in (a) do not apply, then (b) if the following conditions are fulfilled: (1) more than 30% by mass of its fibrous content is made up of fibres (excluding chemically digested vegetable fibres) with a length to diameter ratio greater than 300 and (2) its density is less than 0.40 g/cm^3' [1].

A revised definition was subsequently proposed by EDANA and INDA, the major trade associations representing the nonwovens industry in 2010 as [2] *a sheet of fibres, continuous filaments, or chopped yarns of any nature or origin, that have been formed into a web by any means, and bonded together by any means, with the exception of weaving or knitting. Felts obtained by wet-milling are not nonwovens. Furthermore, Wetlaid webs are nonwovens provided they contain a minimum of 50% of manmade fibres or other fibres of nonvegetable origin with a length to diameter ratio equals or superior to 300, or a minimum of 30% of man-made fibres with a length to diameter ratio equals or superior to 600, and a maximum apparent density of 0.40 g/cm^3. Composite structures are considered nonwovens provided their mass is constituted of at least 50% of nonwoven as per to the above definitions, or if the nonwoven component plays a prevalent role.*

Nonwovens were defined in ISO 9092:2011 as *structures of textile materials, such as fibres, continuous filaments, or chopped yarns of any nature or origin, that have been formed into webs by any means, and bonded together by any means, excluding the interlacing of yarns as in woven fabric, knitted fabric, laces, braided fabric or tufted fabric (Film and paper structures are not considered as nonwovens)* [3].

In the ISO standard (ISO 9092:2019), a nonwoven is defined as *an engineered fibrous assembly, primarily planar, which has been given a designed level of structural integrity by physical and/or chemical means, excluding weaving, knitting or papermaking* [4]. The physical and/or chemical means here refer to *bonding technologies that result in frictional forces between fibres (through entanglement) or adhesive forces between fibres (with or without the use of binders)* [4].

The ISO standard also defines other specific types of nonwovens and differentiates wetlaid process from the paper making process [4]. Paper making refers to the *process of producing a thin material by pressing together, short, refined cellulose fibres formed on a screen from a water suspension of these fibres, and drying them, with hydrogen bonding as the predominant mechanism holding the web together*, and the wetlaid process that is not paper making refers to the *process where cellulose or other fibres are engineered to a level of structural integrity primarily by physical and/or chemical means other than hydrogen bonding.*

Films cast, blown or extruded from polymers, which then through physical or chemical means are made into fibrous assemblies, are considered nonwovens if the length to diameter (L/D) ratio of the fibrous elements is over 30. Stitch-bonded nonwovens are fabrics *engineered to a given level of integrity by physical means for specific applications and the warp or circular knit stitching is the additional bonding technology.* Wadding refers to *high-loft assemblies, primarily fibrous, engineered to a given level of integrity by physical means for specific applications.*

To summarise, nonwoven structures are different from other textiles because the fabrics:

1. Principally consist of individual fibres or layers of fibrous webs rather than yarns.
2. Are anisotropic both in terms of structure and properties due to the fibre orientation distribution and the diversity of bonding point arrangements.
3. Are highly porous and permeable.
4. Are usually not entirely uniform in either fabric weight per unit areas and/or fabric thickness, or both.

Although nonwovens share some characteristics with textile fabrics, paper, and plastics in terms of composition and structures, they are particularly diverse, and specific functionalities and performance characteristics can be easier to achieve than with traditional textile materials.

The structural architecture of nonwoven fabrics is influenced by raw materials selection, web formation, bonding, and fabric finishing processes. Therefore, adjusting processing conditions during fabric manufacture enables structural features to be modulated and physical properties to be engineered. In addition to the structural architecture of the fabric, key parameters in the engineering of nonwovens include fibre properties, the type of bonding elements, and the nature of the bonding interfaces between fibres and binder elements (if present). Specific examples of dimensional and structural parameters relating to nonwovens may be listed as follows:

1. *Fibre dimensions and properties*: e.g., fibre diameter and its distribution, fibre length and its distribution, fibre cross-sectional shape and aspect ratio, crimp frequency and amplitude, density and the detailed mechanical, physical, and chemical properties of the fibres.
2. *Fibre orientation*: usually considered in terms of the fibre orientation distribution.
3. *Fabric dimensional properties*: e.g., weight per unit area and its variation, thickness and its variation, porosity, pore size, pore size distribution, pore shape, density, and dimensional stability.
4. Bond point structural properties: e.g., bond type, shape, size, bonding area, bonding density, bond strength, bond point distribution, geometrical arrangement, degree of liberty of fibre movement within and between the bonding points, binder–fibre interface properties, and the surface properties of bond points.

By careful selection of the polymers, fibres, structural and dimensional properties of the nonwoven, key performance attributes can be engineered, for example:

1. *Mechanical properties*: fabric tensile properties (Young's modulus, tenacity, strength and elasticity, elastic recovery, work of rupture), compression and compression recovery, bending and shear rigidity, tear resistance, bursting strength, crease resistance, abrasion resistance, frictional properties (smoothness, roughness, friction coefficient), energy absorption.
2. *Fluid handling properties*: permeability, liquid absorption (liquid absorbency), penetration/inlet time, wicking rate, re-wet, liquid retention, particle collection, liquid repellency and barrier properties, run-off, strike time, water vapour transport, and breathability.
3. *Physical properties*: thermal and acoustic insulation and conductivity, electrostatic properties, dielectric constant and electrical conductivity, opacity, and many others.
4. *Chemical properties*: surface wetting angle, oleophobicity and hydrophobicity, interface compatibility with binders and resins, chemical resistance and stability, flame resistance, dyeing capability, flammability, soiling resistance.
5. *Application specific performance*: linting (particle generation), aesthetics and handle, filtration efficiency, biocompatibility, sterilisation compatibility, dispersiblity (flushability), biodegradability and environmental stability, regulatory as well as health and safety compliance.

12.2 Characterisation of fabric bond structure

Nonwoven fabrics contain bond structures that hold the fibrous network together, the type, shape, rigidity, size, and density of which may be characterised. The bonds are typically frictional, cohesive, or adhesive. In nonwovens, the bonds can be grouped into two categories: (1) rigid, solid bonds and (2) flexible, elastic joints. These result from specific nonwoven manufacturing process. The bond points in mechanically bonded nonwovens, for example, needlepunched and hydroentangled fabrics, are formed by either interlacing individual fibres or loose fibre bundles to increase frictional resistance to slippage. These bonds are flexible, and under tension, component fibres are able to slip or move within the bonding points. By contrast, the bonds in thermally bonded and chemically bonded fabrics are formed by cohesion or adhesion between polymer surfaces, in which a small portion of the fibrous network is firmly bonded, and the fibres have little freedom to move within the bond points. For example, the bonds in thermoplastic spunbond and through-air bonded fabrics are formed by melting polymer surfaces to produce bonding at fibre cross-over points and the fibres associated with these bonds cannot move individually. In meltblown fabrics, the fibres are usually not as well bonded together as in spunbond fabrics, although in some applications, the large fibre surface area is sufficient to give the web acceptable cohesion without the need for additional bonding. Stitch-bonded fabrics are stabilised by knitting fibres or yarns through the web and the bonding points are flexible despite the knitted connections.

The size of bond structures in nonwovens is influenced by fabric manufacturing parameters. For example, the needle barb depth in relation to the fibre diameter, punch density and number of barbs that penetrate the batt (needlepunching); water jet diameter, specific energy and number of injectors (hydroentanglement); the land area and bond point area, pressure and the temperature (thermal bonding) and the method of binder application, for example full saturation, spray or printing and binder viscosity (chemical bonding).

The stability of bond structures in most nonwoven materials can be physically characterised in terms of measured tensile properties, for example, stress and strain at break, while the degree of bonding can also be directly determined by measurement of local frictional resistance or microscopic analysis of bond point structures within the fabric cross-section. In mechanically bonded fabrics, specifically needlepunched and hydroentangled structures, the depth of bent fibre loops in the bonding points can be determined and based on the depth of these fibre segment loops, a simple but limited estimate of bonding intensity can be derived [5].

12.2.1 Needlepunched fabrics

Needlepunched fabrics have characteristic periodicities in their structural architecture that result from the interaction of fibres with the needle barbs. Fibre segments (or bundles of fibre segments) caught by the barbs of a needle are reoriented and migrated from the surface of the web along the direction of fabric thickness towards the interior of the fabric forming pillars of fibre oriented approximately perpendicular to the plane (see Fig. 12.1C and D). The entangled fibre pillars act as bonding points inside the needlepunched nonwoven fabric structure and influence a range of mechanical properties. On the fabric surface,

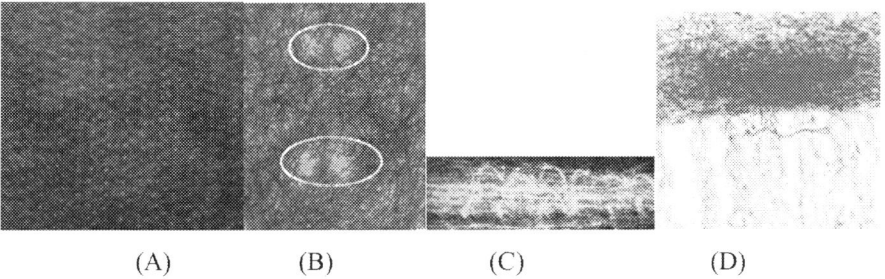

(A)   (B)   (C)   (D)

FIG. 12.1 Structure of needlepunched fabrics. (A) Needle marks in the surface of a needlepunched fabric; (B) Two pairs of needle marks in the fabric surface; (C) Cross-section of needle marks [6]; (D) 3D pillar structures in the fabric cross-section revealed by X-ray microcomputed tomography [7].

needle marking is frequently visible (see Fig. 12.1), which is associated with the punch hole locations, and fibres within the vertical pillars formed in the cross-section during the process, may also be joined in the fabric plane running in the machine direction [6]. The three-dimensional structure of the vertical pillars and their development under different needlepunching conditions have been visualised using various techniques, including X-ray microcomputed tomography (XMT) (Fig. 12.1D).

On a microstructural scale, needlepunched fabrics consist of at least two different regions. The first, between the impact areas associated with the needle marks, is not directly disturbed by the needles and retains a similar structure as the original unbonded web. The second region, the needle-marked area, contains fibre segments that are orientated approximately perpendicular to the fabric plane to produce pillars. Some fibre segments are realigned in the machine direction. This rearrangement of fibre segments induced by the process effectively increases the structural anisotropy as compared with the original web and therefore the structure of needlepunched fabrics, like many other nonwovens, is not homogeneous. Both the number of needle marks and the depth of fibre penetration are related to the degree of bonding and fabric tensile strength. The shape and number of the pillars depends mainly on the number of needles in the needle board, the size of the needles, the needle throat depth, the fibre type and dimensions relative to the barb dimensions, the advance per stroke, and the punch density.

The depth of needle penetration, the number of barbs that pass through the web, and the distance each barb and its attached fibres travel are important variables influencing the microstructure. Previously, the effects of changes in penetration depth and the number of barbs [6,8,9] on the fabric structure have been investigated. These experiments demonstrated that fabric strength is influenced by changes in barb position as the needle passes through the web. Maximum fabric tenacity may be obtained with only three barbs per apex if the depth of penetration is adjusted accordingly. Although needlepunched fabrics have some fibre segments aligned in the transverse direction [8], the majority remain aligned in-plane and the fabrics have a greater porosity and a larger number of curved inter-connected pore channels than woven fabrics. Hearle et al. [6] observed that the punched loops of fibre do not protrude from the lower surface of the fabric when the needle penetration is small and the resulting fabric appearance and needle marks are illustrated in Fig. 12.2A. A pseudo-knitted appearance

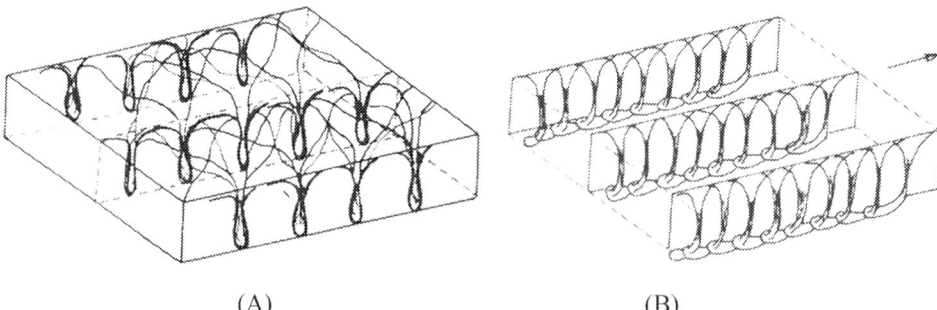

(A)   (B)

FIG. 12.2 Needled fabric structures [6]. (A) Low level of needling density and low needle penetration; (B) high level of needling density and high needle penetration.

resulting from linked loops of fibre tufts [6] produced by the needle barbs can be detected on the fabric surface when the needle penetration is large, as illustrated in Fig. 12.2B.

12.2.2 Hydroentangled fabrics

The microstructure of hydroentangled fabrics is quite different from needlepunched fabrics in that the formation of discrete pillars of fibre in the fabric cross-section is absent. However, the incident high-speed water jets locally migrate, fibre segments, both in the transverse and in-plane machine directions. Some fibre segments impacted by the water jets are bent and formed into 'U' shape configurations visible in the fabric cross-section (see Fig. 12.3). The number of fibre segments deformed in this way and their maximum penetration depth in

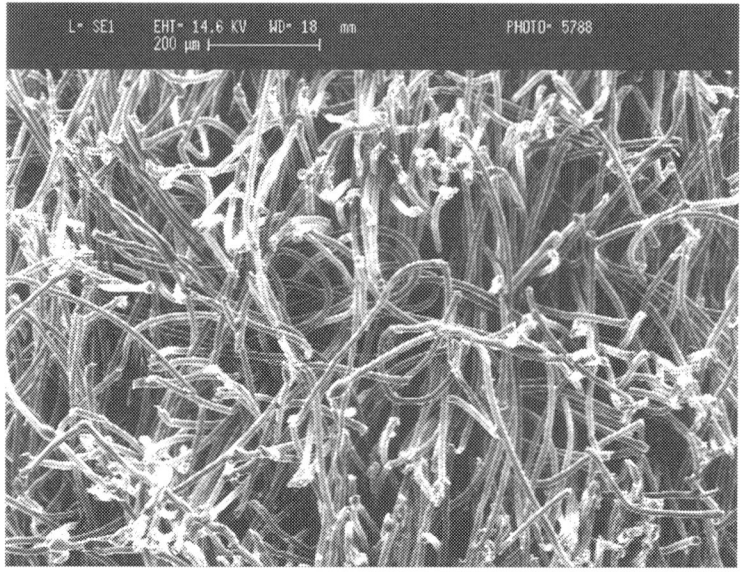

FIG. 12.3 Example of curved, 'U'-shaped fibre segments in the cross-section of a hydroentangled fabric [5].

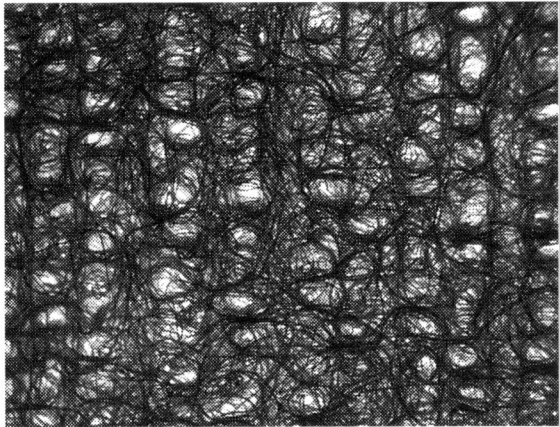

FIG. 12.4 Jet marks in the surface of a thin hydroentangled fabric ($30\,\mathrm{g\,m^{-2}}$).

the fabric cross-section can be linked to the specific energy consumed in the hydro-entanglement process, and this together with the level of the intertwining of fibres relates to the fabric bonding intensity [5]. The fabric strength mainly depends on the degree to which fibres are intertwined in the process.

Hydroentangling produces local density variations and texture in the fabric that can influence tensile and fluid flow properties as well as creating variations in local fibre segment orientation. The fabrics are consolidated mainly in the areas where the water jets impact and jet marks are formed on the fabric surface, which appear as visible 'lines' on the jet-side of the fabric running in the machine direction (Fig. 12.4). Jet marking becomes less pronounced as the number of injectors increases and as more of the surface is impacted. Where the support surface is three-dimensional, fibres can be displaced to form apertures and other structural patterns. This also produces local density variations in the fabric. Therefore, even if the original web is isotropic, structural anisotropy is introduced during hydroentanglement that may be of a periodic nature.

Various patterned structures including apertures (see Fig. 12.5) can be introduced into hydroentangled fabrics by changing the design of the support surface and adjusting the

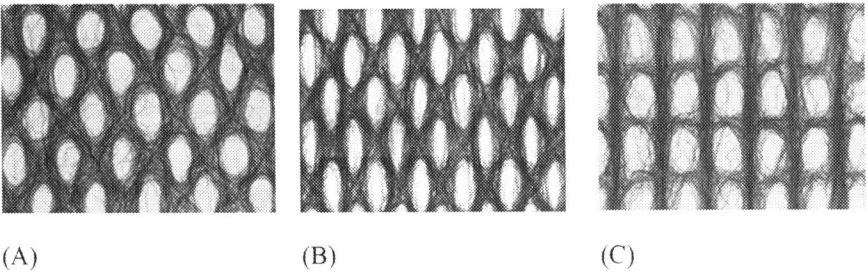

(A)   (B)   (C)

FIG. 12.5 Examples of apertured hydroentangled fabrics. (A) Elliptical holes, (B) Extended-elliptical holes, (C) Pseudo-rectangular holes.

degree of prebonding in the web and the water jet operating conditions can influence the clarity and definition of these structural features.

The structure of hydroentangled fabrics depends on process parameters and fibre properties. At low water jet pressure, only a small portion of fibre segments in the surface of the web are entangled and intertwined. At high water jet pressure, some fibre segments are reoriented towards the reverse side of the web and some fibre ends project. Fibre rigidity and bending recovery influence the ability of the jet to produce fibre entanglements during hydroentanglement and therefore the structural features of hydroentangled fabrics can differ according to fibre type. An example is fabrics made from polypropylene and viscose rayon using the same process conditions. The specific flexural rigidity of polypropylene fibre (0.51 mN mm/tex^2) is higher than viscose rayon (0.35 mN mm/tex^2), and polypropylene fibre has higher compression recovery, bending recovery and tensile recovery [10] compared to viscose rayon fibre [11]. In a polypropylene hydroentangled fabric, when the water pressure is low, only the surface fibres are effectively bonded, and the fibres inside the fabric are poorly entangled. The surface is therefore more compact than the fabric core. In contrast, a viscose rayon fabric is more consistently bonded through the cross-section and the compaction is greater than in the corresponding polypropylene fabric. This is just one illustration of how a combination of fibre properties and process settings combine to allow the nonwoven fabric structure to be engineered according to the requirements of the intended end-use or product.

12.2.3 Stitch-bonded fabrics

Stitch-bonded fabrics are formed by stitching a fibrous web together by using either a system of additional yarns (filaments) or the fibres in the web using a warp-knitting action. Because the formation of a stitch-bonded fabric is basically a hybrid of warp-knitting and sewing, it is reflected in the fabric structure. The fabric: (a) integrates stitching and a fibrous web; (b) has a clear stitching pattern on at least one side of the fabric and; (c) the stitches hold the fibres in the fibrous web together. There are three basic types of stitch-bonded fabric structure: (i) fibres bonded with the constituent fibres in which the stitches are observed on one side of the fabric (Malivlies); (ii) stitches of yarns on one surface and a projecting pile of pleated fibres on the reverse surface (Kunit); and (iii) stitches of yarns on both surfaces (Multiknit, Maliwatt). The basic structural characteristics of these different types of stitch-bonded fabric structure may be summarised as follows [12]:

- Malivlies: Fabrics are bonded by knitting fibres in the web rather than by additional yarns (filaments); therefore, the fabric consists of staple fibres. They have a warp knitted-loop stitch pattern on one side of the fabric, and the intensity of stitch bonding depends on the number of fibres carried in the needle hook. The carrying capacity depends on the dimensions of the hook and the fibre fineness.
- Kunit: Fabrics consist of a three-dimensional pile structure made from 100% fibres stitched using the constituent fibres in the web. Fibres on one side of the fabric are formed into stitches, while the other side of the fabric has a pile loop structure with fibres arranged with an almost perpendicular orientation with respect to the plane. The fabric has very good air permeability because of the high-loft structure, and excellent compression elasticity because of the vertical pile loop structure.

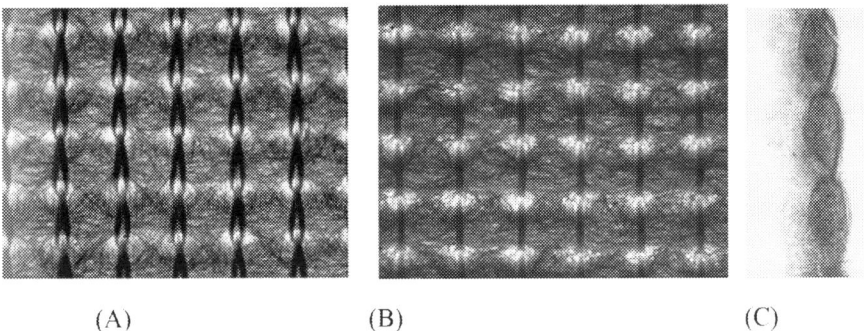

(A)   (B)   (C)

FIG. 12.6 Example of the lengthwise bond structure of a Maliwatt fabric. (A) Surface, (B) back, (C) cross-section.

- Multiknit (Malimo): These constructions are formed from Kunit pile loop fabric, and both sides are formed into a closed surface by stitched loops of fibre. The two sides of the fabric are joined together by fibres orientated almost perpendicular to the plane. The fabric is stitched using the fibres in the original web rather than by additional yarns and therefore, a three-dimensional fabric composed of 100% staple fibres is formed.
- Maliwatt: Fabrics (see Figs 12.6 and 12.7) are fibrous webs stitched through with one or two stitch-forming yarns. Both sides of the fabric have a yarn stitch pattern and the fabric weight per unit area ranges from 15 to 3000 g/m^2 with a fabric thickness of up to 20 mm, and a stitching yarn linear density in the range 44–4400 dtex.

In stitch-bonded fabrics, yarn stitches are usually aligned in the fabric plane, while the fibre piles or the fibre pile loops are fixed by the stitches and are generally orientated perpendicular to the stitched fabric surface. Stitch-bonded fabric structure is determined by the warp-knitting action applied by the machine, fibre properties and dimensions, web density and structure, stitching yarn structure, stitch density, machine gauge (number of needles per 25 mm), stitching yarn tension, and stitch length. Both the stitch holes

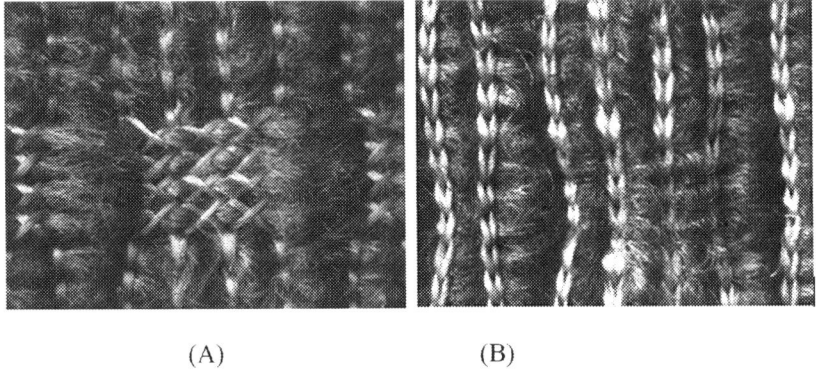

(A)   (B)

FIG. 12.7 Example of the complex bond structure in a Maliwatt fabric. (A) Surface, (B) back.

and the pile formed in the fabric surface are two unique structural characteristics of stitch-bonded fabrics. The number and size of the stitch holes depends on the properties of the stitching yarn, the properties of the fibrous web, machine gauge (number of needles per 25 mm), intermeshing intensity, interlacing, and stitching yarn tension. The pile height, visible in certain stitch-bonded fabrics, ranges from 2 to 20 mm, and depends on how the oscillating element is set at the stitch-bonding position. Both the stitch holes and the piles formed in the fabric surface influence fabric properties. The warp-knitted structure in stitch-bonded nonwovens has an open fabric construction and short underlaps; it is dimensionally extensible in the cross direction (CD), as well as in the machine direction (MD). To increase the fabric tensile strength in the MD, a specific stitch construction is used (pillar stitch). To increase the width-wise stability, the underlaps are lengthened (e.g. satin stitch) and a three-dimensionally stable structure is achieved by combining these two types of stitch construction (e.g. pillar-satin).

12.2.4 Thermal-bonded fabrics

The types of bonding structure formed in thermally bonded fabrics depend on the method used to introduce heat to the fibres, as well as the web structure and the type of binder fibre present. In calendered thermal point-bonded fabrics (see Fig. 12.8), the fibres are compressed together locally, and heat is introduced by conduction. This produces deformation of the fibres and polymer flow around the bond points. Around the immediate vicinity of the bond points, the heating of surrounding fibres can also introduce interfacial bonding at the cross-over points of uncompressed fibres. This is known as secondary bonding and is particularly noticeable when bicomponent fibres are present as the binder component.

In through-air thermal-bonded fabrics (see Fig. 12.9), core-sheath bicomponent fibres are commonly utilised and convected heat introduced during the process produces the interfacial bonding at the fibre cross-over points as the polymer softens and flows. There is no associated deformation of the fibres at these locations and therefore the resulting fabric density is lower as compared to a calendered thermally bonded fabric. Calendered thermal bond structures are found in a host of nonwoven materials both in monolithic and multilayer fabrics including SMS and other spunbond–meltblown nonwoven structures.

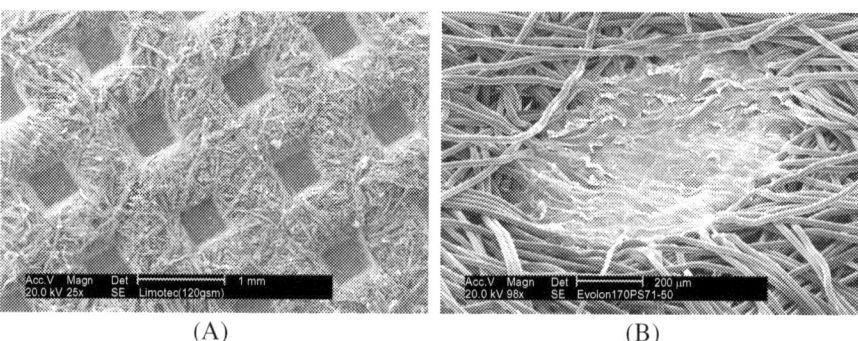

(A)   (B)

FIG. 12.8   Examples of thermally point-bonded fabrics. (A) Thermal bonding points, (B) a thermal bond point showing a change in fibre morphology within the bond point.

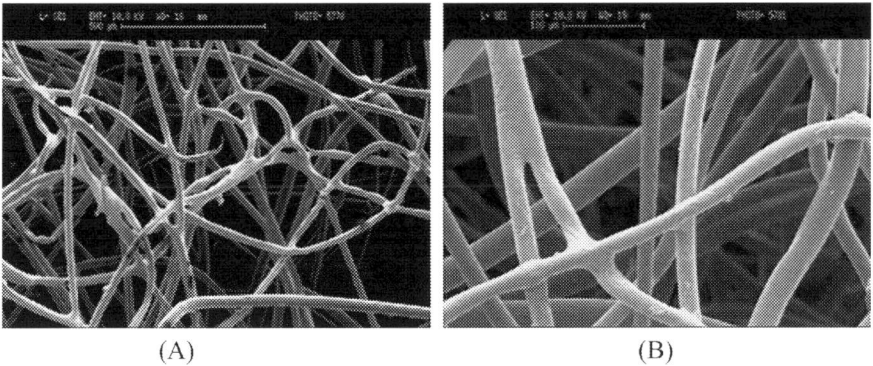

FIG. 12.9 (A) Through-air thermally bonded fabric structure. (B) Bond points in a through-air thermally bonded fabric.

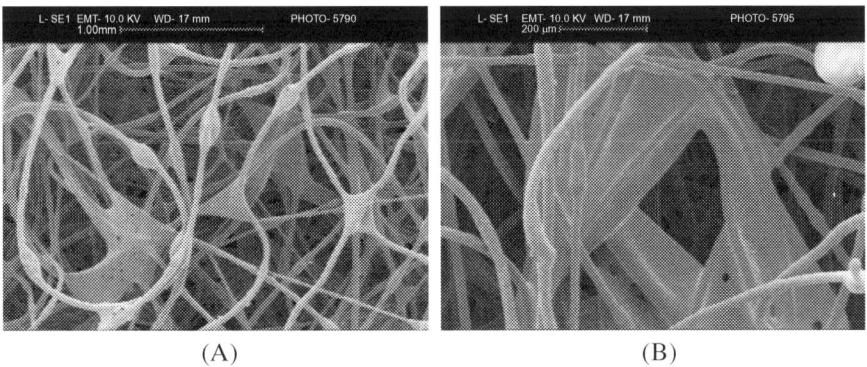

FIG. 12.10 Examples of bond points in a chemically bonded fabric.

12.2.5 Chemically bonded fabrics

Chemically bonded fabrics are produced by the application of a binder (often termed a 'resin'), which takes the form of an emulsion, made of, for example, acrylic, polyvinyl acetate, or other suitable chemical binder. The binder is applied to the web and then dried and cured. The distribution of the binder in the fabric is largely governed by its method of application to the web and the flow properties of the binder between fibres (see Fig. 12.10). Large numbers of fibres may be enveloped by a film of the binder connecting both fibre cross-over points and interfibre spaces. Large segments of these film binders are visible in such structures, spanning adjacent fibres. Alternatively, the binder may be concentrated at the fibre cross-over points, producing localised bonding in these regions and either rigid or flexible bonds, depending on the polymer composition.

12.3 General standards for testing nonwovens

Various testing methods are in use for the measurement of nonwoven fabric properties and performance. These test methods can be grouped as follows:

• Standard test methods defined by standard authorities (e.g. ISO, CEN/BSI, ASTM, and ANSI).

- Test methods established by industry associations (e.g. INDA, EDANA, AATCC, etc.), or by individual companies.
- Nonstandard tests, mainly designed for research purposes or to measure properties not yet covered by standard or industry test methods.

Standard test methods are orderly procedures that are carried out under controlled conditions and frequently in a standard testing environment. They are designed to provide reliable measurements with a certain degree of precision to enable the specification and trading of nonwovens and associated products. Industrial test methods are usually established for routine measurements linked to the evaluation, benchmarking, and quality control of semi-finished or final end products. In addition to these standard tests, numerous techniques are available to characterise nonwoven materials for either research purposes or monitoring of nonwoven production processes.

Various national and international standard systems (ISO (BS, EN, ERT) and ASTM (ITS, AATCC) standards) apply to nonwoven materials and the products into which they are incorporated. Seven major standards in Europe and North America (ISO, BS, EN with ERT, ASTM with ITS and AATCC) are summarised in this section. Many of the test standards proposed by the two nonwoven industrial organisations in Europe (EDANA) and in North America (INDA), ERT (by EDANA), and ITS (by INDA) have become part of either ISO (BS and EN) or ASTM standards. EDANA and INDA have also worked together to produce a harmonised set of nonwoven test standards, as Worldwide Strategic Partners (WSP). These nonwoven test standards became Nonwovens Standard Procedures (NWSP) in 2015 and many have been adopted by ISO or CEN. A summary of the standards relating to nonwovens is given in Table 12.1.

12.3.1 Standards for nonwoven wound dressings (ISO, BP, ASTM, and BS)

There are also standard testing methods for medical devices that relate to wound dressings in the ASTM standards, which include general practice for medical devices [13,14], analysis of medical materials [15–18], methods for medical packages [19,20], fluid penetration [21], sterilisation, and disinfection [22,23]. The BP [24] has defined a series of test methods for surgical dressings. These methods include: fibre identification, yarn number, threads per stated length (unstretched, fully stretched), weight per unit area (nonadhesive dressings, adhesive dressings, weight of adhesive mass), minimum breaking load, elasticity, extensibility, adhesiveness, water–vapour permeability (tapes, foam dressings), waterproofness, absorbency (sinking time, water holding capacity), water-soluble substances, ether-soluble substances, colour fastness, content of antiseptics, content of zinc oxide in the adhesive mass, X-ray opacity, sulphated ash of surgical dressings, and water retention capacity. Other standards relevant to wound dressings, such as the test of sterility [25], test of microbial contamination [26], efficacy of antimicrobial preservation [27], and methods of sterilisation [28] are also available.

A series of standard testing methods for wound dressings was introduced in British and European Standards BS EN 13726. These methods include aspects of absorbency [29], moisture vapour transmission rate of permeable film dressings [30], waterproofness [31], conformability [32], and bacterial barrier properties [33] and antibacterial properties [34]. Other standards related to medical fabrics include the specification for spinal and abdominal

TABLE 12.1 Summary of test standards (ASTM, BS, EN, ISO, ERT, IST, WSP, NWSP) in relation to nonwovens.

	ASTM, AATCC and MIL	EDANA and INDA (NWSP, WSP, IST, and ERT)				ISO, BS, and EN
		NWSP	WSP	IST	ERT	
Glossary of Terms (Vocabulary)	ASTM D123–19	NWSP 001.0.R0(15)	WSP 1.0(05)	IST 1	ERT 1.4-02	BS ISO 11224:2003
Nonwovens. Vocabulary					ERT 0.0-89	BS EN 29092:1992; BS EN ISO 9092:2019
How to Write a Test Method		NWSP 002.0.R0 (15)	WSP 2.0 (05)			
Sample and Laboratory Conditioning	ASTM D1776–20	NWSP 003.0.R0 (15)	WSP 3.0(05)		ERT 60.2–99	BS EN ISO 139:2005
Worldwide Associations		NWSP 004.0.R0 (15)	WSP 4.0(05)		ERT Useful Addresses	
Sampling		NWSP 005.0.R0 (15)	WSP 5.0(05)		ERT 130.2–99	BS EN 12751:1999; BS EN ISO 186:2002; ISO 2859-1: 1999; ISO 3951-1: 2013
List of Vendors		NWSP 006.0.R0 (15)	WSP 6.0(05)	IST Useful Vendor's List		
Guideline Test Methods for Nonwoven Fabrics		NWSP 007.0.R0 (15)	WSP 7.0(05)	IST GL		
Guidance to High-loft Test Methods		NWSP 008.0.R0 (15)	WSP 8.0(05)	IST GL		
Guideline Test Methods for Evaluating Nonwoven Felts				IST GL FELTS		
Safety requirements of nonwoven machinery						BS EN ISO 22291
Noise emission of nonwoven machinery						BS EN ISO 9902–3:2001

Continued

	ASTM, AATCC and MIL	EDANA and INDA (NWSP, WSP, IST, and ERT)				ISO, BS, and EN
		NWSP	WSP	IST	ERT	
Absorption						
Nonwoven Absorption		NWSP 010.1.R0 (15)	WSP 10.1(05)	IST 10.1	ERT 10.4–02	BS EN ISO 9073-6:2003
Rate of Sorption of Wiping Materials		NWSP 010.2.R1 (15)	WSP 10.2(05)	IST 10.2		
Demand Absorbency		NWSP 010.3.R0 (15)	WSP 10.3 (05)	IST 10.3	ERT 230.1–02	BS ISO 9073-12:2002
Oil/Fatty Liquids Absorption		NWSP 010.4.R0 (15)				BS EN ISO 9073-6:2003
Abrasion resistance						
Inflated Diaphragm	ASTM D3886–99	NWSP 020.1.R0 (15)	WSP 20.1 (05)	IST 20.1		
Flexing and Abrasion	ASTM D3885–07	NWSP 020.2.R0 (15)	WSP 20.2 (05)	IST 20.2		
Oscillatory Cylinder	ASTM D4157–13			IST 20.3		
Rotary Platform, double head method	ASTM D3884–09	NWSP 020.4.R0 (15)	WSP 20.4 (05)	IST 20.4		
Modified Martindale	ASTM D4966–12	NWSP 020.5.R0 (15)	WSP 20.5 (05)	IST 20.5		BS EN ISO 12947-1:1998 (BS 5690:1991)
Uniform Abrasion Method	ASTM D4158–08			IST 20.6		
Bursting strength						
Diaphragm	ASTM D3786–18	NWSP 030.1.R0 (15)	WSP 30.1 (05)	IST 30.1		
Nonwoven Burst		NWSP 030.2.R0 (15)	WSP 30.2 (05)	IST 30.2	ERT 80.4–02	BS EN ISO 13938-1:2019; 13,938-2:2019;

Electrostatic properties						
Surface Resistivity	ASTM D257–14; FED-STD-191A, Method 5930	NWSP 040.1.R1 (15)	WSP 40.1 (05)	IST 40.1		(EN 1149–1: 1995; EN 1149–2: 1995)
Decay	MIL-STD-3010, Method 4046; NFPA-99-92 Sec. 12–4.1.3.8(f)	NWSP 040.2.R0 (15)	WSP 40.2 (05)	IST 40.2		
Optical properties						
Opacity (INDA)		NWSP 060.1.R0 (15)	WSP 60.1 (05)	IST 60.1		BS ISO 2471: 2008
Brightness (INDA)	TAPPI T452 om-92	NWSP 060.2.R0 (15)	WSP 60.2 (05)	IST 60.2		
Brightness (EDANA)		NWSP 060.3.R1 (15)	WSP 60.3 (05)		ERT 100.1–78	BS ISO 2469: 2014
Opacity (EDANA)		NWSP 060.4.R0 (15)	WSP 60.4 (05)		ERT 110.1–78	BS ISO 2469: 2014
Permeability						
Air Permeability	ASTM D737–18	NWSP 070.1.R0 (15)	WSP 70.1 (05)	IST 70.1	ERT 140.2–99	BS EN ISO 9073-15:2008; BS EN ISO 9237:1995
Water Vapour transmission (Multiple Tests)	ASTM D6701–16			IST 70.2		
Liquid Strike-Through Time (Simulated Urine)		NWSP 070.3.R1 (19)	WSP 70.3 (05)	IST 70.3	ERT 150.5–02	BS EN ISO 9073-8:1998
Water Vapour Transmission (Mocon) (rates of 500–100,000 g/m^2/day)		NWSP 070.4.R0 (15)	WSP 70.4 (05)	IST 70.4		
Water Vapour Transmission (Mocon) (Relative Humidity)		NWSP 070.5.R0 (15)	WSP 70.5 (05)		ERT New Method part 1	
Water Vapour Transmission Rate (Lyssy)		NWSP 070.6.R0 (15)	WSP 70.6 (05)		ERT New Method part 2	

Continued

TABLE 12.1 Summary of test standards (ASTM, BS, EN, ISO, ERT, IST, WSP, NWSP) in relation to nonwovens—cont'd

	ASTM, AATCC and MIL	EDANA and INDA (NWSP, WSP, IST, and ERT)				ISO, BS, and EN
		NWSP	WSP	IST	ERT	
Repeated Liquid Strike-Through Time		NWSP 070.7.R1 (19)	WSP 70.7 (05)		ERT 153.0–02	BS EN ISO 9073-13:2007
Wetback After Repeated Strike-Through Time (Simulated Urine)		NWSP 070.8.R1 (19)	WSP 70.8 (05)			(BS EN ISO 9073-14:1998)
Incontinence products: Rate of Acquisition and Re-Wet Test		NWSP 070.9.R1 (15)	WSP 070.9.R1 (12)			
Incontinence products: Centrifugal Liquid Retention Capacity Test		NWSP 070.10.R2 (19)	WSP 070.10.R1 (12)			
Repellency						
Repellency						
Surface Wetting Spray Test	AATCC 22–17	NWSP 080.1.R0 (15)	WSP 80.1 (05)	IST 80.1		(BS EN ISO 4920:2012)
Penetration by water (Rain Test)	AATCC 35–18; FED-STD-191A method 5524	NWSP 080.2.R0 (15)	WSP 80.2 (05)	IST 80.2		
Penetration by Water (Spray Impact Penetration Test)	AATCC 42–17; AAMI PB70:2012	NWSP 080.3.R1 (19)	WSP 80.3 (05)	IST 80.3		BS EN ISO 9073-17: 2008; (ISO18695:2007)
Penetration by Saline Solution (Automated Mason Jar End Point Detector)		NWSP 080.5.R0 (15)	WSP 80.5 (05)	IST 80.5		
Water Resistance (Hydrostatic Pressure Test)	AATCC 127–17	NWSP 080.6.R0 (15)	WSP 80.6 (05)	IST 80.6 / IST 80.4	ERT 120.2–02	BS EN ISO 811:2018; BS EN ISO 9073-16: 2007
Penetration by Oil (Hydrocarbon Resistance)		NWSP 080.7.R0 (15)	WSP 80.7 (05)	IST 80.7		(BS EN ISO 14419:2010)
Alcohol Repellency of Nonwoven Fabrics	AATCC 193–17	NWSP 080.8.R0 (15)	WSP 80.8 (05)	IST 80.8		BS ISO 23232:2009
Nonwovens Run-Off		NWSP 080.9.R1 (19)	WSP 80.9 (05)	IST 80.9	ERT 152.1–02	BS EN ISO 9073-11: 2002

Property	ASTM/TAPPI/MIL	NWSP	WSP	IST	ERT	BS EN ISO
Nonwovens Coverstock Wetback		NWSP 080.10.R1 (19)	WSP 80.10 (05)		ERT 151.3–02	
Nonwoven Wet Barrier Mason Jar	MIL-F-36901A, SECTION 4.3.3	NWSP 080.11.R0 (15)	WSP 80.11 (05)	(IST 80.5)	ERT 170.1–02	
Stiffness						
Cantilever	ASTM D1388–18	NWSP 090.1.R0 (15)	WSP 90.1 (05)	IST 90.1		BS EN ISO 9073-7: 1998
Gurley	TAPPI T 543	NWSP 090.2.R0 (15)	WSP 90.2 (05)	IST 90.2		
Handle-O-Meter	TAPPI T498; ASTM D 6828–02	NWSP 090.3.R0 (15)	WSP 90.3 (05)	IST 90.3		
Drape		NWSP 090.4.R0 (15)	WSP 90.4 (05)	IST 90.4	ERT 90.4-99	BS EN ISO 9073-9: 2008
Bending Length		NWSP 090.5.R0 (15)	WSP 90.5 (05)		ERT 50.6–02	BS EN ISO 9073-7:1998 BS 3356: 1990
Drapeability Including Drape Coefficient		NWSP 090.6.R0 (15)	WSP 090.6 (12)			ISO 9073-9: 2008
Tear strength						
Falling-Pendulum (Elmendorf)	TAPPI T414	NWSP 100.1.R0 (15)	WSP 100.1 (05)	IST 100.1		BS EN ISO 13937-1:2000
Trapezoid	ASTM D4533–15	NWSP 100.2.R1 (15)	WSP 100.2 (05)	IST 100.2	ERT 70.4-99	BS EN ISO 9073-4:1997
Tongue (Single Rip)		NWSP 100.3.R0 (15)	WSP 100.3 (05)	IST 100.3		BS EN ISO 13937-4:2000
Tensile						
Grab	ASTM D5034-09	NWSP 110.1.R0 (15)	WSP 110.1 (05)	IST 110.1		ISO 9073-18:2007
Seam Strength	ASTM D1683-17			IST 110.2		
Internal Bond Strength		NWSP 110.3.R0 (15)		IST 110.3219		

Continued

TABLE 12.1 Summary of test standards (ASTM, BS, EN, ISO, ERT, IST, WSP, NWSP) in relation to nonwovens—cont'd

	ASTM, AATCC and MIL	EDANA and INDA (NWSP, WSP, IST, and ERT)				ISO, BS, and EN
		NWSP	WSP	IST	ERT	
Strip	ASTM D5035–11	NWSP 110.4.R0 (15)	WSP 110.4 (05)	IST 110.4	ERT 20.2-89	BS EN 29073–3:1992; ISO 9073-3:1992
Ball burst	ASTM D6797–15	NWSP 110.5.R0 (15)	WSP 110.5 (05)			BS EN ISO 9073-5:2008
Thickness						
Thickness of Nonwoven Fabrics		NWSP 120.1.R0 (15)	WSP 120.1 (05)	IST 120.1		BS EN ISO 9073-2:1997
High-loft Nonwovens	ASTM D5736–95 (withdrawn)	NWSP 120.2.R0 (15)	WSP 120.2 (05)	IST 120.2		
High-loft Compression and Recovery	ASTM D6571–01 (withdrawn)	NWSP 120.3.R0 (15)	WSP 120.3 (05)	IST 120.3		
High-loft Compression and Recovery (at room temperature)		NWSP 120.4.R1 (15)	WSP 120.4 (05)	IST 120.4		
High-loft Compression and Recovery (at High Temperature and Humidity)		NWSP 120.5.R1 (15)	WSP 120.5 (05)	IST 120.5		
Thickness of Nonwoven Fabrics		NWSP 120.6.R0 (15)	WSP 120.6 (05)		ERT 30.5-99	BS EN ISO 9073-2:1997
Weight						
Nonwovens Mass Per Unit Area	ASTM D3776–09a	NWSP 130.1.R0 (15)	WSP 130.1 (05)	IST 130.1	ERT 40.3-90	BS EN 29073–1:1992; ISO 9073-1:1989
Friction						
Static and Kinetic				IST 140.1		
Binder						
Resin Binder Distribution and Penetration		NWSP 150.1.R0 (15)	WSP 150.1 (05)	IST 150.1		
Appearance & Integrity of High-loft Batting		NWSP 150.2.R0 (15)	WSP 150.2 (05)	IST 50.2		

Linting

	ASTM	NWSP	WSP	IST	ERT	BS EN ISO / Other
Particulate Shedding (Dry)	ASTM F51–20	NWSP 160.1.R0 (15)	WSP 160.1 (05)	IST 160.1	ERT 220.1–02	BS EN ISO 9073-10:2004
Particulate Shedding (Wet)		NWSP 160.2.R0 (15)	WSP 160.2 (05)	IST 160.2		
Fibrous Debris from Nonwovens		NWSP 160.3. R1 (19)	WSP 160.3 (05)	IST 160.3		
Fibrous Debris from Hydrophobic Nonwovens		NWSP 160.4.R0 (15)	WSP 160.4 (05)	IST 160.4		
Surface linting		NWSP 400.0.R0 (15)	WSP 400.0 (05)		ERT 300.0–84	

Fibre identification

	ASTM	NWSP	WSP	IST	ERT	BS EN ISO / Other
Identification of Fibres in Textiles	ASTM D276–12			IST 170.1		(ISO/TR 11827:2012)

Geotextiles

	ASTM	NWSP	WSP	IST	ERT	BS EN ISO / Other
Geotextiles—Vocabulary						BS EN ISO 10318:2015
Guidelines on durability						ISO/TR 13434:2008
Sampling	ASTM D5251–10			IST 180.1		(BS EN ISO 9862:2005)
Mass per unit area						BS EN ISO 9864:2005
Thickness						BS EN ISO 9863-1:2016 BS EN ISO 9863-2:1996
Breaking (Grab Strength)	ASTM D5C34–09			IST 110.1		
Tensile strength (joints/seams)						BS EN ISO 10321:2008
Trapezoid Tear	ASTM D4533–15			IST 180.3		
Puncture Strength	ASTM D4833–07			IST 180.4		BS EN ISO 12236:2006

Continued

TABLE 12.1 Summary of test standards (ASTM, BS, EN, ISO, ERT, IST, WSP, NWSP) in relation to nonwovens—cont'd

	ASTM, AATCC and MIL	EDANA and INDA (NWSP, WSP, IST, and ERT)				ISO, BS, and EN
		NWSP	WSP	IST	ERT	
Dynamic perforation (Cone drop test)						BS EN ISO 13433:2006
Bursting Strength	ASTM D3786–18			IST 30.1		
Pore Size	ASTM D4751–20; (ASTM D6767–20)			IST 180.6		BS EN ISO 12956:2020
Permittivity	ASTM D4491–17; (D5493–06)			IST 180.7		(BS EN ISO 11058:2019)
In-plane Transmissivity	ASTM D6574-13e1					BS EN ISO 12958:2010
Thermoplastic Fabrics in Roofing/Waterproofing	ASTM D4830–98			IST 180.8		
Wide-width tensile test	ASTM D4595–17			IST 180.9		BS EN ISO 10319:2015
Abrasion damage simulation (sliding block test)	ASTM D4886–18					BS EN ISO 13427:2014
Degradable nonwoven fabrics						
Guide to Assess the Compostability of Nonwoven Fabrics				IST 190.1		BS EN ISO 14855-1:2012; BS EN ISO 14855-2:2018
Superabsorbent materials						
pH of Polyacrylate (PA) Powders		NWSP 200.0.R2 (19)	WSP 200.2 (05)		ERT 400.2–02	BS ISO 17190-1:2001
Residual Monomers		NWSP 210.0.R2 (19)	WSP 210.2 (05)		ERT 410.2–02	BS ISO 17190-2:2001
Particle Size Distribution		NWSP 220.0.R2 (19)	WSP 220.2 (05)		ERT 420.2–02	BS ISO 17190-3:2001

Mass Loss Upon Heating		NWSP 230.0.R2 (19)	WSP 230.2 (05)	ERT 430. 2–02	BS ISO 17190-4:2001
Free Swell Capacity in Saline, Gravimetric Determination		NWSP 240.0.R2 (19)	WSP 240.2 (05)	ERT 440. 2–02	BS ISO 17190-5:2001
Fluid Retention Capacity in Saline, After Centrifugation		NWSP 241.0.R2 (19)	WSP 241.2 (05)	ERT 441.2–02	BS ISO 17190-6:2001
Absorption Under Pressure, Gravimetric Determination		NWSP 242.0.R2 (19)	WSP 242.2 (05)	ERT 442.2–02	BS ISO 17190-7:2001
Permeability Dependent Absorption under Pressure		NWSP 243.0.R2 (19)	WSP 243.1 (05)		
Gravimetric Determination of Flow rate and bulk density		NWSP 251.0.R2 (19)	WSP 250.2 (05); WSP 260.2 (05)	ERT 450.2–02; ERT 460.2–02	BS ISO 17190-8:2001 + BS ISO 17190-9:2001
Extractable		NWSP 270.0.R2 (19)	WSP 270.2 (05)	ERT 470.2–02	BS ISO 17190-10:2001
Respirable Particles			WSP 280.2 (05)	ERT 480.2–02	BS ISO 17190-11:2001
Dust in Collection, Sodium Atomic Absorption/Emission Spectrometry			WSP 290.2 (05)	ERT 490.2–02	BS ISO 17190-12:2002
Bacterial					
Bacterial Filtration Efficiency	ASTM F2100–19; F2101–9; MIL-M-36954C-1975	NWSP 300.0.R0 (15)	WSP 300.0 (05)	ERT 180.0. (89)	EN 14683:2019
Dry Bacterial Penetration	ASTM F1670–17; F1671–13; F1819–19; F1862–17.	NWSP 301.0.R0 (15)	WSP 301.0 (05)	ERT 190.1–02	ISO 22612:2005
Wet Bacterial Penetration		NWSP 302.0.R0 (15)	WSP 302.0 (05)	ERT 200.1–02	ISO 22610:2018
Toxicity					
Free formaldehyde—I (water extraction method)		NWSP 310.1.R1 (15)	WSP 310.1 (05)	ERT 210.1–99	EN ISO 14184-1:2011
Free formaldehyde—II (under stressed conditions)		NWSP 311.1.R1 (15)	WSP 311.0 (05)	ERT 211.1–99	

Continued

TABLE 12.1 Summary of test standards (ASTM, BS, EN, ISO, ERT, IST, WSP, NWSP) in relation to nonwovens—cont'd

	ASTM, AATCC and MIL	EDANA and INDA (NWSP, WSP, IST, and ERT)				ISO, BS, and EN
		NWSP	WSP	IST	ERT	
Free formaldehyde—III (determination by HPLC)	ASTM D5910–05	NWSP 312.0.R1 (15)	WSP 312.0 (05)		ERT 212.0-96	
Free formaldehyde—IV (in drying conditions)		NWSP 313.1.R1 (15)	WSP 313.0 (05)		ERT 213.0-99	EN ISO 14184-2:2011
Absorbent hygiene products						
Syngina Method (Tampons)		NWSP 350.1.R1 (15)	WSP 350.1 (05)		ERT 350.0-02	
Ethanol—Extractable Organotin 1		NWSP 351.0.R0 (15)	WSP 351.0 (05)		ERT 360.0-02	
Synthetic Urine—Extractable Organotin II		NWSP 352.0.R0 (15)	WSP 352.0 (05)		ERT 361.0-02	
Acetone Extractable Finish		NWSP 353.0.R0 (15)				
Absorption Before Leakage (ABL) Using an Adult Mannequin		NWSP 354.0.R1 (15)				ISO/CD 19331 (deleted)
Other methods						
Surface Linting		NWSP 400.0.R1 (15)				
Composites Lamination Strength		NWSP 401.0.R0 (15)	WSP 401.0 (05)			
Cup Crush		NWSP 402.0.R0 (15)				
Flame Resistance and Thermal Transfer Properties		NWSP 403.0.R0 (15)				
Homogenization of Absorbent Hygiene Products Using a Laboratory Cutting Mill		NWSP 404.0.R0 (15)				
Polyacrylate Superabsorbent Powders- Determining the Content of Respirable Particles		NWSP 405.0.R0 (15)				
Polyacrylate Superabsorbent Powders- Determination of Dust in Collection Cassettes by Sodium Atomic Absorption/Emission Spectrometry		NWSP 406.0.R0 (15)				
Fibre Orientation Distribution of Nonwoven Fabrics		NWSP 407.0.R0 (15)				

Interlinings

Interlinings	ASTM D2724-19	BS 4973-1:1973 BS 4973-2:1973 International Water Services Flushability Group (IWSFG)	
Fusible interlinings	ASTM D2724-19		
Flushability			
Toilet Bowl and Drainline Clearance	FG501.R1(18)	FG501.R1(18)	WSP 510.1 (09)
Dispersability Shake Flask			WSP 511.1 (09)
Dispersability Tipping Tube			WSP 511.2 (09)
Dispersability Vortex			WSP 511.3 (09)
Column Settling	FG504.R1(18)	FG504.R1(18)	WSP 512.1 (09)
Aerobic Biodisintegration	FG505A.R1(18)		WSP 513.1 (09)
Aerobic Biodegradation			WSP 513.2 (09)
Anaerobic Biodisintegration	FG506A.R1(18)	FG506A.R1(18)	WSP 514.1 (09)
Anaerobic Biodegradation			WSP 514.2 (09)
Toilet Bowl & Drainline Clearance		FG501.R1(18)	WSP 520.1 (09)
Laboratory Household Pump-Compatibility			WSP 521.1 (09)
Laboratory Household Pump-Dispersability	FG503.R1(18)		WSP 522.1 (09)
Slosh Box Disintegration	FG502.R1(18)	IWSFG 2018 PAS3	WSP 522.2 (09)
Septic Tank Retention			WSP 523.1 (09)
Activated Sludge Porous Pot Disintegration			WSP 524.1 (09)
Lab. Untreated Discharge Disintegration			WSP 525.1 (09)
Lab. Soil Biodisintegration			WSP 525.2 (09)
Home Use Toilet and Drainline			WSP 530.1 (09)

Continued

TABLE 12.1 Summary of test standards (ASTM, BS, EN, ISO, ERT, IST, WSP, NWSP) in relation to nonwovens—cont'd

	ASTM, AATCC and MIL	EDANA and INDA (NWSP, WSP, IST, and ERT)				ISO, BS, and EN
		NWSP	WSP	IST	ERT	
Household Sewage Pump		FG507.R1(18) Municipal Sewage Pump Test	WSP 531.1 (09)			
Sewer Conveyance and Municipal Wastewater			WSP 532.1 (09)			
	TAPPI/ANSI Test Method T401 Fibre Analysis					TAPPI/ANSI Test Method T401 Fibre Analysis of Paper and Paperboard.
Fibre glass mats			Pending to revision by TAPPI/INDA			
Sampling and Lot Acceptance, Stiffness, Tear Resistance, and Thickness	TAPPI T-1006		WSP 600.0 (08)			
Sample Location for Fibre Glass Mat Sheets	TAPPI T-1007		WSP 601.0 (08)			
Test conditions for Fibre Glass Mat	TAPPI T-1008		WSP 602.0 (08)			
Tensile Strength and Elongation at Break	TAPPI T-494		WSP 603.0 (08)			
Basis Weight	TAPPI T-410		WSP 604.0 (08)			
Moisture Content	TAPPI T-1012		WSP 605.0 (08)			
Loss on Ignition	TAPPI T-1013		WSP 606.0 (08)			
Moisture Sensitivity	TAPPI T-1014		WSP 607.0 (08)			
Fibre Glass Mat Uniformity (visual defects)	TAPPI T-1015		WSP 608.0 (08)			
Average Fibre Diameter	TAPPI T-1016		WSP 609.0 (08)			

fabric supports [35], the specification for the elastic properties of fabric bandages [36], and medical nonwoven compresses [37,38].

The packaging for terminally sterilised medical devices has been defined in BS EN 868 and BS EN ISO 11607 as follows: General requirements and test methods [39], validation requirements for forming, sealing, and assembly processes [40], sterilisation wrap [41], paper for use in the manufacture of paper bags and in the manufacture of pouches and reels [42], paper bags [43], sealable pouches and reels of porous materials and plastic film construction [44], paper for low-temperature sterilisation processes [45], adhesive coated paper for low-temperature sterilisation processes [46], reusable sterilisation containers for steam sterilisers conforming to EN 285 [47], uncoated nonwoven materials of polyolefins [48], and adhesive coated nonwoven materials of polyolefins [49].

12.3.2 Standards for air filtration

The following international and industry standards specify the filtration performance for various applications:

(a) *National and international standards*:
ISO (International Standardisation Organisation), IEC (International Electrotechnical Commission), CEN (European Committee for Standardisation), CENELEC (European Committee for Electrotechnical Standardisation), BS (British Standard), ANSI (American National Standard Institute), and ASTM (American Society for Testing Methods).

(b) *Standards for specific industrial filtration products are also available in the absence of a specific international standard*:
ASHRAE (American Society of Heating and Refrigerating and Air-conditioning Engineers), SAE (Society for Automotive Engineers), ISIAQ (International Society of Indoor Air Quality and Climate), UL (Underwriters Laboratories), AHAM (Association for Home Appliance Manufactures), and IEST (Institute of Environmental Sciences and Technology).

12.3.2.1 Standards for heating, ventilation, air conditioning (HVAC), efficient particulate air (EPA), HEPA (high-efficiency particulate air), ULPA (ultra low penetration air) filters

For HVAC, BS EN ISO 16890 (all parts) (Table 12.2) refers to particulate air filter elements for general ventilation having an ePM 1 efficiency less than or equal to 99% and an ePM 10 efficiency greater than 20% when tested as per the procedures defined within BS EN ISO 16890 (all parts). It is applicable for air flow rates between $0.25\,m^3/s$ ($900\,m^3/h$) and $1.5\,m^3/s$ ($5400\,m^3/h$), referring to a test rig with a nominal face area of $610\,mm \times 610\,mm$.

BS EN ISO 16890 classifies filter groups in relation to WHO recommendations; it evaluates a filter's performance by its arrestance of particles ranging from 0.3 to 10 μm in size. Filter Group PM 1 has the capacity to arrest particulates in the sizes ≤1 μm, PM 2.5 includes the capacity to capture particles in the particulate sizes ≤2.5 μm and PM 10 covers particulate sizes ≤10 μm.

TABLE 12.2   Classification of filter groups in BS EN ISO 16890.

Filter group	Particulate size (µm)	Classification criteria (ePM[a])
ISO ePM$_1$	$0.3 \leq x \leq 1$	Minimum efficiency ePM $\geq 50\%$
ISO ePM$_{2.5}$	$0.3 \leq x \leq 2.5$	Minimum efficiency ePM $\geq 50\%$
ISO ePM$_{10}$	$0.3 \leq x \leq 10$	Average efficiency ePM $\geq 50\%$
ISO Coarse	$0.3 \leq x \leq 10$	Average efficiency ePM $< 50\%$

[a] *ePM = efficiency Particulate Matter.*

For EPA, HEPA, and ULPA filters, BS EN ISO 29463 parts 2–5 have defined a series of methods for testing air filters based on particle arrestance at MPPS (most penetrating particle size) and have defined 13 different filter classes ranging from ISO 15E to ISO 75U in BS EN ISO 29463-1 as follows:

EPA filter: ISO 15E–ISO 30E
HEPA Filter: ISO 35H–ISO 45H
ULPA Filter: ISO 50U–ISO 75U

In the EU, while all the filters are also tested according to BS EN ISO 29463 parts 2–5, the EU has its own air filter classification system defined in BS EN 1822–1 including EPA (for E10–E12), HEPA (for H13–H14), and ULPA (for U15–U17) filters. Note that BS EN 1822–1 requires only one leakage tests, while BS EN ISO 29463-1 requires three leakage tests (Table 12.3).

BS EN 13142:2013 Ventilation for buildings. Components/products for residential ventilation. Required and optional performance characteristics
BS EN 13053:2019 Ventilation for buildings. Air handling units. Ratings and performance for units, components and sections
BS EN ISO 16890-1:2016 Air filters for general ventilation. Technical specifications, requirements and classification system based upon particulate matter efficiency (ePM)
BS EN ISO 16890-2:2016 Air filters for general ventilation. Measurement of fractional efficiency and air flow resistance
BS EN ISO 16890-3:2016 Air filters for general ventilation. Determination of the gravimetric efficiency and the air flow resistance versus the mass of test dust captured
BS EN ISO 16890-4:2016 Air filters for general ventilation. Conditioning method to determine the minimum fractional test efficiency
BS EN 1822–1:2019 High-efficiency air filters (HEPA and ULPA). Classification, performance testing, marking
ISO 29463-1:2017 High-efficiency filters and filter media for removing particles from air— Part 1: Classification, performance, testing, and marking
BS EN ISO 29463-2:2018, High-efficiency filters and filter media for removing particles in air. Aerosol production, measuring equipment and particle-counting statistics
BS EN ISO 29463-3:2018 High-efficiency filters and filter media for removing particles in air. Testing flat sheet filter media

TABLE 12.3 EN 1822–1:2019 vs ISO 29463-1:2017.

BS EN 1822-1:2019			BS EN ISO 29463-1:2017		
Filter class and group	Overall value		Filter class and group	Overall value	
	Efficiency (%)	Penetration (%)		Efficiency (%)	Penetration (%)
E10	≥85	≤15			
E11	≥95	≤5	ISO 15E	≥95	≤5
			ISO 20E	≥99	≤1
E12	≥99.5	≤0.5	ISO 25E	≥99.5	≤0.5
			ISO 30E	≥99.9	≤0.1
H13	≥99.95	≤0.05	ISO 35H	≥99.95	≤0.05
			ISO 40H	≥99.99	≤0.01
H14	≥99.995	≤0.005	ISO 45H	≥99.995	≤0.005
			ISO 50U	≥99.999	≤0.001
U15	≥99.9995	≤0.0005	ISO 55U	≥99.9995	≤0.0005
			ISO 60U	≥ 99.9999	≤0.0001
U16	≥ 99.99995	≤0.00005	ISO 65U	≥99.99995	≤0.00005
			ISO 70U	≥99.99999	≤0.00001
U17	≥99.999995	≤0.000005	ISO 75U	≥99.999995	≤0.000005

For group H filters, local penetration is given for reference to the MPPS particle scanning method. Alternate limits may be specified when photometer or oil thread leak testing is used

BS EN ISO 29463-4:2018 High-efficiency filters and filter media for removing particles in air. Test method for determining leakage of filter elements. Scan method
BS EN ISO 29463-5:2018 High-efficiency filters and filter media for removing particles in air. Test method for filter elements
ASHRAE 52.2 Method of testing general ventilation air-cleaning devices for removal efficiency by particle size
MIL-STD-282 Filter units, protective clothing, gas-mask components and related products: performance test methods
IEST-RP-CC001.6 HEPA and ULPA Filters
IEST-RP-CC007.3 Testing ULPA Filters
IES RP-CC021.4 Testing HEPA and ULPA Filter Media
EST-RP-CC034.4 HEPA and ULPA Filter Leak Tests

12.3.2.2 Standards for filters used in the automotive industry

ISO/TS 11155–1: Road vehicles—Air filters for passenger compartments—Part 1: Test for particulate filtration (DIN 71460-1:2001)

ISO/TS 11155-2: Road vehicles—Air filters for passenger compartments—Part 2: Test for gaseous filtration (DIN 71460-2: 2003)

12.3.2.3 Standards for filters used in appliances: Vacuum cleaners, room air cleaners, and room air purifiers

IEC 60312: Vacuum cleaners for Household use-Methods of measuring the performance (2001 – 11).
ASTM 1977–99: Standard Test Method for Determining Initial, Fractional Efficiency of a Vacuum Cleaner System.
ANSI AHAM AC-1-1988: Method for Measuring Performance of Portable Household Electric Cord-Connected Room Air Cleaners (RAC).
ASTM D 6830-02 Standard Test Method for Characterizing the Pressure Drop and Filtration Performance of Cleanable Filter Media
ASHRAE Standard 52.1-1992, Gravimetric and Dust Spot Procedures for Testing Air Cleaning Devices Used in General Ventilation for Removing Particulate Matter
ANSI/ASHRAE standard 52.2-1999, Method of Testing General Ventilation Air-Cleaning Devices for Removal Efficiency by Particle Size are the accepted test methods for air filter

12.3.3 Standards for masks and respirators

Nonwoven filter media are key components of masks and respirators used for workwear and in medical and healthcare settings. The performance is measured in terms of filter efficiency (e.g. aerosols of defined compositions and sizes), breathing resistance, and CO_2 residuals, with requisite threshold values defined in a variety of regulations and standards. In the USA, occupational face masks and respirators are regulated by the National Institute for Occupational Safety and Health (NIOSH) (e.g. NIOSH 42 CFR Part 84). Medical masks and respirators used in preventing and treating disease (e.g. N95 surgical respirators) are also subject to regulation under the device provisions of the Federal Food, Drug, and Cosmetic Act and overseen by the Food and Drug Administration (FDA).

NIOSH-approved respirators (Table 12.4) have three classes of filters, N-(Not oil resistant), R-(Resistant to Oil), and P-(Oil Proof) series, each of the classes has three levels of filter efficiency, 95%, 99%, and 912.97% at the most penetrating aerosol particle size (MPPS), i.e., 0.3 µm aerodynamic mass median diameter. The N-series are tested against a mildly degrading aerosol of sodium chloride (NaCl). The R- and P-series filters are tested against a highly degrading aerosol of dioctylphthalate (DOP).

In the EU, both surgical face masks and filtering facepiece (FFP) respirators are regulated by both Directive 93/42/EEC as medical devices and Regulation (EU) 2016/425 as personal protective equipment, which sets out requirements for the testing procedure, classification, safety performance, and other performance attributes, as defined in BS EN 149:2001, BS EN 529:2005, and BS EN 14683:2019. These requirements together with those in other standards are summarised in Tables 12.5–12.7.

The major CEN, NOISH, and ASTM standards for testing masks and respirators are summarised below.

TABLE 12.4 N-, R-, and P-series surgical respirators defined in NIOSH 42 CFR Part 84.

Filter designation	Minimum efficiency 95%	Minimum efficiency 99%	Minimum efficiency 912.97%	Test agent	Maximum test challenge loading
N	N95	N99	N100	NaCl	Filter loading: 200 mg of aerosol particles, in an average 0.075 μm mass median diameter
R	R95	R99	R100	DOP	
P	P95	P99	P100		Maximum filter degradation

N-series filters should be restricted to use in those workplaces free of oil or other severely degrading aerosols. The R-series filters would not have similar aerosol-use restrictions.
The N-series and P-series filters will be tested with DOP until no further decrease in filter efficiency is observed. The P-series filters have neither aerosol-use nor time-use limitations.

TABLE 12.5 Comparison of filter efficiency of various types of face masks.

Mask Type		Filtration efficiency		
Single use face masks	YYT0969	3.0 μm: ≥95% 0.1 μm: N/A		
Surgical masks	YY0469	3.0 μm: ≥95% 0.1 μm: ≥30%		
	ASTM 2100	**Level 1**	**Level 2**	**Level 3**
		3.0 μm: ≥95% 0.1 μm: ≥95%	3.0 μm: ≥98% 0.1 μm: ≥98%	3.0 μm: ≥98% 0.1 μm: ≥98%
	EN 14683	**Type 1**	**Type 2**	**Type 3**
		3.0 μm: ≥95% 0.1 μm: N/A	3.0 μm: ≥98% 0.1 μm: N/A	3.0 μm: ≥98% 0.1 μm: N/A
Respirator masks	NIOSH 42 CFR Part 84	**N95**	**N99**	**N100**
		0.3 μm: ≥95%	0.3 μm: ≥99%	0.3 μm: ≥99.97%
	EN 149:2001	**FFP1**	**FFP2**	**FFP3**
		0.3 μm: >80%	0.3 μm: >94%	0.3 μm: ≥99%

TABLE 12.6 Other performance requirements of surgical masks defined in BS EN 14683.

Test	Type I	Type II	Type IIR[a]
Splash resistance, mmHg	—	—	120 kPal
Bacterial filtration efficiency (BFE), %	≥95%	≥98%	≥98%
Differential pressure, mmH$_2$O/cm^2	<3.0	<3.0	<5.0
Microbial cleanliness	≤30	≤30	≤30

[a] 'R' indicates that the mask offers protection for liquid splash, while other performances such as BFE remain the same as Type II.

TABLE 12.7 Comparison of the testing conditions and performance requirements of surgical respirators defined in six different standards.

Standards	N95 (NIOSH 42 CFR Part 84)	FPP2 (EN 149:2001)	P2 (AS/NZ 1716:2012)	DS2 (Japan JMHLW-Notification 214, 2018)	Korea 1st class (Korea KMOEL-2017-64)	KN95 (China GB2626-2006)
Minimum filter efficiency	≥95% (at 85 L/min)	≥94% (at 95 L/min)	≥94% (at 95 L/min)	≥95% (at 85 L/min)	≥94% (at 95 L/min)	≥95% (at 85 L/min)
Test agent	NaCl	NaCl and DOP	NaCl	NaCl	NaCl and DOP	NaCl
Inhalation resistance[a]	≤343 Pa (at 85 L/min)	≤70 Pa (at 30 L/min); ≤240 Pa (at 95 L/min); ≤500 Pa (clogging)	≤70 Pa (at 30 L/min); ≤240 Pa (at 95 L/min)	≤70 Pa (w/valve); ≤50 Pa (no valve) (at 40 L/min)	≤70 Pa (at 30 L/min); ≤240 Pa (at 95 L/min)	≤350 Pa (at 85 L/min)
Exhalation resistance[a]	≤245 Pa (at 85 L/min)	≤300 Pa (at 160 L/min)	≤120 Pa (at 85 L/min)	≤70 Pa (w/valve); ≤50 Pa (no valve) (at 40 L/min)	≤300 Pa (at 160 L/min)	≤250 Pa
Exhalation valve leakage requirement	Leak rate ≤30 mL/min	N/A	Leak rate ≤30 mL/min	Depressurization to 0 Pa ≥15 s	Visual inspection after 300 L/min for 30 s	Depressurization to 0 Pa ≥20 s
Total inward leakage (TIL)[b]	N/A	≤8% leakage (arithmetic mean)	≤8% leakage (arithmetic mean)	Inward Leakage (IL)[c] measured and included in User Instructions	≤8% leakage (individual & arithmetic mean)	≤8% leakage (arithmetic mean)

[a] Resistance—the pressure drop of the air flow is subjected to as it moves through a respirator filter.

[b] Total inward leakage (TIL)—the amount of a specific aerosol that enters the tested respirator facepiece via both filter penetration and face seal leakage, while a wearer performs a series of exercises in a test chamber.

[c] Inward leakage (IL)—the amount of a specific aerosol that enters the tested respirator facepiece, while a wearer performs a normal breathing for 3 min in a test chamber. The test aerosol size (count median diameter) is about 0.5 μm.

BS EN 143:2000 Respiratory protective devices. Particle filters. Requirements, testing, marking

BS EN 149:2001 Respiratory protective devices. Filtering half masks to protect against particles. Requirements, testing, marking

BS EN 529:2005 Respiratory protective devices. Recommendations for selection, use, care, and maintenance. Guidance document

BS EN ISO 23328-1:2008 Breathing system filters for anaesthetic and respiratory use. Salt test method to assess filtration performance

BS EN ISO 23328-2:2009 Breathing system filters for anaesthetic and respiratory use. Nonfiltration aspects

BS EN 14683:2019 Medical face masks. Requirements and test methods

ASTM F2100–19 Standard Specification for Performance of Materials Used in Medical Face Masks

NIOSH 42 CFR Part 84 Respiratory Protective Devices

21 CFR 878.4040—Surgical apparel

29 CFR 1910.134 Appendix B-1 Respiratory protection. Including Fit test and Assigned Protection Factors (APF)

12.4 Measurement of basic nonwoven structural parameters

The structural and dimensional attributes of nonwoven fabrics are frequently characterised in terms of fabric weight per unit area, thickness, density, fabric uniformity, fabric porosity, pore size and pore size distribution, fibre orientation distribution, bonding segment structure, and the distribution.

Nonwoven fabric weight (fabric mass, or basis weight) is defined as the mass per unit area of the fabric and is usually measured in g/m^2 (or gsm) [50]. Fabric thickness is defined as the distance between the two outermost fabric surfaces under a specified applied pressure [51], which varies if the fabric is very bulky (or compressible). The fabric weight and thickness determine the fabric packing density, which influences the freedom of movement of the fibres and determines the porosity (the proportion of void volume) in a nonwoven structure. The freedom of movement of the fibres plays an important role in nonwoven mechanical properties and the void structure affects fabric porosity, pore size distribution, as well as the permeability of the fabric. Fabric density, or bulk density, is the weight per unit volume of the nonwoven fabric (kg/m^3). It equals the measured weight per unit area (kg/m^2) divided by the measured thickness of the fabric (m). Fabric bulk density together with fabric porosity are important because of their influence on how easily fluids, heat, and sound transport through the structure.

12.4.1 Standard test method for resin binder distribution and binder penetration analysis of polyester nonwoven fabrics

NWSP 150.1 (similar to ITS50.1 and WSP150.1) is a test method designed for analysing resin binder distribution and binder penetration in polyester nonwovens. A specimen of the fabric (in full width) is dyed by using C.I. Basic Red 14 dyes in a 60-l solution with a concentration of 0.2% at 49–60°C for 15 min. After drying, the stained specimen is examined and rated for binder distribution on the nonwoven fabric surface and binder penetration through the fabric thickness by comparison to photographic rating standards on a scale of 1–5.

12.4.2 Fabric thickness [52,53]

Testing of nonwoven fabric thickness and fabric weight is similar to other textile fabrics but due to the greater compressibility and unevenness, a different sampling procedure is adopted. The thickness of a nonwoven fabric is defined as the distance between the face and back of the fabric and is measured as the distance between a reference plate on which the nonwoven rests and a parallel presser-foot that applies a pressure to the fabric (see BS EN ISO 9073-2:1997, NWSP120.6). Nonwoven fabrics with a high specific volume, i.e., bulky fabrics, require a special procedure. In this context, bulky fabrics are defined as those that are compressible by 20% or more when the pressure applied changes from 0.1 to 0.5 kPa. Three procedures are defined in the test standard (BS EN ISO 9073-2:1997, and NWSP120.6) as summarised in Table 12.8.

Three NWSP test methods (i.e. NWSP 120.1, 120.2, and 120.3) (Table 12.9) are defined for the measurement of the thickness, compression, and recovery of conventional nonwovens and high-loft nonwovens (defined, in the NWSP, as a low-density fibre network structure characterised by a high ratio of thickness to mass per unit area. High-loft batts have no more than a 10% solid volume and are greater than 3 mm in thickness). Two more test standards (NWSP 120.4 and 120.5) are defined for rapid measurement of the compression and recovery of high-loft nonwovens.

12.4.3 Fabric weight per unit area

This involves a specific sampling procedure, with specific dimensions for the test samples, and a greater weighing balance accuracy than is needed for conventional textiles. According to the ISO standards (BS EN 29073–1:1992, ISO 9073-1:1989, NWSP 130.1), the measurement of nonwoven fabric weight per unit area requires each fabric specimen to be at least 50,000 mm^2. The mean value is calculated in grams per square metre and the coefficient of variation is expressed as a percentage.

12.4.4 Fabric uniformity

Nonwoven fabric uniformity refers to the magnitude of variations in local fabric dimensions, which can include fabric weight, thickness, and density, along and across the nonwoven fabric, but is usually expressed as the variation of weight per unit area. The variations are frequently of a periodic nature with a recurring wavelength due to the mechanics of different web formation and bonding processes. For example, the uniformity of spunbond and meltblown fabrics is influenced by the ratio of filament spinning velocity to the velocity of conveyor belt as well as by the oscillations prior to deposition [54]. Persistent cross-machine variation in web or fabric weight is commonly encountered in nonwovens, which is one

TABLE 12.8 Summary of testing method in BS EN ISO 9073-2:1997 and NWSP120.6.

	Area of presser foot plate	Area of lower reference plate	Measurement accuracy (mm)	Orientation of reference plate	Pressure applied	Size of samples (mm²)	Number of test samples
Normal fabric	2500 mm²	>19,216 mm²	±1.0	Horizontal, circular	0.5 kPa	2500	10
Bulky fabrics — Maximum thickness up to 20 mm	2500 mm²	1000 mm²	±0.1	Vertical, circular/square	0.02 kPa	130 mm × 80 mm	Measurement time duration: 10 s
Maximum thickness from 20 mm to 100 mm	200 mm × 200 mm	300 mm × 300 mm	±0.5	Horizontal, square	0.02 kPa	200 mm × 200 mm	

TABLE 12.9 Summary of testing methods NWSP 120.1, 120.2, and 120.3.

		Size of presser foot plate (mm)	Pressure applied (kPa)	Size of samples	Number of samples	Test duration time	Report
Thickness	Conventional treated or untreated fabrics,	Ø25.4±0.02	4.14±0.21	20% greater than presser foot	10	5 s	Thickness, SD, CV
	High-loft nonwovens[a]	300 mm by 300 mm	0.03	130 mm by 80 mm	5	9–10 s	
Compression and recovery of high-loft nonwovens[a]	Repeated compression and recovery (Weight-plate, NWSP120.4; NWSP120.5)	230 mm by 230 mm by 6.4 mm	1.83	200 mm by 200 mm by (min) 100 mm	Applied and removed at a series of time intervals.	10 min to 56 h	Compression resistance; Elastic loss; Immediate recovery; Long-term recovery

[a] High-loft nonwovens are defined as porosity >90% and thickness >=3 mm.

reason for edge trimming. Variations in either thickness and/or weight per unit area also give rise to variations in local fabric packing density, local fabric porosity, and pore size distribution, and therefore influence the uniformity of appearance, tensile properties, permeability, thermal insulation, sound insulation, filtration, liquid barrier and penetration properties, energy absorption, light opacity, and conversion behaviour of nonwoven products.

$$\text{Standard deviation: } \sigma^2 = \frac{\sum_{i=1}^{n}(w_i - \bar{w})^2}{n} \tag{12.1}$$

$$\text{Coefficient of variation: } CV = \frac{\sigma}{\bar{w}} \tag{12.2}$$

$$\text{Index of dispersion [55]: } I_{\text{dispersion}} = \frac{\sigma^2}{\bar{w}} \tag{12.3}$$

where n is the number of test samples, $\bar{w}$ is the average of the measured parameter, and w_i is the local value of the measured parameter. Usually, the fabric uniformity is characterised by the percentage coefficient of variation ($CV\%$).

Nonwoven fabrics are normally anisotropic, i.e., measured properties vary in different directions (e.g. in the MD and CD) in the fabric structure. The index of dispersion, $I_{\text{dispersion,}}$ is used to determine if the distribution is a random distribution, uniform distribution or if there is a presence of patterns and clusters [55]. The ratio of the index of dispersion has been used to represent the anisotropy of uniformity [56]. The local anisotropy of weight uniformity in a nonwoven has also been defined by Scharcanski and Dodson [57] in terms of the 'local dominant orientations of fabric weight'.

Both subjective and objective techniques are used to evaluate fabric uniformity. In subjective assessment, visual inspection can distinguish nonuniform areas as small as about 10 mm^2 from a distance of about 30 cm. Qualitative assessments of this type can be used to produce ratings of nonwoven fabric samples by a group of experts against benchmark standards. The consensual benchmark standards are usually established by an observer panel using paired comparison, graduated scales, or similar voting techniques; these standard samples are then used to grade future samples.

To enable on-line determination of weight variation during nonwoven manufacturing, the fabric uniformity is objectively measured. Approaches include measuring the variation in the intensity of optical or infrared light [58,59], the optical and grey-level intensity of fabric images [60,61], the transmission and reflection of beta rays, gamma rays (CO-60), lasers, and electromagnetic rays absorbed by the fabric [62,63]. With optical light scanning methods, the fabrics are evaluated for uniformity using an optical electronic method, which screens the nonwoven to register 32 different shades of grey [59,64]. The intensity of the points in the different shades of grey provides a measure of the uniformity. A statistical analysis of the optical transparency and the fabric uniformity is then produced. This method is suitable for light-weight nonwovens of 10–50 g/m^2. Optical light measurements are commonly coupled with image analysis to determine the coefficient of variation of grey-level intensities from scanned images of nonwoven fabrics [56].

In practice, nonwoven fabric uniformity depends on fibre properties, fabric weight, and nonwoven manufacturing conditions. It is usually true that the variation in fabric thickness and fabric weight decrease as the mean fabric weight per unit area increases. Reducing the mean fibre diameter can have a similar effect in terms of improving uniformity. Wetlaid nonwovens are usually among the most uniform of any nonwoven fabric, particularly at low basis weights, because of the way in which the fibres are deposited during web formation. Short-fibre airlaid fabrics are commonly more uniform than those made from carded cross-laid and parallel-laid webs, and spunbond and meltblown fabrics are often more uniform than fabrics produced from staple fibres.

12.5 Fibre orientation distribution (FOD) and its measurement using image analysis

The fibres in nonwoven fabrics are rarely completely randomly oriented in three-dimensions, rather, individual fibres are aligned in various directions, mostly in-plane. Therefore, fibre orientation in nonwoven webs and fabrics is usually anisotropic, i.e., the number of fibres oriented in each direction in the fabric is not equal. Significant differences in the in-plane fibre orientation between the machine and cross directions of nonwovens are common and give rise to directional property variations. A further consideration is the difference between the fibre orientation in the fabric plane and in the direction perpendicular to the fabric plane, i.e., the transverse direction or fabric thickness direction. In most nonwovens with the exception of vertically lapped fabrics, the majority of fibres are preferentially aligned in the fabric plane rather than in the fabric thickness. However, a small proportion of fibres in nonwoven structures can be oriented out of plane, and through thickness, depending on the method of web formation and bonding conditions.

The structural anisotropy can be characterised in terms of the fibre orientation distribution function. The preferential fibre orientation distributions in 2D or 3D anisotropic fibrous assemblies have been modelled using a fibre orientation vector [65]. Preferential fibre (either staple fibre or continuous filament) orientation in one or multiple directions can be introduced during web formation [66] and during mechanical bonding processes [67]. A simplified example of an anisotropic nonwoven structure is a unidirectional fibrous bundle in which fibres are aligned in one direction only. Parallel-laid or cross-laid carded webs are usually anisotropic, with a highly preferential direction of fibre orientation. Fibre orientation in airlaid structures is usually more isotropic than in other drylaid webs, both in two and three dimensions. In perpendicular-laid webs, such as vertically lapped nonwovens, e.g., V-lap, fibres are preferentially oriented in the direction of the fabric thickness. Spunlaid nonwovens composed of filaments are less anisotropic in the fabric plane than layered carded webs [68]; however, the anisotropy of continuous filament webs depends on the way in which the webs are collected and tensioned.

The structural anisotropy of nonwoven webs and fabrics is important because it influences the anisotropy of mechanical and physical properties including, tensile, bending, thermal insulation, acoustic absorption, dielectric behaviour, liquid transport, and permeability. The ratio of physical properties obtained in the machine and cross directions in the fabric (MD/CD) is a well-established means of expressing the anisotropy. The MD/CD ratio of tensile strength is most commonly encountered, although the same approach may be used to express directional in-plane differences in elongation, liquid wicking distance, liquid transport rate, dielectric constant, permeability, and other properties. However, the MD/CD ratio characterises the anisotropy based on property measurements in just two specific directions in the fabric plane, which can misrepresent the true anisotropy of a nonwoven structure.

12.5.1 Fibre orientation distribution

The fibre segment orientations in a nonwoven fabric are in two and three dimensions and the orientation angle can be determined as shown in Fig. 12.11 [69].

Although the fibre segment orientation in a nonwoven is potentially in any three-dimensional direction, measurement in three dimensions is complex [70,71]. In certain nonwoven structures, the majority of fibres can be aligned in the fabric plane, but with a small proportion nearly vertical to the fabric plane. The structure of a needlepunched fabric is frequently simplified in this way. In this case, the structure of a three-dimensional nonwoven may be thought of as a combination of two-dimensional layers connected by fibres orientated perpendicular to the plane (Fig. 12.12). The fibre orientation in such a three-dimensional fabric can be described by measuring the fibre orientation in two dimensions in the fabric plane [72].

In the two-dimensional fabric plane, fibre orientation is measured by the fibre orientation angle, which is defined as the relative directional position of individual fibres in the structure relative to the machine direction as shown in Fig. 12.13. The orientation angles of individual fibres or fibre segments can be determined by evaluating photomicrographs of the fabric, or directly by means of microscopy or X-ray microcomputed tomography and image analysis.

The frequency distribution (or statistical function) of the fibre orientation angles in a nonwoven fabric is called fibre orientation distribution (FOD) or orientation distribution function

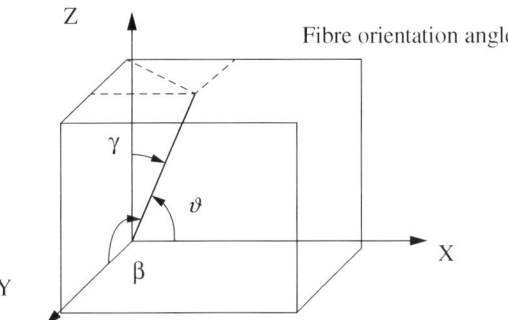

FIG. 12.11 Fibre orientation angle in three-dimensional nonwoven fabrics [69].

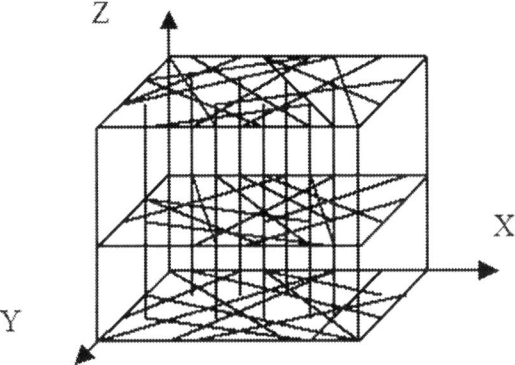

FIG. 12.12 Simplified three-dimensional nonwoven structures [72].

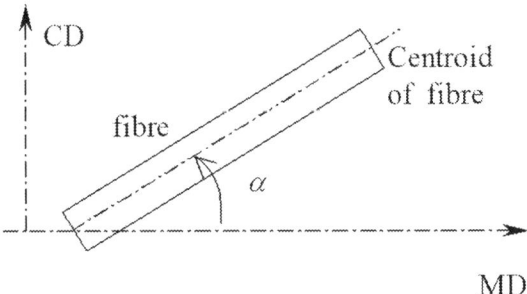

FIG. 12.13 Fibre orientation and the orientation angle.

(ODF). Frequency distributions are obtained by determining the fraction of the total number of fibres (fibre segments) falling within a series of predefined ranges of orientation angle. Discrete frequency distributions are used to estimate continuous probability density functions. The following general relationship is proposed for the fibre orientation distribution in a two-dimensional web or fabric [8]:

$$\int_0^\pi \Omega(\alpha)d\alpha = 1(\Omega(\alpha) \geq 0) \text{ or} \tag{12.4}$$

$$\sum\nolimits_{\alpha=0}^{\pi} \Omega(\alpha)\Delta\alpha = 1\,(\Omega(\alpha) \geq 0) \tag{12.5}$$

where α is the fibre orientation angle, and $\Omega(\alpha)$ is the fibre orientation distribution function in the examined area. The numerical value of the orientation distribution indicates the number of observations that fall in the direction α which is the angle relative to the examined area.

Attempts have been made to fit the fibre orientation distribution frequency with mathematical functions including uniform, normal, and exponential distribution density functions. The following two functions in combination with the constraints in the equation $\int_0^{\pi} \Omega(\alpha)d\alpha = 1$ have been suggested by Petterson [73] and Hansen [74], respectively.

$$\text{Petterson: } \Omega(\alpha) = A + B\cos\alpha + C\cos^3\alpha + D\cos^8\alpha + E\cos^{16}\alpha \tag{12.6}$$

$$\text{Hansen: } \Omega(\alpha) = A + B\cos^2(2\alpha) \tag{12.7}$$

12.5.2 Measurement of fibre orientation distribution using image analysis

In modelling the properties of nonwoven fabrics, and particularly in any quantitative analysis of the anisotropic properties of nonwoven fabrics, it is important to obtain an accurate measure of the fibre orientation distribution (FOD). A number of measuring techniques have been developed.

A direct visual and manual method of measurement was first described by Petterson [73]. Hearle and co-workers [75,76] found that visual methods produce accurate measurements and it is the most reliable way to evaluate the fibre orientation. Manual measurements of fibre segment angles relative to a given direction were conducted and the lengths of segment curves were obtained within a given range. Chuleigh [77] developed an optical processing method in which an opaque mask was used in a light microscope to highlight fibre segments that are oriented in a known direction. However, the application of this method is limited by the tedious and time-consuming work required in visual examinations. To increase the speed of assessment, various indirect-measuring techniques have been introduced including both the zero span [78,79] and short span [80] tensile analysis for predicting the fibre orientation distribution. Stenemur [81] devised a computer system to monitor fibre orientation on running webs based on the light differentiation phenomenon. Methods that employ X-ray-diffraction analysis and X-ray differentiation patterns of fibre webs have also been studied [82,83]. In this method, the distribution of the diffraction peak of the fibre to X-ray is directly related to the distribution of the fibre orientation. X-ray microcomputed tomography (XMT) or micro-CT techniques have also been successfully employed to enable three-dimensional fibre orientation to be visualised. Microscopic techniques based on scanning electron microscopy (SEM) or optical microscopy coupled with image analysis software is also common. To enable the three-dimensional structure to be visualised, serial sectioning of fabric specimens (usually after resin embedding) followed by stacking of the resulting 2D images can be conducted. Digital volumetric imaging (DVI) to study the three-dimensional microstructure of nonwovens is also a proven technique. Other methods include the use of microwaves [84], ultrasound [85], light diffraction methods [86], light reflection and light refraction [87], electrical measurements [88,89], and liquid-migration-pattern analysis [90,91]. Following microscopy or other visualisation techniques, image analysis has been widely employed to quantify fibre orientation [92–95] in nonwoven fabrics. In addition to direct measurement of fibre orientation in real nonwoven fabrics, a range of computer simulation techniques have come into use for the creation of simulated nonwoven structures, which can be particularly useful for performance modelling of nonwovens [96–99].

Huang and Bressee [92] developed a random sampling algorithm and software to analyse fibre orientation in thin webs. In this method, fibres are randomly selected and traced to estimate the orientation angles; test results showed excellent agreement with results from visual measurements. Xu and Ting [99] used image techniques to measure structural characteristics of fibre or fibre bundle segments in a thin nonwoven fabric. The structural characteristics measured included length, thickness, curl, and the orientation of fibre segments. Pourdeyhimi et al. [100–103] completed a series of studies on the fibre orientation of nonwovens by using an image analyser to determine the fibre orientation in which image processing techniques such as computer simulation, fibre tracking, Fourier transforms, and flow field techniques were employed. In contrast to two-dimensional imaging techniques suitable only for thin nonwoven fabrics, the theory of Hilliard–Komori–Makishima [104] and the visualisations made by Gilmore et al. [71] using X-ray tomographic techniques have provided a means of analysing the three-dimensional orientation. Image analysis is a computer-based means of converting the visual qualitative features of a particular image into quantitative data. The measurement of the fibre orientation distribution in nonwoven fabrics using image analysis is based on the assumption that in thin materials, a two-dimensional structure can be assumed, although in reality, the fibres in a nonwoven are arranged in three dimensions. Characterising the fabric structure in terms of the three-dimensional geometry is generally more complex. The fabric geometry is reduced to two dimensions by evaluation of the planar projections of the fibres within the fabric. The assumption of a two-dimensional fabric structure is adequate to describe thin fabrics. The image analysis system in the measurement of the fibre orientation distribution is based on a computerised image capture system operating with an integrated image analysis software package in which numerous functions can be performed [105]. A series of sequential operations is required to perform image analysis, and in a simple system, the following procedures are carried out [105]: Production of a grey image of the sample fabric, processing the grey image, detection of the grey image and conversion into binary form, storage and processing of the binary image, measurement of the fibre orientation, and output of results. Nonlinear statistical equations for estimating the parametric orientation distribution of fibres, as well as a numerical algorithm in a digital greyscale image of a fibrous structure have also been proposed [106] based on scaled variograms of grey levels along the line transects in a few fixed directions.

12.6 Measuring porosity, pore size, and pore size distribution

The pore structure of a nonwoven fabric may be characterised in terms of the total pore volume (or porosity), the pore size, pore size distribution, the pore connectivity, and tortuosity. Porosity provides information on the overall pore volume of a fibrous structure and is defined as the ratio of the nonsolid volume (voids) to the total volume of the nonwoven fabric.

The volume fraction of solid material is defined as the ratio of solid fibre material to the total volume of the fabric. While the fibre density is related to the weight of a given volume of the solid component only (i.e. not containing other materials), the porosity can be calculated as follows using the fabric bulk density and the fibre density:

$$\varphi(\%) = \frac{\rho_{\text{fabric}}}{\rho_{\text{fibre}}} \times 100\% \tag{12.8}$$

$$P\,(\%) = (1 - \varphi) \times 100\% \tag{12.9}$$

where P is the fabric porosity (%), φ is the volume fraction of solid material (%), ρ_{fabric} (kg/m^3) is the fabric bulk density, and ρ_{fibre} (kg/m^3) is the fibre density.

In resin coated, impregnated or laminated nonwoven composites, a small proportion of the pores in the fabric is not accessible i.e., they are not connected to the fabric surface. The definition of porosity as shown above refers to the so-called total porosity of the fabric. Thus, the open porosity (or effective porosity) is defined as the ratio of accessible pore volume to total fabric volume, which is a component part of the total fabric porosity. The majority of nonwoven fabrics have porosities >50% and usually above 80%. A fabric with a porosity of 100% is a totally open fabric and cannot realistically be made, while a fabric with a porosity of 0% is a solid polymer without any pore volume, and so cannot be described as a nonwoven. High-loft nonwoven fabrics usually have a low bulk density because they have more pore space than a heavily compacted nonwoven fabric; the porosity of high-loft nonwovens can reach >98%.

Individual pores in a nonwoven structure are difficult to define, as almost all of the pores are connected to each other, and pore size and pore size distributions calculated or measured by a specific test method are normally based on certain assumptions and theoretical models. Usually, the geometric shapes of the pores are modelled as 2D circular holes, 2D polygon holes, 3D spherical holes, 3D cylindrical capillary tubes, 3D conical capillary tubes, nonuniform 3D cylindrical capillary tubes with bottlenecks, 3D capillary grooves, networks of 3D spherical holes, and networks of 3D cylindrical/conical capillary tubes.

Pore sizes in nonwoven fabrics are inherently characterised by statistical distributions, which vary considerably depending on the local uniformity of the fabric. In addition, the orientation of different pores is related to fibre orientation and the fibre orientation distribution. If fibres are predominantly oriented in one or more directions, the pores formed in the fabric are likely to be dominantly oriented in corresponding directions. Pore connectivity, which gives the geometric pathway between pores cannot be readily quantified and described. If the total pore area responsible for liquid transport across any distance along the direction of liquid transport is known, its magnitude and change in magnitude are believed to indicate the combined characteristics of the pore structure and connectivity.

Tortuosity is another useful quantity relating to the pore structure of a nonwoven, and in simple terms can be defined as the ratio of the actual flow path length (average) to the thickness of the fabric. The least tortuous pore would therefore be one that followed a straight line through the fabric thickness. As the pore waviness and therefore tortuosity increases, the length of the flow path through the fabric increases. Tortuosity influences the resistance to fluid flow through the fabric, as well as properties such as filter efficiency, thermal and electrical conductivity, as well as sound absorption.

12.6.1 Test methods for measuring fabric porosity, pore size, and pore size distribution

Porosity can be simply obtained from the ratio of the fabric density and the fibre density. In addition to the direct method of determination for resin impregnated nonwovens, the fabric porosity can be determined by measuring densities using liquid buoyancy or gas expansion porosimetry [107]. Other methods include small angle neutron, small angle X-ray scattering, and quantitative image analysis for total porosity. Open porosity may be obtained from xylene and water impregnation techniques [107], liquid metal (mercury) impregnation, nitrogen adsorption, and air or helium penetration.

Existing definitions of pore geometry and the size of pores in a nonwoven are based on various physical models of fabrics for specific applications. In general, cylindrical-, spherical-, or convex-shaped pores are assumed with a distribution of pore diameters. Three groups of pore size are defined: (i) the near-largest pore size (known as apparent opening pore size, or opening pore size), (ii) the constriction pore size (known as the pore-throat size), and (iii) the pore volume size. Pore size and the pore size distribution of nonwoven fabrics can be measured using optical methods, density methods, gas expansion and adsorption, electrical resistance, image analysis, porosimetry, and porometry. The apparent pore opening (or opening pore) size is determined by the passage of spherical solid glass beads of different sizes (50–500 μm) through the largest pore size of the fabric under specified conditions. The pore size can be measured using sieving test methods (dry sieving, wet sieving, and hydrodynamic sieving).

The opening pore sizes are important for determining the filtration and clogging performance of nonwoven geosynthetics, and it enables the determination of the absolute rating of filter fabrics. The constriction pore size, or pore-throat size, is different from the apparent pore opening size. The constriction pore size is the dimension of the smallest part of the flow channel in a pore, and it is important for fluid flow transport in nonwoven fabrics. The largest pore-throat size is called the bubble point pore size, which is related to the degree of clogging of geotextiles and the performance of filter fabrics. The pore-throat size distribution and the bubble point pore size can be obtained by liquid expulsion methods. However, it is found that wetting fluid, air pressure, and equipment type affects the measured constriction pore size [108,109]. A summary of the test methods for the determination of pore size distribution is given in Table 12.10.

12.6.2 Dry sieving (ASTM D4751[110,111])

Dry sieving involves passing spherical glass beads (or sand particles) through a nonwoven fabric to determine the fraction of bead sizes for which 5% (or 10%, 50%) or less, by weight, passes through the fabric. The apparent opening size (AOS) or O_{95} (or O_{90}, O_{50}) of the fabric is determined. However, the test accuracy for pore opening sizes smaller than 90 μm is questionable due to various problems in the testing procedure [112,113].

12.6.3 Wet sieving [114,115]

Wet sieving is based on the dry sieving method, and the primary differences are that a continuous water spray is applied to the glass beads and the test fabric during shaking. The

TABLE 12.10   Summary of test methods for the determination of pore size distribution.

Testing methods	Types of materials	Mechanism	Pore size distribution	Porosity	Range of pore sizes
Adsorption	Surface of Pores	Multilayer molecular adsorption on solids		specific surface (0.1–1000 m^2/g), pore volume	0.3–200 nm
Pycnometry	Surface of Pores	Liquid–gas—He or a set of gases with known molecular size and adsorption on the sample	Pore size and distribution	total pore volume, pore volume, density of the solid sample	0.2–1 nm
Calorimetry	Pore surfaces	Thermal effect of wetting liquid penetration into the pores	pore size and distribution	specific surface,	0.5–1 nm
Porosimetry	Pore volume	Filling up the volume of the pores, weight or volume of gas/liquid		porosity, pore volume and size, specific surface	1 nm–1000 μm
Porometry	Constriction pore size	Rate of gas flow when liquid was expulsed from pores	Filter flow pore size distribution		2–1000 μm
Small angle X-rays or neutron dissipation (0–2 degrees)	Closed pores		pore size and distribution		0.5–700 nm.
Sieving test	Largest apparent opening size (AOS)	Particles passing through opening pores	Largest apparent opening size (AOS)		≥90 μm
Bubble point	Largest constriction size of pores	Gases passing through liquid occupied pores	Largest constriction size of pores		N/A
Image analysis	Apparent opening size	Pores in 2D image	Pore size distribution		N/A

continuous water spray reduces electrostatic charging associated with the glass bead particles; mixtures of many different glass bead sizes are used in testing rather than size fractions.

12.6.4 Hydrodynamic sieving method [116–118]

Hydrodynamic sieving is based on hydrodynamic filtration as proposed by Fayoux [116]. Glass bead mixtures are sieved through nonwovens by introducing a water flow with repeated immersion of the fabric in water. Hydrodynamic sieving is usually used to determine the O_{95}, known as the filtration opening size (FOS), of geotextile fabrics. An example of the hydrodynamic sieving testing procedure was proposed by Mlynarek et al. [117] Four testing chambers are used each consisting of a 140 mm diameter cylinder. At the base of each testing cylinder, a fabric sample is supported by two perpendicular supports, each of which is 12.7 mm wide by 55 mm long and has nine equally spaced holes with a diameter of 9 mm. The spherical glass bead mixtures used for hydrodynamic sieving tests are similar to those used in wet sieving, but their diameters range from 25 to 250 μm.

In summary, sieving test methods (dry, hydrodynamic, and wet) are [112]:

- All based on the probability of the spherical particles of a certain diameter passing through an opening during shaking or cycles of immersion.
- Methods that provide arbitrary results, because random probability governs whether a bead meets an opening size through which it can pass.
- Limited because they measure only the largest pore sizes in the fabric.

12.6.5 Image analysis

Image analysis can be used to determine the pore size and pore size distribution of non-woven fabrics [119–125]. Thin specimens of fabric are required. This can necessitate epoxy-resin impregnation of the sample, cutting, grinding, lapping, and polishing. Measurements are performed following optical microscopy or SEM imaging. The pore size distribution obtained from image analysis is different from sieving test results because in the former, the pore dimensions are measured in a two-dimensional plane and measuring accuracy depends on the quality of the cross-section taken. For geosynthetics, the apparent opening sizes (AOS) of the pores in the nonwoven, O_{95} and O_{50}, can be obtained by image analysis techniques. It has been established that [126] the image-based O_{95} pore opening sizes obtained for nonwoven geosynthetics are comparable to dry sieving results based on AOS, while the image-based O_{50} pore opening sizes are lower than those obtained by the dry sieving test (AOS) (O_{50}).

12.6.6 Bubble point test method [127–129]

Bubble point refers to the pressure at which the first flow of air through a liquid saturated fabric sample occurs, and it is a measure of the largest pore-throat in a sample [130]. The bubble point method is based on the principle that the critical pressure of an airflow applied across the thickness of a fabric evacuates the fluid trapped in the pore with the largest pore-throat. Therefore, the applied pressure must exceed the capillary pressure of the fluid in the largest pore-throat. In testing, a nonwoven fabric specimen is saturated with a liquid. The gas pressure on the upstream face of the saturated fabric is then slowly increased to a critical pressure when the first air bubble passes through the largest pore-throat in the saturated fabric. Based on the Laplace equation of capillary pressure, the diameter of the largest pore-throat can then be calculated.

12.6.7 Liquid expulsion porometry [130–133]

Both the pore-throat size distribution and the largest pore-throat size can be determined by means of porometry, which is based on liquid expulsion. First, the relationship between the airflow rate through a liquid saturated fabric and the applied pressure when the liquid is expelled from the saturated fabric sample is determined. In the test, the airflow pressure is applied across the saturated fabric to force liquid out of the pores. With an increase in the applied pressure, the trapped liquid in the pores of the fabric is gradually forced out. According to the Laplace theory of capillary pressure, the smaller the pore diameter, the greater the applied pressure needed to overcome the capillary pressure and to push the liquid out of the pore. The relationship between the applied pressure, the pore sizes, and the airflow rate through the pores can be established. However, to quantify the airflow rate through pores of different sizes, the relationship between the airflow rate through the pores in the dry fabric sample and the applied pressure should be established. By comparing the flow rates for both a dry and a saturated sample at the same applied pressure, the percentage of flow passing through pores larger than or equal to a certain size can be calculated, and the pore size distribution between the pore diameters corresponding to any pressure interval l to h from flows at l and h in terms of air flow rate (not in terms of the number of pores in the fabric) can be defined:

$$Q = \left(\frac{wet\ flow\ h}{dry\ flow\ h} - \frac{wet\ flow\ l}{dry\ flow\ l} \right) \times 100\% \tag{12.10}$$

The pore size corresponding to the applied pressure can be determined as follows:

$$d = \frac{4\sigma \times 10^6}{p} \tag{12.11}$$

where, d is the pore diameter (mm); σ is the surface tension (N/m) of the liquid, where the contact angle between the liquid and the pore wall is assumed to be zero and p is the capillary pressure (equivalent to the applied pressure) (Pa). An example of the flow rate against applied pressure for wet and dry runs, performed on a nonwoven fabric is given in Fig. 12.14.

12.6.8 Pore volume distribution and mercury porosimetry [134–139]

Unlike porometry where the measurement of the pore-throat size distribution is based on measurement of the airflow rate through a fabric sample, the pore volume distribution is determined by liquid porosimetry, which is based on the liquid uptake concept proposed by Haines [140]. A fabric sample (either dry or saturated) is placed on a perforated plate and connected to a liquid reservoir. The liquid having a known surface tension and contact angle is gradually forced into or out of the pores in the fabric by an external applied pressure. Porosimetry is grouped into two categories based on the liquid used, which is either nonwetting (e.g. mercury) or wetting (e.g. water). Each is used for intrusion porosimetry and extrusion porosimetry where the advancing contact angle and receding contact angle are applied in liquid intrusion and extrusion porosimetry, respectively.

Mercury has a high surface tension and is strongly nonwetting on most fabrics at room temperature. In a typical mercury porosimetry measurement, a nonwoven fabric is evacuated to remove moisture and impurities and then immersed in mercury. A gradually increasing pressure is applied to the sample forcing mercury into increasingly smaller

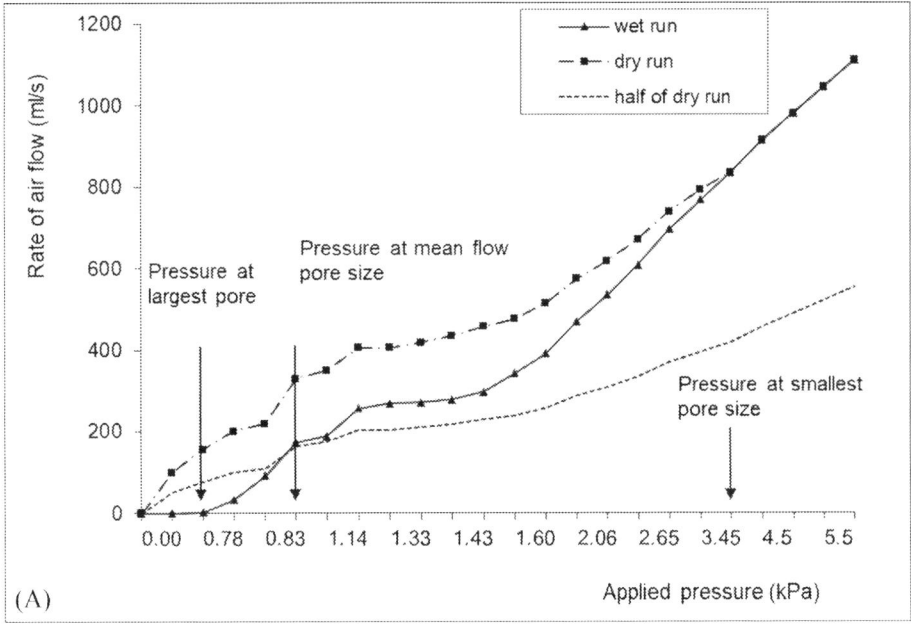

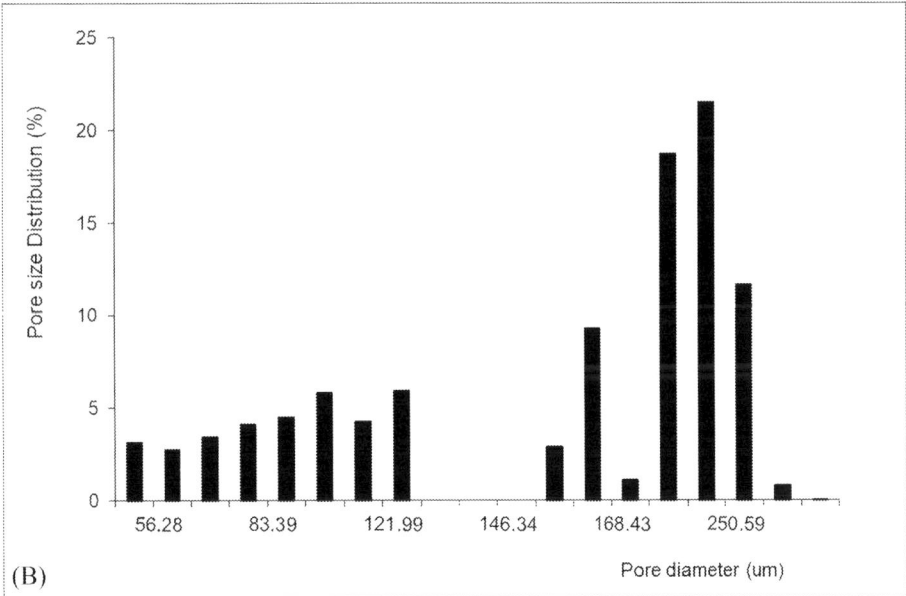

FIG. 12.14 Examples of differential flow pore size distributions for a nonwoven fabric measured by liquid expulsion porometry. (A) The rate of airflow against applied pressure for wet and dry runs; (B) Differential flow pore size distribution of the nonwoven fabric.

'pores' in the fabric. The pressure P required to force a nonwetting fluid into a circular cross-section capillary of diameter d is given by:

$$P = \frac{4\sigma_{Hg} \cos \gamma_{Hg}}{d} \tag{12.12}$$

where σ_{Hg} is the surface tension of the mercury (0.47 N/m) [141], and γ_{Hg} is the contact angle of the mercury on the material being intruded (the contact angle ranges from 135° to 180°), and d is the diameter of a cylindrical pore. The incremental volume of mercury is recorded as a function of the applied pressure to obtain a mercury intrusion curve. The pore size distribution of the sample can be estimated in terms of the volume of the pores intruded for a given cylindrical pore diameter d.

The pressure can be increased incrementally or continuously (scanning porosimetry). The process is reversed by lowering the pressure to allow the mercury to extrude from the pores in the fabric to generate a mercury extrusion curve. Analysis of the data is based on a model that assumes the pores in the fabric are a series of parallel nonintersecting cylindrical capillaries of random diameters (capillary tube model) [142]. However, as a consequence of the nonwetting behaviour of mercury in mercury intrusion porosimetry, relatively high pressure is needed to force mercury into the smaller pores; therefore, compressible nonwoven fabrics are not suitable for testing using the mercury porosimetry method.

Liquids other than mercury are also used in porosimetry [143–149]. Test procedures are similar to that of mercury porosimetry, but any liquid that wets the sample, such as water, organic liquids, or solutions may be utilised. The cumulative and differential pore volume distribution, total pore volume, porosity, average, main, effective, and equivalent pore size can be obtained.

12.6.9 Specific surface area by using gas adsorption [150–152]

For nanofibrous materials and nonwoven fabrics containing nanoporous materials such as nanoparticles and nanoporous fibres, characterisation of the specific surface area is often required to determine suitability for their intended applications. The characterisation methods are usually based on measurement of the amount of gaseous molecules that are adsorbed on the surface of micropores, mesopores, and macropores within the structure. The number of gas molecules adsorbed depends on both the gas pressure and the temperature. An experimental adsorption isotherm plot of the incremental increases in weight of the fabric due to absorption against the gas pressure can be obtained in isothermal conditions. Prior to measurement, the sample needs to be pretreated at an elevated temperature in a vacuum or flowing gas to remove contaminants. In physical gas adsorption, when an inert gas (such as nitrogen or argon) is used as an absorbent gas, the adsorption isotherm indicates the surface area and/or the pore size distribution of the material by applying experimental data to the theoretical adsorption isotherm for gas adsorption on the polymer surface. In chemical gas adsorption, the chemical properties of a polymeric surface are revealed if the absorbent is acidic or basic. In some experiments, a liquid absorbent such as water is used in the same manner.

12.6.9.1 Physical gas adsorption

In physical gas adsorption, an inert gas such as nitrogen (or argon, krypton, carbon dioxide) is adsorbed on the fibre surfaces within the fabric. Usually the Brunauer–Emmett–Teller (BET) [153] multilayer adsorption isotherm theory is used based on the following hypotheses:

(i) gas molecules are physically adsorbed on a solid polymer surface in layers infinitely; (ii) there is no interaction between each adsorption layer; and (iii) the Langmuir theory for monolayer adsorption can be applied to each layer. The BET equation is therefore shown as follows:

$$\frac{1}{V_{adsorption}\left[\left(p_0/p\right)-1\right]} = \frac{1}{V_{monolayer}c}\left(\frac{p}{p_0}\right) + \frac{1}{V_{monolayer}} \tag{12.13}$$

where c is the BET constant, $c = e^{\left(\frac{E_1-E_L}{RT}\right)}$, E_1 is the heat of adsorption for the first layer, E_L is that for the second and additional layers and is equal to the heat of liquefaction, p and p_0 are the equilibrium and the saturation pressure of gases at the temperature of adsorption, $V_{adsorption}$ is the adsorbed gas quantity (for example, in units of volume), and $V_{monolayer}$ is the monolayer adsorbed gas quantity. R is gas constant and T is temperature. There is linear relationship of the adsorption isotherm between, $\frac{1}{V_{adsorption}\left[\left(p_0/p\right)-1\right]}$ and p/p_0 when $0.05 < p/p_0 < 0.35$, and the monolayer adsorbed gas quantity $V_{monolayer}$ and the BET constant c can be obtained from the slope and the y-intercept of the straight line respectively in the plot. The total surface area of the nonwoven, S_{total}, and the specific surface area, S, can therefore be obtained as follows:

$$S_{total} = \frac{V_{monolayer}Ns}{M} \tag{12.14}$$

$$S = \frac{S_{total}}{a} \tag{12.15}$$

where N is Avogadro's number (6.022×10^{23}), s is the adsorption cross-section of the fibre polymer material to specific gases, M is the molecular weight of the fibre polymer materials, and a is the weight of the fabric sample.

12.6.9.2 Chemical gas adsorption

In chemical gas adsorption, a reactive gas such as hydrogen [154], ammonia [155], or carbon monoxide is used to obtain information on the active properties of the porous material [156–158], and is frequently used in the characterisation of nanoscale pores in polymer membranes and metal materials but not usually for nonwoven fabrics containing bigger pores.

12.6.9.3 Helium porosity analysis using pycnometry

Helium-pycnometry gives information on the true density of solids (or skeletal density) by means of helium [159], which is able to enter the smallest voids or pores (up to 1 Å) in the surface to measure the volume per unit weight [160].

12.7 Measuring gas and liquid permeability

Intrinsic permeability, which also called the specific permeability or absolute permeability of a nonwoven fabric, is a characteristic feature of the fabric structure and relates to the void capacity through which a fluid can flow. The specific permeability k is defined by Darcy [161] as:

$$v = -\frac{k\,dp}{\eta\,dx} \tag{12.16}$$

where v is the volumetric flow rate of the fluid in a unit flow area (m/s), η is the liquid viscosity (Pa s), dp is the difference in hydraulic pressure (Pa), dx is the conduit distance (m), and k is the specific permeability (m^2). In practical engineering applications of Darcy's law, sometimes the preference is to use the permeability coefficient, K, which is also referred to as conductivity or Darcy's coefficient. This characterises a fluid flowing through the porous medium at a superficial flow rate. The permeability coefficient K is defined in Darcy's law as: $v = K * i$. Where v is the volumetric flow rate of the fluid in a unit flow area (m/s); i is the hydraulic gradient, i.e., the differential hydraulic head per conduit distance (m/m); and K is the permeability coefficient (m/s). The relationship between k and K is given as: $k = \frac{K\eta}{\rho g}$ (m^2), where ρ is the liquid density ($\frac{kg}{m^3}$) and g is the gravity accelerator constant ($\frac{m}{s^2}$). When the liquid is water at a temperature of 20^{0}C, the constant becomes, $k(m^2) = 1.042 \times 10^{-7} K(\frac{m}{s})$.

In permeability testing, the fluids used are either air or water, and the volumetric rate of the fluid flow per unit cross-sectional area are measured and recorded against specific differential pressures to obtain the air permeability or water permeability. The testing of air permeability in nonwoven fabrics is defined by the ASTM [162], ISO [163], and NWSP [164] (IST and ERT) standards. The testing equipment includes the Frazier air permeability tester, the liquid expulsion porometer, and the water permeability tester for geosynthetics. In air permeability tests, the volumetric airflow rate through a nonwoven fabric of unit cross-sectional area at a unit differential pressure under laminar flow conditions is the fabric permeability.

In water permeability tests, the volumetric flow rate of water flow through a fabric of unit cross-sectional area at a unit differential pressure under laminar flow conditions is measured as the hydraulic conductivity or permittivity under standard conditions (also frequently called the permeability coefficient) [128,165–168]. Two procedures are utilised: the constant hydraulic pressure head method and the falling hydraulic pressure head method. In the falling hydraulic head test, a column of water is introduced to the fabric to induce flow through its structure and both the water flow rate and the pressure change against time are taken. The constant head test is used when the fabric is so highly porous that the flow rate becomes very large and it is difficult to obtain a relationship of the pressure change against time during the falling hydraulic test.

The intrinsic permeability may be obtained by dividing this fluid flow rate by both the fabric thickness and the viscosity of air (or water). However, the nonwoven fabric is usually compressed by the applied pressure during permeability testing, which makes it impossible to use the nominal thickness of the fabric if an accurate assessment of specific permeability is to be obtained. In the water permeability test, the in-plane permeability of nonwoven fabrics [169–172] is also defined and has been studied for many applications including RTM for composites, geosynthetics, and medical textiles. A test standard for measuring the in-plane permeability is defined by the ASTM for geosynthetics.

Adams and Rebenfeld [171,172] developed a method to quantify the directional specific permeability of anisotropic fabrics using an image analysis apparatus that allowed flow visualisation of in-plane radial flow movement. Montgomery [173] studied the directional in-plane permeability of geosynthetics and gave methods for obtaining the maximum and minimum principal specific permeabilities and the resulting degree of anisotropy in the fabric. In the test, viscous liquid is forced by gas pressure to flow within a fabric sample. A mirror is positioned just below the apparatus so that the shape and the position of the radially advancing liquid front can be measured by means of an image analysis system. In this way, the local and dynamic anisotropy of liquid transport through a fabric can be evaluated and the specific permeabilities can be calculated. Capacitance methods [174,175] have also been designed to measure the in-plane directional permeability in which separate capacitance segments are arranged radially around a central point to enable directional measurements of liquid volume to be measured in real time.

12.8 Measuring water vapour transmission rate [176]

The water vapour transmission rate through a nonwoven refers to the mass of the water vapour (or moisture) at a steady state flow through a thickness of unit area per unit time. This is taken at a unit differential pressure across the fabric thickness under specific conditions of temperature and humidity (g/Pa s m^2). It can be tested by two standard methods: the Desiccant method and the Water method. In the Desiccant method, the specimen is sealed to an open mouth of a test dish containing a desiccant, and the assembly is placed in a controlled atmosphere. Periodic weighing determines the rate of water vapour movement through the specimen into the desiccant. In the Water method, the dish contains distilled water, and repeated weighing determines the rate of water vapour movement through the specimen to the controlled atmosphere. The vapour pressure difference is nominally the same in both methods except when testing conditions involve extremes of humidity on opposite sides.

12.9 Measuring wetting and liquid absorption

There are two main modes of liquid transport in nonwovens. One is liquid absorption, driven by the capillary pressure, where the liquid is taken up by means of a negative capillary pressure gradient. The other mode is forced flow, where the liquid is driven through the fabric by an external pressure gradient. The liquid absorption that takes place when one edge of a dry fabric is dipped into a liquid so that it is absorbed primarily in the fabric plane is referred to as wicking. When the liquid front enters into the fabric from one side to the other side of the fabric, it is referred to as demand absorbency or spontaneous uptake.

12.9.1 Wettability and contact angle

The wettability of a nonwoven fabric refers to its ability to be wetted by liquid [177], and is determined by the balance of surface energies in the interface between the air, liquid, and solid material, i.e., the fibres or filaments in the fabric. Wetting is concerned with the initial behaviour of the fabric when it is first brought into contact with the liquid [178], and involves the displacement of a solid–air (vapour) interface with a solid–liquid interface. Thus, the wettability of a nonwoven fabric depends on the chemical nature of the constituent fibre surfaces [179], the fibre geometry [180] (especially surface roughness [181]), and the nonwoven fabric structure. The wettability of a fibre is determined by the fibre–liquid contact angle [182] and that of the associated fabric containing a single fibre type might be considered to be the same as the constituent single fibres [179].

However, wetting of a porous nonwoven fabric is a much more complex process than wetting of a fibre because other wetting mechanisms, such as liquid spreading, immersion, adhesion, and capillary penetration are at play [183]. Because nonwovens are porous, heterogeneous, and anisotropic in structure, the reliability of contact angle measurement is debatable, especially when the fabric is hydrophilic. Although the contact angle of nonwovens can be measured by a goniometer or other indirect methods, there is no standard procedure [184], and it is usually obtained according to standards applicable for sheet materials and paper [185–192]. The contact angle is usually evaluated by two types of technique (i) direct measurement of the contact angle by observation or, (ii) optical techniques, including the goniometer and the direct imaging sessile drop method, as well as wetting force measurement by methods such as the Wilhelmy technique [193,194] and other methods [195–197]. This particular group of measurement methods does not give the contact angle (θ) directly but usually requires either a force measurement or compensation of a capillary force to show $\gamma \cos \theta$ (where γ is the liquid surface tension that needs to be known, or determined independently). Methods used in testing the wettability of other porous materials [198,199] can also provide a good reference.

12.9.2 Wettability and liquid strike time (areal wicking spot test)

The areal wicking spot test method is based on the modification of two existing standards, BS4554 (1970), Determination of wettability of textile fabrics and AATCC method (79–2018), Test method for absorbency of textiles. The 'spot' test attempts to measure the in-plane wicking behaviour, or the capability of a liquid drop to penetrate and spread inside the fabric. In the test, a liquid droplet of either distilled water or, for highly wettable fabrics, a 50% sugar solution, is delivered from a height of approximately 6 mm onto a flat preconditioned nonwoven fabric. A beam of light illuminates the fabric to create bright reflections from the droplet surface as it contacts the fabric. The elapsed time between the droplet reaching the fabric surface and the disappearance of the reflection from the liquid surface is measured. The disappearance of the reflection is assumed to indicate that the liquid has spread over and wetted the fabric surface. The elapsed time is taken as a direct measure of the fabric wettability. The shorter the time, the more wettable the fabric. In some cases, the wetted area of the fabric at the moment reflection ceases is also recorded [200]. An alternative approach is to replace the droplet by a continuous supply of liquid delivered by a capillary tube or a saturated fabric 'wick' in contact with the test specimen and to measure the rate of increase in diameter of the wetted region [201]. For the single drop test, the results are dependent on the local fabric structure, and therefore the measurements are subject to marked variation even within the same fabric.

12.9.3 Liquid absorbency

The capacity of a nonwoven fabric to absorb (absorption capacity) and retain liquid is defined in INDA, EDANA, and ISO test methods. In contrast to the strip test to measure the amount of liquid wicking into a nonwoven fabric in the direction of the fabric plane, the demand absorbency test (also referred to as the demand wettability test or the transverse

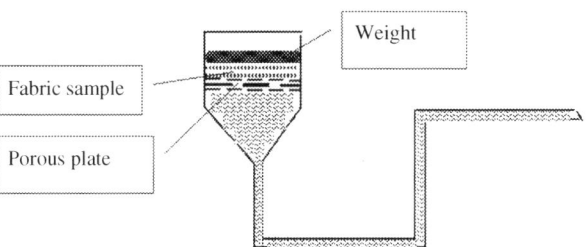

FIG. 12.15 Instrument for measuring demand wettability.

wicking 'plate' test) [202,203] measures liquid wicking into the nonwoven fabric driven by the capillary pressure in the direction of fabric thickness. In demand absorbency tests, the liquid will only enter into the fabric when the sample demands it. These tests involve contacting the dry sample with a liquid in such a way that absorption occurs under a zero or slightly negative hydrostatic head. No standards for this test method are currently available. A classic example of this type of tester [202] is shown in Fig. 12.15. The device consists of a filter funnel fitted with a porous glass plate that is connected to flexible tubing and to a horizontal length of capillary glass tube. The horizontal porous plate is fed from below with water from a horizontal capillary tube, the level of which can be set so that the upper surface of the plate is filled with an uninterrupted column of test liquid and kept damp. This is often used to simulate a sweating skin surface. A disc of test nonwoven fabric is placed on the plate and held in contact with it under a defined pressure achieved by placing weights on top of it. The position of the meniscus along the capillary tube is recorded at various time intervals as water is wicked into the fabric. Given the diameter of the capillary tube, the wicking rate of the water absorbed into the fabric can be obtained. The method can be modified to integrate with an electronic balance and a computer to improve the measurement accuracy and to indicate the dynamic variance of the liquid uptake process against time.

When modified and combined with the electronic balance method, the transverse porous plate method is called the gravimetric absorbency testing system (GATS) [204]. The GATS system is based on an obsolete ASTM standard [205] where the amount of liquid absorbed is determined gravimetrically. The liquid introduction method is modified in the GATS system. Instead of horizontal tubing or a burette with an air bleed, the liquid source rests on top of an electronic balance via a coil spring, which has a known Hooke constant and is capable of compensating the weight loss (due to absorption of liquid by the fabric) or weight gain (due to exsorption) of the liquid source so that the liquid level can be maintained constant. The amount of liquid absorbed is measured continuously by the electronic balance and is recorded continuously against time via a computer. Several test cells allowing different modes of contacting the absorbent sample and the liquid (including the porous plate and a point source) can be used with this equipment. Also, the system may incorporate a sample thickness measuring device which allows continuous monitoring of the change in bulk volume under a constant load. The load can be programmed to allow cyclic loading tests.

A problem related to this method is that the wicking rate is strongly dependent on the applied weight on top of the test fabric, particularly for bulky nonwovens. The structure of nonwoven fabrics can change considerably under compression, but very low compression may not provide uniform porous plate to fabric contact. Another criticism of the method is that the resistance to flow imposed by the capillary tube decreases during the course of the test, as

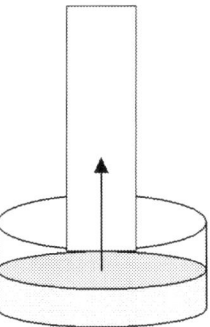

FIG. 12.16   Vertical upwards wicking strip test.

water is withdrawn from the tube, although this can be improved by replacing the capillary tube with an air bleed. A further limitation is that the hydrostatic head in some GATS systems, which is set at a low level at the start of the experiment, decreases during the test as water wicks up through the fabric sample. This can be a particular problem with thick fabrics.

12.9.4 Liquid wicking rate

12.9.4.1 One-dimensional liquid wicking rate (wicking strip test)

The liquid wicking rate can be measured in terms of the linear rate of advance of the liquid in a strip of nonwoven fabric in a strip test. In an upward wicking strip test, the nonwoven fabric is first conditioned at 20°C, 65% relative humidity for 24 h. A strip of the test fabric is suspended vertically with its lower end immersed in a reservoir of distilled water (or other liquid). After a fixed time has elapsed, the height reached by the water in the fabric above the water level in the reservoir is measured (Fig. 12.16). Both the wicking rate and the ultimate height the water reaches are taken as direct indications of the wicking behaviour of the test fabric. Liquid wicking in both the MD and CD of the fabric is tested to determine the anisotropy of liquid wicking properties. The main standard test methods are:

1. BS3424-18, method 21(1986): Methods 21A and 21B. Methods for determination of the resistance to wicking and lateral leakage.
2. NWSP 010.1R$_0$ (2015) Three Standard Test Methods for Nonwoven Absorption
3. NWSP 010.2.R1 (15) Rate of Sorption of Wiping Materials
4. NWSP 010.3.R0 (15) and BS ISO 9073-12:2002 Nonwoven Demand Absorbency
5. NWSP 010.4.R0 (2015): Evaluation of Oil and Fatty Liquids Absorption
6. BS EN ISO 9073-6 (2003) Textiles. Test methods for nonwovens. Absorption

There are some differences in these test procedures. The BS3424–18 (Method 21) specifies a very long test period (24 h) and is intended for coated fabrics with very slow wicking rates. In contrast, other test methods (e.g. NWSP 010.1, NWSP 010.2, and BS EN ISO 9073-6) specify a much shorter test time (maximum 5 min) and apply to fabrics that exhibit rapid wicking. The upward wicking strip test method can be connected with a computer-integrated image analyser to obtain dynamic wicking measurements or it can be modified for integration with an electronic balance to monitor the mass of water absorbed. A downward wicking strip test is also reported to enable the wicking rate and capillary pressure to be obtained [206]. Other test methods include the horizontal strip test.

When the strip test method is used for determining the rate of advance of the liquid front, the position of the advancing front might not be obvious because of the so-called finger-effect [207]. This occurs in nonwoven fabrics, which are usually of high porosity and heterogeneous with local variations in fabric density. However, comparison of the strip test results in fabrics having large differences in fabric structure needs caution. The effect of liquid evaporation cannot be ignored in a strip test performed for a long time and the influence of the fabric structure on the gravity effect in the strip test needs to be considered.

12.9.4.2 Two-dimensional liquid wicking rate

The demand absorbency 'plate' test method can be converted into a two-dimensional radial dynamic wicking measurement method [204] when the liquid is introduced from a point source into the nonwoven fabric (also known as the point source demand wettability test). One example of this method is the GATS system mentioned in Section 12.9.3 when a point source liquid introduction cell was used.

Modified laser-Doppler anemometry (MLDA) [208] is another alternative method to monitor liquid wicking in the two-dimensional fabric plane based on the Doppler principle. When a laser beam is passed through a flowing liquid, light is scattered by the particles suspended in the liquid. The scattered light is subject to a frequency shift and contains information about the velocity of the particles which can then be examined by electro-optical techniques. In order to obtain the velocity of liquid flow in a nonwoven fabric using this method, it is required that the flow medium be partly transparent and contain particles that scatter light used in the measurement.

Electrical capacitance techniques have been used to monitor the liquid absorption in multidirections in a nonwoven fabric plane [174,175]. The principle of the method is based on the fact that the dielectric constant of water is about 15–40 times higher than that of normal fibres and fabrics, and therefore the capacitance of a transducer in a measuring system will be very sensitive to the amount of liquid absorbed by a fabric. The device is computerised and is able to provide both dynamic (real time) and multidirectional measurements of the wicking rate in terms of the volume of liquid absorbed. Problems in testing nonwovens may arise from the influence of the significant geometric deformations in saturated fabrics, the liquid evaporation, and limitations in the size of the capacitance transducers. Also, different types of fibrous material may have different dielectric constants, which can lead to difficulties when comparing different materials.

Similar to measuring the permeability and the anisotropy of liquid transport in the nonwoven fabric plane [171], the image analysis method has been used to track the in-plane radial liquid advancing front to determine the rate of capillary spreading in the two-dimensional fabric plane. Kawase et al. [209,210] used a video camera to determine the capillary spreading of liquids in the fabric plane. The apparatus used in his studies comprised a desiccator with a 200 mm diameter. The cover had an orifice for inserting a micropipette. The liquid used was n-decane which was placed at the bottom of the desiccator to minimise volatilisation of the liquid spreading in the fabrics. To aid observation of the liquid spreading in test fabrics, the n-decane was dyed with a 0.1% solution of Sudan IV or acid blue 9. The fabric was mounted on a 12.0 cm wooden ring (embroidery hoop) and placed into the desiccator along with a stopwatch. The cell was covered, and the fabric was left for at least 2 h. A measured amount (0.05–0.20 mL) of liquid was introduced onto the fabric by a micropipette. The area of the spreading liquid and the reading on the stopwatch were recorded simultaneously using a video camera. The spreading area was copied onto film, cut out, and weighed. A calibration curve was determined by recording areas of several known sizes and weighing the copied film in every experiment in order to determine the actual spreading area. The correlation coefficients of the calibration curves were reported to be higher than 0.99. A camera coupled with an image analyser allows capillary spreading in the fabric plane to be quantified as a distribution of brightness levels in an image. The profile of the distribution of the liquid concentration in the fabric can be obtained by calibrating the brightness or intensity values with liquid concentration levels in a fabric.

12.9.5 Liquid drainage rate (syphon test)

The syphon test [201,211] measures the rate of drainage under external pressure rather than the wicking rate. In this test, a rectangular strip of saturated fabric is used as a syphon, by immersing one end in a reservoir of water or saline solution and allowing the liquid to drain from the other end into a collecting beaker. The amount of liquid transmission at successive time intervals is recorded. Because the saturated fabric has a lower resistance to flow than a dry fabric, the rate of drainage is usually greater than the wicking rate.

12.10 Measuring thermal conductivity and insulation

The thermal resistance and the thermal conductivity of flat nonwoven fabrics, fibrous slabs, and mats can be measured with a guarded hot plate apparatus [212–214] according to BS 4745: 2005, ISO 5085-1:1989, ISO 5085-2:1990. For testing the thermal resistance of quilt [215], the testing standard is defined in BS 5335 Part 1:1991. The heat transfer in the measurement of thermal resistance and thermal conductivity in current standard methods is the overall heat transfer by conduction, radiation, and by convection where applicable.

The core components of the guarded hot plate apparatus consist of one cold plate and a guarded hot plate. A sample of the fabric or insulating wadding to be tested, 330 mm in diameter and disc shaped, is placed over the heated hot metal plate. The sample is heated by the hot plate, and the temperature on both sides of the sample is recorded using thermocouples. The apparatus is encased in a fan-assisted cabinet and the fan ensures enough air movement to prevent heat build-up around the sample and also isolates the test sample from external influences. The test takes approximately 8 h including warm-up time. The thermal resistance is calculated based on the surface area of the plate and the difference in temperature between the inside and outside surfaces.

When the hot and cold plates of the apparatus are in contact and a steady state has been established, the contact resistance, R_c ($m^2\ KW^{-1}$), is given by the equation:

$$\frac{R_c}{R_s} = \frac{\theta_2 - \theta_3}{\theta_1 - \theta_2} \tag{12.17}$$

R_s is the thermal resistance of the 'standard'.

θ_1 is the temperature registered by the thermocouple, T_1, θ_2 is the temperature registered by T_2, θ_3 is the temperature registered by T_3. Thus, the thermal resistance of the test specimen, R_f ($m^2 KW^{-1}$), is given by the equation:

$$\frac{R_f}{R_s} = \frac{\theta_2' - \theta_3'}{\theta_1' - \theta_2'} - \frac{\theta_2 - \theta_3}{\theta_1 - \theta_2} \tag{12.18}$$

where θ_1' is the temperature registered by T_1, θ_2' is the temperature registered by T_2, and θ_3' is the temperature registered by T_3. Since R_s (m^2KW^{-1}) is a known constant and can be calibrated for each specific apparatus, $R_f(m^2KW^{-1})$ can thus be calculated. Then the thermal conductivity of the specimen, k $(Wm^{-1}K^{-1})$ can be calculated from the equation:

$$k = \frac{d(mm)*10^{-3}}{R_f(m^2KW^{-1})} \tag{12.19}$$

The conditioning and testing atmosphere is usually the standard atmosphere for testing textiles defined in ISO139, i.e., a relative humidity (RH) of 65%$\pm$2% and a temperature of 20°C$\pm$2°C.

12.11 Measuring tensile properties

Some of the most important fabric properties governing the functionality of nonwoven fabrics are mechanical properties, e.g., tensile, compression, bending, and stiffness. Tensile properties of nonwoven fabrics are usually tested in both the MD and CD, and other in-plane directions if required. Several test methods are available for tensile testing of nonwovens, chief among these are the strip and grab test methods. In the grab test, the central section across the fabric width is clamped by jaws a fixed distance apart. The edges of the sample therefore extend beyond the width of the jaws. In the standard grab tests for nonwoven fabrics [216], the width of the nonwoven fabric strip is 100 mm, and the clamping width in the central section of the fabric is 25 mm. The fabric is extended at a rate of 100 mm/min (according to the ISO standards) or 300 mm/min (according to the ASTM standards) and the separation distance (gauge length) of the two clamps is 200 mm (ISO standards) or 75 mm (ASTM standards). Nonwoven fabrics usually give a maximum breaking force before rupture. In the strip test [217,218], the full width of the fabric specimen is gripped between the two clamps. The width of the fabric strip is 50 mm (ISO standard) or either 25 or 50 mm (ASTM standards). Both the rate of extension and gauge length in a strip test are the same as they are in the grab test. The gauge length of the two clamps is 200 mm (ISO standards) or 75 mm (ASTM standards). Note that the observed force at break for a 50 mm specimen is not necessarily double that of a 25 mm wide specimen.

12.12 Measuring fabric tactile properties

The tactile perception of nonwoven fabrics is important for products that are to be worn next to the skin. It is usually determined by subjective 'touch-feel' of the fabric, which depends on the complex interplay of rubbing, squeezing, stretching, and sliding on the surface of fabrics. It involves multidimensional sensory perception. Subjective assessment of fabric touch-feel frequently results in disagreement between different user groups. Therefore, various fabric objective measurement (FOM) methods have been developed to quantify relevant fabric properties and to try to correlate them with human touch perceptions of the skin–fabric interface. The criteria for evaluating tactile comfort of woven and nonwoven fabrics have been previously reviewed [219].

Fabric softness perception of touch-feel properties can relate to many objective fabric properties such as fabric stiffness, flexibility, and compression properties in the direction of fabric thickness. Peirce [220] pioneered the objective measurement of fabric stiffness by determining the flexural rigidity of a fabric strip based on a cantilever test. This test method is now designated as a standard method for determining nonwoven fabric bending length, flexural rigidity, and bending modulus (see, Stiffness of Nonwoven Fabrics Using the Cantilever Test and Nonwovens Bending Length (i.e. NWSP 090.1.R0 (15), NWSP 090.5.R0 (15), and ISO 9073-7: 1995)). The fabric strip is held flat at its back end as the front end bends under its own weight and the force of gravity. When the front end of the fabric strip intersects the plane of a slope at an angle of $\theta = 43°$ relative to the horizontal plane, the projecting length of the fabric strip is defined as the bending length l (see Fig. 12.17).

Other mechanisms are used to test fabric stiffness and softness. In the Gurley Tester method (NWSP 090.2.R0 (15)), nonwoven fabric stiffness is characterised during a forced bending deformation exerted by an external force. In the Handle-O-Meter stiffness tester for nonwoven fabrics (i.e. NWSP 090.3.R0 (15)), the nonwoven sample is deformed through a narrow slit opening by a plunger, and the force required is measured to reflect fabric stiffness. However, the measured force not only depends on fabric stiffness but also relies on fabric surface friction.

One method for measuring fabric softness is based on the measurement of compression and recovery properties in the fabric thickness direction [222]. Dutkiewicz et al. [223] described another method of measuring the softness of absorbent sheets by placing a strip of material between two plates and applying force from both sides (Fig. 12.18). A semicircular clamp served to hold the sheet, whose shape then mimicked the anatomy

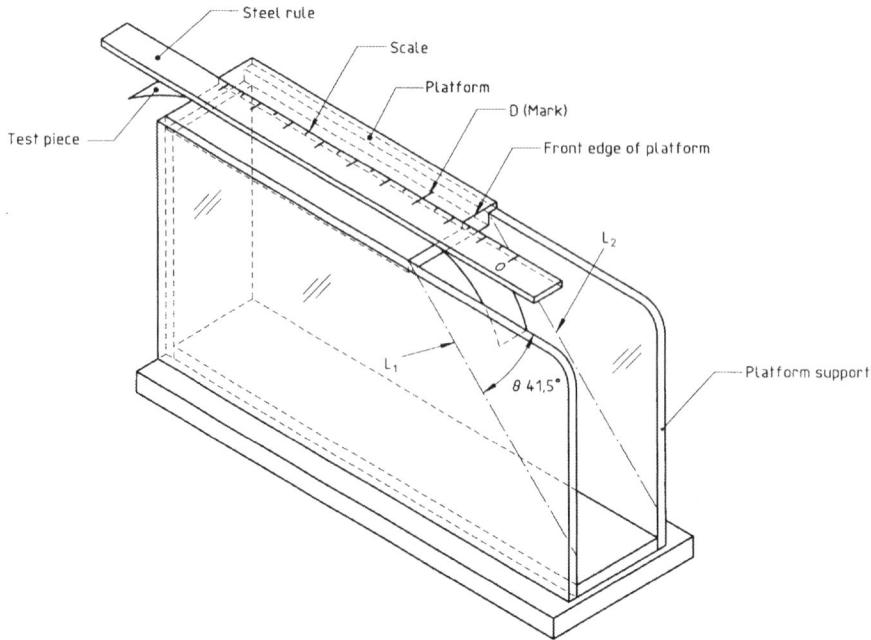

FIG. 12.17   Peirce bending length tester [221].

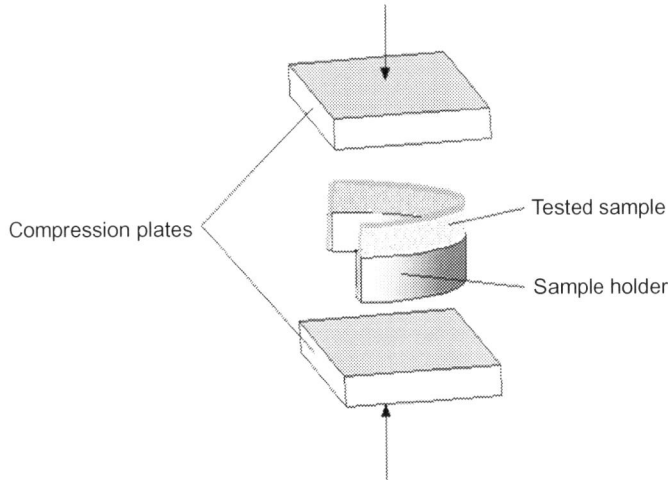

Compression plates

Tested sample

Sample holder

FIG. 12.18   Measurement of fabric softness according to reference [223].

of the human body. The softness was defined to be the inverse of the energy needed to compress the sheet to half of the sample's width, the latter being constant for all measurements. This energy, Edmax, (expressed in J), consumed at the maximum deflection, was calculated to indicate the fabric softness.

While fabric properties obtained from many FOM methods are objective, the conclusions from such instrumental evaluation methods, including the Kawabata Evaluation System for Fabrics (KES-F) [224] and Fabric Assurance by Simple Testing (FAST), [225] are not necessarily objective because of their dependency on the relationships between measured fabric mechanical properties and the subjective hand preferences of human panels with respect to reference fabrics [224,226,227]. Furthermore, both the KES-F and FAST methods examine fabric tactile properties in a one-dimensional approach, and the standard fabrics are based on 100% woven wool fabrics. Consequently, neither the measurement conditions required, e.g., pretension, nor the conclusions obtained from the data are ideally suitable for thin, easily deformable nonwoven fabrics.

In fabric handle assessment methods based on extraction [228], a fabric is forced through a narrow opening, such as a slit opening or a nozzle, creating fabric deformations in multiple directions. In commercial instruments utilising this mechanism, e.g., the PhabrOmeter [228] and the Wool HandleMeter [229], the resultant quantitative assessment of fabric handle is based upon statistical Principal Component Analysis (PCA) of the dynamic force–displacement relationship obtained. The handle values are expressed as a relative index to those of standard fabrics, including the relative hand value, stiffness, softness, smoothness, wrinkle recovery rate, and drape index. However, the subjective determination of initial weight added to a target fabric as well as random fabric deformations occurring in the fabric extraction process can lead to problems of reproducibility. In addition, standard fabrics are usually required in those tests to differentiate fabric stiffness and softness in relation to subjective assessment results. It is notable that in all of the aforementioned existing approaches, the recovery of the fabric following initial deformation is not determined, despite the fact that it is known to affect tactile perception in subjective fabric handle evaluations [227].

To bridge the gap between objective and subjective handle evaluations, the Leeds University Fabric Handle Evaluation System (LUFHES) [230] was developed to mimic the multidimensional fabric buckling deformations that occur in fabric–human interactions, which includes rubbing, squeezing, and stretching. In the LUFHES method, fabric deformations are quantified by determining the energy consumed in elastic, plastic, and permanent deformation during cyclic deformation and recovery of a fabric cylindrical shell subjected to compression buckling and shear. Fabric tactile indices for softness, stiffness, flexibility, sponginess, crispiness, and formability are thus obtained to objectively discriminate and differentiate the tactile properties of fabrics without reference to any standard fabrics. Fabric surface smoothness and textural features are obtained during analysis of the fabric–fabric dynamic friction process. The application of LUFHES to differentiate and discriminate differences in tactile properties of nonwoven fabrics containing different components has been demonstrated [231,232] and subtle differences can be detected, including before and after washing using different detergents [233,234].

Nonwoven fabric drape and drape coefficient are characterised using quite similar methods by means of instruments such as the Nonwovens Cusick Drape test (NWSP 090.4.R0 (15) and ISO 9073-9: 2008) and the Evaluation of Drapeability, including the Drape Coefficient of Nonwoven Fabrics (NWSP 090.6.R0 (15) and ISO 9073-9: 2008). In LUFHES, fabric formability defined by Lindberg [235] can be obtained by using the critical buckling force measured in this system. Other tactile indices to represent fabric drape properties can therefore be derived. Cool/warm touch-feel properties of nonwoven fabrics can also be tested using instruments such as the KES-F7 [236], C-Therm [237,238], and the Fabric Touch Tester [239]. Either the amount of heat absorption or fabric thermal emissivity, as well as thermal conductivity is measured to indicate the level of fabric thermal touch comfort.

12.13 Modelling tensile strength

Backer and Petterson [240] pioneered a fibre network theory from work on needlepunched fabrics. This estimates the tensile properties of nonwovens based on the fibre orientation, fibre tensile properties, and the assumption that fibre segments between bonds are straight. Hearle and Stevenson [241] expanded this theory by taking account of the effects of fibre curl. They indicated that the stress–strain properties of a nonwoven fabric were dictated by the orientation distribution of fibre segments. Later, Hearle and Ozsanlav [242] developed a further theoretical model to incorporate binder deformation into the model. The fibre orientation distribution is an essential parameter in constructing these models. When the fibres in a nonwoven are assumed to lie in layers parallel to the two-dimensional fabric plane, the prediction of the stress–strain curve of the fabric under uniaxial extension can be established based on three models: the orthotropic models, the force analysis method in a small strain model, and the energy analysis method in an elastic energy absorption model.

12.13.1 Orthotropic models of tensile strength [237]

12.13.1.1 Predicting fabric tensile properties based on tensile properties in different principal directions

It is assumed that the deformation of a two-dimensional nonwoven fabric is analogous to that of a two-dimensional orthotropic woven fabric where stress (σ)–strain (ε) relationships are known for the two principal directions of the nonwoven fabric. It is assumed that the following properties are known: elastic modulus (EX, EY) in two principal directions respectively; shear modulus (G_{XY}) between the two principal directions and the Poisson's ratio ($v_{XY} = \frac{1}{v_{YX}} = \frac{\varepsilon_X}{\varepsilon_Y}$) in two principal directions. For a unidirectional force on the fabric with a small strain ($\frac{\sigma(\theta)}{\varepsilon(\theta)} = E(\theta)$), the fabric modulus and the Poisson's ratio in the direction θ are as follows:

$$\frac{1}{E(\theta)} = \frac{\varepsilon(\theta)}{\sigma(\theta)} = \frac{\cos^4\theta}{E_X} + \left(\frac{1}{G_{XY}} - \frac{2v_{XY}}{E_X}\right)\cos^2\theta\,\sin^2\theta + \frac{\sin^4\theta}{E_Y} \tag{12.20}$$

$$v(\theta) = -\frac{\varepsilon(\theta)}{\varepsilon\left(\theta + \frac{\pi}{2}\right)} = -\frac{\left(\frac{1}{E_X} + \frac{1}{E_Y} - \frac{1}{G_{XY}}\right)\cos^2\theta\,\sin^2\theta - v_{YX}\frac{\left(\cos^4\theta + \sin^4\theta\right)}{E_X}}{\frac{\cos^4\theta}{E_X} + \left(\frac{1}{G_{XY}} - \frac{2v_{XY}}{E_X}\right)\cos^2\theta\,\sin^2\theta + \frac{\sin^4\theta}{E_Y}} \tag{12.21}$$

12.13.1.2 Predicting fabric tensile properties based on fibre orientation distribution and fibre properties

For a nonwoven fabric, it is assumed that (i) fibres in the fabric are straight and cylindrical with no buckling, (ii) the bond strength between fibres in the fabric is considerably higher than the fibre strength, i.e., nonwoven rupture results from fibre failure, and (iii) the shear stress and shear strain are negligible. Thus, we have the following equations for nonwoven fabric tensile properties, where the fibre orientation distribution function in the fabric is $\Omega(\beta)$.

$$\sigma(\theta) = E_f\varepsilon_x \int_{-\pi/2}^{\pi/2} \left(\cos^4\beta - v(\theta)\sin^2\beta\,\cos^2\beta\right)\Omega(\beta)d\beta \tag{12.22}$$

$$\sigma\left(\theta + \frac{\pi}{2}\right) = E_f\varepsilon_x \int_{-\pi/2}^{\pi/2} \left(\sin^2\beta\,\cos^2\beta - v(\theta)\sin^4\beta\right)\Omega(\beta)d\beta \tag{12.23}$$

$$v(\theta) = \frac{\int_{-\pi/2}^{\pi/2} \left(\sin^2\beta\,\cos^2\beta\right)\Omega(\beta)d\beta}{\int_{-\pi/2}^{\pi/2} \left(\sin^4\beta\right)\Omega(\beta)d\beta} \tag{12.24}$$

$$E(\theta) = \frac{\sigma(\theta)}{\varepsilon(\theta)} = E_f \int_{-\pi/2}^{\pi/2} \left(\cos^4\beta - \frac{\int_{-\pi/2}^{\pi/2} \left(\sin^2\beta\,\cos^2\beta\right)\Omega(\beta)d\beta}{\int_{-\pi/2}^{\pi/2} \left(\sin^4\beta\right)\Omega(\beta)d\beta}\sin^2\beta\,\cos^2\beta\right)\Omega(\beta)d\beta \tag{12.25}$$

When the nonwoven fabric is isotropic, i.e., $\Omega(\beta) = \frac{1}{\pi}$, then from the above equations, we have $v(\theta) = \frac{1}{3}$.

12.13.2 Force analysis method in the small strain model

For this model, the following assumptions are made about the nonwoven structure:

1. The fibres are assumed to lie in layers parallel to the two-dimensional fabric plane.
2. The fabric is subjected to a small strain.
3. The nonwoven fabric is a pseudo-elastic material and Hooke's law applies.
4. No lateral contraction of the material takes place.
5. No transverse force exists between fibres.
6. No curl is present in the fibres.

The stress–strain relationship can be established by using the analysis of the components of force in the fibre elements in a nonwoven fabric is given as follows [237,238]:

$$\left(1+\varepsilon_j\right)^2 = \left(1+\varepsilon_L\right)^2 \cos^2\theta_j + \left[1+\varepsilon_T + \left(1+\varepsilon_L\right)\cot\theta_j\ \tan\beta\right]^2 \sin^2\theta_j \qquad (12.26)$$

where β is the shear undertaken by the fabric, ε_j is the fibre strain in the jth fibre element, ε_L and ε_T are the fabric strains in longitudinal direction and transverse direction respectively; θ_j is the fibre orientation angle of the jth fibre element. If there is no shear in the fabric plane, then,

$$\left(1+\varepsilon_j\right)^2 = \left(1+\varepsilon_L\right)^2 \cos^2\theta_j + \left(1+\varepsilon_T\right)^2 \sin^2\theta_j \qquad (12.27)$$

12.13.3 Energy analysis method [243]

In the previous models, the fibres are assumed to be straight cylindrical rods. In fact, fibres in real nonwoven fabrics usually have various degrees of curl, thus energy analysis methods have been adopted to improve tensile property modelling. The nonwoven structure is treated as a network of energy absorbing fibrous elastic elements, where elastic energy in reversible deformation can be solely determined by changes in fibre length. The deformation geometry is defined by minimum energy criteria. The applied stresses and strain are used in the analysis rather than the applied forces and displacement used in the first method. The following assumptions are made:

1. The fabric is a two-dimensional planar sheet.
2. The sheet consists of networks of fibre elements between bond points.
3. The bond points move in a way that corresponds to the overall fabric deformation.
4. Stored energy is derived from changes solely in fibre length (i.e. no contribution of binder, each point is freely jointed, fibres are free to move independently between bonds).

When a unidirectional force is applied, we have:

$$\varepsilon_j' = \frac{1}{C_j}\left[\left(1+\varepsilon_L^2\right)\cos^2\theta_j + \left(1-v_{XY}\sin^2\theta_j\right)^2 \sin^2\theta_j\right]^{\frac{1}{2}} - 1 \qquad (12.28)$$

$$\sigma_L = \frac{\sum_{j=1}^{N} \mu_j\sigma_j\left(\dfrac{\cos^2\theta_j}{C_j^2\left(1+\varepsilon_j'\right)}\right)}{\sum_{j=1}^{N}\mu_j} \qquad (12.29)$$

where ε_j'=strain in the j_{th} fibre element, ε_L=overall fabric strain, σ_j=stress in the j_{th} fibre element, σ_L=overall fabric stress, v_{XY}=fabric contraction factor, which defined as the contraction in the Y direction due to a force in the X direction, which is equal to the ratio of strain in the Y direction to the strain in the X direction. θ_j=orientation angle of the j_{th} fibre element, C_j=curl factor of the j_{th} fibre element, μ_j=mass of the j_{th} fibre element, and N=total number of fibre elements.

12.14 Modelling bending rigidity [244]

The bending rigidity or flexural rigidity of adhesive-bonded nonwovens has been modelled by Freeston and Platt [244]. A nonwoven fabric is assumed to be composed of unit cells and the bending rigidity of the fabric is the sum of the bending rigidities of all the unit cells in the fabric, defined as the bending moment times the radius of curvature of a unit cell. The analytical equations for bending rigidity were established in the two cases of 'no freedom' and 'complete freedom' of relative motion of the fibres inside a fabric. The following assumptions about the nonwoven structure are made for modelling the bending rigidity.

1. The fibre cross-section is cylindrical and constant along the fibre length.
2. The shear stresses in the fibre are negligible.
3. The fibres are initially straight and the axes of the fibres in the bent cell follow a cylindrical helical path.
4. The fibre diameter and fabric thickness are small compared to the radius of curvature; the neutral axis of bending is in the geometric centreline of the fibre.
5. The fabric density is high enough that the fibre orientation distribution density function is continuous.
6. The fabric is homogeneous in the fabric plane and in the fabric thickness.

The general unit cell bending rigidity, (EI) cell, is therefore as follows:

$$(EI)_{cell} = N_f \int_{-\pi/2}^{\pi/2} \left[E_f I_f \cos^4\theta + GI_p \sin^2\theta \, \cos^2\theta\right]\Omega(\theta)d\theta \tag{12.30}$$

where N_f=number of fibres in the unit cell, $E_f I_f$=fibre bending rigidity around the fibre axis, G=shear modulus of the fibre, I_p=polar moment of the inertia of the fibre cross-section, a torsion term, $\Omega(\theta)$=the fibre orientation distribution in the direction, θ. The bending rigidities of a nonwoven fabric in two specific cases of fibre mobility are as follows:

1. 'Complete freedom' of relative fibre motion. If the fibres are free to twist during fabric bending, e.g., in a needlepunched fabric, the torsion term ($GI_p \sin^2\theta \cos^2\theta$) will be zero. Therefore,

$$(EI)_{cell} = \frac{\pi d_f^4 N_f E_f}{64} \int_{-\pi/2}^{\pi/2} \Omega(\theta)\cos^4\theta \, d\theta \tag{12.31}$$

where d_f=fibre diameter and E_f=Young's modulus of the fibre.

2. 'No freedom" of relative fibre motion. In chemically bonded nonwovens, the freedom of relative fibre motion is severely restricted. It is assumed in this case that there is no freedom of relative fibre motion, and the unit cell bending rigidity, $(EI)_{cell}$, is therefore as follows:

$$(EI)_{cell} = \frac{\pi N_f E_f d_f^2 h}{48} \int_{-\pi/2}^{\pi/2} \Omega(\theta)\cos^4\theta \; d\theta \tag{12.32}$$

where h = fabric thickness and d_f = fibre diameter.

12.15 Modelling pore size and pore size distribution

In the design and engineering of nonwoven fabrics, it is desirable to make property predictions based on the structural parameters of the fabric. Extensive work has been conducted to simulate isotropic nonwoven structural geometries by modulating parameters such as the number of fibre contact points [245] and interfibre cross distances [246]. Before manufacturing the actual nonwoven, it is then possible to predict how different structures will behave in terms of, for example, their mechanical or fluid flow properties. Another important structural characteristic is pore size. In this section, we summarise some of the approaches that have been developed for predicting pore size in nonwovens.

12.15.1 Models of pore size

Although it is arguable if the term 'pore' accurately describes the voids in a highly connective, low density nonwoven fabric, it is still helpful to use this term in modelling porous nonwoven structures. The pore size in simplified nonwoven structures can be approximately estimated by Wrotnowski's model [247,248] (Fig. 12.19), although the assumptions that are made for the fabric structure are based on fibres that are circular in cross-section, straight, parallel, equidistant, and arranged in a square pattern. The radius of a pore in Wrotnowski's model is shown as follows,

$$r = \left(0.075737\sqrt{\frac{Tex}{\rho_{fabric}}}\right) - \frac{d_f}{2} \tag{12.33}$$

where Tex = fibre linear density (tex), ρ_{fabric} is the fabric density (g/cm^3) and d_f is the fibre diameter (m).

Several other models relating pore size and fibre size by researchers can also be used for nonwoven fabrics. For example, both the largest pore size and the mean pore size can be predicted as follows, using Goeminne's equation [249].

Largest pore size ($2r_{max}$): $r\frac{d_f}{2(1-\varepsilon)}$max.

Mean pore size ($2r$) (porosity <0.9): $r = \frac{d_f}{4(1-\varepsilon)}$where ε is the fabric porosity. In addition, pore size ($2r$) can also be obtained based on Hagen–Poiseuille's law in a cylindrical tube,

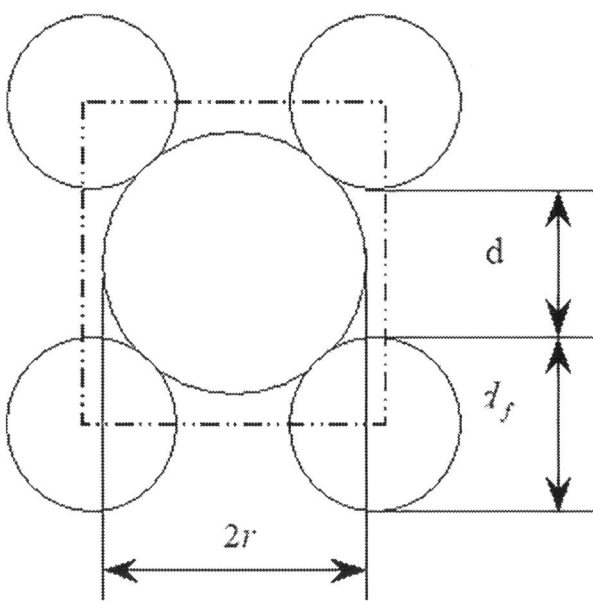

FIG. 12.19 Wrotnowski's model for pore size in a bundle of paralleled cylindrical fibres arranged in a square pattern.

$$r = \sqrt[4]{\frac{8k}{\pi}} \qquad (12.34)$$

where k is the specific permeability (m^2) in Darcy's law.

12.15.2 Models of pore size distribution

If it is assumed that the fibres are randomly aligned in a nonwoven fabric following Poisson's law, then the probability, $P(r)$, of a circular pore of known radius, r, is distributed as follows [119],

$$P(r) = -(2\pi v')exp(-\pi r^2 v') \qquad (12.35)$$

where $v' = \frac{0.36}{r^2}$, and it is defined as the number of fibres per unit area.

In geosynthetics, a series of critical pore sizes have been defined, i.e., the fabric apparent opening size (AOS, or O_{95}) and the fabric filtration opening size (FOS). The apparent opening size (AOS, or O_{95}) indicates the approximate largest particle that would effectively pass through the structure. In the dry sieving method, it is defined as the bead size at which 5% or less of the weight of the beads pass through the nonwoven fabric.

Giroud [116] proposed a theoretical equation for calculating the filtration pore size of nonwoven geosynthetics. The equation is based on the fabric porosity, fabric thickness, and fibre diameter:

$$O_f = \left[\frac{1}{\sqrt{1-\varepsilon}} - 1 + \frac{\xi \varepsilon d_f}{(1-\varepsilon)h} \right] d_f \qquad (12.36)$$

where d_f=fibre diameter; ε=porosity; h=fabric thickness; ξ=an unknown dimensionless parameter to be obtained by calibration with test data to account for the further influence of fabric porosity, and $\xi=10$ for particular experimental results; and O_f=filtration opening size, usually given by the near largest constriction size of the fabric (e.g. O_{95}).

Lambard [250] and Faure [251] applied Poissonian line network theory to establish a theoretical model of the 'opening sizing' of nonwoven fabrics. In this model, the fabric thickness is assumed to rely on randomly stacked elementary fibre layers, where each layer has a thickness T_e and is simulated by two-dimensional straight lines (a Poissonian line network). Faure et al. [252] and Gourc and Faure [253] also presented a theoretical technique for determining constriction size based on the Poissonian polyhedra model. In Faure's approach, epoxy-impregnated nonwoven specimens were sliced and the structure was modelled as a pile of elementary layers, in which fibres were randomly distributed in planar images of the fabric. The cross-sectional images were obtained by slicing at a thickness of fibre diameter df and the statistical distribution of pores was modelled by inscribing a circle into each polygon defined by the fibres (Fig. 12.20).

The pore size distribution, which is obtained from the probability of passage of different spherical particles (similar to glass beads in the dry sieving test) through the layers forming the fabric, can thus be determined theoretically using the following equation [250],

$$Q(d) = (1-\varphi)\left(\frac{2+\lambda(d+d_f)}{2+\lambda d_f}\right)^{2N} e^{-\lambda Nd} \tag{12.37}$$

where $\lambda = \frac{4}{\pi}\frac{(1-\varphi)}{d_f}$, and $N = \frac{T}{d_f}$

$Q(d)$=probability of a particle with a diameter d passing through a pore channel in the fabric; φ=fraction of solid fibre material in the fabric; λ=total length of straight lines per unit area in a planar surface (also termed the specific length); and N=number of slices in the cross-sectional image. Because of the assumption in Faure's approach that the constriction size in a relatively thick fabric tends to approach zero, Faure's model generally produces lower values than other methods [113]. The use of this method is therefore not recommended for fabrics with a porosity of 50% or less [120].

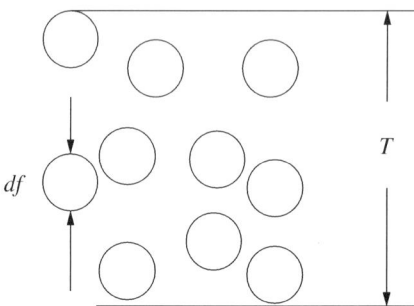

FIG. 12.20   Model for constriction pore size in a nonwoven fabric consisting of randomly stacked elementary layers of fibres.

12.16 Modelling specific permeability

The specific permeability of a nonwoven fabric is solely determined by fabric structure and is defined based on Darcy's Law [254], which may be written as follows:

$$Q = -\frac{k}{\eta}\frac{\Delta p}{h}$$

(12.38)

where Q (m^3/m^2/s) is the volumetric flow rate of the fluid flow through a unit cross-sectional area in the porous structure, η (Pa.s) is the viscosity of the fluid, Δp is the pressure drop (Pa) along the conduit length of the fluid flow, h (m), and k (m^2) is the specific permeability of the fabric.

12.16.1 Theoretical models of specific permeability

Numerous theoretical models describing laminar flow through porous media have been proposed to predict permeability. The existing theoretical models of permeability applied in nonwoven fabrics can be grouped into two main categories based on:

1. Capillary channel theory, e.g., Kozeny [266], Carman [256], Davies [257], Piekaar and Clarenburg [258], and Dent [259].
2. Drag force theory, e.g., Emersleben [260], Brinkman [261], Iberall [262], Happel [272], Kuwabara [255], Cox [264], and Sangani and Acrivos [265].

Many permeability models established for textile fabrics are based on capillary channel theory or the hydraulic radius model, which is based on the work of Kozeny [266] and Carman [256]. The flow through a nonwoven fabric is treated as a conduit flow between cylindrical capillary tubes. The Hagen–Poiseuille equation for fluid flow through such a cylindrical capillary tube structure is as follows:

$$q = \frac{\pi r^4}{8\eta}\frac{\Delta P}{h}$$

(12.39)

where r is the radius of the hydraulic cylindrical tube. However, it has been argued that models based on capillary channel theory are suitable only for fabrics having a low porosity and are unsuitable for highly porous structures where the porosity is greater than 0.8, see, for example, Carman [256].

In drag force theory, the walls of the pores in the structure are treated as obstacles to an otherwise straight flow of the fluid. Drag force theory is believed to be more applicable to highly porous fibrous assemblies, such as nonwoven fabrics, where the single fibres can be regarded as elements within the fluid that cannot be displaced, see Scheidegger [274]. The drag of the fluid acting on each portion of the wall is estimated from the Navier–Stokes equations, and the sum of all the individual 'drags' is assumed to be equal to the total resistance to flow in the fabric.

Iberall [262] adopted the drag force models obtained by Emersleben [260] and Lamb [267] and established a model of permeability for a structure having a random distribution of cylindrical fibres of circular cross-section and identical fibre diameter. The model accounted for the permeability on the basis of the drag forces acting on individual elements in the structure. It was assumed that the flow resistivity of all random distributions of fibres per unit volume does not differ. The resistivity was obtained by assuming the fabric has an isotropic structure in which the number of fibres in each axis is equal and one of the axes is along the direction of macroscopic flow.

Happel [263], Kuwabara [255], Sparrow and Loeffler [268], as well as Drummond and Tahir [269] have given detailed analyses of the permeability in unidirectional fibrous structures using a so-called 'unit cell' theory, or 'free surface' theory. In these models, the fibres are assumed to be unidirectionally aligned in a periodic pattern such as a square, triangular, or hexagonal array. The permeability is then solved using the Navier–Stokes equation in the unit cell with appropriate boundary conditions. These models have shown good agreement with experimental results when the fabric porosity is greater than 0.5 [253,260,270]. Unlike capillary flow theory, drag force theory and the unit cell model clearly demonstrate the relationship between permeability and the internal structural architecture of the fabric.

12.16.2 Summary of permeability models

Theoretical models of permeability and empirical equations for fibrous structures are based on the fabric being homogeneous and either isotropic, unidirectional [257,261,263,265,266,271,272] or, anisotropic [72,161,273]. There are distinct differences between the three types of permeability model. The permeability in isotropic nonwovens is identical in all directions throughout the entire structure, while the permeabilities in the three principal directions in homogeneous unidirectional nonwoven structures are obtained parallel and perpendicular to the orientation of the fibres. The permeabilities in anisotropic fabrics vary in all directions throughout the fabric structure. Various empirical permeability models for nonwoven fabrics have also been obtained. A comparison of the available models for 2D and 3D nonwovens is shown in Figs 12.21 and 12.22 and Table 12.11.

As shown in Fig. 12.21A, the Kozeny equation [266] and its derivations [274], which are based on capillary channel theory, agree with experimental data very well when the fabric porosity is low (<0.8), but agreement is poorer for high porosity (>0.8) fabrics, see Fig. 12.21B. Iberall's equation gives predicted permeability results that are higher than those obtained from empirical models in low porosity fabrics (where porosity <0.8) but agrees well with the empirical equations when porosity approaches 1 as shown in Fig. 12.22B. In Fig. 12.21, Rushton's equation, which is based on woven fabric, agrees well with Shen's experimental results [279], which are based on both the transverse permeability and the in-plane permeability of needlepunched nonwoven fabrics, for porosities ranging from 0.3 to 0.8. Davies' equation [253], which was obtained from experimental results of air permeability in fabrics composed of glass fibres having a porosity of 0.70–0.99, appears to provide reasonable predictions for structures with higher porosities of 0.70–0.99.

The Mao–Russell equation for isotropic fibrous structures (denoted as M_R_ISO in Fig. 12.21) shows good agreement with capillary theory at low porosity and is also in reasonable agreement with the results of empirical models at high porosity. Predicted results from the M_R_ISO model are in close agreement with the empirical data from both Shen's equation at low porosity (0.3–0.8) and with Davies' equation at higher porosity (0.85–0.99). It would appear that the M_R_ISO model is applicable for both low and high porosity fibrous structures.

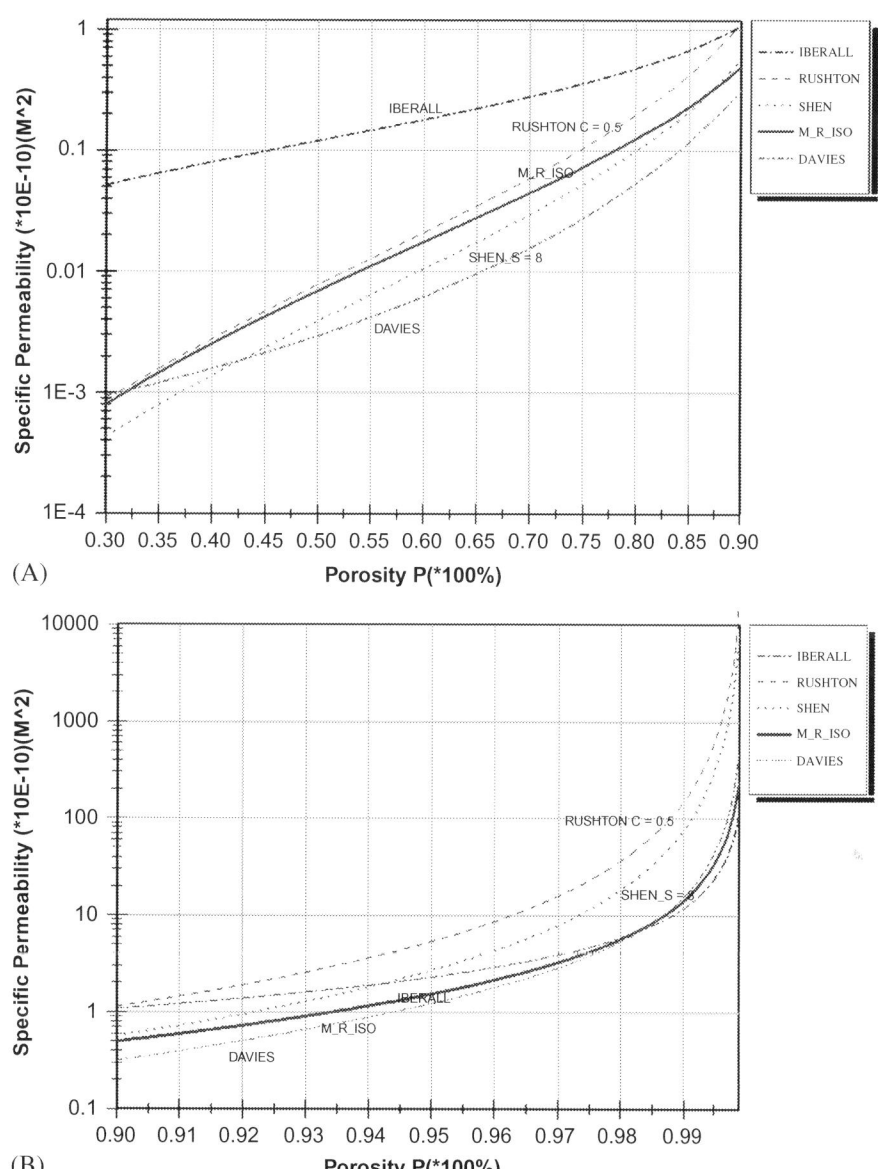

FIG. 12.21 Comparison of existing permeability models for homogeneous isotropic two-dimensional nonwoven fabrics. (A) $\varepsilon = 0.30$–0.90; (B) $\varepsilon = 0.90$–1.0. (Note: In Rushton's equation, the product of the roughness factor and Kozeny constant ϕk_0 is taken as 0.5.) [161,273].

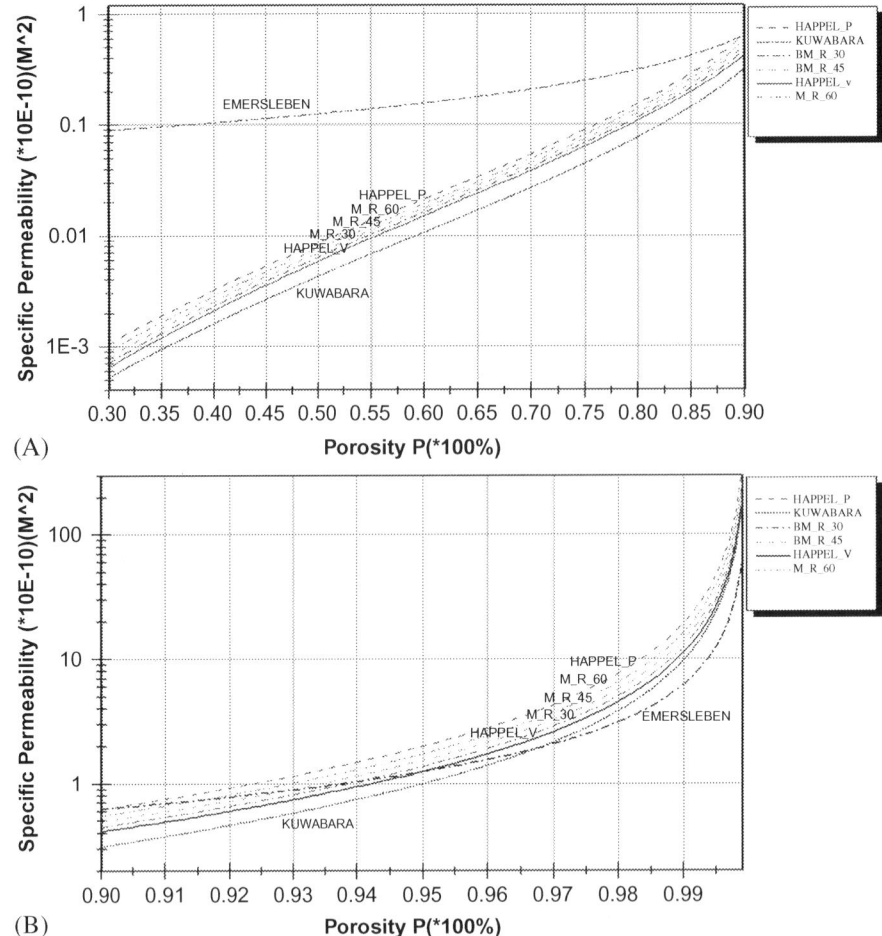

FIG. 12.22 Comparison of existing permeability models for homogeneous unidirectional fibrous structures [270,271]. (A) ($\varepsilon = 0.30$–0.9); (B) ($\varepsilon = 0.90$–1.00). (Note: M_R_30, M_R_45 and M_R_60 are referred to as the directional permeabilities in unidirectional fabrics $k(\theta) = -\frac{1}{32}\frac{d_f^2}{\varphi}\{\frac{ST}{Tsin^2\theta + Scos^2\theta}\}$ when $\theta = \frac{\pi}{6}, \frac{\pi}{4}$ and $\theta = \frac{\pi}{3}$ respectively.)

12.16.3 Directional permeability in anisotropic nonwovens [69]

Nonwovens are three-dimensional anisotropic structures in which fibres are oriented in preferred directions, and the directional permeability in such structures, is closely related to fibre orientation [174,267]. In Fig. 12.23, fluid flow through a three-dimensional nonwoven fabric in the direction $\vec{\tau}$ and $\vec{f}$ is linked to the fibre orientation. When the two direction vectors $\vec{f}$ and $\vec{\tau}$ are represented by angles with three principal axes in a spherical coordinate system, $\vec{f} = (\beta, \frac{\pi}{2} - \beta, \alpha)$ and $\vec{\tau} = (\beta_\tau, \frac{\pi}{2} - \beta_\tau, \alpha_\tau)$, the angle between the two direction vectors $\vec{f}$ and $\vec{\tau}$, $\psi \left(\psi = \angle \left(\vec{f}, \vec{\tau} \right) \right)$ is as follows [69],

TABLE 12.11 Existing permeability models for isotropic and unidirectional fibrous structures.

Name of theory		Permeability (m²)		Notes
Ferrandon's theory k_n' [274]		$\frac{1}{k_n} = \frac{1}{k_1}\cos^2\theta + \frac{1}{k_2}\sin^2\theta$		For anisotropic nonwoven fabrics
Directional permeability of 2D nonwoven fabric having unidirectional fibre alignments $k(\theta)$ (Mao-Russell_Uni [161])		$k(\theta) = -\frac{1}{32}\frac{d_f^2}{\varphi}\{\frac{ST}{T\sin^2\theta + S\cos^2\theta}\}$		Directional permeability of unidirectional fibrous bundles
Drag force theory	Emersleben's equation [260]	$k = \frac{1}{C}\frac{d_f^2}{\varphi}$	$C = 16$	
	Happel's model [263]		$C_{//} = -\frac{32}{S}$ $C_\perp = -\frac{32}{T}$	
	Kuwabara's model [255]		$C_\perp = -\frac{64}{S}$	
	Langmuir model [275]	$k_\perp = -\frac{S}{16\varphi}d_f^2$		
	Van der Westhuizen & du Plessis [276]	$k_{//} = \frac{(3.142 + 2.157\varphi)(1-\varphi)^2}{192\varphi^2}d_f^2$ $k_\perp = \frac{\pi(1-\varphi)(1-\sqrt{\varphi})^2}{96\varphi^{1.5}}d_f^2$		
	Miao	$k_\perp = \frac{S}{9\varphi}d_f^2$		
	Permeability of 2D isotropic nonwoven fabric, k (Mao–Russell_ISO [273])		$C = -16\{\frac{S+T}{ST}\}$	Permeability of isotropic structures
	Permeability of 3D isotropic nonwoven fabric, k (Mao–Russell_ISO3D [72]		$C = -\frac{32}{3}\{\frac{2S+T}{ST}\}$	
	Iberall's Model [262]		$C = \frac{16}{3}\frac{(4-lnRe)}{(2-lnRe)}\frac{1}{(1-\varphi)}$	
	Hagen–Poiseuille Equation [280,281]	$k = \frac{\pi r^4}{8}$	N/A	
Capillary channel theory	Kozeny–Carman's equation for structure of capillary channels [256]	$k = \frac{1}{C}\frac{(1-\varphi)^3}{\varphi^2}$	$C = k_0 S_0^2$	
	Kozeny–Carman's equation for fibrous materials [289]		$C = \frac{k_0}{d_f^2}$	
	Rushton's Equation for woven fabrics [278]		$C = \frac{16\tau k_0}{d_f^2}$	
	Sullivan's Equation k [270]		$C = \frac{32}{\xi d_f^2}$	Permeability of anisotropic structures
	Davies' Model [257]	$k = \frac{1}{64\varphi^{\frac{3}{2}}[1 + 56\varphi^3]}d_f^2 (\varphi = 0.16 \sim 0.30)$		Permeability of isotropic structures
Empirical models	Shen's model [279]	$k = \frac{1}{128}\frac{(1-\varphi)^3}{\varphi^2}d_f^2$		
	Rollin's Model [119]	$k = 7.376*10^{-6}\frac{d_f}{\sqrt{\varphi}}$		

Where (1) τ—Roughness factor, k_0—the Kozeny constant, ξ the orientation factor.
(2) S_i—Specific internal surface area, and S_0—the specific surface area, where $S_0 = \frac{S_i}{(1-\varphi)}$, r is the radius of a cylindrical capillary tube.
(3) k_1 and k_2 are the two principal permeabilities, respectively, in the fabric plane.
(4) $S = [2\ln\varphi + \varphi^2 - 4\varphi + 3]$ and $T = \left[\ln\varphi + \frac{1-\varphi^2}{1+\varphi^2}\right]$
(5) $C_{//}$ and $C_\perp$ are the coefficient for permeability in the direction parallel and perpendicular to the fibre orientation, respectively, in

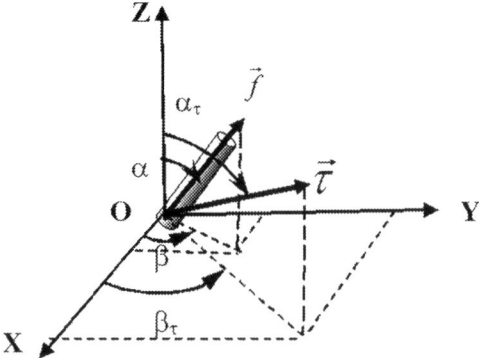

FIG. 12.23 Liquid flows through three-dimensional fabric structures [69].

$$\cos(\psi) = \cos(\alpha)\cos(\alpha_\tau) + \sin(\alpha)\sin(\alpha_\tau)\cos(\beta - \beta_\tau) \quad (12.40)$$

The Mao–Russell equations were developed to enable modelling of directional permeability in nonwovens based on measurable parameters of fabric structure. The following assumptions are made:

(i)   The distance between fibres and the length of individual fibres is much larger than the fibre diameter, i.e., the structure has high porosity. The disturbance of the flow due to adjacent fibres is assumed to be negligible.

(ii) The fibres constituting the nonwoven fabric are of the same diameter and are distributed horizontally in-plane and in two-dimensions. No fibres are aligned in the Z-direction.

(iii) The flow resistivity of the fibres per unit volume in the entire structure of the fabric is equal, i.e., the fabric is homogeneous.

(iv) The number of fibres oriented in each direction is not the same, but obeys the function of the fibre orientation distribution, $\Omega(\beta, \alpha)$, in the fibre orientation direction $\vec{f}$ (β, α).

(v)   The inertial forces of the fluid are negligible, i.e., the fluid has a low local Reynolds number Re, and the pressure drop between planes perpendicular to the direction of the macroscopic flow is equal to the drag force on all elements between the planes.

(vi) The pressure drop necessary to overcome the viscous drag is linearly additive for the various fibres, whether they are arranged parallel, perpendicular or in any other direction relative to the flow.

The directional permeability, $k(\vec{\tau})$, in the direction $\vec{\tau}$ of a three-dimensional nonwoven fabric can be written as follows [69]:

$$k(\vec{\tau}) = -\frac{\delta^2}{32\varphi}\left\{\frac{ST}{\int_0^\pi \int_0^\pi [T\cos^2(\psi) + S\sin^2(\psi)]\Omega(\beta, \alpha)d\beta d\alpha}\right\} \quad (12.41)$$

where, $S = [\varphi^2 + 2\ln\varphi - 4\varphi + 3]$ and $T = \left[\ln\varphi + \frac{1-\varphi^2}{1+\varphi^2}\right]$, $\vec{f}$ is the fibre orientation, φ is the volume fraction of the solid material, d_f is the fibre diameter, $\Omega(\beta, \alpha)$ is the fibre orientation distribution function in direction $\vec{f}$.

Nonwoven fabrics can be considered to be three-dimensional anisotropic structures, containing mostly planar, preferentially oriented fibres (Table 12.12). The planar fibre orientation can be readily determined by microscopy coupled with image analysis. To calculate the permeability in nonwovens where a proportion of fibres are oriented out of plane, i.e., in the Z-direction or fabric thickness, the structure can be simplified as shown in Fig. 12.12. In this structure:

1. Fibres are either aligned in the fabric plane or perpendicular to it.
2. The fibre distribution in the Z-direction and in the fabric plane is homogeneous and uniform.
3. The number of fibres perpendicular to the fabric plane represents a fraction z of the total number of fibres.
4. The fluid flow is laminar and in-plane, i.e., the flow along the Z-axis is ignored.

The directional in-plane permeability $k(\theta)$ of the fabric plane based on the two-dimensional fibre orientation distribution function, can be written as follows [72,174,273]:

$$k(\theta) = -\frac{1}{32}\frac{d_f^2}{\varphi}\left\{\frac{ST}{zS + (1-z)\int_0^\pi \Omega(\alpha)\left[T\cos^2(\theta-\alpha) + S\sin^2(\theta-\alpha)\right]d\alpha}\right\} \quad (12.42)$$

where $S = [\varphi^2 + 2\ln\varphi - 4\varphi + 3]$ and $T = \left[\ln\varphi + \frac{1-\varphi^2}{1+\varphi^2}\right]$, θ is the flow direction, α is the fibre orientation in each direction in the fabric plane, φ is the volume fraction of the solid material, $k(\theta)$ is the directional permeability of the fabric, d_f is the fibre diameter, $\Omega(\alpha)$ is the fibre orientation distribution function, and z is the fraction of fibres aligned perpendicular to the fabric plane. The permeability perpendicular to the fabric plane, k_Z, can be written as:

$$k_Z = -\frac{1}{32}\frac{d_f^2}{\varphi}\left\{\frac{ST}{(1-z)S + zT}\right\} \quad (12.43)$$

12.17 Modelling capillary wicking

Liquid wicking in nonwovens can be studied as steady state fluid flow in a porous medium, although in many practical situations, the liquid is in fact an unsteady state flow, where the nonwoven fabric is heterogeneous and not uniformly and completely saturated. There is usually a saturation gradient in the fabric along the direction of flow, and this saturation gradient changes with time as the wicking and absorption process continues. Wicking processes can be divided into four categories:

1. Pure wicking of a liquid without diffusion into the interior of the fibres.
2. Wicking accompanied by diffusion of the liquid into the fibres or into chemical finishes or coatings on the fibre surfaces.
3. Wicking accompanied by adsorption by fibres.
4. Wicking involving adsorption and diffusion into fibres.

TABLE 12.12 Directional permeabilities in various two- and three-dimensional nonwoven structures.

Fibrous Structure model	FOD in the fabric plane	Directional permeability, $k(\theta)$
Ferrandon's equation [161]	$\dfrac{1}{k(\vartheta,\beta,\gamma)} = \dfrac{\cos^2\vartheta}{k_X} + \dfrac{\cos^2\beta}{k_Y} + \dfrac{\cos^2\gamma}{k_Z}$ $\dfrac{1}{k(\theta)} = \dfrac{\cos^2(\theta-\phi)}{k_X} + \dfrac{\sin^2(\theta-\phi)}{k_Y}$ $\dfrac{1}{k(\beta_\tau,\alpha_\tau)} = \dfrac{\cos^2(\beta_\tau-\phi)}{k_X} + \dfrac{\cos^2\left(\frac{\pi}{2}-\beta_\tau+\phi\right)}{k_Y} + \dfrac{\cos^2\alpha_\tau}{k_Z}$	
Generalised permeability model for 3D nonwoven structures [69]	$\Omega(\alpha)$	$k(\vec{\tau}) = -\dfrac{\delta^2}{32\varphi}\left\{\dfrac{ST}{\int_0^\pi\int_0^\pi[T\cos^2(\psi)+S\sin^2(\psi)]\Omega(\beta,\alpha)d\beta d\alpha}\right\}$ $\cos(\psi) = \cos(\alpha)\cos(\alpha_\tau)+\sin(\alpha)\sin(\alpha_\tau)\cos(\beta-\beta_\tau)$
Generalised permeability model for simplified 3D nonwoven structures [69]	$\Omega(\alpha)$	$k(\vec{\tau}) = -\dfrac{\delta^2}{32\varphi}\left\{\dfrac{ST}{zT+(1-z)S+(1-z)(T-S)\int_0^\pi[\cos^2(\beta-\beta_\tau)\sin^2(\alpha_\tau)]\Omega(\beta)d\beta}\right.$ $\dfrac{4(T+S)}{T-S}\cos(2\beta)-\cos(4\phi)\int_0^\pi[\Omega(\beta)\cos(2\beta)]d\beta+\sin(4\phi)$ $\left.\int_0^\pi[\Omega(\beta)\sin(2\beta)]d\beta = 3\int_0^\pi[\Omega(\beta)\cos(2\beta)]d\beta\right\}$
3D isotropic nonwovens [72]	Constant	$k = k_X = k_Y = k_Z = -\dfrac{3}{32}\dfrac{d_f^2}{\varphi}\left\{\dfrac{ST}{2S+T}\right\}$
3D fabric with isotropic fibre alignment in the fabric plane [72,273]	$\Omega(\alpha) = \dfrac{1}{\pi}$	$k_X = k_Y = -\dfrac{1}{16(1-z)}\dfrac{d_f^2}{\varphi}\left\{\dfrac{ST}{S+T}\right\}$ $k_Z = -\dfrac{d_f^2}{32\varphi}\left\{\dfrac{ST}{(1-z)S+zT}\right\}$

3D nonwoven fabric having layers of unidirectional fibre alignment in the fabric plane [72,273]

$$k_X = -\frac{d_f^2}{32\phi}\left(\frac{ST}{zT+(1-z)S}\right)$$

$$k_Y = -\frac{d_f^2}{32\phi}\left(\frac{ST}{zS+(1-z)T}\right)$$

$$k_Z = -\frac{d_f^2}{32\phi}\left\{\frac{ST}{(1-z)S+zT}\right\}$$

where

$$\Omega(\alpha) = \begin{cases} 1; & when\ \alpha = \dfrac{\pi}{2} \\ 0; & when\ \alpha \neq \dfrac{\pi}{2} \end{cases}$$

3D fabric having layers of fibres aligned in two orthogonal directions in the fabric plane [72]

$$k_X = -\frac{d_f^2}{32\phi}\left\{\frac{ST}{((1-z)(1-X)+z)S+X(1-z)T}\right\},$$

$$k_Y = -\frac{d_f^2}{32\phi}\left\{\frac{ST}{((1-z)+z)S+(1-X)(1-z)T}\right\}$$

$$k_Z = -\frac{d_f^2}{32\phi}\left\{\frac{ST}{(1-z)S+zT}\right\}$$

where

$$\Omega(\alpha) = \begin{cases} X; & when\ \alpha = 0 \\ 1-X; & when\ \alpha = \dfrac{\pi}{2} \end{cases}$$

Where, (1) $S = [\varphi^2 + 2ln\varphi - 4\varphi + 3]$ and $T = \left[ln\varphi + \dfrac{1-\varphi^2}{1+\varphi^2}\right]$

(2) X and z are the fraction of fibres aligned in the X–Y fabric plane and Z-directions, respectively. The Z-direction is perpendicular to the fabric plane.

(3) k_X, k_Y, and k_Z are the three principal permeabilities in the equation, respectively.

12.17.1 Capillary pressure (Laplace's equation)

Liquid wicking into a nonwoven fabric is driven by the capillary pressure in the void spaces between adjacent fibres in the fabric. It is known that capillary pressure in a cylindrical capillary tube is given by Laplace's equation,

$$P_{cap} = \frac{2\sigma \cos\gamma}{r} \tag{12.44}$$

where r is the radius of the capillary tube, γ is the contact angle between the liquid and the capillary tube surface, and σ is the surface tension of the liquid. Nonwoven fabrics containing capillary pores having an average diameter of $2r$ are frequently modelled as an equivalent system of parallel cylindrical capillary tubes having the same diameter $2r$.

12.17.2 Hagen–Poiseuille equation

From a consideration of the laws of hydrodynamic flow through capillary channels, Poiseuille [280] first deduced the relation between the volume of fluid flowing through a narrow tube and the pressure difference across its ends. The Hagen–Poiseuille equation [281] for a laminar fluid flow through a cylindrical capillary tube is as follows:

$$\frac{dh}{dt} = \frac{r^2}{8\eta}\frac{\Delta P}{h} \tag{12.45}$$

where, h is the distance through which the fluid flows in time t and η is the viscosity of the fluid.

12.17.3 Lucas–Washburn equation

Based on Poiseuille's equation, Lucas [282] and Washburn [142] calculated the distance along which turbulence occurs and, by converting the volume-flow in Poiseuille's equation into linear-flow in uniform cylindrical tubes, they developed the Lucas–Washburn equation as follows:

$$h = Ct^{\frac{1}{2}} \tag{12.46}$$

where h is the distance through which the fluid flows in time t and C is a constant related to both the liquid properties and nonwoven fabric structure.

A typical example of the application of the Lucas–Washburn equation is in liquid absorption in the upward vertical strip test, in which the capillary pores in a nonwoven fabric are modelled as a series of vertically supported parallel capillary tubes and liquid is absorbed from one end upwards into the tube. The upward driving pressure is as follows:

$$\Delta P = P_{cap} - \rho g h \tag{12.47}$$

where P_{cap} is the capillary pressure in the tube, h is the rising height of liquid in the tube, g is the gravitation acceleration, and ρ is the liquid density.

Substituting Eqs (12.46) and (12.44) into Hagen–Poiseuille's equation (Eq. 12.45), the rising height of the liquid capillary flow is as follows:

$$\frac{dh}{dt} = \left(\frac{r}{4\eta h} \sigma \cos\gamma - \frac{r^2}{8\eta} \rho g \right) \tag{12.48}$$

The solution of this equation is then:

$$t = -\frac{8\eta h}{r^2 \rho g} - \frac{16\eta \sigma \cos\gamma}{r^3 \rho^2 g^2} \log_e \left\{ 1 - \frac{\rho g r h}{2\sigma \cos\gamma} \right\} \tag{12.49}$$

To obtain a simplified form of the relationship between t and h, Laughlin [283] rewrote the above equation as follows:

$$bt = -h_m \log_e \left\{ 1 - \frac{h}{h_m} \right\} - h \tag{12.50}$$

where $h_m = \frac{a}{b}$ with $a = \frac{r\sigma\cos\gamma}{4\eta}$ and $b = \frac{r^2 \rho g}{8\eta}$

Using Taylor's expansion [284], and when the effect of gravity is negligible, e.g., $h \ll h_m$, then the equation can be reduced to the form of the Lucas–Washburn equation [142,282,285] as follows:

$$h = \left\{ \frac{r\sigma \cos\gamma}{2\eta} \right\}^{\frac{1}{2}} t^{\frac{1}{2}} \tag{12.51}$$

These equations can be applied to approximate the horizontal and vertical liquid wicking in nonwoven fabrics, respectively, where $2r$ would be the average pore size in the nonwoven fabric. To calculate the mass transmission of water absorbed during the upwards vertical strip test, Law [286] developed an equation as follows:

$$m_v t = \int_0^t B_v \left\{ \frac{1}{x(t)} - \frac{1}{h_m} \right\} dt \tag{12.52}$$

where $B_v = \frac{\rho \pi r^3 \sigma \cos\gamma}{4\eta}$ (Kg·m/s)

Many authors [277,282] have shown that the flow of liquids through textile fabrics obeys the Lucas–Washburn equation. This equation is reported to accurately characterise the water penetration in other fibrous structures, such as paper [287].

12.17.4 Directional capillary pressure and anisotropic liquid wicking in nonwovens (Mao–Russell equations) [69,291]

According to the capillary channel theory as shown in both the Poiseuille equation and the Lucas–Washburn equation, capillary wicking is determined by the geometric structure of the pores in a nonwoven fabric. However, there are difficulties in quantifying the average equivalent capillary radii [288] of pores in nonwovens because the capillary channels differ in size and shape. They are also inter-connected as well as interdependent, forming a three-dimensional network system. Additionally, the capillary channels in real nonwoven fabrics do not have circular cross-sections and are not necessarily uniform along their lengths. Also, because most of the volume of a nonwoven fabric consists of pore voids (the porosity of high-loft nonwovens may be as high as 0.99) and the spaces between fibres are very large and widely distributed, any attempt to give an exact analysis of the fabric based on theoretical capillary channels would be impracticable because of the complex nature of the nonwoven structure.

12.17.4.1 *Two-dimensional models of capillary pressure in a nonwoven fabric*

The liquid flow over a fibre during wicking in a two-dimensional nonwoven fabric plane is shown in Fig. 12.24.

The fibre of a defined unit length is aligned in the direction α and the liquid flow over the fibre driven by capillary pressure is in the direction θ. The capillary wicking model in a nonwoven two-dimensional plane is based on both the hydraulic radius theory proposed by Kozeny [266] and Carman [256] (see also Collins [289]) and has the following assumptions:

1. The fibres in the nonwoven fabric are of the same diameter and are distributed horizontally in-plane and in two dimensions. No fibres are aligned in the Z-direction.
2. The distance between fibres and the length of individual fibres is much larger than the fibre diameter, i.e., the structure has high porosity. The disturbance of the flow due to adjacent fibres is assumed to be negligible.
3. The flow resistivity of the fibres per unit volume in the entire structure of the fabric is equal, i.e., the fabric is homogeneous.
4. The number of fibres orientated in each direction is not the same, but obeys the function of the fibre orientation distribution, $\Omega(\alpha)$, where α is the fibre orientation angle.
5. The inertial forces of the fluid are negligible, i.e., the fluid has a low local Reynolds number Re, and the pressure drop between planes perpendicular to the direction of the macroscopic flow is equal to the drag force on all elements between the planes.
6. The pressure drop necessary to overcome the viscous drag is linearly additive for the various fibres, whether they are arranged parallel, perpendicular or in any other direction relative to the flow.

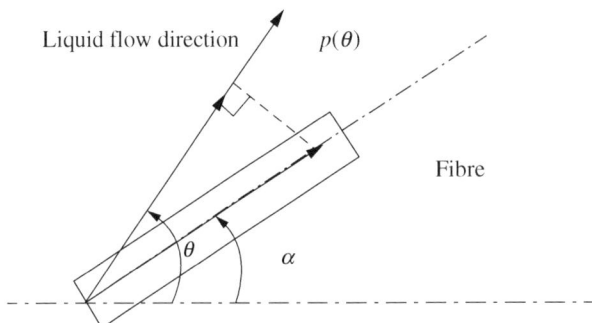

FIG. 12.24   Liquid flow over a fibre during liquid absorption [161].

7. The capillary pressure in direction θ in the fabric plane is hydraulically equivalent to a capillary tube assembly in which there are N capillary cylindrical capillary tubes of the same hydraulic diameter. The basic requirements are therefore [161]:
   - **(i)** The wetted specific area in the direction θ in the fabric plane, $S_0(\theta)$, should be identical to the capillary tube assembly.
   - **(ii)** The porosity of the capillary tube assembly should be the same as the nonwoven fabric.
8. The capillary pressure occurs only between the longitudinal axes of fibres and varies in different directions in the fabric plane, depending on the fibre orientation distribution. This assumption is based on the fact that the capillary phenomenon only occurs along the direction of the fibre orientation and the drag resistance exists perpendicular to the fibre orientation. This resistance force was modelled by Princen [290].

12.17.4.2 Directional capillary pressure [291]

For a nonwoven fabric having the fibre orientation distribution function in the fabric plane, $\Omega(\alpha)$ (where α is the fibre orientation angle), the capillary pressure in the direction θ in the fabric, $p(\theta)$, is as follows [291],

$$p(\theta) = \frac{4\varphi \sum_{\alpha=0}^{\pi} \Omega(\alpha)|\cos(\theta - \alpha)|}{d_f(1 - \varphi)} \sigma \cos\gamma \tag{12.53}$$

where γ is the contact angle of the fibre, σ is the surface tension of the liquid, d_f is the fibre diameter, and φ is the fraction of the solid fibre in the fabric.

12.17.4.3 Anisotropic liquid wicking

The volumetric rate of liquid wicking in the direction θ can be described by Darcy's law, rewritten as follows:

$$Q(\theta) = -\frac{k(\theta)}{\eta} \frac{p(\theta)}{x(\theta)} \tag{12.54}$$

where, $Q(\theta) =$ volumetric flow rate through a unit area in the fabric cross-section [m^3/s* m^2] in the direction θ, $p(\theta) =$ capillary pressure difference [Pa] in the direction θ, $x(\theta) =$ conduit distance [m] in the direction θ, $k(\theta) =$ specific permeability of the fabric [m^2] in the direction θ.

With reference to the capillary pressure, $p(\theta)$ as given above, the directional permeability, $k(\theta)$, has been established in previous work (see Sections 12.16.2 and 12.16.3) as follows [273]:

$$k(\theta) = -\frac{1}{32} \frac{d_f^2}{\varphi} \left\{ \frac{ST}{\int_0^\pi \Omega(\alpha)\left[T\cos^2(\theta - \alpha) + S\sin^2(\theta - \alpha)\right]d\alpha} \right\} \tag{12.55}$$

where d_f is the fibre diameter and φ is the volume fraction of the solid material (defined as $\varphi = \frac{\rho_{fabric}}{\rho_{fibre}}$, where ρ_{fabric} is the density of a fibrous pack in a vacuum and ρ_{fibre} is the fibre density). $\Omega(\alpha)$ is the fibre orientation distribution function, which defines the arrangement of fibres within the nonwoven fabric structure and $T = \left[\ln\varphi + \frac{1-\varphi^2}{1+\varphi^2}\right]$ and $S = [2\ln\varphi - 4\varphi + 3 + \varphi^2]$.

Substituting $k(\theta)$ and $p(\theta)$ into Darcy's law, the rate of liquid absorption (or the wicking rate) in direction θ, $V(\theta)$, can be written as follows [291]:

$$V(\theta) = -\frac{d_f}{8(1-\varphi)x(\theta)}\left\{\frac{ST\sum_{\alpha=0}^{\pi}\Omega(\alpha)|\cos(\theta-\alpha)|}{\sum_{\alpha=0}^{\pi}\Omega(\alpha)\left[T\cos^2(\theta-\alpha)+S\sin^2(\theta-\alpha)\right]}\right\}\frac{\sigma\cos\gamma}{\eta} \quad (12.56)$$

There are two important conclusions to be drawn from this equation. Firstly, it has been shown that the liquid absorption velocity depends on the fibre diameter, the fibre orientation distribution and the fabric porosity, in addition to liquid properties (viscosity and surface tension) and conditions at the fibre surface (liquid contact angle). Secondly, it is evident that the anisotropy of liquid wicking in nonwoven fabrics largely depends on the combination of the fibre orientation distribution and the fabric porosity. Additionally, the relationship between the distance wicked by the liquid and the wicking time in the form of the Lucas–Washburn equation can be obtained from Eq. (12.56) as follows:

$$x(\theta) = Ct^{\frac{1}{2}} \quad (12.57)$$

where, $C = \frac{1}{2}\left[-\frac{d_f}{(1-\varphi)}\left\{\frac{ST\sum_{\alpha=0}^{\pi}\Omega(\alpha)|\cos(\theta-\alpha)|}{\sum_{\alpha=0}^{\pi}\Omega(\alpha)[T\cos^2(\theta-\alpha)+S\sin^2(\theta-\alpha)]}\right\}\frac{\sigma\cos\gamma}{\eta}\right]^{\frac{1}{2}}$

For wicking in a generalised three dimensional nonwoven fabric, the directional capillary pressure $p(\vec{\tau})$ in the direction $\vec{\tau}$ of the structure (see Fig. 12.21) can be written as follows [69],

$$p(\vec{\tau}) = \frac{\phi S_0\displaystyle\int_0^{\pi}\int_0^{\pi}|\cos(\alpha)\cos(\alpha_\tau)+\sin(\alpha)\sin(\alpha_\tau)\cos(\beta-\beta_\tau)|\Omega(\beta,\alpha)d\alpha d\beta}{(1-\phi)}\sigma\cos\gamma \quad (12.58)$$

where $\Omega(\beta,\alpha)$ is the fibre orientation distribution function in the fabric structure in direction $\vec{f}$. Combining the directional permeability in the direction $\vec{\tau}$, both the wicking rate, $V(\vec{\tau})$, and wicking distance, $L(\vec{\tau})$ during horizontal wicking can be determined in direction $\vec{\tau}$ as follows [69],

$$V(\vec{\tau}) = V(\beta_\tau,\alpha_\tau) = -\frac{\delta}{8x(\vec{\tau})(1-\phi)}\left\{\frac{ST\displaystyle\int_0^{\pi}\int_0^{\pi}|\cos(\psi)|\Omega(\beta,\alpha)d\alpha d\beta}{\displaystyle\int_0^{\pi}\int_0^{\pi}[T\cos^2(\psi)+S\sin^2(\psi)]\Omega(\beta,\alpha)d\alpha d\beta}\right\}\frac{\sigma\cos\gamma}{\eta} \quad (12.59)$$

$$L(\vec{\tau}) = L(\beta_\tau,\alpha_\tau) = Ct^{\frac{1}{2}} \quad (12.60)$$

where, $\cos(\psi) = \cos(\alpha)\cos(\alpha_\tau)+\sin(\alpha)\sin(\alpha_\tau)\cos(\beta-\beta_\tau)$

$$C = \frac{1}{2}\left\{-\frac{\delta}{(1-\phi)}\left\{\frac{ST\displaystyle\int_0^{\pi}\int_0^{\pi}|\cos(\psi)|\Omega(\beta,\alpha)d\alpha d\beta}{\displaystyle\int_0^{\pi}\int_0^{\pi}[T\cos^2(\psi)+S\sin^2(\psi)]\Omega(\beta,\alpha)d\alpha d\beta}\right\}\frac{\sigma\cos\gamma}{\eta}\right\}^{\frac{1}{2}}$$

It is therefore apparent that in addition to liquid viscosity, surface tension, and the liquid contact angle, the wicking rate depends on the fibre diameter, the fibre orientation distribution, and the fabric porosity.

12.17.5 Liquid diffusion into nonwovens containing absorbent materials

For hygroscopic or absorbent fibres, it is believed that the spontaneous flow of liquid within capillary spaces is accompanied by a simultaneous diffusion of the liquid into the interior of the fibre or a film on the fibre surface [200]. The sorption of the liquid into fibres can cause fibre swelling [292–294], thereby reducing the capillary spaces between fibres, and complicating the kinetics [183]. Liquid absorption into nonwoven structures containing either absorbent materials or membranes usually involves both intra-fibre liquid diffusion and interfibre liquid absorption. In many cases, liquid saturation gradients have been observed in studies of paper, soil physics, and water resources. Based on the liquid diffusion phenomenon described by Fick's Law [180,295,296], many researchers [297,298] have studied the constant diffusivity coefficient in the one-dimensional strip test by combining the law of conservation of mass and the differential form of Darcy's law. A one-dimensional general equation for the saturation rate has been derived [294,295]:

$$\frac{\partial s}{\partial t} = \frac{\partial}{\partial t}\left[F(s)\frac{\partial s}{\partial x}\right] \tag{12.61}$$

In this equation, the diffusivity factor $F(s)$ is related to the specific permeability k (m^2) via $F(s) = \dfrac{\left(\dfrac{k}{\eta}\right)\left(\dfrac{dp}{ds}\right)}{\varphi}$ [180]. Where s (%) is fabric saturation, t (seconds) is time, x (m) is the liquid conduit distance, p (Pa) is the hydraulic pressure, and η (N * s/m^2) is the viscosity of the liquid. It is also observed [294] that $\frac{x}{\sqrt{t}}$ is a constant and related to saturation $s(x)$, which agrees qualitatively with the Washburn equation [142] with respect to the proportionality between the wicking distance x and $\sqrt{t}$.

12.18 Modelling liquid absorbency and retention

Liquid absorbency (or liquid absorption capacity), C, is defined as the weight of liquid absorbed at equilibrium by a unit weight of nonwoven fabric. Thus, liquid absorbency is based on determining the total interstitial space available for holding fluid per unit dry mass of fibre. The equation is shown as follows [299,300]:

$$C = A\frac{T}{W_f} - \frac{1}{\rho_f} + (1-\alpha)\frac{V_d}{W_f} \tag{12.62}$$

where, A is the area of the fabric, T is the fabric thickness, W_f is the mass of the dry fabric, ρ_f is the density of the dry fibre, V_d is the amount of fluid diffused into the structure of the fibres, α is the ratio of increase in volume of a fibre upon wetting to the volume of fluid diffused into the fibre. In the above equation, the second term is negligible compared to the first, and the third term is nearly zero if the fibre is assumed to swell strictly by replacement of fibre volume with fluid volume [301]. Thus, the dominant factor that controls the fabric absorbent capacity is the fabric thickness per unit mass on a dry basis (T/W_f).

In a given fabric and fluid system, only the mean pore radius r and thickness per unit mass (T/Wf) in the above equation are not constant. The value of r is predicted by the following equation based on the assumption that a capillary is bound by three fibres, orientated parallel or randomly, and the specific volume of the capillary unit cell is equal to that of the parent fabric [302].

$$r = \left[\frac{1}{6\pi\xi} \left(A \frac{T}{W_f} \times \frac{\rho_1\rho_2}{f_1\rho_2 + f_2\rho_1} - 1 \right) \left(\frac{d_1 n_1}{\rho_1} + \frac{d_2 n_2}{\rho_2} \right) \right]^{\frac{1}{2}}$$ (12.63)

for $n_1 = 3 - n_2, n_2 = \frac{3f_2 d_1}{f_1 d_2 + f_2 d_1}$

where the subscripts 1 and 2 represent different fibre types and ξ is a constant with a value of 9×10^5, d is fibre denier, ρ is fibre density (g/cm^3) and f is the mass fraction of the fibre in a blend ($f_1 + f_2 = 1$).

12.19 Modelling thermal resistance and thermal conductivity

Heat is transferred in three ways [303,304]:

(i)   By conduction arising from the vibration of particles (molecules, atoms, and electrons). Here, some of the energy in the high-temperature region of a solid, gas, or liquid is transmitted to the adjacent lower-temperature regions through particle interaction.
(ii) By convection, arising from a fluid flow process.
(iii) By emittance of electromagnetic radiation.

Conduction is heat transfer through a material or through several types of material in direct contact. In nonmetallic solids, the primary mechanism is by lattice-vibration wave propagation. In the case of conduction in gases, the interchange of kinetic energy by molecules colliding is the predominant mechanism. Higher temperatures are associated with higher molecular energies, and when neighbouring molecules collide, a transfer of energy from the more energetic to the less energetic molecules occurs.

Convection is heat transfer due to the internal movement of fluid particles in a fluid flow. The fluid is frequently a gas such as air. The flow can be caused by an external force and induced by buoyancy in the fluid that arises from fluid density variations caused by temperature variations in the fluid. Convection depends on the conditions in the boundary layer, which are influenced by surface geometry, the nature of the fluid motion, and the fluid thermodynamic and transport properties. Convective heat transfer can occur between solids and fluids.

Thermal radiation is heat transfer between two bodies by means of electromagnetic waves, i.e., the radiation or propagation of a collection of particles termed photons or quanta. Radiation is a surface phenomenon. The heat flow in the 'steady state' condition passing through a flat plate material is found to be proportional to the area and to the temperature difference between the two faces, and inversely proportional to the thickness of the plate material. The heat transfer can be expressed as a one-dimensional form of Fourier's heat conduction equation for steady state heat flow through the flat plate as follows:

$$Q = k\frac{\Delta T}{AL} \tag{12.64}$$

where Q is the heat flow rate in unit area (Wm^{-2}), k is the thermal conductivity $(Wm^{-1} K^{-1})$, A is the meter area normal to heat flow (m^2), ΔT is the temperature difference across the plate (K), and L is the fabric thickness (m).

Based on the above equation, the thermal conductivity and thermal resistance have been defined to quantify the thermal properties of materials as follows:

1. Thermal conductivity. The thermal conductivity, k, is a material property which defines the capacity of the material to conduct heat through its mass. It is defined as the amount of heat/energy in watts (W) that will flow through a unit area of (m^2) and unit thickness of (1 m) of the material when a temperature difference of 1 K is established between its surfaces. Thermal conductivity can be expressed in $Wm^{-1} K^{-1}$.
2. Thermal resistance. The thermal resistance, $Rf(m^2 KW^{-1})$, is related to the thermal conductivity of the material and its thickness by the relationship:
3. The thermal resistance, $R_f(m^2 KW^{-1})$, is related to the thermal conductivity of the material, k, and its thickness, d(mm), by the relationship:

$$R_f = \frac{d(mm)*10^{-3}}{k(Wm^{-1}K^{-1})} \tag{12.65}$$

The thermal conductivities of some solid polymer materials and fabrics are listed in Tables 12.13 and 12.14. Steam has the lowest thermal conductivity $(0.016\ Wm^{-1} K^{-1})$, whereas silver has the highest thermal conductivity $(406\ Wm^{-1} K^{-1})$. Among polymers, polypropylene has a low thermal conductivity $(0.10\ Wm^{-1}K^{-1})$ and high-density polyethylene (HDPE) has a comparatively high thermal conductivity $(0.52\ Wm^{-1}K^{-1})$.

TABLE 12.13 Thermal conductivity of polymer materials commonly used in the textile industry.

Materials	Density (kg/m³)	Thermal conductivity (W/m*K)	Source of data
Nylon 6 fibre	1140	0.25	Morton and Hearle [11]
PET fibre	1390	0.14	
PP fibre	910	0.12	
PE fibre	920	0.34	
PVC fibre	1360	0.16	
Wool keratin	1300	0.20	Baxter [305]
Silver	10,490	406	Kreider [306]
Air (25°C)	1.29	0.024	
Water	1000	0.58	

TABLE 12.14 Thermal conductivity of some textile fabrics.

Materials	Density (kg/m^3)	Thermal conductivity (W/m*K)	Source of data
Silk pad	500	0.05	Morton and Hearle [11]
Cotton pad	500	0.071	
Wool pad	500	0.054	
Wool felt	320	0.047	ASHRAE [307]
Wool felt	300	0.071	Kreider [303]
Loose Wool	100	0.03	
Human hair felt	100	0.05	

Nonwoven fabrics are a mixture of fibre and air (and chemical binders if present), and the thermal conductivity of air and solid materials (e.g. fibres) needs to be considered when heat transfers through the fabric. The overall heat transfer in nonwovens is the sum of the contributions of the fibre and the air, which can involve multiple transport mechanisms. Therefore, the thermal conductivity of a nonwoven fabric, k, is the sum of all the individual conductivities of the structural components. The heat transfer value includes all the conduction, convection, and radiation values of air, solid fibre (and binders if any). The thermal conductivities of a nonwoven fabric can therefore be expressed as follows:

$$k = k_{\text{air conduction}} + k_{f \text{ conduction}} + k_{\text{convection}} + k_{\text{radiation}} + k_{\text{fibre}-\text{air}} \qquad (12.66)$$

where $k_{\text{air conduction}}$ = heat conduction via air, $k_{f \text{ conduction}}$ = heat conduction through fibres, $k_{\text{convection}}$ = heat transfer by convection, $k_{\text{radiation}}$ = heat transfer by radiation, $k_{\text{fibre}-\text{air}}$ = heat interaction between the air and fibre.

The heat transfer of a nonwoven fabric is influenced by the thermal conductivity of the fibre components, the fabric structure, fabric dimensions, and the environmental temperature. The fabric bulk density, porosity, and the fibre arrangement are particularly important. Conduction occurs in the solid fibre material and in the pore spaces or void volume between the fibres. Free convection takes place in the presence of a gravity field and there is radiation from the surface of the fabric as well as internally among the fibres. Both convection and radiation can be reduced by increasing the bulk density of the fabric, whereas there may be an increase in the thermal conduction by the fibres. With the increases of the fabric density, the increase in thermal conduction gradually outweighs the decrease in radiation, there is a minimal value in the thermal conductivity curve for a given fabric density and temperature. Therefore, in order to maximise the thermal insulation properties of a nonwoven fabric, i.e., to minimise the effective thermal conductivity of the fabric, it is necessary to find a balance between minimising fibre content to reduce conduction, while providing sufficient fibre surface area to prevent convection and to decrease radiation effects.

For modelling purposes, it is assumed that (i) the nonwoven structure can be approximated as a homogeneous medium of conductivity k; (ii) the interaction of fibres influencing

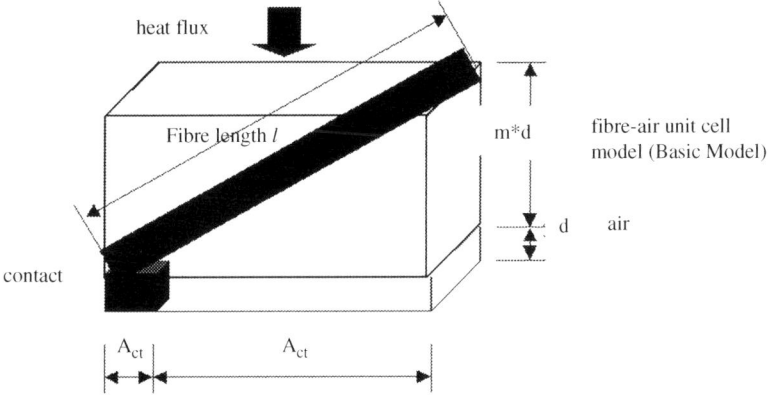

FIG. 12.25   Modified fibre–air unit cell model for the thermal conductivity of fabrics [308].

k can be averaged over a unit volume; (iii) any individual fibre can be assumed to be a spheroid whose major axis is very large compared with the minor axis, and (iv) the fibres are perpendicular to the heat flow.

By considering the fibre orientation and the thermal resistance of air based on the basic fibre–air unit cell (Fig. 12.25), Stark and Fricke [308,309] established a model to determine the thermal conductivity, k^{BM} as follows,

$$k^{BM} = k_s \left(1 + \frac{\beta - 1}{1 + \alpha(1 + Z(\beta - 1)/(\beta + 1))} \right) \tag{12.67}$$

where $\alpha = \frac{v_s}{v_a}$, $\beta = \frac{k_a}{k_s}$ and v_a and v_s are the fractional volumes of the medium of thermal conductivity k_a, and k_s, respectively, and $v_a + v_s = 1$. The term Z is the fraction of fibres arranged perpendicularly to the macroscopic heat flow ($Z = 1$ when fibres are aligned perpendicular to the heat flux, $Z = 0.66$ when randomly arranged and $Z = 0.83$ when arranged parallel to the heat flux).

The thermal resistance in the contact area between adjacent fibres (Fig. 12.24) is modelled using statistical probability and the coupling effects of the fibre and air. The thermal conductivity of the fabric based on a modified model, k^{MMC}, may be calculated using the diagram in Fig. 12.26 [309].

$$k^{MMC} = (m + 1) \left\{ \frac{m}{k^{BM}} + \frac{\xi + 1}{\xi} \frac{1}{k_a + \dfrac{2k_s A d_f}{\xi \pi a_{ct}}} \right\}^{-1} \tag{12.68}$$

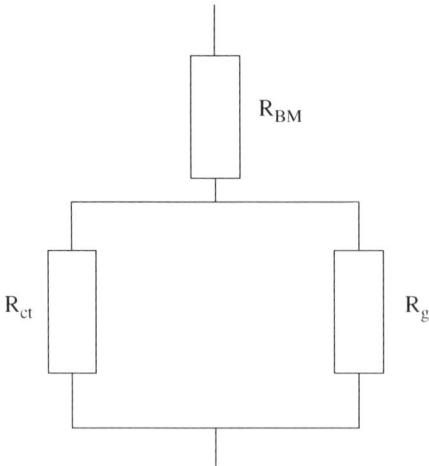

FIG. 12.26   Circuit diagram of the modified model based on the thermal resistances in the basic model (R_{BM}), fibre–fibre contact area (R_{ct}) and air (R_g) [309].

where,

$$m = \frac{l}{d_f} \cos \vartheta_0 - 1$$

$$\xi = \left(\frac{2l}{d_f}\right)^{\frac{2}{3}} \frac{(0.5 \sin \vartheta_0)^{\frac{1}{3}}}{\pi \left(1.5 \left(1 - \mu_0^2\right) p_{ext}/E\right)^{\frac{2}{3}}} - 1$$

$$\frac{l}{d_f} = \frac{1}{2} \left(\frac{\pi \left(1 + \frac{1}{\alpha}\right)}{0.5 \sin^2 \vartheta_0 \cos \vartheta_0}\right)^{\frac{1}{2}}$$

Here, a_{ct} is the contact radius, A is the connection parameter and $A = 0.611$, A_{ct} is the area of one contact, d_f is the fibre diameter, E is the Young's modulus, l is the fibre length in the basic model of the fibre–air unit cell (diagonally across the cell), k^{BM} is the thermal conductivity based on the basic model of the fibre–air unit cell, k^{MMC} is the thermal conductivity based on the modified model of the fibre–air unit cell, m is the height of the fibre–air unit cell in units of the fibre diameter (d_f), p_{ext} is the external pressure, μ_0 is the Poisson's number of the fibre material, $\xi + 1$ is the area of a fibre–air cell in units of the contact area (A_{ct}), and ϑ_0 is the mean fibre orientation angle.

Empirical equations for the relationship between the thermal conductivity of a homogeneous and isotropic nonwoven fabric at a certain regain, k, and the proportion of its component fibres has been given by Schuhmeister [310] as follows:

$$k = \frac{1}{3}(k_1 v_1 + k_2 v_2) + \frac{2}{3}\left(\frac{k_1 k_2}{k_1 v_2 + k_2 v_1}\right) \tag{12.69}$$

where v_1 and v_2 are the fractional volumes of the media of thermal conductivity $k1$, and $k2$, respectively, and $v_1 + v_2 = 1$. When nonwovens are composed of wool fibres, Baxter [305] expanded Schuhmeister's equation and found that the thermal conductivity of the wool fabric is as follows:

$$k_m = x(k_1 v_1 + k_2 v_2) + y\left(\frac{k_1 k_2}{k_1 v_2 + k_2 v_1}\right) \tag{12.70}$$

where $x = 0.21$, $y = 0.79$, k_1 is the thermal conductivity of air ($k_1 = 0.0264$), k_2 is the thermal conductivity of wool fibre at a certain regain ($k_2 = 0.2226$ at a regain of 0.7%, $k_2 = 0.1933$ at a regain of 10.7%) and v_1 and v_2 are the fractional volumes of air and the wool fibre, respectively.

For many nonwoven fabrics, the bulk density of the fabric is the primary factor contributing to the heat transfer through the fabrics [311]. However, the thermal conductivity of a fabric is not linearly related to the fabric density. There is a range of very low fabric densities in which the thermal conductivity of nonwovens decreases with an increase in the fabric density. The thermal conductivity of the constituent fibres dominates the heat transfer of the fabric because the increased fabric density blocks a large proportion of the radiation, convection, and air infiltration effects. When fabric density increases to a threshold point, the thermal conductivity of the fabric then increases with an increase in the fabric density. The thermal conductivity of a nonwoven fabric also increases with an increase in the environmental temperature [307] because the contribution of radiation, convection, and conduction to the thermal conductivity of a fabric increases significantly with an increase in the heating temperature.

12.20 Modelling acoustic impedance [312–315]

Sound is concerned with the transmission of energy through a medium in the form of mechanical vibrations. Sound waves are characterised by the wavelength, λ, the sound frequency, f, and the velocity of sound propagation, v. These terms are related as follows: $v = \lambda * f$. Sound energy is characterised by sound intensity and sound pressure. Sound propagation in air-filled, high porosity nonwoven materials involves the elastic response of the fibre segments as well as thermal and viscous effects at the fibre–air boundaries.

There are three acoustical effects in nonwoven fabrics when sound waves transport through the structure: reflection, transmission, and absorption. The latter two depend on the interaction between sound waves and fibres in the fabrics, and they are the two main mechanisms of sound isolation. Nonwoven fabrics are good sound absorption materials but are relatively poor sound barriers. Sound absorption in nonwoven structures involves the ability of the fibres in the nonwoven to absorb sound energy by converting it into heat. Sound isolation is also possible in a dense nonwoven composed of stiff fibres, which tend to reflect sound waves to stop the propagation of the sound through the fabric, while nonwovens of low density tend to be poor sound insulators because sound waves propagate easily through the fabric. Nonwoven fabrics can be used in a wide range of sound transmission control applications, including wall claddings, acoustic barriers, and acoustic ceilings. Understanding sound propagation through nonwoven fabrics is of prime importance for evaluating their noise absorption capacities.

12.20.1 Theoretical models

Most of the present theoretical analyses of sound propagation in nonwoven fabrics are based on the assumption that nonwovens are two-phase media containing both solid (rigid or flexible) fibres and air. In fact, the flexibility of fibres and deformability of the bonded structure in nonwovens makes sound transmission in such structures quite complicated. Therefore, the existing theoretical models introduced in this section are designed to help give some basis for the design and engineering of nonwoven structures for noise insulation. However, reliable predictive models are yet to be established. In this section, we focus on the reflection and absorption of small amplitude, air-borne sound waves from nonwoven fabric surfaces in the audio-frequency range.

In general, propagation of sound in an isotropic homogeneous material is determined by two complex quantities, the characteristic impedance $Z_0(f, d_f, \varepsilon) = z_0(R) - jz_0(I)$ and the propagation coefficient per metre $\gamma(f, d_f, \varepsilon) = \alpha + j\beta$. The normal-incidence energy absorption coefficient is defined as follows:

$$\alpha = 1 - \left| \frac{Z - \rho_0 c_0}{Z + \rho_0 c_0} \right|^2 \tag{12.71}$$

where P_0 and c_0 are the density of free air and the speed of sound in free air, respectively. When a nonwoven of small thickness, h (up to 10 cm), is fixed on a rigid wall, the impedance Z of the nonwoven fabric may be calculated from the following equation:

$$Z_0 = Z \coth \gamma \left(f, d_f, \varepsilon \right) h \tag{12.72}$$

Here $z_0(R)$ and $z_0(I)$ are the real and imaginary parts of Z_0 (f, df, E) respectively, and the absorption coefficient of the fabric given in Eq. (12.72) can be rewritten as follows,

$$\alpha = \frac{4z_0(R)\rho_0 c_0}{(z_0(R) + \rho_0 c_0)^2 + z_0^2(I)} \tag{12.73}$$

The absorption coefficients of a nonwoven fabric are based on the theory of sound propagation through the structure. Three types of model are available for sound transport in nonwoven fabrics [316]: parallel capillary pore models, parallel fibre models, and semiempirical models based on the preceding two models. In the parallel capillary pore models, the fabric is modelled as a medium containing identical parallel cylindrical capillary pores running normal to the fabric surface. There are two groups of parallel fibre models. In the first, the fibres are parallel to each other and to the fabric surface. In the second, the fibres are parallel to each other and normal to the fabric surface.

Zwikker and Kosten [317] introduced a method to decouple the influence of the viscous and thermal effects of each cylindrical fibre in a 'Raleigh type' model. The bulk acoustic properties of a nonwoven are modelled using a combination of the effective dynamic (complex) density and bulk modulus of individual cylindrical fibres. Although the separation of the viscous and thermal effects [298,318] places limitations upon the validity of solutions for sound propagation in some porous materials under certain conditions, it has been shown that this decoupling has little significant influence on the accuracy for sound in the low frequency range assuming a parallel fibre model. A schematic of a parallel fibre microstructure model [309] is shown in Fig. 12.27.

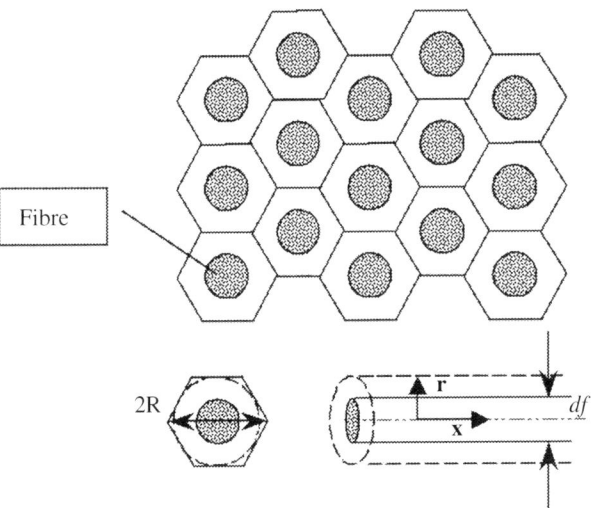

FIG. 12.27 Model for sound propagation in a bundle of parallel fibres. The direction of the sound propagation is along the fibre surface [309].

The sound absorption in the parallel fibre model in the 'open face' case is defined by the following system of equations [319]:

The equation of motions for a fibre:

$$\frac{\partial p_{fibre}(x,t)}{\partial x} + \frac{\partial u_{fibre}(x,t)}{\partial t}\rho_{fabric} = D\left(u_{air}(x,t) - u_{fibre}(x,t)\right) \tag{12.74}$$

The equation of motions for air:

$$\frac{\partial p_{air}(x,t)}{\partial x} + \frac{\partial u_{air}(x,t)}{\partial t}\rho_{air} = -D\left(u_{air}(x,t) - u_{fibre}(x,t)\right) \tag{12.75}$$

The continuity equations for fibre:

$$-\frac{\partial p_{fibre}(x,t)}{\partial t} + \frac{\varphi_{fibre}}{\varepsilon}\frac{\partial p_{air}(x,t)}{\partial t} = K_{fibre}\frac{\partial u_{fibre}(x,t)}{\partial x} \tag{12.76}$$

The continuity equations for air:

$$-\frac{\partial p_{air}(x,t)}{\partial t} = K_{air}\varepsilon\frac{\partial u_{air}(x,t)}{\partial x} + (K_{air} - P_0)\varphi_{fibre}\frac{\partial u_{fibre}(x,t)}{\partial x} \tag{12.77}$$

Then, the following boundary conditions associated with the above equations are:

$$u_{air}(h,t) = 0$$

$$u_{fibre}(h,t) = 0$$

$$\varepsilon p_{fibre}(0,t) = \varphi_{fibre}p_{air}(0,t)$$

$$x \in [0,h]$$

$$t \in [0,T]$$

where ρ_0 and ρ_p are the densities of free air and the fibre polymeric material respectively; ε and φ_{fibre} are the porosity of the nonwoven fabric and the volume fraction of fibre, respectively, (i.e. $\varphi_{fibre}=1-\varepsilon$). In Fig. 12.25, $\varepsilon=1-\left(\dfrac{d_f}{2R}\right)^2$), ρ_{air} and ρ_{fabric} are the bulk densities of the air and fibre components in a unit volume of the fabric respectively; i.e., $\rho_{air}=\rho_0\varepsilon$ and $\rho_{fabric}=\rho_{fibre}\varphi_{fibre}$; h is the thickness of the fabric, x is the distance, t is the time, p is the pressure and $u(x,t)$ is the sound velocity. K_{fibre} is the fibre bulk modulus, K_{air} is the air bulk modulus and P_0 is the air pressure.

The coupling parameter, $D(u_{air}(x,t)-u_{fibre}(x,t))$, representing the drag force between the solid fibre material and air is given by [314]:

$$D\left(u_{air}(x,t)-u_{fibre}(x,t)\right)=i\varpi\rho_{air}(m-1)+\varepsilon^2\sigma \tag{12.78}$$

where $\omega=2\pi f$ is the angular frequency, m is the structure constant and σ is the resistance constant. The characteristic impedance at the front face ($x=0$) of a nonwoven fabric is given as follows [316]:

$$\frac{1}{Z_0}=\frac{\left(u_{fibre}+\alpha_1 u_{air}\right)\left(u_{air}-\beta_2 u_{fibre}\right)}{(\beta_1-\beta_2)z_{f1}\coth{(ik_1 h)}}-\frac{\left(u_{fibre}+\alpha_2 u_{air}\right)\left(u_{air}-\beta_1 u_{fibre}\right)}{(\beta_1-\beta_2)z_{f2}\coth{(ik_2 h)}} \tag{12.79}$$

where $\alpha_j=\dfrac{u_{air}^{k_j}}{u_{fibre}^{k_j}}=\dfrac{u_{air}\left(1-c_f^2\left(\dfrac{k_j}{\omega}\right)^2\right)-iB\theta}{\left(u_{fibre}-i\theta\right)B}$ $\beta_j=\dfrac{p_{air}^{k_j}}{p_{fibre}^{k_j}}=\dfrac{(1-i\theta)\left(1-c_f^2\left(\dfrac{k_j}{\omega}\right)^2\right)-iB\theta}{u_{fibre}(1-i\theta)(1+B))-i\theta u_{air}c_f^2\left)\left(\dfrac{k_j}{\omega}\right)^2 u_{air}}$

$$z_{fj}=\frac{p_{fibre}^{k_j}}{u_{fibre}^{k_j}}=\frac{\omega}{k_j}\frac{u_{fibre}(1-i\theta(1+B))-i\theta u_{air}c_f^2\left(\dfrac{k_j}{\omega}\right)^2}{u_{fibre}-i\theta}P_{fabric}$$

where $c_f^2=\dfrac{K_{fibre}}{\rho_{fabric}}$ is the square of the fibre wave velocity, $\theta=\dfrac{D\left(u_{air}(x,t)-u_{fibre}(x,t)\right)}{\rho_a\omega}$ is a dimensionless coupling coefficient, $B=\dfrac{\rho_{air}}{\rho_{fabric}}$, and the subscripts air and $fibre$ refer to air and fibre phases, respectively. Thus, the sound absorption coefficient, α, can be obtained from the characteristic impedance, Z_0, as shown above.

12.20.2 Empirical models

Given that the real part of the characteristic impedance $z_0(R)$ is the most sensitive to the influence of fibrous structure, the structural characteristic [319] (Q) is usually defined as $Q=z_0(R)-1$. The other two dimensionless parameters, the structural factor (K_s) and the air-flow resistance per unit thickness (σ) in complex form are also defined as follows [320,321],

$$\frac{K_s}{\varepsilon}=\frac{(z_0(R)\beta-z_0(I)\alpha)}{k} \tag{12.80}$$

$$\sigma=\lim_{k\to 0}(z_0(R)\alpha-z_0(I)\beta)$$

where, $z_0(R)$ and $z_0(I)$ are the real and imaginary parts of Z_0 and α and β are the attenuation and phase constants.

12.20.2.1 Voronina models [322]

Since the real part of the characteristic impedance, $z_0(R)$, is the most sensitive to the influence of fibrous structure, the structural characteristic [301] (Q) is usually defined as $Q = z_0(R) - 1$. Voronina also defined other two dimensionless parameters, a structural factor (K_s) and the airflow resistance per unit thickness (σ) in complex forms as follows [317,318],

$$\frac{K_s}{\varepsilon} = \frac{(z_0(R)\beta - z_0(I)\alpha)}{k} \tag{12.81}$$

$$\sigma = \lim_{k \to 0} (z_0(R)\alpha - z_0(I)\beta)$$

where $z_0(R)$ and $z_0(I)$ are the real and imaginary parts of Z_0, α and β are the attenuation and phase constants, and ε is the fabric porosity.

Based on the sound absorption performance (sound frequencies $f = 250$–2000 Hz) absorption performance of a range of nonwoven materials composed of glass fibre, silica fibre, mineral wool, and basalt fibre, respectively ($d_f = 2$–8 μm, $\varepsilon = 0.996$–0.92), Voronina established approximate empirical models of sound propagation through nonwoven fibrous media as follows,

$$Z(f, d_f, \varepsilon) = (1 + Q) - jQ \tag{12.82}$$

$$\gamma(f, d_f, \varepsilon) = \frac{kQ(2 + Q)}{(1 + Q)} + jk(1 + Q) \tag{12.83}$$

$$\frac{k_s}{\varepsilon} = 1 + 2Q \tag{12.84}$$

$$\sigma = 2kQ \tag{12.85}$$

To predict the sound absorption properties based on nonwoven fabric structural parameters, Voronina established the following empirical equations for structural characteristics (Q) and sound resistance (σ) in nonwoven sound absorption materials [317,318]. Based on experimental results:

$$Q = \frac{(1 - \varepsilon)\left(1 + 0.25*10^{-4}(1 - \varepsilon)^{-2}\right)}{\varepsilon d_f} \sqrt{\frac{8\eta}{\omega \rho_0 c_0}} \tag{12.86}$$

$$\sigma = \frac{16(1 - \varepsilon)^2 \left(1 + 0.25*10^{-4}(1 - \varepsilon)^{-2}\right)\eta}{\varepsilon^2 d_f^2 \rho_0 c_0} \tag{12.87}$$

where $\eta = 1.85*10^{-5}$ is dynamic viscosity of air (Pa*s), ρ_0 is the air density (kg/m^3); c_0 is the sound velocity in air (m/s); $\omega = \frac{2\pi f}{c_0}$ is wave number (m^{-1}).

12.20.2.2 Delany–Bazley equations [323]

For nonwoven fabrics of large thickness (up to 2 m) and when the sound waves are in the range of $10 \leq \frac{f}{\sigma} \leq 1000$, the Delany–Bazley equations are as follows,

$$\frac{z_0(R)}{\rho_0 c_0} = 1 + 9.08 \left(\frac{f}{\sigma}\right)^{-0.75} \tag{12.88}$$

$$\frac{z_0(I)}{\rho_0 c_0} = -11.9 \left(\frac{f}{\sigma}\right)^{-0.73} \tag{12.89}$$

$$\alpha = 10.3 \left(\frac{2\pi f}{c_0}\right) \left(\frac{f}{\sigma}\right)^{-0.59} \tag{12.90}$$

$$\beta = \left(\frac{2\pi f}{c_0}\right) \left[1 + 10.8 \left(\frac{f}{\sigma}\right)^{-0.70}\right] \tag{12.91}$$

12.21 Modelling air filtration properties

The filtration process can be considered in terms of dry filtration (air filtration, aerosol filtration), wet filtration (mist filtration), and liquid filtration. The prime objective of modelling filtration processes using nonwoven filter fabrics is to improve their engineering design. Improving particle capture efficiency and minimising pressure drop across the filter thickness are particularly important. A good filter usually has a high filter efficiency in terms of particle capture and removal combined with a minimal pressure drop, and performance should ideally be consistent over the full service life of the filter. In practice, this is very difficult to achieve as the filter medium progressively accumulates particles and is subjected to different environmental or physical conditions during use that can affect its performance. Nonwoven filter fabrics are highly permeable and compared to other types of porous media, generally exhibit low flow resistance with potential for high particle capture capacity, across a range of particle sizes due to the large fibre surface area. Many nonwovens operate as depth filters, in which particles gradually penetrate through the thickness of the structure and are captured by individual fibres within. The filtration mechanism is influenced by the structure of the filter medium and the properties of the particles, as well as the fluid within which they are suspended. Consequently, nonwoven filter fabric performance cannot be modelled using a universal model. For brevity, this section focuses on models relating to air filtration and depth filtration.

12.21.1 Filtration mechanism

Filtration mechanisms and evaluation of filter performance have been thoroughly described [324,325]. The most frequently encountered mechanisms of particle–fibre–fluid interactions in generalised filtration models are straining, Brownian diffusion, direct interception, and inertial impaction. The effects of electrostatic forces and of gravity sedimentation are also important in certain filtration processes. The relative importance of the various mechanisms depends in part on the particle size and the surface area of the fibres, among other factors:

Straining: This refers to the entrapment of particles between adjacent fibres when the particle is larger than the pore opening in the fabric. Straining is the dominant method for removing large particles in low efficiency air filters.

Inertial impaction (*impingement*): Impaction occurs when the particle inertia is so high that it breaks free of the airflow streamlines and impacts fibres within the nonwoven structure. The streamline of the airflow circumvents a fibre in a curve and a large particle travelling at high velocity does not follow the streamline around the fibre periphery. Instead, it proceeds in a straight line due to the effect of its inertia and collides with the fibre. This inertial effect is proportional to the square of the particle size and the velocity of the fluid flow, and the temperature effect is reflected by the viscosity of the fluid.

Direct interception: A particle can be caught by a fibre if it approaches the fibre within a small distance. It usually assumes that the distance is equal to, or smaller than, half the particle diameter. Direct interception occurs when a particle following a normal airflow streamline is carried to within contact range of a fibre, at which point it becomes attached. The effect is therefore proportional to particle size, and there is usually a high possibility that any particles of 1 μm or larger in the airflow can be intercepted in a typical high-efficiency nonwoven filter. This process is of importance for normal filter velocities and small particle filtration including microbial filtration.

Brownian diffusion: Brownian motion is the random movement of a particle in the fluid flow caused by collisions with the surrounding fluid medium at a molecular scale. Because of Brownian motion, the particles do not follow fluid flow streamlines around fibres before being deposited on fibre surfaces within the nonwoven. Diffusion is important for particles smaller than about 0.1 μm [326,327]. The effect of Brownian motion is influenced by the particle size, the fibre diameter, the fluid flow velocity and the temperature of the fluid flow.

Each of these basic filtration mechanisms depends on the nature of the particles, the particle size, the structure of the nonwoven fabric, airflow velocity, and the mechanical and physiochemical interactions of the particle–fluid–filter fabric system.

Electrostatic charging of fibre surfaces in nonwoven air filter media is frequently necessary to achieve high filter efficiency combined with low pressure drop (or low breathing resistance) when attempting to remove submicron diameter particles. This group of air filter media are referred to as nonwoven electrets. Here, the electrostatic charge provides an additional particle capture mechanism, depending on the nature of the charge present on the incoming submicron particles. Removal of the electrostatic charge in the nonwoven electret by for example, liquid immersion or other charge leakage mechanisms during use, results in a major reduction in filter efficiency.

The major criteria characterising the performance of a nonwoven air filter include the filter efficiency, pressure drop, filter quality coefficient, filter loading, filter clogging, and filter cleaning and filtration cycling time. They are defined as follows:

Filter efficiency

The filter efficiency, E, is the ability of the filter to retain particles and is defined as the percentage of particles of a given size retained by the filter. It can be calculated from the ratio of the particle concentrations in the upstream (P_{in}) and downstream (P_{out}) fluid flows, respectively.

$$E = 1 - \frac{P_{out}}{P_{in}} \qquad (12.92)$$

Pressure drop

The pressure drop refers to the difference in pressures in the upstream (p_{in}) and downstream (p_{out}) fluid flows across the filter thickness.

$$\Delta p = p_{in} - p_{out} \qquad (12.93)$$

Filter quality performance

The filter quality performance, or the filter quality coefficient, is defined as the ratio of the filter efficiency to the pressure drop across the filter thickness.

$$Q = \frac{-\ln(1-E)}{\Delta p} \tag{12.94}$$

Dust holding capacity

Dust-holding capacity, or filter capacity, is defined as the amount of deposited particles that the filter is capable of accumulating before reaching a certain pressure drop for particles of a given size.

12.21.2 Filter efficiency in dry air filtration

Nonwoven filters can be modelled as layered, two-dimensional fibrous networks of high porosity. During the filtration process, a small proportion of the particles in the air or fluid flow may penetrate through the fabric, but most of are gradually deposited on the fibre surfaces. As the fibres become covered with particles, a filter cake forms and the permeability of the fabric structure gradually decreases. Therefore, the overall filtration process is not constant. The progressive deposition of particles leads to an increase in pressure drop, but there may also be an increase in the filter efficiency as smaller particles are retained. The formation of the filter cake during filtration makes modelling the filtration process complex because the filter cake largely influences both the particle capture capacity and the pressure drop across the filter thickness. A key performance attribute is the filter quality coefficient. In this section, we focus on the nonwoven filter fabric design, based on models of the relationship between the filter quality coefficient i.e., the filter efficiency and pressure drop, as well as the fabric structure.

12.21.2.1 Single-fibre collection efficiency theory

The filter efficiency of a nonwoven fabric filter is modelled based on the single-fibre collection efficiency theory, and the analysis relates mainly to depth air filtration. Before considering the theory, it is useful to summarise some key terms:

$C_C = Cunningham$ slip factor $C_C = 1 + \left(\frac{\lambda}{d_p}\right)\left(2.492 + 0.84e^{\frac{-0.435d_p}{\lambda}}\right)$

$D_d =$ particle diffusion coefficient, (m^2/s), which is a measure of the degree of diffusion motion and is a function of the mean free path of air molecules, $D_d = \frac{C_C k_B T}{3\pi\eta d_p}$

$d_f =$ fibre diameter.

$d_p =$ particle diameter.

$e_f =$ effective fibre length factor, which is the ratio of theoretical pressure drop for a Kuwabara flow field to the experimental pressure drop [327,328],

$$\left(e_f = {16\eta u_i \varphi h}\Big/{\left(Ku d_f^2 \Delta P_O\right)}\right)$$

E = single-fibre collection efficiency is the fraction of particles that can be collected by a fibre from a normal cross-sectional area of the air stream, equal to the frontal area of the fibre.

E_j = single-fibre collection efficiency for size subrange j of a polydisperse aerosol.

G = gravitation parameter, $G = \frac{d_p \rho_p g}{18 \eta u_i}$

Ku = Kuwabara hydrodynamic factor, $Ku = -\frac{3}{4} - \frac{1}{2} \ln\varphi + \varphi - \frac{1}{4}\varphi^2$

g = gravitational acceleration.

h = thickness of filter fabric.

k_B = Boltzmann's constant, 1.3708×10^{-23} J/K.

n = aerosol number concentration leaving the filter fabric.

n_o = aerosol number concentration entering the filter fabric.

N'_{cap} = simplified capillary number, $N'_{cap} = \frac{u\eta}{\sigma \cos\gamma}$

Pe = Peclet number characterises the intensity of diffusional deposition. An increase in the Peclet number decreases the single-fibre diffusion efficiency, $P_e = \frac{U_0 d_f}{D_d}$

ΔP_0 = pressure drop across the dry filter fabric.

R = interception parameter, $R = \frac{d_p}{d_f}$

R_{ef} = Reynolds number, $R_{ef} = \frac{\rho u_i d_f}{\eta}$

Stk = Stokes number, which is the ratio of the particle's kinetic energy to work done against viscous drag over a distance of one fibre radius, $Stk = \frac{\rho_p u_i d_p^2}{9 \mu d_f}$

T = temperature (K).

u = superficial gas velocity (m/s).

u_i = interstitial gas velocity (m/s).

Y = filter efficiency (%).

ε = porosity of filter fabric.

φ = packing density (fibre volume fraction) of the dry filter fabric.

η = gas absolute viscosity (Ns/m^2).

λ = mean free path of gas molecules at NTP (0.067 μm [329]), which is inversely proportional to the air pressure.

γ = contact angle between the liquid and fibre.

ρ = density of gas.

ρ_p = density of particle.

σ = surface tension of liquid (N/m).

Filter efficiency based on a single-fibre collection efficiency

Nonwoven filter fabrics are composed of many individual fibres each of which provides solid surface onto which incoming particles can be deposited. Understanding the particle collection efficiency of a single fibre is therefore important. The equation defining overall filter efficiency of a nonwoven fabric, $Y(df)$, for any particle size, d_p, and set of conditions is as follows [257,328,330–332].

$$Y(d_f) = 1 - \exp\left(\frac{-4\varphi E h}{\pi(1 - \varphi)d_f e_f}\right) \tag{12.95}$$

The particle collection efficiency of a single fibre, E, depends on the particle size, air velocity, and fibre properties, based on the six primary mechanisms that operate in filtration: impaction (EI), direct interception (ER), diffusion (ED), enhanced interception due to diffusion (ED_r), gravitational settling (EG), and electrostatic attractions (E_q). Several equations have been proposed for predicting E from these different collection mechanisms; in particular, Davies' equation [257] as well as Friederlandler and Stenhouse.

Davies [257]:

$$E = E_{DRI} = \left(R + (0.25 + 0.4R)\left(Stk + 2P_e^{-1}\right) - 0.0263R\left(Stk + 2P_e^{-1}\right)^2\right)\left(0.16 + 10.9\varphi - 17\varphi^2\right)$$

(12.96)

Friederlandler [333,334]

$$E = E_{DRI} = \frac{1}{RP_e}\left(6\left(RP_e^{\frac{1}{3}}R_e^{\frac{1}{6}}\right) + 3\left(RP_e^{\frac{1}{3}}R_e^{\frac{1}{6}}\right)^3\right)$$

(12.97)

Stenhouse [341]

$$E = E_D + E_R + E_{Dr} + E_I + E_G$$

(12.98)

Each of the component collection efficiencies for the different mechanisms is as follows:
Diffusion [336]:

$$E_D = 2.9\left(\frac{1-\varphi}{Ku}\right)^{-\frac{1}{3}}P_e^{-\frac{2}{3}} + 0.62P_e^{-1}$$

(12.99)

(Valid for $0.005 < \varphi < 0.2$, $0.1 < U_0 < \frac{2m}{s}$; $0.1 < d_f < 50\mu m$; $R_{ef} < 1$)
Interception [326,328,337]:

$$E_R = \frac{(1+R)}{2Ku}\left(2\ln(1+R) - 1 + \varphi + \left(\frac{1}{1+R}\right)^2\left(1 - \frac{\varphi}{2}\right) - \frac{\varphi}{2}(1+R)^2\right) = \frac{(1-\varphi)R^2}{2Ku(1+R)^{\frac{2}{3(1-\varphi)}}}$$

(12.100)

Impaction [338]:

$$E_I = \frac{(Stk)J}{2K_u^2}$$

(12.101)

where $Stk = \frac{\rho_d d_p^2 C_c U_0}{18\eta d_f}$, $J = (29.6 - 28\varphi^{0.62})R^2 - 27.5R^{2.8}$ for $0.01 < R < 0.4$ and $0.0035 < \varphi < 0.111$, $J = 2$ for $R > 0.4$.
Enhanced diffusion due to interception of diffusing particles:

$$E_{Dr} = \frac{1.24R^{\frac{2}{3}}}{(KuPe)^{\frac{1}{2}}} \text{ for Pe} > 100$$

(12.102)

Gravitational settling [335]:
$$E_G \cong (1+R)G \text{ for } V_{TS} \text{ and } U_0 \text{ in the same direction.}$$
$$E_G \cong -(1+R)G, \text{ for } V_{TS} \text{ and } U_0 \text{ in the opposite direction.}$$

$$E_G \cong -G^2, \text{ for } V_{TS} \text{ and } U_0 \text{ in the orthogonal direction.} \tag{12.103}$$

where $G = \frac{V_{TS}}{U_0} = \frac{\rho_d d_p^2 C_c g}{18\eta U_0}$

V_{TS} and U_0 are the particle terminal setting velocity and face air velocity, respectively. *Electrostatic attraction* [327]:

$$E_q = \left(\frac{\ni - 1}{\ni + 1}\right)^{\frac{1}{2}} \left(\frac{q^2}{3\pi \eta d_p d_f^2 U_0 (2 - lnRe_f)}\right) \tag{12.104}$$

where $\ni$ is the dielectric constant of the particle and q is the charge on the particle.

12.21.2.2 *Filter efficiency of nonwovens containing multiple fibre diameters*

When there are multiple sizes of particles in the fluid flow and the nonwoven consists of fibres having the same diameter, the filter efficiency can be obtained from the above model of collection efficiency for a single fibre, E, by subdividing the size range of the particles into several subranges, E_j. The value of E_j is obtained for each subrange j of average particle diameter d_{pj} from the above equations for a single fibre. The filter efficiency Y is then calculated from the following equations.

$$Y = 1 - \sum_j a_j \left(\frac{n}{n_0}\right)_j \tag{12.105}$$

where $\left(\frac{n}{n_0}\right)_j = \exp\left(\frac{4\varphi E_j h}{\pi(1-\varphi)d_f e_f}\right)$, $\left(\frac{n}{n_0}\right)_j$ and a_j are the number penetration and mass fraction of the j_{th} size range of particles, respectively.

In respect of predicting the filter efficiency of a nonwoven having fibres of the same diameter, poor filter efficiency will be observed for particles of certain sizes. This is unavoidable unless the fibre diameters in the nonwoven are different. An example of the variation in filter efficiency linked to particle size is illustrated in Fig. 12.28. Here, it is evident that for very

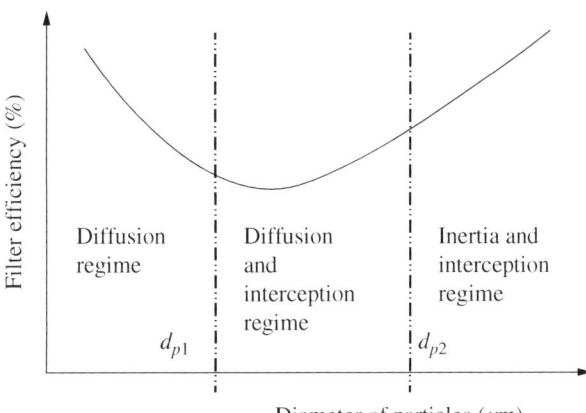

FIG. 12.28   Filter efficiency of a nonwoven fabric against the particle size in an airflow [339].

small particles less than d_{p1} in diameter, the primary filtration mechanism is diffusion. For particle sizes between d_{p1} and d_{p2}, the filter is less efficient because the particles are too large for diffusion effects and too small for a large interception effect. For particles of diameter above d_{p2}, the filter is very efficient because the interception, along with inertial impaction effects, are predominant during filtration. The relatively low filter efficiency for particles of diameters between d_{p1} and d_{p2} is therefore a weakness of this nonwoven filter, but is inevitable in cases where the particles are of multiple sizes in the airflow being filtered.

To design high filter efficiency media that avoids this limitation requires the fabric to contain fibres of different diameter. Of course, this is straightforward in nonwoven filter media based on, for example, meltblown and electrospun media, but requires fibre blending in wetlaid and drylaid processes.

For example, high-efficiency and HEPA filters consist of fibre diameters ranging from 0.65 to 6.5 µm, usually in three nominal diameter groups [340]. If a nonwoven filter is composed of multiple fibre components, the filter efficiency when dealing with a fluid containing particles of multiple diameters will be as follows:

$$Y = 1 - \sum_j a_j \left(\frac{n}{n_0}\right)_j \quad where$$

$$\left(\frac{n}{n_0}\right)_j = \exp\left(\frac{4\phi \sum_{d_f} E_j(d_f) h}{\pi (1 - \phi) d_f e_f}\right) \tag{12.106}$$

$$E_j(d_f) = E_D(d_f)_j + E_R(d_f)_j + E_{Dr}(d_f)_j + E_I(d_f)_j + E_G(d_f)_j + E_e(d_f)_j$$

where $\left(\frac{n}{n_0}\right)_j$ and a_j are the number penetration and mass fraction of the j_{th} size range of particles respectively, and $E_j(d_f)$ is the collection efficiency of a single fibre having the diameter of d_f against a particle of the diameter of d_{pj}.

12.21.3 Pressure drop in dry air filtration

The pressure drop across a dry fibrous filter, ΔP_0, can be predicted using the expression developed by Davies [327].

$$\Delta P_0 = \frac{U_0 \eta h}{d_f^2} \left(64 \varphi^{1.5} + \left(1 + 56 \varphi^3\right)\right) \tag{12.107}$$

12.21.4 Wet-operating filters [341]

For nonwoven fabrics intended for mist filtration or the filtration of liquid particles, the specified collection efficiencies can be obtained with various combinations of filter thickness, fibre diameter, packing density and gas velocity. For a specified efficiency of 90%, the required filter thickness varies according to the approximate relation $h = 5\varphi^{-1.5} d_f^{2.5}$. The corresponding pressure drop at constant filtration efficiency is insensitive to d_f but varies approximately according to the relation:

$$\Delta P_{wet} \propto \varphi^{0.6} U^{0.3} \quad when \; \varphi > 0.01 \tag{12.108}$$

12.21.5 Liquid filtration

Particle capture in liquid filtration is much more complicated and less efficient than in air filtration. For example, captured particles may easily reenter the liquid flow, and this particle reentrainment is partly responsible for low filter efficiency. The pressure drop through the fabric is determined using the following expression:

$$\Delta P = \Delta P_{H-P} + \Delta P_{B-P} \tag{12.109}$$

where ΔP_{H-P} is the pressure drop for a Hagen–Poiseuille fluid, and ΔP_{B-P} is the pressure drop due to the flow resistance of particles. They can be calculated based on the assumption that the captured dust particles are spherical [342] in the following equations:

$$\Delta P_{H-P} = \frac{f\eta hUA}{(OA)^2} \tag{12.110}$$

$$\Delta P_{B-P} = \frac{k\eta hU(1-\varepsilon)^2}{d_p^2\varepsilon^2} \tag{12.111}$$

where, U is the face velocity, A is the filter area, and OA is the open filter area for flow, f is the correction factor for the clean filter fraction as determined experimentally, h is the filter thickness, d_p is the mean particle diameter, k is the Carman–Kozeny constant, and ε is the porosity of the filter containing deposited particles. The porosity, ε, should be calculated by considering the total area of the filter and the area covered by the deposited particles, thus ε will change with time.

12.22 Influence of fibre orientation distribution on the properties of thermal-bonded nonwoven fabrics

Nonwovens, regardless of the process utilised, are assemblies of fibres bonded together by chemical, mechanical or thermal means. In most nonwovens, the overwhelming majority of fibres are planar x, y stacks of fibres having little or no orientation through the plane (z-direction). Some airlaid processes make an attempt to create a third dimension in the orientation of webs they produce. It may be argued that needlepunching and perhaps hydroentangling also result in some fibres lying in the z direction. However, the ratio of fibres in the z-direction is a small fraction of the total number of fibres and that the planar x-y orientation is still responsible for the performance of the nonwoven. It may be argued therefore, that the x, y planar fibre orientation is the most important structural characteristic in any nonwoven. Clearly, the properties of a nonwoven fabric will depend on the nature of the component fibres as well as the way in which the fibres are arranged and bonded [343–348]. Modelling and predicting the performance of nonwovens cannot be separated from the fibre orientation distribution and the structure anisotropy it brings about. Equally important is the so-called basis weight uniformity of a nonwoven. This refers to the degree of mass variation in a nonwoven normally measured over a certain scale. Local variation of the mass results in an unattractive, appearance, but more importantly will lead to, and potentially dictate, the failure point of a nonwoven. For example, tensile failure may be initiated and propagated first in areas that are fibre-poor (regions with low mass), or barrier properties are lost because of the existence of fibre-poor regions in the fabric. Another characteristic of a nonwoven may be the extent to which the fibre diameter varies. This is particularly important in spunbonded and meltblown structures, where the fibre diameter variation may come about as a result of roping (fibres sticking together to form bundles), or because of the process, which leads to thick and thin places along the length of the fibres. This becomes significant at the micro scale and can lead to failure in a similar manner to the variations in basis weight. In addition to weight variation, the fibre orientation distribution function is of particular importance in governing fabric properties and the directional variation in properties within a fabric. The rest of this section discusses fibre orientation and its role on performance by examining some case studies.

12.22.1 Fibre orientation distribution

The definition of Folgar and Tucker [340] best describes the fibre orientation distribution function (ODF) in a nonwoven. The orientation distribution function [ODF] Ψ is a function of the angle α. The integral of the function Ψ from an angle α_1 to α_2 is equal to the probability that a fibre will have an orientation between the angles α_1 and α_2. The function Ψ must additionally satisfy the following conditions:

$$\psi(\alpha + \pi) = \psi(\alpha)$$
$$\int_0^\pi \psi(\alpha)d\alpha = 1 \tag{12.112}$$

The peak direction mean is at an angle $\bar{\alpha}$ given by [341]:

$$\bar{\alpha} = \frac{1}{2}\tan^{-1}\frac{\sum_{i=1}^{N}f(\alpha_i)\sin 2\alpha_i}{\sum_{i=1}^{N}f(\alpha_i)\cos 2\alpha_i} \tag{12.113}$$

while the standard deviation about this mean is given by [341]:

$$\sigma(\alpha) = \left[\frac{1}{2N}\sum_{i=1}^{N}f(\alpha_i)(1 - \cos 2(\alpha_i - \bar{\alpha}))\right]^{1/2} \tag{12.114}$$

Anisotropy is often described by the ratio of the maximum to the minimum frequency of the ODF. For uni-modal distributions, in the range 0–180 degrees, the degree of anisotropy can also be characterised by the width of the orientation distribution peak given above. These definitions have to be reinterpreted for bimodal distributions in the range 0–180 degrees such as are obtained from cross-lapped webs or for crimped fibre webs viewed at short segment lengths. A more general approach would be to use the so-called $\cos^2$ anisotropy parameter, H_t, given by [349]

$$H_t = 2cos^2\varphi - 1,\tag{12.115}$$

$$\text{where } \langle cos^2\varphi \rangle = \int_{-\frac{\pi}{2}}^{\frac{\pi}{2}} f_t(\varphi)cos^2(\varphi)d\varphi$$

The average cos^2 anisotropy parameters can range between -1 and 1. A value of 1 indicates a perfect alignment of the fibres parallel to a reference direction and a value of -1 indicates a perfect perpendicular alignment to that direction. A uniform ODF (random ODF) would yield a zero value. It is customary to set the reference direction to the machine direction. More appropriately, the peak direction should be used as the reference instead of the machine direction. A direct experimental method for measuring fibre orientation extends back several decades when orientation was measured manually [345]. Other indirect methods explored since include short span tensile analysis [350–352], microwaves (used primarily for paper) [353], ultrasound [354], diffraction methods [346], and more recently, image analysis [355,356] methods. In a series of publications, the present author evaluated various optical means and methods for determining the fibre orientation distribution in nonwovens [63,357–361]. There are various commercial systems now available for measuring fibre orientation distribution [362].

12.22.2 The influence of the production method on anisotropy

Today, nonwovens are made by a variety of processes, alone or together. The final structure and its anisotropy are therefore, a function of the ODF, the bonding and the layering of various webs to form a consolidated web. Orientation anisotropies are induced by various nonwoven processes. Most thermally bonded nonwoven fabrics are made by hot calendering a carded web of short staple fibres. A typical thermal bonding line has an opening section, a carding section, and a subsequent calender-bonding section. The opening and carding processes have a significant impact on the orientation of the resultant web. The primary goal of the opening section is to separate fibres and provide a uniform feed to the cards. Openability is affected by fibre crimp and finish level. Each of these properties must be carefully controlled if the opener is to provide a uniform batt to the card. A high crimp value provides more cohesion, but it also makes the fibre opening more difficult. A low-crimp fibre opens easily and yields a high-quality web, but it is more difficult to process. The opening properties of the fibre must be balanced with its cohesive properties to have an efficient bonding line. The carding process, by nature, imparts a high degree of orientation to the fibres in the machine direction. The main cylinder and the workers in the card align the fibres parallel to the machine direction. Inadequate opening of the fibres creates a blotchy, nonuniform fabric that has a tendency to break easily during processing. A fabric formed from a web with fibres mostly aligned in the machine direction is expected to have high strength in the machine direction and relatively low strength in the cross direction. Other properties follow the same pattern.

To improve the cross direction strength requires the rearrangement of the fibres so as to have a higher degree of orientation in the cross direction. This can be achieved by several mechanical methods. One method involves stretching the web in the cross direction prior to the consolidation or bonding step. When the web is stretched in the cross direction, fibres are pulled away from the machine direction and realigned in the cross direction. Of course, the web must be cohesive enough to prevent too much fibre slippage, which could tear the web. The second method of imparting cross direction orientation to fibres involves a randomising doff mechanism at the exit of the card. This randomising is accomplished by buckling the web as it is doffed.

Another method commonly employed is a cross-lapper that takes a card feed and cross-laps it into a uniform batt before consolidation or bonding. Most cross-lapped webs have a bimodal fibre orientation distribution. The ODF in the wet lay process can also have a machine direction dependency. Here, the ODF can be adjusted by controlling the relative throughput and the speed of the belt. Unlike the systems above, most air-lay systems have a tendency to create a more randomised web. The spunmelt, spunbonded, and meltblown variety of webs also often have a machine direction dependency. Some spunbonded products also have a bimodal distribution. Here, the aspirator and the lay-down system are responsible for the lay-down of the webs. What is perhaps significant is that most nonwovens are anisotropic and machine direction dependency and that the web anisotropy typically increases with machine (belt) speed. This also implies that the properties of most nonwovens are also anisotropic. Also significant is that the ODF is typically symmetrical around the machine or cross directions. The symmetry is lost at any other direction.

12.22.3 The role of ODF on mechanical performance

When a simple tensile deformation is applied along a direction around which the initial orientation distribution is symmetric, it will remain symmetric through the deformation process. However, if it is applied along a different direction, the symmetry could be lost with respect to the *initial* symmetry direction, but develop progressively with regard to the test direction. The changes in ODF that occur as a result of fabric strain can be followed by the following three average anisotropy parameters and an asymmetry parameter.

1. Overall average anisotropy parameter, H_t, given above

$$H_t = 2 < cos^2 \varphi > -1 \qquad (12.116)$$

where $<cos^2 \varphi> = \int_{-\frac{\pi}{2}}^{\frac{\pi}{2}} f_t(\varphi) cos^2(\varphi) d\varphi$

We define a left-quadrant average anisotropy parameter, H_t^L, as

$$H_t^L = 2 < cos^2 \varphi >_L - 1, \qquad (12.117)$$

where $<cos^2 \varphi >_L = \dfrac{\int_{-\frac{\pi}{2}}^{0} f_t(\varphi) cos^2(\varphi) d\varphi}{\int_{-\frac{\pi}{2}}^{0} f_t(\varphi) d\varphi}$ and a right-quadrant average anisotropy parameter, H_t^R, as:

$$H_t^R = 2 < cos^2 \varphi >_R - 1, \qquad (12.118)$$

where $<\cos^2\varphi>_R = \dfrac{\int_0^{\frac{\pi}{2}} f_t(\varphi)\cos^2(\varphi)d\varphi}{\int_0^{\frac{\pi}{2}} f_t(\varphi)d\varphi}$

We can, therefore, define an asymmetry parameter, $A_t^{(m)}$, as:

$$A_t^{(m)} = 4\left(\left(\int_0^{\frac{\pi}{2}} f_t(\varphi)d\varphi\right)<\cos^2\varphi\sin^2\varphi>_R - \left(\int_{-\frac{\pi}{2}}^0 f_t(\varphi)d\varphi\right)<\cos^2\varphi\sin^2\varphi>_L\right) \qquad (12.119)$$

Each of the average anisotropy parameters can range between -1 and 1. A value of 1 indicates a perfect alignment of the fibres parallel to a reference direction and a value of -1 indicates a perfect perpendicular alignment to that direction. A uniform ODF (random ODF) would yield a zero value. The asymmetry parameter, $A_t^{(m)}$, will govern the magnitude of the moment that can arise around the tensile test direction and also its direction, with $A < 0$ and $A > 0$ leading to clockwise and anticlockwise moments, respectively. The factor, 4, has been introduced in the definition of $A_t^{(m)}$ only to limit its range from -1 to 1. These limiting values represent conditions that would lead, respectively, to maximum clockwise and anticlockwise moments when a tensile stress is applied along the reference (test) direction.

Let us examine the behaviour of a carded, calendered nonwoven under tension in various directions. Tensile testing was performed at 0 degrees (machine direction), $\pm$ 34 degrees (bond pattern stagger angles), and 90 degrees (cross direction). The choice of these three specific test directions was based on the goal of exploring the anisotropic mechanical properties of the fabric and the requirement that the repeating unit of the bond pattern is easily identifiable with respect to the test direction. The nonwoven sample strips, 25.4mm (1in) wide, were tested at a gauge length of 101.6mm (4in). The tensile tests were carried out at a 100%/min extension rate. Five strips were tested at each angle; the average values are used in the plots. From the images digitised during tensile testing at 0°, +34°, 90°, and − 34° directions, the fibre orientation distribution function (ODF) and the shear deformation angle of the unit cell were measured. The deformation parameters are described in Fig. 12.29.

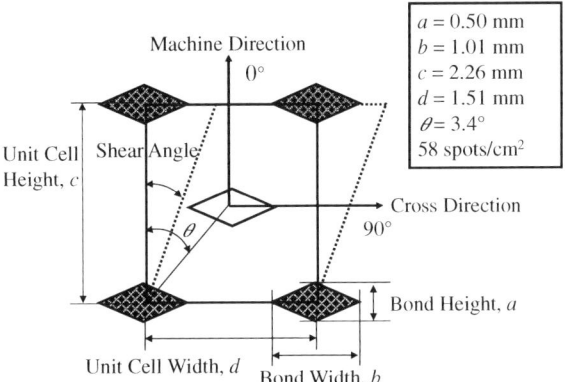

FIG. 12.29   Unit cell.

⊢ 1mm

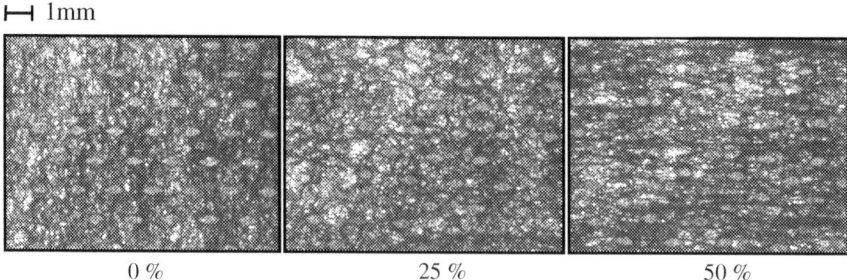

0 %   25 %   50 %

FIG. 12.30   Images at 0%, 25%, and 50% strain.

Fig. 12.30 shows a typical sequence of images captured during tensile testing, in this case in the 90-degree direction (cross direction). The ODF was measured from a series of such images captured at regular intervals of deformation in each test direction.

The ODF results are summarised in Figs 12.31–12.34. The loading direction is defined with respect to the sample axis (i.e. orientation angle). As may be noted from Fig. 12.31, when the samples are tested in the cross direction (90°), the fibres reorient significantly and the dominant orientation angle changes from its initially preferred machine direction towards the loading direction.

In the case of samples tested in the machine direction (0°), where the initially preferred orientation coincides with the loading direction, the deformation-induced effect is, as expected, primarily to increase this preference of fibres (Fig. 12.32).

Because of the anisotropy of the initial structure, it is expected that, when the samples are tested in different directions, the relative contributions to the total deformation from structural reorientations and fibre deformations would be different. The reorientations due to the test deformations imposed at 34 degrees and −34 degrees also show similar changes in the dominant orientation angle (Figs 12.33 and 12.34), but of a much smaller magnitude than that obtained at 90 degrees.

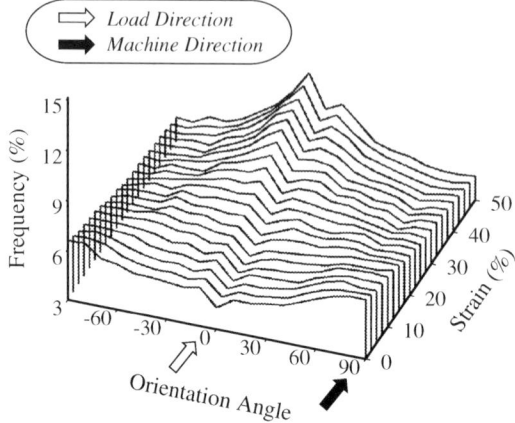

FIG. 12.31   Reorientation when tested in cross direction.

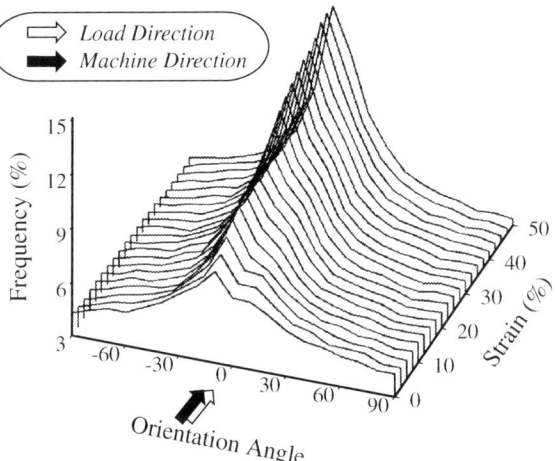

FIG. 12.32   Reorientation when tested in machine direction.

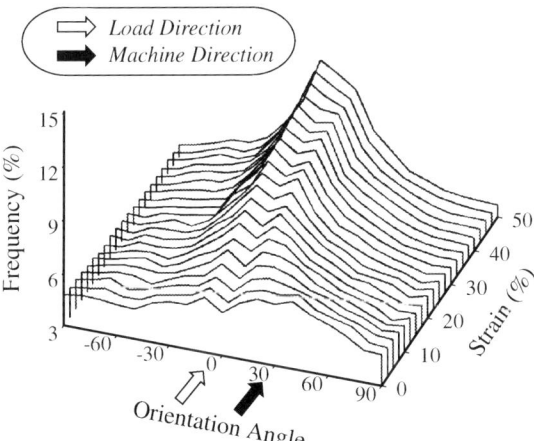

FIG. 12.33   Reorientation when tested in +34 direction.

When the samples are tested in the cross direction, the nonwoven structure undergoes significant reorientation before the fibres themselves are strained. This is reflected in a high failure strain. In this case, reorientation is due to bending of fibres at their interfaces with the bonds. This would obviously lead to highly localised stress concentrations and high shear stresses at the fibre–bond interface, leading to a relatively low failure stress. In contrast, if the samples are tested in the machine direction, which is the direction of initial preferred orientation, there can only be a limited extent of fibre–reorientation facilitated deformation of the nonwoven material. This is reflected in a low strain, but high stress, at failure, occurring

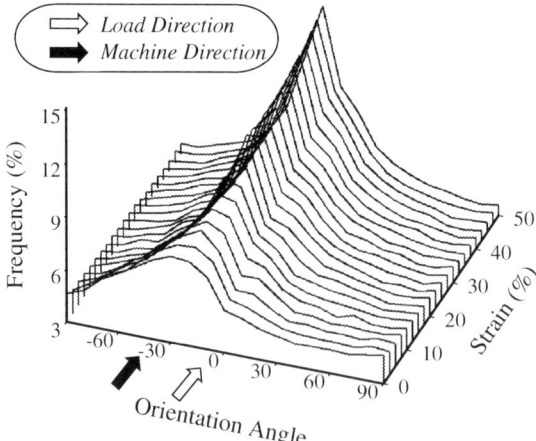

FIG. 12.34 Reorientation when tested in −34 direction.

predominantly due to tensile failure of the fibres. If the bonding is optimal, failure can be initiated at the fibre–bond interface, or any other position in the path of the fibres that traverse between bonds. As can be seen in Fig. 12.35, the samples tested in the 34-degree and −34-degree directions fall between the two cases of 'low stress—high strain' and 'high stress—low strain' failure along the cross and machine directions, respectively. Also, the failures are dominated by shear when the fabrics are tested at 34 degrees and −34 degrees. The fracture edges are shown for each case in Fig. 12.35. As expected, failure tends to propagate *along* the dominant orientation angle.

As expected, failure tends to propagate *along* the dominant orientation angle. The propensity for shear deformation along the direction of preferred fibre orientation is clearly

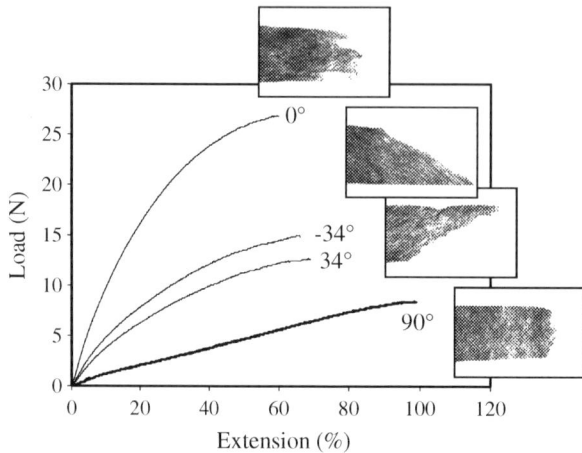

FIG. 12.35 Stress–strain behaviour and fracture surfaces.

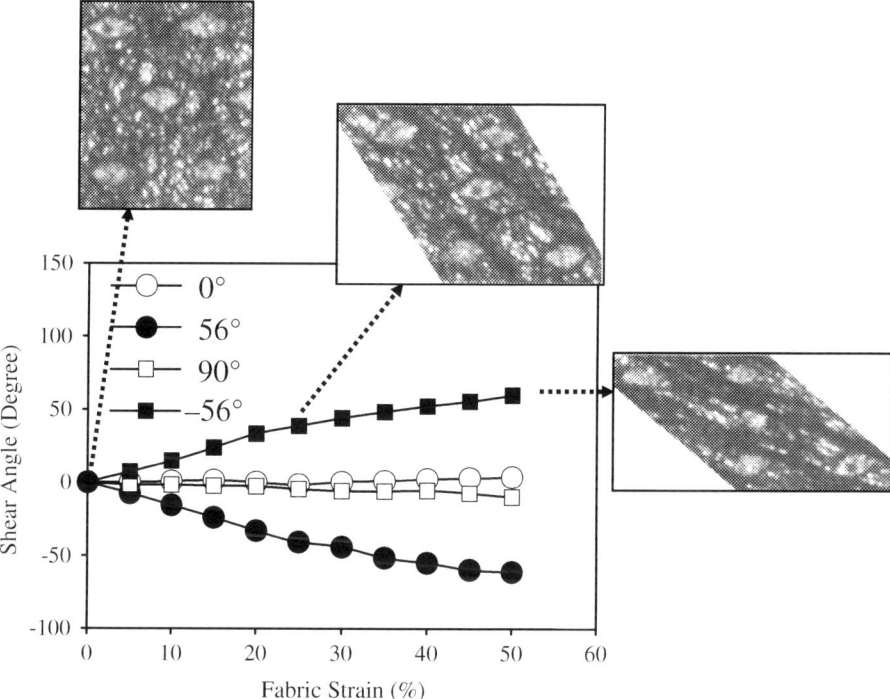

FIG. 12.36   Shear angle as a function of strain.

manifested in these tests. The unit-cell shear deformation results are shown in Fig. 12.36. It is clear that application of a macroscopic tensile strain produces a significant shear deformation along the initially preferred direction in fibre ODF, except when the two directions are either parallel or normal to each other.

The degree of asymmetry in the structure is shown in Fig. 12.37. As may be noted, the moments are greatest when the test is performed in directions other than the two principal directions (machine and cross).

12.22.4 Concluding remarks with respect to ODF

A fundamental link that can serve to identify the appropriate structural parameters, and to establish relationships between them and properties of interest, pertains to quantitative relationships between macroscopic stress fields, or deformation parameters, and the consequent structural changes. It has been shown clearly that the fabric performance is a function of its structure or the manner in which the fibres are arranged within the structure. It has also been revealed that, while failure can follow different modes, it is likely to be dictated, under most conditions, by shear along the preferred direction of fibre orientation. Regardless of bonding conditions (a most important processing parameter), the structural changes brought about in the structure and the microscopic deformations are driven by the initial orientation

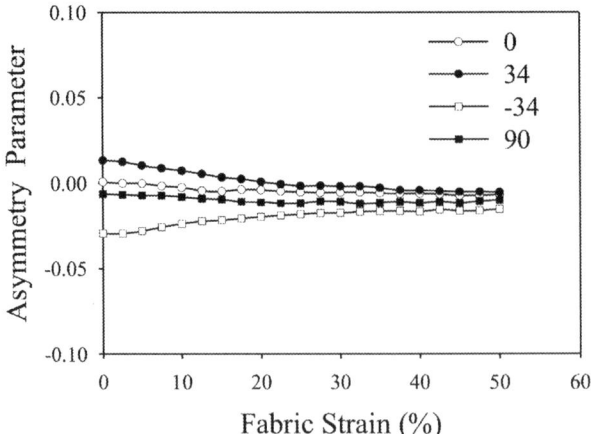

FIG. 12.37   Asymmetry parameter as a function of strain.

distribution function (ODF) of the fibres and are similar for all structures with the same initial ODF. The bonding conditions only dictate the point of failure. The magnitude of the moment during the deformation process that can arise around the tensile test direction and also its direction can be determined by the asymmetry parameter. It is confirmed that from the asymmetry parameter values, the moments are greatest when the test is performed in directions other than the two principal directions (machine and cross).

References

[1] ISO 9092:1988, BS EN 29092:1992 Textiles. Nonwovens. Definition.

[2] ISO 9092:2010 Textiles. Nonwovens. Definition.

[3] BS EN ISO 9092:2011 Textiles. Nonwovens. Definition.

[4] BS EN ISO 9092:2019 Nonwovens—Vocabulary.

[5] N. Mao, S.J. Russell, A framework for determining the bonding intensity in hydroentangled nonwoven fabrics, Compos. Sci. Technol. 66 (1) (2006) 66–81.

[6] J.W.S. Hearle, M.A.J. Sultan, A study of needled fabrics. Part 2: effect of needling process, J. Text. Inst. 59 (1968) 103–116.

[7] T. Ishikawa, K. Kim, Y. Ohkoshi, Visualization of a pillar-shaped fiber bundle in a model needle-punched nonwoven fabric using X-ray micro-computed tomography, Text. Res. J. 87 (11) (2017) 1387–1393, https://doi.org/10.1177/0040517516652351.

[8] R. Krcma, Manual of Nonwoven Textiles, Textile Trade Press, Manchester, 1972.

[9] J.W.S. Hearle, M.A.J. Sultan, A study of needled fabrics. Part 1: experimental methods and properties, J. Text. Inst. 58 (1967) 251–265.

[10] A. Smolen, Polypropylene (BSc Dissertation), Department of Textile Industries, University of Leeds, 1967.

[11] W.E. Morton, J.W.S. Hearle, Physical Properties of Textile Fibres, The Textile Institute, London, 1993.

[12] S. Raz, Mayer Textile Machine Corporation, The Karl Mayer Guide to Technical Textiles, Karl Mayer/Mayer Textile Machine Corp., Obertshausen, Deutschland/Clifton, New Jersey, 1988.

[13] ISO 20753:2018 Plastics—Test Specimens.

[14] ISO 10079-1:2015 Medical Suction Equipment—Part 1: Electrically Powered Suction Equipment; ISO 10079-2:2014 Medical Suction Equipment—Part 2: Manually Powered Suction Equipment; ISO 10079-3:2014 Medical Suction Equipment—Part 3: Suction Equipment Powered From a Vacuum or Positive Pressure Gas Source.

[15] ASTM F561-19 Standard Practice for Retrieval and Analysis of Medical Devices, and Associated Tissues and Fluids.
[16] ASTM F619-14 Standard Practice for Extraction of Medical Plastics.
[17] ASTM F997-18 Standard Specification for Polycarbonate Resin for Medical Applications.
[18] ASTM F1855-00(2019) Standard Specification for Polyoxymethylene (Acetal) for Medical Applications.
[19] ASTM F1886/F1886M-16 Standard Test Method for Determining Integrity of Seals for Medical Packaging by Visual Inspection.
[20] ASTM F1929-15 Standard Test Method for Detecting Seal Leaks in Porous Medical Packaging by Dye Penetration.
[21] ASTM F1862/F1862M-17 Standard Test Method for Resistance of Medical Face Masks to Penetration by Synthetic Blood (Horizontal Projection of Fixed Volume at a Known Velocity).
[22] ASTM E1766-15 Standard Test Method for Determination of Effectiveness of Sterilization Processes for Reusable Medical Devices.
[23] ASTM E1837-96 Standard Test Method to Determine Efficacy of Disinfection Processes for Reusable Medical Devices (Simulated Use Test).
[24] Appendix XX, Methods of Test for Surgical Dressings (A~T), British Pharmacopoeia, 1993, p. A214.
[25] Appendix XVIA, Test of Sterility, British Pharmacopoeia, 1993, p. A180.
[26] Appendix XVIB, Test of Microbial Contamination, British Pharmacopoeia, 1993, p. A184.
[27] Appendix XVIC, Efficacy of Antimicrobial Preservation, British Pharmacopoeia, 1993, p. A191.
[28] Appendix XVIII, Methods of Sterilisation, British Pharmacopoeia, 1993, p. A197.
[29] BS EN 13726-1:2002. Test Methods for Primary Wound Dressings. Part 1. Aspects of Absorbency.
[30] BS EN 13726-2:2002. Test Methods for Primary Wound Dressings. Part 2. Moisture Vapour Transmission Rate of Permeable Film Dressings.
[31] BS EN 13726-3:2003. Test Methods for Primary Wound Dressings. Part 3. Waterproofness.
[32] BS EN 13726-4:2003. Test Methods for Primary Wound Dressings. Part 4. Conformability.
[33] BS EN 13726-5:2003. Test Methods for Primary Wound Dressings. Part 5. Bacterial Barrier Properties.
[34] BS EN 16756. Antimicrobial Wound Dressings. Requirements and Test Methods.
[35] BS 8432:2005 Spinal Orthoses. Guide to Design.
[36] BS 7505:1995 Specification for the Elastic Properties of Flat, Non-adhesive, Extensible Fabric Bandages.
[37] BS EN 1644-1:1997 Test Methods for Nonwoven Compresses for Medical Use: Nonwovens Used in the Manufacture of Compresses.
[38] BS EN 1644-2:2000 Test Methods for Nonwoven Compresses for Medical Use: Finished Compresses.
[39] BS EN ISO 11607-1:2020, Packaging for Terminally Sterilized Medical Devices. Requirements for Materials, Sterile Barrier Systems and Packaging Systems.
[40] BS EN ISO 11607-2:2020, Packaging for Terminally Sterilized Medical Devices. Validation Requirements for Forming, Sealing and Assembly Processes.
[41] BS EN 868-2:2017, Packaging for Terminally Sterilized Medical Devices. Sterilization Wrap. Requirements and Test Methods.
[42] BS EN 868-3:2017, Packaging for Terminally Sterilized Medical Devices. Paper for Use in the Manufacture of Paper Bags and in the Manufacture of Pouches and Reels. Requirements and Test Methods.
[43] BS EN 868-4:2017, Packaging for Terminally Sterilized Medical Devices. Paper Bags. Requirements and Test Methods.
[44] BS EN 868-5:2018, Packaging for Terminally Sterilized Medical Devices. Sealable Pouches and Reels of Porous Materials and Plastic Film Construction. Requirements and Test Methods.
[45] BS EN 868-6:2017, Packaging for Terminally Sterilized Medical Devices. Paper for Low Temperature Sterilization Processes. Requirements and Test Methods.
[46] BS EN 868-7:2017, Packaging for Terminally Sterilized Medical Devices. Adhesive Coated Paper for Low Temperature Sterilization Processes. Requirements and Test Methods.
[47] BS EN 868-8:2018, Packaging for Terminally Sterilized Medical Devices. Re-Usable Sterilization Containers for Steam Sterilizers Conforming to EN 285. Requirements and Test Methods View Details.
[48] BS EN 868-9:2018, Packaging for Terminally Sterilized Medical Devices. Uncoated Nonwoven Materials of Polyolefines. Requirements and Test Methods.

[49] BS EN 868-10:2018, Packaging for Terminally Sterilized Medical Devices. Adhesive Coated Nonwoven Materials of Polyolefines. Requirements and Test Methods.

[50] BS EN 29073-1:1992, ISO 9073-1: 1989, Textiles. Methods of Test for Nonwovens. Part 1: Determination of Mass per Unit Area.

[51] BS EN ISO 9073-2:1997 Textiles. Test Methods for Nonwovens. Part 2: Determination of Thickness.

[52] NWSP 120.1.R0 (15) Thickness of Nonwoven Fabrics.

[53] NWSP 120.2.R0 (15) Thickness of Highloft Nonwoven Fabrics.

[54] F. Battocchio, M.P.F. Sutcliffe, Modelling fibre laydown and web uniformity in nonwoven fabric, Model. Simul. Mater. Sci. Eng. 25 (2017), 035006.

[55] J.M.R. Perry, On the power of the index of dispersion test to detect spatial pattern, Biometrics 35 (1979) 613–622.

[56] R. Chhabra, Nonwoven uniformity—measurements using image analysis, Int. Nonwovens J. 12 (1) (2003) 43–50.

[57] J. Scharcanski, C.T. Dodson, Texture analysis for estimating spatial variability and anisotropy in planar stochastic structures, Opt. Eng. 35 (8) (1996) 2302–2309.

[58] H.J. Chen, D.K. Huang, Online Measurement of Nonwoven Weight Evenness Using Optical Methods, ACT Paper, 1999.

[59] W.H. Pound, Real world uniformity measurement in nonwoven coverstock, Int. Nonwovens J. 10 (1) (2001) 35–39.

[60] X. Huang, R.R. Bresee, Characterizing nonwoven web structure using image analysis techniques, part III: web uniformity analysis, Int. Nonwovens J. 5 (3) (1993) 28–38.

[61] E. Amirnasr, E. Shim, B. Yeom, B. Pourdeyhimi, Basis weight uniformity analysis in nonwovens, J. Text. Inst. 105 (4) (2014) 444–453, https://doi.org/10.1080/00405000.2013.820017.

[62] R.K. Aggarwal, W.R. Kennon, I. Porat, A scanned-laser technique for monitoring fibrous webs and nonwoven fabrics, J. Text. Inst. 83 (3) (1992) 386–398.

[63] P.A. Boeckerman, Meeting the special requirements for on-line basis weight measurement of lightweight nonwoven fabrics, TAPPI J. 75 (12) (1992) 166–172.

[64] Hunter Lab Color Scale. http://www.hunterlab.com/appnotes/an08_96a.pdf.

[65] B. Neckář, D. Das, Modelling of fibre orientation in fibrous materials, J. Text. Inst. 103 (3) (2012) 330–340, https://doi.org/10.1080/00405000.2011.578357.

[66] T. Ishikawa, Y. Ishii, Y. Ohkoshi, K.H. Kim, Microstructural analysis of melt-blown nonwoven fabric by X-ray micro computed tomography, Text. Res. J. 89 (9) (2019) 1734–1747, https://doi.org/10.1177/0040517518779255.

[67] T. Ishikawa, Y. Ishii, K. Nakasone, Y. Ohkoshi, K. Kyoung Hou, Structure analysis of needle-punched nonwoven fabrics by X-ray computed tomography, Text. Res. J. 89 (1) (2019) 20–31, https://doi.org/10.1177/0040517517736470.

[68] D. Groitzsch, Ultrafine Microfiber Spunbond for Hygiene and Medical Application. http://www.technica.net/NT/NT2/eedana.htm.

[69] N. Mao, S.J. Russell, Capillary pressure and liquid wicking in three-dimensional nonwoven materials, J. Appl. Phys. 104 (3) (2008), 034911.

[70] M. Tausif, B. Duffy, S. Grishanov, H. Carr, S. Russell, Three-dimensional Fiber segment orientation distribution using X-ray microtomography, Microsc. Microanal. 20 (4) (2014) 1294–1303, https://doi.org/10.1017/S1431927614000695.

[71] T. Gilmore, H. Davis, Z. Mi, Tomographic approaches to nonwovens structure definition, in: National Textile Centre Annual Report, 1993. USA, September 1993.

[72] N. Mao, S.J. Russell, Modelling of permeability in homogeneous three- dimensional nonwoven fabrics, Text. Res. J. 91 (2003) 243–258.

[73] D.R. Petterson, The Mechanics of Nonwoven Fabrics (Sc D. thesis), MIT, Cambridge, MA, 1958.

[74] S.M. Hansen, Nonwoven engineering principles, in: A.F. Turbak (Ed.), Nonwovens—Theory, Process, Performance & Testing, Tappi Press, Atlanta, 1993.

[75] J.W.S. Hearle, P.J. Stevenson, Nonwoven fabric studies, part 3: the anisotropy of nonwoven fabrics, Text. Res. J. 33 (1963) 877–888.

[76] J.W.S. Hearle, V. Ozsanlav, Nonwoven fabric studies, part 5: studies of adhesive-bonded nonwoven fabrics part 3: the determination of fibre orientation and curl, J. Text. Inst. 70 (1979) 487–497.

[77] P.W. Chuleigh, Image formation by fibres and fibre assemblies, Text. Res. J. 54 (1983) 813.

[78] O.J. Kallmes, Techniques for determining the fibre orientation distribution throughout the thickness of a sheet, TAPPI J. 52 (1969) 482–485.

[79] A. Votava, Practical method–measuring paper asymmetry regarding fibre orientation, TAPPI J. 65 (1982) 67.

[80] W.F. Cowan, E.J.K. Cowdrey, Evaluation of paper strength components by short span tensile analysis, TAPPI J. 57 (2) (1973) 90.

[81] B. Stenemur, Method and device for monitoring fibre orientation distributions based on light diffraction phenomenon, Int. Nonwovens J. 4 (1992) 42–45.

[82] G.L. Clark, Comparative degree of preferred orientation in nineteen wood pulps as evaluated from X-ray diffraction patterns, TAPPI J. 33 (1950) 384.

[83] R.E. Prud'homme, N.V. Hien, J. Noah, R.H. Marchessault, Determination of fiber orientation of cellulosic samples by X-ray diffraction, J. Appl. Polym. Sci. 19 (1975) 2609–2620.

[84] S. Osaki, Dielectric anisotropy of nonwoven fabrics by using the microwave method, TAPPI J. 72 (1989) 171.

[85] S. Lee, Effect of fibre orientation on thermal radiation in fibrous media, Int. J. Heat Mass Transfer 32 (2) (1989) 311.

[86] S.H. McGee, R.L. McCullough, Characterization of fibre orientation in short-fibre composites, J. Appl. Phys. 55 (1) (1983) 1394.

[87] G.A. Orchard, The measurement of fibre orientation in card webs, J. Text. Inst. 44 (1953) T380.

[88] P.P. Tsai, R.R. Bresse, Fibre orientation distribution from electrical measurements. Part 1, theory, Int. Nonwovens J. 3 (3) (1991) 36.

[89] P.P. Tsai, R.R. Bresse, Fibre orientation distribution from electrical measurements. Part 2, instrument and experimental measurements, Int. Nonwovens J. 3 (4) (1991) 32.

[90] M.M. Chaudhray, M.Sc. dissertation, University of Manchester, 1972.

[91] S.M. Judge, M.Sc. dissertation, University of Manchester, 1973.

[92] X.C. Huang, R.R. Bressee, Characteristizing nonwoven web structure using image analysing techniques, part 2: fibre orientation analysis in thin webs, Int. Nonwovens I. (2) (1993) 14–21.

[93] B. Pourdeyhimi, A. Nayernouri, Assessing fibre orientation in nonwoven fabrics, INDA J. Nonw. Res. 5 (1993) 29–36.

[94] B. Pouredyhimi, B. Xu, Charaterizing pore size in nonwoven fabrics: shape considerations, Int. Nonwovens J. 6 (1) (1993) 26–30.

[95] R.H. Gong, A. Newton, Image analysis techniques. Part II: the measurement of fibre orientation in nonwoven fabrics, Text. Res. J. 87 (1996) 371.

[96] P.N. Britton, A.J. Sampson Jr., C.F. Elliot, H.W. Grabben, W.E. Gettys, Computer simulation of the technical properties of nonwoven fabrics, part 1: the method, Text. Res. J. 53 (1983) 363–368.

[97] T.H. Grindstaff, S.M. Hansen, Computer model for predicting point-bonded nonwoven fabric strength, part 1, Text. Res. J. 56 (1986) 383–388.

[98] O. Jirsak, D. Lukas, R. Charrat, A two-dimensional model of mechanical properties of textiles, J. Text. Inst. 84 (1993) 1–14.

[99] B. Xu, Y. Ting, Measuring structural characteristics of fibre segments in nonwoven fabrics, Text. Res. J. 65 (1995) 41–48.

[100] B. Pourdeyhimi, R. Dent, H. Davis, Measuring fibre orientation in nonwovens. Part 3: Fourier transform, Text. Res. J. 67 (1997) 143–151.

[101] B. Pourdeyhimi, R. Ramanathan, R. Dent, Measuring fibre orientation in nonwovens. Part 2: direct tracking, Text. Res. J. 66 (1996) 747–753.

[102] B. Pourdeyhimi, R. Ramanathan, R. Dent, Measuring fibre orientation in nonwovens. Part 1: simulation, Text. Res. J. 66 (1996) 713–722.

[103] B. Pourdeyhimi, R. Dent, Measuring fibre orientation in nonwovens. Part 4: flow field analysis, Text. Res. J. 67 (1997) 181–187.

[104] T. Komori, K. Makishima, Number of fibre-to-fibre contacts in general fibre assemblies, Text. Res. J. 47 (1977) 13–17.

[105] Leica Microsystems Imaging Solutions, Manual of Quantimet 570, Leica Microsystems Imaging Solutions, Cambridge, UK, 1993.

[106] S. Kärkkäinnen, A. Penttinen, N. Ushakov, A. Ushakova, Estimation of orientation characteristic of fibrous material, Adv. Appl. Probab. 33 (3) (2001) 559–575, https://doi.org/10.1239/aap/1005091352.

[107] BS EN 993-1:2018, Methods of Test for Dense Shaped Refractory Products. Determination of Bulk Density, Apparent Porosity and True Porosity.

[108] S.K. Bhatia, J.L. Smith, Application of the bubble point method to the characterization of the pore size distribution of geotextile, Geotech. Test. J. 18 (1) (1995) 94–105.

[109] S.K. Bhatia, J.L. Smith, Geotextile characterization and pore size distribution, part II: a review of test methods and results, Geosynth. Int. 3 (2) (1996) 155–180.

[110] ASTM D4751, Test Method for Determining Apparent Opening Size of a Geotextile.

[111] E. Blond, O. Veermersch, R.A. Diederich, Comprehensive analysis of the measurement techniques used to determine geotextile opening size: AOS, FOS, O90, and 'bubble point', in: Proceedings of Geosynthetics 2015, February 15–18, Portland, Oregon, 2015.

[112] L. Van der Sluys, W. Dierickx, Comparative studies of different porometry determination methods for geotextiles, Geotext. Geomembr. 9 (1991) 183–198.

[113] J.P. Giroud, Granular filters and geotextile filters, in: Proc., Geo-filters'96, Montréal, 1996, pp. 565–680.

[114] F. Saathoff, S. Kohlhase, Research at the Franzius-Institut on geotextile filters in hydraulic engineering, in: Proceedings of the Fifth Congress Asian and Pacific Regional Division, ADP/IAHR, Seoul, Korea, 1986, pp. 9–10.

[115] BS EN ISO 12956:2020 Geotextiles and Geotextile-Related Products: Determination of the Characteristic Opening Size.

[116] D. Fayoux, Filtration Hydrodynamique des Sols par des Textiles, in: Proceedings of the International Conference on the Use of Fabrics in Geotechnics, 2, Paris, France, April 1977, 1977, pp. 329–332 (in French).

[117] J. Mlynarek, J. Lafleur, R. Rollin, G. Lombard, Filtration opening size of geotextiles by hydrodynamic sieving, Geotech. Test. J. 16 (1) (1993) 61–69.

[118] CAN/CGSB-148.1-10..

[119] A.L. Rollin, R. Denis, L. Estaque, J. Masounave, Hydraulic behaviour of synthetic nonwoven filter fabrics, Can. J. Chem. Eng. 60 (1982) 226–234.

[120] A.H. Aydilek, S.H. Oguz, T.B. Edil, Constriction size of geotextile filters, J. Geotech. Geoenviron. 131 (1) (2005) 28–38.

[121] W. Dierickx, Opening size determination of technical textiles used in agricultural applications, Geotext. Geomembr. 17 (4) (1999) 231–245.

[122] S.K. Bhatia, Q. Huang, J.L. Smith, Application of digital image processing in morphological analysis of geotextiles, in: Proc. Conf. on Digital Image Processing: Techniques and Applications in Civil Engineering, vol. 1, ASCE, New York, 1993, pp. 95–108.

[123] V.K. Kothari, G. Agarwal, Determination of pore size parameters and its distribution of hydroentangled fabrics by image processing, J. Text. Inst. 99 (4) (2008) 317–324, https://doi.org/10.1080/00405000701414824.

[124] A. Rawal, Structural analysis of pore size distribution of nonwovens, J. Text. Inst. 101 (4) (2010) 350–359, https://doi.org/10.1080/00405000802442351.

[125] R.A. e Silva, R.G. Negri, V.D. de Mattos, A new image-based technique for measuring pore size distribution of nonwoven geotextiles, Geosynth. Int. 26 (3) (2019) 261–272.

[126] A.H. Aydilek, S.H. Oguz, T.B. Edil, Digital image analysis to determine pore opening size distribution of nonwoven geotextiles, J. Comput. Civ. Eng. 16 (4) (2002) 280–290.

[127] ASTM F316-03, Standard Test Methods for Pore Size Characteristics of Membrane Filters by Bubble Point and Mean Flow Pore Test.

[128] ASTM E128-99 Standard Test Method for Maximum Pore Diameter and Permeability of Rigid Porous Filters for Laboratory Use.

[129] BS 3321:1986 Method for Measurement of the Equivalent Pore Size of Fabrics (Bubble Pressure Test).

[130] BS 7591-4:1993 Porosity and Pore Size Distribution of Materials. Method of Evaluation by Liquid Expulsion.

[131] ASTM D6767-14 Standard Test Method for Pore Size Characteristics of Geotextiles by Capillary Flow Test.

[132] ASTM F2450-18, Standard Guide for Assessing Microstructure of Polymeric Scaffolds for Use in Tissue-Engineered Medical Products.

[133] TenCate Geosynthetics Americas, Understanding Porometer Versus AOS Testing of a Geotextile, 2018. www.tencategeo.com.

[134] ASTM D4284-12 Standard Test Method for Determining Pore Volume Distribution of Catalysts and Catalyst Carriers by Mercury Intrusion Porosimetry.

[135] ASTM D4404-10 Standard Guide for Assessing Microstructure of Polymeric Scaffolds for Use in Tissue-Engineered Medical Products.

[136] BS 7591-1:1992 Porosity and Pore Size Distribution of Materials. Method of Evaluation by Mercury Porosimetry (Withdrawn).

[137] BS 1902-3.16:1990 Methods of Testing Refractory Materials, General and Textural Properties: Determination of Pore Size Distribution (Method 1902-316).

[138] BS ISO 15901-1:2016 Evaluation of Pore Size Distribution and Porosity of Solid Materials by Mercury Porosimetry and Gas Adsorption. Mercury Porosimetry.

[139] ASTM D 4404-18, Standard Test Method for Determination of Pore Volume and Pore Volume Distribution of Soil and Rock by Mercury Intrusion Porosimetry.

[140] W.B. Haines, J. Agric. Sci. 20 (1930) 97–116.

[141] P.M. Whelan, M.J. Hodgson, Essential Principles of Physics, John Murray, London, 1978.

[142] E. Washburn, The dynamics of capillary flow, Phys. Rev. 17 (3) (1921) 273–283.

[143] ASTM E 1294-89 Standard Test Method for Pore Size Characteristics of Membrane Filters Using Automated Liquid Porosimeter (Withdrawn in 2008).

[144] B. Miller, I. Tyomkin, J.A. Wehner, Quantifying the porous structure of fabrics for filtration applications, in: R.R. Raber (Ed.), Fluid Filtration: Gas, 1, ASTM Special Technical Publication 975, Proceedings of a Symposium Held in Philadelphia, Pennsylvania, USA, 1986, pp. 97–109.

[145] B. Miller, I. Tyomkin, An extended range liquid extrusion method for determining pore size distributions, Text. Res. J. 56 (1) (1994) 35–40.

[146] . https://www.micromeritics.com/product-showcase/autopore-v.aspx.

[147] . https://mpmandp.com/porosimeter.

[148] . https://www.corelab.com/cli/routine-rock/ultrapore-porosimeter.

[149] . https://www.anton-paar.com.

[150] BS ISO 15901-2:2006 Pore Size Distribution and Porosity of Solid Materials by Mercury Porosimetry and Gas Adsorption. Analysis of Mesopores and Macropores by Gas Adsorption.

[151] BS ISO 15901-3:2007, Pore Size Distribution and Porosity of Solid Materials by Mercury Porosimetry and Gas Adsorption. Analysis of Micropores by Gas Adsorption.

[152] ASTM D1993-18 Standard Test Method for Precipitated Silica-Surface Area by Multipoint BET Nitrogen Adsorption.

[153] S. Brunauer, P.H. Emmett, E. Teller, Adsorption of gases in multimolecular layers, J. Am. Chem. Soc. 60 (2) (1938) 309.

[154] ASTM D3908-03 Standard Test Method for Hydrogen Chemisorption on Supported Platinum Catalysts by Volumetric Vacuum Method.

[155] ASTM D4824-13: Standard Test Method for Determination of Catalyst Acidity by Ammonia Chemisorption.

[156] ASTM D5160-95: Standard Guide for Gas-Phase Adsorption Testing of Activated Carbon.

[157] ASTM D5228-92: Standard Test Method for Determination of Butane Working Capacity of Activated Carbon.

[158] ASTM D5742-95 (2015): Standard Test Method for Determination of Butane Activity of Activated Carbon.

[159] ASTM B923-10: Standard Test Method for Metal Powder Skeletal Density by Helium or Nitrogen Pycnometry.

[160] ASTM D6226-15: Standard Test Method for Open Cell Content of Rigid Cellular Plastics.

[161] N. Mao, S.J. Russell, Directional permeability of homogeneous anisotropic fibrous material, part 1, J. Text. Inst. 91 (2000) 235–243.

[162] ASTM D737-96 Test Method for Air Permeability of Textile Fabrics.

[163] BS EN ISO 9237:1995 Textiles. Determination of the Permeability of Fabrics to Air; ISO/CD 9073–15:2005 Textiles—Test Methods for Nonwovens—Part 15: Evaluation of Air Permeability.

[164] WSP 70.1–19 (IST 70.1, ERT 140.2).

[165] ASTM F2952-14 Standard Guide for Determining the Mean Darcy Permeability Coefficient for a Porous Tissue Scaffold.

[166] BS EN 14150:2019 Geosynthetic Barriers. Determination of Permeability to Liquids.

[167] BS EN ISO 11058:2019 Geotextiles and Geotextile-Related Products. Determination of Water Permeability Characteristics Normal to the Plane, without Load.

[168] BS EN ISO 10776:2012 Geotextiles and Geotextile-Related Products. Determination of Water Permeability Characteristics Normal to the Plane, under Load.

[169] BS EN ISO 12958:2010 Geotextiles and Geotextile-Related Products. Determination of Water Flow Capacity in their Plane.

[170] R.V. Zantam, Geotextile and Geomembrance in Civil Engineering, John Wiley, New York, 1986, pp. 181–192.

[171] K.L. Adams, et al., Radial penetration of a viscous liquid into a planar anisotropic porous medium, Int. J. Multiphase Flow 14 (2) (1988) 203–215.

[172] K.L. Adams, et al., In plane flow of fluids in fabrics structure, flow characterization, Text. Res. J. 57 (1987) 647–654.

[173] S.M. Montagomery, Directional in-plane permeabilities of geotextile, Geotext. Geomembr. 7 (1988) 275–292.

[174] N. Mao, Effect of Fabric Structure on the Liquid Transport Characteristics of Nonwoven Wound Dressings (Ph. D. thesis), University of Leeds, 2000.

[175] S.J. Russell, N. Mao, Apparatus and method for the assessment of in-plane anisotropic liquid absorption in nonwoven fabrics, J. Autex 1 (2001) 47–53.

[176] IST70.2..

[177] E. Kissa, Wetting and wicking, Text. Res. J. 66 (1996) 660.

[178] P.R. Harnett, P.N. Mehta, A survey and comparison of laboratory test methods for measuring wicking, Text. Res. J. 54 (1984) 471–478.

[179] Y.L. Hsieh, B. Yu, Wetting and retention properties of fibrous materials, part 1: water wetting properties of woven fabrics and their constituent single fibres, Text. Res. J. 62 (1992) 677–685.

[180] P.K. Chatterjee, Absorbency, Elsevier, New York, 1985.

[181] R.E. Johnson, R.H. Dettre, in: R.F. Gould (Ed.), Contact Angle, Wettability and Adhesion, Advances in Chemistry Series, vol. 43, American Chemistry Society, Washington, DC, 1964, p. 112.

[182] B. Miller, I. Tymokin, Spontaneous transplanar uptake of liquids by fabrics, Text Res. J. 54 (1983) 706–712.

[183] E. Kissa, in: Cutler, E. Kissa (Eds.), Detergency, Theory and Technology, Surfactant Science Series, vol. 20, Marcel Dekker, New York, 1987, p. 193.

[184] A.W. Newman, R.J. Good, Techniques of measuring contact angles, in: R.J. Good, P.R. Stromberg (Eds.), Surface and Colloid Science, vol. 11, Plenum Press, New York, 1977, p. 31.

[185] ASTM 5725 Standard Test Method for Surface Wettability and Absorbency of Sheeted Materials Using an Automated Contact Angle Tester.

[186] ASTM 724 Standard Method for Surface Wettability of Paper (Angle-of-Contact Method).

[187] ASTM D 1590 Standard Test Method for Surface Tension of Water.

[188] ASTM D 7490 Standard Test Method for Measurement of the Surface Tension of Solid Coatings.

[189] BS EN ISO 19403-6:2017 Paints and Varnishes—Wettability—Part 6: Measurement of Dynamic Contact Angle.

[190] BS EN ISO 19403-2:2017 Paints and Varnishes—Wettability—Part 2: Determination of the Surface Free Energy of Solid Surfaces by Measuring the Contact Angle.

[191] ISO 15989:2004 Plastics—Film and Sheeting—Measurement of Water-Contact Angle of Corona-Treated Films.

[192] BS ISO 27448:2009 Fine Ceramics (Advanced Ceramics, Advanced Technical Ceramics)—Test Method for Self-Cleaning Performance of Semiconducting Photocatalytic Materials—Measurement of Water Contact Angle.

[193] B. Miller, in: M.J. Schick (Ed.), Surface Characterization of Fibres and Textiles, Part II, Marcel Dekker, NY, 1977, p. 47.

[194] M. Tagawa, K. Gotoh, A. Yasukawa, M. Ikuta, Estimation of surface free from energies & Hawaker constants for fibrous solids by wetting force measurements, Colloid Polym. Sci. 268 (1990) 689.

[195] R.V. Dyba, B. Miller, Dynamic measurements of the wetting of single filaments, Text. Res. J. 40 (1970) 884.

[196] R.V. Dyba, B. Miller, Dynamic wetting of filaments in solutions, Text. Res. J. 41 (1971) 978.

[197] Y.K. Kamath, C.J. Dansizer, S. Hornby, H.D. Weigmann, Surface wettability scanning of long filaments by a liquid emmbrane method, Text. Res. J. 57 (1987) 205.

[198] H.G. Bruil, J.J. Van Aartsen, The determination of contact angles of aqueous surfacant solutions on powders, J. Colloid Polym. Sci. (1979) 32. 252.

[199] T. Gillespie, T. Johnson, The penetration of aqueous surfactant solutions and non-Newtonian polymer solutions into paper by capillary action, J. Colloid Interface Sci. 36 (1971) 282–285.

[200] J.J. DeBoer, The wettability of scoured and dried cotton fabrics, Text. Res. J. 50 (1980) 624–631.

[201] P.L. Lennox-Kerr, Super-absorbent acrylic from Italy, Text. Inst. Ind. 19 (1981) 83–84.

[202] E.M. Buras, et al., Measurement and theory of absorbency of cotton fabrics, Text. Res. J. 20 (1950) 239–248.

[203] W. Korner, New results on the water comfort of the absorbent synthetic fibre Dunoua, Chemiefasern/Textilind 31 (1981) 112–116.

[204] . https://mksystems.com/mk-gats-liquid-absorbency-system.

[205] ASTM D5802-95 Standard Test Method for Sorption of Bibulous Paper Products (Sorptive Rate and Capacity Using Gravimetric Principles) (Withdrawn 2009).

[206] B. Miller, Critical evaluation of upward wicking tests, Int. Nonwovens J. 9 (1) (2000) 35–40.

[207] S.M. Montgomery, B. Miller, L. Rebenfeld, Spatial distribution of local permeability in fibrous networks, Text. Res. J. 62 (1992) 151–161.

[208] M. Howaldt, A.P. Yoganathan, Laser-Doppler anemometry to study fluid transport in fibrous assemblies, Text. Res. J. 53 (9) (1983) 544–551.

[209] T. Kawase, Y. Morimoto, T. Fujii, M. Minagawa, Spreading of liquids in textile assemblies, I. spreading of liquids in textile assemblies. I. Capillary spreading of liquids, Text. Res. J. 56 (7) (1986) 409–414.

[210] T. Kawase, Y. Morimoto, T. Fujii, M. Minagawa, Spreading of liquids in textile assemblies, III. Application of an image analyser system to capillary spreading of liquids, Text. Res. J. 58 (5) (1988) 306–308.

[211] D. Tanner, Development of textile yarns based on customer performance, in: Symposium on Yarns and Yarn Manufacturing, University of Manchester, 1979.

[212] BS 4745:2005 Determination of the Thermal Resistance of Textiles. Two-Plate Method: Fixed Pressure Procedure, Two-Plate Plate Method: Fixed Opening Procedure, and Single-Plate Method.

[213] ISO 5085-1:1989 Textiles—Determination of Thermal Resistance—Part 1: Low Thermal Resistance.

[214] ISO 5085-2:1990 Textiles—Determination of Thermal Resistance—Part 2: High Thermal Resistance.

[215] BS 5335-1:1991 Continental Quilts. Specification for Quilts Containing Fillings Other than Feather and/or Down.

[216] ASTM D5034-09, IST110.1 Standard Test Method for Breaking Strength and Elongation of Textile Fabrics (Grab Test).

[217] ASTM D5035-11 Standard Test Method for Breaking Force and Elongation of Textile Fabrics (Strip Method).

[218] BS EN 29073-3:1992, ISO 9073-3:1992 Methods of Test for Nonwovens. Methods of Test for Nonwovens. Determination of Tensile Strength and Elongation.

[219] E. Kamalha, Y. Zeng, J. Mwasiagi, S. Kyatuheire, The Comfort Dimension: A Review of Perception in Clothing, Wiley, Hoboken, NJ, USA, 2013, p. 8.

[220] F.T. Peirce, The "handle" of cloth as a measurable quantity, J. Text. Inst. 21 (1930) T337–T416.

[221] ISO 9073-7: 1995..

[222] Measuring Compression/Softness of Nonwovens Using the Vantage^{NX} Universal Testing Machine and MAP4 Software. https://www.thwingalbert.com/media/wysiwyg/CompressionSoftness.pdf.

[223] J. Dutkiewicz, Some advances in nonwoven structures for absorbency, comfort and aesthetics, AUTEX Res. J. 2 (3) (2002). http://www.autexrj.com/cms/zalaczone_pliki/6-02-3.pdf.

[224] S. Kawabata, The Standardization and Analysis of Hand Evaluation, second ed., Textile Machinery Society of Japan, Osaka, Japan, 1980, p. 96.

[225] P. Giorgio Minazio, FAST—fabric assurance by simple testing, Int. J. Cloth. Sci. Technol. 7 (2/3) (1995) 43–48, https://doi.org/10.1108/09556229510087146.

[226] N. Yaman, F.M. Şenol, P. Gurkan, Applying artificial neural networks to total hand evaluation of disposable diapers, J. Eng. Fibers Fabr. 6 (2011) 38–43.

[227] N. Mao, Towards objective discrimination & evaluation of fabric tactile properties: quantification of biaxial fabric deformations by using energy methods, in: Proceedings of the 14th AUTEX World Textile Conference, Bursa, Turkey, 26–28 May, 2014.

[228] . http://www.phabrometer.com/.

[229] Anonymous, Wool Handlemeter, Available from: http://www.woolcomfortandhandle.com/index.php/wool-handlemeter. (Accessed 26 February 2021).

[230] N. Mao, M. Taylor, Evaluation Apparatus and Method. WO2012104627, 13 October 2012.

[231] J.V. Edwards, N. Mao, S. Russell, E. Carus, B. Condon, D. Hinchliffe, L. Gary, E. Graves, A. Bopp, Y. Wang, Fluid handling and fabric handle profiles of hydroentangled greige cotton and spunbond polypropylene nonwoven topsheets, Proc. IMechE L J. Mater. Des. Appl. (2015) 1–13, https://doi.org/10.1177/1464420715586020.

[232] M. Easson, J. Vincent Edwards, N. Mao, C. Carr, D. Marshall, J. Qu, E. Graves, M. Reynolds, A. Villalpando, B. Condon, Structure/function analysis of nonwoven cotton topsheet fabrics: multi-fiber blending effects on fluid handling and fabric handle mechanics, Materials 11 (11) (2018) 2077, https://doi.org/10.3390/ma11112077.

[233] C. Carr, N. Mao, J. Qu, C. Boardman, S. Mjornstedt, N. Miller, Discriminatory mechanical and sensory properties for the study of fabric conditioners in the laundry cycle, in: C. Carr (Ed.), Proceedings of the 91st Textile Institute World Conference, TBC. The Textile Institute World Conference, 24–26 Jul 2018, University of Leeds, UK, 2018.

[234] N. Mao, C. Carr, J. Qu, A. Williams, L. Connell, H. Rieley, Evaluation of methods for characterising fabric deformations, in: C. Carr (Ed.), TBC, Proceedings of the 91st Textile Institute World Conference, 24–26 Jul 2018, University of Leeds, UK, 2018.

[235] J. Lindberg, L. Waesterberg, R. Svenson, Wool fabrics as garment construction materials, J. Text. Inst. Trans. 51 (1960) T1475–T1493.

[236] . https://english.keskato.co.jp/archives/products/kes-f7.

[237] Thermal Effusivity Tester. https://ctherm.com/.

[238] ASTM D7984, Standard Test Method for Measurement of Thermal Effusivity of Fabrics Using a Modified Transient Plane Source (MTPS) Instrument.

[239] J. Hu, Characterization of Sensory Comfort of Apparel Products, Hong Kong Polytechnic University, Hong Kong, China, 2006.

[240] S. Backer, D.R. Petterson, Some principles of nonwoven fabrics, Text. Res. J. 30 (12) (1960) 704–711.

[241] J.W.S. Hearle, P.J. Stevenson, Studies in nonwoven fabrics: prediction of tensile properties, Text. Res. J. 34 (1964) 181–191.

[242] J.W.S. Hearle, V. Ozsanlav, Nonwoven fabric studies, part 1: a theoretical model of tensile response incorporating binder deformation, J. Text. Inst. 70 (1979) 19–28.

[243] J.W.S. Hearle, A. Newton, Nonwoven fabric studies, part XIV: derivation of generalized mechanics by the energy method, Text. Res. J. 37 (9) (1967) 778.

[244] W.D. Freeston, M.M. Platt, Mechanics of elastic performance of textile materials, part XVI: bending rigidity of nonwoven fabrics, Text. Res. J. 35 (1) (1965) 48–57.

[245] C.T.J. Dodson, Fibre crowding, fibre contacts and fibre flocculation, TAPPI J. 79 (9) (1996) 211–216.

[246] Dodson CTJ. http://www.ma.umist.ac.uk/kd/pslec/node9.html.

[247] A.C. Wrotnowski, Nonwoven filter media, Chem. Eng. Prog. 58 (12) (1962) 61–67.

[248] A.C. Wrotnowski, Felt filter media, Filtr. Sep. (September/October) (1968) 426–431.

[249] H. Goeminne, The geometrical and filtration characteristics of metal-fiber filters—a comparative study, Filtr. Sep. (August) (1974) 350–355.

[250] G. Lambard, et al., Theoretical and experimental opening size of heat-bonded geotextiles, Text. Res. J. (April) (1988) 208–217.

[251] Y.H. Faure, et al., Theoretical and experimental determination of the filtration opening size of geotextiles, in: 3rd International Conference on Geotextiles, Vienna, Austria, 1989, pp. 1275–1280.

[252] Y.H. Faure, J.P. Gourc, P. Gendrin, Structural study of porometry and filtration opening size of geotextiles, in: I.D. Peggs (Ed.), Geosynthetics: Microstructure and Performance, ASTM STP 1076, ASTM, Philadelphia, 1990, pp. 102–119.

[253] J.P. Gourc, Y.H. Faure, Soil particle, water, and fiber—A fruitful interaction now controlled, in: Proc., 4th Int. Conf. on Geotextiles, Geomembranes and Related Products, The Hague, The Netherlands, 1990, pp. 949–971.

[254] H. Darcy, Les Fontaines Publiques de la Ville de Dijon, Victor Valmont, Paris, 1856.

[255] S.J. Kuwabara, The forces experienced by randomly distributed parallel circular cylinder or spheres in a viscous flow at small Reynolds numbers, J. Phys. Soc. Jpn. 14 (1959) 527.

[256] P.C. Carman, Flow of Gases through Porous Media, Academic Press, New York, 1956.

[257] C.N. Davies, The separation of airborne dust and particles, Proc. Inst. Mech. Eng. B (1952) 185–213.

[258] H.W. Piekaar, L.A. Clarenburg, Aerosol filters: pore size distribution in fibrous filters, Chem. Eng. Sci. 22 (1967) 1399.

[259] R.W. Dent, The air permeability of nonwoven fabrics, J. Text. Inst. 67 (1976) 220–223.

[260] V.O. Emersleben, Das darcysche filtergesetz, Phsikalische Z. 26 (1925) 601.

[261] H.C. Brinkman, On the permeability of media consisting of closely packed porous particles, Appl. Sci. Res. A1 (1948) 81.

[262] A.S. Iberall, Permeability of glass wool and other highly porous media, J. Res. Natl. Bur. Stand. 45 (1950) 398.

[263] J. Happel, Viscous flow relative to arrays of cylinders, AICHE J. 5 (1959) 174–177.

[264] R.G. Cox, The motion of long slender bodies in a viscous fluid, part 1, J. Fluid Mech. 44 (1970) 791–810.

[265] A.S. Sangani, A. Acrivos, Slow flow past periodic arrays of cylinders with applications to heat transfer, Int. J. Multiphase Flow 8 (1982) 193–206.

[266] J. Kozeny, Royal Academy of Science, Vienna, Proceedings Class 1, 136, 1927, p. 271.

[267] H. Lamb, Hydrodynamics, Cambridge University Press, 1932.

[268] E.M. Sparrow, A.L. Loeffler Jr., Longitudinal laminar flow between cylinders arranged in regular array, AICHE J. 5 (1959) 325–330.

[269] J.E. Drummond, M.I. Tahir, Laminar viscous flow through regular arrays of parallel solid cylinders, Int. J. Multiphase Flow 10 (1983) 515–540.

[270] R.R. Sullivan, Specific surface measurements on compact bundles of parallel fibres, J. Appl. Phys. 13 (1942) 725–730.

[271] T. Nogai, M. Ihara, Study on air permeability of fibre assemblies oriented unidirectionary, J. Text. Mach. Soc. Jpn. 26 (1980) 10.

[272] J. Happel, H. Brenner, Low Reynolds's Number Hydrodynamics, Prentice Hall, 1965.

[273] N. Mao, S.J. Russell, Directional permeability of homogeneous anisotropic fibrous material, part 2, J. Text. Inst. 91 (2000) 244–258.

[274] A.E. Scheidegger, The Physics of Flow through Porous Media, University of Toronto Press, Toronto, 1972.

[275] I. Langmuir, Report on Smokes and Filters, Section I. U.S. Office of Scientific Research and Development, No. 865, Part IV, 1942.

[276] J. Van der Westhuizen, J. Prieur Du Plessis, An attempt to quantify fibre bed permeability utilizing the phase average Navier Stokes equation, Compos. A: Appl. Sci. Manuf. 27 (1996) 263–269, https://doi.org/10.1016/1359-835X(95)00039-5.

[277] F.W. Minor, A.M. Schwartz, L.C. Buckles, E.A. Wulkow, The migration of liquids in textile assemblies, Text. Res. J. 29 (1959) 931.

[278] A. Rushton, The analysis of textile filter media, Sep. Filtr. (November/December) (1968) 516.

[279] X. Shen, An Application of Needle-Punched Nonwovens in the Press Casting of Concrete (Ph.D. thesis), University of Leeds, 1996.

[280] J.L. Poiseuille, CR Acad. Sci. Paris 11 (1840) 961. 961, 1041; 12, p 112, 1841.

[281] G.G. Stokes, On the theories of the internal friction of fluids in motion, and of the equilibrium and motion of elastic solids, Trans. Camb. Philos. Soc. 8 (1845) 287–341.

[282] R. Lucas, Ueber das Zeitgesetz des Kapillaren Aufstiegs von Flussigkeiten, Kolloid Z. 23 (1918) 15.

[283] R.D. Laughlin, J.E. Davies, Some aspects of capillary absorption in fibrous textile wicking, Text. Res. J. 31 (1961) 904.

[284] T.B. Bahder, Mathematica for Scientists and Engineers, Addison-Wesley Pub. Co., Reading, 1995.

[285] B.S. Gupta, L.C. Wadsworth, Differentially absorbent cotton-surfaced spunbond copoplyester and spunbond PP with wetting agent, in: Proceedings of 7th Nonwovens Conference at 2004 Beltwide Cotton Conferences, San Antonio, TX, January 5–9, 2004.

[286] W.M. Law, Water Transport in Fabric (Ph.D. thesis), Department of Textiles, University of Leeds, 1988.

[287] R.L. Peek, D.A. McLean, Ind. Eng. Chem. Anal. Ed. 6 (1934) 85.

[288] G.D. Robinson, A Study of the Voids within the Interlock Structure and their Influence on Thermal Properties of Fabric (Ph.D. thesis), Department of Textile Industries, University of Leeds, 1982.

[289] R.E. Collins, Flow of Fluids through Porous Materials, Reinhold Publishing Corporation, New York, 1961.

[290] H.M. Princen, J. Colloid Interface Sci. 30 (1969) 359.

[291] N. Mao, S.J. Russell, Anisotropic liquid absorption in homogeneous two-dimensional nonwoven structures, J. Appl. Phys. 94 (6) (2003) 4135–4138.

[292] S.P. Rowland, D.J. Stanonis, W.D. King, Penetration-sorption of cotton fibres measured by immersed weight, J. Appl. Polym. Sci. 25 (1980) 2229.

[293] S.P. Rowland, N.R. Bertoniere, Some interactions of water-soluble solutes with cellulose and Sephadex, Text. Res. J. 46 (1976) 770.

[294] D.J. Stanonis, S.P. Rowland, Interactions of carbamates and their N-Methylol derivatives with cotton and Sephadex, measured by gel filtration, Text. Res. J. 49 (1979) 72.

[295] H.S. Carslaw, J.C. Jaeger, Conduction of Heat in Solids, second ed., Clarendo Press, Oxford, 1959.

[296] J. Crank, Mathematics of Diffusion, Clarendon Press, London, 1956, p. 148.

[297] D.F. Rudd, J. Phys. Chem. 64 (1960) 1254.

[298] R.B. Bird, W.E. Stewart, E.N. Lightfoot, Transport Phenomena, John Wiley, New York, 1960.

[299] B.S. Gupta, C.J. Hong, Changes in dimensions of web during fluid uptake and its impact on absorbency, TAPPI J. 77 (1994) 181–188.

[300] B.S. Gupta, H.S. Whang, Capillary Absorption Behaviors of Hydroentangled and Needlepunched webs of cellulosic Fibers, in: Proceedings of INDA-TEC 96: International nonwovens conference, September 11–13, 1996, Hyatt Regency Crystal City, Crystal City, Virginia, USA, 1996.

[301] B.S. Gupta, D.K. Smith, Nonwovens in Absorbent Materials, Text. Sci. Technol. 13 (2002) 349–388.

[302] B.S. Gupta, The effect of structural factors on Absorbent characteristics of nonwovens, TAPPI J. 71 (1988) 147–152.

[303] C. Bankvall, Heat transfer in fibrous material, J. Test. Eval. 1 (1973) 235–243.

[304] M. Bomberg, S. Klarsfeld, Semi-empirical model of heat transfer in dry mineral Fiber insulations, J. Therm. Insul. 6 (1) (1983) 157–173.

[305] S. Baxter, The thermal conductivity of textiles, Proc. Phys. Soc. 58 (1946) 105–118.

[306] J.F. Kreider, Handbook of Heating, Ventilation, and Air Conditioning, CRC Press LLC, London, 2001.

[307] American Society of Heating, 1993 ASHRAE Handbook, Fundamentals, I-P edition, Refrigerating and Air-Conditioning Engineers, Inc., Atlanta, 1993.

[308] J. Fricke, D. Büttner, R. Caps, J. Gross, O. Nilsson, Solid conductivity of loaded fibrous insulation, in: D.L. McElroy, J.F. Kimpflen (Eds.), Insulation Materials, Testing, and Applications, ASTM STP 1030, American Society for Testing and Materials, Philadelphia, 1990, pp. 66–78.

[309] C. Stark, J. Fricke, Improved heat-transfer models for fibrous insulations, Int. J. Heat Mass Transf. 36 (3) (1993) 617–625.

[310] J. Schuhmeister, Ber. K. Akad. Wien (Math.-Naturw. Klasse) 76 (1877) 283.

[311] R.S. Grewal, P. Banks-Lee, Development of thermal insulation for textile wet processing machinery using Needlepunched nonwoven fabrics, Int. Nonwovens J. 2 (1999) 121–129.

[312] R. Kirby, A. Cummings, Prediction of the bulk acoustic properties of fibrous materials at low frequencies, Appl. Acoust. 56 (2) (1999) 101–125.

[313] S.H. Burns, Propagation constant and specific impedance of airborne sound in metal wool, J. Acoust. Soc. Am. 49 (1971) 1–8.

[314] F.P. Mechel, Eine Modelltheorie zum Faserabsorber, Teil I: Regulare Faseranordnung; Teil II: Absorbermodell aus Elementarzellen und numerische Ergelnisse, Acustica 36 (1976/1977) 53–89.

[315] A. Cummings, I.-J. Chang, Acoustic propagation in porous media with internal mean flow, J. Sound Vib. 114 (1987) (1982) 565–581.

[316] Y. Attenborough, Acoustical characteristics of porous materials, Phys. Rep. 82 (3) (1982) 179–227.

[317] C. Zwikker, C.W. Kosten, Sound Absorbing Materials, Elsevier, Amsterdam, 1949.

[318] H. Tijdeman, On the propagation of sound waves in cylindrical tubes, J. Sound Vib. 39 (1975) 1–33.

[319] Y. Shoshani, Y. Yakubov, Numerical assessment of maximal absorption coefficients for nonwoven fibrewebs, Appl. Acoust. 59 (1) (2000) 77–87.

[320] N.N. Voronina, Empirical equations for a calculation of acoustic parameters of fibrous materials in terms their structural characteristic, Tr./NIISF, in: Building Acoustics, 1976, pp. 20–27.

[321] N.N. Voronina, Influence of fibrous materials structure on their acoustic properties, Acoustic J. 29 (1983) 598–602.

[322] N.N. Voronina, Acoustic properties of fibrous materials, Appl. Acoust. 42 (1994) 165–174.

[323] M.E. Delany, E.N. Bazley, Acoustical properties of fibrous absorbent materials, Appl. Acoust. 3 (1970) 105–116.

[324] BS EN 779:2002, Particulate Air Filters for General Ventilation—Determination of the Filtration Performance.

[325] BS ISO 19438:2003, Diesel Fuel and Petrol Filters for Internal Combustion Engines—Filtration Efficiency Using Particle Counting and Contaminant Retention Capacity.

[326] R.C. Brown, Air Filtration—An Integrated Approach to the Theory and Applications of Fibrous Filters, Pergamon Press, Oxford, UK, 1988.

[327] C.N. Davies (Ed.), Air Filtration, Academic Press, London, 1973.

[328] A.A. Krish, I.B. Stechkina, The theory of aerosol filtration with fibrous filters, in: D.T. Shaw (Ed.), Fundamentals of Aerosol Science, Wiley, 1978.

[329] P.C. Reist, Aerosol Science and Technology, McGraw-Hill, New York, 1993.

[330] A.A. Kirsh, N.A. Fuchs, Investigation of fibrous filters: diffusional deposition of aerosols in fibrous filters, Colloid J. 30 (1968) 630.

[331] I.B. Stechkina, A.A. Kirsh, N.A. Fuchs, Effect of inertia on the captive coefficient of aerosol particles by cylinders at low Stokes' numbers, Kolloidn. Zh. 32 (1970) 467.

[332] I.B. Stechkina, A.A. Kirsh, N.A. Fuchs, Studies on fibrous aerosol filters. IV. Calculation of aerosol deposition in model filters in the range of maximum penetration, Ann. Occup. Hyg. 12 (1969) 1–8.

[333] S.K. Friedlander, Theory of aerosol filtration, Ind. Eng. Chem. 30 (1958) 1161–1164.

[334] S.K. Friedlander, Aerosol filtration by fibrous filters, in: Blakebrough (Ed.), Biochemical and Biological Engineering, vol. 1, Academic Press, London, 1967 (Chapter 3).

[335] W.C. Hinds, Aerosol Technology: Properties, Behaviour and Measurements of Airborne Particles, John Wiley and Sons, New York, 1999.

[336] I.B. Steckina, N.A. Fuchs, Studies on fibrous aerosol filters I: calculation of diffusional deposition of aerosols in fibrous filters, Ann. Occup. Hyg. 9 (1966) 59–64.

[337] K.W. Lee, J.A. Gieseke, Note on the approximation of interceptional collection efficiencies, J. Aerosol Sci. 11 (1980) 335–341.

[338] H.C. Yeh, B.Y.H. Liu, Aerosol filtration by fibrous filters, J. Aerosol Sci. 5 (1974) 191–217.

[339] . http://www.tsi.com/AppNotes/appnotes.aspx?Pid=33&lid=439&file=iti_041.

[340] N.P. Vaughan, R.C. Brown, Observations of the microscopic structure of fibrous filters, Filtr. Sep. 9 (1996) 741–748.

[341] J.I.T. Stenhouse, Filtration of air by fibrous filters, Filtr. Sep. 12 (May/June) (1975) 268–274.

[342] R.B. Bird, W.E. Steward, E.N. Lightfood, Transport Phenomena, John Wiley and Sons, 2002, pp. 196–200.

[343] F. Folgar, C. Tucker III, J. Reinf. Plast. Compos. 3 (1984) 98–119.

[344] H.S. Kim, A. Deshpande, B. Pourdeyhimi, A.S. Abhiraman, P. Desai, Characterizing structural changes in point-bonded nonwoven fabrics during load-deformation experiments, Text. Res. J. 71 (2) (2001) 157–164.

[345] H.S. Kim, B. Pourdeyhimi, A.S. Abhiraman, P. Desai, Angular mechanical properties in thermally point-bonded nonwovens, part I: experimental observations, Text. Res. J. 71 (11) (2001) 965–976.

[346] S.M. Lee, A.S. Argon, The mechanics of the bending of nonwoven fabrics, part I: spunbonded fabric (Cerex), J. Text. Inst. 1 (1983) 1–11.

[347] S.M. Lee, A.S. Argon, The mechanics of the bending of nonwoven fabrics, part II: spunbonded fabric with spot bonds (Fibretex), J. Text. Inst. 1 (1983) 12–18.

[348] S.M. Lee, A.S. Argon, The mechanics of the bending of nonwoven fabrics, part III: print-bonded fabric (Masslinn), J. Text. Inst. 1 (1983) 19–30.

[349] B. Pourdeyhimi, R. Dent, A. Jerbi, S. Tanaka, A. Deshpande, Measuring fibre orientation in nonwovens, part V: real fabrics, Text. Res. J. 69 (1999) 185–192.

[350] S.M. Lee, A.S. Argon, The mechanics of the bending of nonwoven fabrics, part IV: print-bonded fabric with a pattern of elliptical holes (Keybak), J. Text. Inst. 1 (1983) 31–37.

[351] S.V. Patel, S.B. Warner, Modeling the bending stiffness of point bonded nonwoven fabrics, Text. Res. J. 64 (9) (1994) 507–513.

[352] B. Pourdeyhimi, B. Xu, Characterizing pore size in nonwoven fabrics: shape considerations, Int. Nonwovens J. 6 (1) (1994) 26–30.

[353] B. Pourdeyhimi, R. Ramanathan, R. Dent, Measuring fibre orientation in nonwovens, part ii: direct tracking, Text. Res. J. 66 (1996) 747–753.

[354] B. Pourdeyhimi, R. Dent, H. Davis, Measuring fibre orientation in nonwovens, part III: Fourier transform, Text. Res. J. 67 (1997) 43–151.

[355] S.I. Guceri, J.W. Gillespie, R. Shanker, Polym. Eng. Sci. 31 (3) (1991) 161.

[356] . http://www.allasso-industries.com.

[357] H. Xuan-chao, R. Bresee, Characterizing nonwoven web structure using image analysis techniques, part III: web uniformity analysis, INDA J. Nonwovens Res. 5 (3) (1994) 28–38.

[358] B. Drouin, R. Gagnon, C. Cheam, J. Silvy, A new way for testing paper sheet formation, Compos. Sci. Technol. 61 (2001) 389–393.

[359] O.J. Kallmes, Techniques for determining the Fiber orientation distribution throughout the thickness of a sheet, TAPPI J. 52 (1969) 482.

[360] O.J. Kallmes, H. Corte, Formation and structure of paper, 1, technical section, in: British Paper and Board Maker's Association, William Clowes & Sons, Ltd., London, 1962, pp. 13–46.

[361] R.G. Weigert, The selection of an optimum quadrant size for sampling the standard crop of grasses and Forbes, Ecology 43 (1962) 125–129.

[362] B. Pourdeyhimi, L. Kohel, Area based strategy for determining web uniformity, Text. Res. J. 72 (12) (2002) 1065–1072.

Index

Note: Page numbers followed by *f* indicate figures and *t* indicate tables.

Printed in the United States
by Baker & Taylor Publisher Services